# CLINICAL ANATOMY

## SYSTEMS

# CLINICAL ANATOMY

# BY SYSTEMS

## Richard S. Snell, MD, PhD

Emeritus Professor of Anatomy
George Washington University
School of Medicine and Health Sciences
Washington, DC

LIPPINCOTT WILLIAMS & WILKINS
A **Wolters Kluwer** Company
Philadelphia • Baltimore • New York • London
Buenos Aires • Hong Kong • Sydney • Tokyo

*Acquisitions Editor:* Crystal Taylor
*Managing Editor:* Kathleen Scogna
*Marketing Manager:* Valerie Shannahan
*Production Editor:* Jenn Glazer
*Designer:* Doug Smock
*Compositor:* Maryland Composition
*Printer:* R.R. Donnelley and Sons

**Library of Congress Cataloging-in-Publication Data**

Snell, Richard S.
    Clinical anatomy by systems / Richard S. Snell
        p. ; cm.
    Includes bibliographical references and index.
    ISBN-13: 978-0-7817-9164-9
    ISBN-10: 0-7817-9164-2
    1. Human anatomy.    I. Title.
    [DNLM:    1. Anatomy.    QS 4 S671c 2006]
  QM23.2.S64 2006
  611–dc22

2005027302

07 08 09 10 11
3 4 5 6 7 8 9 10

During the last three decades, the emphasis in many Anatomy Departments has been directed toward the study of cellular and molecular structures. This new information has greatly enhanced our understanding of the physiologic and biochemical mechanisms and their relationship to disease and drug treatment. The unquestionable fact remains, however, that each patient that comes to your clinic with medical problems is composed of gross anatomical structures, which may or may not exhibit functional deficits. It is thus the clear responsibility of medical school faculty to provide all students entering medicine with a basic knowledge of anatomy that is clinically relevant.

The explosion in the knowledge of medical disease and the technological advances associated with the diagnosis and treatment of diseases have dictated a complete restructuring of the curriculum for medical students, dental students, allied health students, and nursing students. Many medical schools outside the United States still enjoy the luxury of offering students the possibility of completely dissecting the human cadaver region by region. Here, in the United States, students in many schools are now being offered programs in gross anatomy in which only part of the body is dissected, and this is supplemented by the use of prosected specimens, plastinated specimens, and computer imagery.

To further streamline the basic science and clinical programs, many schools are introducing the organ approach or are teaching the entire medical curriculum system by system. The student is thus exposed to the basic science of each system of the body, together with the pathology, medicine, and surgery of the system. It is hoped that this approach will result in considerable streamlining of the curriculum.

This new text of Clinical Anatomy has been written to accommodate these new approaches to the teaching of Anatomy in a modern medical curriculum. The book is predicated on the fact that medical professionals require a detailed knowledge of certain systems of the body, whereas a superficial knowledge of other regions is quite adequate. For example, the anatomy of the upper and lower airways of the respiratory system is of paramount importance, while the anatomy of the sole of the foot is of less importance. At certain sites in the body where disease is common and likely to involve many systems, a brief regional overview has been provided in the Appendix. Tables have been used wherever possible to reduce the size of the text. This includes tables that give important dimensions and capacities of various anatomic structures; these are incorporated in an Appendix at the end of the CD-ROM.

The book is heavily illustrated. Most figures have been kept simple, and color has been used extensively. Colored photographs of surface anatomy models have been used, and a number of colored photographs of prosected specimens have been included.

Each chapter focuses the student on the material that is most important to learn and understand. It emphasizes the basic structures in the area being studied so that once mastered, the student is easily able to build up his or her knowledge base.

Throughout the book, each chapter is constructed in a similar manner to provide easy access to the material. Each chapter includes:

1. **Basic Clinical Anatomy.** This section provides useful basic information to assist medical professionals in making diagnoses and instructing patients in treatment; it also provides a ready source of anatomic material to assist them in performing many basic medical procedures. Numerous examples of normal radiographs, CT scans, MRI studies, and sonograms are also provided. Labeled photographs of cross-sectional anatomy are included to stimulate students to think in terms of three-dimensional anatomy, which is so important in the interpretation of imaging studies.

2. **Physiologic Notes.** These notes are interspersed amongst the basic anatomical material to emphasize to the reader the functional significance of the material.

3. **Embryologic Notes.** The development of many organs is briefly considered where such knowledge significantly contributes to the understanding of the structure and relationships of the organs. A detailed description

of the common congenital anomalies that may be seen in the clinic is found in the CD-ROM.

4. **Surface Anatomy.** This section provides surface landmarks of important anatomic structures, many of which are located some distance beneath the skin. This section is important because most practicing medical personnel seldom explore tissues to any depth beneath the skin.

5. **Review Questions.** The purpose of the questions is threefold—to focus attention on areas of importance, to enable students to assess their areas of weakness, and to provide a form of self-evaluation for questions asked under examination conditions. Some of the questions are centered around a clinical problem that requires an anatomic answer. Solutions are provided at the end of the section.

The CD-ROM contains all the clinical material relevant to each chapter. This is followed by clinical problem solving questions with answers and explanations.

R.S.S.

## Publisher's Note

The reader will note that on occasion, some figures are cited out of order. Such a writing strategy was used when a relevant aspect from a figure also discussed in another context clarifies an aspect of the topic under discussion. To avoid confusion, the editors placed the word "see" in front of any out-of-order reference, whether it refers to a figure already discussed or still to be detailed.

# Acknowledgments

I am greatly indebted to many faculty members of the Department of Radiology at the George Washington University School of Medicine and Health Sciences for the loan of radiographs, CT scans, and MRIs that have been reproduced in different sections of this book. I am also grateful to Dr. Carol Lee, Dr. Gordon Sze, and Dr. Robert Smith of the Department of Radiology at Yale University Medical Center for supplying examples of mammograms, CT scans of the vertebral column, and MRIs of the limbs. My thanks are also owed to Dr. Michael Remetz of the Department of Cardiology at Yale for providing examples of coronary arteriograms.

My special thanks are owed to Larry Clark, who, as a senior technician in the Department of Anatomy at George Washington University, greatly assisted me in the preparation of anatomical specimens for photography and for the preparation of plastinated specimens of many different organs. His enthusiasm for the many projects was contagious and greatly helped in the final production of outstanding specimens, many of which are illustrated in the text.

I wish also to express my sincere thanks to Terry Dolan, Virginia Childs, Myra Feldman, and Ira Grunther for preparation of the artwork.

Finally, I wish to express my deep gratitude to the staff of Lippincott, Williams & Wilkins for their great help and support in the preparation of this new textbook.

# Contents

# THE RESPIRATORY SYSTEM

# THE CARDIOVASCULAR SYSTEM

# THE LYMPHATIC SYSTEM

# THE MUSCULOSKELETAL SYSTEM

# THE NERVOUS SYSTEM

# THE DIGESTIVE SYSTEM

# THE URINARY SYSTEM

# THE REPRODUCTIVE SYSTEM

# THE ENDOCRINE SYSTEM

**NOTE: The CD-ROM contains all the clinical material relevant to each chapter.**

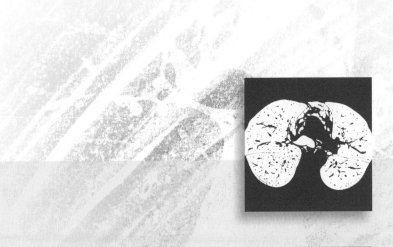

# 1 Introduction to Clinical Anatomy

All clinical material relevant to this chapter can be found on the CD-ROM.

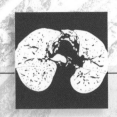

# Chapter Outline

I t is essential that students understand the terms used for describing the structure and function of different parts of gross anatomy. Without these terms, it is impossible to describe in a meaningful way the composition of the body. Moreover, the medical professional needs these terms so that anatomic abnormalities found on clinical examination of a patient can be accurately recorded.

This chapter also introduces some of the basic structures that compose the body, such as skin, fascia, muscles, bones, blood vessels, lymphatic system, nervous system, and mucous membranes and serous membranes.

# BASIC ANATOMY

**Anatomy** is the science of the structure and function of the body. **Clinical anatomy** is the study of the macroscopic structure and function of the body as it relates to the practice of medicine and other health sciences. **Basic anatomy** is the study of the minimal amount of anatomy consistent with the understanding of the overall structure and function of the body.

# Descriptive Anatomic Terms

Medical personnel must understand the basic anatomic terms. With the aid of a medical dictionary, you will find that understanding anatomic terminology greatly assists you in the learning process.

The accurate use of anatomic terms enables you to communicate with your colleagues both nationally and internationally. Without anatomic terms, one cannot accurately discuss or record the abnormal functions of joints, the actions of muscles, the alteration of position of organs, or the exact location of swellings or tumors.

## Terms Related to Position

All descriptions of the human body are based on the assumption that the person is standing erect, with the upper limbs by the sides and the face and palms of the hands directed forward (Fig. 1-1). This is called the **anatomic position.** The various parts of the body are then described in relation to certain imaginary planes.

### Median Sagittal Plane

This is a vertical plane passing through the center of the body, dividing it into equal right and left halves (Fig. 1-1). Planes situated to one or the other side of the median plane and parallel to it are termed **paramedian.** A structure situated nearer to the median plane of the body than another is said to be **medial** to the other. Similarly, a struc-

ture that lies farther away from the median plane than another is said to be **lateral** to the other.

### Coronal Planes

These are imaginary vertical planes at right angles to the median plane (Fig. 1-1).

### Horizontal or Transverse Planes

These planes are at right angles to both the median and the coronal planes (Fig. 1-1). The terms **anterior** and **posterior** are used to indicate the front and back of the body, respectively (Fig. 1-1). To describe the relationship of two structures, one structure is said to be anterior or posterior to the other insofar as it is closer to the anterior or posterior body surface.

In describing the hand, the terms **palmar** and **dorsal surfaces** are used in place of anterior and posterior, and in describing the foot, the terms **plantar** and **dorsal surfaces**

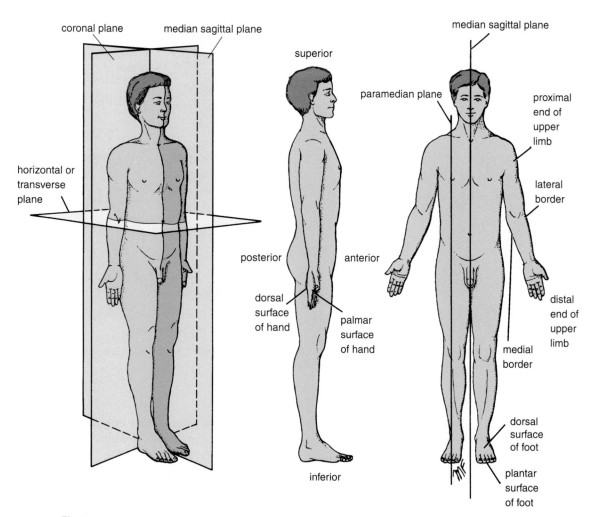

**Figure 1-1**   Anatomic terms used in relation to position. Note that the subjects are standing in the anatomic position

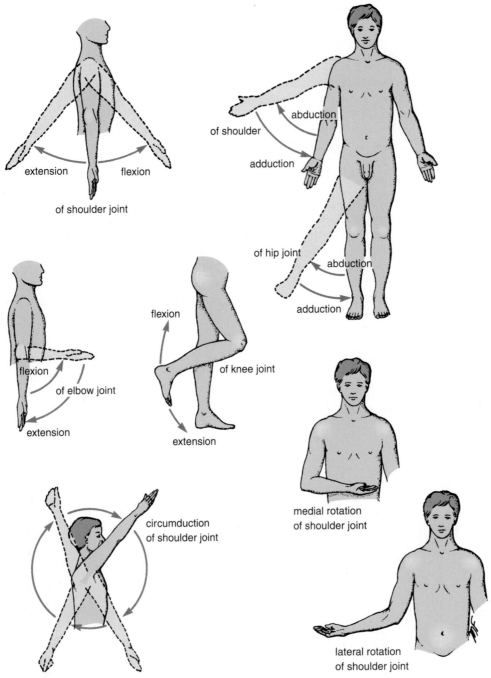

**Figure 1-2**  Some anatomic terms used in relation to movement. Note the difference between flexion of the elbow and that of the knee.

are used instead of lower and upper surfaces (Fig. 1-1). The terms **proximal** and **distal** describe the relative distances from the roots of the limbs; for example, the arm is proximal to the forearm, and the hand is distal to the forearm.

The terms **superficial** and **deep** denote the relative distances of structures from the surface of the body, and the terms **superior** and **inferior** denote levels relatively high or low with reference to the upper and lower ends of the body.

The terms **internal** and **external** are used to describe the relative distance of a structure from the center of an organ or cavity; for example, the internal carotid artery is found inside the cranial cavity, and the external carotid artery is found outside the cranial cavity.

The term **ipsilateral** refers to the same side of the body; for example, the left hand and left foot are ipsilateral. **Contralateral** refers to opposite sides of the body; for example, the left biceps brachii muscle and the right rectus femoris muscle are contralateral.

The **supine** position of the body is lying on the back. The **prone** position is lying face downward.

## Terms Related to Movement

A site where two or more bones come together is known as a **joint**. Some joints have no movement (sutures of the skull), some have only slight movement (superior tibiofibular joint), and some are freely movable (shoulder joint).

■ **Flexion:** Movement that takes place in a sagittal plane. For example, flexion of the elbow joint approximates the anterior surface of the forearm to the anterior surface of the arm. It is usually an anterior movement, but it is occasionally posterior, as in the case of the knee joint (Fig. 1-2).

■ **Extension:** Movement involving the straightening of the joint and that usually takes place in a posterior direction (Fig. 1-2).

■ **Lateral flexion:** Movement of the trunk in the coronal plane (Fig. 1-3).

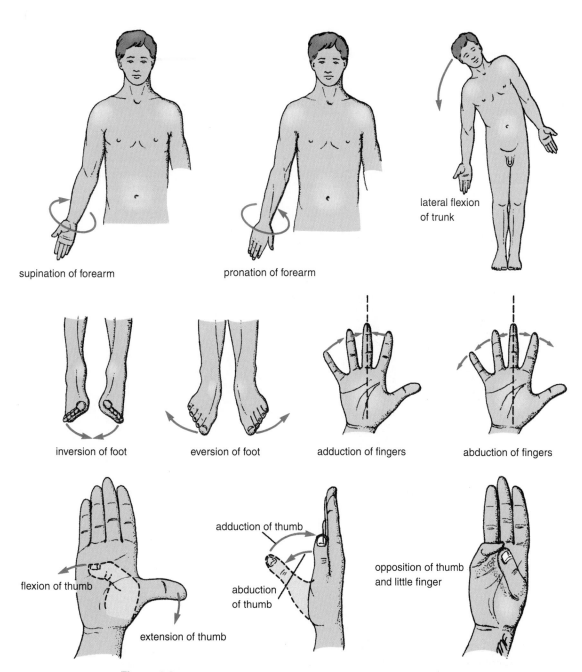

supination of forearm

pronation of forearm

lateral flexion of trunk

inversion of foot

eversion of foot

adduction of fingers

abduction of fingers

flexion of thumb

extension of thumb

adduction of thumb

abduction of thumb

opposition of thumb and little finger

**Figure 1-3** Additional anatomic terms used in relation to movement.

▦ **Abduction:** Movement of a limb away from the midline of the body in the coronal plane (Fig. 1-2).

▦ **Adduction:** Movement of a limb toward the body in the coronal plane (Fig. 1-2). In the fingers and toes, abduction is applied to the spreading of these structures, and adduction is applied to the drawing together of these structures (Fig. 1-3). The movements of the thumb (Fig. 1-3), which are a little more complicated, are described on page 403.

▦ **Rotation:** Movement of a part of the body around its long axis.

   ▦ **Medial rotation:** Movement that results in the anterior surface of the part facing medially.

▦ **Lateral rotation:** Movement that results in the anterior surface of the part facing laterally.

   ▦ **Pronation of the forearm:** Medial rotation of the forearm in such a manner that the palm of the hand faces posteriorly (Fig. 1-3).

   ▦ **Supination of the forearm:** Lateral rotation of the forearm from the pronated position so that the palm of the hand comes to face anteriorly (Fig. 1-3).

▦ **Circumduction:** The combination in sequence of the movements of flexion, extension, abduction, and adduction (Fig. 1-2).

▦ **Protraction:** Moving forward.

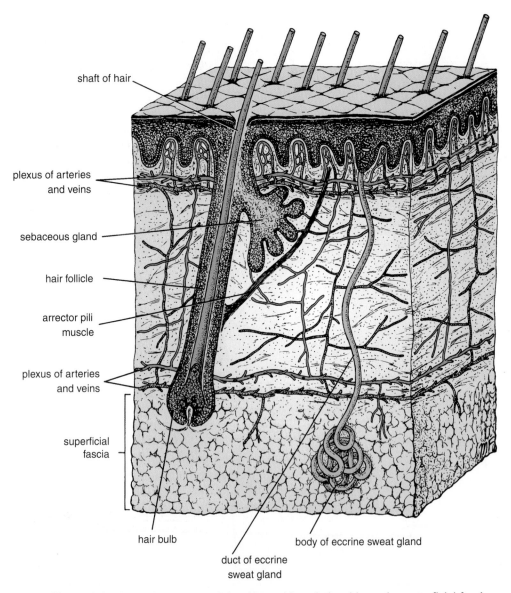

shaft of hair

plexus of arteries and veins

sebaceous gland

hair follicle

arrector pili muscle

plexus of arteries and veins

superficial fascia

hair bulb

duct of eccrine sweat gland

body of eccrine sweat gland

**Figure 1-4** General structure of the skin and its relationship to the superficial fascia. Note that hair follicles extend down into the deeper part of the dermis or even into the superficial fascia, whereas sweat glands extend deeply into the superficial fascia.

- **Retraction:** Moving backward (used to describe the forward and backward movement of the jaw at the temporomandibular joints).
- **Inversion:** Movement of the foot so that the sole faces in a medial direction (Fig. 1-3).
- **Eversion:** The opposite movement of the foot so that the sole faces in a lateral direction (Fig. 1-3).

# Basic Structures

## Skin

The skin is divided into two parts: the superficial part, the **epidermis,** and the deep part, the **dermis** (Fig. 1-4). The epidermis is a stratified epithelium whose cells become flattened as they mature and rise to the surface. On the palms of the hands and the soles of the feet, the epidermis is extremely thick to withstand the wear and tear that occurs in these regions. In other areas of the body, for example, on the anterior surface of the arm and forearm, the epidermis is thin. The dermis is composed of dense connective tissue containing many blood vessels, lymphatic vessels, and nerves. It shows considerable variation in thickness in different parts of the body, tending to be thinner on the anterior than on the posterior surface. It is thinner in women than in men. The dermis of the skin is connected to the underlying deep fascia or bones by the **superficial fascia,** otherwise known as **subcutaneous tissue.**

The skin over joints always folds in the same place, the **skin creases** (Fig. 1-5). At these sites, the skin is thinner than elsewhere and is firmly tethered to underlying structures by strong bands of fibrous tissue.

The appendages of the skin are the **nails, hair follicles, sebaceous glands,** and **sweat glands.**

## Nails

The **nails** are keratinized plates on the dorsal surfaces of the tips of the fingers and toes. The proximal edge of the plate is the **root of the nail** (Fig. 1-5). With the exception of the distal edge of the plate, the nail is surrounded and overlapped by folds of skin known as **nail folds.** The surface of skin covered by the nail is the **nail bed** (Fig. 1-5).

## Hairs

**Hairs** grow out of **follicles,** which are invaginations of the epidermis into the dermis (Fig. 1-4). The follicles lie obliquely to the skin surface, and their expanded extremities, called **hair bulbs,** penetrate to the deeper part of the dermis. Each hair bulb is concave at its end, and the concavity is occupied by vascular connective tissue called the **hair papilla.** A band of smooth muscle, the **arrector pili,** connects the undersurface of the follicle to the superficial part of the dermis (Fig. 1-4). The muscle is innervated by sympathetic nerve fibers, and its contraction causes the hair to move into a more vertical position; it also compresses the sebaceous gland and causes it to extrude some of its secretion. The pull of the muscle also causes dimpling of the skin surface, or so-called **gooseflesh.** Hairs are distributed in various numbers over the whole surface of the body, except on the lips, the palms of the hands, the sides of the fingers, the glans penis and clitoris, the labia minora and the internal surface of the labia majora, and the soles and sides of the feet and the sides of the toes.

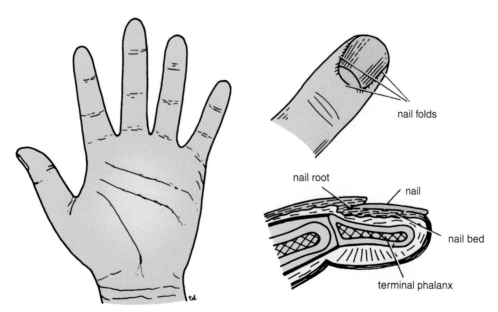

**Figure 1-5** The various skin creases on the palmar surface of the hand and the anterior surface of the wrist joint. The relationship of the nail to other structures of the finger is also shown.

## Sebaceous Glands

**Sebaceous glands** pour their secretion, the sebum, onto the shafts of the hairs as they pass up through the necks of the follicles. They are situated on the sloping undersurface of the follicles and lie within the dermis (Fig. 1-4). **Sebum** is an oily material that helps preserve the flexibility of the emerging hair. It also oils the surface epidermis around the mouth of the follicle.

## Sweat Glands

**Sweat glands** are long, spiral, tubular glands distributed over the surface of the body, except on the red margins of the lips, the nail beds, and the glans penis and clitoris (Fig. 1-4). These glands extend through the full thickness of the dermis, and their extremities may lie in the superficial fascia. Therefore, the sweat glands are the most deeply penetrating structures of all the epidermal appendages.

## Fasciae

The fasciae of the body can be divided into **superficial** and **deep fascia**, and they lie between the skin and the underlying muscles and bones.

## The Superficial Fascia

The **superficial fascia,** or subcutaneous tissue, is a mixture of loose areolar and adipose tissue that unites the dermis of the skin to the underlying deep fascia (Fig. 1-6). In the scalp, the back of the neck, the palms of the hands, and the soles of the feet, it contains numerous bundles of collagen fibers that hold the skin firmly to the deeper structures. In the eyelids, auricle of the ear, penis and scrotum, and clitoris, the superficial fascia is devoid of adipose tissue.

## The Deep Fascia

The **deep fascia** is a membranous layer of connective tissue that clothes the muscles and other deep structures (Fig. 1-6). In the thorax and abdomen, it is just a thin film of areolar tissue covering the muscles and aponeuroses. In the limbs, the deep fascia forms a definite sheath around the muscles and other structures, holding them in place. Fibrous septa extend from the deep surface of the membrane, between the groups of muscles, and, in many places, divide the interior of the limbs into compartments (Fig. 1-6). In the region of joints, the deep fascia may be considerably thickened to form restraining bands called **retinacula** (Fig. 1-7). Their function is to hold underlying tendons in position or to serve as pulleys around which the tendons may move.

# Muscle

The three types of muscle are skeletal, smooth, and cardiac.

## Skeletal Muscle

Skeletal muscles produce the movements of the skeleton; they are sometimes called **voluntary muscles** and are made up of striped muscle fibers. A skeletal muscle has two or more attachments. The attachment that moves the least is referred to as the **origin,** and the one that moves the most is called the **insertion** (Fig. 1-8). Under varying circumstances, the degree of mobility of the attachments may be reversed; therefore, the terms *origin* and *insertion* are interchangeable.

The fleshy part of the muscle is referred to as its **belly** (Fig. 1-8). The ends of a muscle are attached to bones, cartilage, or ligaments by cords of fibrous tissue called **tendons** (Fig. 1-9). Occasionally, flattened muscles are attached by a thin but strong sheet of fibrous tissue called an **aponeurosis**

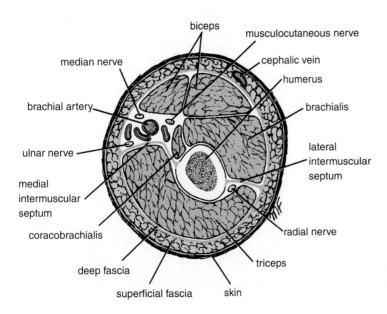

Figure 1-6 Section through the middle of the right arm showing the arrangement of the superficial and deep fascia. Note how the fibrous septa extend between groups of muscles, dividing the arm into fascial compartments.

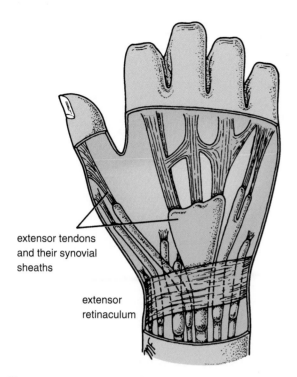

**Figure 1-7** Extensor retinaculum on the posterior surface of the wrist holding the underlying tendons of the extensor muscles in position.

(Fig. 1-9). A **raphe** is an interdigitation of the tendinous ends of fibers of flat muscles (Fig. 1-9).

## Smooth Muscle

Smooth muscle consists of long, spindle-shaped cells closely arranged in bundles or sheets.

### PHYSIOLOGIC NOTE

#### Function of Smooth Muscle

In the tubes of the body, smooth muscle provides the motive power for propelling the contents through the lumen. In the digestive system, it also causes the ingested food to be thoroughly mixed with the digestive juices. A wave of contraction of the circularly arranged fibers passes along the tube, milking the contents onward. By their contraction, the longitudinal fibers pull the wall of the tube proximally over the contents. This method of propulsion is referred to as **peristalsis.**

In storage organs such as the urinary bladder and the uterus, the fibers are irregularly arranged and interlaced with one another. Their contraction is slow and sustained and brings about expulsion of the contents of the organs. In the walls of the blood vessels, the smooth muscle fibers are arranged circularly and serve to modify the caliber of the lumen.

Depending on the organ, smooth muscle fibers may be made to contract by local stretching of the fibers, by nerve impulses from autonomic nerves, or by hormonal stimulation.

## Cardiac Muscle

Cardiac muscle consists of striated muscle fibers that branch and unite with each other. It forms the myocardium of the heart. The fibers of cardiac muscle tend to be arranged in whorls and spirals, and they have the property of spontaneous and rhythmical contraction. Specialized cardiac muscle fibers form the **conducting system of the heart.**

Cardiac muscle is supplied by autonomic nerve fibers that terminate in the nodes of the conducting system and in the myocardium.

## Joints

A site where two or more bones come together, whether or not movement occurs between them, is called a **joint.** Joints are classified according to the tissues that lie between the bones: fibrous joints, cartilaginous joints, and synovial joints.

## Fibrous Joints

The articulating surfaces of the bones are joined by fibrous tissue (Fig. 1-10 ), and thus, very little movement is possible. The sutures of the vault of the skull and the inferior tibiofibular joints are examples of fibrous joints.

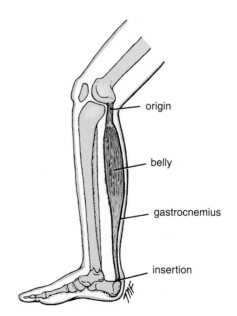

**Figure 1-8** Origin, insertion, and belly of the gastrocnemius muscle.

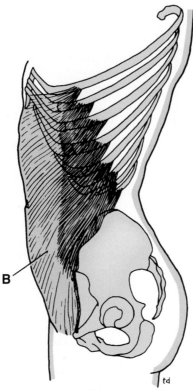

common tendon for the insertion of the gastrocnemius and soleus muscles

external oblique aponeurosis

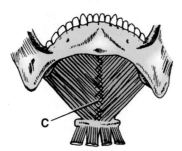

raphe of mylohyoid muscles

**Figure 1-9** Examples of (**A**) a tendon, (**B**) an aponeurosis, and (**C**) a raphe.

## Cartilaginous Joints

### Primary Cartilaginous Joint

A **primary cartilaginous joint** is one in which the bones are united by a plate or bar of hyaline cartilage. Thus, the union between the **epiphysis** and the **diaphysis** of a growing bone and the union between the first rib and the manubrium sterni are examples of such a joint. No movement is possible.

### Secondary Cartilaginous Joint

A **secondary cartilaginous joint** is one in which the bones are united by a plate of fibrocartilage, and the articular surfaces of the bones are covered by a thin layer of hyaline cartilage. Examples are the joints between the vertebral bod-

ies (Fig. 1-10) and the **symphysis pubis**. A small amount of movement is possible.

## Synovial Joints

The articular surfaces of the bones are covered by a thin layer of hyaline cartilage separated by a joint cavity (Fig. 1-10). This arrangement permits a great degree of freedom of movement. The cavity of the joint is lined by **synovial membrane**, which extends from the margins of one articular surface to those of the other. The synovial membrane is protected on the outside by a tough fibrous membrane referred to as the **capsule** of the joint. The articular surfaces are lubricated by a viscous fluid called **synovial fluid**, which is produced by the synovial membrane. In certain synovial joints, for example, in the knee joint, discs or wedges of fibrocartilage are interposed between the articular surfaces of the bones. These are referred to as **articular discs**.

**Fatty pads** are found in some synovial joints lying between the synovial membrane and the fibrous capsule or bone. Examples are found in the hip (Fig. 1-10) and knee joints.

The degree of movement in a synovial joint is limited by the shape of the bones participating in the joint, the coming together of adjacent anatomic structures (e.g., the thigh against the anterior abdominal wall on flexing the hip joint), and the presence of fibrous **ligaments** uniting the bones. Most ligaments lie outside the joint capsule, but in

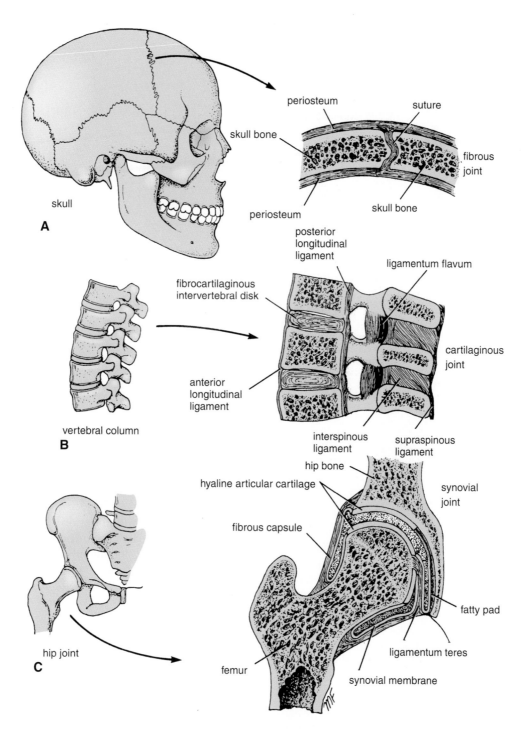

**Figure 1-10** Examples of three types of joints. **A.** Fibrous joint (coronal suture of skull). **B.** Cartilaginous joint (joint between two lumbar vertebral bodies). **C.** Synovial joint (hip joint).

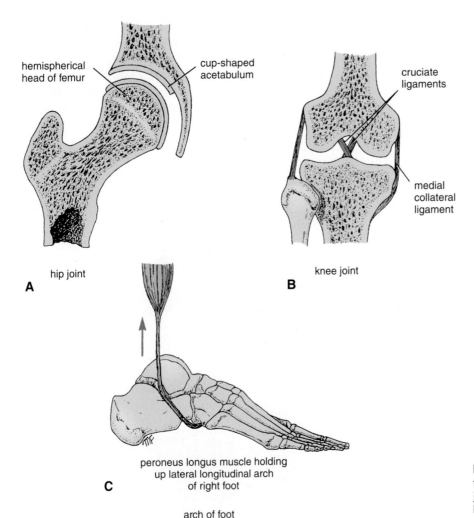

A   hip joint

B   knee joint

peroneus longus muscle holding
up lateral longitudinal arch
of right foot

C

arch of foot

**Figure 1-11**   The three main factors responsible for stabilizing a joint. **A.** Shape of articular surfaces. **B.** Ligaments. **C.** Muscle tone.

the knee, some important ligaments, the **cruciate ligaments,** lie within the capsule (Fig. 1-11).

## Ligaments

A ligament is a cord or band of connective tissue uniting two structures (Fig.1-11). Commonly found in association with joints, ligaments are of two types. Most are composed of dense bundles of collagen fibers and are unstretchable under normal conditions (e.g., the iliofemoral ligament of the hip joint and the collateral ligaments of the elbow joint). The second type is composed largely of elastic tissue and can, therefore, regain its original length after stretching (e.g., the ligamentum flavum of the vertebral column and the calcaneonavicular ligament of the foot).

## Bursae

A bursa is a lubricating device consisting of a closed fibrous sac lined with a delicate smooth membrane. Its walls are separated by a film of viscous fluid. Bursae are found wher-

ever tendons rub against bones, ligaments, or other tendons. They are commonly found close to joints where the skin rubs against underlying bony structures, for example, the prepatellar bursa (Fig. 1-12). Occasionally, the cavity of a bursa communicates with the cavity of a synovial joint. For example, the suprapatellar bursa communicates with the knee joint (Fig. 1-12), and the subscapularis bursa communicates with the shoulder joint.

## Synovial Sheath

A synovial sheath is a tubular bursa that surrounds a tendon. The tendon invaginates the bursa from one side so that the tendon becomes suspended within the bursa by a **mesotendon** (Fig. 1-12). The mesotendon enables blood vessels to enter the tendon along its course.

## Blood Vessels

Blood vessels are of three types: arteries, veins, and capillaries (Fig. 1-14).

## Arteries

**Arteries** transport blood from the heart and distribute it to the various tissues of the body by means of their **branches** (Figs. 1-13 and 1-14). The smallest arteries, <0.1 mm in diameter, are called **arterioles**. The joining of branches of arteries is called an **anastomosis**. Arteries do not have valves.

**Anatomic end arteries** (Fig. 1-14) are vessels whose terminal branches do not anastomose with branches of arteries supplying adjacent areas. **Functional end arteries** are vessels whose terminal branches do anastomose with those of adjacent arteries, but the caliber of the anastomosis is insufficient to keep the tissue alive should one of the arteries become blocked.

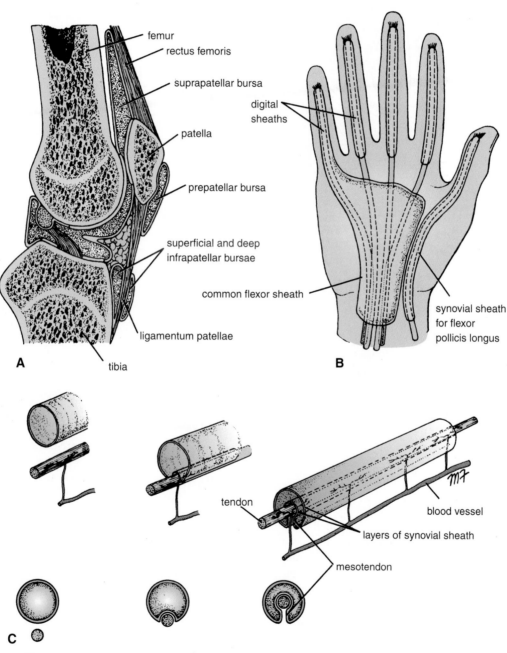

**Figure 1-12    A.** Four bursae related to the front of the knee joint. Note that the suprapatellar bursa communicates with the cavity of the joint. **B.** Synovial sheaths around the long tendons of the fingers. **C.** How tendon indents synovial sheath during development, and how blood vessels reach the tendon through the mesotendon.

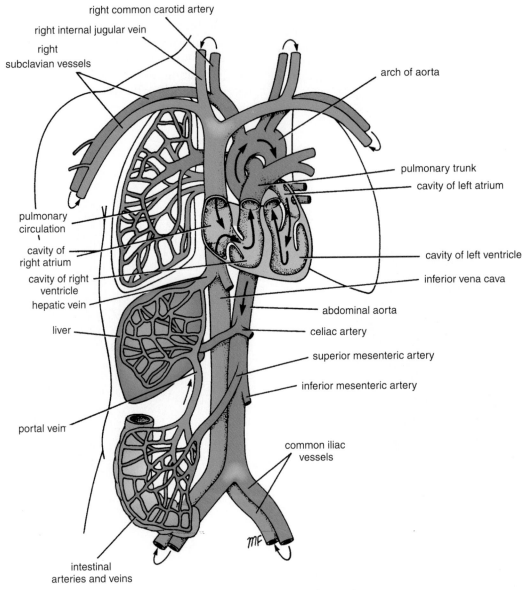

right common carotid artery

right internal jugular vein

right subclavian vessels

arch of aorta

pulmonary trunk

cavity of left atrium

pulmonary circulation

cavity of right atrium

cavity of right ventricle

hepatic vein

liver

cavity of left ventricle

inferior vena cava

abdominal aorta

celiac artery

superior mesenteric artery

inferior mesenteric artery

portal vein

common iliac vessels

intestinal arteries and veins

**Figure 1-13** General plan of the blood vascular system.

## Veins

**Veins** are vessels that transport blood back to the heart; many of them possess valves. The smallest veins are called **venules** (Fig. 1-14). The smaller veins, or **tributaries**, unite to form larger veins, which may join with one another to form **venous plexuses**. Medium-size deep arteries are often accompanied by two veins, one on each side, called **venae comitantes**.

Veins leaving the gastrointestinal tract do not go directly to the heart but converge on the **portal vein**; this vein enters the liver and breaks up again into veins of diminishing size, which ultimately join capillary-like vessels, termed **sinusoids,** in the liver (Fig. 1-14). A **portal system** is thus a system of vessels interposed between two capillary beds.

## Capillaries

**Capillaries** are microscopic vessels in the form of a network connecting the arterioles to the venules (Fig. 1-14).

### Sinusoids

**Sinusoids** resemble capillaries in that they are thin-walled blood vessels, but they have an irregular cross diameter and are wider than capillaries. They are found in the bone marrow, spleen, liver, and some endocrine glands. In some areas of the body, principally the tips of the fingers and toes, direct connections occur between the arteries and veins without the intervention of capillaries. The sites of such connections are referred to as **arteriovenous anastomoses** (Fig. 1-14).

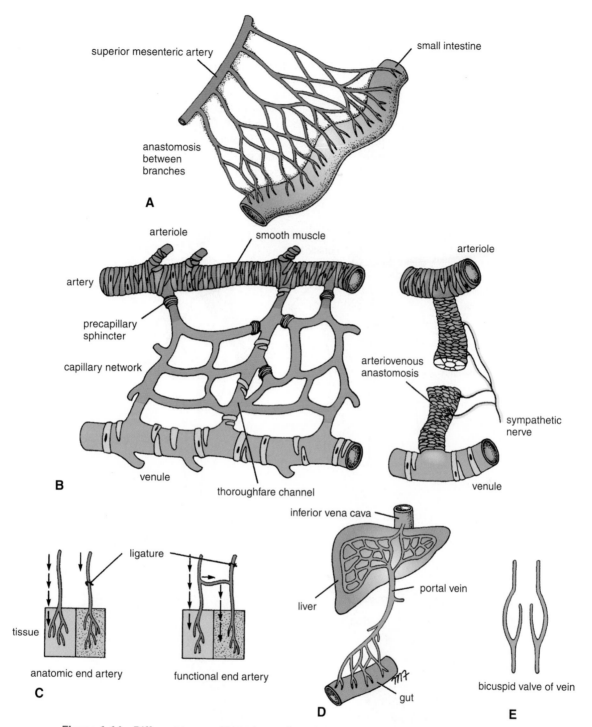

**Figure 1-14** Different types of blood vessels and their methods of union. **A.** Anastomosis between the branches of the superior mesenteric artery. **B.** A capillary network and an arteriovenous anastomosis. **C.** Anatomic end artery and functional end artery. **D.** A portal system. **E.** Structure of the bicuspid valve in a vein.

## Lymphatic System

The lymphatic system consists of lymphatic tissues and lymphatic vessels (Fig. 1-15).

### Lymphatic Tissues

**Lymphatic tissues** are a type of connective tissue that contains large numbers of lymphocytes. Lymphatic tissue is organized into the following organs or structures: the thymus, the lymph nodes, the spleen, and the lymphatic nodules. Lymphatic tissue is essential for the immunologic defenses of the body against bacteria and viruses.

### Lymphatic Vessels

**Lymphatic vessels** are tubes that assist the cardiovascular system in the removal of tissue fluid from the tissue spaces of the body; the vessels then return the fluid to the blood. The lymphatic system is essentially a drainage system, and **there is no circulation.** Lymphatic vessels are found in all tissues and organs of the body except the central nervous

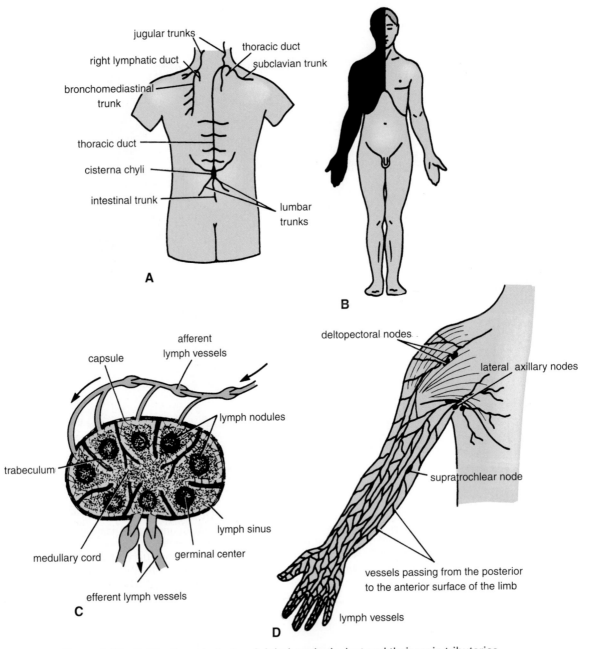

**Figure 1-15    A.** The thoracic duct and right lymphatic duct and their main tributaries. **B.** The areas of body drained into thoracic duct (*clear*) and right lymphatic duct (*black*). **C.** General structure of a lymph node. **D.** Lymph vessels and nodes of the upper limb.

system, the eyeball, the internal ear, the epidermis of the skin, the cartilage, and the bone.

### Lymph

**Lymph** is the name given to tissue fluid once it has entered a lymphatic vessel. **Lymph capillaries** are a network of fine vessels that drain lymph from the tissues. The capillaries are, in turn, drained by small lymph vessels, which unite to form large lymph vessels. Lymph vessels have a beaded appearance because of the presence of numerous valves along their course.

Before lymph is returned to the bloodstream, it passes through at least one **lymph node** and often through several. The lymph vessels that carry lymph to a lymph node are referred to as **afferent** vessels (Fig. 1-15); those that transport it away from a node are **efferent** vessels. The lymph reaches the bloodstream at the root of the neck by large lymph vessels called the **right lymphatic duct** and the **thoracic duct** (Fig. 1-15).

## Nervous System

The nervous system is divided into two main parts: the **central nervous system**, which consists of the brain and spinal cord, and the **peripheral nervous system**, which consists of 12 pairs of cranial nerves and 31 pairs of spinal nerves and their associated ganglia.

Functionally, the nervous system can be further divided into the **somatic nervous system**, which controls voluntary activities, and the **autonomic nervous system**, which controls involuntary activities.

---

### PHYSIOLOGIC NOTE

### Functions of the Nervous System

The nervous system, together with the endocrine system, controls and integrates the activities of the different parts of the body.

## Central Nervous System

The central nervous system is composed of large numbers of nerve cells and their processes, supported by specialized tissue called **neuroglia.** **Neuron** is the term given to the nerve cell and all its processes. The nerve cell has two types of processes, called **dendrites** and an **axon.** Dendrites are the short processes of the cell body; the axon is the longest process of the cell body (Fig. 1-16).

The interior of the central nervous system is organized into gray and white matter. **Gray matter** consists of nerve cells embedded in neuroglia. **White matter** consists of nerve fibers (axons) embedded in neuroglia.

## Peripheral Nervous System

The peripheral nervous system consists of the cranial and spinal nerves and their associated ganglia. On dissection, the cranial and spinal nerves are seen as grayish white cords. They are made up of bundles of nerve fibers (axons) supported by delicate areolar tissue.

### Cranial Nerves

There are 12 pairs of cranial nerves that leave the brain and pass through foramina in the skull. All the nerves are distributed in the head and neck except the Xth (vagus) nerve, which also supplies structures in the thorax and abdomen. The cranial nerves are described in Chapter 15.

### Spinal Nerves

A total of 31 pairs of spinal nerves leave the spinal cord and pass through intervertebral foramina in the vertebral column (Figs. 1-17 and 1-18). The spinal nerves are named according to the region of the vertebral column with which they are associated: 8 **cervical**, 12 **thoracic**, 5 **lumbar**, 5 **sacral**, and 1 **coccygeal**. Note that there are 8 cervical nerves and only 7 cervical vertebrae and that there is 1 coccygeal nerve and 4 coccygeal vertebrae. The spinal nerves are described in Chapter 17.

---

### EMBRYOLOGIC NOTE

### Spinal Cord Growth during Development

During development, the spinal cord grows in length more slowly than the vertebral column. In the adult, when growth ceases, the lower end of the spinal cord reaches inferiorly only as far as the lower border of the first lumbar vertebra. To accommodate for this disproportionate growth in length, the length of the roots increases progressively from above downward. In the upper cervical region, the spinal nerve roots are short and run almost horizontally, but the roots of the lumbar and sacral nerves below the level of the termination of the cord form a vertical bundle of nerves that resembles a horse's tail and is called the **cauda equina** (Fig. 1-17).

---

Each spinal nerve is connected to the spinal cord by two roots: the **anterior root** and the **posterior root** (Figs. 1-16 and 1-18). The anterior root consists of bundles of nerve fibers carrying nerve impulses away from the central nervous system (Fig. 1-16). Such nerve fibers are called **efferent** fibers. Those efferent fibers that go to skeletal muscle and cause them to contract are called **motor fibers.** Their cells of origin lie in the anterior gray horn of the spinal cord.

The posterior root consists of bundles of nerve fibers that carry impulses to the central nervous system and are called **afferent** fibers (Fig. 1-16). Because these fibers are

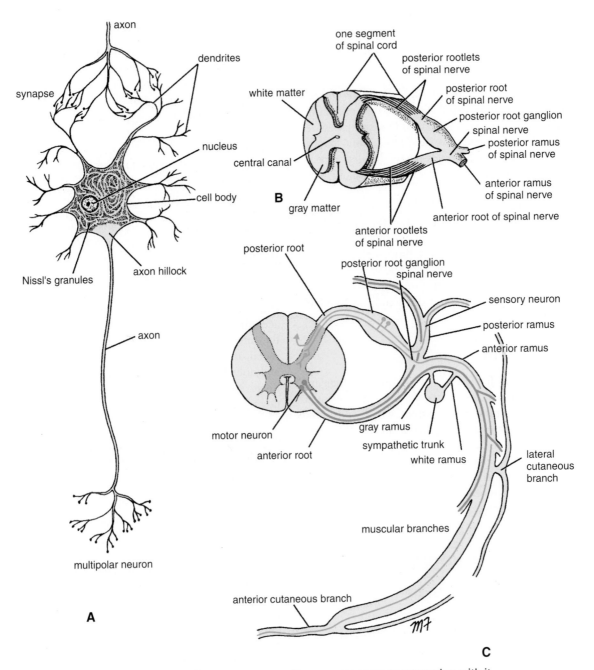

**Figure 1-16**  **A.** Multipolar motor neuron with connector neuron synapsing with it. **B.** Section through thoracic segment of spinal cord with spinal roots and posterior root ganglion. **C.** Cross section of thoracic segment of spinal cord showing roots, spinal nerve, and anterior and posterior rami and their branches.

concerned with conveying information about sensations of touch, pain, temperature, and vibrations, they are called **sensory fibers**. The cell bodies of these nerve fibers are situated in a swelling on the posterior root called the **posterior root ganglion** (Figs. 1-16 and 1-18).

At each intervertebral foramen, the anterior and posterior roots unite to form a spinal nerve (Fig. 1-18). Here, the motor and sensory fibers become mixed together, so that a

spinal nerve is made up of a mixture of motor and sensory fibers (Fig. 1-16). On emerging from the foramen, the spinal nerve divides into a large **anterior ramus** and a smaller **posterior ramus**. The posterior ramus passes posteriorly around the vertebral column to supply the muscles and skin of the back (Figs. 1-16 and 1-18). The anterior ramus continues anteriorly to supply the muscles and skin over the anterolateral body wall and all the muscles and skin of the limbs.

In addition to the anterior and posterior rami, spinal nerves give a small **meningeal branch** that supplies the vertebrae and the coverings of the spinal cord (the meninges). Thoracic spinal nerves also have branches, called **rami communicantes** that are associated with the sympathetic part of the autonomic nervous system (see p. 20 and 21).

## Plexuses

At the root of the limbs, the anterior rami join one another to form complicated nerve plexuses (Fig. 1-17). The **cervical** and **brachial plexuses** are found at the root of the upper limbs, and the **lumbar** and **sacral plexuses** are found at the root of the lower limbs.

The classic division of the nervous system into central and peripheral parts is purely artificial and one of descriptive convenience because the processes of the neurons pass freely between the two. For example, a motor neuron located in the anterior gray horn of the first thoracic segment of the spinal cord gives rise to an axon that passes through the anterior root of the first thoracic nerve (Fig. 1-19), passes through the brachial plexus, travels down the arm and forearm in the ulnar nerve, and finally reaches the motor end plates on several muscle fibers of a small muscle of the hand—a total distance of about 3 ft (90 cm).

As another example, consider the sensation of touch felt on the lateral side of the little toe. This area of skin is supplied by the first sacral segment of the spinal cord (S1).

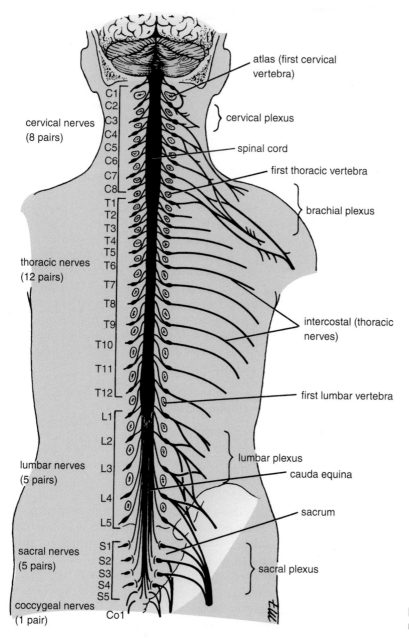

cervical nerves (8 pairs)

thoracic nerves (12 pairs)

lumbar nerves (5 pairs)

sacral nerves (5 pairs)

coccygeal nerves (1 pair)

C1
C2
C3
C4
C5
C6
C7
C8
T1
T2
T3
T4
T5
T6
T7
T8
T9
T10
T11
T12
L1
L2
L3
L4
L5
S1
S2
S3
S4
S5
Co1

atlas (first cervical vertebra)

cervical plexus

spinal cord

first thoracic vertebra

brachial plexus

intercostal (thoracic nerves)

first lumbar vertebra

lumbar plexus

cauda equina

sacrum

sacral plexus

**Figure 1-17** Brain, spinal cord, spinal nerves, and plexuses of limbs.

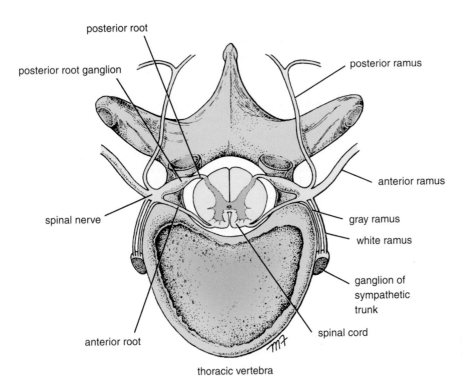

posterior root

posterior root ganglion

posterior ramus

anterior ramus

gray ramus

white ramus

spinal nerve

ganglion of sympathetic trunk

anterior root

spinal cord

thoracic vertebra

**Figure 1-18**   The association between spinal cord, spinal nerves, and sympathetic trunks.

The fine terminal branches of the sensory axon, called **dendrites**, leave the sensory organs of the skin and unite to form the axon of the sensory nerve. The axon passes up the leg in the sural nerve (Fig. 1-19) and then in the tibial and sciatic nerves to the lumbosacral plexus. It then passes through the posterior root of the first sacral nerve to reach the cell body in the posterior root ganglion of the first sacral nerve. The central axon now enters the posterior white column of the spinal cord and passes up to the nucleus gracilis in the medulla oblongata—a total distance of about 5 ft (1.5 m). Thus, a single neuron extends from the little toe to the inside of the skull. Both these examples illustrate the possible great length of a single neuron.

## Autonomic Nervous System

The autonomic nervous system is the part of the nervous system concerned with the innervation of involuntary structures such as the heart, smooth muscle, and glands throughout the body and is distributed throughout the central and peripheral nervous system. The autonomic system may be divided into two parts—the **sympathetic** and the **parasympathetic**—and both parts have afferent and efferent nerve fibers.

### PHYSIOLOGIC NOTE

### Functions of the Autonomic Nervous System

The activities of the sympathetic part of the autonomic system prepare the body for an emergency. It accelerates the heart rate, causes constriction of the peripheral blood vessels, and raises the blood pressure. The sympathetic part of the autonomic system brings about a redistribution of the blood so that it leaves the areas of the skin and intestine and becomes available to the brain, heart, and skeletal muscle. At the same time, it inhibits peristalsis of the intestinal tract and closes the sphincters.

The activities of the parasympathetic part of the autonomic system aim at conserving and restoring energy. They slow the heart rate, increase peristalsis of the intestine and glandular activity, and open the sphincters.

The hypothalamus of the brain controls the autonomic nervous system and integrates the activities of the autonomic and neuroendocrine systems, thus preserving homeostasis in the body.

### Sympathetic System

#### Efferent Fibers

The gray matter of the spinal cord, from the first thoracic segment to the second lumbar segment, possesses a lateral horn, or column, in which are located the cell bodies of the sympathetic connector neurons (Fig. 1-20). The myelinated axons of these cells leave the spinal cord in the anterior nerve roots and then pass via the **white rami communicantes** to the **paravertebral ganglia** of the **sympathetic trunk** (Figs. 1-18, 1-20, and 1-21). The connector cell fibers are called preganglionic as they pass to a peripheral ganglion. Once the preganglionic fibers reach the ganglia in the sympathetic trunk, they may pass to the following destinations. (1) They may terminate in the ganglion they have entered by synapsing with an excitor cell in the ganglion (Fig. 1-20). A **synapse** can be defined as the site where two neurons come into close proximity but not into anatomic continuity. The gap between the two neurons is bridged by a neurotransmitter substance, **acetylcholine**. The axons of the excitor neurons leave the ganglion and are nonmyelinated. These postganglionic nerve fibers now pass to the thoracic spinal nerves as **gray rami communicantes** and are distributed in the branches of the spinal nerves to supply the smooth muscle in the walls of blood vessels, the sweat glands, and the arrector pili muscles of the skin.

(2) Those fibers entering the ganglia of the sympathetic trunk high up in the thorax may travel up in the sympathetic trunk to the ganglia in the cervical region, where they synapse with excitor cells (Figs. 1-20 and 1-21). Here, again, the postganglionic nerve fibers leave the sympathetic trunk as gray rami communicantes, and most of them join the cervical spinal nerves. Many of the preganglionic fibers entering the lower part of the sympathetic trunk from the lower thoracic and upper two lumbar segments of the spinal cord travel down to ganglia in the lower lumbar and sacral regions, where they synapse with excitor cells (Fig. 1-21). The postganglionic fibers leave the sympathetic trunk as gray rami communicantes that join the lumbar, sacral, and coccygeal spinal nerves.

(3) The preganglionic fibers may pass through the ganglia on the thoracic part of the sympathetic trunk without synapsing. These myelinated fibers form the three **splanchnic nerves** (Fig. 1-21). The **greater splanchnic nerve** arises from the 5th to the 9th thoracic ganglia, pierces the diaphragm, and synapses with excitor cells in the ganglia of the celiac plexus. The **lesser splanchnic nerve** arises from the 10th and 11th ganglia, pierces the diaphragm, and synapses with excitor cells in the ganglia of the lower part of the celiac plexus. The **lowest splanchnic nerve** (when

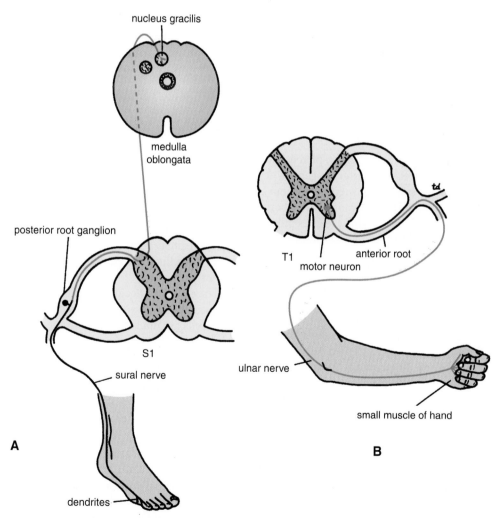

**Figure 1-19** Two neurons that pass from the central to the peripheral nervous system. **A.** Afferent neuron that extends from little toe to brain. **B.** Efferent neuron that extends from the anterior gray horn of the first thoracic segment of spinal cord to the small muscle of the hand.

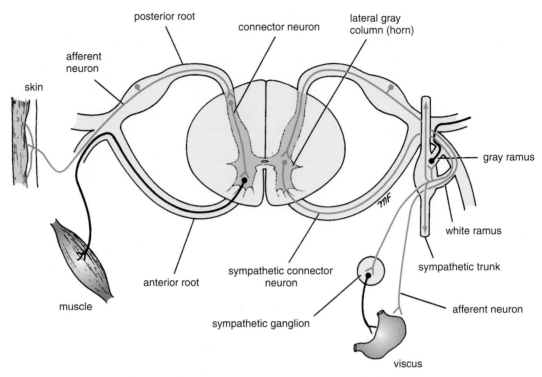

**Figure 1-20**   General arrangement of somatic part of nervous system (left) compared to autonomic part of nervous system (right).

present) arises from the 12th thoracic ganglion, pierces the diaphragm, and synapses with excitor cells in the ganglia of the renal plexus. Therefore, splanchnic nerves are composed of preganglionic fibers. The postganglionic fibers arise from the excitor cells in the peripheral plexuses previously noted and are distributed to the smooth muscle and glands of the viscera. A few preganglionic fibers traveling in the greater splanchnic nerve end directly on the cells of the suprarenal medulla. These medullary cells may be regarded as modified sympathetic excitor cells.

Sympathetic trunks are two ganglionated nerve trunks that extend the whole length of the vertebral column (Fig. 1-21). There are 3 ganglia in each trunk of the neck, 11 or 12 ganglia in the thorax, 4 or 5 ganglia in the lumbar region, and 4 or 5 ganglia in the pelvis. The two trunks lie close to the vertebral column and end below by joining together to form a single ganglion, the **ganglion impar**.

### Afferent Fibers

The afferent myelinated nerve fibers travel from the viscera through the sympathetic ganglia without synapsing (Fig. 1-20). They enter the spinal nerve via the white rami communicantes and reach their cell bodies in the posterior root ganglion of the corresponding spinal nerve. The central axons then enter the spinal cord and may form the afferent component of a local reflex arc. Others may pass up to higher autonomic centers in the brain.

### Parasympathetic System

### Efferent Fibers

The connector cells of this part of the system are located in the brain and the sacral segments of the spinal cord (Fig. 1-21). Those in the brain form parts of the nuclei of origin of cranial nerves III, VII, IX, and X, and the axons emerge from the brain contained in the corresponding cranial nerves.

The sacral connector cells are found in the gray matter of the second, third, and fourth sacral segments of the cord. These cells are not sufficiently numerous to form a lateral gray horn, as do the sympathetic connector cells in the thoracolumbar region. The myelinated axons leave the spinal cord in the anterior nerve roots of the corresponding spinal nerves. They then leave the sacral nerves and form the **pelvic splanchnic nerves**.

All of the efferent fibers described so far are preganglionic, and they synapse with excitor cells in peripheral ganglia, which are usually situated close to the viscera they innervate. The cranial preganglionic fibers relay in the **ciliary, pterygopalatine, submandibular,** and **otic ganglia** (Fig. 1-21). The preganglionic fibers in the pelvic splanchnic nerves relay in ganglia in the hypogastric plexuses or in the walls of the viscera. Characteristically, the postganglionic fibers are nonmyelinated and are relatively short compared with sympathetic postganglionic fibers.

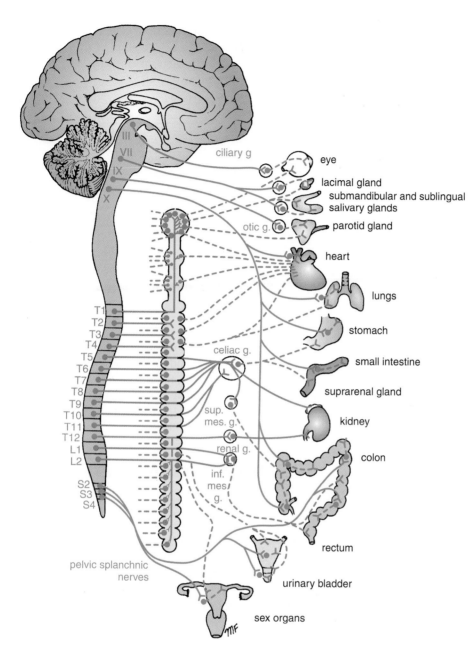

**Figure 1-21** Efferent part of autonomic nervous system. Preganglionic parasympathetic fibers are shown in solid blue; postganglionic parasympathetic fibers are shown in interrupted blue. Preganglionic sympathetic fibers are shown in solid red; postganglionic sympathetic fibers are shown in interrupted red.

### Afferent Fibers

The afferent myelinated fibers travel from the viscera to their cell bodies located either in the sensory ganglia of the cranial nerves or in the posterior root ganglia of the sacrospinal nerves. The central axons then enter the central nervous system and take part in the formation of local reflex arcs or pass to higher centers of the autonomic nervous system.

**PHYSIOLOGIC NOTE**

### Afferent Component of the Autonomic Nervous System

The afferent component of the autonomic system is identical to the afferent component of somatic nerves and forms part of the general afferent segment of the entire nervous system. The nerve endings in the autonomic afferent component may not be activated by such sensations as heat or touch but instead by stretch or lack of oxygen. Once the afferent fibers gain entrance to the spinal cord or brain, they are thought to travel alongside, or are mixed with, the somatic afferent fibers.

## Mucous Membranes

A mucous membrane is the lining of organs or passages that communicate with the surface of the body. It consists essentially of a layer of epithelium supported by a layer of connective tissue, the **lamina propria**. Smooth muscle, called the **muscularis mucosa**, is sometimes present in the connective tissue. A mucous membrane may or may not secrete mucus on its surface.

## Serous Membranes

Serous membranes line the cavities of the trunk and are reflected onto the mobile viscera lying within these cavities (Fig. 1-22). They consist of a smooth layer of mesothelium supported by a thin layer of connective tissue. The serous membrane lining the wall of the cavity is referred to as the **parietal layer**, and the membrane covering the viscera is called the **visceral layer**. The narrow, slitlike interval that separates these layers forms the **pleural, pericardial,** and **peritoneal cavities** and contains a small amount of serous liquid, the **serous exudate**.

### PHYSIOLOGIC NOTE

#### Serous Exudate

The serous exudate lubricates the surfaces of the serous membranes and allows the two layers to slide easily on each other.

The mesenteries, omenta, and serous ligaments are described in Chapter 19. The parietal layer of a serous membrane is developed from the somatopleure (inner cell layer of mesoderm) and is richly supplied by spinal nerves. It is therefore sensitive to all common sensations such as touch and pain. The visceral layer is developed from the splanchnopleure (inner cell layer of mesoderm) and is supplied by autonomic nerves. It is insensitive to touch and temperature but very sensitive to stretch.

## Bone

Bone is a living tissue capable of changing its structure as the result of the stresses to which it is subjected. Like other

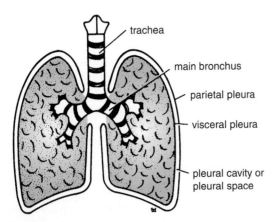

**Figure 1-22** Arrangement of pleura within the thoracic cavity. Note that under normal conditions the pleural cavity is a slitlike space; the parietal and visceral layers of pleura are separated by a small amount of serous fluid.

trachea

main bronchus

parietal pleura

visceral pleura

pleural cavity or pleural space

connective tissues, bone consists of cells, fibers, and matrix. It is hard because of the calcification of its extracellular matrix and possesses a degree of elasticity because of the presence of organic fibers.

### PHYSIOLOGIC NOTE

#### Bone Functions

Bone has a protective function; the skull and vertebral column, for example, protect the brain and spinal cord from injury; the sternum and ribs protect the thoracic and upper abdominal viscera (Fig. 1-23). Bone serves as a lever, as seen in the long bones of the limbs, and as an important storage area for calcium salts. It houses and protects within its cavities the delicate blood-forming bone marrow.

Bone exists in two forms: **compact** and **cancellous**. Compact bone appears as a solid mass; cancellous bone consists of a branching network of **trabeculae** (Fig. 1-24). The trabeculae are arranged in such a manner as to resist the stresses and strains to which the bone is exposed.

### EMBRYOLOGIC NOTE

#### Development of Bone

Bone is developed by two processes: **membranous and endochondral**. In the first process, the bone is developed directly from a connective tissue membrane; in the second process, a cartilaginous model is first laid down and is later replaced by bone. For details of the cellular changes involved, a textbook of histology or embryology should be consulted.

The bones of the vault of the skull are developed rapidly by the membranous method in the embryo, and this serves to protect the underlying developing brain. At birth, small areas of membrane persist between the bones. This is important clinically because it allows the bones a certain amount of mobility, so that the skull can undergo molding during its descent through the female genital passages.

The long bones of the limbs are developed by endochondral ossification, which is a slow process that is not completed until the 18th to 20th year or even later. The center of bone formation found in the shaft of the bone is referred to as the **diaphysis**; the centers at the ends of the bone are referred to as the **epiphyses**. The plate of cartilage at each end, lying between the epiphysis and diaphysis in a growing bone, is called the **epiphyseal plate**. The **metaphysis** is the part of the diaphysis that abuts onto the epiphyseal plate.

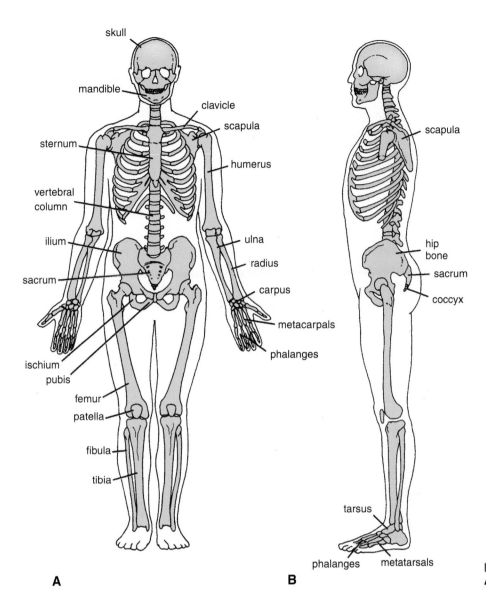

**A**                                    **B**

**Figure 1-23** The skeleton.
**A.** Anterior view. **B.** Lateral view.

## Cartilage

Cartilage is a form of connective tissue in which the cells and fibers are embedded in a gel-like matrix, the latter being responsible for its firmness and resilience. Except on the exposed surfaces in joints, a fibrous membrane called the **perichondrium** covers the cartilage. There are three types of cartilage: **hyaline cartilage**, **fibrocartilage**, and **elastic cartilage**.

# EFFECTS OF SEX, RACE, AND AGE ON STRUCTURE

Descriptive anatomy tends to concentrate on a fixed descriptive form. Medical personnel must always remember that sexual and racial differences exist and that the body's structure and function change as a person grows and ages.

The adult male tends to be taller than the adult female and to have longer legs; his bones are bigger and heavier, and his muscles are larger. He has less subcutaneous fat, which makes his appearance more angular. His larynx is larger and his vocal cords are longer so that his voice is deeper. He has a beard and coarse body hair. He possesses axillary and pubic hair, the latter extending to the region of the umbilicus.

The adult female tends to be shorter than the adult male and to have smaller bones and less bulky muscles. She has more subcutaneous fat and fat accumulations in the breasts, buttocks, and thighs, giving her a more rounded appearance. Her head hair is finer, and her skin is smoother in appearance. She has axillary and pubic hair, but the latter does not extend up to the umbilicus. The adult female has larger breasts and a wider pelvis than the male. She has a wider carrying angle at the elbow, which results in a greater lateral deviation of the forearm on the arm.

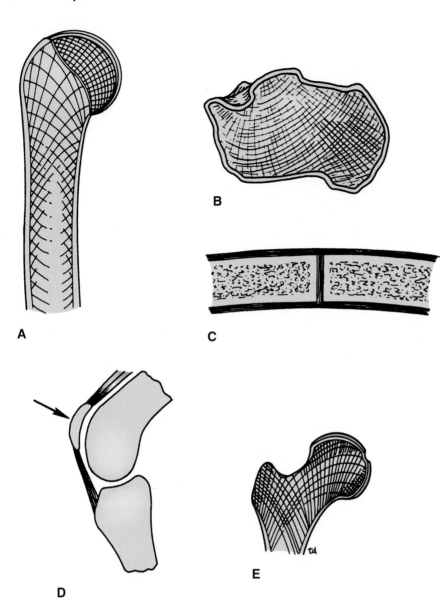

**Figure 1-24** Sections of different types of bones. **A.** Long bone (humerus). **B.** Irregular bone (calcaneum). **C.** Flat bone (two parietal bones separated by the sagittal suture). **D.** Sesamoid bone (patella). **E.** Note arrangement of trabeculae to act as struts to resist both compression and tension forces in the upper end of the femur.

Until the age of approximately 10 years, boys and girls grow at about the same rate. Around 12 years, boys often start to grow faster than girls, so that most males reach a greater adult height than females.

Puberty begins between ages 10 and 14 in girls and between 12 and 15 in boys. In the girl at puberty, the breasts enlarge and the pelvis broadens. At the same time, a boy's penis, testes, and scrotum enlarge; in both sexes, axillary and pubic hair appear.

Racial differences may be seen in the color of the skin, hair, and eyes and in the shape and size of the eyes, nose, and lips. Africans and Scandinavians tend to be tall, as a result of long legs, whereas Asians tend to be short, with short legs. The heads of central Europeans and Asians also tend to be round and broad.

## PHYSIOLOGIC NOTE

### Age and Functional Efficiency

After birth and during childhood, the bodily functions become progressively more efficient, reaching their maximum degree of efficiency during young adulthood. During late adulthood and old age, many bodily functions become less efficient.

## EMBRYOLOGIC NOTE

### Embryology and Clinical Anatomy

Embryology provides a basis for understanding anatomy and an explanation of many of the congenital anomalies that are seen in clinical medicine. A very brief overview of the development of the embryo follows.

Once the ovum has been fertilized by the spermatozoon, a single cell is formed, called the **zygote**. This undergoes a rapid succession of mitotic divisions with the formation of smaller cells. The centrally placed cells are called the **inner cell mass** and ultimately form the tissues of the embryo. The outer cells, called the **outer cell mass**, form the trophoblast, which plays an important role in the formation of the **placenta** and the **fetal membranes**.

The cells that form the embryo become defined in the form of a bilaminar **embryonic disc**, composed of two germ layers. The upper layer is called the **ectoderm**, and the lower layer is called the **entoderm**. As growth proceeds, the embryonic disc becomes pear shaped, and a narrow streak appears on its dorsal surface formed of ectoderm, called the **primitive streak**. The further proliferation of the cells of the primitive streak forms a layer of cells that will extend between the ectoderm and the entoderm to form the third germ layer, called the **mesoderm**.

## Ectoderm

Further thickening of the ectoderm gives rise to a plate of cells on the dorsal surface of the embryo called the **neural plate**. This plate sinks beneath the surface of the embryo to form the **neural tube**, which ultimately gives rise to the **central nervous system**. The remainder of the ectoderm forms the **cornea, retina**, and **lens** of the **eye** and the **membranous labyrinth** of the **inner ear**. The ectoderm also forms the **epidermis** of the **skin**; the **nails** and **hair**; the **epithelial cells** of the **sebaceous, sweat**, and **mammary glands**; the **mucous membrane** lining the **mouth, nasal cavities**, and **paranasal sinuses**; the **enamel** of the **teeth**; the **pituitary gland** and the **alveoli** and **ducts** of the **parotid salivary glands**; the mucous membrane of the lower half of the **anal canal**; and the terminal parts of the **genital tract** and the **male urinary tract**.

## Entoderm

The entoderm eventually gives origin to the following structures: the **epithelial lining** of the **alimentary tract** from the **mouth cavity** down to halfway along the anal canal and the **epithelium** of the glands that develop from it—namely, the **thyroid, parathyroid, thymus, liver**, and **pancreas**—and the epithelial linings of the **respiratory tract, pharyngotympanic tube** and **middle ear, urinary bladder**, parts of the female and male **urethras, greater vestibular glands, prostate gland, bulbourethral glands**, and **vagina**.

## Mesoderm

The mesoderm becomes differentiated into the paraxial, intermediate, and the lateral mesoderms.

### Paraxial Mesoderm

The **paraxial mesoderm** is situated initially on either side of the midline of the embryo. It becomes segmented and forms the **bones, cartilage**, and **ligaments** of the **vertebral column** and part of the base of the **skull**. The lateral cells form the **skeletal muscles** of their own segment, and some of the cells migrate beneath the ectoderm and take part in the formation of the **dermis** and **subcutaneous tissues** of the skin.

### Intermediate Mesoderm

The **intermediate mesoderm** is a column of cells on either side of the embryo that is connected medially to the paraxial mesoderm and laterally to the lateral mesoderm. It gives rise to portions of the **urogenital system**.

### Lateral Mesoderm

The **lateral mesoderm** splits into a **somatic layer** and a **splanchnic layer** associated with the ectoderm and the entoderm, respectively. It encloses a cavity within the embryo called the **intraembryonic coelom**. The coelom eventually forms the **pericardial, pleural**, and **peritoneal cavities.**

The embryonic mesoderm, in addition, gives origin to **smooth, voluntary**, and **cardiac muscle**; all forms of **connective tissue**, including cartilage and bone; **blood vessel walls** and **blood cells; lymph vessel walls** and **lymphoid tissue**; the **synovial membranes** of **joints** and **bursae**; and the **suprarenal cortex**.

When appropriate, a more detailed account of the development of different organs is given in the chapters to follow.

# RADIOGRAPHIC ANATOMY

As a medical professional, you will be frequently called on to study normal and abnormal anatomy as seen on radiographs. Familiarity with normal radiographic anatomy permits one to recognize abnormalities quickly, such as fractures and tumors.

The most common form of radiographic anatomy is studied on a **radiograph** (x-ray film), which provides a two-dimensional image of the interior of the body (Fig. 1-25). To produce such a radiograph, a single barrage of x-rays is passed through the body and exposed to the film. Tissues of differing densities show up as images of differing densities on the radiograph (or fluorescent screen). A tissue that is relatively dense absorbs (stops) more x-rays than tissues that are less dense. A very dense tissue is said to be **radiopaque**, but a less dense tissue is said to be **radiolucent**. Bone is very dense, and fat is moderately dense; other soft tissues are less dense.

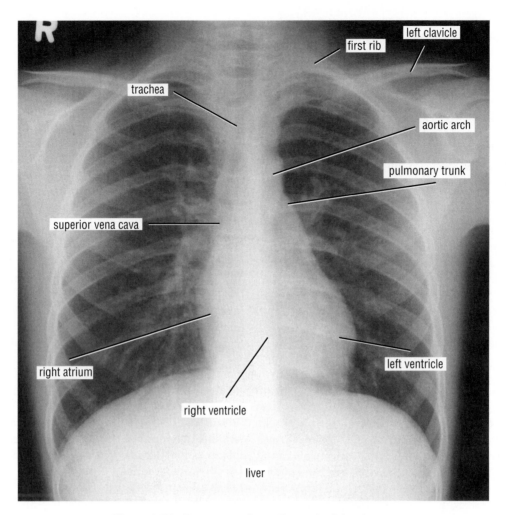

**Figure 1-25**    Posteroanterior radiograph of the thorax.

Unfortunately, an ordinary radiograph shows the images of the different organs superimposed onto a flat sheet of film. This overlap of organs and tissues often makes it difficult to visualize them. This problem is overcome to some extent by taking films at right angles to one another or by making stereoscopic films.

# Computed Tomography

**Computed tomography** (CT) scanning or **computerized axial tomography** (CAT) scanning permits the study of tissue slices so that tissues with minor differences in density can be recognized. CT scanning relies on the same physics as conventional x-rays but combines it with computer technology. A source of x-rays moves in an arc around the part of the body being studied and sends out a beam of x-rays. The x-rays, having passed through the region of the body, are collected by a special x-ray detector. Here, the x-rays are converted into electronic impulses that produce readings of the density of the tissue in a 1-cm slice of the body. From these readings, the computer is able to assemble a picture of the body called a **CT scan**, which can be viewed on a fluorescent screen and then photographed for later examination (Fig. 1-26). The procedure is safe and quick, lasts only a few seconds for each slice, and, for most patients, requires no sedation.

# Magnetic Resonance Imaging

**Magnetic resonance imaging** (MRI) is a technique that uses the magnetic properties of the hydrogen nucleus excited by radiofrequency radiation transmitted by a coil surrounding the body part. The excited hydrogen nuclei emit a signal that is detected as induced electric currents in a receiver coil. MRI is absolutely safe to the patient, and because it provides better differentiation between different soft tissues, its use can be more revealing than a CT scan. The reason for this is that some tissues contain more hydrogen in the form of water than other tissues (Fig. 1-27).

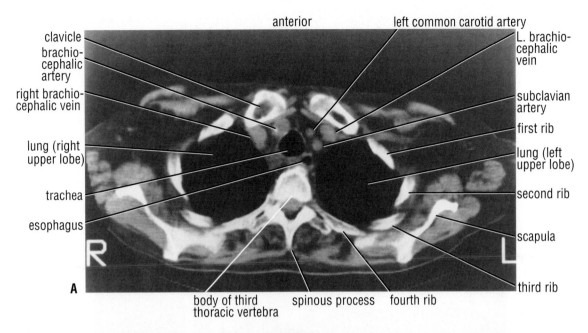

anterior — left common carotid artery

clavicle
brachio-cephalic artery
right brachio-cephalic vein
lung (right upper lobe)
trachea
esophagus

L. brachio-cephalic vein
subclavian artery
first rib
lung (left upper lobe)
second rib
scapula
third rib

R  L

**A**

body of third thoracic vertebra — spinous process — fourth rib

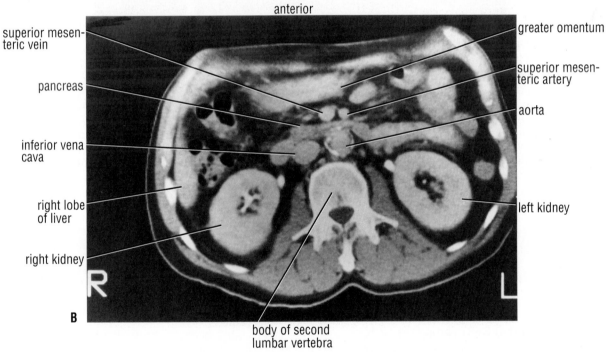

anterior

superior mesen-teric vein
pancreas
inferior vena cava
right lobe of liver
right kidney

greater omentum
superior mesen-teric artery
aorta
left kidney

R  L

**B**

body of second lumbar vertebra

**Figure 1-26** CT scans. **A.** The upper thorax at the level of the third thoracic vertebra. **B.** The upper abdomen at the level of the second lumbar vertebra. All CT scans are viewed from below. Thus, the right side of the body appears on the left side of the figure.

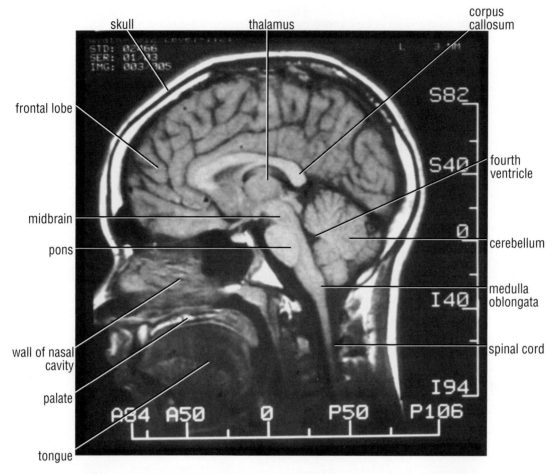

**Figure 1-27**    MRI study of the head in a sagittal plane showing different parts of the brain.

# Review Questions

## Completion Questions

**Select the phrase that BEST completes each statement.**

1. A patient who is standing in the anatomic position is
   A. facing laterally.
   B. has the palms of the hands directed medially.
   C. has the ankles several inches apart.
   D. is standing on his or her toes.
   E. has the upper limbs by the sides of the trunk.

2. A patient is performing the movement of flexion of the hip joint when she
   A. moves the lower limb away from the midline in the coronal plane.
   B. moves the lower limb posteriorly in the paramedian plane.
   C. moves the lower limb anteriorly in the paramedian plane.
   D. rotates the lower limb so that the anterior surface faces medially.
   E. moves the lower limb toward the median sagittal plane.

## Matching Questions

**Match each structure listed below with a structure or occurrence with which it is most closely associated. Each lettered answer may be used more than once.**

3. Superficial fascia

4. Deep fascia

5. Skeletal muscle
   A. Divides up interior of limbs into compartments
   B. Adipose tissue
   C. Voluntary movement
   D. None of the above

**For each joint listed below, indicate with which type of movement it is associated.**

6. Elbow joint

7. Temporomandibular joint

8. Hip joint
   A. Flexion
   B. Extension
   C. Both A and B
   D. Protraction
   E. Flexion, extension, and abduction.

**For each joint listed below, give the most appropriate classification.**

9. Joints between vertebral bodies

10. Inferior tibiofibular joint

11. Sutures between bones of vault of skull

12. Knee joint
    A. Synovial joint
    B. Cartilaginous
    C. Fibrous
    D. None of the above

**For each type of blood vessel listed below, select an appropriate definition.**

13. Arteriole

14. Portal vein

15. Anatomic end artery

16. Venule
    A. A vessel that connects two capillary beds
    B. A vessel whose terminal branches do not anastomose with branches or arteries supplying adjacent areas
    C. A vessel that connects large veins to capillaries
    D. An artery <0.1 mm in diameter

E. A thin-walled vessel that has an irregular cross diameter

**For each of the lymphatic structures listed below, select an appropriate structure or function.**

17. Lymph capillary

18. Thoracic duct

19. Right lymphatic duct

20. Lymph node
    A. Present in the central nervous system
    B. Drains lymph directly from the tissues
    C. Contains lymphatic tissue and has both afferent and efferent vessels
    D. Drains lymph from the right side of the head and neck, the right upper limb, and the right side of the thorax
    E. Drains lymph from the right side of the abdomen

## Multiple Choice Questions

**Read the case histories and select the BEST answer to the question following them.**

The surgical notes of a patient state that she had a right infraumbilical paramedian incision through the skin of the anterior abdominal wall.

21. Where exactly was this incision made?
    A. In the midline below the umbilicus
    B. In the midline above the umbilicus
    C. To the right of the midline above the umbilicus
    D. To the right of the midline below the umbilicus
    E. Just below the xiphoid process in the midline

After an attack of pericapsulitis of the left shoulder joint, a patient finds that a particular movement of the joint is restricted.

22. Which of the joint movements is restricted and by how much?
    A. Abduction is limited to 30°.
    B. Lateral rotation is limited to 45°.
    C. Medial rotation is limited to 55°.
    D. Flexion is limited to 90°.
    E. Extension is limited to 45°.

# Answers and Explanations

1. **E** is correct. The patient is standing erect, with the upper limbs by the sides and the face and palms of the hands directed forward (Fig. 1-1).

2. **C** is correct. The patient is performing the movement of flexion of the hip joint when she moves the lower limb anteriorly in the paramedian plane.

3. **B** is correct. Superficial fascia, or subcutaneous tissue, is a mixture of loose areolar and adipose tissue that unites the dermis of the skin to the underlying deep fascia (Fig. 1-6).

4. **A** is correct. Deep fascia is a membranous layer of connective tissue that invests the muscles and other deep structures (Fig. 1-6). Fibrous septa extend from the membrane between groups of muscles and may divide the interior of the limbs into compartments.

5. **C** is correct. Skeletal or voluntary muscles produce the movements of the skeleton.

6. **C** is correct. The elbow joint can perform the movement of flexion and extension (Fig. 1-2).

7. **D** is correct. The temporomandibular joint can perform the movement of protraction in which the jaw moves forward on the temporal bone

8. **E** is correct. The hip joint can perform the movements of flexion, extension, medial rotation and lateral rotation, and adduction and abduction (Fig. 1-2), as well as circumduction.

9. **B** is correct. The joints between vertebral bodies are cartilaginous (Fig. 1-10).

10. **C** is correct. The joint between the lower end of the tibia and fibular bones (inferior tibiofibular joint) is fibrous.

11. **C** is correct. The sutures (joints) between the bones of the vault of the skull are fibrous (Fig. 1-10).

12. **A** is correct. The knee joint is a synovial joint.

13. **D** is correct. An arteriole is a small artery that is less than 0.1 mm in diameter.

14. **A** is correct. The portal vein is a blood vessel that connects two capillary beds.

15. **B** is correct. An anatomic end artery is a blood vessel whose terminal branches do not anastomose with branches or arteries supplying adjacent areas (Fig. 1-14).

16. **C** is correct. A venule is a vessel that connects large veins to capillaries (Fig. 1-14).

17. **B** is correct. A lymph capillary drains lymph directly from the tissues.

18. **E** is correct. The thoracic duct drains lymph from the right side of the abdomen and, in fact, drains lymph from the entire body, except the right side of the head and neck, the right upper limb, and the right side of the thorax, which are drained by the right lymphatic duct (Fig. 1-15).

19. **D** is correct. The right lymphatic duct drains lymph from the right side of the head and neck, the right upper limb, and the right side of the thorax (Fig. 1-15).

20. **C** is correct. A lymph node contains lymphatic tissue and has both afferent and efferent lymph vessels.

21. **D** is correct. A right infraumbilical paramedian incision through the skin of the anterior abdominal wall is located to the right of the midline below the umbilicus.

22. **A** is correct. Patients with pericapsulitis of the shoulder joint experience limitation of the movement of abduction, sometimes to as little as 30°.

# The Respiratory System

# 2

# The Upper and Lower Airway and Associated Structures

All clinical material relevant to this chapter can be found on the CD-ROM.

# Chapter Outline

No medical emergency quite produces the drama, urgency, and anxiety of the compromised airway. The medical professional not only has to make a rapid diagnosis but also has to institute almost immediate treatment. All the techniques of airway management, from manual manipulation to endotracheal intubation to cricothyroidotomy, require a detailed knowledge of anatomy.

Epistaxis and nasal lacerations and diseases of the paranasal sinuses and the salivary glands may also have to be considered. Since anatomic structures change as one matures from infancy to childhood to adulthood, it is necessary that these changes be recognized.

 # BASIC ANATOMY

The airway extends from the nostrils of the nose and the lips of the mouth to the alveoli of the lungs.

# The Nose

The nose consists of the external nose and the nasal cavity, both of which are divided by a septum into right and left halves.

## External Nose

The external nose has two elliptical orifices called the **nostrils,** which are separated from each other by the **nasal septum** (Fig. 2-1). The lateral margin, the **ala nasi,** is rounded and mobile.

The framework of the external nose is made up above by the nasal bones, the frontal processes of the maxillae, and the nasal part of the frontal bone. Below, the framework is formed of plates of hyaline cartilage (Fig. 2-1).

## Blood Supply of the External Nose

The skin of the external nose is supplied by branches of the ophthalmic and the maxillary arteries. The skin of the ala and the lower part of the septum are supplied by branches from the facial artery.

## Sensory Nerve Supply of the External Nose

The infratrochlear and external nasal branches of the ophthalmic nerve (V CN) and the infraorbital branch of the maxillary nerve (V CN) supply the external nose.

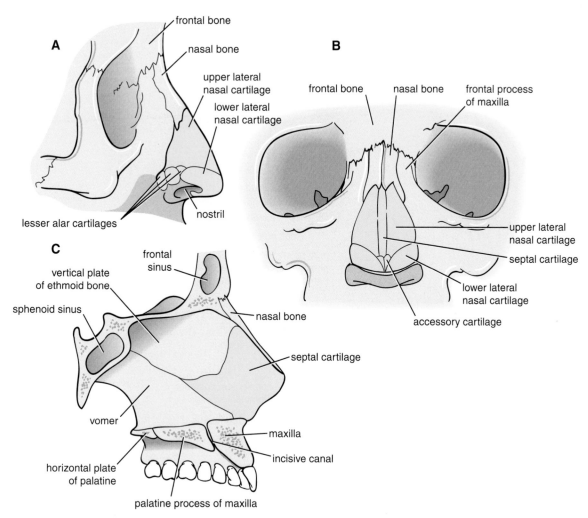

**Figure 2-1** External nose and nasal septum. **A.** Lateral view of bony and cartilaginous skeleton of external nose. **B.** Anterior view of bony and cartilaginous skeleton of external nose. **C.** Bony and cartilaginous skeleton of nasal septum.

# Nasal Cavity

The nasal cavity extends from the nostrils in front to the **posterior nasal apertures** or **choanae** behind, where the nose opens into the nasopharynx. **The nasal vestibule is the area** of the nasal cavity lying just inside the nostril (Fig. 2-2). The nasal cavity is divided into right and left halves by the **nasal septum** (Fig. 2-1). The septum is made up of the **septal cartilage**, the **vertical plate of the ethmoid**, and the **vomer**.

## Walls of the Nasal Cavity

Each half of the nasal cavity has a floor, a roof, a lateral wall, and a medial or septal wall.

### Floor

The floor is the palatine process of the maxilla and the horizontal plate of the palatine bone (Fig. 2-1).

### Roof

The roof is narrow and is formed anteriorly beneath the bridge of the nose by the nasal and frontal bones, in the middle by the cribriform plate of the ethmoid, located beneath the anterior cranial fossa, and posteriorly by the downward sloping body of the sphenoid (Fig. 2-2).

### Lateral Wall

The lateral wall has three projections of bone called the **superior, middle,** and **inferior nasal conchae** (Fig. 2-2). The space below each concha is called a **meatus.**

#### Sphenoethmoidal Recess

The sphenoethmoidal recess is a small area above the superior concha. It receives the opening of the **sphenoid air sinus.**

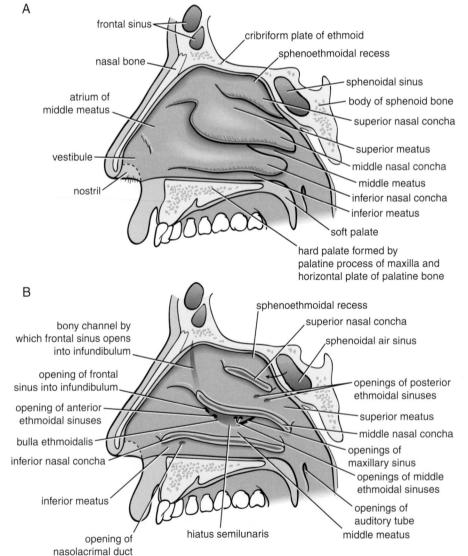

**A**

frontal sinus

nasal bone

atrium of middle meatus

vestibule

nostril

cribriform plate of ethmoid

sphenoethmoidal recess

sphenoidal sinus

body of sphenoid bone

superior nasal concha

superior meatus

middle nasal concha

middle meatus

inferior nasal concha

inferior meatus

soft palate

hard palate formed by palatine process of maxilla and horizontal plate of palatine bone

**B**

bony channel by which frontal sinus opens into infundibulum

opening of frontal sinus into infundibulum

opening of anterior ethmoidal sinuses

bulla ethmoidalis

inferior nasal concha

inferior meatus

opening of nasolacrimal duct

hiatus semilunaris

sphenoethmoidal recess

superior nasal concha

sphenoidal air sinus

openings of posterior ethmoidal sinuses

superior meatus

middle nasal concha

openings of maxillary sinus

openings of middle ethmoidal sinuses

openings of auditory tube

middle meatus

**Figure 2-2  A.** Lateral wall of the right nasal cavity. **B.** Lateral wall of the right nasal cavity; the superior, middle, and inferior conchae have been partially removed to show openings of the paranasal sinuses and the nasolacrimal duct into the meati.

*Superior Meatus*

The superior meatus lies below the superior concha (Fig. 2-2). It receives the openings of the **posterior ethmoid sinuses**.

*Middle Meatus*

The middle meatus lies below the middle concha. It has a rounded swelling called the **bulla ethmoidalis** that is formed by the **middle ethmoidal air sinuses**, which open on its upper border. A curved opening, the **hiatus semilunaris,** lies just below the bulla (Fig. 2-2). The anterior end of the hiatus leads into a funnel-shaped channel called the **infundibulum,** which is continuous with the **frontal sinus**. The **maxillary sinus** opens into the middle meatus through the hiatus semilunaris.

*Inferior Meatus*

The inferior meatus lies below the inferior concha and receives the opening of the lower end of the **nasolacrimal duct,** which is guarded by a fold of mucous membrane (Fig. 2-2).

**Medial Wall**

The medial wall is formed by the nasal septum. The upper part is formed by the vertical plate of the ethmoid and the vomer (Fig. 2-1). The anterior part is formed by the septal cartilage. The septum rarely lies in the midline, thus increasing the size of one half of the nasal cavity and decreasing the size of the other half.

## Mucous Membrane of the Nasal Cavity

The vestibule is lined with modified skin and has coarse hairs. The area above the superior concha is lined with olfactory mucous membrane and contains nerve endings sensitive to the reception of smell. The lower part of the nasal cavity is lined with respiratory mucous membrane. A large plexus of veins in the submucous connective tissue is present in the respiratory region.

**P H Y S I O L O G I C   N O T E**

**Function of Warm Blood and Mucus of Mucous Membrane**

The presence of warm blood in the venous plexuses serves to heat up the inspired air as it enters the respiratory system. The presence of mucus on the surfaces of the conchae traps foreign particles and organisms in the inspired air, which are then swallowed and destroyed by gastric acid.

## Nerve Supply of the Nasal Cavity

The olfactory nerves from the olfactory mucous membrane ascend through the cribriform plate of the ethmoid bone to the olfactory bulbs (Fig. 2-3). The nerves of ordinary sensation

are branches of the ophthalmic division (V1) and the maxillary division (V2) of the trigeminal nerve (Fig. 2-3).

## Blood Supply to the Nasal Cavity

The arterial supply to the nasal cavity is from branches of the maxillary artery, one of the terminal branches of the external carotid artery. The most important branch is the sphenopalatine artery (Fig. 2-4). The sphenopalatine artery anastomoses with the septal branch of the superior labial branch of the facial artery in the region of the vestibule. The submucous venous plexus is drained by veins that accompany the arteries.

## Lymph Drainage of the Nasal Cavity

The lymph vessels draining the vestibule end in the submandibular nodes. The remainder of the nasal cavity is drained by vessels that pass to the upper deep cervical nodes.

**E M B R Y O L O G I C   N O T E**

**Development of the Nose**

The roof of the nose is formed from the lateral nasal processes, from which the lateral walls are also formed, with the assistance of the maxillary processes (Fig. 2-5). The anterior openings of the nose begin as olfactory pits in the frontonasal process. Each olfactory pit is bounded medially by the medial nasal process, laterally by the lateral nasal process, and inferiorly by the maxillary process. As these processes fuse, the olfactory pits become deeper and form well-defined blind sacs, the opening into each of which is the nostril.

The floor of the nose at first is very short and consists of the medial nasal process and the anterior part of the maxillary process on each side. At this stage, the floors of the olfactory pits rupture so that the nasal cavities communicate with the developing mouth (Fig. 2-6). Meanwhile, the nasal septum is forming as a downgrowth from the medial nasal process (Fig. 2-6). Later, the palatal processes of the maxilla grow medially and fuse with each other and with the nasal septum, thus completing the floor of the nose. Therefore, each nasal cavity communicates anteriorly with the exterior through the nostril and posteriorly through the choana with the nasopharynx. In the early stages of development, the nose is a much flattened structure and gains its recognizable form only after the facial development is complete.

# The Paranasal Sinuses

The paranasal sinuses are cavities found in the interior of the maxilla, frontal, sphenoid, and ethmoid bones

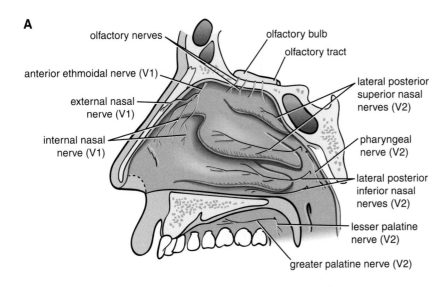

**A**

olfactory nerves

olfactory bulb

olfactory tract

anterior ethmoidal nerve (V1)

external nasal nerve (V1)

internal nasal nerve (V1)

lateral posterior superior nasal nerves (V2)

pharyngeal nerve (V2)

lateral posterior inferior nasal nerves (V2)

lesser palatine nerve (V2)

greater palatine nerve (V2)

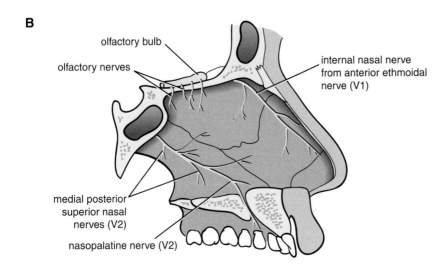

**B**

olfactory bulb

olfactory nerves

internal nasal nerve from anterior ethmoidal nerve (V1)

medial posterior superior nasal nerves (V2)

nasopalatine nerve (V2)

**Figure 2-3** **A.** Lateral wall of nasal cavity showing sensory innervation of mucous membrane. **B.** Nasal septum showing sensory innervation of mucous membrane.

(Fig. 2-7). They are lined with mucoperiosteum and filled with air; they communicate with the nasal cavity through relatively small apertures. The maxillary and sphenoidal sinuses are present in a rudimentary form at birth; they enlarge appreciably after the 8th year and become fully formed in adolescence.

---

### PHYSIOLOGIC NOTE

#### Drainage of Mucus and Function of the Paranasal Sinuses

The mucus produced by the mucous membrane is moved into the nose by ciliary action of the columnar cells.

Drainage of the mucus is also achieved by the siphon action created during the blowing of the nose. The function of the sinuses is to act as resonators to the voice; they also reduce the weight of the skull. When the apertures of the sinuses are blocked, or they become filled with fluid, the quality of the voice is markedly changed.

### Maxillary Sinus

The maxillary sinus is pyramidal in shape and located within the body of the maxilla behind the skin of the cheek (Fig. 2-7). The roof is formed by the floor of the orbit, and the floor is related to the roots of the premolars and molar

**A**

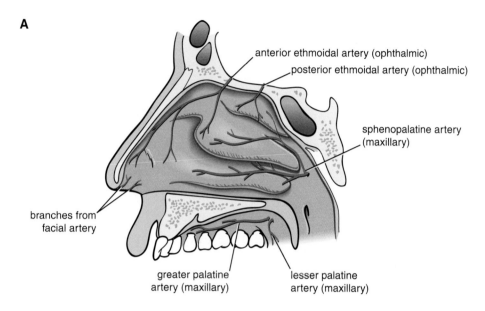

anterior ethmoidal artery (ophthalmic)

posterior ethmoidal artery (ophthalmic)

sphenopalatine artery (maxillary)

branches from facial artery

greater palatine artery (maxillary)

lesser palatine artery (maxillary)

**B**

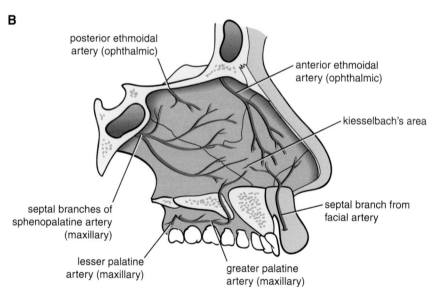

posterior ethmoidal artery (ophthalmic)

anterior ethmoidal artery (ophthalmic)

kiesselbach's area

septal branch from facial artery

septal branches of sphenopalatine artery (maxillary)

lesser palatine artery (maxillary)

greater palatine artery (maxillary)

**Figure 2-4** **A.** Lateral wall of nasal cavity showing the arterial supply of the mucous membrane. **B.** Nasal septum showing the arterial supply of the mucous membrane.

teeth. The maxillary sinus opens into the middle meatus of the nose through the hiatus semilunaris (Fig. 2-7).

## Frontal Sinuses

The two frontal sinuses are contained within the frontal bone (Fig. 2-7). They are separated from each other by a bony septum. Each sinus is roughly triangular, extending upward above the medial end of the eyebrow and backward into the medial part of the roof of the orbit.

Each frontal sinus opens into the middle meatus of the nose through the infundibulum (Fig. 2-7).

## Sphenoidal Sinuses

The two sphenoidal sinuses lie within the body of the sphenoid bone (Fig. 2-7). Each sinus opens into the sphenoethmoidal recess above the superior concha.

## Ethmoid Sinuses

The ethmoidal sinuses are anterior, middle, and posterior, and they are contained within the ethmoid bone, between the nose and the orbit (Fig. 2-7). They are separated from the latter by a thin plate of bone so that infection can readily spread from

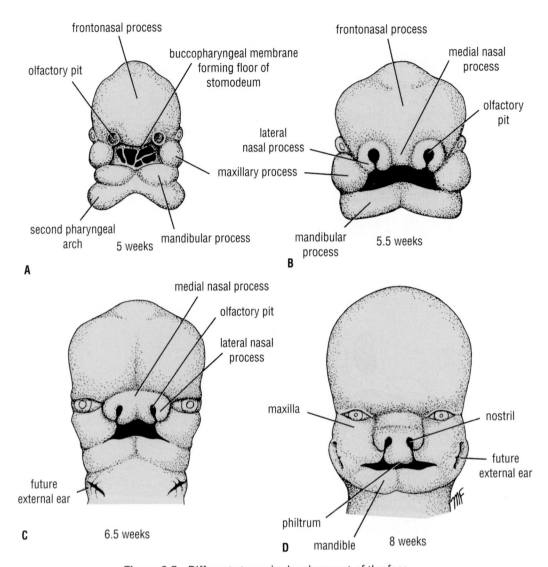

**Figure 2-5**   Different stages in development of the face.

the sinuses into the orbit. The anterior sinuses open into the infundibulum; the middle sinuses open into the middle meatus, on or above the bulla ethmoidalis; and the posterior sinuses open into the superior meatus. The various sinuses and their openings into the nose are summarized in Table 2-1.

# The Mouth

## The Lips

The lips are two fleshy folds that surround the oral orifice (Fig. 2-8). They are covered on the outside by skin and are lined on the inside by mucous membrane. The substance of the lips is made up by the orbicularis oris muscle and the muscles that radiate from the lips into the face (Fig. 2-9). Also included are the labial blood vessels and nerves, connective tissue, and many small salivary glands. The philtrum is the shallow vertical groove seen in the midline on the outer surface of the upper lip. Median folds of mucous membrane—the labial frenulae—connect the inner surface of the lips to the gums.

## The Mouth Cavity

The mouth extends from the lips to the pharynx. The entrance into the pharynx, the **oropharyngeal isthmus**, is formed on each side by the palatoglossal fold (Fig. 2-8).

The mouth is divided into the vestibule and the mouth cavity proper.

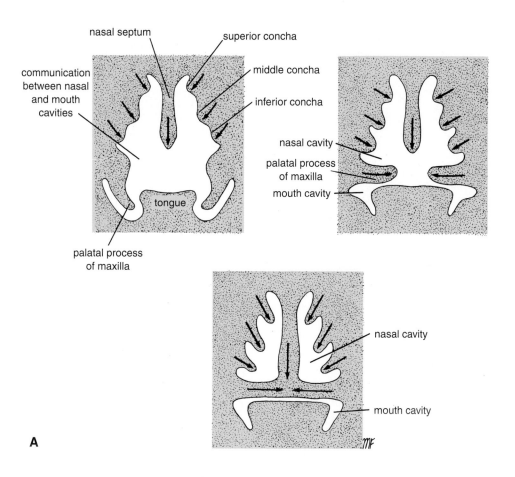

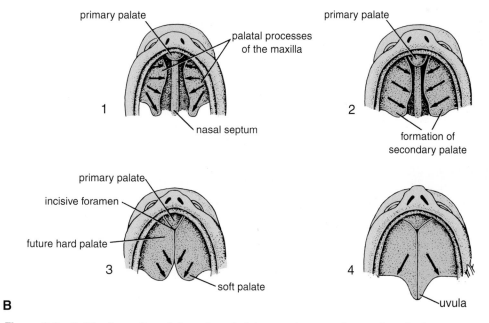

**Figure 2-6   A.** The formation of the palate and the nasal septum (coronal section).
**B.** The different stages in the formation of the palate.

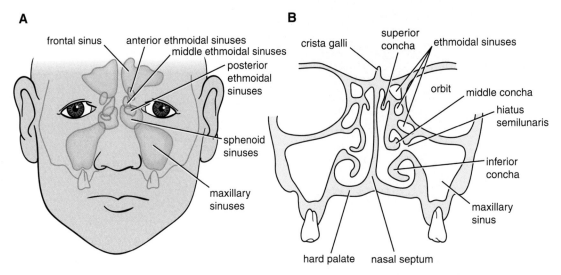

**Figure 2-7**   **A.** The position of the paranasal sinuses in relation to the face. **B.** Coronal section through the nasal cavity showing the ethmoidal and the maxillary sinuses.

## Vestibule

The vestibule lies between the lips and the cheeks externally and the gums and the teeth internally. This slit-like space communicates with the exterior through the oral fissure between the lips. When the jaws are closed, it communicates with the mouth proper behind the third molar tooth on each side. The vestibule is limited above and below by the reflection of the mucous membrane from the lips and cheeks to the gums.

The lateral wall of the vestibule is formed by the cheek, which is made up by the buccinator muscle and lined with mucous membrane. The tone of the buccinator muscle and that of the muscles of the lips keeps the walls of the vestibule in contact with one another. The **duct of the parotid salivary gland** opens on a small papilla into the vestibule opposite the upper second molar tooth (Fig. 2-8).

## Mouth Proper

The mouth proper has a roof and a floor.

## Roof of Mouth

The roof of the mouth is formed by the hard palate in front and the soft palate behind (Fig. 2-8).

## Floor of Mouth

The floor is formed largely by the anterior two thirds of the tongue and by the reflection of the mucous membrane from

| Table 2-1 | Paranasal Sinuses and Their Site of Drainage into the Nose[a] |
|---|---|
| **Sinus** | **Site of Drainage** |
| Maxillary sinus | Middle meatus through hiatus semilunaris |
| Frontal sinuses | Middle meatus via infundibulum |
| Sphenoidal sinuses | Sphenoethmoidal recess |
| Ethmoidal sinuses | |
|    Anterior group | Infundibulum and into middle meatus |
|    Middle group | Middle meatus on or above bulla ethmoidalis |
|    Posterior group | Superior meatus |

[a]Note that maxillary and sphenoidal sinuses are present in rudimentary form at birth, enlarge appreciably after the 8th year, and are fully formed in adolescence.

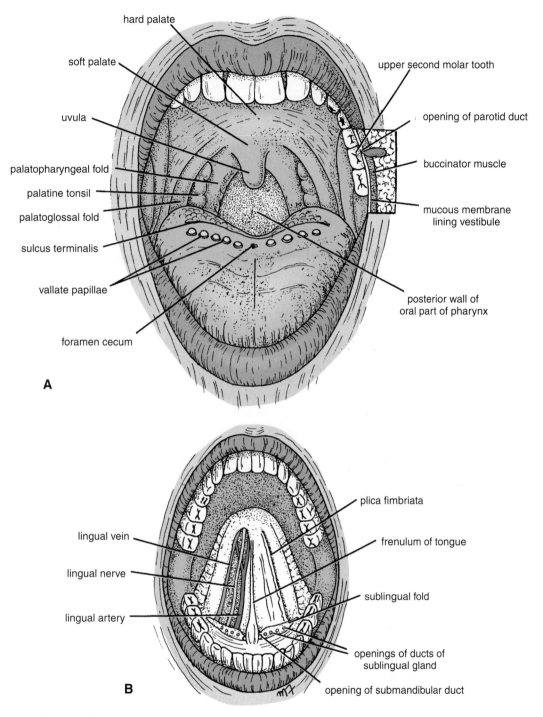

**Figure 2-8** **A.** Cavity of the mouth. Cheek on the left side of the face has been cut away to show the buccinator muscle and the parotid duct. **B.** Undersurface of the tongue.

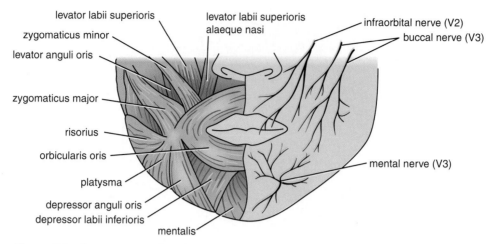

**Figure 2-9**   Arrangement of the facial muscles around the lips; the sensory nerve supply of the lips is shown.

the sides of the tongue to the gum of the mandible. A fold of mucous membrane called the **frenulum of the tongue** connects the undersurface of the tongue in the midline to the floor of the mouth (Fig. 2-8). Lateral to the frenulum, the mucous membrane forms a fringed fold, the **plica fimbriata** (Fig. 2-8).

The submandibular duct of the submandibular gland opens onto the floor of the mouth on the summit of a small papilla on either side of the frenulum of the tongue (Fig. 2-8). The sublingual gland projects up into the mouth, producing a low fold of mucous membrane, the sublingual fold. Numerous ducts of the gland open on the summit of the fold.

## Mucous Membrane of the Mouth

In the vestibule, the mucous membrane is tethered to the buccinator muscle by elastic fibers in the submucosa that prevent redundant folds of mucous membrane from being bitten between the teeth when the jaws are closed. The mucous membrane of the gingiva, or gum, is strongly attached to the alveolar periosteum.

## Sensory Innervation of the Mouth

**Roof:** The greater palatine and nasopalatine nerves (Fig. 2-10) from the maxillary division of the trigeminal nerve.

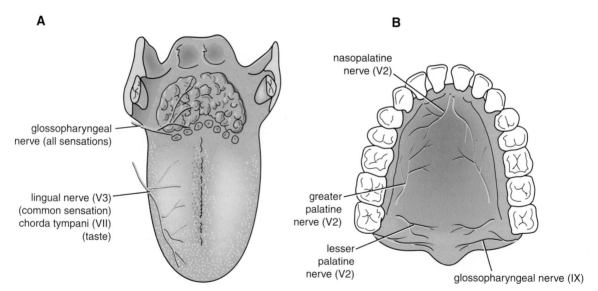

**Figure 2-10   A.** Sensory nerve supply to the mucous membrane of the tongue. **B.** Sensory nerve supply to the mucous membrane of the hard and soft palate; taste fibers run with branches of the maxillary nerve (V2) and join the greater petrosal branch of the facial nerve.

**Floor:** The lingual nerve (common sensation), a branch of the mandibular division of the trigeminal nerve. The taste fibers travel in the chorda tympani nerve, a branch of the facial nerve.

**Cheek:** The buccal nerve, a branch of the mandibular division of the trigeminal nerve (the buccinator muscle is innervated by the buccal branch of the facial nerve).

---

### E M B R Y O L O G I C   N O T E

### Development of the Mouth

The cavity of the mouth is formed from two sources: a depression from the exterior, called the **stomodeum,** which is lined with ectoderm, and a part immediately posterior to the stomodeum, derived from the cephalic end of the foregut and lined with entoderm. These two parts at first are separated by the **buccopharyngeal membrane,** but this breaks down and disappears during the 3rd week of development (Fig. 2-11). If this membrane were to persist into adult life, it would occupy an imaginary plane extending obliquely from the region of

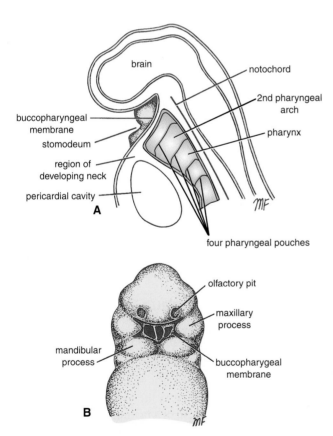

**Figure 2-11   A.** Sagittal section of the embryo showing the position of the buccopharyngeal membrane. **B.** The face of the developing embryo showing the buccopharyngeal membrane breaking down.

the body of the sphenoid, through the soft palate, and down to the inner surface of the mandible inferior to the incisor teeth. This means that the structures that are situated in the mouth anterior to this plane are derived from ectoderm. Thus, the epithelium of the hard palate, sides of the mouth, lips, and enamel of the teeth are ectodermal structures. The secretory epithelium and cells lining the ducts of the parotid salivary gland also are derived from ectoderm. On the other hand, the epithelium of the tongue, the floor of the mouth, the palatoglossal and palatopharyngeal folds, and most of the soft palate are entodermal in origin. The secretory and duct epithelia of the sublingual and submandibular salivary glands also are believed to be of entodermal origin.

# The Teeth

## Deciduous Teeth

There are 20 **deciduous teeth**: four incisors, two canines, and four molars in each jaw. They begin to erupt about 6 months after birth and have all erupted by the end of 2 years. The teeth of the lower jaw usually appear before those of the upper jaw.

## Permanent Teeth

There are 32 **permanent teeth**, including four incisors, two canines, four premolars, and six molars in each jaw (Fig. 2-12). They begin to erupt at 6 years of age. The last tooth to erupt is the third molar, this may happen between the ages of 17 and 30. The teeth of the lower jaw appear before those of the upper jaw.

# The Tongue

The tongue is a mass of striated muscle covered with mucous membrane (Fig. 2-8). The muscles attach the tongue to the styloid process and the soft palate above and to the mandible and the hyoid bone below. The tongue is divided into right and left halves by a median **fibrous septum**.

## Mucous Membrane of the Tongue

The mucous membrane of the upper surface of the tongue can be divided into anterior and posterior parts by a V-shaped sulcus, the **sulcus terminalis** (Fig. 2-13) The apex of the sulcus projects backward and is marked by a small pit, the foramen cecum. The sulcus serves to divide the tongue into the anterior two thirds, or oral part, and the posterior third, or pharyngeal part. The foramen cecum is an embryologic remnant and marks the site of the upper end of the thyroglossal duct.

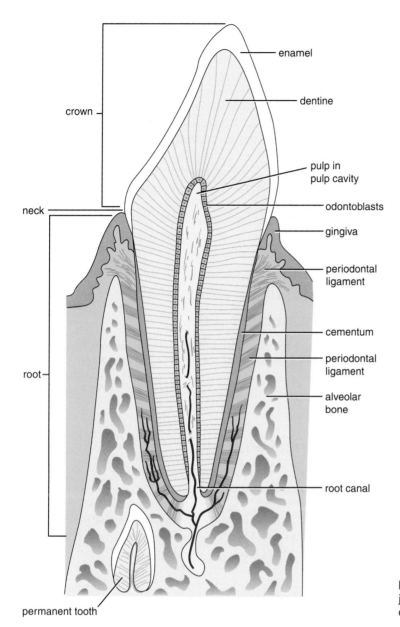

crown

neck

root

enamel

dentine

pulp in
pulp cavity

odontoblasts

gingiva

periodontal
ligament

cementum

periodontal
ligament

alveolar
bone

root canal

permanent tooth

**Figure 2-12** Sagittal section through the lower jaw and gum showing an erupted temporary incisor tooth and a developing permanent tooth.

Three types of papillae are present on the upper surface of the anterior two thirds of the tongue: the **filiform papillae**, the **fungiform papillae**, and the **vallate papillae**. The mucous membrane covering the posterior third of the tongue is devoid of papillae but has a nodular irregular surface (Fig. 2-13) caused by the presence of underlying lymph nodules, the **lingual tonsil.**

The mucous membrane on the inferior surface of the tongue is reflected from the tongue to the floor of the mouth. In the midline anteriorly, the undersurface of the tongue is connected to the floor of the mouth by a fold of mucous membrane, the **frenulum of the tongue.** On the lateral side of the frenulum, the deep lingual vein can be seen through the mucous membrane. Lateral to the lingual vein, the mucous membrane forms a fringed fold called the **plica fimbriata** (Fig. 2-8).

## Muscles of the Tongue

The muscles of the tongue are divided into two types: intrinsic and extrinsic.

### Intrinsic Muscles

These muscles are confined to the tongue and are not attached to bone. They consist of longitudinal, transverse, and vertical fibers.

**Nerve supply:** Hypoglossal nerve.
**Action:** Alter the shape of the tongue.

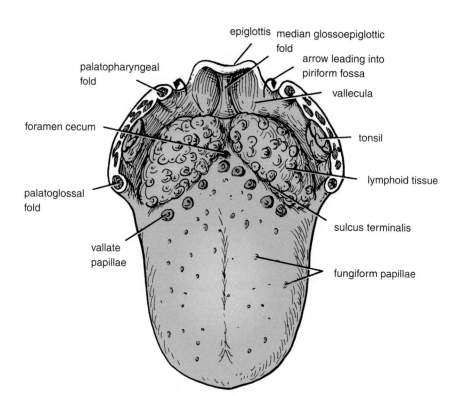

**Figure 2-13** Dorsal surface of the tongue showing the valleculae, the epiglottis, and the entrance into the piriform fossa on each side (arrows).

## Extrinsic Muscles

These muscles are attached to bones and the soft palate. They are the genioglossus, the hyoglossus, the styloglossus, and the palatoglossus.

**Nerve supply:** Hypoglossal nerve.
**Action:** Alter the position of the tongue within the mouth.

The origin, insertion, nerve supply, and action of the tongue muscles are summarized in Table 2-2.

## Blood Supply

The lingual artery, the tonsilar branch of the facial artery, and the ascending pharyngeal artery supply the tongue. The veins drain into the internal jugular vein.

## Lymph Drainage

**Tip:** Submental lymph nodes.
**Sides of the anterior two thirds:** Submandibular and deep cervical lymph nodes.
**Posterior third:** Deep cervical lymph nodes.

## Sensory Innervation

**Anterior two thirds:** Lingual nerve branch of mandibular division of trigeminal nerve (general sensation) and chorda tympani branch of the facial nerve (taste).

**Posterior third:** Glossopharyngeal nerve (general sensation and taste).

## Movements of the Tongue

**Protrusion:** The genioglossus muscles on both sides acting together (Fig. 2-14).
**Retraction:** Styloglossus and hyoglossus muscles on both sides acting together.
**Depression:** Hyoglossus muscles on both sides acting together.
**Retraction and elevation of the posterior third:** Styloglossus and palatoglossus muscles on both sides acting together.
**Shape changes:** Intrinsic muscles.

### E M B R Y O L O G I C   N O T E

### Development of the Tongue

At about the 4th week, a median swelling called the **tuberculum impar** appears in the entodermal ventral wall or floor of the pharynx (Fig. 2-15). A little later, another swelling, called the **lateral lingual swelling** (derived from the anterior end of each first pharyngeal arch), appears on each side of the tuberculum impar. The lateral lingual swellings now enlarge, grow medially, and fuse with each other and the tuberculum impar. Thus, the

| Table 2-2 | Muscles of Tongue | | | |
|---|---|---|---|---|
| **Muscle** | **Origin** | **Insertion** | **Nerve Supply** | **Action** |
| **Intrinsic Muscles** | | | | |
| Longitudinal<br><br>Transverse<br>Vertical | Median septum and submucosa | Mucous membrane | Hypoglossal nerve | Alters shape of tongue |
| **Extrinsic Muscles** | | | | |
| Genioglossus | Superior genial spine of mandible | Blends with other muscles of tongue | Hypoglossal nerve | Protrudes apex of tongue through mouth |
| Hyoglossus | Body and greater cornu of hyoid bone | Blends with other muscles of tongue | Hypoglossal nerve | Depresses tongue |
| Styloglossus | Styloid process of temporal bone | Blends with other muscles of tongue | Hypoglossal nerve | Draws tongue upward and backward |
| Palatoglossus | Palatine aponeurosis | Side of tongue | Pharyngeal plexus | Pulls roots of tongue upward and backward, narrows oropharyngeal isthmus |

lingual swellings form the anterior two thirds of the body of the tongue, and because they are derived from the first pharyngeal arches, the mucous membrane on each side will be innervated by the lingual nerve, a branch of the mandibular division of the fifth cranial nerve (common sensation). The chorda tympani from the seventh cranial nerve (taste) also supplies this area.

Meanwhile, a second median swelling, called the **copula,** appears in the floor of the pharynx behind the tuberculum impar. The copula extends forward on each side of the tuberculum impar and becomes V shaped. At about this time, the anterior ends of the second, third, and fourth pharyngeal arches are entering this region. The anterior ends of the third arch on each side overgrow the other arches and extend into the copula, fusing in the midline. The copula now disappears. Thus, the mucous membrane of the posterior third of the tongue is formed from the third pharyngeal arches and is innervated by the ninth cranial nerve (common sensation and taste).

The anterior two thirds of the tongue is separated from the posterior third by a groove, the **sulcus terminalis,** which represents the interval between the lingual swellings of the first pharyngeal arches and the anterior ends of the third pharyngeal arches. Around the edge of the anterior two thirds of the tongue, the ectodermal cells proliferate and grow inferiorly into the underlying mesenchyme. Later, these cells degenerate so that this part of the tongue becomes free. Some of entodermal cells remain in the midline and help form the frenulum of the tongue.

Remember that the **circumvallate papillae** are situated on the mucous membrane just anterior to the sulcus terminalis and that their taste buds are innervated by the ninth cranial nerve. It is presumed that during development the mucous membrane of the posterior third of the tongue becomes pulled anteriorly slightly, so that fibers of the ninth cranial nerve cross the sucus terminalis to supply these taste buds (Fig. 2-15).

The muscles of the tongue are derived from the occipital myotomes, which, at first, are closely related to the developing hind brain and later migrate inferiorly and anteriorly around the pharynx and enter the tongue. The migrating myotomes carry with them their innervation, the twelfth cranial nerve, and this explains the long curving course taken by the twelfth cranial nerve as it passes downward and forward in the carotid triangle of the neck (see p. 573).

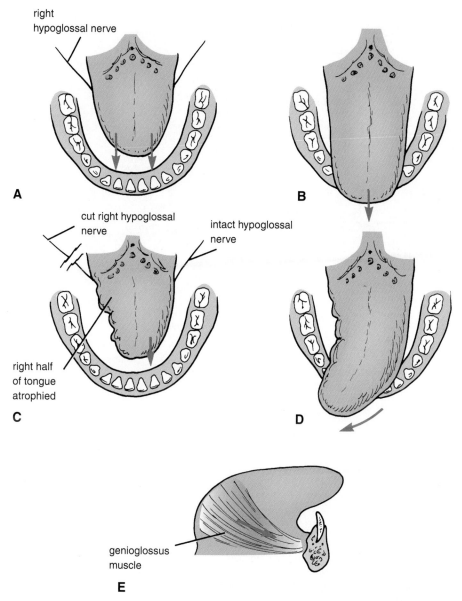

**Figure 2-14** Diagrammatic representation of the action of the right and left genioglossus muscles of the tongue. **A.** The right and left muscles contract equally together, and as a result (**B**), the tip of the tongue is protruded in the midline. **C.** The right hypoglossa nerve (which innervates the genioglossus muscle and the intrinsic tongue muscles on the same side) is cut, and as a result, the right side of the tongue is atrophied and wrinkled. **D.** When the patient is asked to protrude the tongue, the tip points to the side of the nerve lesion. **E.** The origin and insertion and direction of pull of the genioglossus muscle.

# The Palate

The palate forms the roof of the mouth and the floor of the nasal cavity. It is divided into two parts: the hard palate in front and the soft palate behind.

## Hard Palate

The hard palate is formed by the palatine processes of the maxillae and the horizontal plates of the palatine bones (Fig. 2-16). It is continuous behind with the soft palate.

## The Soft Palate

The soft palate is a mobile fold attached to the posterior border of the hard palate (Fig. 2-17). Its free posterior border presents in the midline a conical projection called the **uvula**. The soft palate is continuous at the sides with the lateral wall of the pharynx. The soft palate is composed of mucous membrane, palatine aponeurosis, and muscles.

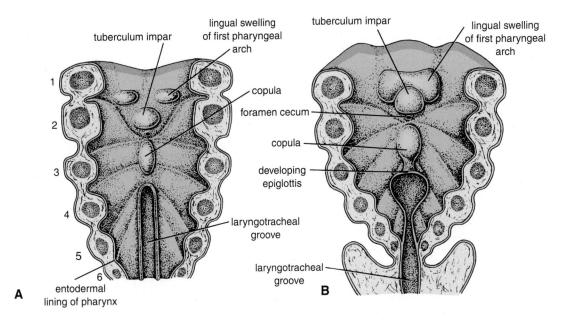

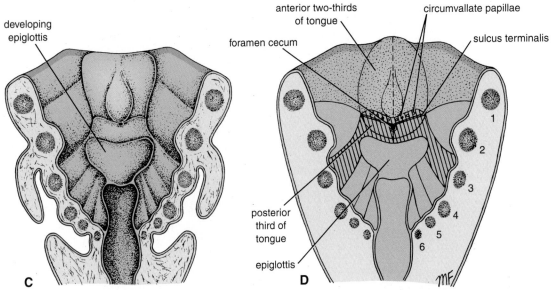

**Figure 2-15** The floor of the pharynx showing the stages in the development of the tongue.

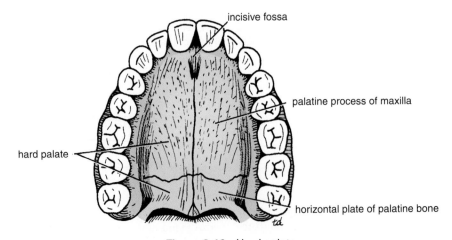

**Figure 2-16** Hard palate.

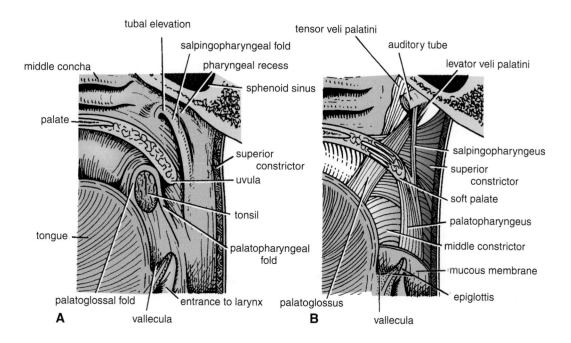

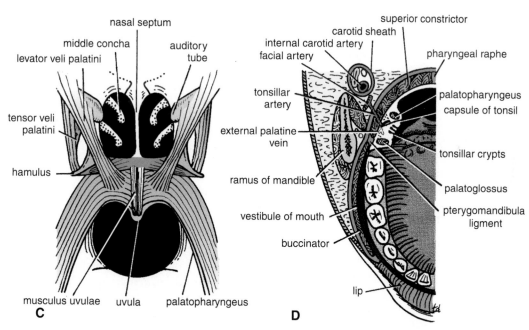

**Figure 2-17** **A.** Junction of the nose with the nasal part of the pharynx and the mouth with the oral part of the pharynx. Note the position of the tonsil and the opening of the auditory tube. **B.** Muscles of the soft palate and the upper part of the pharynx. **C.** Muscles of the soft palate seen from behind. **D.** Horizontal section through the mouth and the oral part of the pharynx showing the relations of the tonsil.

## The Mucous Membrane

The mucous membrane covers the upper and lower surfaces of the soft palate.

## Palatine Aponeurosis

The palatine aponeurosis is a fibrous sheet attached to the posterior border of the hard palate. It is the expanded tendon of the tensor veli palatini muscle.

## Muscles of the Soft Palate

The muscles of the soft palate are the tensor veli palatini, the levator veli palatini, the palatoglossus, the palatopharyngeus, and the musculus uvulae (Fig. 2-17). The muscle fibers of the tensor veli palatini converge as they descend from their origin to form a narrow tendon, which turns medially around the pterygoid hamulus. The tendon, together with the tendon of the opposite side, expands to form the

| | **Table 2-3** | **Muscles of the Soft Palate** | | | |
|---|---|---|---|---|---|
| **Muscle** | **Origin** | **Insertion** | **Nerve Supply** | **Action** |
| Tensor veli palatini | Spine of sphenoid, auditory tube | With muscle of other side, forms palatine aponeurosis | Nerve to medial pterygoid from mandibular nerve | Tenses soft palate |
| Levator veli palatini | Petrous part of temporal bone, auditory tube | Palatine aponeurosis | Pharyngeal plexus | Raises soft palate |
| Palatoglossus | Palatine aponeurosis | Side of tongue | Pharyngeal plexus | Pulls root of tongue upward and backward, narrows oropharyngeal isthmus |
| Palato-pharyngeus | Palatine aponeurosis | Posterior border of thyroid cartilage | Pharyngeal plexus | Elevates wall of pharynx, pulls palatopharyngeal folds medially |
| Musculus uvulae | Posterior border of hard palate | Mucous membrane of uvula | Pharyngeal plexus | Elevates uvula |

palatine aponeurosis. When the muscles of the two sides contract, the soft palate is tightened so that the soft palate may be moved upward or downward as a tense sheet. The muscles of the soft palate, and their origins, insertions, nerve supply, and actions are summarized in Table 2-3.

## Nerve Supply of the Palate

The greater and lesser palatine nerves from the maxillary division of the trigeminal nerve enter the palate through the greater and lesser palatine foramina (Fig. 2-10). The nasopalatine nerve, also a branch of the maxillary nerve, enters the front of the hard palate through the incisive foramen. The glossopharyngeal nerve also supplies the soft palate.

## Blood Supply of the Palate

The blood supply of the palate consists of the greater palatine branch of the maxillary artery, the ascending palatine branch of the facial artery, and the ascending pharyngeal artery.

## Lymph Drainage of the Palate

Lymph is drained from the palate by the deep cervical lymph nodes.

## Palatoglossal Arch

The palatoglossal arch is a fold of mucous membrane containing the **palatoglossus muscle,** which extends from the soft palate to the side of the tongue (Fig. 2-8). **The palatoglossal arch marks where the mouth becomes the pharynx.**

## Palatopharyngeal Arch

The palatopharyngeal arch is a fold of mucous membrane behind the palatoglossal arch (Fig. 2-8) that runs downward and laterally to join the pharyngeal wall. The muscle contained within the fold is the **palatopharyngeus muscle.** The **palatine tonsils,** which are masses of lymphoid tissue, are located between the palatoglossal and palatopharyngeal arches (Fig. 2-8).

## Movements of the Soft Palate

The pharyngeal isthmus (the communicating channel between the nasal and oral parts of the pharynx) is closed by raising the soft palate. Closure occurs during the production of explosive consonants in speech.

The soft palate is raised by the contraction of the levator veli palatini on each side. At the same time, the upper fibers of the superior constrictor muscle contract and pull the posterior pharyngeal wall forward. The palatopharyngeus muscles on both sides also contract so that the palatopharyngeal arches are pulled medially, like side curtains. By this means, the nasal part of the pharynx is closed off from the oral part.

## Development of the Palate

In early fetal life, the nasal and mouth cavities are in communication, but later they become separated by the development of the palate (Fig. 2-6). The **primary palate**, which carries the four incisor teeth, is formed by the medial nasal process. Posterior to the primary palate, the maxillary process on each side sends medially a horizontal plate called the **palatal process**; these plates fuse to form the **secondary palate** and also unite with the primary palate and the developing nasal septum. The fusion takes place from the anterior to the posterior region. The primary and secondary palates later will form the **hard palate.** Two folds grow posteriorly from the posterior edge of the palatal processes to create the **soft palate**, so that the **uvula** is the last structure to be formed (Fig. 2-6). The union of the two folds of the soft palate occurs during the 8th week. The two parts of the uvula fuse in the midline during the 11th week. The interval between the primary palate and secondary palate is represented in the midline by the **incisive foramen**.

# The Salivary Glands

## Parotid Gland

The parotid gland is the largest salivary gland and is composed mostly of serous acini. It lies in a deep hollow below the external auditory meatus, behind the ramus of the mandible (Fig. 2-18) and in front of the sternocleidomastoid muscle. The facial nerve divides the gland into **superficial** and **deep lobes.** The **parotid duct** emerges from the anterior border of the gland and passes forward over the lateral surface of the masseter. It enters the vestibule of the mouth upon a small papilla opposite the upper second molar tooth (Fig. 2-18).

**A**

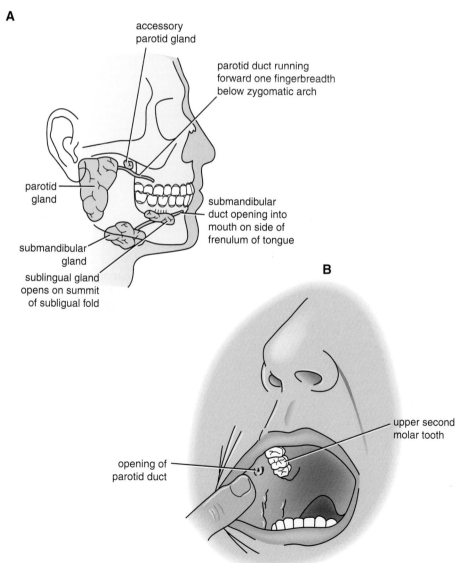

accessory
parotid gland

parotid duct running
forward one fingerbreadth
below zygomatic arch

parotid
gland

submandibular
duct opening into
mouth on side of
frenulum of tongue

submandibular
gland

sublingual gland
opens on summit
of subligual fold

**B**

upper second
molar tooth

opening of
parotid duct

**Figure 2-18　A.** The general position of the major salivary glands and their ducts. **B.** The interior of the mouth showing the opening of the right parotid duct into the vestibule opposite the upper second molar tooth.

## Nerve Supply

Parasympathetic secretomotor supply arises from the glossopharyngeal nerve. The nerves reach the gland via the tympanic branch, the lesser petrosal nerve, the otic ganglion, and the auriculotemporal nerve.

## Submandibular Gland

The submandibular gland consists of a mixture of serous and mucous acini. It lies beneath the lower border of the body of the mandible (Fig. 2-18) and is divided into superficial and deep parts by the mylohyoid muscle. The deep part of the gland lies beneath the mucous membrane of the mouth on the side of the tongue. The **submandibular duct** emerges from the anterior end of the deep part of the gland and runs forward beneath the mucous membrane of the mouth. It opens into the mouth on a small papilla, which is situated at the side of the frenulum of the tongue (Figs. 2-8 and 2-18).

## Nerve Supply

Parasympathetic secretomotor supply is from the facial nerve via the chorda tympani and the submandibular ganglion. The postganglionic fibers pass directly to the gland.

## Sublingual Gland

The sublingual gland lies beneath the mucous membrane (sublingual fold) of the floor of the mouth, close to the frenulum of the tongue. It has both serous and mucous acini, with the latter predominating. The **sublingual ducts** (8 to 20 in number) open into the mouth on the summit of the sublingual fold (Figs. 2-8 and 2-18).

## Nerve Supply

Parasympathetic secretomotor supply is from the facial nerve via the chorda tympani and the submandibular ganglion. Postganglionic fibers pass directly to the gland.

# The Pharynx

The pharynx is situated behind the nasal cavities, the mouth, and the larynx (Fig. 2-19) and may be divided into **nasal, oral,** and **laryngeal parts.** The pharynx is funnel-shaped, with its upper, wider end lying under the skull, and its lower, narrow end becoming continuous with the esophagus opposite the sixth cervical vertebra. The pharynx has a musculomembranous wall, which is deficient anteriorly. Here, it is replaced by the posterior openings into the nose

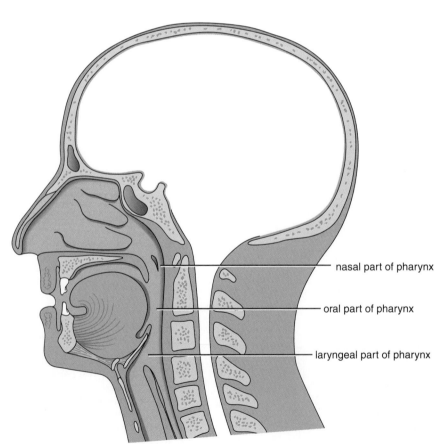

nasal part of pharynx

oral part of pharynx

laryngeal part of pharynx

**Figure 2-19** Sagittal section through the nose, mouth, pharynx, and larynx to show the subdivisions of the pharynx.

(choanae), the opening into the mouth, and the inlet of the larynx. By means of the auditory tube, the mucous membrane is also continuous with the mucous membrane of the tympanic cavity.

## Muscles of the Pharynx

The muscles in the wall of the pharynx consist of the **superior, middle,** and **inferior constrictor muscles** (Fig. 2-20), whose fibers run in a somewhat circular direction, and the **stylopharyngeus** and **salpingopharyngeus muscles,** whose fibers run in a somewhat longitudinal direction.

The three constrictor muscles extend round the pharyngeal wall to be inserted into a fibrous band or raphe that extends from the pharyngeal tubercle on the basilar part of the occipital bone of the skull down to the esophagus. The three constrictor muscles overlap each other so that the middle constrictor lies on the outside of the lower part of the superior constrictor and the inferior constrictor lies outside the lower part of the middle constrictor (Fig. 2-21).

The lower part of the inferior constrictor, which arises from the cricoid cartilage, is called the **cricopharyngeus muscle** (Fig. 2-21). The fibers of the cricopharyngeus pass horizontally round the lowest and narrowest part of the pharynx and act as a sphincter. **Killian's dehiscence** is the area on the posterior pharyngeal wall between the upper propulsive part of the inferior constrictor and the lower sphincteric part, the cricopharyngeus.

The details of the origins, insertions, nerve supply, and actions of the pharyngeal muscles are summarized in Table 2-4.

## Interior of Pharynx

The pharynx is divided into three parts: the nasal pharynx, the oral pharynx, and the laryngeal pharynx.

### Nasal Pharynx

The nasal pharynx lies above the soft palate and behind the nasal cavities (Fig. 2-19). In the submucosa of the roof is a collection of lymphoid tissue called the **pharyngeal tonsil** (Fig. 2-22). The pharyngeal isthmus is the opening in the floor between the soft palate and the posterior pharyngeal wall. On the lateral wall is the opening of the **auditory tube,** the elevated ridge of which is called the **tubal elevation**

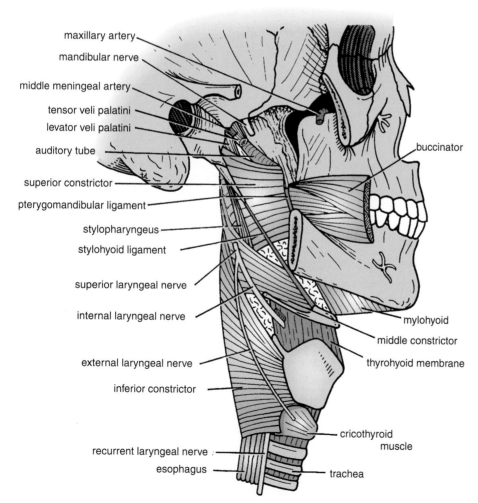

**Figure 2-20** The three constrictors of the pharynx; the superior and recurrent laryngeal nerves are also shown.

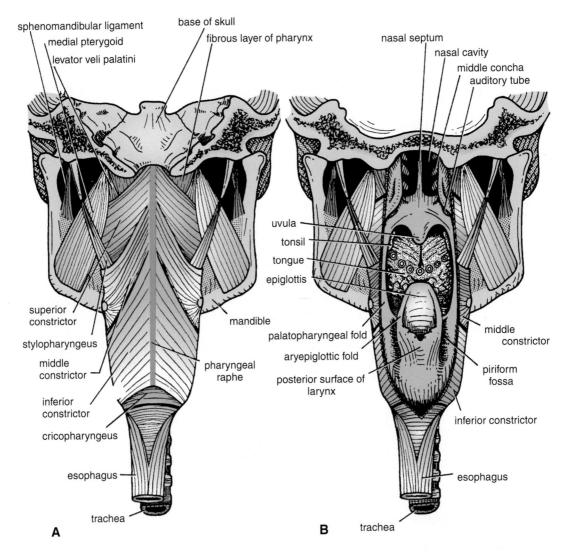

**Figure 2-21**    The pharynx seen from behind. **A.** Note the three constrictor muscles and the position of the stylopharyngeus muscles. **B.** The greater part of the posterior wall of the pharynx has been removed to display the nasal, oral, and laryngeal parts of the pharynx.

(Fig. 2-22). The **pharyngeal recess** is a depression in the pharyngeal wall behind the tubal elevation. The **salpingopharyngeal fold** is a vertical fold of mucous membrane covering the salpingopharyngeus muscle.

## Oral Pharynx

The oral pharynx lies behind the oral cavity (Figs. 2-19 and 2-23). The floor is formed by the posterior one third of the tongue and the interval between the tongue and epiglottis. In the midline is the **median glossoepiglottic fold** (Fig. 2-13), and on each side is the **lateral glossoepiglottic fold.** The depression on each side of the median glossoepiglottic fold is called the **vallecula** (Fig. 2-13).

On the lateral wall on each side are the palatoglossal and the palatopharyngeal arches or folds and the palatine tonsils between them (Fig. 2-22). The palatoglossal arch is a fold of mucous membrane covering the palatoglossus muscle. The interval between the two palatoglossal arches is called the **oropharyngeal isthmus** and marks the boundary between the mouth and pharynx. The palatopharyngeal arch is a fold of mucous membrane covering the palatopharyngeus muscle. The recess between the palatoglossal and palatopharyngeal arches is occupied by the **palatine tonsil.**

## Laryngeal Pharynx

The laryngeal pharynx lies behind the opening into the larynx (Fig. 2-19). The lateral wall is formed by the thyroid cartilage and the thyrohyoid membrane. The **piriform fossa** is a depression in the mucous membrane on each side of the laryngeal inlet (Fig. 2-21).

### Table 2-4    Muscles of the Pharynx

| Muscle | Origin | Insertion | Nerve Supply | Action |
|---|---|---|---|---|
| Superior constrictor | Medial pterygoid plate, pterygoid hamulus, pterygomandibular ligament, mylohyoid line of mandible | Pharyngeal tubercle of occipital bone, raphe in midline posteriorly | Pharyngeal plexus | Aids soft palate in closing off nasal pharynx, propels bolus downward |
| Middle constrictor | Lower part of stylohyoid ligament, lesser and greater cornu of hyoid bone | Pharyngeal raphe | Pharyngeal plexus | Propels bolus downward |
| Inferior constrictor | Lamina of thyroid cartilage, cricoid cartilage | Pharyngeal raphe | Pharyngeal plexus | Propels bolus downward |
| Cricopharyngeus | Lowest fibers of inferior constrictor muscle | | | Sphincter at lower end of pharynx |
| Stylopharyngeus | Styloid process of temporal bone | Posterior border of thyroid cartilage | Glossopharyngeal nerve | Elevates larynx during swallowing |
| Salpingopharyngeus | Auditory tube | Blends with palatopharyngeus | Pharyngeal plexus | Elevates pharynx |
| Palatopharyngeus | Palatine aponeurosis | Posterior border of thyroid cartilage | Pharyngeal plexus | Elevates wall of pharynx, pulls palatopharyngeal arch medially |

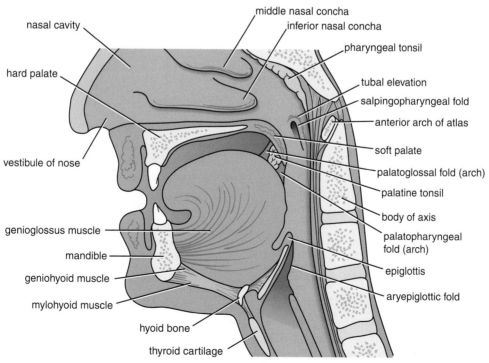

**Figure 2-22**   Sagittal section of the head and neck showing the relations of the nasal cavity, mouth, pharynx, and larynx.

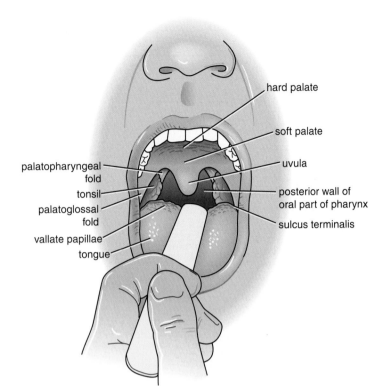

hard palate

soft palate

palatopharyngeal fold

uvula

tonsil

posterior wall of oral part of pharynx

palatoglossal fold

sulcus terminalis

vallate papillae

tongue

**Figure 2-23**   The oral pharynx as seen through the open mouth.

## Sensory Nerve Supply of the Pharyngeal Mucous Membrane

**Nasal pharynx:** The maxillary nerve (V2).
**Oral pharynx:** The glossopharyngeal nerve.
**Laryngeal pharynx** (around the entrance into the larynx): The internal laryngeal branch of the vagus nerve.

## Blood Supply of Pharynx

Ascending pharyngeal arteries, tonsillar branches of facial arteries, and branches of maxillary and lingual arteries compose the blood supply of the pharynx.

## Lymph Drainage of Pharynx

Lymph drains from the pharynx directly into the deep cervical lymph nodes or indirectly via the retropharyngeal or paratracheal nodes into the deep cervical nodes.

### PHYSIOLOGIC NOTE

### The Process of Swallowing (Deglutition)

Masticated food is formed into a ball or bolus on the dorsum of the tongue and voluntarily pushed upward and backward against the undersurface of the hard palate. This is brought about by the contraction of the styloglossus muscles on both sides, which pull the root of the tongue upward and backward. The palatoglossus muscles then squeeze the bolus backward into the pharynx. From this point onward, the process of swallowing becomes an involuntary act.

The nasal part of the pharynx is now shut off from the oral part of the pharynx by the elevation of the soft palate, the pulling forward of the posterior wall of the pharynx by the upper fibers of the superior constrictor muscle, and the contraction of the palatopharyngeus muscles. This prevents the passage of food and drink into the nasal cavities.

The larynx and the laryngeal part of the pharynx are pulled upward by the contraction of the stylopharyngeus, salpingopharyngeus, thyrohyoid, and palatopharyngeus muscles. The main part of the larynx is thus elevated to the posterior surface of the epiglottis, and the entrance into the larynx is closed. The laryngeal entrance is made smaller by the approximation of the aryepiglottic folds, and the arytenoid cartilages are pulled forward by the contraction of the aryepiglottic, oblique arytenoid, and thryoarytenoid muscles.

The bolus moves downward over the epiglottis, the closed entrance into the larynx, and reaches the lower part of the pharynx as the result of the successive contraction of the superior, middle, and inferior constrictor muscles. Some of the food slides down the groove on either side of the entrance into the larynx (i.e., down through the **piriform fossae**). Finally, the lower part of the pharyngeal wall (the cricopharyngeus muscle) relaxes, and the bolus enters the esophagus.

## Palatine Tonsils

The palatine tonsils are two masses of lymphoid tissue, each located in the depression on the lateral wall of the oral part of the pharynx between the palatoglossal and palatopharyngeal arches (Fig. 2-24). Each tonsil is covered by mucous membrane, and its free medial surface projects into the pharynx. The surface is pitted by numerous small openings that lead into the **tonsillar crypts.**

The tonsil is covered on its lateral surface by a **fibrous capsule** (Fig. 2-24). The capsule is separated from the superior constrictor muscle by loose areolar tissue (Fig. 2-24), and the external palatine vein descends from the soft palate in this tissue to join the pharyngeal venous plexus. Lateral to the superior constrictor muscle lie the styloglossus muscle, the loop of the facial artery, and the internal carotid artery.

The tonsil reaches its maximum size during early childhood, but after puberty, it diminishes considerably in size.

### Blood Supply of the Tonsil

The blood supply of the tonsil is the tonsillar branch of the facial artery. The veins pierce the superior constrictor muscle and join the external palatine, pharyngeal, or facial veins.

### Lymph Drainage of the Tonsil

Lymph drains from the tonsil into the upper deep cervical lymph nodes, just below and behind the angle of the mandible.

## Waldeyer's Ring of Lymphoid Tissue

The lymphoid tissue that surrounds the opening into the respiratory and digestive systems forms a ring. The lateral part of the ring is formed by the palatine tonsils and tubal tonsils (lymphoid tissue around the opening of the auditory tube in the lateral wall of the nasopharynx). The pharyngeal tonsil in the roof of the nasopharynx forms the upper part, and the lingual tonsil on the posterior third of the tongue forms the lower part.

# The Larynx

The larynx is an organ that provides a protective sphincter at the inlet of the air passages and is responsible for voice production. It is situated below the tongue and hyoid bone and between the great blood vessels of the neck and lies at the level of the fourth, fifth, and sixth cervical vertebrae

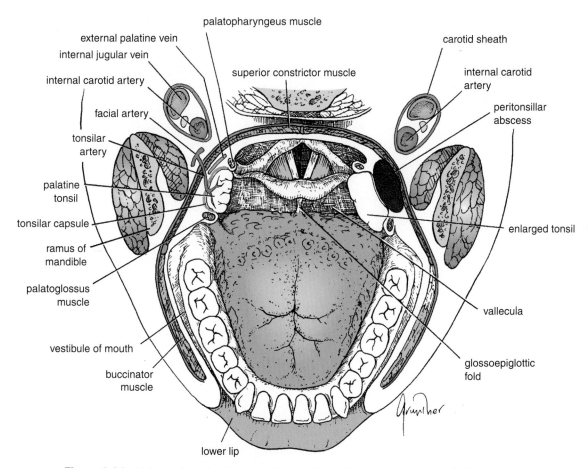

**Figure 2-24** Hoizontal section through the mouth and the oral pharynx. Left, the normal palatine tonsil and its relationships. Right, the position of a peritonsillar abscess. Note the relationship of the abscess to the superior constrictor muscle and the carotid sheath. The opening into the larynx can also be seen below and behind the tongue.

(Fig. 2-25). Above, the larynx opens into the laryngeal part of the pharynx, and below, it is continuous with the trachea. The larynx is covered in front by the infrahyoid strap muscles and is covered at the sides by the thyroid gland.

The framework of the larynx is formed of cartilages that are held together by ligaments and membranes, moved by muscles, and lined by mucous membrane.

## Cartilages of the Larynx
### Thyroid Cartilage

The thyroid cartilage is the largest cartilage of the larynx (Fig. 2-26) and consists of two laminae of hyaline cartilage that meet in the midline in the prominent V angle (the so called Adam's apple). The posterior border extends upward

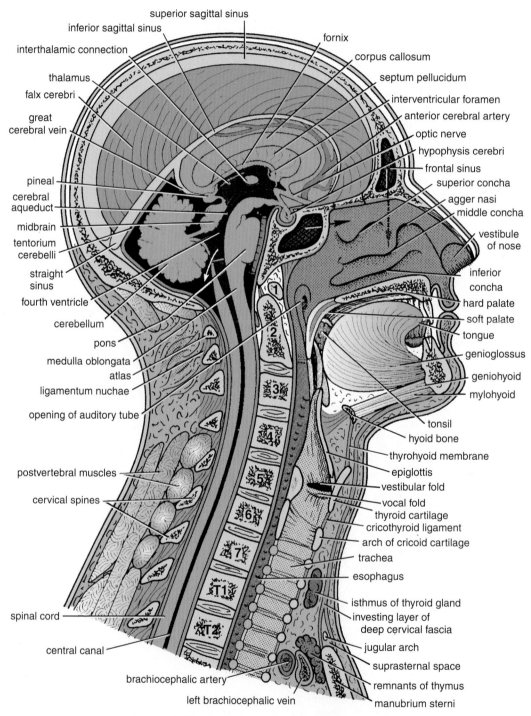

**Figure 2-25**   Sagittal section of the head and neck.

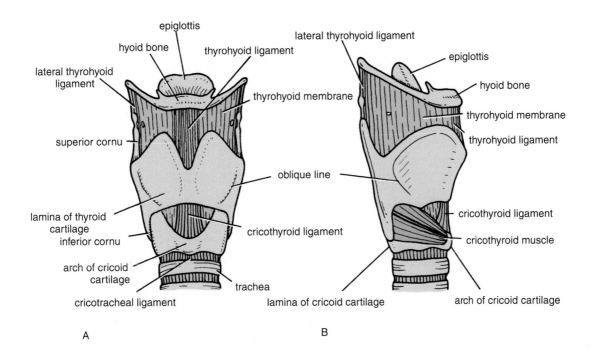

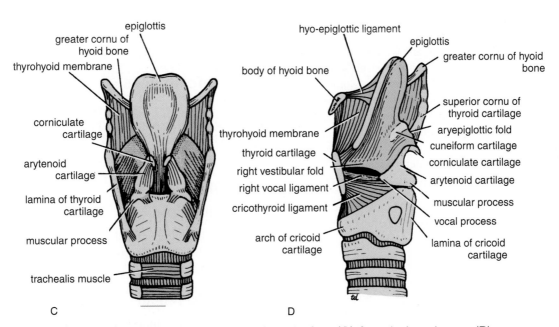

**Figure 2-26** The larynx and its ligaments from the front (**A**), from the lateral aspect (**B**), and from behind (**C**). **D.** The left lamina of the thyroid cartilage has been removed to display the interior of the larynx.

into a **superior cornu** and downward into an **inferior cornu**. On the outer surface of each lamina is an **oblique line** for the attachment of muscles.

## Cricoid Cartilage

The cricoid cartilage is formed of hyaline cartilage and shaped like a signet ring, having a broad plate behind and a shallow arch in front (Fig. 2-26). The cricoid cartilage lies below the thyroid cartilage, and on each side of the lateral surface is a facet for articulation with the inferior cornu of the thyroid cartilage. Posteriorly, the lamina has on its upper border on each side a facet for articulation with the arytenoid cartilage. All these joints are synovial.

## Arytenoid Cartilages

There are two arytenoid cartilages; they are small and pyramidal shaped and located at the back of the larynx (Fig. 2-26). They articulate with the upper border of the lamina of the cricoid cartilage. Each cartilage has an **apex** above that articulates with the small corniculate cartilage, a **base** below that articulates with the lamina of the cricoid cartilage, and a **vocal process** that projects forward and gives attachment to the vocal ligament. A **muscular process** that projects laterally gives attachment to the posterior and lateral cricoarytenoid muscles.

## Corniculate Cartilages

Two small conical-shaped cartilages articulate with the arytenoid cartilages (Fig. 2-27). They give attachment to the aryepiglottic folds.

## Cuneiform Cartilages

These two small rod-shaped cartilages are found in the aryepiglottic folds and serve to strengthen them (Fig. 2-27).

## Epiglottis

This leaf-shaped lamina of elastic cartilage lies behind the root of the tongue (Fig. 2-26). Its stalk is attached to the back of the thyroid cartilage. The sides of the epiglottis are attached to the arytenoid cartilages by the aryepiglottic folds of mucous membrane. The upper edge of the epiglottis is free. The covering of mucous membrane passes forward onto the posterior surface of the tongue as the **median glossoepiglottic fold**; the depression on each side of the fold is called the **vallecula** (Fig. 2-24). Laterally, the mucous membrane passes onto the wall of the pharynx as **the lateral glossoepiglottic fold.**

# Membranes and Ligaments of the Larynx

## Thyrohyoid Membrane

The thyrohyoid membrane connects the upper margin of the thyroid cartilage to the hyoid bone (Fig. 2-26). In the midline, it is thickened to form the **median thyrohyoid ligament.** The membrane is pierced on each side by the superior laryngeal vessels and the internal laryngeal nerve, a branch of the superior laryngeal nerve (Fig. 2-20).

## Cricotracheal Ligament

The cricotracheal ligament connects the cricoid cartilage to the first ring of the trachea (Fig. 2-26).

## Quadrangular Membrane

The quadrangular membrane extends between the epiglottis and the arytenoid cartilages (Fig. 2-27). Its thickened inferior margin forms the **vestibular ligament,** and the vestibular ligaments form the interior of the **vestibular folds** (Fig. 2-27).

## Cricothyroid Ligament

The lower margin of the cricothyroid ligament is attached to the upper border of the cricoid cartilage (Fig. 2-27). Instead of being attached to the thyroid cartilage, the superior margin of the ligament ascends on the medial surface of the thyroid cartilage. Its upper free margin, which is composed almost entirely of elastic tissue, forms the important **vocal ligament** on each side. The vocal ligaments form the interior of the **vocal folds (vocal cords)** (Fig. 2-27). The anterior end of each vocal ligament is attached to the thyroid cartilage, and the posterior end is attached to the vocal process of the arytenoid cartilage.

# Inlet of the Larynx

The inlet of the larynx looks backward and upward into the laryngeal part of the pharynx (Fig. 2-21). The opening is wider in front than behind and is bounded in front by the epiglottis, laterally by the aryepiglottic fold of mucous membrane, and posteriorly by the arytenoid cartilages with the corniculate cartilages. The cuneiform cartilage lies within and strengthens the aryepiglottic fold and produces a small elevation on the upper border.

# The Piriform Fossa

The piriform fossa is a recess on either side of the fold and inlet (Fig. 2-27). It is bounded medially by the aryepiglottic fold and laterally by the thyroid cartilage and the thyrohyoid membrane.

# Laryngeal Folds

## Vestibular Fold

The vestibular fold is a **fixed** fold on each side of the larynx (Fig. 2-26). It is formed by mucous membrane covering the vestibular ligament and is vascular and **pink** in color.

## Vocal Fold (Vocal Cord)

The vocal fold is a **mobile** fold on each side of the larynx and is concerned with voice production. It is formed by mucous membrane covering the vocal ligament and is avascular and **white** in color. **The vocal fold moves with respiration, and its white color is easily seen when viewed with a laryngoscope (Fig. 2-27).**

The gap between the vocal folds is called the **rima glottidis** or **glottis** (Fig. 2-27). The glottis is bounded in front by the vocal folds and behind by the medial surface of the arytenoid cartilages. The glottis is the narrowest part of the larynx and measures about 2.5 cm from front to back in the male adult and less in the female. In children, the lower part of the larynx within the cricoid cartilage is the narrowest part.

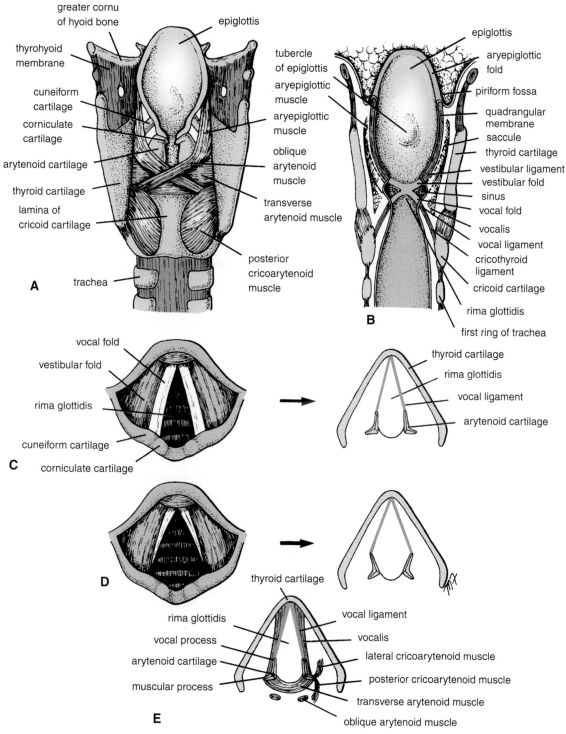

**Figure 2-27** **A.** Muscles of the larynx seen from behind. **B.** Coronal section through the larynx. **C.** Rima glottidis partially open as in quiet breathing. **D.** Rima glottidis wide open as in deep breathing. **E.** Muscles that move vocal ligaments.

# Cavity of the Larynx

The cavity of the larynx extends from the inlet to the lower border of the cricoid cartilage, where it is continuous with the cavity of the trachea. It is divided into three regions:

- The **vestibule**, which is situated between the inlet and the vestibular folds.
- The **middle region**, which is situated between the vestibular folds above and the vocal folds below.
- The **lower region,** which is situated between the vocal folds above and the lower border of the cricoid cartilage below.

# Sinus of the Larynx

The sinus of the larynx is a small recess on each side of the larynx situated between the vestibular and vocal folds. It is lined with mucous membrane (Fig. 2-27).

# Saccule of the Larynx

The saccule of the larynx is a diverticulum of mucous membrane that ascends from the sinus (Fig. 2-27). The mucous secretion lubricates the vocal cords.

# Muscles of the Larynx

The muscles of the larynx may be divided into two groups: extrinsic and intrinsic.

## Extrinsic Muscles

Extrinsic muscles move the larynx up and down during swallowing. Note that many of these muscles are attached to the hyoid bone, which is attached to the thyroid cartilage by the thyrohyoid membrane. It follows that movements of the hyoid bone are accompanied by movements of the larynx.

- **Elevation:** The digastric, the stylohyoid, the mylohyoid, the geniohyoid, the stylopharyngeus, the salpingopharyngeus, and the palatopharyngeus muscles.
- **Depression:** The sternothyroid, the sternohyoid, and the omohyoid muscles.

## Intrinsic Muscles

Two muscles modify the laryngeal inlet (Fig. 2-27).

- **Narrowing the inlet:** The oblique arytenoid muscle.
- **Widening the inlet:** The thyroepiglottic muscle.

Five muscles move the vocal folds (cords) (Fig. 2-27).

- **Tensing the vocal cords:** The cricothyroid muscle.
- **Relaxing the vocal cords:** The thyroarytenoid (vocalis) muscle.
- **Adducting the vocal cords:** The lateral cricoarytenoid muscle.
- **Abducting the vocal cords:** The posterior cricoarytenoid muscle.
- **Approximates the arytenoid cartilages:** The transverse arytenoid muscle.

The details of the origins, insertions, nerve supply, and actions of intrinsic muscles are given in Table 2-5.

# Movements of the Vocal Folds (Cords)

The movements of the vocal folds depend on the movements of the arytenoid cartilages, which rotate and slide up and down on the sloping shoulder of the superior border of the cricoid cartilage.

The rima glottidis is opened by the contraction of the posterior cricoarytenoid, which rotates the arytenoid cartilage and abducts the vocal process (Fig. 2-27). The elastic tissue in the capsules of the cricoarytenoid joints keeps the arytenoid cartilages apart so that the posterior part of the glottis is open.

The rima glottidis is closed by contraction of the lateral cricoarytenoid, which rotates the arytenoid cartilage and adducts the vocal process (Fig. 2-27). The posterior part of the glottis is narrowed when the arytenoid cartilages are drawn together by contraction of the transverse arytenoid muscles.

The vocal folds are stretched by contraction of the cricothyroid muscle (Fig. 2-28). The vocal folds are slackened by contraction of the vocalis, a part of the thyroarytenoid muscle (Fig. 2-27).

# Movements of the Vocal Folds with Respiration

On quiet inspiration, the vocal folds are abducted, and the rima glottidis is triangular in shape with the apex in front (Fig. 2-27). On expiration, the vocal folds are adducted, leaving a small gap between them (Fig. 2-27).

On deep inspiration, the vocal folds are maximally abducted, and the triangular shape of the glottis becomes a diamond shape because of the maximal lateral rotation of the arytenoid cartilages (Fig. 2-27).

---

**PHYSIOLOGIC NOTE**

### Sphincteric Function of the Larynx

There are two sphincters in the larynx, one at the inlet and another at the rima glottidis. The sphincter at the inlet is used only during swallowing. As the bolus of food is passed backward between the tongue and the hard palate, the larynx is pulled up beneath the back of the tongue. The inlet of the larynx is narrowed by the action of the oblique arytenoid and aryepiglottic muscles. The epiglottis is pulled backward by the tongue and serves as a cap over the laryngeal inlet. The bolus of food, or fluids, then enters the esophagus by passing over the epiglottis

| Table 2-5 | Intrinsic Muscles of the Larynx | | | |
|---|---|---|---|---|
| **Muscle** | **Origin** | **Insertion** | **Nerve Supply** | **Action** |
| **Muscles Controlling the Laryngeal Inlet** | | | | |
| Oblique arytenoid | Muscular process of arytenoid cartilage | Apex of opposite arytenoid cartilage | Recurrent laryngeal nerve | Narrows the inlet by bringing the aryepiglottic folds together |
| Thyroepiglottic | Medial surface of thyroid cartilage | Lateral margin of epiglottis and aryepiglottic fold | Recurrent laryngeal nerve | Widens the inlet by pulling the aryepiglottic folds apart |
| **Muscles Controlling the Movements of the Vocal Folds (Cords)** | | | | |
| Cricothyroid | Side of cricoid cartilage | Lower border and inferior cornu of thyroid cartilage | External laryngeal nerve | Tenses vocal cords |
| Thyroarytenoid (vocalis) | Inner surface of thyroid cartilage | Arytenoid cartilage | Recurrent laryngeal nerve | Relaxes vocal cords |
| Lateral cricoarytenoid | Upper border of cricoid cartilage | Muscular process of arytenoid cartilage | Recurrent laryngeal nerve | Adducts the vocal cords by rotating arytenoid cartilage |
| Posterior cricoarytenoid | Back of cricoid cartilage | Muscular process of arytenoid cartilage | Recurrent laryngeal nerve | Abducts the vocal cords by rotating arytenoid cartilage |
| Transverse arytenoid | Back and medial surface of arytenoid cartilage | Back and medial surface of opposite arytenoid cartilage | Recurrent laryngeal nerve | Closes posterior part of rima glottidis approximating arytenoid cartilages |

or moving down the grooves on either side of the laryngeal inlet, the piriform fossae.

In coughing or sneezing, the rima glottidis serves as a sphincter. After inspiration, the vocal folds are adducted, and the muscles of expiration are made to contract strongly. As a result, the intrathoracic pressure rises, and the vocal folds are suddenly abducted. The sudden release of the compressed air will often dislodge foreign particles or mucus from the respiratory tract and carry the material up into the pharynx, where the material is either swallowed or expectorated.

In the Valsalva maneuver, forced expiration takes place against a closed glottis. In abdominal straining associated with micturition, defecation, and parturition, air is often held temporarily in the respiratory tract by closing the rima glottidis. After deep inspiration, the rima glottidis is closed. The muscles of the anterior abdominal wall now contract, and the upward movement of the diaphragm is prevented by the presence of compressed air within the respiratory tract. After a prolonged effort, the person often releases some of the air by momentarily opening the rima glottidis, producing a grunting sound.

## PHYSIOLOGIC NOTE

### Voice Production in the Larynx

The intermittent release of expired air between the adducted vocal folds results in their vibration and in the production of sound. The **frequency, or pitch,** of the sound is determined by changes in the length and tension of the vocal ligaments. The quality of the voice depends on the resonators above the larynx,

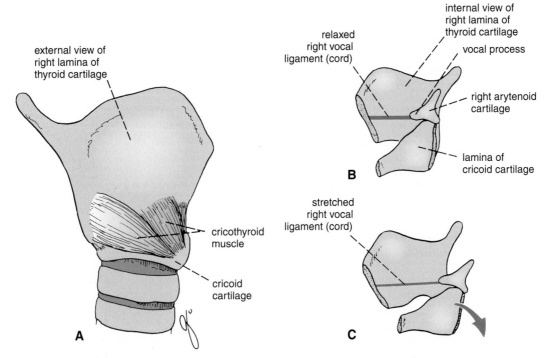

**external view of right lamina of thyroid cartilage**

cricothyroid muscle

cricoid cartilage

**A**

**relaxed right vocal ligament (cord)**

**internal view of right lamina of thyroid cartilage**

vocal process

right arytenoid cartilage

lamina of cricoid cartilage

**B**

**stretched right vocal ligament (cord)**

**C**

**Figure 2-28** Diagrams showing the attachments and actions of the cricothyroid muscle. **A.** Right lateral view of the larynx and the cricothyroid muscle. **B.** Interior view of the larynx showing the relaxed right vocal ligament. **C.** Interior view of the larynx showing the right vocal ligament stretched as a result of the cricoid and arytenoid cartilages tilting backward by contraction of the cricothyroid muscles.

namely the pharynx, mouth, and paranasal sinuses. The **quality** of the voice is controlled by the muscles of the soft palate, tongue, floor of the mouth, cheeks, lips, and jaws. Normal speech depends on the modification of the sound into recognizable consonants and vowels by the use of the tongue, teeth, and lips. Vowel sounds are usually purely oral, with the soft palate raised so that the air is channeled through the mouth rather than the nose.

**Speech** involves the intermittent release of expired air between the adducted vocal folds. **Singing** a note requires a more prolonged release of the expired air between the adducted vocal folds. In **whispering**, the vocal folds are adducted, but the arytenoid cartilages are separated; the vibrations are given to a constant stream of expired air that passes through the posterior part of the rima glottidis.

## Mucous Membrane of the Larynx

The mucous membrane of the larynx lines the cavity and is covered with ciliated columnar epithelium. On the vocal cords, however, where the mucous membrane is subject to repeated trauma during phonation, the mucous membrane is covered with stratified squamous epithelium.

## Nerve Supply of the Larynx

### Sensory Nerves

**Above the vocal cords:** The internal laryngeal branch of the superior laryngeal branch of the vagus.
**Below the level of the vocal cords:** The recurrent laryngeal nerve (Fig. 2-29).

### Motor Nerves

All the intrinsic muscles of the larynx except the cricothyroid muscle are supplied by the recurrent laryngeal nerve. The cricothyroid muscle is supplied by the external laryngeal branch of the superior laryngeal branch of the vagus.

## Blood Supply of the Larynx

**Upper half of the larynx:** The superior laryngeal branch of the superior thyroid artery.
**Lower half of the larynx:** The inferior laryngeal branch of the inferior thyroid artery.

## Lymph Drainage of the Larynx

The lymph vessels drain into the deep cervical group of nodes.

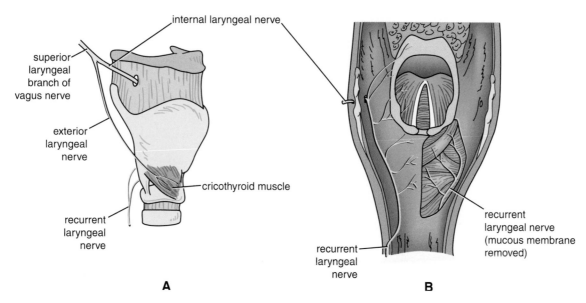

**Figure 2-29  A.** Lateral view of larynx showing the internal and external laryngeal branches of the superior laryngeal branch of the vagus nerve. **B.** The distribution of the terminal branches of the internal and recurrent laryngeal nerves. The larynx is viewed from above and posteriorly.

# The Trachea

The trachea is a mobile cartilaginous and membranous tube (Fig. 2-30). It begins as a continuation of the larynx at the lower border of the cricoid cartilage at the level of the sixth cervical vertebra. It descends in the midline of the neck. In the thorax, the trachea ends at the **carina** by dividing into right and left principal (main) bronchi at the level of the sternal angle (opposite the disc between the fourth and fifth thoracic vertebrae), where it lies slightly to the right of the midline. During expiration, the bifurcation rises by about one vertebral level, and during deep inspiration, the bifurcation may be lowered as far as the sixth thoracic vertebra, a distance of about 3 cm.

In adults, the trachea is about 4½ in (11.25 cm) long and 1 in (2.5 cm) in diameter. In infants, the trachea is about 1.6 to 2 in (4 to 5 cm) long and may be as small as 3 mm in diameter. As children grow, the diameter in millimeters corresponds approximately to their age in years. The fibroelastic tube is kept patent by the presence of U-shaped cartilaginous bars (rings) of hyaline cartilage embedded in its wall. The posterior free ends of the cartilage are connected by smooth muscle, the **trachealis muscle.**

The mucous membrane of the trachea is lined with pseudostratified ciliated columnar epithelium (Fig. 2-31) and contains many goblet cells and tubular mucous glands.

## Relations of the Trachea in the Neck (Fig. 2-32)

**Anteriorly:** Skin, fascia, isthmus of the thyroid gland (in front of the second, third, and fourth rings), inferior thyroid vein, jugular arch, thyroidea ima artery (if present), and the left brachiocephalic vein in children, overlapped by the sternothyroid and sternohyoid muscles.

**Posteriorly:** Right and left recurrent laryngeal nerves and the esophagus.

**Laterally:** Lobes of the thyroid gland and the carotid sheath and contents.

## Relations of the Trachea in the Superior Mediastinum of the Thorax (Fig. 2-33).

**Anteriorly:** Sternum, thymus, left brachiocephalic vein, origins of brachiocephalic and left common carotid arteries, and arch of aorta.

**Posteriorly:** Esophagus and left recurrent laryngeal nerve.

**Right Side:** Azygos vein, right vagus nerve, and the pleura.

**Left Side:** Arch of aorta, left common carotid, left subclavian arteries, left vagus nerve, left phrenic nerve, and pleura.

## Nerve Supply of the Trachea

The sensory nerve supply is from the vagi and the recurrent laryngeal nerves.

## Blood Supply of the Trachea

The upper two thirds of the trachea is supplied by the inferior thyroid arteries, and the lower third is supplied by the bronchial arteries.

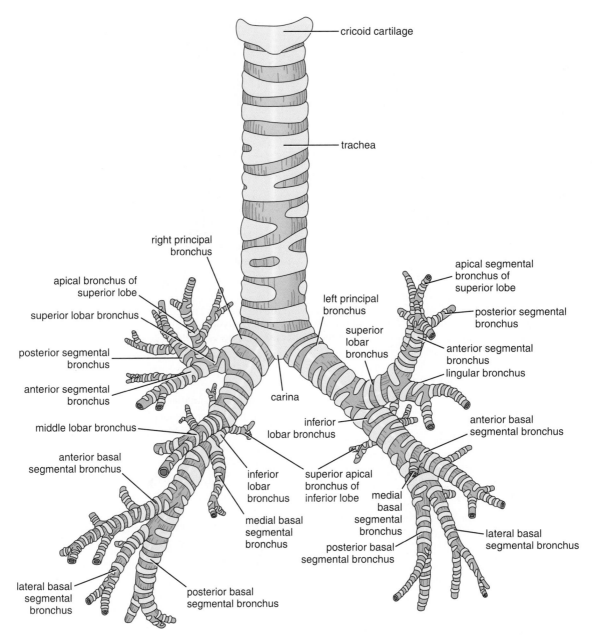

**Figure 2-30**   The trachea and the bronchi.

## Lymph Drainage of the Trachea

The lymph drains into the pretracheal and paratracheal lymph nodes and the deep cervical nodes.

# The Bronchi

The trachea bifurcates behind the arch of the aorta into the **right and left principal (primary, or main) bronchi** (Fig. 2-30). The right bronchus leaves the trachea at an angle 25 degrees from the vertical, and the left bronchus leaves the trachea at an angle 45 degrees from the vertical. In children younger than 3 years, both bronchi arise from the trachea at equal angles.

The bronchi divide dichotomously, giving rise to several million terminal bronchioles that terminate in one or more respiratory bronchiole. Each respiratory bronchiole divides into 2 to 11 alveolar ducts that enter the alveolar sacs. The alveoli arise from the walls of the sacs as diverticula.

## Right Principal Bronchus

The right principal bronchus is wider, shorter, and more vertical than the left bronchus and is about 1 in (2.5 cm) long (Fig. 2-30). The azygos vein arches over its superior border. The **superior lobar bronchus** arises within 2 cm of the commencement and is level with the carina. The right principal bronchus then enters the hilum of the right lung,

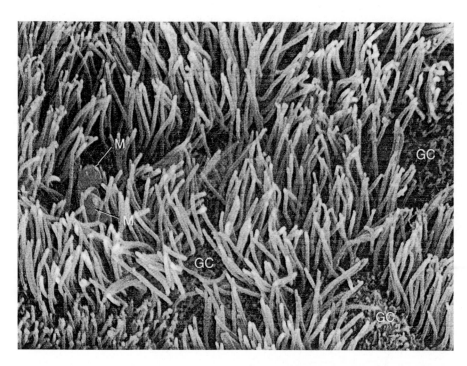

**Figure 2-31**  Scanning electron micrograph of surface epithelial cells of the mucous membrane of the trachea showing ciliated columnar cells and goblet cells (GC). M, droplet of mucus. (Courtesy of M. Koering.)

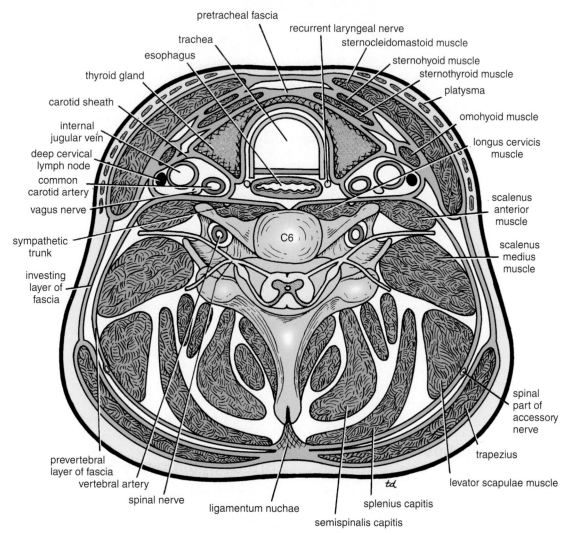

**Figure 2-32**  Cross section of the neck at the level of the sixth cervical vertebra.

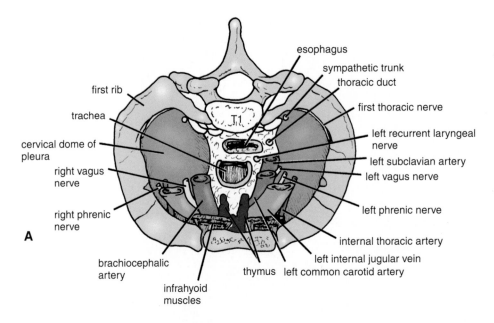

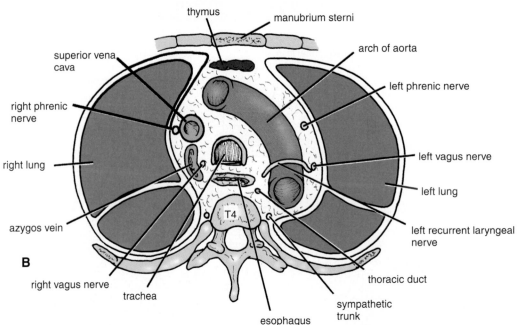

**Figure 2-33** Cross section of the chest. **A.** At the level of the first thoracic vertebra, as seen from above. **B.** At the level of the fourth thoracic vertebra, as seen from below.

where it divides into a **middle** and an **inferior lobar bronchus**.

## Left Principal Bronchus

The left principal bronchus is narrower, longer, and more horizontal than the right bronchus and is about 2 in (5 cm) long (Fig. 2-30). It passes to the left below the arch of the aorta and in front of the esophagus. On entering the hilum of the left lung, it divides into a **superior** and an **inferior lobar bronchus**.

## Bronchopulmonary Segments

The bronchopulmonary segments will be discussed with the structure of the lungs in Chapter 3.

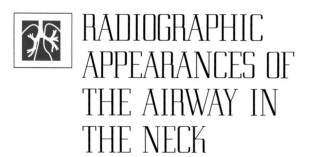

# RADIOGRAPHIC APPEARANCES OF THE AIRWAY IN THE NECK

The radiographic appearances of the larynx and trachea are shown in a lateral radiograph of the neck.

# SURFACE ANATOMY OF THE AIRWAY IN THE NECK

Preferably, the neck should be palpated while the patient is in a supine position, when the muscles overlying the deeper structures are relaxed and the structures become easier to feel.

In the midline anteriorly, the following structures may be palpated from above downward.

The **symphysis menti** may be felt where the two halves of the body of the mandible unite in the midline (Fig. 2-34).

The **body of the hyoid bone** lies opposite the body of the third cervical vertebra (Figs. 2-34 and 2-35). The hyoid bone is a horseshoe-shaped structure, and the greater cornua can be felt on each side of the neck between the finger and thumb. The hyoid bone moves superiorly when the patient swallows.

The **thyrohyoid membrane** occupies the interval between the hyoid bone and the thyroid cartilage (Fig. 2-35).

The **notched upper border of the thyroid cartilage** lies opposite the fourth cervical vertebra (Figs. 2-34 and 2-35). The anterior border of the thyroid cartilage is more prominent in male adults than in female adults.

The **cricothyroid ligament** fills in the interval between the cricoid cartilage and the thyroid cartilage (Fig. 2-35).

The **cricoid cartilage** lies at the level of the sixth cervical vertebra at the junction of the larynx with the trachea (Figs. 2-34 and 2-35). It is not as prominent as the thyroid cartilage, but it can be identified with gentle palpation when the patient is asked to swallow and the cartilage rises in the neck. **In the unresponsive patient, it can be identified as the first cartilaginous ring below the thyroid cartilage.**

The **cricotracheal ligament** fills in the interval between the cricoid cartilage and the first ring of the trachea (Fig. 2-35). It can be recognized with gentle palpation.

The **first ring of the trachea** can usually be identified by careful palpation.

The **isthmus of the thyroid gland** can be recognized as a soft structure crossing in front of the second, third, and fourth rings of the trachea (Figs. 2-34 and 2-35).

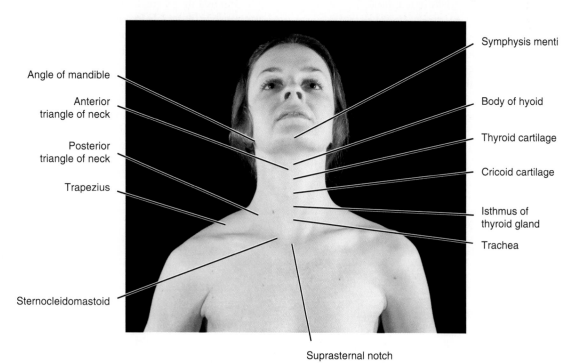

Angle of mandible

Anterior triangle of neck

Posterior triangle of neck

Trapezius

Sternocleidomastoid

Symphysis menti

Body of hyoid

Thyroid cartilage

Cricoid cartilage

Isthmus of thyroid gland

Trachea

Suprasternal notch

**Figure 2-34** Anterior view of the head and neck showing important surface landmarks.

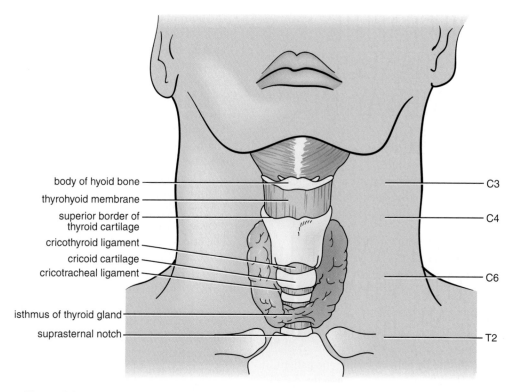

**Figure 2-35**    Head and neck of an adult showing the vertebral levels of different parts of the larynx.

The **inferior thyroid veins** and the **thyroidea ima artery** (when present), although not palpable, lie in front of the fifth, sixth, and seventh rings of the trachea. The trachea is now receding from the surface as it approaches the root of the neck.

The **brachiocephalic artery**, the **left brachiocephalic vein**, the **thymus gland**, and even the **upper margin of the arch of the aorta** are sometimes present in front of the trachea just above the suprasternal notch in young children.

The **jugular arch** connects the two anterior jugular veins just above the suprasternal notch.

The **suprasternal notch**, which is the upper margin of the manubrium sterni, can be felt between the anterior

ends of the clavicles (Figs. 2-34 and 2-35). It lies opposite the lower border of the body of the second thoracic vertebra in the midrespiratory position.

The **trachea** in the neck lies in the midline, and the examiner should be able to confirm this by inserting the index and middle finger into the grooves on either side of the trachea, between the trachea and the sternocleidomastoid muscles.

In infants, many of the important anatomical structures in the neck, referred to above, lie at different vertebral levels from those found in adults (see CD Fig. 2-6). Also, in adults, the trachea may measure as much as 1 in (2.5 cm) in diameter, whereas in infants, it is much smaller (3 mm).

# Review Questions

## Completion Questions

**Select the phrase that BEST completes each statement.**

1.  The hyoglossus muscle
    A.  changes the shape of the tongue.
    B.  elevates the tongue.
    C.  depresses the tongue.
    D.  protrudes the tongue.
    E.  retracts the tongue upward and backward.

2.  The maxillary sinuses drain into the
    A.  middle meatus of the nose.
    B.  superior meatus of the nose.
    C.  sphenoethmoidal recess.
    D.  inferior meatus of the nose.
    E.  nasolacrimal duct.

3. The frontal sinus drains into the
   A. inferior meatus of the nose.
   B. lacrimal sac.
   C. middle meatus of the nose.
   D. sphenoethmoidal recess.
   E. superior meatus of the nose.

## Multiple Choice Questions

**Select the BEST answer for each question.**

4. The following statements concerning the trachea are correct **except** which?
   A. It lies anterior to the esophagus in the neck.
   B. The left principal bronchus is wider than the right principal bronchus.
   C. The sensory innervation of the mucous membrane lining the trachea is derived from branches of the vagi and the recurrent laryngeal nerves.
   D. It begins at the level of the sixth cervical vertebra.
   E. In the adult, it measures about 4½ in (11.25 cm) long.

5. The following statements concerning the parotid salivary gland are correct **except** which?
   A. The facial nerve passes through it between the superficial and deep parts.
   B. The parotid duct pierces the buccinator muscle in the cheek and opens into the mouth.
   C. The duct passes forward in the face superficial to the masseter muscle.
   D. The parotid salivary gland is the largest salivary gland.
   E. The papilla of the parotid duct opens into the vestibule of the mouth opposite the upper third molar tooth.

6. Which of the following muscles elevates the soft palate during swallowing?
   A. Tensor veli palatini
   B. Palatoglossus
   C. Palatopharyngeus
   D. Levator veli palatini
   E. Salpingopharyngeus

7. The following processes are responsible for closing off the nasal cavity from the oropharynx during swallowing **except** which?
   A. The soft palate is made taught and rigid by the contraction of the tensor veli palatini muscles.
   B. The soft palate is raised by the contraction of the levator veli palatini muscles.
   C. The posterior wall of the pharynx is pulled forward by the contraction of the upper fibers of the superior constrictor muscles.
   D. The palatopharyngeal folds are pulled medially by the contraction of the palatopharyngeus muscles.
   E. The palatoglossal folds are pulled medially by the contraction of the palatoglossus muscles.

8. Sustained tension of the vocal cords (folds) is best achieved through the action of which of the following muscles?
   A. The thyroarytenoid
   B. The posterior cricoarytenoid
   C. The cricothyroid
   D. The lateral cricoarytenoid
   E. The transverse arytenoid

9. The following structures take part in the formation of the lateral wall of the external nose **except** which?
   A. The ethmoid bone
   B. The nasal part of the frontal bone
   C. The frontal process of the maxilla
   D. The upper lateral nasal cartilage
   E. The nasal bone

10. The following structures take part in the formation of the lips of the mouth **except** which?
    A. The skin
    B. The masseter muscle
    C. The orbicularis oris muscle
    D. The zygomaticus major
    E. The mucous membrane
    F. The branches of the facial artery and vein

11. A patient having lunch accidentally bites the side of her tongue. To which lymph nodes are infecting bacteria likely to spread?
    A. Submental nodes
    B. Submandibular nodes
    C. Superficial cervical nodes
    D. Parotid nodes
    E. Mastoid nodes

**Read the case history and select the BEST answer to the question following it.**

A 4-week-old baby boy was examined by a pediatrician because of failure to gain weight and difficulty with feeding. The mother said that the child was breast-fed and eagerly accepted the milk when it was manually expressed from the breast but obviously was having difficulty in sucking at the nipple. The physician carefully examined the baby and then made a diagnosis and advised appropriate treatment.

12. The following statements about this case are correct **except** which?
    A. The condition is often associated with a cleft upper lip.
    B. The baby had a median cleft palate.
    C. The cleft in the palate involved the hard palate but not the soft palate or the uvula.
    D. The difficulty with the feeding was that the cleft palate prevented the child from actively sucking milk from the breast.
    E. Surgical repair of a cleft palate should be undertaken at or before 18 months.

# Answers and Explanations

1. **C** is correct. The hyoglossus muscle depresses the tongue.

2. **A** is correct. The maxillary sinus drains through the hiatus semilunaris into the middle meatus of the nose (Fig. 2-7).

3. **C** is correct. The frontal sinus drains into the middle meatus of the nose via the infundibulum (Fig. 2-2).

4. **B** is the incorrect statement. The right principal bronchus is wider than the left principal bronchus.

5. **E** is the incorrect statement. The papilla of the parotid duct opens into the vestibule of the mouth opposite the upper second molar tooth (Fig. 2-8).

6. **D** is correct. The levator veli palatini muscle elevates the soft palate during swallowing.

7. **E** is the incorrect statement. The contraction of the palatoglossus muscles does not take part in the closing off of the nasal cavity during swallowing. They assist the tongue in moving the bolus of food from the mouth backward into the oropharynx.

8. **C** is correct. The cricothyroid muscle tilts the cricoid cartilage and the arytenoid cartilages backward and thus tenses the vocal cords (see p 66).

9. **A** is correct. The ethmoid bone does not form part of the lateral wall of the external nose (see p 37).

10. **B** is the incorrect statement. The masseter is a muscle of mastication and does not form part of the lips.

11. **B** is correct. The side of the tongue is drained into the submandibular lymph nodes (see p 49).

12. **C** is the incorrect statement. During development, the palatal processes of the maxilla grow medially and fuse with each other and the nasal septum; the fusion of the processes takes place from anterior to posterior so that the uvula is the last part of the palate to fuse and this occurs at about the 11th week. If the pediatrician had made a more thorough examination in a good light, he would have seen that the cleft in the hard palate extended all the way posteriorly to the tip of the uvula.

# 3

# The Chest Wall, Chest Cavity, Lungs, and Pleural Cavities

All clinical material relevant to this chapter can be found on the CD-ROM.

# Chapter Outline

An understanding of the structure of the chest wall and the diaphragm is essential if one is to understand the normal movements of the chest wall in the process of aeration of the lungs.

The thoracic cage also performs a protective function, not only for the lungs, but also for other life-sustaining organs such as the heart and major blood vessels. In addition, the lower part of the cage overlaps the upper abdominal organs, such as the liver, stomach, and spleen, and offers them considerable protection. Although the chest wall is strong, blunt or penetrating wounds can injure the soft organs beneath it. This is especially so in an era in which automobile accidents, stab wounds, and gunshot wounds are commonplace.

 BASIC ANATOMY

# Chest Wall

The chest wall is formed by the sternum, ribs, and the costal cartilages (Fig. 3-1).

## Sternum

The sternum lies in the midline of the anterior chest wall. It is a flat bone that is divided into three parts: manubrium sterni, body of the sternum, and xiphoid process.

The **manubrium** forms the upper part of the sternum. It articulates with the body of the sternum at the manubriosternal joint, and it also articulates with the clavicles and with the first costal cartilage and the upper part of the second costal cartilages on each side (Fig. 3-1). It lies opposite the third and fourth thoracic vertebrae.

The **body of the sternum** articulates above with the manubrium at the **manubriosternal joint** and below with the xiphoid process at the **xiphisternal joint.** On each side, it articulates with the second to the seventh costal cartilages (Fig. 3-1).

The **xiphoid process** (Fig. 3-1) is a thin plate of cartilage that becomes ossified at its proximal end during adult life. No ribs or costal cartilages are attached to it.

## Ribs

There are 12 pairs of ribs, all of which are attached posteriorly to the thoracic vertebrae (Figs. 3-1 and 3-2). The ribs are divided into the following three categories:

**True ribs:** The upper seven pairs are attached anteriorly to the sternum by their costal cartilages.
**False ribs:** The eighth, ninth, and tenth pairs of ribs are attached anteriorly to each other and to the seventh rib by means of their costal cartilages and small synovial joints.
**Floating ribs:** The eleventh and twelfth pairs have no anterior attachment.

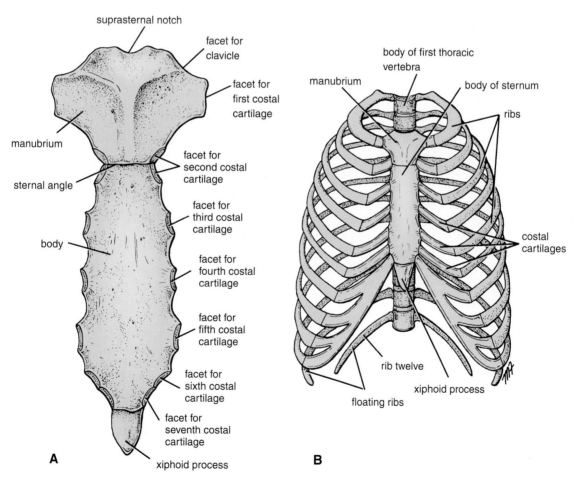

**Figure 3-1** **A.** Anterior view of the sternum. **B.** Sternum, ribs, and costal cartilages forming the thoracic skeleton.

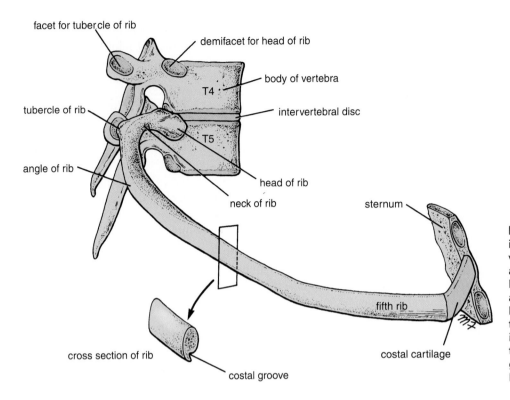

facet for tubercle of rib

demifacet for head of rib

body of vertebra

T4

intervertebral disc

tubercle of rib

T5

angle of rib

head of rib

neck of rib

sternum

cross section of rib

fifth rib

costal cartilage

costal groove

**Figure 3-2** Fifth right rib as it articulates with the vertebral column posteriorly and the sternum anteriorly. Note that the rib head articulates with the vertebral body of its own number and that of the vertebra immediately above. Note also the presence of the costal groove along the inferior border of the rib.

## Typical Rib

The typical rib is a long, twisted, flat bone having a rounded superior border and a grooved inferior border (the **costal groove**), which accommodates the intercostal vessels and nerve. The anterior end of each rib is attached to the corresponding costal cartilage (Fig. 3-2).

A rib has a **head, neck, tubercle, shaft,** and **angle** (Fig. 3-2). The head has two facets for articulation with the numerically corresponding vertebral body and that of the vertebra immediately above (Fig. 3-2). The neck is a constricted portion situated between the head and the tubercle. The tubercle is a prominence on the outer surface of the rib and has a facet for articulation with the transverse process of the numerically corresponding vertebra (Fig. 3-2). The shaft is thin and flattened and twisted on its long axis. Its inferior border has the costal groove. The angle is where the shaft of the rib bends sharply forward.

## Atypical Rib

The **first rib** is important clinically because of its close relationship to the lower nerves of the brachial plexus and the main vessels to the arm, namely, the subclavian artery and vein (Fig. 3-3). This rib is small and flattened from above downward. The scalenus anterior muscle is attached to its upper surface and inner border. Anterior to the scalenus anterior, the subclavian vein crosses the rib; posterior to the muscle attachment, the subclavian artery and lower trunk of the brachial plexus cross the rib and lie in contact with the bone.

## Costal Cartilages

The costal cartilages are bars of cartilage connecting the upper seven ribs to the lateral edge of the sternum and the eighth, ninth, and tenth ribs to the cartilage immediately above them (Fig. 3-1). The cartilages of the eleventh and twelfth ribs end in the abdominal musculature (Fig. 3-1).

The costal cartilages contribute significantly to the elasticity and mobility of the thoracic walls. In old age, the costal cartilages tend to lose some of their flexibility as the result of superficial calcification.

## Joints of the Chest Wall

### Joints of the Sternum

The manubriosternal joint is a cartilaginous joint between the manubrium and the body of the sternum (Fig. 3-1). A small amount of angular movement is possible during respiration.

The xiphisternal joint is a cartilaginous joint between the xiphoid process (cartilage) and the body of the sternum (Fig. 3-1). The xiphoid process usually fuses with the body of the sternum during middle age.

### Joints of the Ribs

#### The Joints of the Heads of the Ribs

From the second to the ninth ribs, the head articulates by a synovial joint with the corresponding vertebral body and that of the vertebra above it (Fig. 3-2). There is a strong **intra-articular ligament** that connects the head to the

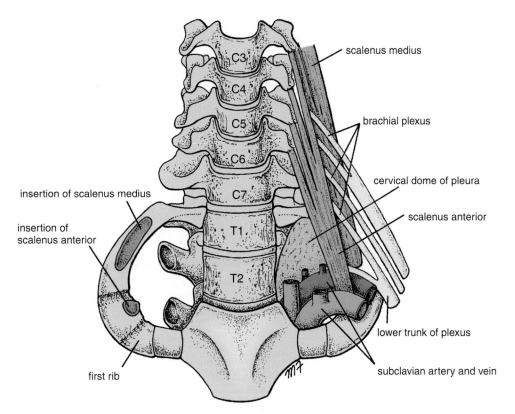

**Figure 3-3** Thoracic outlet showing the cervical dome of pleura on the left side of the body and its relationship to the inner border of the first rib. Note the presence of the brachial plexus and subclavian vessels.

intervertebral disc. The heads of the first rib and the lowest three ribs have a single synovial joint with their corresponding vertebral body.

### The Joints of the Tubercles of the Ribs

The tubercle of a rib articulates by means of a synovial joint with the transverse process of the corresponding vertebra (Fig. 3-2). (This joint is absent on the eleventh and twelfth ribs.)

## Joints of the Ribs and Costal Cartilages

These joints are cartilaginous joints. No movement is possible.

## Joints of the Costal Cartilages with the Sternum

The first costal cartilages articulate with the manubrium by cartilaginous joints that permit no movement (Fig. 3-1). The second to the seventh costal cartilages articulate with the lateral border of the sternum by synovial joints. In addition, the sixth, seventh, eighth, ninth, and tenth costal cartilages articulate with one another along their borders by small synovial joints. The cartilages of the eleventh and twelfth ribs are embedded in the abdominal musculature.

### Movements of the Ribs and Costal Cartilages

The first ribs and their costal cartilages are fixed to the manubrium and are immobile. The raising and lowering of the ribs during respiration are accompanied by movements in both the joints of the head and the tubercle, permitting the neck of each rib to rotate around its own axis.

## Thoracic Outlet

The chest cavity communicates with the root of the neck through an opening called the **thoracic outlet.** It is called an outlet because important vessels and nerves emerge from the thorax here to enter the neck and upper limbs (Fig. 3-3). The opening is bounded posteriorly by the first thoracic vertebra, laterally by the medial borders of the first ribs and their costal cartilages, and anteriorly by the superior border of the manubrium sterni. The opening is obliquely placed facing upward and forward. Through this small opening pass the esophagus and trachea and many vessels and nerves. Because of the obliquity of the opening, the apices of the lung and pleurae project upward into the neck (Fig. 3-3).

## Intercostal Spaces

The spaces between the ribs contain three muscles of respiration: the external intercostal, the internal intercostal, and the innermost intercostal muscles. The innermost intercostal muscle is lined internally by the **endothoracic fascia,** which is lined internally by the parietal pleura. The intercostal nerves and blood vessels run between the intermediate and deepest layers of muscles (Fig. 3-4). They are arranged in the following order from above downward: intercostal vein, intercostal artery, and intercostal nerve (i.e., **VAN**).

## Intercostal Muscles

The **external intercostal muscle** forms the most superficial layer. Its fibers are directed downward and forward from the inferior border of the rib above to the superior border of the rib below (Fig. 3-4). The muscle extends forward to the costal cartilage, where it is replaced by an aponeurosis, the **anterior (external) intercostal membrane** (Fig. 3-5).

The **internal intercostal muscle** forms the intermediate layer. Its fibers are directed downward and backward from the subcostal groove of the rib above to the upper border of the rib below (Fig. 3-4). The muscle extends backward from the sternum in front to the angles of the ribs behind, where the muscle is replaced by an aponeurosis, the **posterior (internal) intercostal membrane** (Fig. 3-5).

The **innermost intercostal muscle** forms the deepest layer (Fig. 3-4) and corresponds to the transversus abdominis muscle in the anterior abdominal wall. It is an incomplete muscle layer and crosses more than one intercostal space within the ribs. It is related internally to fascia

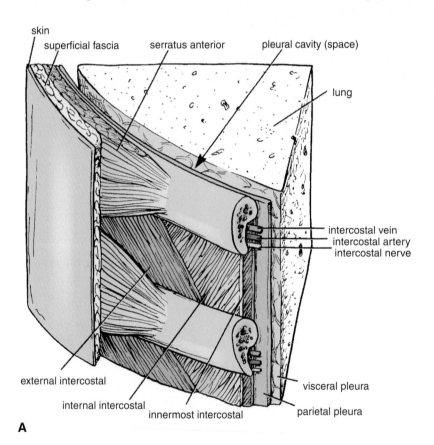

A

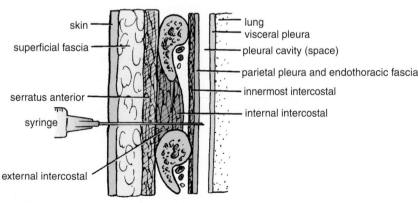

B

**Figure 3-4   A.** Section through an intercostal space. **B.** Structures penetrated by a needle when it passes from skin surface to pleural cavity. Depending on the site of penetration, the pectoral muscles will be pierced in addition to the serratus anterior muscle.

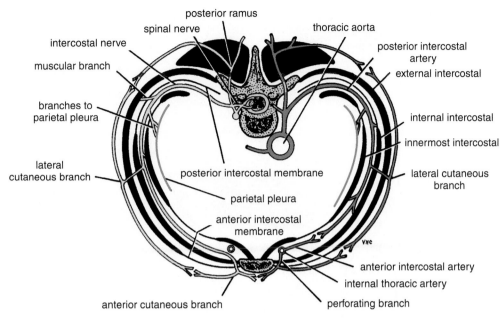

**Figure 3-5** Cross section of the thorax showing distribution of a typical intercostal nerve and a posterior and an anterior intercostal artery.

(endothoracic fascia) and parietal pleura and externally to the intercostal nerves and vessels. The innermost intercostal muscle can be divided into three portions (Fig. 3-5), which are more or less separate from one another.

## Action

When the intercostal muscles contract, they all tend to pull the ribs nearer to one another. If the first rib is fixed by the contraction of the muscles in the root of the neck, namely, the scaleni muscles, the intercostal muscles raise the second to the twelfth ribs toward the first rib, as in inspiration. If, conversely, the twelfth rib is fixed by the quadratus lumborum muscle and the oblique muscles of the abdomen, the first to the eleventh ribs will be lowered by the contraction of the intercostal muscles, as in expiration. In addition, the tone of the intercostal muscles during the different phases of respiration serves to strengthen the tissues of the intercostal spaces, thus preventing the sucking in or the blowing out of the tissues with changes in intrathoracic pressure. For further details concerning the action of these muscles, see **Mechanics of Respiration** on page 108.

## Nerve Supply

Intercostal nerves.

## Intercostal Arteries and Veins

Each intercostal space possesses a large, single **posterior intercostal artery** and two small **anterior intercostal arteries.**

▦ The **posterior intercostal arteries** of the first two spaces are branches from the superior intercostal artery, a branch of the costocervical trunk of the subclavian artery.

The posterior intercostal arteries of the lower nine spaces are branches of the descending thoracic aorta (Figs. 3-5 and 3-6).

▦ The **anterior intercostal arteries** of the first six spaces are branches of the internal thoracic artery (Figs. 3-5 and 3-6), which arises from the first part of the subclavian artery. The anterior intercostal arteries of the lower spaces are branches of the musculophrenic artery (one of the terminal branches of the internal thoracic artery).

Each intercostal artery gives off branches to the muscles, skin, and parietal pleura. In the region of the breast in the female, the branches to the superficial structures are particularly large.

The corresponding **posterior intercostal veins** drain backward into the azygos or hemiazygos veins (Figs. 3-6 and 3-7), and the **anterior intercostal veins** drain forward into the internal thoracic and musculophrenic veins.

## Intercostal Nerves

The intercostal nerves are the anterior rami of the first 11 thoracic spinal nerves (Fig. 3-8). The anterior ramus of the twelfth thoracic nerve lies in the abdomen and runs forward in the abdominal wall as the **subcostal nerve.** Refer to Figure 17-25.

Each intercostal nerve enters an intercostal space between the parietal pleura and the posterior intercostal membrane (Figs. 3-4 and 3-5). It then runs forward inferiorly to the intercostal vessels in the subcostal groove of the corresponding rib, between the innermost intercostal and internal intercostal muscles. The first six nerves are distributed within their intercostal spaces. The seventh to ninth intercostal

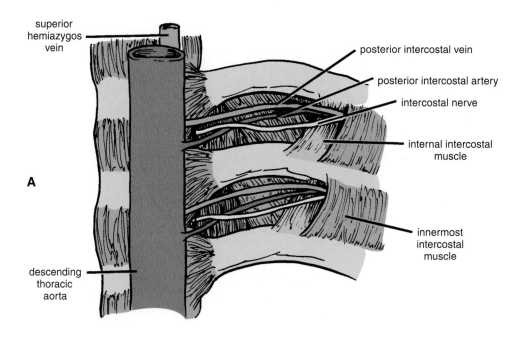

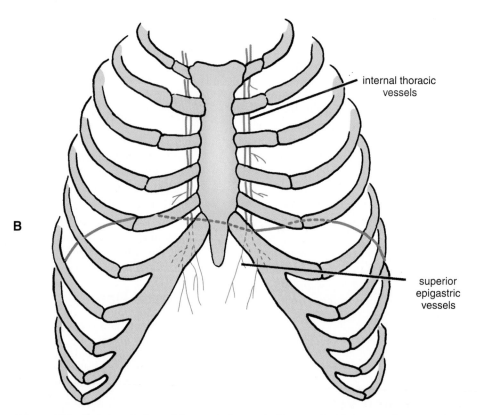

**Figure 3-6   A.** Internal view of the posterior end of two typical intercostal spaces; the posterior intercostal membrane has been removed for clarity. **B.** Anterior view of the chest showing the courses of the internal thoracic vessels. These vessels descend about one fingerbreadth from the lateral margin of the sternum.

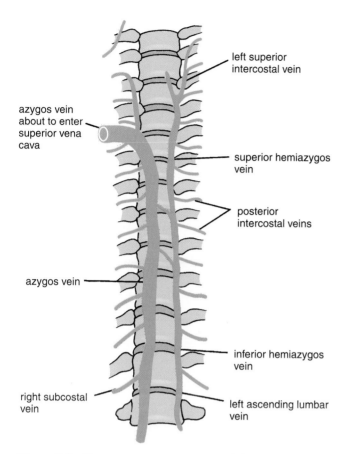

**Figure 3-7** The common arrangement of the azygos vein, the superior hemiazygos (accessory hemiazygos) vein, and the inferior hemiazygos (hemiazygos) vein.

The **first intercostal nerve** is joined to the brachial plexus by a large branch that is equivalent to the lateral cutaneous branch of typical intercostal nerves. The remainder of the first intercostal nerve is small, and there is no anterior cutaneous branch.

The **second intercostal nerve** is joined to the medial cutaneous nerve of the arm by a branch called the **intercostobrachial nerve,** which is equivalent to the lateral cutaneous branch of other nerves. Therefore, the second intercostal nerve supplies the skin of the armpit and the upper medial side of the arm. **In coronary artery disease, pain is referred along this nerve to the medial side of the arm.**

Thus, with the exceptions noted, the first six intercostal nerves supply the skin and the parietal pleura covering the outer and inner surfaces of each intercostal space, respectively, and the intercostal muscles of each intercostal space and the levatores costarum and serratus posterior muscles.

In addition, the seventh to the eleventh intercostal nerves supply the skin and the parietal peritoneum covering the outer and inner surfaces of the abdominal wall, respectively, and the anterior abdominal muscles, which include the external oblique, internal oblique, transversus abdominis, and rectus abdominis muscles.

## Suprapleural Membrane

Superiorly, the thorax opens into the root of the neck by a narrow aperture, the **thoracic outlet.** The outlet transmits structures that pass between the thorax and the neck (esophagus, trachea, blood vessels, etc.) and, for the most part, lie close to the midline. On either side of these structures, the outlet is closed by a dense fascial layer called the **suprapleural membrane** (see Fig. 3-15). This tent-shaped fibrous sheet is attached laterally to the medial border of the first rib and costal cartilage. It is attached at its apex to the tip of the transverse process of the seventh cervical vertebra and medially to the fascia investing the structures passing from the thorax into the neck. It protects the underlying cervical pleura and resists the changes in intrathoracic pressure occurring during respiratory movements.

# Diaphragm

The diaphragm is a thin, muscular, and tendinous septum that separates the chest cavity above from the abdominal cavity below (Fig. 3-9). It is pierced by the structures that pass between the chest and the abdomen.

The diaphragm is the most important muscle of respiration. It is dome shaped and consists of a peripheral muscular part and a centrally placed tendon. The origin of the diaphragm is divided into three parts:

- A **sternal part** arising from the posterior surface of the xiphoid process.
- A **costal part** arising from the deep surfaces of the lower six ribs and their costal cartilages (Fig. 3-9).

nerves leave the anterior ends of their intercostal spaces by passing deep to the costal cartilages to enter the anterior abdominal wall. The tenth and eleventh nerves pass forward directly into the abdominal wall.

## Branches

See Figure 3-8.

- **Rami communicantes** connect the intercostal nerve to a ganglion of the sympathetic trunk (Fig. 1-18). The gray ramus joins the nerve medial to the point at which the white ramus leaves it.
- **Collateral branch** runs forward below the main nerve.
- **Lateral cutaneous branch** reaches the skin on the side of the chest and divides into an anterior and a posterior branch.
- **Anterior cutaneous branch** forms the terminal part of the main nerve. It reaches the skin near the midline. It divides into a medial and a lateral branch.
- **Muscular branches** run to the intercostal muscles.
- **Pleural sensory branches** run to the parietal pleura.
- **Peritoneal sensory branches** (seventh to eleventh intercostal nerves only) run to the parietal peritoneum.

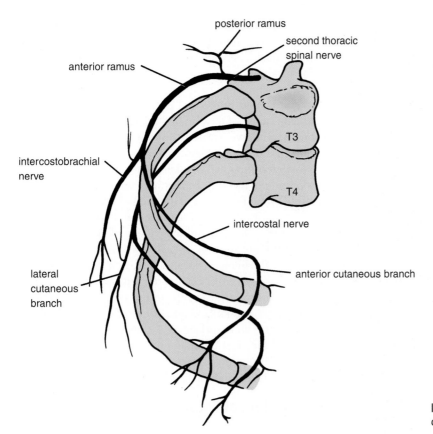

**Figure 3-8**   The distribution of two intercostal nerves relative to the rib cage.

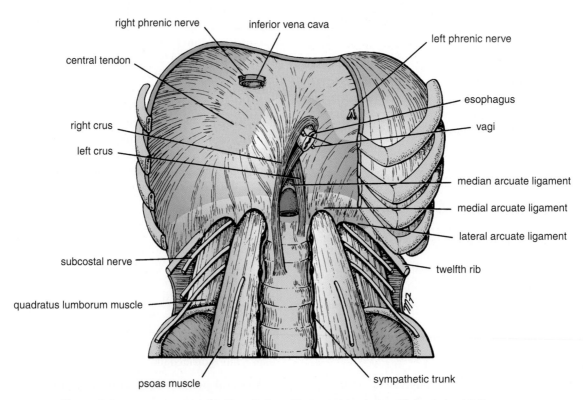

**Figure 3-9**   Diaphragm as seen from below. The anterior portion of the right side has been removed. Note the sternal, costal, and vertebral origins of the muscle and the important structures that pass through it.

■ A **vertebral part** arising by vertical columns or **crura** and from the arcuate ligaments.

The **right crus** arises from the sides of the bodies of the first three lumbar vertebrae and the intervertebral discs; the **left crus** arises from the sides of the bodies of the first two lumbar vertebrae and the intervertebral disc (Fig. 3-9 ). Lateral to the crura, the diaphragm arises from the medial and lateral arcuate ligaments (Fig. 3-9). The **medial arcuate ligament** extends from the side of the body of the second lumbar vertebra to the transverse process of the first lumbar vertebra, and the **lateral arcuate ligament** extends from the transverse process of the first lumbar vertebra to the twelfth rib. The medial borders of the two crura are connected by a **median arcuate ligament,** which crosses over the anterior surface of the aorta (Fig. 3-9). The diaphragm is inserted into a flat **central tendon.** The superior surface of the tendon is partially fused with the inferior surface of the fibrous pericardium.

## Shape and Structure of the Diaphragm

As seen from in front, the diaphragm curves up in the form of thin muscular sheets to form the **right and left domes.** The right dome reaches as high as the upper border of the fifth rib, and the left dome may reach the lower border of the fifth rib. (The right dome lies at a higher level because of the large size of the right lobe of the liver.) The central tendon lies at the level of the xiphisternal joint. The domes support the right and left lungs, whereas the central tendon supports the heart. The levels of the diaphragm vary with the phase of respiration, the posture, and the degree of distention of the abdominal viscera. The diaphragm is lower when a person is sitting or standing; it is higher in the supine position and after a large meal.

When seen from the side, the diaphragm has the appearance of an inverted J, the long limb extending up from the vertebral column and the short limb extending forward to the xiphoid process (see Fig. 3-17).

## Nerve Supply of the Diaphragm

■ **Motor nerve supply:** The right and left phrenic nerves (C3, 4, and 5).
■ **Sensory nerve supply:** The parietal pleura and peritoneum covering the central surfaces of the diaphragm is from the phrenic nerve, and the periphery of the diaphragm is from the lower six intercostal nerves.

> **P H Y S I O L O G I C   N O T E**

### Action of the Diaphragm

■ **Muscle of inspiration:** On contraction, the diaphragm pulls its central tendon down and increases the vertical diameter of the thorax.

■ **Muscle of abdominal straining:** The contraction of the diaphragm assists the contraction of the muscles of the anterior abdominal wall in raising the intra-abdominal pressure for micturition, defecation, and the giving birth to a child.
■ **Weight-lifting:** In a person taking a deep breath and holding it (fixing the diaphragm), the diaphragm assists the muscles of the anterior abdominal wall in raising the intra-abdominal pressure to such an extent that it helps support the vertebral column and prevent its flexion.
■ **Thoracoabdominal pump:** The descent of the diaphragm decreases the intrathoracic pressure and increases the intra-abdominal pressure. This mechanism assists the return of venous blood in the inferior vena cava to the right atrium and the passage of lymph upward in the thoracic duct.

## Openings in the Diaphragm

The diaphragm has three main openings:

■ The **aortic opening** lies anterior to the body of the twelfth thoracic vertebra between the crura (Fig. 3-9) and transmits the aorta, the thoracic duct, and the azygos vein.
■ The **esophageal opening** lies at the level of the tenth thoracic vertebra in a sling of muscle fibers derived from the right crus (Fig. 3-9). It transmits the esophagus, the right and left vagus nerves, the esophageal branches of the left gastric vessels, and the lymphatics from the lower third of the esophagus.
■ The **caval opening** lies at the level of the eighth thoracic vertebra in the central tendon (Fig. 3-9) and transmits the inferior vena cava and terminal branches of the right phrenic nerve.

In addition to these openings, the sympathetic splanchnic nerves pierce the crura; the sympathetic trunk passes posterior to the medial arcuate ligament on each side; and the superior epigastric vessels pass between the sternal and costal origins of the diaphragm on each side.

> **E M B R Y O L O G I C   N O T E**

### Development of the Diaphragm

The diaphragm is formed from the following structures: (1) the **septum transversum,** which forms the muscle and central tendon; (2) the two **pleuroperitoneal membranes,** which are largely responsible for the peripheral areas of the diaphragmatic pleura and peritoneum that cover its upper and lower surfaces, respectively; and (3) the **dorsal mesentery of the esophagus,** in which the crura develop.

The septum transversum is a mass of mesoderm that is formed in the neck by the fusion of the myotomes of the third, fourth, and fifth cervical segments. With the descent of the heart from the neck to the thorax, the septum is pushed caudally, pulling its nerve supply with it; thus its motor nerve supply is derived from the 3rd, 4th, and 5th cervical nerves, which are contained within the phrenic nerve.

The pleuroperitoneal membranes grow medially from the body wall on each side until they fuse with the septum transversum anterior to the esophagus and with the dorsal mesentery posterior to the esophagus. During the process of fusion, the mesoderm of the septum transversum extends into the other parts, forming all the muscles of the diaphragm.

# Accessory Muscles of Respiration

The small levator costarum muscles and the small, thin serratus posterior muscles play an insignificant role in the movements of the chest wall. A summary of the muscles of the chest wall, their nerve supply, and their actions is given in Table 3-1.

# Internal Thoracic Artery

The internal thoracic artery supplies the anterior wall of the body from the clavicle to the umbilicus. It is a branch of the

| Table 3-1 | Muscles of the Thorax | | | |
|---|---|---|---|---|
| **Name of Muscle** | **Origin** | **Insertion** | **Nerve Supply** | **Action** |
| External intercostal muscle (11) (fibers pass downward and forward) | Inferior border of rib | Superior border of rib below | Intercostal nerves | With first rib fixed, they raise ribs during inspiration and thus increase anteroposterior and transverse diameters of thorax; with last rib fixed by abdominal muscles, they lower ribs during expiration |
| Internal intercostal muscle (11) (fibers pass downward and backward) | Inferior border of rib | Superior border of rib below | Intercostal nerves | |
| Innermost intercostal muscle (incomplete layer) | Adjacent ribs | Adjacent ribs | Intercostal nerves | Assist external and internal intercostal muscles |
| Diaphragm (most important muscle of respiration) | Xiphoid process; lower six costal cartilages, first three lumbar vertebrae | Central tendon | Phrenic nerve | Very important muscle of inspiration; increases vertical diameter of thorax by pulling central tendon downward and assists in raising lower ribs; also used in abdominal straining and weight lifting |
| Levatores costarum (12) | Tip of transverse process of C7 and T1–11 vertebrae | Rib below | Posterior rami of thoracic spinal | Raise ribs and therefore inspiratory muscles |
| Serratus posterior superior | Lower cervical and upper thoracic spines | Upper ribs | Intercostal nerves | Raise ribs and therefore inspiratory muscles |
| Serratus posterior inferior | Upper lumbar and lower thoracic spines | Lower ribs | Intercostal nerves | Depresses ribs and therefore expiratory muscles |

first part of the subclavian artery in the neck. It descends vertically on the pleura behind the costal cartilages, a fingerbreadth lateral to the sternum, and ends in the sixth intercostal space by dividing into the superior epigastric and musculophrenic arteries (Fig. 3-6).

## Branches

- Two **anterior intercostal arteries** for the upper six intercostal spaces.
- **Perforating arteries,** which accompany the terminal branches of the corresponding intercostal nerves.
- The **pericardiacophrenic artery,** which accompanies the phrenic nerve and supplies the pericardium.
- **Mediastinal arteries** to the contents of the anterior mediastinum (e.g., the thymus).
- The **superior epigastric artery,** which enters the rectus sheath of the anterior abdominal wall and supplies the rectus muscle as far as the umbilicus.
- The **musculophrenic artery,** which runs around the costal margin of the diaphragm and supplies the lower intercostal spaces and the diaphragm.

# Internal Thoracic Vein

The internal thoracic veins drain into the brachiocephalic vein on each side.

# Muscles Connecting the Upper Limb and the Chest Wall

## Pectoralis Major

This thick triangular muscle covers part of the anterior chest wall (Fig. 3-10). Its lower margin forms the anterior axillary fold.

> **Origin:** Medial half of the clavicle, the sternum, and the upper six costal cartilages.
> **Insertion:** Its fibers converge and are inserted into the lateral lip of the bicipital groove of the humerus.
> **Nerve supply:** Medial and lateral pectoral nerves from the medial and lateral cords of the brachial plexus.
> **Action:** Adducts and internally rotates the shoulder joint; the clavicular fibers also flex the shoulder.

## Pectoralis Minor

This thin triangular muscle lies deep to the pectoralis major muscle (Fig. 3-11).

> **Origin:** From the third, fourth, and fifth ribs.
> **Insertion:** The fibers converge to be attached to the tip of the coracoid process of the scapula.
> **Nerve supply:** Medial pectoral nerve, a branch of the medial cord of the brachial plexus.

> **Action:** Pulls the shoulder downward and forward; if the shoulder is fixed, it elevates the ribs of origin.

## Serratus Anterior

This large, thin muscle covers the lateral chest wall (Figs. 3-10 and 3-11).

> **Origin:** From the outer surfaces of the upper eight ribs.
> **Insertion:** The anterior surface of the medial margin of the scapula, especially near the inferior angle.
> **Nerve supply:** The long thoracic nerve from the roots C5, C6, and C7 of the brachial plexus.
> **Action:** Pulls the scapula forward around the thoracic wall and rotates the scapula.

# Muscles of the Root of the Neck Associated with the First Rib

## Scalenus Anterior

The scalenus anterior is a deep muscle on the side of the neck connecting the vertebral column to the first rib. It lies beneath the sternocleidomastoid muscle and runs almost vertically downward (Fig. 3-3).

> **Origin:** Transverse processes of the third, fourth, fifth, and sixth cervical vertebrae.
> **Insertion:** Inner border of the first rib.
> **Nerve supply:** Cervical spinal nerves.
> **Action:** Elevates first rib; laterally flexes and rotates the cervical part of the vertebral column.
> **Posterior relations:** Subclavian artery, brachial plexus, and cervical dome of pleura.

## Scalenus Medius

The scalenus medius is a large muscle connecting the vertebral column to the upper surface of the first rib; it lies posterior to the scalenus anterior (Fig. 3-3).

> **Origin:** Transverse processes of upper six cervical vertebrae.
> **Insertion:** Upper surface of the first rib behind the subclavian artery.
> **Nerve supply:** Cervical spinal nerves.
> **Action:** Elevates the first rib; laterally flexes and rotates the cervical part of the vertebral column.
> **Anterior relations:** Subclavian artery, brachial plexus, and cervical domes of pleura.

# The Clavicle and Its Relationship to the Thoracic Outlet

The clavicle is a long S-shaped bone that lies horizontally at the root of the neck and articulates with the sternum and first

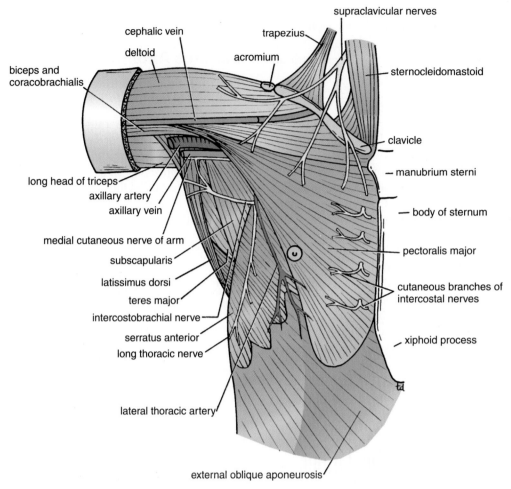

**Figure 3-10**   Pectoral region and axilla.

costal cartilage medially and with the acromion process of the scapula laterally. The clavicle acts as a strut, which holds the upper limb away from the trunk. It transmits forces from the upper limb to the axial skeleton and provides attachment for muscles. The clavicle lies just beneath the skin throughout its length (Fig. 3-10). The clavicle crosses anterior to the apex of the axilla, and thus, it is closely related to the first rib and the underlying brachial plexus and subclavian and axillary vessels (Fig. 3-11).

# The Breasts

## Location and Description

The breasts are specialized accessory glands of the skin that secrete milk (Fig. 3-12). They are present in both sexes. In males and immature females, they are similar in structure. The **nipples** are small and surrounded by a colored area of skin called the **areola.** The breast tissue consists of a system of ducts embedded in connective tissue that does not extend beyond the margin of the areola.

## Puberty

At puberty in females, the breasts gradually enlarge and assume their hemispherical shape under the influence of the ovarian hormones (Fig. 3-12). The ducts elongate, but the increased size of the glands is mainly from the deposition of fat. The base of the breast extends from the second to the sixth rib and from the lateral margin of the sternum to the midaxillary line. The greater part of the gland lies in the superficial fascia. A small part, called the **axillary tail** (Fig. 3-12), extends upward and laterally, pierces the deep fascia at the lower border of the pectoralis major muscle, and enters the axilla. Behind the breasts there is a space filled by loose areolar tissue called the **retromammary space.**

Each breast consists of 15 to 20 **lobes** that radiate out from the nipple. The main duct from each lobe opens separately on the summit of the nipple and possesses a dilated **ampulla** just before its termination. The base of the nipple is surrounded by the **areola** (Fig. 3-12). Tiny tubercles on the areola are produced by the underlying **areolar glands.**

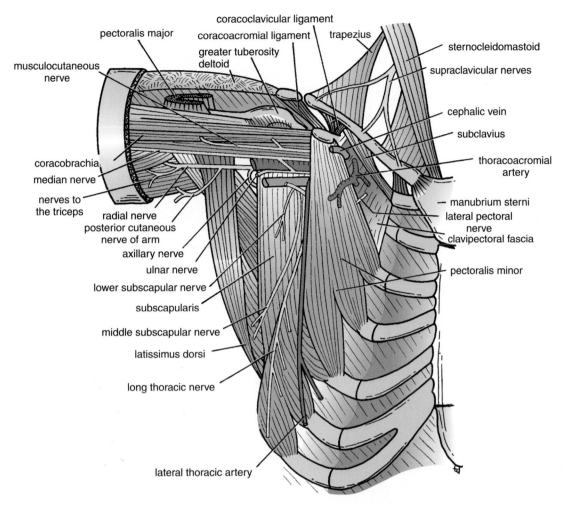

**Figure 3-11** Pectoral region and axilla; the pectoralis major muscle has been removed to display underlying structures.

The lobes of the gland are separated by fibrous septa that serve as **suspensory ligaments** (Fig. 3-12).

## Young Women

In young women, the breasts tend to protrude forward from a circular base.

## Pregnancy

- **Early:** In the early months of pregnancy, there is a rapid increase in length and branching in the duct system (Fig. 3-13). The secretory alveoli develop at the ends of the smaller ducts, and the connective tissue becomes filled with expanding and budding secretory alveoli. The vascularity of the connective tissue also increases to provide adequate nourishment for the developing gland. The nipple enlarges, and the areola becomes darker and more extensive as a result of increased deposits of melanin pigment in the epidermis. The areolar glands enlarge and become more active.

- **Late:** During the second half of pregnancy, the growth process slows. The breasts, however, continue to enlarge, mostly because of the distension of the secretory alveoli with the fluid secretion called **colostrum**.

- **Post-weaning:** Once the baby has been weaned, the breasts return to their inactive state. The remaining milk is absorbed, the secretory alveoli shrink, and most of the alveoli disappear. The interlobular connective tissue thickens. The breasts and the nipples shrink and return nearly to their original size. The pigmentation of the areola fades, but the area never lightens to its original color.

## Postmenopause

After menopause, the breast atrophies (Fig. 3-13). Most of the secretory alveoli disappear, leaving behind the ducts. The amount of adipose tissue may increase or decrease. The breasts tend to shrink in size and become more pendulous. The atrophy after menopause is caused by an absence of ovarian estrogens and progesterone.

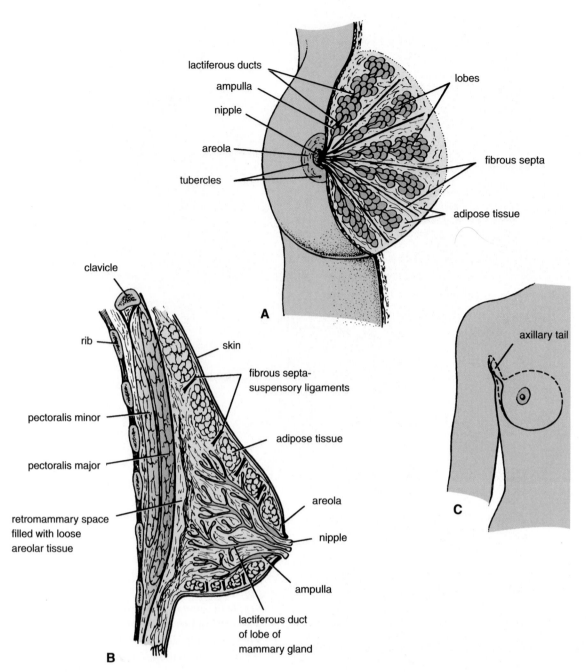

**Figure 3-12** Mature breast in the female. **A.** Anterior view with skin partially removed to show internal structure. **B.** Sagittal section. **C.** The axillary tail, which pierces the deep fascia and extends into the axilla.

## Blood Supply

### Arteries

The branches to the breasts include the perforating branches of the internal thoracic artery and the intercostal arteries, lateral thoracic and thoracoacromial arteries, and branches of the axillary artery.

### Veins

The veins correspond to the arteries.

## Lymph Drainage

The lymph drainage of the mammary gland is of considerable clinical importance because of the frequent development of cancer in the gland and the subsequent dissemination of the malignant cells along the lymph vessels to the lymph nodes.

The lateral quadrants of the breast drain into the anterior axillary or the pectoral nodes (Fig. 3-14) (situated just posterior to the lower border of the pectoralis major muscle). The medial quadrants drain by means of vessels that

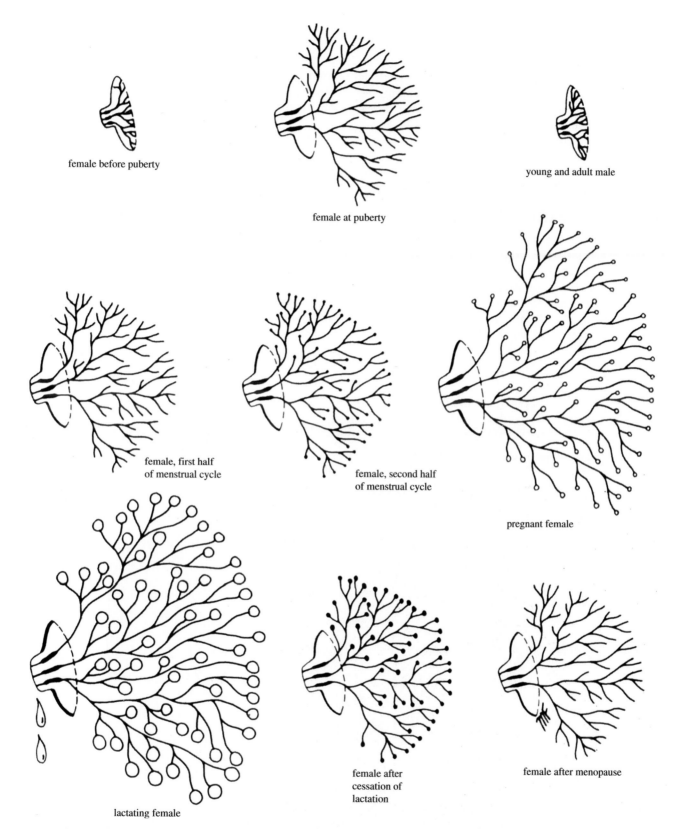

**Figure 3-13** Extent of the development of the ducts and secretory alveoli in the breasts in both sexes at different stages of activity.

pierce the intercostal spaces and enter the internal thoracic group of nodes (situated within the thoracic cavity along the course of the internal thoracic artery). A few lymph vessels follow the posterior intercostal arteries and drain posteriorly into the posterior intercostal nodes (situated along the course of the posterior intercostal arteries); some vessels communicate with the lymph vessels of the opposite breast and with those of the anterior abdominal wall.

---

### E M B R Y O L O G I C   N O T E

### Development of the Breasts

In the young embryo, a linear thickening of ectoderm appears called the **milk ridge,** which extends from the axilla obliquely to the inguinal region. In animals, several mammary glands are formed along this ridge. In the human, the ridge disappears except for a small part in the pectoral region. This localized area thickens, becomes slightly depressed, and sends off 15 to 20 solid cords, which grow into the underlying mesenchyme. Meanwhile, the underlying mesenchyme proliferates, and the depressed ectodermal thickening becomes raised to form the **nipple.** At the 5th month, the **areola** is recognized as a circular pigmented area of skin around the future nipple.

# Chest Cavity

The chest cavity is bounded by the chest wall and below by the diaphragm. It extends upward into the root of the neck about one fingerbreadth above the clavicle on each side (Fig. 3-15).

The diaphragm, which is a very thin muscle, is the only structure (apart from the pleura and peritoneum) that separates the chest from the abdominal viscera. The chest cavity is divided by a midline partition, called the mediastinum, and has laterally placed pleurae and lungs (Figs. 3-16, 3-17, and 3-18).

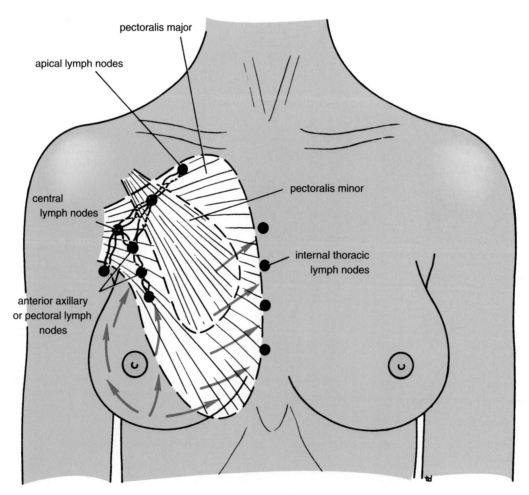

**Figure 3-14   Lymph drainage of the breast.**

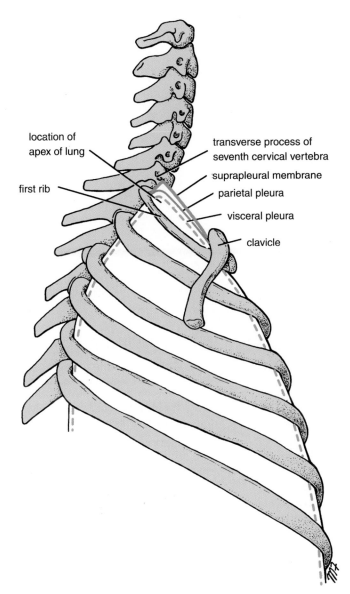

location of
apex of lung

first rib

transverse process of
seventh cervical vertebra

suprapleural membrane

parietal pleura

visceral pleura

clavicle

**Figure 3-15** Lateral view of the upper opening of the thoracic cage (thoracic outlet) showing how the apex of the lung projects superiorly into the root of the neck. The apex of the lung is covered with visceral and parietal layers of pleura and is protected by the suprapleural membrane, which is a thickening of the endothoracic fascia.

## Mediastinum

The mediastinum, although thick, is a movable partition that lies between the pleurae and the lungs (Fig. 3-18 ). It extends superiorly to the thoracic outlet and the root of the neck and inferiorly to the diaphragm. It extends anteriorly to the sternum and posteriorly to the vertebral column. It is divided into the **superior** and **inferior mediastina** by an imaginary plane passing from the sternal angle (joint between the manubrium and the body of the sternum) anteriorly to the lower border of the body of the fourth thoracic vertebra posteriorly (Fig. 3-17). The inferior mediastinum is further subdivided into the **middle mediastinum**, which consists of the pericardium and heart; the **anterior mediastinum**, which is a space between the pericardium and the sternum; and the **posterior mediastinum**, which lies between the pericardium and the vertebral column.

### Superior Mediastinum

The contents of the superior mediastinum, from anterior to posterior, include the remains of the thymus, brachiocephalic veins, the upper part of the superior vena cava, the brachiocephalic artery, the left common carotid artery, the left subclavian artery, the arch of the aorta, both phrenic and vagus nerves, left recurrent laryngeal and cardiac nerves, the trachea and lymph nodes, the esophagus and thoracic duct, and the sympathetic trunks.

### Anterior Mediastinum

The contents of the anterior mediastinum include the sternopericardial ligaments, lymph nodes, and remains of the thymus.

### Middle Mediastinum

The contents of the middle mediastinum include the pericardium, the heart and roots of great blood vessels, phrenic nerves, bifurcation of the trachea, and lymph nodes.

### Posterior Mediastinum

The contents of the posterior mediastinum include the descending thoracic aorta, esophagus, thoracic duct, azygos and hemiazygos veins, vagus nerves, splanchnic nerves, sympathetic trunks, and lymph nodes.

# Pleurae

The pleurae and lungs lie on either side of the mediastinum within the chest cavity (Fig. 3-16). The pleurae are two serous sacs surrounding and covering the lungs. Each pleura has two parts: a **parietal layer,** which lines the thoracic wall, covers the thoracic surface of the diaphragm and the lateral aspect of the mediastinum, and extends into the root of the neck; and a **visceral layer,** which completely covers the outer surfaces of the lungs and extends into the interlobar fissures (Figs. 3-16, 3-18, and 3-19). The parietal layer of each pleura becomes continuous with the visceral layer by means of a cuff of pleura that surrounds the structures entering and leaving the lung at the hilum of each lung (Figs. 3-18 and 3-19). To allow for movement of the pulmonary vessels and large bronchi during respiration, the pleural cuff hangs down as a loose fold called the **pulmonary ligament** (Fig. 3-20).

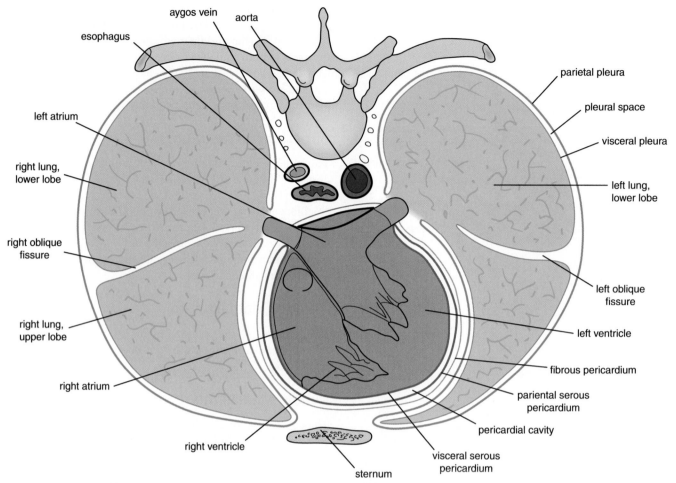

**Figure 3-16**   Cross section of the thorax at the level of the eighth thoracic vertebra. Note the arrangement of the pleura and pleural cavity (space) and the fibrous and serous pericardia.

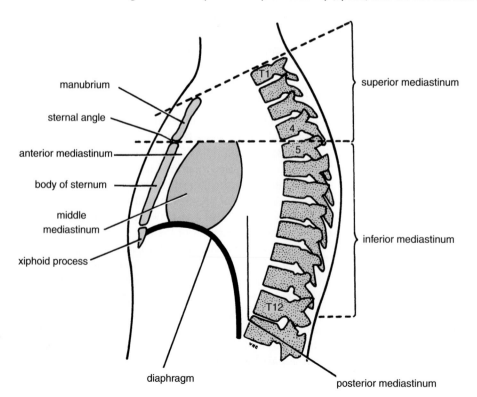

**Figure 3-17**   Subdivisions of the mediastinum.

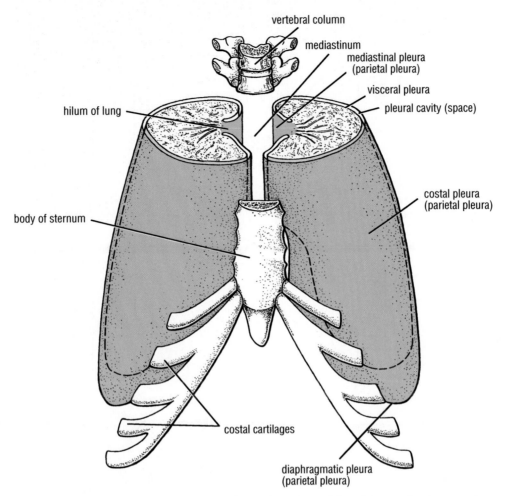

**Figure 3-18** Pleurae from above and in front. Note the position of the mediastinum and hilum of each lung.

The parietal and visceral layers of pleura are separated from one another by a slitlike space, the **pleural cavity** (Figs. 3-18 and 3-19). (Clinicians are increasingly using the term **pleural space** instead of the anatomic term **pleural cavity.** This is probably to avoid confusion between the pleural cavity [slitlike] space and the larger chest cavity.) The pleural cavity normally contains a small amount of tissue fluid, the **pleural fluid,** which lubricates the apposing pleural surfaces.

The **costodiaphragmatic recess** is the lowest area of the pleural cavity into which the lungs expand during inspiration (Figs. 3-19 and 3-20).

## Nerve Supply of the Pleura

**Parietal pleura (Fig. 3-21):** The parietal pleura is sensitive to pain, temperature, touch, and pressure and is supplied as follows:

■ The costal pleura is segmentally supplied by the intercostal nerves.
■ The mediastinal pleura is supplied by the phrenic nerve.

■ The diaphragmatic pleura is supplied over the domes by the phrenic nerve and around the periphery by the lower six intercostal nerves.

**Visceral pleura (Fig. 3-21):** The visceral pleura receives an autonomic supply from the pulmonary plexus. It is sensitive to stretch but is insensitive to common sensations such as pain and touch.

# Trachea and Principal Bronchi

The basic anatomy of these structures (Figs. 3-22 and 3-23) is described in Chapter 2 (p. 69).

# Lungs

The lungs (right and left) are situated on each side of the mediastinum (Fig. 3-16). Between them, in the mediastinum, lie the heart and great blood vessels. The lungs are

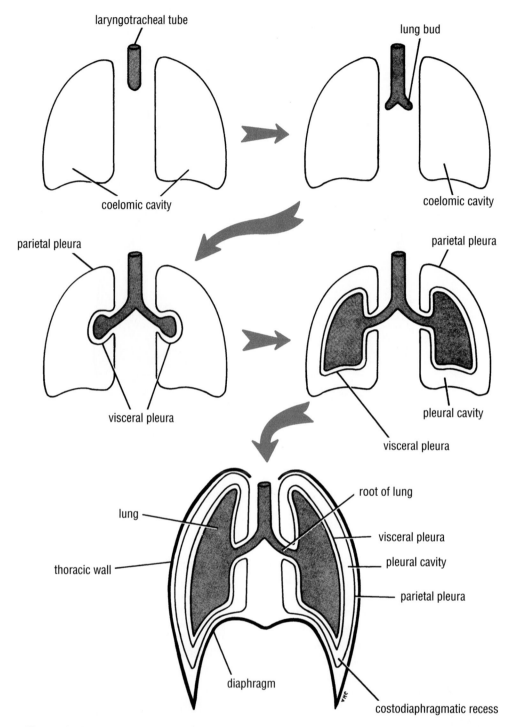

**Figure 3-19** Formation of the lungs. Note that each lung bud invaginates the wall of the coelomic cavity and then grows to fill a greater part of the cavity. Note also that the lung is covered with visceral pleura and the thoracic wall is lined with parietal pleura. The original coelomic cavity is reduced to a slitlike space called the pleural cavity as a result of the growth of the lung.

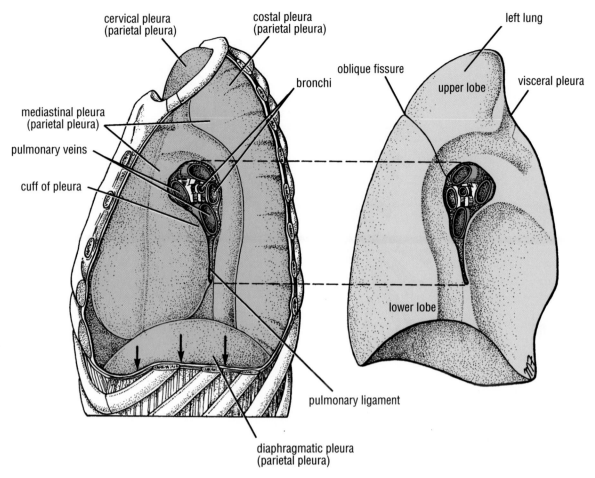

Figure 3-20   Different areas of the parietal pleura. Note the cuff of pleura (*dotted lines*) that surrounds structures entering and leaving the hilum of the left lung. It is here that the parietal and visceral layers of pleura become continuous. *Arrows* indicate the position of the costodiaphragmatic recess.

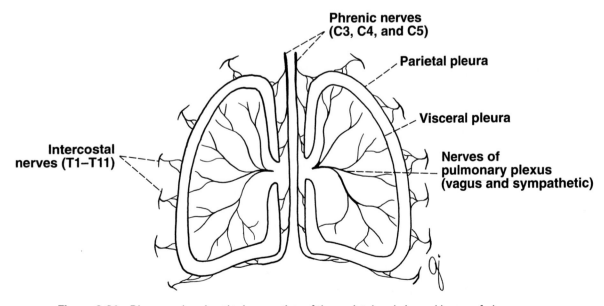

Figure 3-21   Diagram showing the innervation of the parietal and visceral layers of pleura.

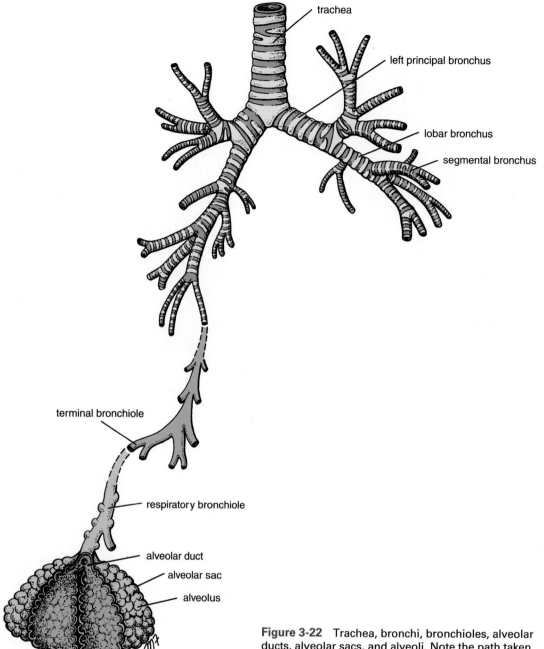

trachea

left principal bronchus

lobar bronchus

segmental bronchus

terminal bronchiole

respiratory bronchiole

alveolar duct

alveolar sac

alveolus

**Figure 3-22**   Trachea, bronchi, bronchioles, alveolar ducts, alveolar sacs, and alveoli. Note the path taken by inspired air from the trachea to the alveoli.

conical in shape and are covered with visceral pleura. The lungs are freely suspended but are attached by their roots to the mediastinum.

Each lung has a blunt **apex,** which projects upward into the neck (see Figs. 3-24 and 3-25) for about 1 in. (2.5 cm) above the clavicle, a concave **base** that sits on the diaphragm, a convex **costal surface** that corresponds to the concave chest wall, and a concave **mediastinal surface** that is molded to the pericardium and other mediastinal structures (see Figs. 3-24, 3-25, and 3-26). At about the middle of the mediastinal surface is the **hilum,** which is a depression where the bronchi, vessels, and nerves enter and leave the lung to form the **root.**

The **anterior border** is thin and overlaps the heart; and here on the left lung is a notch called the **cardiac notch.** The posterior border is thick and lies beside the vertebral column.

## Lobes and Fissures

### Right Lung

The right lung is slightly larger than the left lung, and it is divided into three lobes (the **upper, middle,** and **lower lobes**) by the oblique and the horizontal fissures (Fig. 3-24). The **oblique fissure** runs from the inferior border upward and backward across the medial and costal surfaces until it cuts

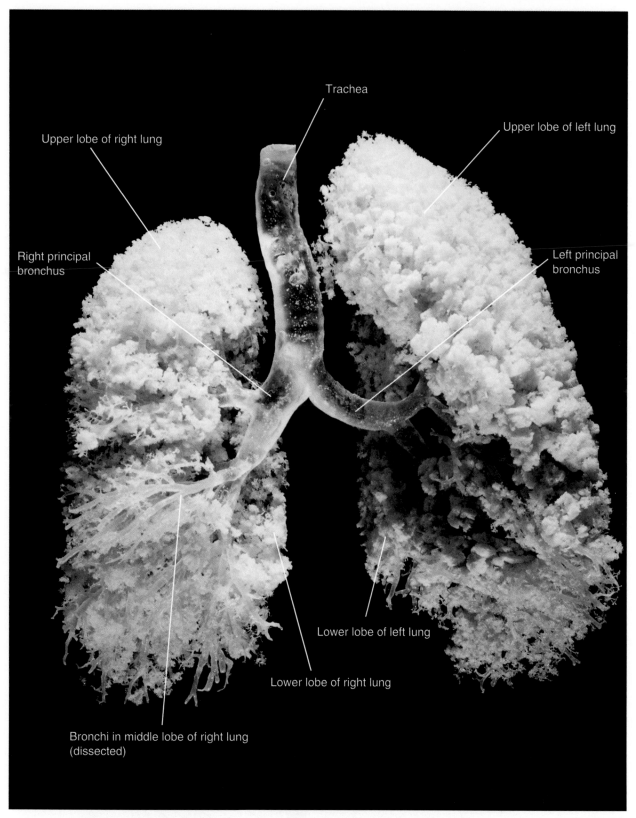

**Figure 3-23** A plastinized specimen of an adult trachea, principal bronchi, and lung; some of the lung tissue has been dissected to reveal the larger bronchi. Note that the right principal bronchus is wider and a more direct continuation of the trachea than the left principal bronchus.

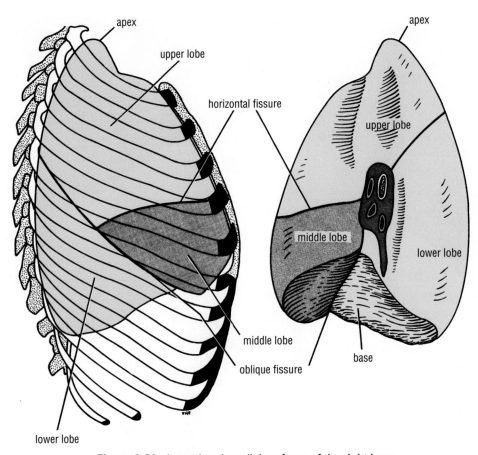

**Figure 3-24**    Lateral and medial surfaces of the right lung.

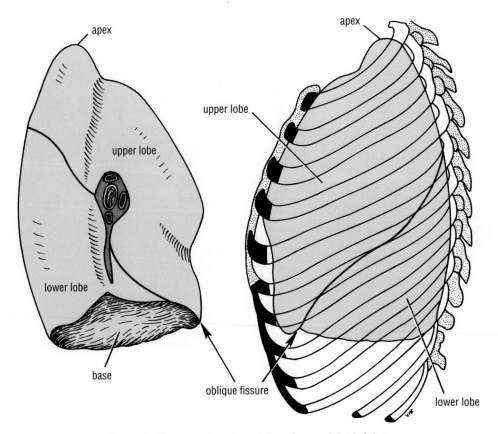

**Figure 3-25**    Lateral and medial surfaces of the left lung.

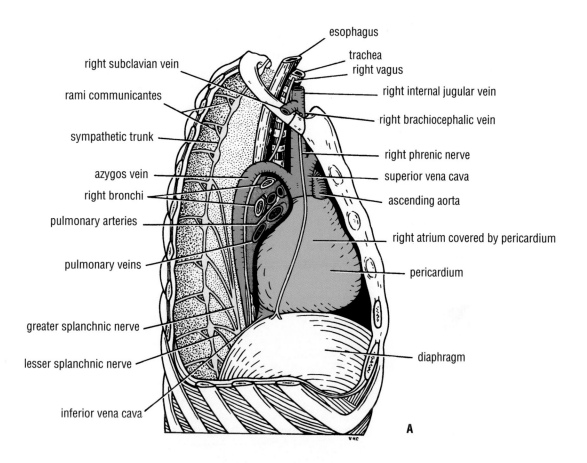

esophagus

trachea
right vagus

right subclavian vein

right internal jugular vein

rami communicantes

right brachiocephalic vein

sympathetic trunk

right phrenic nerve

azygos vein

superior vena cava

right bronchi

ascending aorta

pulmonary arteries

right atrium covered by pericardium

pulmonary veins

pericardium

greater splanchnic nerve

lesser splanchnic nerve

diaphragm

inferior vena cava

A

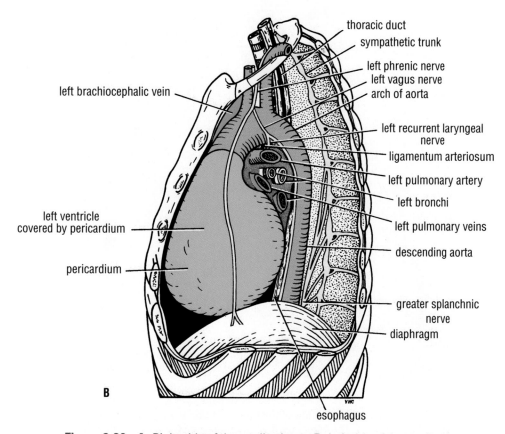

thoracic duct
sympathetic trunk

left phrenic nerve
left vagus nerve
arch of aorta

left brachiocephalic vein

left recurrent laryngeal
nerve

ligamentum arteriosum

left pulmonary artery

left bronchi

left ventricle
covered by pericardium

left pulmonary veins

pericardium

descending aorta

greater splanchnic
nerve

diaphragm

B

esophagus

**Figure 3-26   A.** Right side of the mediastinum. **B.** Left side of the mediastinum.

the posterior border. The **horizontal fissure** runs horizontally across the costal surface to meet the oblique fissure. The middle lobe is thus a small, triangular lobe bounded by the horizontal and oblique fissures.

## Left Lung

The left lung is divided by only one fissure (the oblique fissure) into two lobes: the **upper** and **lower lobes** (Fig. 3-25).

## Bronchopulmonary Segments

The bronchopulmonary segments are the anatomic, functional, and surgical units of the lungs. Each lobar (sec-

ondary) bronchus, which passes to a lobe of the lung, gives off branches called **segmental (tertiary) bronchi** (Fig. 3-22). Each segmental bronchus then enters a **bronchopulmonary segment**. A **bronchopulmonary segment** has the following characteristics:

- It is a subdivision of a lung lobe.
- It is pyramidal in shape, with its apex toward the lung root.
- It is surrounded by connective tissue.
- It has a segmental bronchus, a segmental artery, lymph vessels, and autonomic nerves.
- The segmental vein lies in the connective tissue between adjacent bronchopulmonary segments.

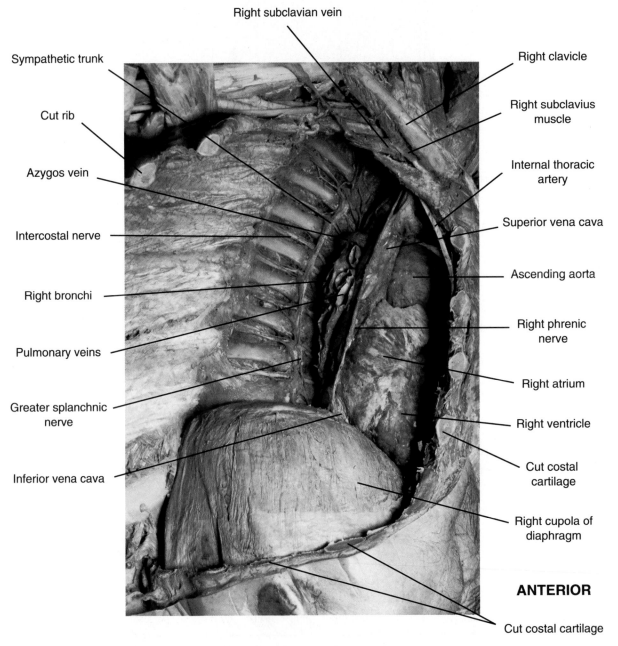

**Figure 3-27** Dissection of the right side of the mediastinum; the right lung and the pericardium have been removed. The costal parietal pleura has also been removed.

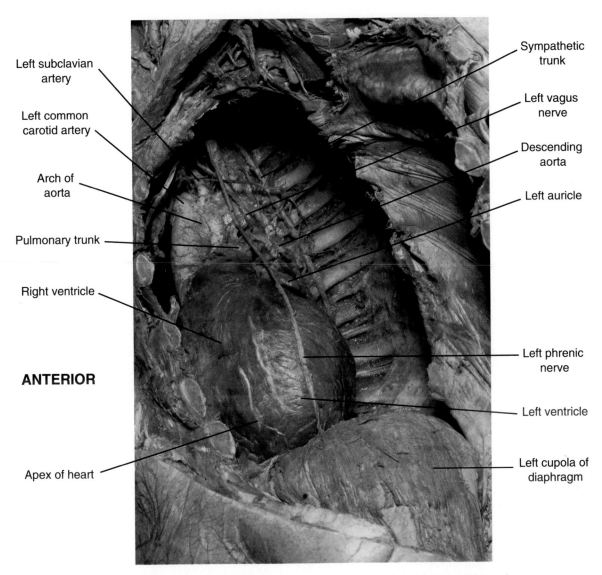

Left subclavian artery

Left common carotid artery

Arch of aorta

Pulmonary trunk

Right ventricle

**ANTERIOR**

Apex of heart

Sympathetic trunk

Left vagus nerve

Descending aorta

Left auricle

Left phrenic nerve

Left ventricle

Left cupola of diaphragm

**Figure 3-28** Dissection of the left side of the mediastinum; the left lung and the pericardium have been removed. The costal parietal pleura has also been removed.

■ Because it is a structural unit, a diseased segment can be removed surgically.

On entering a bronchopulmonary segment, each segmental bronchus divides repeatedly (see Fig. 3-29). As the bronchi become smaller, the U-shaped bars of cartilage found in the trachea are gradually replaced by plates of cartilage, which become smaller and fewer in number. The smallest bronchi divide and give rise to **bronchioles,** which are less than 1 mm in diameter (see Fig. 3-29). Bronchioles possess no cartilage in their walls and are lined with columnar ciliated epithelium. The submucosa possesses a complete layer of circularly arranged smooth-muscle fibers.

The bronchioles then divide and give rise to **terminal bronchioles** (see Fig. 3-29), which show delicate outpouchings from their walls. Gaseous exchange between blood and air takes place in the walls of these outpouchings, which

explains the name **respiratory bronchiole.** The diameter of a respiratory bronchiole is about 0.5 mm. The respiratory bronchioles end by branching into **alveolar ducts,** which lead into tubular passages with numerous thin-walled outpouchings called **alveolar sacs.** The alveolar sacs consist of several alveoli opening into a single chamber (Figs. 3-29 and 3-30). Each alveolus is surrounded by a rich network of blood capillaries. Gaseous exchange takes place between the air in the alveolar lumen through the alveolar wall into the blood within the surrounding capillaries.

The main bronchopulmonary segments (see Figs. 3-31 and 3-32) are as follows:

■ **Right lung**
**Superior lobe** Apical, posterior, anterior
**Middle lobe** Lateral, medial
**Inferior lobe** Superior (apical), medial basal, anterior basal, lateral basal, and posterior basal

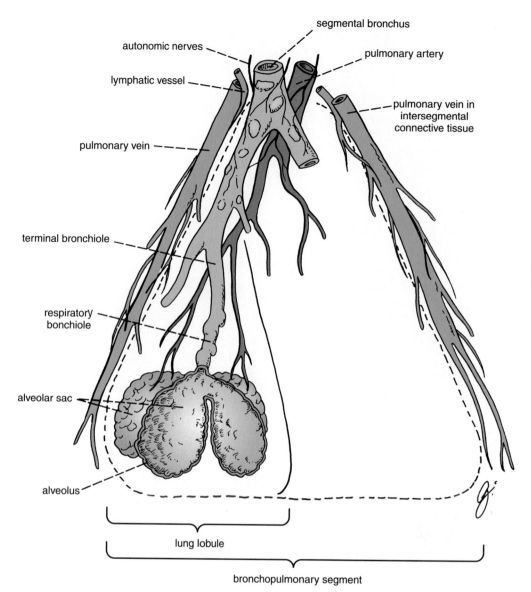

**Figure 3-29** A bronchopulmonary segment and a lung lobule. Note that the pulmonary veins lie within the connective tissue septa that separate adjacent segments.

■ Left lung
**Superior lobe** Apical, posterior, anterior, superior lingular, inferior lingular
**Inferior lobe** Superior (apical), medial basal, anterior basal, lateral basal, and posterior basal

Although the general arrangement of the bronchopulmonary segments is of clinical importance, it is unnecessary to memorize the details unless one intends to specialize in pulmonary medicine or surgery.

## Root of the Lung

The root of the lung is formed of structures that are entering or leaving the lung. It is made up of the bronchi, pulmonary artery and veins, lymph vessels, bronchial vessels, and nerves. The root is surrounded by a tubular sheath of pleura, which joins the mediastinal parietal pleura to the visceral pleura covering the lungs (Figs. 3-20 and 3-26).

## Blood Supply of the Lungs

The bronchi, the connective tissue of the lung, and the visceral pleura are supplied by the bronchial arteries, which are branches of the descending aorta. The bronchial veins drain into the azygos and hemiazygos veins.

The alveoli receive deoxygenated blood from the terminal branches of the pulmonary arteries. The oxygenated blood leaving the alveolar capillaries eventually drains into two pulmonary veins. Two pulmonary veins leave each lung

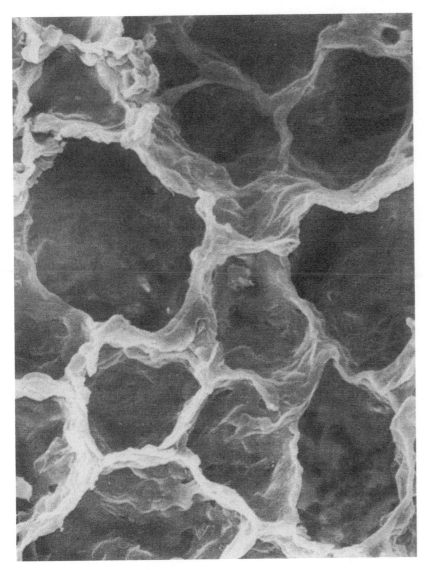

**Figure 3-30**   Scanning electron micrograph of the lung showing numerous alveolar sacs. The alveoli are the depressions, or alcoves, along the walls of the alveolar sac. (Courtesy of Dr. M. Koering.)

root (see Figs. 3-26, 3-27, and 3-28) to empty into the left atrium of the heart.

## Lymph Drainage of the Lungs

The lymph vessels originate in superficial and deep plexuses (Fig. 3-33) and are not present in the alveolar walls. The **superficial (subpleural) plexus** lies beneath the visceral pleura and drains over the surface of the lung toward the hilum, where the lymph vessels enter the **bronchopulmonary nodes.** The **deep plexus** travels along the bronchi and pulmonary vessels toward the hilum of the lung, passing through **pulmonary nodes** located within the lung substance; the lymph then enters the bronchopulmonary nodes in the hilum of the lung. All the lymph from the lung leaves the hilum and drains into the tracheobronchial nodes and then into the **bronchomediastinal lymph trunks.**

## Nerve Supply of the Lungs

At the root of each lung is a **pulmonary plexus** (Fig. 3-21). The plexus is formed from branches of the sympathetic trunk and receives parasympathetic fibers from the vagus nerve.

The sympathetic efferent fibers produce bronchodilatation and vasoconstriction. The parasympathetic efferent fibers produce bronchoconstriction, vasodilatation, and increased glandular secretion. Afferent impulses derived from the bronchial mucous membrane and from stretch receptors in the alveolar walls pass to the central nervous system in both sympathetic and parasympathetic nerves.

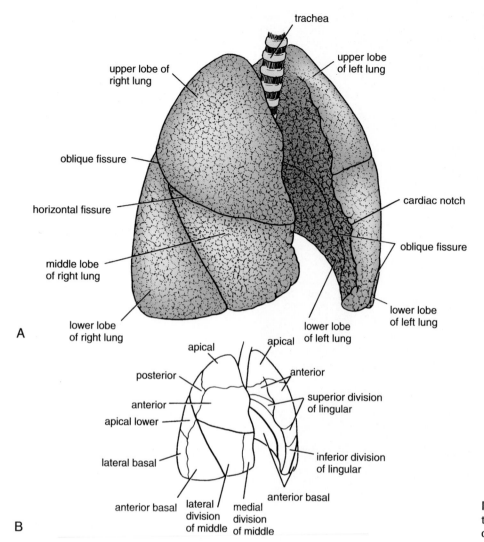

trachea

upper lobe of
right lung

upper lobe
of left lung

oblique fissure

cardiac notch

horizontal fissure

oblique fissure

middle lobe
of right lung

lower lobe
of left lung

A

lower lobe
of right lung

lower lobe
of left lung

apical

apical

posterior

anterior

anterior

superior division
of lingular

apical lower

lateral basal

inferior division
of lingular

anterior basal

anterior basal

B

lateral
division
of middle

medial
division
of middle

**Figure 3-31**   Lungs viewed from
the right. **A.** Lobes. **B.** Bronchopulmonary segments.

---

### Mechanics of Respiration

Respiration consists of two phases—inspiration and
expiration—which are accomplished by the alternate
increase and decrease of the capacity of the thoracic
cavity. The rate varies between 16 and 20 breaths per
minute in normal resting patients and is faster in children
and slower in the elderly.

# Inspiration

## Quiet Inspiration

Compare the thoracic cavity to a box with a single entrance
at the top, which is a tube called the trachea (Fig. 3-34). The
capacity of the box can be increased by elongating all of its
diameters, and this results in air under atmospheric pressure
entering the box through the tube.

Consider now the three diameters of the thoracic cavity
and how they may be increased (Figs. 3-34 and 3-35).

**Vertical diameter:** Theoretically, the roof could
be raised and the floor lowered. The roof is formed
by the suprapleural membrane and is fixed. Conversely,
the floor is formed by the mobile diaphragm. When the
diaphragm contracts, the domes become flattened and
the level of the diaphragm is lowered (Fig. 3-34).

**Anteroposterior diameter:** If the downward-sloping ribs
were raised at their sternal ends, the anteroposterior diameter of the thoracic cavity would be increased and the
lower end of the sternum would be thrust forward (Fig. 3-
34). This can be brought about by fixing the first rib by
the contraction of the scaleni muscles of the neck and
contracting the intercostal muscles (Fig. 3-35). By this
means, all the ribs are drawn together and raised toward
the first rib.

**Transverse diameter:** The ribs articulate in front with the
sternum via their costal cartilages and behind with the
vertebral column. Because the ribs curve downward as

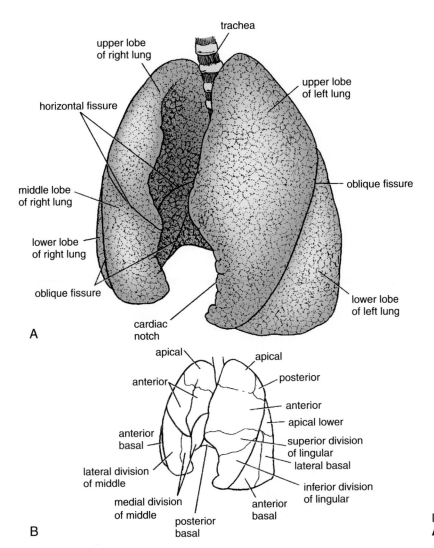

**Figure 3-32** Lungs viewed from the left. **A.** Lobes. **B.** Bronchopulmonary segments.

well as forward around the chest wall, they resemble bucket handles (Fig. 3-34). Therefore, it follows that if the ribs are raised (like bucket handles), the transverse diameter of the thoracic cavity will be increased. As described previously, this can be accomplished by fixing the first rib and raising the other ribs to it by contracting the intercostal muscles (Fig. 3-35).

An additional factor that must not be overlooked is the effect of the descent of the diaphragm on the abdominal viscera and the tone of the muscles of the anterior abdominal wall. As the diaphragm descends on inspiration, intra-abdominal pressure rises. This rise in pressure is accommodated by the reciprocal relaxation of the abdominal wall musculature. However, a point is reached when no further abdominal relaxation is possible, and the liver and other upper abdominal viscera act as a platform for further diaphragmatic descent. On further contraction, the diaphragm will now have its central tendon supported from below, and its shortening muscle fibers will assist the intercostal muscles in raising the lower ribs (Fig. 3-35).

## Forced Inspiration

In deep forced inspiration, a maximum increase in the capacity of the thoracic cavity occurs. Every muscle that can raise the ribs is brought into action, including the scalenus anterior and medius and the sternocleidomastoid. In respiratory distress, the action of all the muscles already engaged becomes more violent, and the scapulae are fixed by the trapezius, levator scapulae, and rhomboid muscles, enabling the serratus anterior and pectoralis minor to pull up the ribs. If the upper limbs can be supported by grasping a chair back or table, the sternal origin of the pectoralis major muscles can also assist the process.

## Lung Changes in Inspiration

In inspiration, the root of the lung descends, and the level of the bifurcation of the trachea may be lowered by as much as two vertebrae. The bronchi elongate and dilate, and the alveolar capillaries dilate, thus assisting the pulmonary

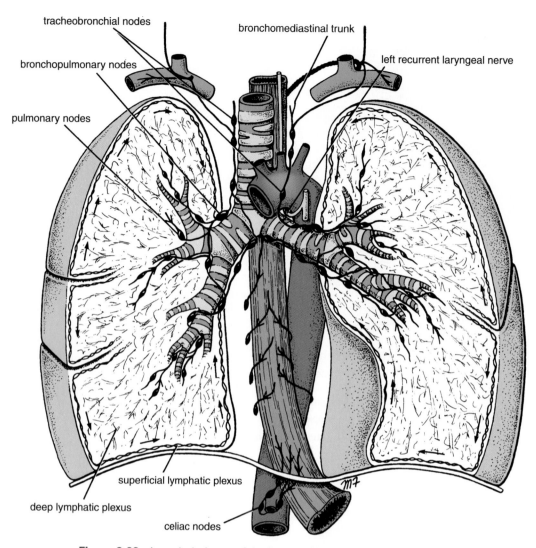

**Figure 3-33** Lymph drainage of the lung and lower end of the esophagus.

circulation. Air is drawn into the bronchial tree as the result of the positive atmospheric pressure exerted through the upper part of the respiratory tract and the negative pressure on the outer surface of the lungs brought about by the increased capacity of the thoracic cavity. With expansion of the lungs, the elastic tissue in the bronchial walls and connective tissue are stretched. As the diaphragm descends, the costodiaphragmatic recess of the pleural cavity opens, and the expanding sharp lower edges of the lungs descend to a lower level.

# Expiration

## Quiet Expiration

Quiet expiration is largely a passive phenomenon and is brought about by the elastic recoil of the lungs, the relaxation of the intercostal muscles and diaphragm, and an increase in tone of the muscles of the anterior abdominal wall, which forces the relaxing diaphragm upward.

## Forced Expiration

Forced expiration is an active process brought about by the forcible contraction of the musculature of the anterior abdominal wall. The quadratus lumborum also contracts and pulls down the twelfth ribs. It is conceivable that under these circumstances some of the intercostal muscles may contract, pull the ribs together, and depress them to the lowered twelfth rib (Fig. 3-35). The serratus posterior inferior and the latissimus dorsi muscles may also play a minor role.

## Lung Changes on Expiration

In expiration, the roots of the lungs ascend along with the bifurcation of the trachea. The bronchi shorten and contract. The elastic tissue of the lungs recoils, and the lungs become reduced in size. With the upward movement of the diaphragm, increasing areas of the diaphragmatic and costal parietal pleura come into apposition, and the costodiaphragmatic recess becomes reduced in size. The lower margins of the lungs shrink and rise to a higher level.

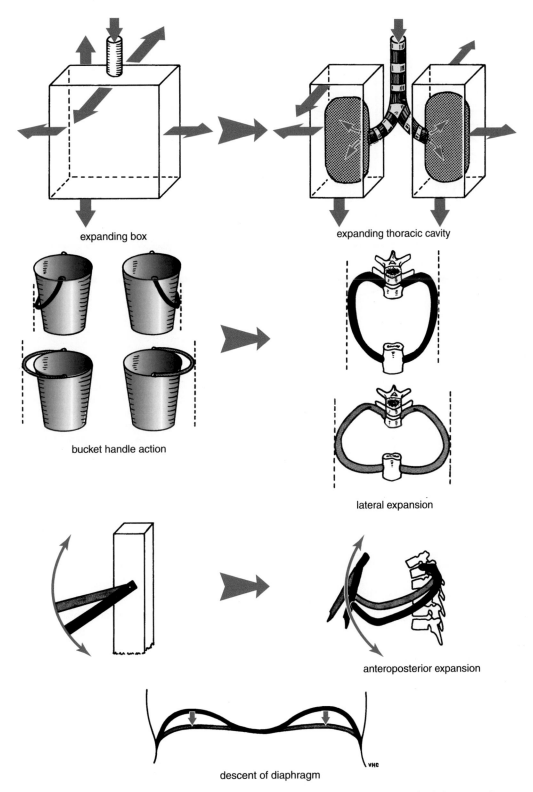

expanding box

expanding thoracic cavity

bucket handle action

lateral expansion

anteroposterior expansion

descent of diaphragm

**Figure 3-34**   The different ways in which the capacity of the thoracic cavity is increased during inspiration.

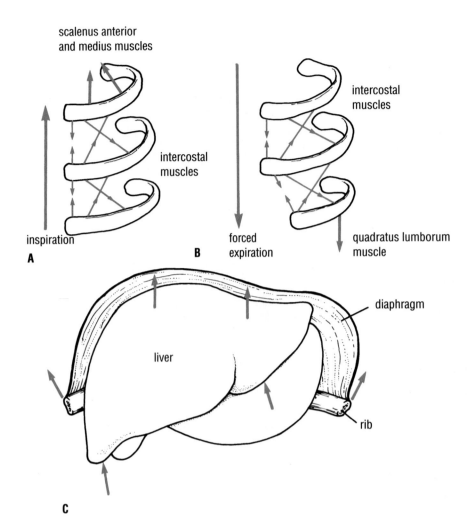

**Figure 3-35  A.** How the intercostal muscles raise the ribs during inspiration. Note that the scaleni muscles fix the first rib or, in forced inspiration, raise the first rib. **B.** How the intercostal muscles can be used in forced expiration, provided that the twelfth rib is fixed or is made to descend by the abdominal muscles. **C.** How the liver provides the platform that enables the diaphragm to raise the lower ribs.

## EMBRYOLOGIC NOTE

### Development of the Lungs and Pleura

A longitudinal groove develops in the entodermal lining of the floor of the pharynx. This groove is known as the **laryngotracheal groove.** The lining of the larynx, trachea, and bronchi and the epithelium of the alveoli develop from this groove. The margins of the groove fuse and form the **laryngotracheal tube** (Fig. 3-36). The fusion process starts distally so that the lumen becomes separated from the developing esophagus. Just behind the developing tongue, a small opening persists that will become the permanent opening into the larynx. The laryngotracheal tube grows caudally into the splanchnic mesoderm and will eventually lie anterior to the esophagus. The tube divides distally into the right and left **lung buds.** Cartilage develops in the mesenchyme surrounding the tube, and the upper part of the tube becomes the **larynx,** whereas the lower part becomes the **trachea.**

Each lung bud consists of an entodermal tube surrounded by splanchnic mesoderm; from this, all the tissues of the corresponding lung are derived. Each bud grows laterally and projects into the pleural part of the embryonic coelom (Fig. 3-36). The lung bud divides into three lobes and then into two, corresponding to the number of **main bronchi** and **lobes** found in the fully developed lung. Each main bronchus then divides repeatedly in a dichotomous manner, until eventually the **terminal bronchioles** and **alveoli** are formed. The division of the terminal bronchioles, with the formation of additional bronchioles and alveoli, continues for some time after birth.

Each lung will receive a covering of **visceral pleura** derived from the splanchnic mesoderm. The **parietal pleura** will be formed from somatic mesoderm. By the 7th month, the capillary loops connected with the pulmonary circulation have become sufficiently well developed to support life should premature birth take place. With the onset of respiration at birth, the lungs expand, and the alveoli become dilated. However, it is only after 3 or 4 days of postnatal life that the alveoli in the periphery of each lung become fully expanded.

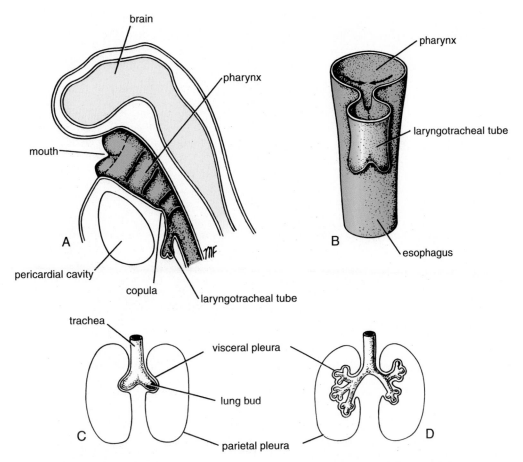

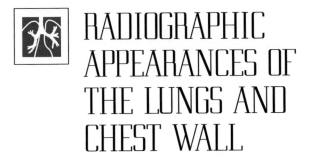

Figure 3-36 The development of the lungs. **A.** The laryngotracheal groove and tube have been formed. **B.** The margins of the laryngotracheal groove fuse to form the laryngotracheal tube. **C.** The lung buds invaginate the wall of the intra-embryonic coelom. **D.** The lung buds divide to form the main bronchi.

# RADIOGRAPHIC APPEARANCES OF THE LUNGS AND CHEST WALL

The more important features seen in standard posteroanterior and oblique lateral radiographs of the chest are shown in Figures 3-37 through 3-42. A bronchogram is also shown (Fig. 3-43). In this specialized study, iodized oil or other contrast medium is introduced into a particular bronchus or bronchi usually under a fluoroscopic control. Examples of CT scans of the chest are shown in Figures 3-44 and 3-45. Examples of cross sections of the chest viewed from below to assist in the interpretation of CT scans are shown in Figure 3-46.

# SURFACE ANATOMY OF THE TRACHEA, LUNGS, AND PLEURA

Before studying the surface anatomy of the trachea, lungs, and pleura, the surface anatomy of the front and back of the chest wall should be reviewed. This is summarized in Figures 3-47 through 3-50.

## Trachea

The trachea extends from the lower border of the cricoid cartilage (opposite the body of the sixth cervical vertebra) in the neck to the level of the sternal angle in the chest

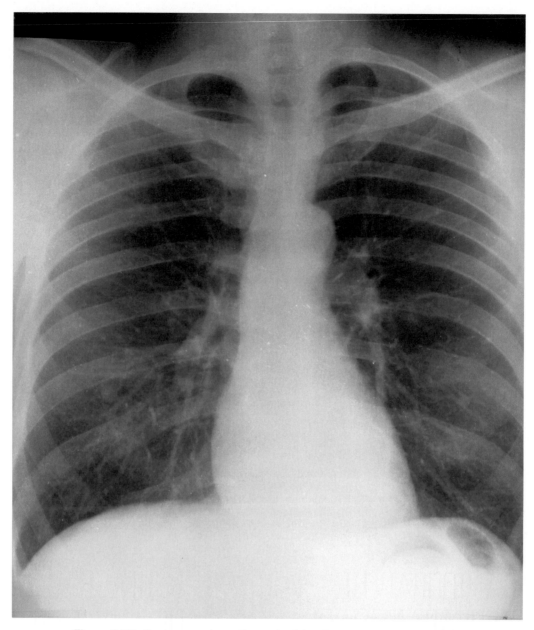

**Figure 3-37**   Posteroanterior radiograph of the chest of a normal adult man.

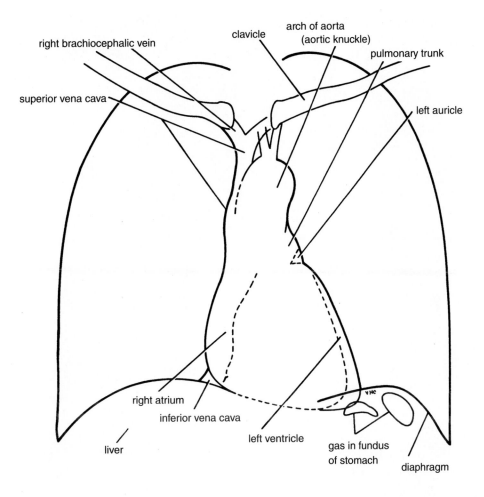

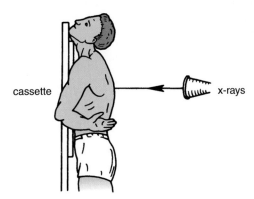

**Figure 3-38** Main features observable in the posteroanterior radiograph of the chest shown in Figure 3-37. Note the position of the patient in relation to the x-ray source and cassette holder.

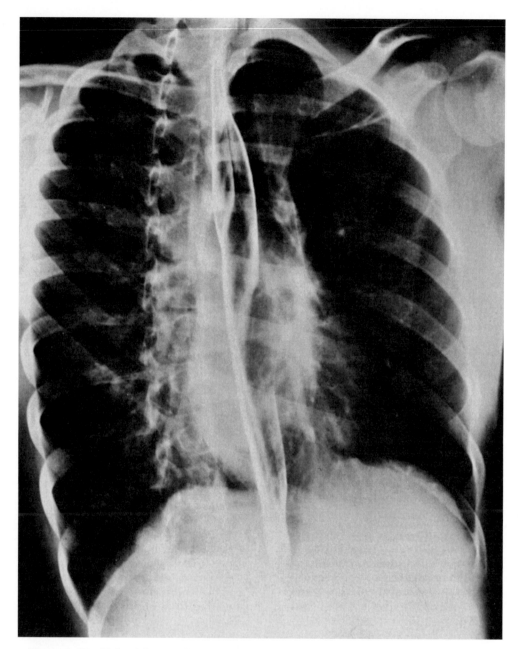

**Figure 3-39**    Right oblique radiograph of the chest of a normal adult man after a barium swallow.

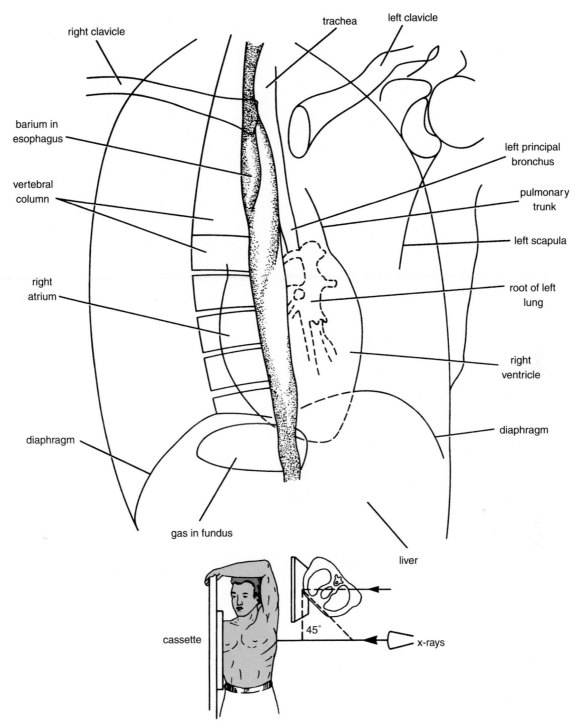

**Figure 3-40**  Main features observable in the right oblique radiograph of the chest shown in Figure 3-39. Note the position of the patient in relation to the x-ray source and cassette holder.

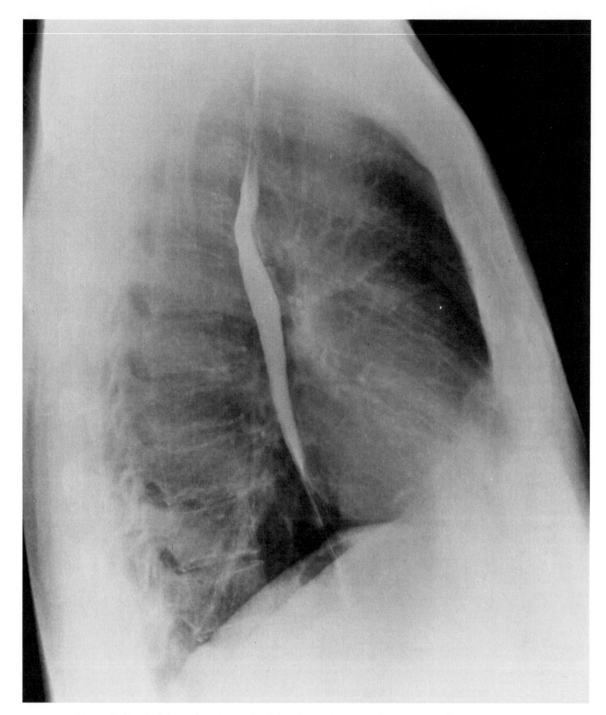

**Figure 3-41**    Left lateral radiograph of the chest of a normal adult man after a barium swallow.

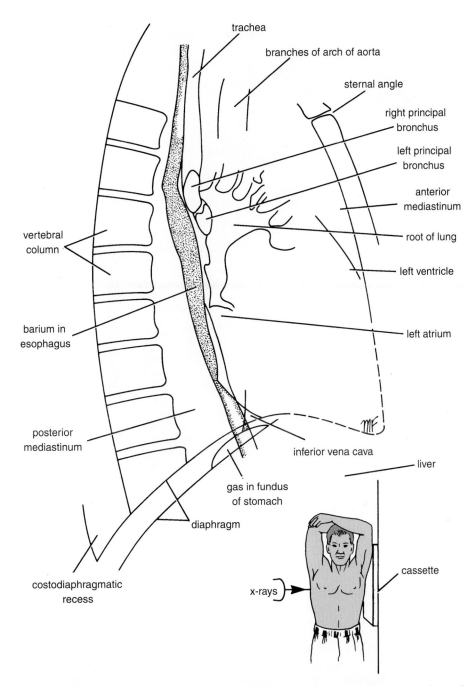

trachea

branches of arch of aorta

sternal angle

right principal bronchus

left principal bronchus

anterior mediastinum

root of lung

left ventricle

left atrium

vertebral column

barium in esophagus

posterior mediastinum

inferior vena cava

liver

gas in fundus of stomach

diaphragm

costodiaphragmatic recess

cassette

x-rays

**Figure 3-42**  Main features observable in a left lateral radiograph of the chest shown in Figure 3-41. Note the position of the patient in relation to the x-ray source and cassette.

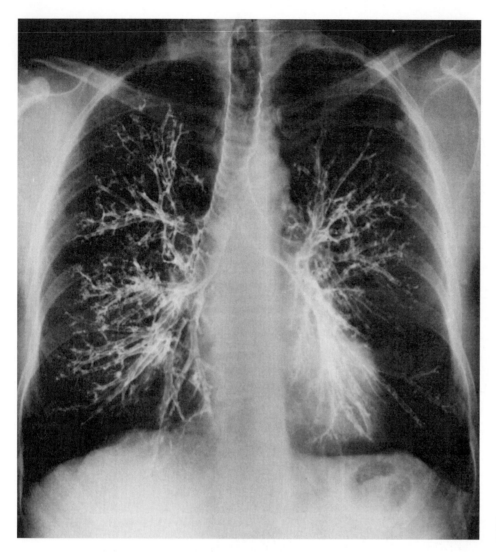

**Figure 3-43**  Posteroanterior bronchogram of the chest.

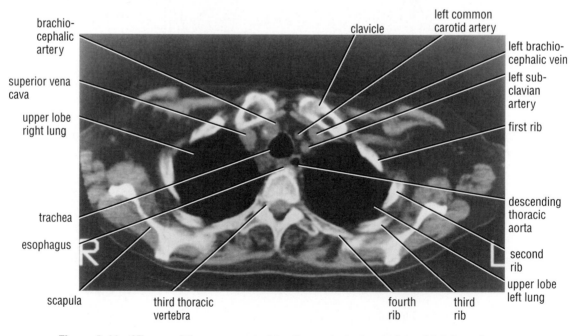

brachio-cephalic artery

superior vena cava

upper lobe right lung

trachea

esophagus

scapula

third thoracic vertebra

clavicle

left common carotid artery

left brachio-cephalic vein

left sub-clavian artery

first rib

descending thoracic aorta

second rib

upper lobe left lung

fourth rib

third rib

**Figure 3-44**  CT scan of the upper part of the thorax at the level of the third thoracic vertebra. The section is viewed from below.

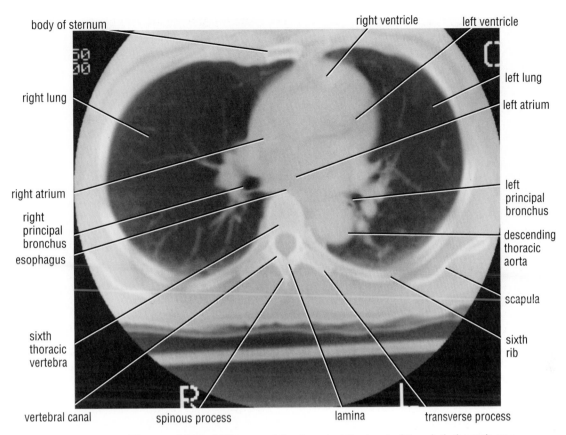

Figure 3-45   CT scan of the middle part of the thorax at the level of the sixth thoracic vertebra. The section is viewed from below.

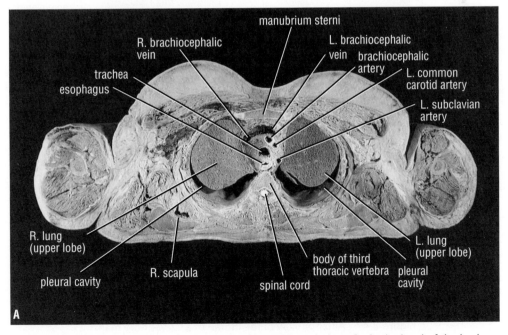

Figure 3-46   Cross sections of the thorax viewed from below. **A.** At the level of the body of the third thoracic vertebra.

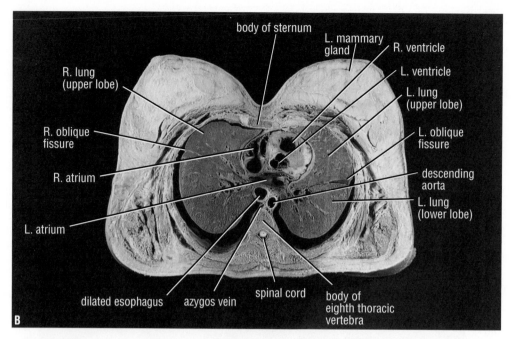

**Figure 3-46** *(continued)* **B.** At the level of the eighth thoracic vertebra. Note that, in the living, the pleural cavity is only a potential space. The large space seen here is an artifact and results from the embalming process.

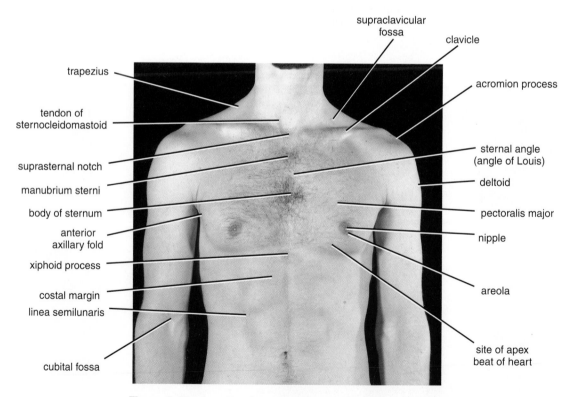

**Figure 3-47** Anterior view of the thorax of a 27-year-old man.

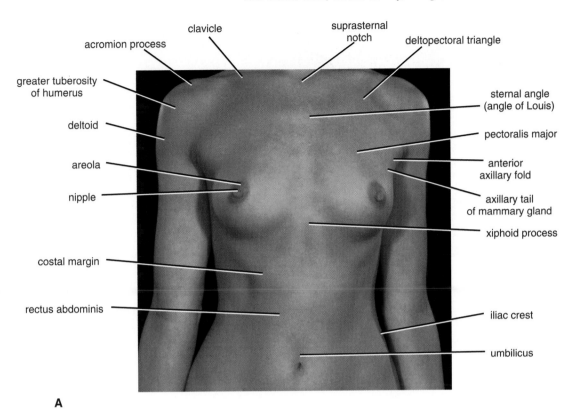

**A**

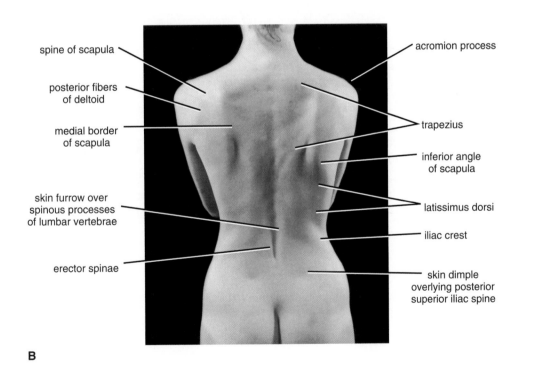

**B**

**Figure 3-48**   **A.** Anterior view of the thorax and abdomen of a 29-year-old woman.
**B.** Posterior view of the thorax of a 29-year-old woman.

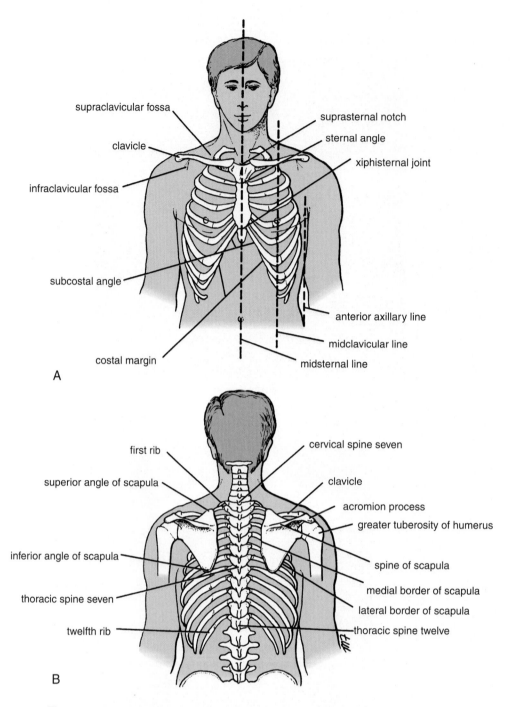

**Figure 3-49**    Surface landmarks of anterior (**A**) and posterior (**B**) thoracic walls.

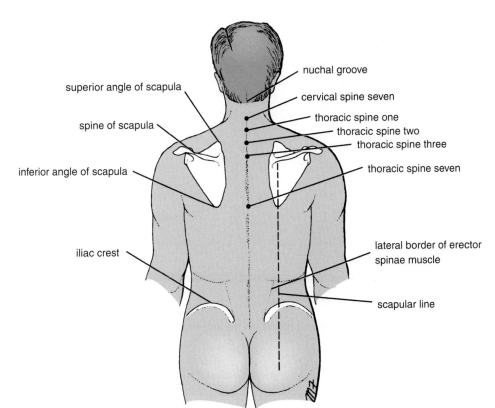

superior angle of scapula

spine of scapula

inferior angle of scapula

iliac crest

nuchal groove

cervical spine seven

thoracic spine one

thoracic spine two

thoracic spine three

thoracic spine seven

lateral border of erector spinae muscle

scapular line

**Figure 3-50** Surface landmarks of the posterior thoracic wall.

(Fig. 3-51). It commences in the midline and ends just to the right of the midline by dividing into the right and left principal bronchi. At the root of the neck, it may be palpated in the midline in the suprasternal notch.

# Lungs

The **apex of the lung** projects up into the neck. It can be mapped out on the anterior surface of the body by drawing a curved line, convex upward, from the sternoclavicular joint to a point 1 in. (2.5 cm) above the junction of the medial and intermediate thirds of the clavicle (Fig. 3-51).

The **anterior border of the right lung** begins behind the sternoclavicular joint and runs downward, almost reaching the midline behind the sternal angle. It then continues downward until it reaches the xiphisternal joint (Fig. 3-51). The **anterior border of the left lung** has a similar course, but at the level of the fourth costal cartilage, it deviates laterally and extends for a variable distance beyond the lateral margin of the sternum to form the **cardiac notch** (Fig. 3-51). This notch is produced by the heart displacing the lung to the left. The anterior border then turns sharply downward to the level of the xiphisternal joint.

The **lower border of the lung** in midinspiration follows a curving line, which crosses the sixth rib in the midclavicular line and the eighth rib in the midaxillary line, and reaches the tenth rib adjacent to the vertebral column posteriorly (Figs. 3-51, 3-52, and 3-53). Of course, the level of the inferior border of the lung changes during inspiration and expiration.

The **posterior border of the lung** extends downward from the spinous process of the seventh cervical vertebra to the level of the tenth thoracic vertebra and lies about 1.5 in. (4 cm) from the midline (Fig. 3-52).

The **oblique fissure** of the lung can be indicated on the surface by a line drawn from the root of the spine of the scapula obliquely downward, laterally and anteriorly, following the course of the sixth rib to the 6th costochondral junction. In the left lung, the upper lobe lies above and anterior to this line, and the lower lobe lies below and posterior to it (Figs. 3-51 and 3-52).

In the right lung, the **horizontal fissure** may be represented by a line drawn horizontally along the fourth costal cartilage to meet the oblique fissure in the midaxillary line (Figs. 3-51 and 3-52). The upper lobe lies above the horizontal fissure, and the middle lobe lies below it; below and posterior to the oblique fissure lies the lower lobe.

# Pleura

The boundaries of the pleural sac can be marked out as lines on the surface of the chest wall. The lines, which indicate the limits of the parietal pleura where it lies close to the body surface, are referred to as the **lines of pleural reflection.**

The **cervical pleura** bulges upward into the neck and has a surface marking identical to that of the apex of the lung. A curved line may be drawn, convex upward, from the sternoclavicular joint to a point 1 in. (2.5 cm) above the

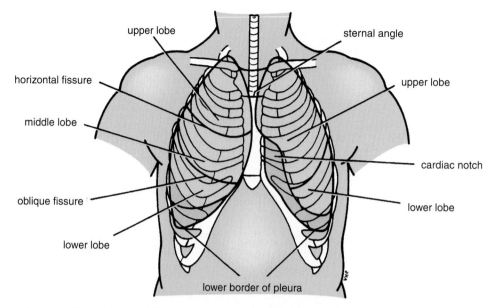

**Figure 3-51**   Surface markings of lungs and parietal pleura on the anterior thoracic wall.

junction of the medial and intermediate thirds of the clavicle (Fig. 3-51).

The **anterior border of the right pleura** runs down behind the sternoclavicular joint, almost reaching the midline behind the sternal angle. It then continues downward until it reaches the xiphisternal joint (Fig. 3-51). The **anterior border of the left pleura** has a similar course, but at the level of the fourth costal cartilage, it deviates laterally and extends to the lateral margin of the sternum to form the cardiac notch. (Note that the pleural cardiac notch is not as large as the cardiac notch of the lung.) It then turns sharply downward to the xiphisternal joint (Fig. 3-51).

The **lower border of the pleura** on both sides follows a curved line, which crosses the eighth rib in the midclavicular line and the tenth rib in the midaxillary line and reaches the twelfth rib adjacent to the vertebral column—that is, at the lateral border of the erector spinae muscle (Figs. 3-51, 3-52, and 3-53). Note that the lower margins of the lungs cross the sixth, eighth, and tenth ribs at the midclavicular lines, the midaxillary lines, and the sides of the vertebral column, respectively; the lower margins of the pleura cross, at the same points, the eighth, tenth, and twelfth ribs, respectively. The distance between the two borders corresponds to the **costodiaphragmatic recess.**

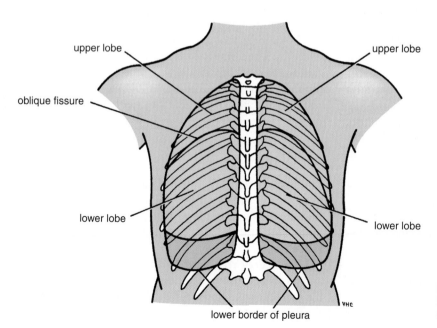

**Figure 3-52**   Surface markings of the lungs and parietal pleura on the posterior thoracic wall.

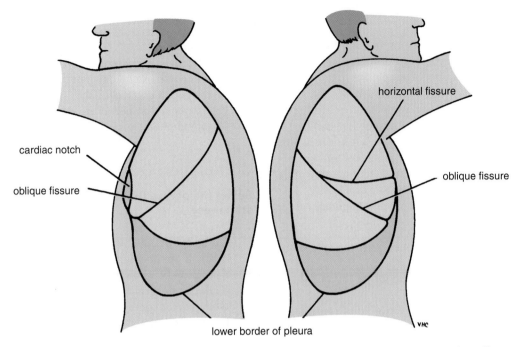

Figure 3-53    Surface markings of the lungs and parietal pleura on the lateral thoracic wall.

 # SURFACE ANATOMY OF THE BLOOD VESSELS

The **internal thoracic vessels** run vertically downward, posterior to the costal cartilages, and a fingerbreadth lateral to the edge of the sternum (Fig. 3-6), as far as the sixth intercostal space.

The **intercostal vessels** and **nerve** ("vein, artery, nerve" or **VAN** is the order from above downward) are situated immediately below their corresponding ribs (Fig. 3-4).

 # MAMMARY GLAND

The mammary gland lies in the superficial fascia covering the anterior chest wall (Figs. 3-12 and 3-48) In the child and in men, it is rudimentary. In the female after puberty, it enlarges and assumes its hemispherical shape. In the young adult female, it overlies the second to the sixth ribs and their costal cartilages and extends from the lateral margin of the sternum to the midaxillary line. In middle-aged multiparous women, the breasts may be large and pendulous. In older women past menopause, the breasts may become reduced in size.The structure of the mammary gland is described fully on page 90.

# Review Questions

## Multiple Choice Questions

**Select the BEST answer for each question.**

1. The following statements concerning structures in the intercostal space are correct **except** which?
   A. The anterior intercostal arteries of the upper six intercostal spaces are branches of the internal thoracic artery.
   B. The intercostal nerves travel forward in an intercostal space between the internal intercostal and innermost intercostal muscles.
   C. The intercostal blood vessels and nerves are positioned in the order of vein, nerve, and artery from superior to inferior in a subcostal groove.
   D. The lower five intercostal nerves supply sensory innervation to the skin of the lateral thoracic and anterior abdominal walls.
   E. The posterior intercostal veins drain backward into the azygos and hemiazygos veins.

2. The following statements concerning the diaphragm are correct **except** which?
   A. The right crus provides a muscular sling around the esophagus and possibly prevents regurgitation of stomach contents into the esophagus.
   B. On contraction, the diaphragm raises the intra-abdominal pressure and assists in the return of the venous blood to the right atrium of the heart.
   C. The level of the diaphragm is higher in the recumbent position than in the standing position.
   D. On contraction, the central tendon descends, reducing the intrathoracic pressure.
   E. The esophagus passes through the diaphragm at the level of the eighth thoracic vertebra.

3. The following statements concerning the intercostal nerves are correct **except** which?
   A. They provide motor innervation to the peripheral parts of the diaphragm.
   B. They provide motor innervation to the intercostal muscles.
   C. They provide sensory innervation to the costal parietal pleura.
   D. They contain sympathetic fibers to innervate the vascular smooth muscle.
   E. The seventh to the eleventh intercostal nerves provide sensory innervation to the parietal peritoneum.

4. To pass a needle into the pleural space (cavity) in the midaxillary line, the following structures will have to be pierced **except** which?
   A. Internal intercostal muscle
   B. Levatores costarum
   C. External intercostal muscle
   D. Parietal pleura
   E. Innermost intercostal muscle

5. The following statements concerning the thoracic outlet are true **except** which?
   A. The manubrium sterni forms the anterior border.
   B. On each side, the lower trunk of the brachial plexus and the subclavian artery emerge through the outlet and pass laterally over the surface of the first rib.
   C. The body of the seventh cervical vertebra forms the posterior boundary.
   D. The first ribs form the lateral boundaries.
   E. The esophagus and trachea pass through the outlet.

6. The following statements concerning the thoracic wall are correct **except** which?
   A. The trachea bifurcates opposite the manubriosternal joint (angle of Louis) in the midrespiratory position.
   B. The superior mediastinum lies behind the body of the sternum.

   C. The apex beat of the heart can normally be felt in the left intercostal space about 3.5 in. (9 cm) from the midline.
   D. The lower margin of the right lung on full inspiration could extend down in the midclavicular line to the eighth costal cartilage.
   E. All intercostal nerves are derived from the anterior rami of thoracic spinal nerves.

## Completion Questions

**Select the phrase that BEST completes each statement.**

7. Clinicians define the thoracic outlet as
   A. the lower opening in the thoracic cage.
   B. the gap between the crurae of the diaphragm.
   C. the esophageal opening in the diaphragm.
   D. the upper opening in the thoracic cage.
   E. the gap between the sternal and costal origins of the diaphragm.

8. The costal margin is formed by
   A. the sixth, eighth, and tenth ribs.
   B. the inner margins of the first ribs.
   C. the edge of the xiphoid process.
   D. the costal cartilages of the seventh, eighth, ninth, and tenth ribs.
   E. the costal cartilages of the seventh to the tenth ribs and the ends of the cartilages of the eleventh and twelfth ribs.

9. The lower margin of the left lung in midrespiration crosses
   A. the sixth, eighth, and tenth ribs.
   B. the seventh, eighth, and ninth ribs.
   C. the tenth, eleventh, and twelfth ribs.
   D. the eighth rib only.
   E. the sixth, eleventh, and twelfth ribs.

10. The breast in the young adult female overlies
    A. the first to the fifth ribs.
    B. the second to the sixth ribs.
    C. the first and second ribs.
    D. the second and third ribs only.
    E. the fourth to the sixth ribs.

11. The parietal pleura
    A. is sensitive only to the sensation of stretch.
    B. is separated from the pleural space by endothoracic fascia.
    C. is sensitive to the sensations of pain and touch.
    D. receives its sensory innervation from the autonomic nervous system.
    E. is formed from splanchnopleuric mesoderm.

## Fill in the Blank Questions

**Fill in the blank with the BEST answer.**

12. The thoracic duct passes through the _____ opening in the diaphragm.

13. The superior epigastric artery passes through the _____ opening in the diaphragm.

14. The right phrenic nerve passes through the _____ opening in the diaphragm.

15. The left vagus nerve passes through the _____ opening in the diaphragm.
    A. aortic
    B. esophageal
    C. caval
    D. none of the above

16. The aortic opening in the diaphragm lies at the level of the _____ thoracic vertebra.

17. The xiphisternal joint lies at the level of the _____ thoracic vertebra.

18. The caval opening in the diaphragm lies at the level of the _____ thoracic vertebra.
    A. tenth
    B. twelfth
    C. eighth
    D. ninth
    E. seventh

## Multiple Choice Questions

**Read the case history and select the BEST answer to the question following it.**

A 35-year-old man complaining of severe pain in the lower part of his left chest was seen by his physician. The patient had been coughing for the last 4 days and was producing blood-stained sputum. He had an increased respiratory rate and had a pyrexia of 104°F. On examination, the patient was found to have fluid in the left pleural space.

19. With the patient in the standing position, the pleural fluid would most likely gravitate down to the
    A. oblique fissure.
    B. cardiac notch.
    C. costomediastinal recess.
    D. horizontal fissure.
    E. costodiaphragmatic recess.

**Select the BEST answer for each question.**

20. The following statements concerning the trachea are true **except** which?
    A. It lies anterior to the esophagus in the superior mediastinum.
    B. In deep inspiration, the carina may descend as far as the level of the sixth thoracic vertebra.
    C. The left principal bronchus is wider than the right principal bronchus.
    D. The arch of the aorta lies on its anterior and left sides in the superior mediastinum.
    E. The sensory innervation of the mucous membrane lining the trachea is derived from branches of the vagi and the recurrent laryngeal nerves.

21. The following statements concerning the root of the right lung are true **except** which?
    A. The right phrenic nerve passes anterior to the lung root.
    B. The azygos vein arches over the superior margin of the lung root.
    C. The right pulmonary artery lies posterior to the principal bronchus.
    D. The right vagus nerve passes posterior to the lung root.
    E. The vessels and nerves forming the lung root are enclosed by a cuff of pleura.

22. The following statements concerning the right lung are true **except** which?
    A. It possesses a horizontal and an oblique fissure.
    B. Its covering of visceral pleura is sensitive to pain and temperature.
    C. The lymph from the substance of the lung reaches the hilum by the superficial and deep lymphatic plexuses.
    D. The pulmonary ligament permits the vessels and nerves of the lung root to move during the movements of respiration.
    E. The bronchial veins drain into the azygos and hemiazygos veins.

23. All of the following statements concerning the mediastinum are correct **except** which?
    A. The mediastinum forms a partition between the two pleural spaces (cavities).
    B. The mediastinal pleura demarcates the lateral boundaries of the mediastinum.
    C. The heart occupies the middle mediastinum.
    D. Should air enter the left pleural cavity, the structures forming the mediastinum are deflected to the right.
    E. The anterior boundary of the mediastinum extends to a lower level than the posterior boundary.

24. All the following statements regarding the mechanics of inspiration are true **except** which?
    A. The diaphragm is the most important muscle of inspiration.
    B. The connective tissue roof of the thorax can be raised.
    C. The sternum moves anteriorly.

D. The ribs are raised superiorly.

E. The tone of the muscles of the anterior abdominal wall is diminished.

25. The following statements concerning the lungs are correct **except** which?

A. Inhaled foreign bodies most frequently enter the right lung.

B. The left lung is in direct contact with the arch of the aorta and the descending thoracic aorta.

C. There are no lymph nodes within the lungs.

D. The structure of the lungs receives its blood supply from the bronchial arteries.

E. The costodiaphragmatic recesses are lined with parietal pleura.

26. The following statements concerning the bronchopulmonary segments are correct **except** which?

A. The veins are intersegmental.

B. The segments are separated by connective tissue septa.

C. The arteries are intrasegmental.

D. Each segment is supplied by a secondary bronchus.

E. Each pyramid-shaped segment has its base pointing toward the lung surface.

# Answers and Explanations

1. **C** is the incorrect statement. The order from superior to inferior is intercostal vein, artery, and nerve (Fig. 3-4).

2. **E** is the incorrect statement. The esophagus passes through the diaphragm at the level of the tenth thoracic vertebra.

3. **A** is the incorrect statement. The intercostal nerves provide sensory innervation to the pleura and peritoneum covering the peripheral parts of the diaphragm.

4. **B** is the incorrect statement. The levator costarum is located on the back away from the area involved.

5. **C** is the incorrect statement. The body of the first thoracic vertebra forms the posterior boundary.

6. **B** is the incorrect statement. The superior mediastinum lies behind the manubrium sterni (Fig. 3-17).

7. **D** is correct. The thoracic outlet is the upper opening in the thoracic cage.

8. **E** is correct. The costal margin is formed by the costal cartilages of the seventh to the tenth ribs and the ends of the cartilages of the eleventh and twelfth ribs (Fig. 3-1).

9. **A** is correct. The lower margin of the left lung in midrespiration crosses the sixth, eighth, and tenth ribs.

10. **B** is correct. The breast in the young adult female overlies the second to the sixth ribs.

11. **C** is correct. The parietal pleura is sensitive to the sensations of pain and touch.

12. **A** is correct. The thoracic duct passes through the aortic opening in the diaphragm.

13. **D** is correct. The superior epigastric artery enters the anterior abdominal wall between the sternal and the costal origins of the diaphragm.

14. **C** is correct. The right phrenic nerve passes through the caval opening in the diaphragm.

15. **B** is correct. The left vagus nerve passes through the esophageal opening in the diaphragm.

16. **B** is correct. The aortic opening in the diaphragm lies at the level of the twelfth thoracic vertebra.

17. **D** is correct. The xiphisternal joint lies at the level of the ninth thoracic vertebra in midrespiration.

18. **C** is correct. The caval opening in the diaphragm lies at the level of the eighth thoracic vertebra in midrespiration.

19. **E** is correct. This patient started his illness with an upper respiratory infection, which he ignored. Now he has left-sided pneumonia complicated by pleurisy. With pleurisy, the inflammatory exudate may remain at the site of the pneumonia. If the pleural fluid is excessive and the patient assumes the upright position, the fluid may gravitate downward to the lowest part of the pleural space—namely, the costodiaphragmatic recess.

20. **C** is incorrect. The right principal bronchus is wider than the left. This is clearly seen in the normal posteroanterior bronchogram shown in Figure 3-43. Also see plastinized specimen in Figure 3-23.

21. **C** is the incorrect statement. The right pulmonary artery lies anterior to the principal bronchus.

22. **B** is the incorrect statement. The visceral pleura is innervated by sympathetic and vagal afferent fibers via the pulmonary plexus and is not sensitive to pain and temperature, but it is sensitive to the sensation of stretch.

23. **E** is the incorrect statement. The anterior boundary of the mediastinum extends down to the xiphisternal joint anteriorly—that is, to the level of the ninth thoracic vertebral body. The posterior boundary extends down farther, to the level of the twelfth thoracic vertebra.

24. **B** is the incorrect statement. The connective tissue roof of the thoracic cavity is attached to the transverse process of the seventh cervical vertebra; it cannot be raised during inspiration.

25. **C** is the incorrect statement. The lungs contain lymph nodes along the course of the bronchi.

26. **D** is the incorrect statement. Each segment of the lung is supplied by a segmental bronchus.

# The Cardiovascular System

# 4

# The Heart, Coronary Vessels, and Pericardium

All clinical material relevant to this chapter can be found on the CD-ROM.

# Chapter Outline

C oronary artery disease, valvular dysfunction, electrical disturbances, myocardiopathies, or problems of the pericardium present the medical professional with a diagnostic and therapeutic challenge. The evaluation of chest pain and the management of disturbances of cardiac rhythm and pump function are common problems facing emergency personnel. In children, congenital heart disease may present as a serious problem.

   The purpose of this chapter is to review the structure of the heart, including its conducting system, its important blood supply, and its surrounding pericardium.

# The Heart

The heart is a hollow muscular organ that is somewhat pyramid shaped and lies within the pericardium in the mediastinum (Figs. 4-1 and 4-2). It is connected at its base to the great blood vessels but otherwise lies free within the pericardium.

## Surfaces of the Heart

The heart has three surfaces: the sternocostal surface (anterior), the diaphragmatic surface (inferior), and the base (posterior). It also has an apex, which is directed downward, forward, and to the left.

The **sternocostal surface** is formed mainly by the right atrium and the right ventricle, which are separated from each other by the vertical atrioventricular groove (Fig. 4-2). The right border is formed by the right atrium; and the left border is formed by the left ventricle and part of the left auricle. The right ventricle is separated from the left ventricle by the anterior interventricular groove.

The **diaphragmatic surface** of the heart is formed mainly by the right and left ventricles separated by the posterior interventricular groove. The inferior surface of the right atrium, into which the inferior vena cava opens, also forms part of this surface.

The **base of the heart,** or the posterior surface, is formed mainly by the left atrium, into which open the four pulmonary veins (Fig. 4-3). The base of the heart lies opposite the apex.

The **apex of the heart,** formed by the left ventricle, is directed downward, forward, and to the left (Figs. 4-1 and 4-2). It lies at the level of the fifth left intercostal space, 3.5 in. (9 cm) from the midline. In the region of the apex, the apex beat can usually be seen and palpated in the living patient.

Note that the base of the heart is called the base because the heart is pyramid shaped; the base lies opposite the apex. The heart does not rest on its base; it rests on its diaphragmatic (inferior) surface.

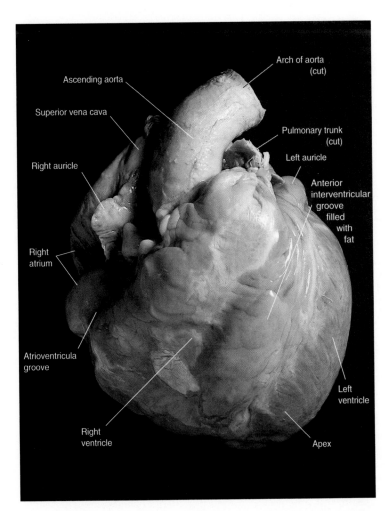

Ascending aorta

Superior vena cava

Right auricle

Right atrium

Atrioventricular groove

Right ventricle

Arch of aorta (cut)

Pulmonary trunk (cut)

Left auricle

Anterior interventricular groove filled with fat

Left ventricle

Apex

**Figure 4-1** The anterior surface of the heart; the fibrous pericardium and the parietal serous pericardia have been removed. Note the presence of fat beneath the visceral serous pericardium in the atrioventricular and interventricular grooves. The coronary arteries are embedded in this fat.

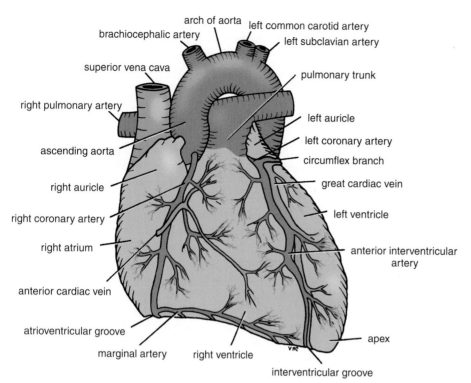

**Figure 4-2**   The anterior surface of the heart and the great blood vessels. Note the course of the coronary arteries and the cardiac veins.

## Borders of the Heart

The right border is formed by the right atrium, and the left border is formed by the left auricle and, below, by the left ventricle (Figs. 4-1 and 4-2). The lower border is formed mainly by the right ventricle but also by the right atrium; the apex is formed by the left ventricle. These borders are important to recognize when examining a radiograph of the heart.

## Structure of the Heart

The heart is divided by vertical septa into four chambers: the right and left atria and the right and left ventricles. The right

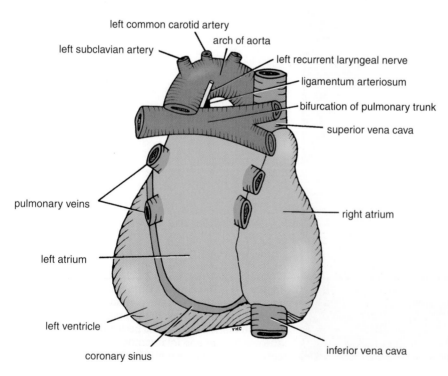

**Figure 4-3**   The posterior surface, or the base, of the heart.

atrium lies anterior to the left atrium, and the right ventricle lies anterior to the left ventricle (Figs. 4-1 and 4-2). The walls of the heart consist of three layers:

■ The outer, visceral layer of serous pericardium (the **epicardium**).
■ The middle, thick layer of cardiac muscle (the **myocardium**).
■ The inner, thin layer (the **endocardium**).

## Skeleton of the Heart

The so-called skeleton of the heart (see Fig. 4-5 ) consists of fibrous rings that surround the atrioventricular, pulmonary, and aortic orifices and are continuous with the membranous upper part of the ventricular septum.

---

### PHYSIOLOGIC NOTE

### Function of the Heart Skeleton

The fibrous rings around the valvular orifices support the bases of the valve cusps and prevent the valves from stretching and becoming incompetent. The fibrous skeleton, although providing attachment for the cardiac muscle fibers, effectively separates the muscular walls of the atria from those of the ventricles. The skeleton of the heart thus forms the basis of electrical discontinuity between the atria and the ventricles.

## Chambers of the Heart

### Right Atrium

The right atrium consists of a main cavity and a small outpouching, the auricle (Figs. 4-2 and 4-4). On the outside of the heart at the junction between the right atrium and the right auricle is a vertical groove, the **sulcus terminalis**, which, on the inside, forms a ridge, the **crista terminalis** (embryologically, this represents the junction between the sinus venosus and the right atrium proper). The main part of the atrium that lies posterior to the ridge is smooth walled (Fig. 4-4), whereas the interior of the auricle is roughened by bundles of muscle fibers called the **musculi pectinati**.

#### Openings into the Right Atrium

The **superior vena cava** (Fig. 4-4) opens into the upper part of the right atrium; there is no valve. It returns the blood to the heart from the upper half of the body. The **inferior vena cava** (larger than the superior vena cava) opens into the lower part of the right atrium; it is guarded by a rudimentary, nonfunctioning valve. It returns the blood to the heart from the lower half of the body.

The **coronary sinus,** which drains most of the blood from the heart wall (Fig. 4-4), opens into the right atrium between the inferior vena cava and the atrioventricular orifice. It is guarded by a rudimentary, nonfunctioning valve.

The **right atrioventricular orifice** lies anterior to the inferior vena caval opening and is guarded by the tricuspid

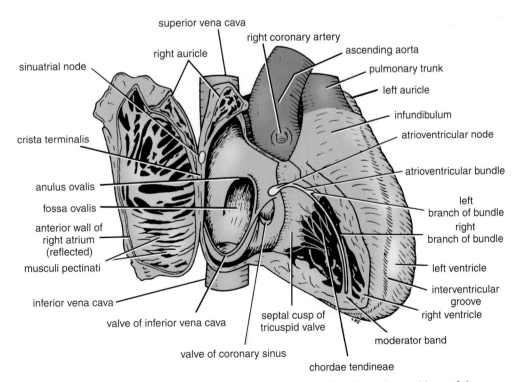

**Figure 4-4** Interior of the right atrium and the right ventricle. Note the positions of the sinoatrial node and the atrioventricular node and bundle.

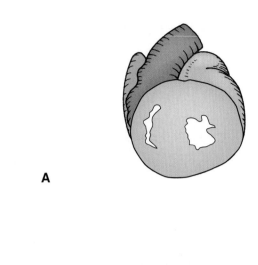

**A**

**B**

**Figure 4-5**    **A.** Cross section of the ventricles of the heart. Note the great thickness of the wall of the left ventricle. **B.** The fibrous skeleton of the heart.

valve (Fig. 4-4). Many small orifices of small veins also drain the wall of the heart and open directly into the right atrium.

### Fetal Remnants in the Right Atrium

In addition to the rudimentary valve of the inferior vena cava are the **fossa ovalis** and **anulus ovalis.** These latter structures lie on the **atrial septum,** which separates the right atrium from the left atrium (Fig. 4-4). The fossa ovalis is a shallow depression, which is the site of the **foramen ovale** in the fetus. (Before birth, oxygenated blood passed through this foramen from the right atrium into the left atrium.) The anulus ovalis forms the upper margin of the fossa.

## Right Ventricle

The right ventricle forms the greater part of the anterior surface of the heart, and it lies anterior to the left ventricle (Fig. 4-4). The right ventricle communicates with the right atrium through the atrioventricular orifice and with the pulmonary trunk through the pulmonary orifice (Fig. 4-4). The approach to the pulmonary orifice is funnel shaped and known as the **infundibulum.**

The walls of the right ventricle are much thicker than the walls of the right atrium. The internal surface shows projecting ridges called **trabeculae carneae.** There are three types of trabeculae carneae:

- The first type comprises the **papillary muscles,** which project inward, being attached by their bases to the

ventricular wall; their apices are connected by fibrous chords (the **chordae tendineae**) to the cusps of the tricuspid valve (Fig. 4-4).

- The second type are attached at their ends to the ventricular wall, being free in the middle. One of these, the **moderator band,** crosses the ventricular cavity from the septal to the anterior wall (Fig. 4-4). It conveys the right branch of the atrioventricular bundle, which is part of the conducting system of the heart.

- The third type is simply composed of prominent ridges.

The **tricuspid valve** guards the atrioventricular orifice (Figs. 4-4, 4-6, and 4-7). It consists of three cusps formed by a fold of endocardium. The cusps are the **anterior, septal,** and **inferior** (posterior) cusps. The anterior cusp lies anteriorly, the septal cusp lies against the ventricular septum, and the inferior or posterior cusp lies inferiorly. The cusps are attached by their bases to the fibrous ring of the skeleton of the heart. To their free edges are attached the chordae tendineae, which connect the cusps to the papillary muscles.

---

### PHYSIOLOGIC NOTE

#### Function of the Papillary Muscles

When the ventricle contracts, the papillary muscles contract and prevent the cusps from being forced into the atrium and turning inside out as the intraventricular pressure rises. To assist in this process, the chordae tendineae of one papillary muscle are connected to the adjacent parts of two cusps.

---

The **pulmonary valve** guards the pulmonary orifice (Fig. 4-7), and the semilunar cusps of this valve are attached by their curved, lower margins to the arterial wall. The open mouths of the cusps are directed upward into the pulmonary trunk. No chordae or papillary muscles are associated with these valve cusps; the attachments of the sides of the cusps to the arterial wall prevent the cusps from prolapsing into the ventricle. At the root of the pulmonary trunk are three dilatations called the **pulmonary sinuses,** and one is situated external to each cusp (see **aortic valve,** p. 143).

The three semilunar cusps are arranged with one posterior (left cusp) and two anterior (anterior and right cusps). (The cusps of the pulmonary and aortic valves are named according to their position in the fetus before the heart has rotated to the left. This, unfortunately, causes a great deal of unnecessary confusion.)

---

### PHYSIOLOGIC NOTE

#### Function of the Pulmonary Valve Cusps

During ventricular systole, the cusps of the valve are pressed against the wall of the pulmonary trunk by the

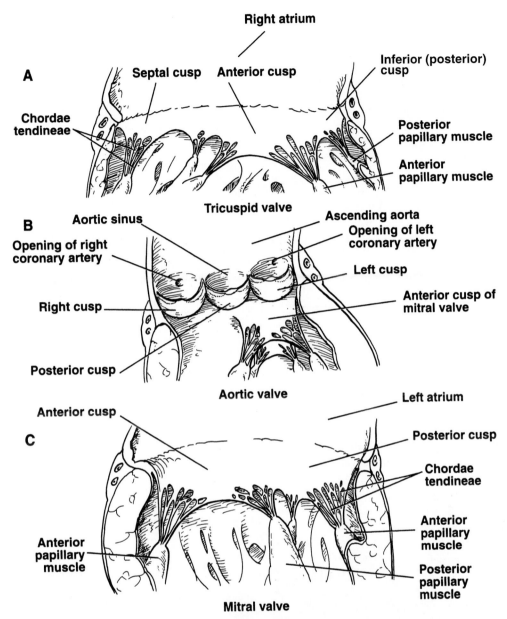

**Figure 4-6** Valves of the heart. **A.** Tricuspid valve showing the septal, anterior, and posterior cusps and their chordae tendineae. **B.** Aortic valve showing the relationship between the cusps and the openings of the coronary arteries. **C.** Mitral valve showing the anterior and posterior cusps and their chordae tendineae.

outrushing blood. During diastole, blood flows back toward the heart and enters the sinuses; the valve cusps fill, come into apposition in the center of the lumen, and close the pulmonary orifice.

## Left Atrium

Similar to the right atrium, the left atrium consists of a main cavity and a left auricle. The left atrium is situated behind the right atrium and forms the greater part of the base or the posterior surface of the heart (Fig. 4-3). Behind it lies the esophagus separated from it by the pericardium (Fig. 4-8).

The interior of the left atrium is smooth, but the left auricle possesses muscular ridges as in the right auricle.

### Openings into the Left Atrium

The four pulmonary veins, two from each lung, open through the posterior wall (Fig. 4-3 ) and have no valves. The left atrioventricular orifice is guarded by the mitral valve.

## Left Ventricle

The left ventricle is situated largely behind the right ventricle (Figs. 4-2 and 4-8). A small portion, however, projects to

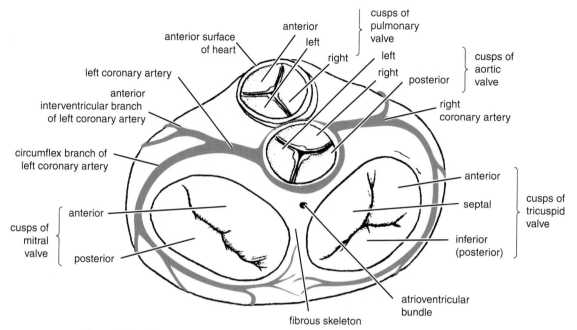

**Figure 4-7**  The valves of the heart and the origin of the coronary arteries, superior view. The atria and the great vessels have been removed.

the left and forms the left margin of the heart and the heart apex. The left ventricle communicates with the left atrium through the atrioventricular orifice and with the aorta through the aortic orifice. The walls of the left ventricle (Fig. 4-5) are three times thicker than the walls of the right ventricle. (The left intraventricular blood pressure is six times higher than that inside the right ventricle.) In cross section, the left ventricle is circular; the right ventricle is crescentic because of the bulging of the ventricular septum into the cavity of the right ventricle (Fig. 4-5). There are

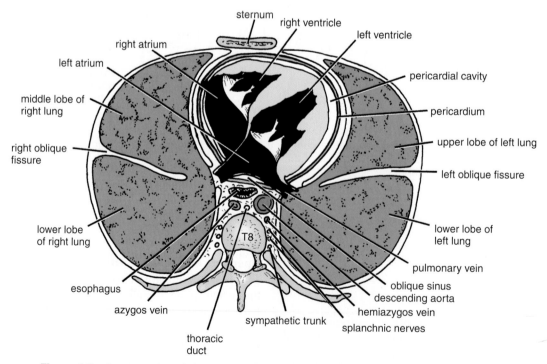

**Figure 4-8**  Cross section of the thorax at the eighth thoracic vertebra, as seen from below. (Note that all CT scans and MRI studies are viewed from below.)

well-developed trabeculae carneae and two large papillary muscles but no moderator band. The part of the ventricle below the aortic orifice is called the **aortic vestibule.**

The **mitral valve** guards the atrioventricular orifice (Figs. 4-6 and 4-7). It consists of two cusps, one anterior and one posterior, which have a structure similar to that of the cusps of the tricuspid valve. The anterior cusp is the larger and intervenes between the atrioventricular and the aortic orifices. The attachment of the chordae tendineae to the cusps and the papillary muscles is similar to that of the tricuspid valve.

The **aortic valve** guards the aortic orifice and is precisely similar in structure to the pulmonary valve (Fig. 4-7). One cusp is situated on the anterior wall (right cusp), and two cusps are located on the posterior wall (left and posterior cusps). Behind each cusp, the aortic wall bulges to form an **aortic sinus.** The anterior aortic sinus gives origin to the right coronary artery, and the left posterior sinus gives origin to the left coronary artery.

## Conducting System of the Heart

**PHYSIOLOGIC NOTE**

### Basic Function of the Conduction System

The normal heart contracts rhythmically at about 70 to 90 beats per minute in the resting adult. The rhythmic contractile process originates spontaneously in the conducting system, and the impulse travels to different regions of the heart, so the atria contract first and together, to be followed later by the contractions of both ventricles together. The slight delay in the passage of the impulse from the atria to the ventricles allows time for the atria to empty their blood into the ventricles before the ventricles contract.

The conducting system of the heart consists of specialized cardiac muscle present in the **sinuatrial node,** the **atrioventricular node,** the **atrioventricular bundle** and its right and left terminal branches, and the subendocardial plexus of **Purkinje fibers** (specialized cardiac muscle fibers that form the conducting system of the heart).

### Sinuatrial Node (Pacemaker)

The sinuatrial node initiates the heart beat. It is located in the wall of the right atrium in the upper part of the sulcus terminalis close to the opening of the superior vena cava (Figs. 4-4 and 4-9). The node spontaneously gives origin to rhythmical electrical impulses that spread in all directions through the cardiac muscle of the atria and cause the muscle to contract.

### Atrioventricular Node

The atrioventricular node is strategically placed on the lower part of the atrial septum just above the attachment of the septal cusp of the tricuspid valve (Figs. 4-4 and 4-9).

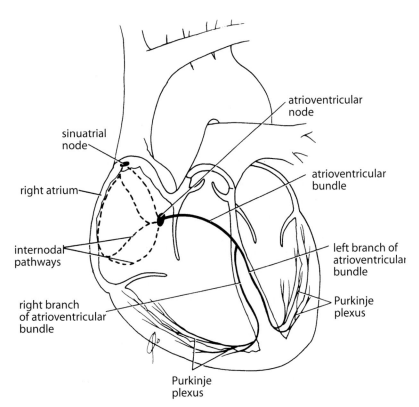

**Figure 4-9** The conducting system of the heart. Note the internodal pathways.

From it, the cardiac impulse is conducted to the ventricles by the atrioventricular bundle. The atrioventricular node is stimulated by the excitation wave as it passes through the atrial myocardium.

---

**PHYSIOLOGIC NOTE**

### Speed of Conduction through the Atrioventricular Node

The speed of conduction of the cardiac impulse through the atrioventricular node (about 0.11 sec) allows sufficient time for the atria to empty their blood into the ventricles before the ventricles start to contract.

---

## Atrioventricular Bundle

The atrioventricular bundle is continuous with the atrioventricular node above and with the fibers of the Purkinje plexus below (Fig. 4-9). The bundle descends through the fibrous skeleton of the heart and then descends behind the septal cusp of the tricuspid valve on the membranous part of the ventricular septum. At the upper border of the muscular part of the septum, it divides into two branches, one for each ventricle. The right bundle branch (RBB) passes down on the right side of the ventricular septum to reach the moderator band, where it crosses to the anterior wall of the right ventricle. Here it becomes continuous with the fibers of the Purkinje plexus (Fig. 4-9).

The left bundle branch (LBB) pierces the septum and passes down on its left side beneath the endocardium. It usually divides into two branches (anterior and posterior), which eventually become continuous with the fibers of the Purkinje plexus of the left ventricle (Fig. 4-9).

---

**PHYSIOLOGIC NOTE**

### Function of the Atrioventricular Bundle

The atrioventricular bundle (bundle of His) is the only pathway of cardiac muscle that connects the myocardium of the atria and the myocardium of the ventricles and, thus, is the only route along which the cardiac impulse can travel from the atria to the ventricles (Fig. 4-9).

Thus, it is seen that the conducting system of the heart is responsible not only for generating rhythmical cardiac impulses but also for conducting these impulses rapidly throughout the myocardium of the heart so that the different chambers contract in a coordinated and efficient manner.

The activities of the conducting system can be influenced by the autonomic nerve supply to the heart. The parasympathetic nerves slow the rhythm and diminish the rate of conduction of the impulse; the sympathetic nerves have the opposite effect.

## Internodal Conduction Paths*

Impulses from the sinuatrial node have been shown to travel to the atrioventricular node more rapidly than they can travel by passing along the ordinary myocardium. This phenomenon has been explained by the description of special pathways in the atrial wall (Fig. 4-9), which have a structure consisting of a mixture of Purkinje fibers and ordinary cardiac muscle cells. The anterior internodal pathway leaves the anterior end of the sinuatrial node and passes anterior to the superior vena caval opening. It descends on the atrial septum and ends in the atrioventricular node. The middle internodal pathway leaves the posterior end of the sinuatrial node and passes posterior to the superior vena caval opening. It descends on the atrial septum to the atrioventricular node. The posterior internodal pathway leaves the posterior part of the sinuatrial node and descends through the crista terminalis and the valve of the inferior vena cava to the atrioventricular node.

# Arterial Supply of the Heart

The arterial supply of the heart is provided by the right and left coronary arteries, which arise from the ascending aorta immediately above the aortic valve (Fig. 4-10). The coronary arteries and their major branches are distributed over the surface of the heart, lying within subepicardial connective tissue.

## The Right Coronary Artery

The right coronary artery arises from the anterior aortic sinus of the ascending aorta (Figs. 4-2, 4-7, and 4-10). It descends in the right atrioventricular groove, and at the inferior border of the heart, it continues posteriorly along the atrioventricular groove to anastomose with the left coronary artery in the posterior interventricular groove. The following branches from the right coronary artery supply the right atrium and right ventricle and parts of the left atrium and left ventricle and the atrioventricular septum.

## Branches of the Right Coronary Artery

- **The right conus artery** supplies the anterior surface of the pulmonary conus (infundibulum of the right ventricle) and the upper part of the anterior wall of the right ventricle.
- **The anterior ventricular branches** are two or three in number and supply the anterior surface of the right ventricle. The **marginal branch** is the largest and runs along the lower margin of the costal surface to reach the apex.

---

*The occurrence of specialized internodal pathways has been dismissed by some researchers, who claim that it is the packaging and arrangement of ordinary atrial myocardial fibers that are responsible for the more rapid conduction.

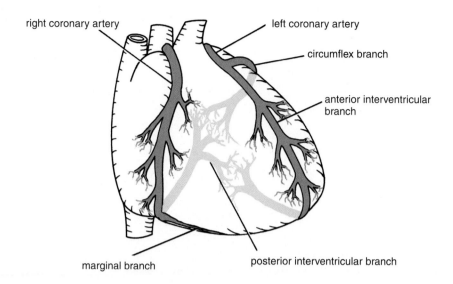

right coronary artery

left coronary artery

circumflex branch

anterior interventricular branch

marginal branch

posterior interventricular branch

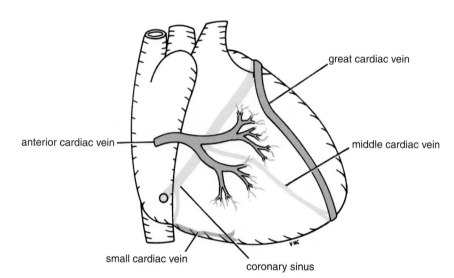

great cardiac vein

anterior cardiac vein

middle cardiac vein

small cardiac vein

coronary sinus

**Figure 4-10**   Coronary arteries and veins.

■ **The posterior ventricular branches** are usually two in number and supply the diaphragmatic surface of the right ventricle.

■ **The posterior interventricular (descending) artery** runs toward the apex in the posterior interventricular groove (Fig 4-10). It gives off branches to the right and left ventricles, including its inferior wall. It supplies branches to the posterior part of the ventricular septum but not to the apical part, which receives its supply from the anterior interventricular branch of the left coronary artery. A large septal branch supplies the **atrioventricular node.** In 10% of individuals, the posterior interventricular artery is replaced by a branch from the left coronary artery.

■ **The atrial branches** supply the anterior and lateral surfaces of the right atrium. One branch supplies the poste-

rior surface of both the right and left atria. The **artery of the sinuatrial node** supplies the node and the right and left atria; in 35% of individuals, it arises from the left coronary artery.

## Left Coronary Artery

The left coronary artery is usually larger than the right coronary artery. It arises from the left posterior aortic sinus of the ascending aorta and passes forward between the pulmonary trunk and the left auricle (Figs. 4-2, 4-7, and 4-11). It then enters the atrioventricular groove and divides into an anterior interventricular branch and a circumflex branch. The left coronary artery supplies the major part of the heart, including the greater part of the left atrium, left ventricle, and ventricular septum.

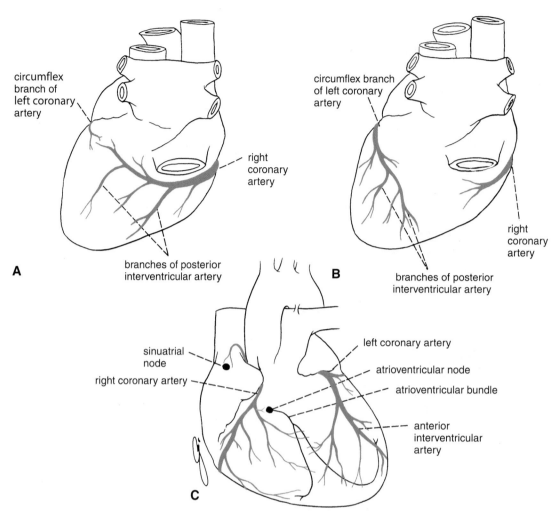

**Figure 4-11    A.** Posterior view of the heart showing the origin and distribution of the posterior interventricular artery in the right dominance. **B.** Posterior view of the heart showing the origin and distribution of the posterior interventricular artery in the left dominance. **C.** Anterior view of the heart showing the relationship of the blood supply to the conducting system.

### Branches of the Left Coronary Artery

- The **anterior interventricular (descending) branch** runs downward in the anterior interventricular groove to the apex of the heart (Fig. 4-10). In most individuals, it then passes around the apex of the heart to enter the posterior interventricular groove and anastomoses with the terminal branches of the right coronary artery. In one third of individuals, it ends at the apex of the heart. The anterior interventricular branch supplies the right and left ventricles with numerous branches that also supply the anterior part of the ventricular septum. One of these ventricular branches (**left diagonal artery**) may arise directly from the trunk of the left coronary artery. A small **left conus artery** supplies the pulmonary conus.
- The **circumflex artery** is the same size as the anterior interventricular artery (Fig. 4-10). It winds around the left margin of the heart in the atrioventricular groove. A

**left marginal artery** is a large branch that supplies the left margin of the left ventricle down to the apex. **Anterior ventricular** and **posterior ventricular branches** supply the left ventricle. **Atrial branches** supply the left atrium.

## Variations in the Coronary Arteries

Variations in the blood supply to the heart do occur, and the most common variations affect the blood supply to the diaphragmatic surface of both ventricles. Here, the origin, size, and distribution of the posterior interventricular artery are variable (Fig. 4-11). In **right dominance**, the posterior interventricular artery is a large branch of the right coronary artery. Right dominance is present in most individuals (90%). In **left dominance**, the posterior interventricular artery is a branch of the circumflex branch of the left coronary artery (10%).

## Coronary Artery Anastomoses

Anastomoses between the terminal branches of the right and left coronary arteries (collateral circulation) exist, but they are usually not large enough to provide an adequate blood supply to the cardiac muscle should one of the large branches become blocked by disease. A sudden block of one of the larger branches of either coronary artery usually leads to myocardial death (myocardial infarction), although sometimes the collateral circulation is enough to sustain the muscle.

## Summary of the Overall Arterial Supply to the Heart in Most Individuals

The **right coronary artery** supplies all of the right ventricle (except for the small area to the right of the anterior interventricular groove), the variable part of the diaphragmatic surface of the left ventricle, the posteroinferior third of the ventricular septum, the right atrium and part of the left atrium, and the sinuatrial node and the atrioventricular node and bundle. The LBB also receives small branches.

The **left coronary artery** supplies most of the left ventricle, a small area of the right ventricle to the right of the interventricular groove, the anterior two thirds of the ventricular septum, most of the left atrium, the RBB, and the LBB.

## Arterial Supply to the Conducting System

The sinuatrial node is usually supplied by the right coronary artery but is sometimes supplied by the left coronary artery. The atrioventricular node and the atrioventricular bundle are supplied by the right coronary artery. The RBB of the atrioventricular bundle is supplied by the left coronary artery; the LBB is supplied by the right and left coronary arteries (Fig. 4-11).

## Venous Drainage of the Heart

Most blood from the heart wall drains into the right atrium through the coronary sinus (Fig. 4-10), which lies in the posterior part of the atrioventricular groove and is a continuation of the great cardiac vein. It opens into the right atrium to the left of the inferior vena cava. The small and middle cardiac veins are tributaries of the coronary sinus. The remainder of the blood is returned to the right atrium by the anterior cardiac vein (Fig. 4-10) and by small veins that open directly into the heart chambers.

### PHYSIOLOGIC NOTE

#### Coronary Circulation

The coronary blood flow in the normal, resting individual is about 225 mL/min and is continuous throughout the cardiac cycle, although about 75% occurs in diastole because of compression of the small branches of the coronary arteries by the cardiac muscle that takes place during systole. Stimulation of the sympathetic nervous system causes slight vasodilatation of the coronary arteries, whereas parasympathetic stimulation causes slight vasoconstriction. Increased coronary flow mainly results from the increased work of the heart muscle and the local effects of the products of metabolism causing vasodilation.

## Nerve Supply of the Heart

The heart is innervated by sympathetic and parasympathetic fibers of the autonomic nervous system via the **cardiac plexuses** situated below the arch of the aorta. The sympathetic supply arises from the cervical and upper thoracic portions of the sympathetic trunks, and the parasympathetic supply comes from the vagus nerves (Fig. 4-12).

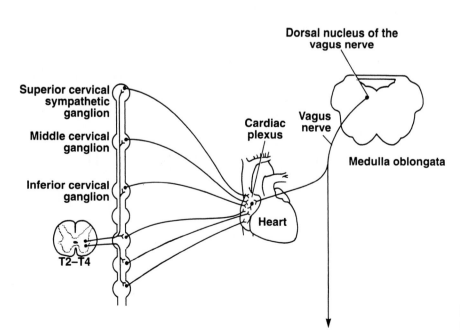

**Figure 4-12** Autonomic innervation of the heart.

The postganglionic sympathetic fibers terminate on the sinuatrial and atrioventricular nodes, on cardiac muscle fibers, and on the coronary arteries. Activation of these nerves results in cardiac acceleration, increased force of contraction of the cardiac muscle, and dilatation of the coronary arteries.

The postganglionic parasympathetic fibers terminate on the sinuatrial and atrioventricular nodes and on the coronary arteries. Activation of the parasympathetic nerves results in a reduction in the rate and force of contraction of the heart and a constriction of the coronary arteries.

Afferent fibers running with the sympathetic nerves carry nervous impulses that normally do not reach consciousness. However, should the blood supply to the myocardium become impaired, pain impulses reach consciousness via this pathway. Afferent fibers running with the vagus nerves take part in cardiovascular reflexes.

### PHYSIOLOGIC NOTE

#### Blood Circulation through the Heart

The heart is a muscular pump. The series of changes that take place within it as it fills with blood and empties is referred to as the **cardiac cycle.** The normal heart beats 70–90 times per minute in the resting adult and 130–150 times per minute in the newborn child.

Blood is continuously returning to the heart; during ventricular systole (contraction), when the atrioventricular valves are closed, the blood is temporarily accommodated in the large veins and atria. Once ventricular diastole (relaxation) occurs, the atrioventricular valves open, and blood passively flows from the atria to the ventricles. When the ventricles are nearly full, atrial systole occurs and forces the remainder of the blood in the atria into the ventricles. The sinuatrial node initiates the wave of contraction in the atria, which commences around the openings of the large veins and milks the blood toward the ventricles. By this means, blood does not reflux into the veins.

The cardiac impulse, having reached the atrioventricular node, is conducted to the papillary muscles by the atrioventricular bundle and its branches. The papillary muscles then begin to contract and take up the slack of the chordae tendineae. Meanwhile, the ventricles start contracting and the atrioventricular valves close. The spread of the cardiac impulse along the atrioventricular bundle (Fig. 4-9) and its terminal branches, including the Purkinje fibers, ensures that myocardial contraction occurs at almost the same time throughout the ventricles.

Once the intraventricular blood pressure exceeds that present in the large arteries (aorta and pulmonary trunk), the semilunar valve cusps are pushed aside, and the blood is ejected from the heart. At the conclusion of ventricular systole, blood begins to move back toward the ventricles and immediately fills the pockets of the semilunar valves. The cusps float into apposition and completely close the aortic and pulmonary orifices.

### PHYSIOLOGIC NOTE

#### Atrial and Ventricular Reflexes

##### Bainbridge Atrial Reflex

Stretch receptors present in the walls of the atria are stimulated by an increase in atrial pressure. The afferent stimuli ascend to the medulla oblongata in the vagus nerves, and the heart rate is increased in response to the diminished activity of the vagi and the increased activity of the sympathetics.

##### Bezold-Jarisch Reflex

Receptors present in the left ventricular walls are stimulated by certain chemicals, such as nicotine. Afferent impulses ascend to the medulla oblongata in the vagus nerves, and the heart rate is slowed in response to increased vagal activity on the heart. It has been postulated that chemicals released by the degenerating tissue in myocardial infarction may initiate this reflex and contribute to the hypotension in this condition.

## Surface Anatomy of the Heart Valves

The surface projection of the heart is seen in Figure 4-13. The surface markings of the heart valves are as follows (Fig. 4-13):

- The **tricuspid valve** lies behind the right half of the sternum opposite the fourth intercostal space.
- The **mitral valve** lies behind the left half of the sternum opposite the fourth costal cartilage.
- The **pulmonary valve** lies behind the medial end of the third left costal cartilage and the adjoining part of the sternum.
- The **aortic valve** lies behind the left half of the sternum opposite the third intercostal space.

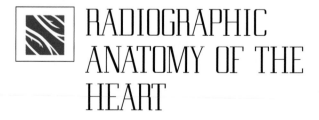

# RADIOGRAPHIC ANATOMY OF THE HEART

The normal radiographic anatomy of the heart in posteroanterior, oblique, and lateral views on radiographs is shown in Figures 3-37 through 3-42.

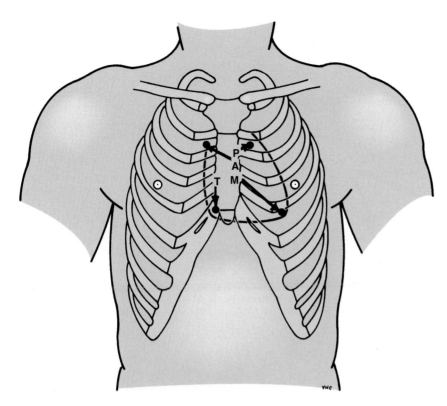

**Figure 4-13**   Position of the heart valves. P = pulmonary valve; A = aortic valve; M = mitral valve; T = tricuspid valve. *Arrows* indicate position where valves may be heard with least interference.

---

### EMBRYOLOGIC NOTE

## Development of the Heart Tube

### Formation of the Heart Tube

Clusters of cells arise in the mesenchyme at the cephalic end of the embryonic disc, cephalic to the site of the developing mouth and the nervous system. These clusters of cells form a plexus of endothelial blood vessels that fuse to form the **right** and **left endocardial heart tubes.** These, too, soon fuse to form a single **median endocardial tube.** As the head fold of the embryo develops, the endocardial tube and the pericardial cavity rotate on a transverse axis through almost 180°, so that they come to lie ventral (in front of) to the esophagus and caudal to the developing mouth.

The heart tube starts to bulge into the pericardial cavity (Fig. 4-14). Meanwhile, the endocardial tube becomes surrounded by a thick layer of mesenchyme, which will differentiate into the **myocardium** and the **visceral layer of the serous pericardium.** The primitive heart has been established, and the cephalic end is the arterial end, and the caudal end is the venous end. The arterial end of the primitive heart is continuous beyond the pericardium with a large vessel, the **aortic sac** (Fig. 4-15). The heart begins to beat during the 3rd week.

### Further Development of the Heart Tube

The heart tube then undergoes differential expansion so that several dilatations, separated by grooves, result.

From the arterial to the venous end, these dilatations are called the **bulbus cordis,** the **ventricle,** the **atrium,** and the **right** and **left horns of the sinus venosus**. The bulbus cordis and ventricular parts of the tube now elongate more rapidly than the remainder of the tube, and because the arterial and venous ends are fixed by the pericardium, the tube begins to bend (Fig. 4-16). The bend soon becomes U shaped and then forms a compound S shape, with the atrium lying posterior to the ventricle; thus the venous and arterial ends are brought close together as they are in the adult. The passage between the atrium and the ventricle narrows to form the **atrioventricular canal.** As these changes are taking place, a gradual migration of the heart tube occurs so that the heart passes from the neck region to what will become the thoracic region.

### Development of the Atria

The primitive atrium becomes divided into two atria—the right and left atria—in the following manner (Fig. 4-17). First, the atrioventricular canal widens transversely. The canal then becomes divided into right and left halves by the appearance of ventral and dorsal atrioventricular cushions, which fuse to form the **septum intermedium**. Meanwhile, another septum, the **septum primum,** develops from the roof of the primitive atrium and grows down to fuse with the septum intermedium. Before fusion occurs, the opening between the lower edge of the septum primum and septum intermedium is referred to as the **foramen primum**. The atrium now is divided into right and left parts. Before complete obliteration of the

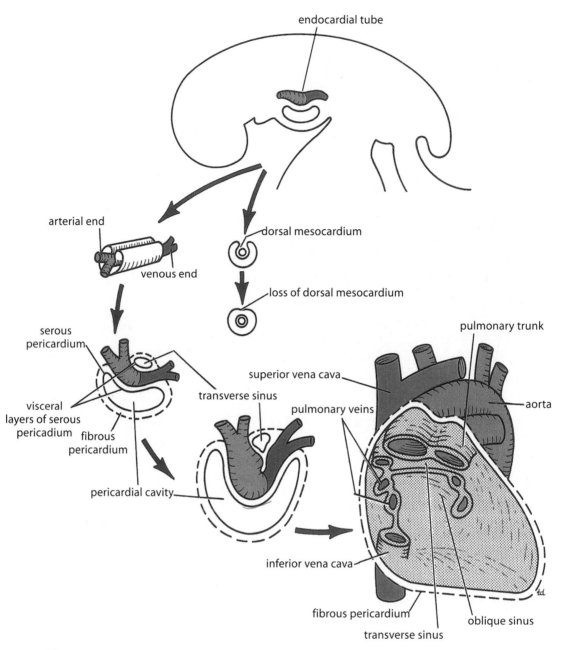

**Figure 4-14**   The development of the endocardial tube in relation to the pericardial cavity.

foramen primum has taken place, degenerative changes occur in the central portion of the septum primum; a foramen appears, the **foramen secundum,** so that the right and left atrial chambers again communicate. Another thicker septum (the **septum secundum**) grows down from the atrial roof on the right side of the septum primum. The lower edge of the septum secundum overlaps the foramen secundum in the septum primum but does not reach the floor of the atrium and does not fuse with the septum intermedium. The space between the free margin of the septum secundum and the septum primum is now known as the **foramen ovale** (Fig. 4-17).

Before birth, the foramen ovale allows oxygenated blood that has entered the right atrium from the inferior vena cava to pass into the left atrium. However, the lower part of the septum primum serves as a flap-like valve to prevent blood moving from the left atrium into the right atrium. At birth, due to raised blood pressure in the left atrium, the septum primum is pressed against the septum secundum and fuses with it, and the foramen ovale is closed. The two atria thus are separated from each other. The lower edge of the septum secundum seen in the right atrium becomes the **anulus ovalis**, and the depression below this is called the **fossa ovalis**. The **right**

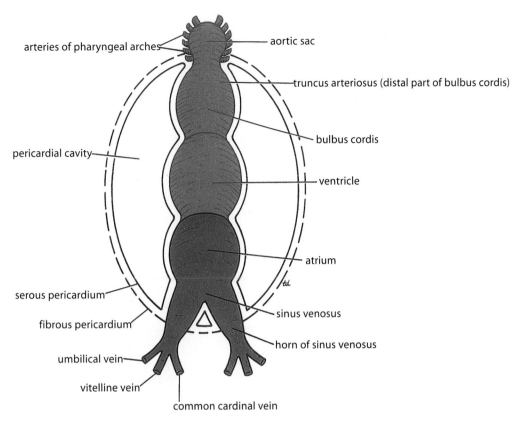

**Figure 4-15** The parts of the endocardial heart tube within the pericardium.

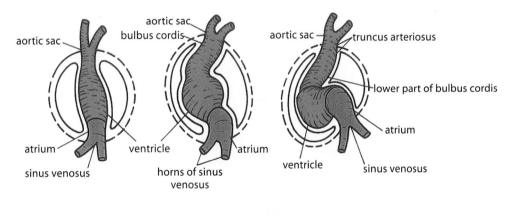

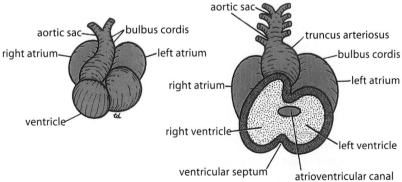

**Figure 4-16** The bending of the heart tube within the pericardial cavity. The interior of the developing ventricles are shown at the bottom right.

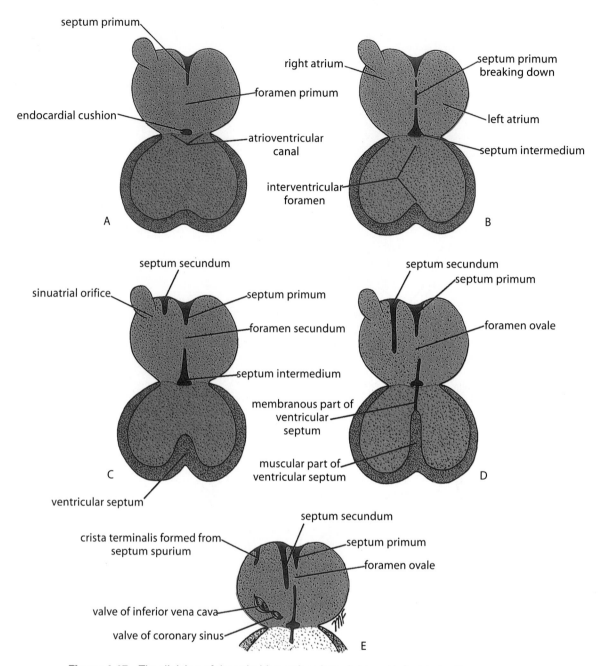

**Figure 4-17** The division of the primitive atrium into right and left atria by the appearance of the septa. The sinuatrial orifice and the fate of the venous valves are shown, as is the appearance of the ventricular septum.

and **left auricular appendages** later develop as small diverticula from the right and left atria, respectively.

### Development of the Ventricles

A muscular partition projects upward from the floor of the primitive ventricle to form the **ventricular septum** (Fig. 4-17). The space bounded by the crescentic upper edge of the septum and the endocardial cushions forms the **interventricular foramen**. Meanwhile, spiral subendocardial thickenings, the **bulbar ridges**, appear in the distal part of the bulbus cordis. The bulbar ridges then grow

and fuse to form a **spiral aorticopulmonary septum** (Fig. 4-17). The interventricular foramen closes as the result of proliferation of the bulbar ridges and the fused endocardial cushions (septum intermedium). This newly formed tissue grows down and fuses with the upper edge of the muscular ventricular septum to form the **membranous part of the septum** (Fig. 4-17). The closure of the interventricular foramen not only shuts off the path of communication between the right and left ventricles but also ensures that the right ventricular cavity communicates with the pulmonary trunk and the left ventricular cavity communicates with the aorta. In addition, the right atrioventricular opening now connects exclusively with the right ventricular cavity; and the left atrioventricular opening connects with the left ventricular cavity.

### Development of the Roots and Proximal Portions of the Aorta and Pulmonary Trunk

The distal part of the bulbus cordis is known as the **truncus arteriosus** (Fig. 4-15). It is divided by the spiral aorticopulmonary septum to form the roots and proximal portions of the aorta and pulmonary trunk (Fig. 4-18). With the establishment of right and left ventricles, the proximal portion of the bulbus cordis becomes incorporated into the right ven-

tricle as the definitive **conus arteriosus** or **infundibulum** and into the left ventricle as the **aortic vestibule**. Just distal to the aortic valves, the two **coronary arteries** arise as outgrowths from the developing aorta.

### Development of the Heart Valves

#### Semilunar Valves of the Aorta and Pulmonary Arteries

After the formation of the aorticopulmonary septum, three swellings appear at the orifices of both the aorta and the pulmonary artery. Each swelling consists of a covering of endothelium over loose connective tissue. Gradually, the swellings become excavated on their upper surfaces to form the semilunar valves.

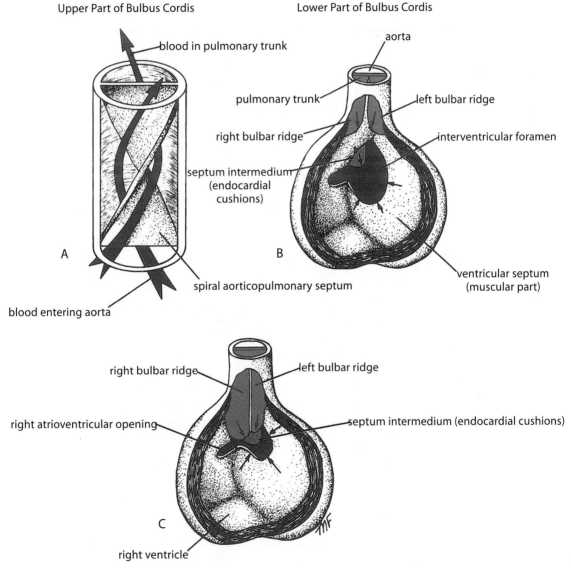

**Figure 4-18**   The division of the bulbus cordis by the spiral aorticopulmonary septum into the aorta and pulmonary trunk. **A.** The spiral septum in the truncus arteriosus (upper part of the bulbus cordis). **B.** The lower part of the bulbus cordis showing the formation of the spiral septum by fusion of the bulbar ridges (red), which then grow down and join the septum intermedium (blue) and the muscular part of the ventricular septum. **C.** The area of the ventricular septum that is formed from the fused bulbar ridges (red), and the septum intermedium (blue) is called the membranous part of the ventricular septum.

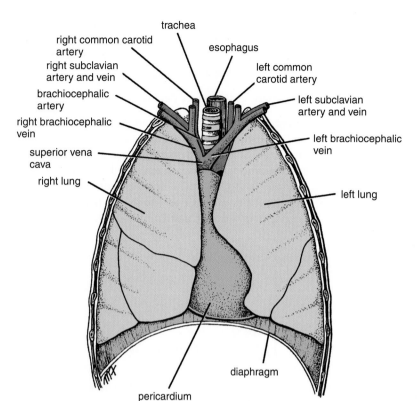

**Figure 4-19** The pericardium and the lungs exposed from in front.

### Atrioventricular Valves

After the formation of the septum intermedium, the atrioventricular canal becomes divided into right and left atrioventricular orifices. Raised folds of endocardium appear at the margins of these orifices. These folds are invaded by mesenchymal tissue that later becomes hollowed out from the ventricular side. Three valvular cusps are formed about the right atrioventricular orifice and constitute the **tricuspid valve**; two cusps are formed about the left atrioventricular orifice to become the **mitral valve**. The newly formed cusps enlarge, and their mesenchymal core becomes differentiated into fibrous tissue. The cusps remain attached at intervals to the ventricular wall by muscular strands. Later, the muscular strands become differentiated into **papillary muscles** and **chordae tendineae**.

# Pericardium

The pericardium is a fibroserous sac that encloses the heart and the roots of the great vessels. Its function is to restrict excessive movements of the heart as a whole and to serve as a lubricated container in which the different parts of the heart can contract. The pericardium lies within the middle mediastinum (Figs. 4-19, 4-20, and 4-21), posterior to the body of the sternum and the second to the sixth costal cartilages and anterior to the fifth to the eighth thoracic vertebrae.

## Fibrous Pericardium

The fibrous pericardium is the strong fibrous part of the sac. It is firmly attached below to the central tendon of the diaphragm. It fuses with the outer coats of the great blood vessels passing through it (Fig. 4-20)—namely, the aorta, the pulmonary trunk, the superior and inferior venae cavae, and the pulmonary veins (Fig. 4-21). The fibrous

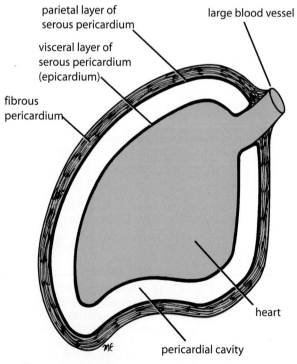

**Figure 4-20** Different layers of the pericardium.

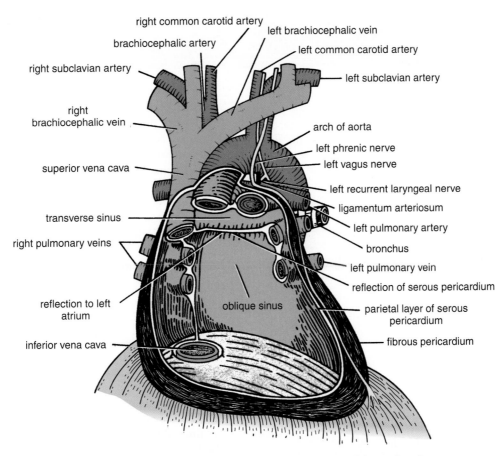

right common carotid artery
brachiocephalic artery
right subclavian artery
right brachiocephalic vein
superior vena cava
transverse sinus
right pulmonary veins
reflection to left atrium
inferior vena cava

left brachiocephalic vein
left common carotid artery
left subclavian artery
arch of aorta
left phrenic nerve
left vagus nerve
left recurrent laryngeal nerve
ligamentum arteriosum
left pulmonary artery
bronchus
left pulmonary vein
reflection of serous pericardium
parietal layer of serous pericardium
fibrous pericardium

oblique sinus

**Figure 4-21**   The great blood vessels and the interior of the pericardium.

pericardium is attached in front to the sternum by the **sternopericardial ligaments.**

## Serous Pericardium

The serous pericardium lines the fibrous pericardium and coats the heart. It is divided into parietal and visceral layers (Fig. 4-20).

The parietal layer lines the fibrous pericardium and is reflected around the roots of the great vessels to become continuous with the visceral layer of serous pericardium that closely covers the heart (Fig. 4-21). The visceral layer is closely applied to the heart and is often called the **epicardium.** The slitlike space between the parietal and visceral layers is referred to as the **pericardial cavity** (Fig. 4-20). Normally, the cavity contains a small amount of tissue fluid (about 50 mL), the **pericardial fluid,** which acts as a lubricant to facilitate movements of the heart.

### Pericardial Sinuses

On the posterior surface of the heart, the reflection of the serous pericardium around the large veins forms a recess called the **oblique sinus** (Fig. 4-21). Also on the posterior surface of the heart is the **transverse sinus,** which is a short passage that lies between the reflection of serous pericardium around the aorta and pulmonary trunk and the re-

flection around the large veins (Fig. 4-21). The pericardial sinuses form as a consequence of the way the heart bends during development. They have no clinical significance.

## Nerve Supply of the Pericardium

The fibrous pericardium and the parietal layer of the serous pericardium are supplied by the phrenic nerves. The visceral layer of the serous pericardium is innervated by branches of the sympathetic trunks and the vagus nerves.

# Principal Relations of the Pericardium and Heart

**Anterior:** Body of the sternum, third to the sixth costal cartilages and the intercostal spaces between them, internal thoracic vessels, anterior borders of the right and left lungs, and the pleural cavities (Fig. 4-8). In young children, the thymus lies anterior to the upper part of the pericardium.
**Posterior:** Fifth to eighth thoracic vertebrae, esophagus, descending thoracic aorta, main bronchi, and the rounded posterior part of each lung.
**Lateral:** Mediastinal parietal pleura, phrenic nerve, and lung and pleural cavities.
**Inferior:** Diaphragm, liver, and fundus of the stomach.

# Review Questions

## Multiple Choice Questions

### Select the BEST answer for each question.

1. The anterior surface of the heart is formed by the following structures **except** which?
   A. Right ventricle
   B. Right atrium
   C. Left ventricle
   D. Left atrium
   E. Right auricle

2. In a posteroanterior radiograph of the thorax, the following structures form the left margin of the heart shadow **except** which?
   A. Left auricle
   B. Pulmonary trunk
   C. Arch of aorta
   D. Left ventricle
   E. Superior vena cava

3. All of the following statements regarding the conducting system of the heart are true **except** which?
   A. The impulse for cardiac contraction spontaneously begins in the sinuatrial node.
   B. The atrioventricular bundle is the sole pathway for conduction of the waves of contraction between the atria and the ventricles.
   C. The sinuatrial node is frequently supplied by the right and left coronary arteries.
   D. The sympathetic nerves to the heart slow the rate of discharge from the sinuatrial node.
   E. The atrioventricular bundle descends behind the septal cusp of the tricuspid valve.

4. The following statements concerning the blood supply to the heart are correct **except** which?
   A. The coronary arteries are branches of the ascending aorta.
   B. The right coronary artery supplies both the right atrium and the right ventricle.
   C. The circumflex branch of the left coronary artery descends in the anterior interventricular groove and passes around the apex of the heart.
   D. Arrhythmias (abnormal heart beats) can occur after occlusion of a coronary artery.
   E. Coronary arteries can be classified as functional end arteries.

5. The following statements concerning the heart are correct **except** which?
   A. The left atrium lies posterior to the right atrium.
   B. The second shorter sound of the heart is dŭp, which is produced by the sharp closure of the aortic and pulmonary valves.
   C. The apex beat of the heart is best felt by asking the patient to sit up and lean forward.
   D. The first sound of the heart is lūb and is produced by the contraction of the ventricles and the closure of the tricuspid and mitral valves.
   E. The pulmonary valve has two semilunar cusps.

6. The following statements concerning the structure of the heart and the pericardium are correct **except** which?
   A. The pericardial cavity is the potential space between the fibrous and the serous pericardia.
   B. The chordae tendineae connect the papillary muscles to the tricuspid and mitral valve cusps in the right and left ventricles.
   C. The trabeculae carneae are internal surface structures of both the left and the right ventricles.
   D. The four pulmonary veins open through the posterior wall of the left atrium, and there are no valves.
   E. The sinuatrial node is supplied by the right and sometimes the left coronary artery.

## Fill in the Blank Questions

### Complete the following statements with the appropriate structure listed below.

7. The left coronary artery gives off the _____ branch, which supplies the right and left ventricles.

8. Both the sinuatrial node and the atrioventricular node are situated in the _____ portion of the heart.

9. The base of the heart is formed mainly by the _____.

10. The coronary arteries are branches of the _____.

11. The coronary sinus drains into the _____.

12. The pericardium has the _____ passing down over its lateral surface.

13. The oblique sinus of the pericardium is related anteriorly to the _____.

14. The anterior cardiac vein drains into the _____.
    A. Right atrium
    B. Left atrium
    C. Ascending aorta
    D. Phrenic nerve
    E. Atrioventricular groove
    F. Anterior interventricular
    G. Posterior interventricular

## Matching Questions

**Match each structure listed below with the region in the heart in which it is found. Each lettered answer may be used more than once.**

15. The inferior vena cava (opening)

16. Moderator band

17. Anulus ovalis

18. Right pulmonary veins (openings)

    A. Left atrium
    B. Right ventricle
    C. Right atrium
    D. Left ventricle
    E. Right auricle

## Multiple Choice Questions

**Read the following case histories and select the best answer to the questions following them.**

On performing a routine examination of a 7-year-old girl, a pediatrician heard a continuous machinery-like murmur in the second left intercostal space. The murmur occupied both systole and diastole. The child was not cyanotic, the heart was of normal size, and there was no clubbing of the fingers. Radiographic examination of the chest revealed slight enlargement of the left atrium, left ventricle, and pulmonary trunk. A diagnosis of patent ductus arteriosus was made.

19. Based on the clinical history and the diagnosis, the following statements concerning the case are correct **except which?**
    A. The patent ductus represents the distal portion of the sixth left aortic arch artery.
    B. The ductus connects the right pulmonary artery to the descending thoracic aorta.
    C. The ductus in fetal life is the normal bypass of blood to the aorta from the pulmonary trunk.
    D. At birth, the ductus arteriosus normally constricts in response to a rise in arterial oxygen.
    E. The ductus arteriosus closes to become the ligamentum arteriosum.

20. The presence of a patent ductus presents the following physiologic and pathologic consequences **except which?**
    A. Aortic blood passes into the pulmonary artery, producing the machinery-like murmur.
    B. The shunting of blood occurs only during systole as the result of the higher blood pressure in the aorta and the lower blood pressure in the pulmonary artery.
    C. The left ventricle shows hypertrophy because of the leak from the aorta.
    D. The pulmonary trunk becomes enlarged and the right ventricle becomes hypertrophied due to the raised pressure in the pulmonary circulation.
    E. Because of the risk of bacterial infection of the wall of the pulmonary artery (bacterial endarteritis) caused by the pulmonary hypertension, the patent ductus should be ligated and divided surgically.

# Answers and Explanations

1. **D** is correct. The left atrium does not form a part of the anterior surface of the heart because it lies behind the right atrium and forms a large portion of the posterior surface or base of the heart (Fig. 4-3).

2. **E** is correct. The superior vena cava forms part of the right margin of the heart shadow in a posteroanterior radiograph of the chest (Fig. 3-37).

3. **D** is correct. The sympathetic nerves to the heart increase the rate of discharge from the sinuatrial node.

4. **C** is correct. The circumflex branch of the left coronary artery descends in the atrioventricular groove and does not pass around the apex of the heart (Fig. 4-10).

5. **E** is correct. The pulmonary valve has three semilunar cusps, similar to that found in the aortic valve (Fig. 4-6).

6. **A** is correct. The pericardial cavity is the potential space between the parietal and visceral layers of serous pericardium (Fig. 4-20).

7.  **F** is correct. The left coronary artery gives off the anterior interventricular branch, which supplies the right and left ventricles (Fig. 4-10).

8.  **A** is correct. Both the sinuatrial and atrioventricular nodes are situated in the right atrium of the heart (Fig. 4-4).

9.  **B** is correct. The left atrium forms the main part of the base of the heart (Fig. 4-3).

10. **C** is correct. The coronary arteries are branches of the ascending aorta (Fig. 4-2).

11. **A** is correct. The coronary sinus drains into the right atrium (Fig. 4-4).

12. **D** is correct. The pericardium has the phrenic nerve passing down over its lateral surface (see Fig. 3-26).

13. **B** is correct. The oblique sinus of the pericardium is related anteriorly to the left atrium of the heart (Fig. 4-8).

14. **A** is correct. The anterior cardiac vein drains into the right atrium of the heart (Fig. 4-10).

15. **C** is correct. The inferior vena cava opens into the lower part of the right atrium (Fig. 4-4).

16. **B** is correct. The moderator band is located in the right ventricle (Fig. 4-4).

17. **C** is correct. The anulus ovalis is located in the right atrium (Fig. 4-4).

18. **A** is correct. The right and left pulmonary veins open into the left atrium (Fig. 4-3).

19. **B** is the incorrect statement. The ductus arteriosus represents the distal portion of the sixth left aortic arch artery and connects the left pulmonary artery at its origin from the pulmonary trunk to the junction of the aortic arch and the descending thoracic aorta.

20. **B** is the incorrect statement. The machinery-like murmur occurs during both systole and diastole and is caused by the shunting of blood from the aorta to the pulmonary artery due to the higher blood pressure in the aorta during both phases of the cardiac cycle.

# 5

# The Blood Vessels
# of the Thorax

All clinical material relevant to this chapter can be found on the
CD-ROM.

# Chapter Outline

The largest blood vessels in the body are located within the chest cavity—the aorta, the pulmonary arteries, the venae cavae, and the pulmonary veins. Trauma to the chest wall can result in the disruption of these vessels, with consequent hemorrhage, rapid exsanguination, and death. Penetrating injuries to the chest can pierce the vessel walls and blunt injuries causing sudden body acceleration or deceleration can tear the vessels. Unfortunately, because these vessels are hidden from view within the thorax, the diagnosis of a major blood vessel injury is often delayed, with disastrous consequences to the patient.

The purpose of this chapter is to familiarize the medical professional with the basic anatomy of the blood vessels of the thorax in order that diagnosis of vascular injury can be made promptly and vascular access can be established quickly and accurately.

# BASIC ANATOMY

## Large Arteries of the Thorax

### Aorta

The aorta is the main arterial trunk that delivers oxygenated blood from the left ventricle of the heart to the tissues of the body (Fig. 5-1). It is divided for purposes of description into the following parts: ascending aorta, arch of the aorta, descending thoracic aorta, and abdominal aorta.

### Ascending Aorta

The ascending aorta begins at the base of the left ventricle and runs upward and forward to come to lie behind the right half of the sternum at the level of the sternal angle, where it becomes continuous with the arch of the aorta. The ascending aorta lies within the fibrous pericardium (Fig. 5-1) and is enclosed with the pulmonary trunk in a sheath of serous pericardium. At its root, it possesses three

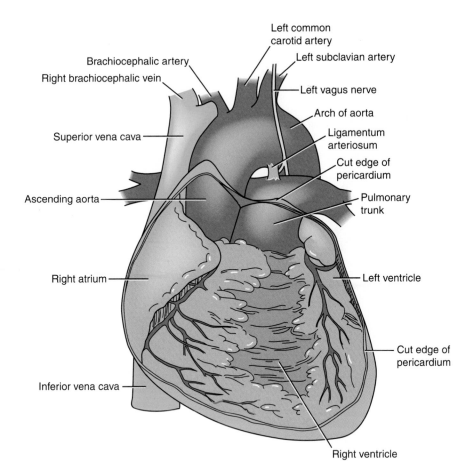

Figure 5-1   Anterior surface of the heart and the great blood vessels. The pericardial sac has been opened to show the ascending aorta and pulmonary trunk.

bulges, the sinuses of the aorta, one behind each aortic valve cusp.

## Important Relations (Figs. 5-1 and 5-2)

- **Anterior:** Pulmonary trunk, right auricle, edge of right pleura and right lung, remains of thymus, and sternum.
- **Posterior:** Left atrium, right pulmonary artery, and right principal bronchus.
- **Right lateral:** Superior vena cava and right atrium.
- **Left lateral:** Left atrium and pulmonary trunk.

## Branches

The **right coronary artery** arises from the anterior aortic sinus, and the **left coronary artery** arises from the left posterior aortic sinus (see Figs. 4- 7 and 4-10).

## Arch of the Aorta

The arch of the aorta is a continuation of the ascending aorta (Fig. 5-2). It lies behind the manubrium sterni and arches upward, backward, and to the left in front of the trachea (its main direction is backward). It then passes downward to the left of the trachea and, at the level of the sternal angle, becomes continuous with the descending thoracic aorta.

---

**P H Y S I O L O G I C   N O T E**

### Ejection Phase of the Cardiac Cycle and the Apex Beat

During the ejection phase of the cardiac cycle, blood is forced under high pressure through the aortic valve of the left ventricle into the aorta. The aortic arch is a flexible curved tube and tends to straighten slightly with the blood pressure, thrusting the heart forward so that the apex contacts the chest wall to form the apex beat. Normally, the beat is felt in the fifth left intercostal space 3½ in. (9 cm) from the midline.

### Important Relations

- **Anterior and to the left** (Fig. 5-3): Left mediastinal pleura, left phrenic nerve, left vagus nerve, cardiac branches of vagus and sympathetic nerves, left superior intercostal vein, left lung, and pleura.
- **Posterior and to the right**: Left recurrent laryngeal nerve, cardiac plexus, trachea, esophagus, and vertebral column.
- **Superior**: Brachiocephalic, left common carotid, and left subclavian arteries take origin from its convexity (Fig. 5-1).

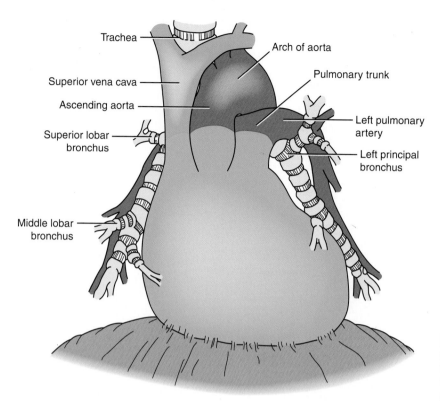

**Figure 5-2** Anterior surface of the heart and the great blood vessels showing their relationship to the bifurcation of the trachea and the main bronchi. The pericardium is intact.

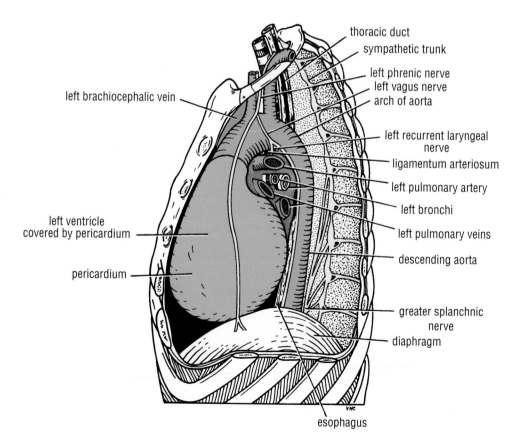

**Figure 5-3** Left side of the mediastinum.

■ **Inferior** (Fig. 5-1): Bifurcation of pulmonary trunk, ligamentum arteriosum, left recurrent laryngeal nerve, and cardiac plexus.

**Branches**

The **brachiocephalic artery** arises from the convex surface of the aortic arch (Figs. 5-1 and 5-4). It passes upward and to the right of the trachea and divides into the right subclavian and right common carotid arteries behind the right sternoclavicular joint.

The **left common carotid artery** arises from the convex surface of the aortic arch on the left side of the brachiocephalic artery (Figs. 5-1 and 5-4). It runs upward and to the left of the trachea and enters the neck behind the left sternoclavicular joint.

The **left subclavian artery arises** from the aortic arch behind the left common carotid artery (Figs. 5-1, 5-3, and 5-4). It runs upward along the left side of the trachea and the esophagus to enter the root of the neck (Fig. 5-3). It arches over the apex of the left lung.

## Descending Thoracic Aorta

The descending thoracic aorta (Fig. 5-4) lies in the posterior mediastinum and begins as a continuation of the arch of the aorta on the left side of the lower border of the body of the fourth thoracic vertebra (i.e., opposite the sternal angle). It runs downward in the posterior mediastinum, inclining forward and medially to reach the anterior surface of the vertebral column (Figs. 5-3 and 5-4). At the level of the twelfth

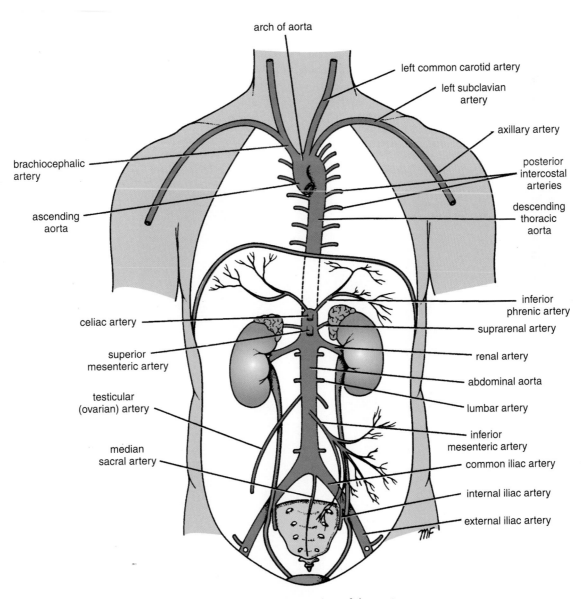

**Figure 5-4**   Major branches of the aorta.

thoracic vertebra, it passes behind the diaphragm (through the aortic opening) in the midline and becomes continuous with the abdominal aorta.

### Important Relations (Fig. 5-3)

- **Anterior:** Hilum of the left lung, the pericardium, esophagus, and the diaphragm.
- **Posterior:** Vertebral column and hemiazygos veins.
- **Right lateral:** Azygos vein, thoracic duct, right pleura, and right lung.
- **Left lateral:** Left pleura and left lung. Note that with respect to the aorta in the posterior mediastinum, the esophagus is right lateral above, anterior lower down, and becomes left anterolateral below. In other words, the aorta and the esophagus cross in the posterior mediastinum.

### Branches

**Posterior intercostal arteries** are given off to the lower nine intercostal spaces on each side (Fig. 5-4). Subcostal arteries are given off on each side and run along the lower border of the twelfth rib to enter the abdominal wall.

   **Pericardial, esophageal,** and **bronchial arteries** are small branches that are distributed to these organs.

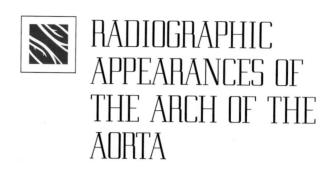

# RADIOGRAPHIC APPEARANCES OF THE ARCH OF THE AORTA

The normal radiographic appearances of the arch of the aorta and its branches are shown in an arteriogram in Figures 5-5 and 5-6 (see also Figs. 3-37 and 3-38).

## Abdominal Aorta

The abdominal aorta is described in Chapter 8.

## Pulmonary Trunk

The pulmonary trunk conveys deoxygenated blood from the right ventricle of the heart to the lungs. It leaves the upper part of the right ventricle and runs upward, backward, and to the left (Fig. 5-1). It is about 2 in. (5 cm) long and terminates in the concavity of the aortic arch by dividing into right and

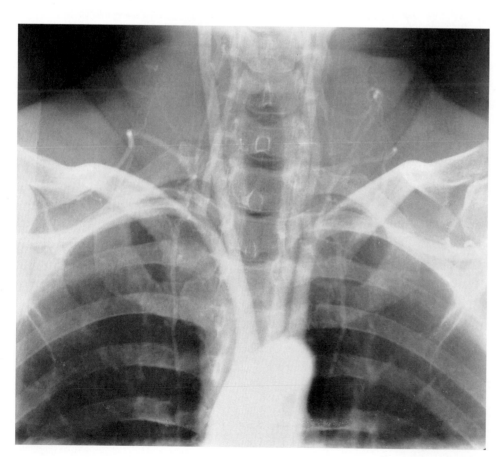

**Figure 5-5**   Aortic arch angiogram showing large arteries at the root of the neck.

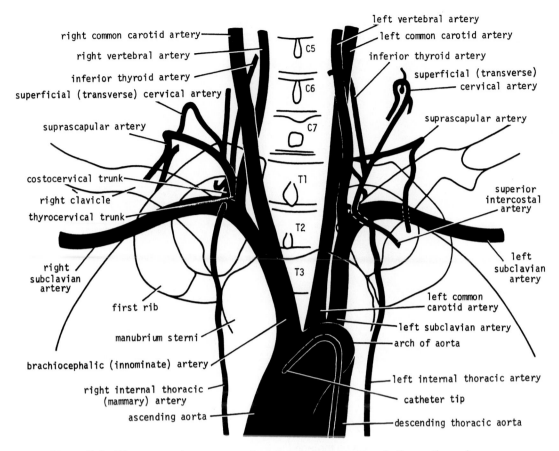

right common carotid artery
right vertebral artery
inferior thyroid artery
superficial (transverse) cervical artery
suprascapular artery
costocervical trunk
right clavicle
thyrocervical trunk
right subclavian artery
first rib
manubrium sterni
brachiocephalic (innominate) artery
right internal thoracic (mammary) artery
ascending aorta

C5
C6
C7
T1
T2
T3

left vertebral artery
left common carotid artery
inferior thyroid artery
superficial (transverse) cervical artery
suprascapular artery
superior intercostal artery
left subclavian artery
left common carotid artery
left subclavian artery
arch of aorta
left internal thoracic artery
catheter tip
descending thoracic aorta

**Figure 5-6** Diagrammatic representation of main features seen in the aortic angiogram.

left pulmonary arteries (Fig. 5-1). Together with the ascending aorta, the pulmonary trunk is enclosed in the fibrous pericardium and a sheath of serous pericardium (see Fig. 4-21).

## Important Relations

- **Anterior:** Sternal end of second left intercostal space, left lung and pleura, and pericardium.
- **Posterior:** Ascending aorta, left coronary artery, and left atrium.

## Branches

The **right pulmonary artery** runs to the right behind the ascending aorta and superior vena cava to enter the root of the right lung (Figs. 5-1 and 5-2).

The **left pulmonary artery** runs to the left in front of the descending aorta to enter the root of the left lung (Figs. 5-1, 5-2, and 5-3).

## Ligamentum Arteriosum

The ligamentum arteriosum is a fibrous band that connects the bifurcation of the pulmonary trunk to the lower concave surface of the aortic arch (Figs. 5-1 and 5-3). The ligamentum arteriosum is the remains of the ductus arteriosus, which in the fetus conducts blood from the pulmonary

trunk to the aorta, thus bypassing the lungs. The left recurrent laryngeal nerve hooks around the lower border of this structure (Figs. 5-1 and 5-3). After birth, the ductus closes.

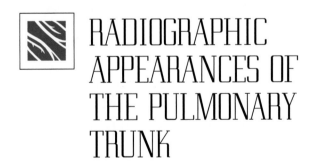

# RADIOGRAPHIC APPEARANCES OF THE PULMONARY TRUNK

The normal radiographic appearance of the pulmonary trunk is shown in the arteriogram in Figure 5- 7.

## EMBRYOLOGIC NOTE

### Development of the Large Arteries of the Thorax

The formation of the single endocardial heart tube and its ultimate differentiation into the definitive heart tube are

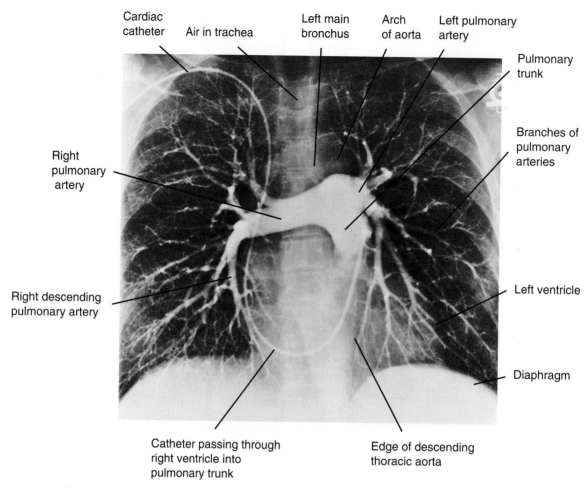

Cardiac catheter    Air in trachea    Left main bronchus    Arch of aorta    Left pulmonary artery    Pulmonary trunk    Branches of pulmonary arteries    Right pulmonary artery    Right descending pulmonary artery    Left ventricle    Diaphragm

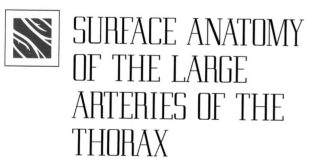

Catheter passing through right ventricle into pulmonary trunk    Edge of descending thoracic aorta

**Figure 5-7**    Angiogram of the pulmonary trunk and pulmonary arteries.

described in Chapter 4. The further development of the large arteries of the thorax from the truncus arteriosus of the bulbus cordis and the pharyngeal arch arteries is described in the CD.

# SURFACE ANATOMY OF THE LARGE ARTERIES OF THE THORAX

## Aorta

The **ascending aorta** lies behind the right half of the sternum just below the sternal angle (Fig.5-3).

The **arch of the aorta** and the roots of the brachiocephalic and left common carotid arteries lie behind the manubrium sterni (Fig. 5-3). The descending thoracic aorta begins at the termination of the arch to the left of the midline at the level of the sternal angle. As it descends, the aorta deviates to the midline and passes through the aortic opening in the diaphragm at the level of the 12th thoracic vertebra.

## Pulmonary Trunk

The pulmonary trunk bifurcates into the right and left pulmonary arteries to the left of the midline at about the level of the sternal angle (Fig. 5-3).

## Large Veins of the Thorax

### Brachiocephalic Veins

The **right brachiocephalic vein** is formed at the root of the neck by the union of the right subclavian and the right internal jugular veins (Figs. 5-8 and 5-9). The **left brachiocephalic vein** has a similar origin (Figs. 5-3 and 5-8). It passes obliquely downward and to the right behind the manubrium sterni and in front of the large branches of the aortic arch. It joins the right brachiocephalic vein to form the superior vena cava.

## Tributaries

Each brachiocephalic vein has the following tributaries: **vertebral vein**, **internal thoracic vein**, **inferior thyroid vein**, and the **first posterior intercostal vein**.

## Superior Vena Cava

The superior vena cava collects all the venous blood from the head and the neck and both upper limbs and is formed by the union of the two brachiocephalic veins (Figs. 5-8 and 5-9). It descends vertically to end in the right atrium of the heart (Fig. 5-1). The azygos vein joins the posterior aspect of the superior vena cava just before it pierces the pericardium (Fig. 5-9).

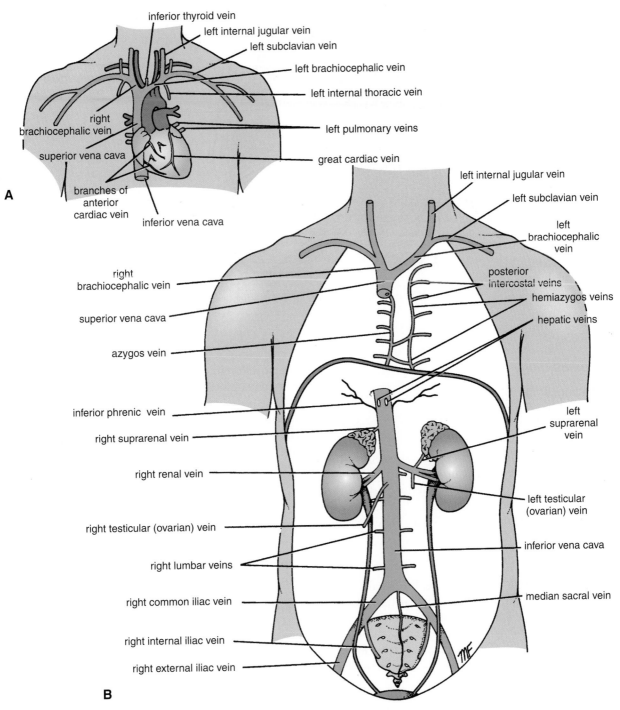

**Figure 5-8   A.** Major veins entering the heart. **B.** Major veins draining into the superior and inferior vena cavae.

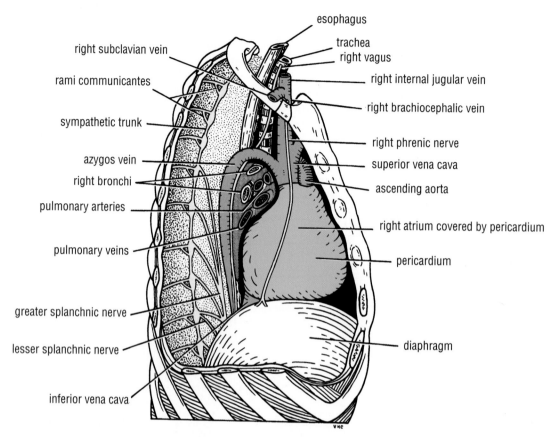

**Figure 5-9**  Right side of the mediastinum.

## Tributaries

The tributaries of the superior vena cava are the right and left brachiocephalic veins.

### Azygos Veins

The azygos veins consist of the main azygos vein, the inferior hemiazygos vein, and the superior hemiazygos vein. They drain blood from the posterior parts of the intercostal spaces, the posterior abdominal wall, the pericardium, the diaphragm, the bronchi, and the esophagus (Fig. 5- 8).

### Azygos Vein

The origin of the azygos vein is variable. It is often formed by the union of the **right ascending lumbar vein** and the **right subcostal vein.** It ascends through the aortic opening in the diaphragm on the right side of the aorta to the level of the fifth thoracic vertebra (Fig. 5-8). Here it arches forward above the root of the right lung to empty into the posterior surface of the superior vena cava (Fig. 5-9).

The azygos vein has numerous tributaries, including the **eight lower right intercostal veins,** the **right superior**

intercostal vein, the **superior** and **inferior hemiazygos veins,** and numerous **mediastinal veins**.

### Inferior Hemiazygos Vein

The inferior hemiazygos vein is often formed by the union of the left ascending lumbar vein and the left subcostal vein. It ascends through the left crus of the diaphragm and, at about the level of the eighth thoracic vertebra, turns to the right and joins the azygos vein (Fig. 5-8). It receives as tributaries some **lower left intercostal veins** and **mediastinal veins**.

### Superior Hemiazygos Vein

The superior hemiazygos vein is formed by the union of the fourth to the eighth intercostal veins. It joins the azygos vein at the level of the seventh thoracic vertebra (Fig. 5- 8).

## Inferior Vena Cava

The inferior vena cava is formed in the abdomen and pierces the central tendon of the diaphragm and the pericardium opposite the eighth thoracic vertebra and almost immediately enters the lowest part of the right atrium (Figs.

5-1, 5-8, and 5- 9). The valve of the inferior vena cava is important in the fetus but rudimentary in the adult.

## Pulmonary Veins

Two pulmonary veins leave each lung carrying oxygenated blood to the left atrium of the heart (Figs. 5-3, 5-8, and 5-9). There are no valves.

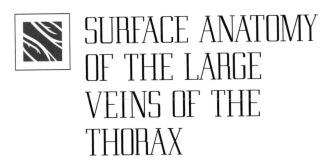

# SURFACE ANATOMY OF THE LARGE VEINS OF THE THORAX

The superior vena cava and the terminal parts of the right and left brachiocephalic veins lie behind the manubrium sterni.

**E M B R Y O L O G I C   N O T E**

### Fetal Circulation

Since the fetus is lying within the uterus of the mother and is surrounded by fluid, it is not surprising that the blood circulatory system of the fetus differs from that of an air-breathing, free-existing individual. In the fetus, the lungs, kidneys, and gastrointestinal tract are nonfunctional; the fetus obtains oxygen and nourishment from the mother's blood through the placenta and gets rid of carbon dioxide and other products of metabolism by the same means.

To understand the fetal circulation, it is a good idea to follow the fetal blood from the placenta through the umbilical vein back to the fetus, then to trace the path taken by the blood through the fetus, and finally to follow its return to the placenta by the two umbilical arteries, which are branches of the fetal internal iliac arteries (see Fig. 5-6 on CD).

For a detailed discussion of the fetal circulation and the changes that take place in the circulation at birth, the reader should consult the CD.

# Review Questions

## Completion Questions

**Select the phrase that BEST completes each statement.**

1. The ascending aorta arises from the
   A. right ventricle.
   B. left atrium.
   C. right auricle.
   D. left ventricle.
   E. left auricle.

2. The ascending aorta has which of the following branches?
   A. Brachiocephalic artery
   B. Left common carotid artery
   C. Right and left coronary arteries
   D. Left subclavian artery
   E. Right subclavian artery

3. The ascending aorta
   A. possesses three bulges, the sinuses of the aorta.
   B. lies outside the fibrous pericardium.
   C. is enclosed in a serous sheath of serous pericardium, which is separate from that surrounding the pulmonary trunk.

D. lies behind the right half of the manubrium sterni.
E. runs upward and backward from the heart.

4. The arch of the aorta begins
   A. at the level of the sternal angle.
   B. behind the right sternoclavicular joint.
   C. behind the left sternoclavicular joint.
   D. at the base of the left ventricle.
   E. behind the third right costal cartilage.

5. The arch of the aorta has which of the following branches?
   A. The right and left coronary arteries
   B. The internal thoracic artery
   C. The brachiocephalic artery
   D. The right common carotid artery
   E. The brachiocephalic, the left common carotid, and the left subclavian arteries

6. The descending thoracic aorta lies in the
   A. superior mediastinum.
   B. middle mediastinum.
   C. anterior mediastinum.
   D. posterior mediastinum.
   E. pericardial cavity.

7. The descending thoracic aorta pierces the diaphragm
   A. at the level of the 8th thoracic vertebra.
   B. at the level of the 10th thoracic vertebra.
   C. at the level of the 12th thoracic vertebra.
   D. through the central tendon.
   E. at the left cupola.

8. The pulmonary trunk
   A. conveys oxygenated blood from the right ventricle of the heart to the lungs.
   B. leaves the upper part of the right ventricle.
   C. terminates at the upper border of the aortic arch.
   D. divides into three right pulmonary arteries and two left pulmonary arteries.
   E. is about 5 ins. (12.5 cm) long.

9. The ligamentum arteriosum is
   A. the remains of the ductus venosus.
   B. a fibrous band that connects the bifurcation of the pulmonary trunk to the ascending aorta.
   C. related to the right recurrent laryngeal nerve.
   D. the remains of the ductus arteriosus.
   E. composed of loose areolar connective tissue.

10. The brachiocephalic veins
    A. are formed from the union of the internal jugular vein and the vertebral vein on each side.
    B. unite to form the superior vena cava.
    C. do not receive blood from the subclavian veins.
    D. receive the superior and middle thyroid veins.
    E. receive the external thoracic veins.

## Multiple Choice Questions

**Select the BEST answer for each question.**

11. The superior vena cava collects venous blood from which of the following areas of the body?
    A. The head
    B. The head and neck
    C. The head and thorax
    D. The head and neck and right upper limb
    E. The head and neck and both upper limbs

12. The superior vena cava drains into which cavity of the heart?
    A. The left atrium
    B. The right ventricle
    C. The right atrium
    D. The right auricle
    E. The left auricle

13. Which of the following statements concerning the superior vena cava is correct?
    A. The entrance of the superior vena cava into the heart is guarded by a functioning valve.
    B. The vena azygos vein joins the posterior aspect of the superior vena cava.
    C. It does not lie within the fibrous pericardium.
    D. It has a thick fibrous wall that resists changing pressures within the thorax.
    E. It is formed by the union of the two internal jugular veins.

# Answers and Explanations

1. **D** is correct. The ascending aorta arises from the left ventricle (Fig. 5-1).

2. **C** is correct. The right coronary artery arises from the anterior aortic sinus, and the left coronary artery arises from the left posterior aortic sinus (see Fig. 4-7). A. The brachiocephalic artery arises from the arch of the aorta (Fig. 5-1). B. The left common carotid artery arises from the arch of the aorta (Fig. 5-1). D. The left subclavian artery also arises from the aortic arch (Fig. 5-1). E. The right subclavian artery arises from the brachiocephalic artery (Fig. 5-4).

3. **A** is correct. At its root, the ascending aorta possesses three bulges, the sinuses of the aorta, one behind each aortic valve cusp (see Fig. 4-6). B. The entire length of the ascending aorta lies within the fibrous pericardium (Fig. 5-1). C. Both the ascending aorta and the pulmonary trunk lie within a serous sheath derived from the serous pericardium (see Fig. 4-21). D. The ascending aorta lies behind the right half of the sternum just below the sternal angle (Fig. 5-9). E. The ascending aorta runs upward and forward from the heart.

4. **A** is correct. The arch of the aorta begins at the level of the sternal angle.

5. **E** is correct. The arch of the aorta gives rise to the brachiocephalic, the left common carotid, and the left subclavian arteries (Fig. 5-1).

6. **D** is correct. The descending thoracic aorta lies in the posterior mediastinum (Fig. 5-3).

7. **C** is correct. The descending thoracic aorta pierces the diaphragm at the level of the 12th thoracic vertebra.

8. **B** is correct. The pulmonary trunk leaves the upper part of the right ventricle (Fig. 5-1).

9. **D** is correct. The ligamentum arteriosum is the remains of the ductus arteriosus. B. The ligamentum arteriosum is a fibrous band that connects the bifurcation of the pulmonary trunk to the lower concave surface of the arch of the aorta (Fig. 5-1). C. The ligamentum arteriosum is related to the left recurrent laryngeal nerve. E. The ligamentum arteriosum is composed of tough fibrous tissue.

10. **B** is correct. The brachiocephalic veins unite to form the superior vena cava (Fig. 5-8). A. The brachiocephalic veins are formed from the union of the internal jugular vein and the subclavian veins on each side (Fig. 5-8). D. The superior and middle thyroid veins drain into the internal jugular veins on each side. E. The internal thoracic veins drain into the brachiocephalic vein.

11. **E** is correct. The superior vena cava collects venous blood from the head and neck and both upper limbs.

12. **C** is correct. The superior vena cava drains into the right atrium of the heart (Fig. 5-8).

13. **B** is correct. The vena azygos vein joins the posterior aspect of the superior vena cava. A. The entrance of the superior vena cava into the right atrium is not guarded by a functioning valve. C. The superior vena cava lies within the fibrous pericardium (see Fig. 4-21). D. The wall of the superior vena cava is thin and easily compressed by external pressures. E. The superior vena cava is formed by the union of the two brachiocephalic veins (Fig. 5-8).

# 6

# The Blood Vessels of the Head and Neck

All clinical material relevant to this chapter can be found on the CD-ROM.

# Chapter Outline

A large number of vital structures are present in the neck. Blunt or penetrating injuries to the neck are potentially life threatening. Vascular structures, including the carotid arteries and their branches and the jugular veins and their tributaries, may be pierced or torn. Approximately 70% of strokes are caused by extracranial arteriosclerosis of the carotid or vertebral arteries.

This chapter provides a review of the anatomy of the blood vessels of the head and neck so that injury or disease of these structures may be more easily assessed and correctly treated.

 # BASIC ANATOMY

## Arteries of the Head and Neck

The **right common carotid** artery arises from the brachiocephalic artery behind the right sternoclavicular joint (Figs. 6-1 and 6-2). The **left artery** arises from the arch of the aorta in the superior mediastinum. The common carotid artery runs upward through the neck under cover of the anterior border of the sternocleidomastoid muscle, from the sternoclavicular joint to the upper border of the thyroid cartilage. Here it divides into the **external and internal carotid arteries** (Figs. 6-3 and 6-4).

At its point of division, the terminal part of the common carotid artery or the beginning of the internal carotid artery shows a localized dilatation, called the **carotid sinus** (Fig. 6-4). The tunica media of the sinus is thinner than elsewhere, but the adventitia is relatively thick and contains numerous nerve endings derived from the glossopharyngeal nerve.

### PHYSIOLOGIC NOTE

**Function of the Carotid Sinus**

The carotid sinus serves as a reflex pressoreceptor mechanism. A rise in blood pressure causes a slowing of the heart rate and vasodilatation of the arterioles.

The carotid body is a small structure that lies posterior to the point of bifurcation of the common carotid artery (Fig. 6-4). It is innervated by the glossopharyngeal nerve.

### PHYSIOLOGIC NOTE

**Function of the Carotid Body**

The carotid body is a chemoreceptor and is sensitive to excess carbon dioxide and reduced oxygen tension in the blood. Such a stimulus reflexly produces a rise in blood pressure and heart rate and an increase in respiratory movements.

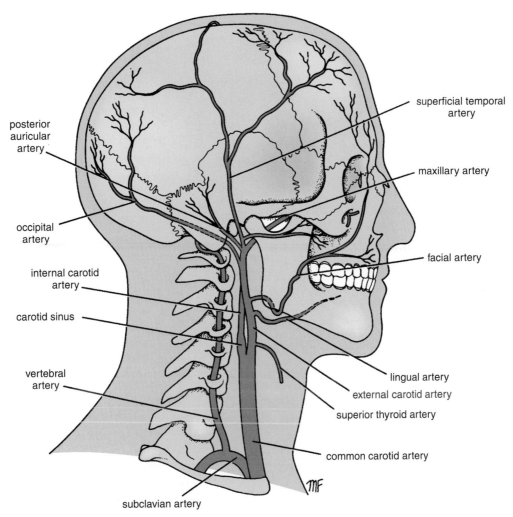

**Figure 6-1** Main arteries of the head and neck. Note that, for clarity, the thyrocervical trunk, the costocervical trunk, and the internal thoracic artery—branches of the subclavian artery—are not shown.

The common carotid artery is embedded in a connective tissue sheath called the **carotid sheath** throughout its course and is closely related to the internal jugular vein and vagus nerve (Fig. 6-5).

## Relations of the Common Carotid Artery

- **Anterolaterally:** The skin, the fascia, the sternocleidomastoid, the sternohyoid, the sternothyroid, and the superior belly of the omohyoid (Fig. 6-3).
- **Posteriorly:** The transverse processes of the lower four cervical vertebrae, the prevertebral muscles, and the sympathetic trunk (Fig. 6-2). In the lower part of the neck are the vertebral vessels.
- **Medially:** The larynx and pharynx and, below these, the trachea and esophagus (Fig. 6-5). The lobe of the thyroid gland also lies medially.
- **Laterally:** The internal jugular vein and, posterolaterally, the vagus nerve (Fig. 6-5).

## Branches of the Common Carotid Artery

Apart from the two terminal branches, the common carotid artery gives off no branches.

## External Carotid Artery

The external carotid artery is one of the terminal branches of the common carotid artery (Fig. 6-1). It supplies structures in the neck, face, and scalp; it also supplies the tongue and the maxilla. The artery begins at the level of the upper border of the thyroid cartilage and terminates in the substance of the parotid gland behind the neck of the mandible by dividing into the superficial temporal and maxillary arteries.

Close to its origin, the artery emerges from undercover of the sternocleidomastoid muscle, where its pulsations can be felt. At first, it lies medial to the internal carotid artery; but as it ascends in the neck, it passes backward and lateral to it. It is crossed by the posterior belly of the digastric and the stylohyoid muscles (Fig. 6-3).

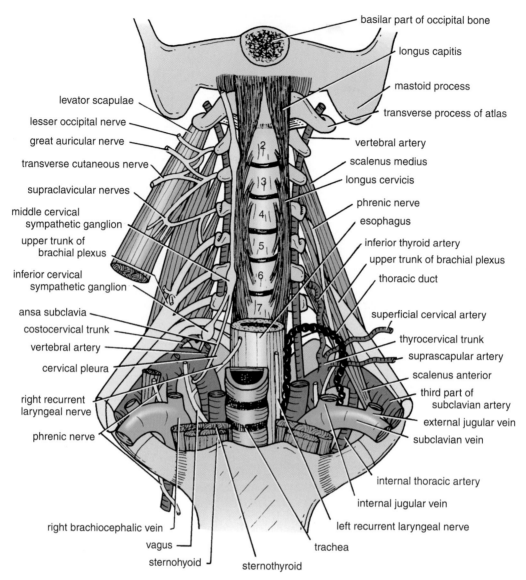

**Figure 6-2** Prevertebral region and the root of the neck.

## Relations of the External Carotid Artery

- **Anterolaterally:** The artery is overlapped at its beginning by the anterior border of the sternocleidomastoid. Above this level, the artery is comparatively superficial, being covered by skin and fascia. It is crossed by the hypoglossal nerve (Fig. 6-3), the posterior belly of the digastric muscle, and the stylohyoid muscles. Within the parotid gland, it is crossed by the facial nerve (Fig. 6-6). The internal jugular vein first lies lateral to the artery and then posterior to it.
- **Medially:** The wall of the pharynx and the internal carotid artery. The stylopharyngeus muscle, the glossopharyngeal nerve, and the pharyngeal branch of the vagus pass between the external and internal carotid arteries (Fig. 6-4).

For the relations of the external carotid artery in the parotid gland, see Fig 6-6.

## Branches of the External Carotid Artery

- Superior thyroid artery.
- Ascending pharyngeal artery.
- Lingual artery.
- Facial artery.
- Occipital artery.
- Posterior auricular artery.
- Superficial temporal artery.
- Maxillary artery.

### Superior Thyroid Artery

The superior thyroid artery curves downward to the upper pole of the thyroid gland (Figs.6-1 and 6-3). It is accompanied by the external laryngeal nerve, which supplies the cricothyroid muscle.

### Ascending Pharyngeal Artery

The ascending pharyngeal artery ascends along and supplies the pharyngeal wall.

### Lingual Artery

The lingual artery loops upward and forward and supplies the tongue (Figs. 6-1 and 6-3).

### Facial Artery

The facial artery loops upward close to the outer surface of the pharynx and the tonsil. It lies deep to the submandibular salivary gland and emerges and bends around the lower border of the mandible. It then ascends over the face close to the anterior border of the masseter muscle. The artery then ascends around the lateral margin of the mouth and terminates at the medial angle of the eye (Figs. 6-1 and 6-3).

**Branches of the facial artery** supply the tonsil, the submandibular salivary gland, and the muscles and the skin of the face.

### Occipital Artery

The artery supplies the back of the scalp (Fig. 6-1).

### Posterior Auricular Artery

The posterior auricular artery supplies the auricle and the scalp (Fig. 6-1).

### Superficial Temporal Artery

The superficial temporal artery ascends over the zygomatic arch, where it may be palpated just in front of the auricle

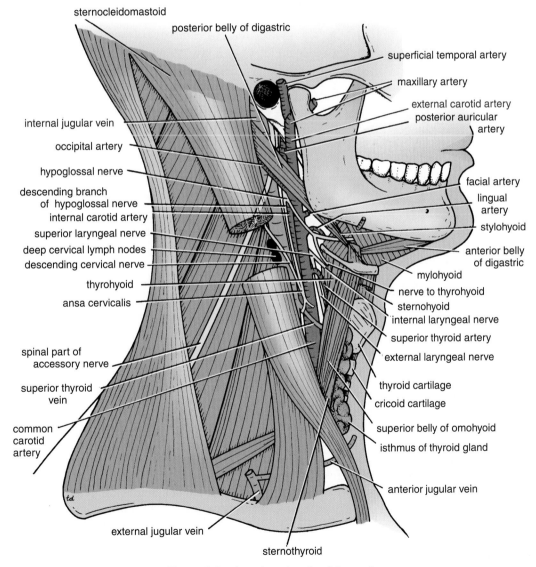

**Figure 6-3**   Anterior triangle of the neck.

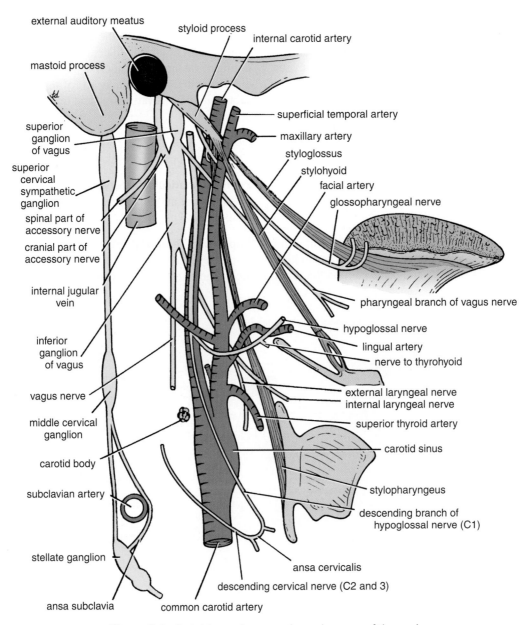

**Figure 6-4** Styloid muscles, vessels, and nerves of the neck.

(Fig. 6-1). It is accompanied by the auriculotemporal nerve, and it supplies the scalp.

### Maxillary Artery

The maxillary artery runs forward medial to the neck of the mandible (Fig. 6-1) and enters the pterygopalatine fossa of the skull.

### Branches of the Maxillary Artery

Branches supply the upper and the lower jaws, the muscles of mastication, the nose, the palate, and the meninges inside the skull.

### Middle Meningeal Artery

The middle meningeal artery enters the skull through the foramen spinosum. It runs laterally within the skull and divides into anterior and posterior branches (Chapter 14, p. 541). The **anterior branch** is important because it lies close to the motor area of the cerebral cortex of the brain. Accompanied by its vein, it grooves (or tunnels) through the upper part of the greater wing of the sphenoid bone of the skull and the thin anteroinferior angle of the parietal bone, where it is prone to damage after a blow to the head.

The origin and distribution of the branches of the **external carotid artery are shown in Figure 6-1.**

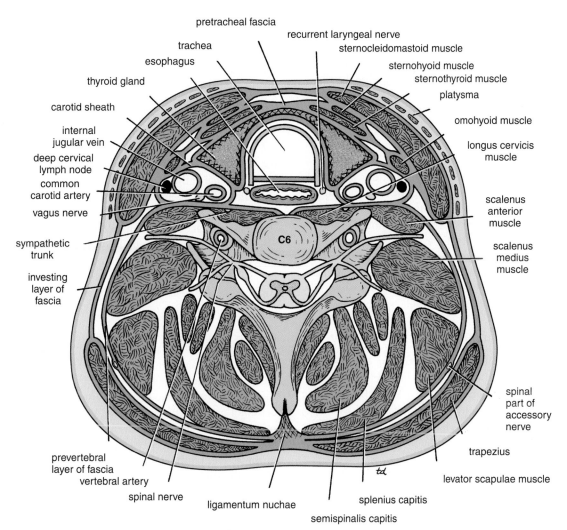

**Figure 6-5** Cross section of the neck at the level of the sixth cervical vertebra.

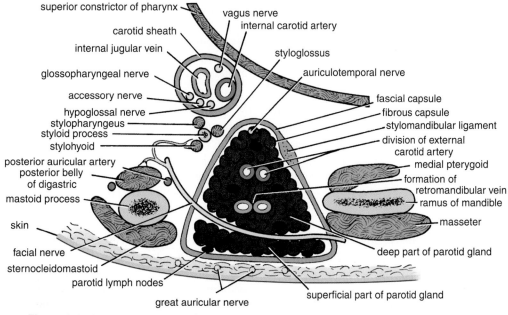

**Figure 6-6** Horizontal section of the parotid gland showing the structures that lie within it.

# Internal Carotid Artery

The internal carotid artery begins at the bifurcation of the common carotid artery at the level of the upper border of the thyroid cartilage (Figs. 6-1 and 6-4). It supplies the brain, the eye, the forehead, and part of the nose. The artery ascends in the neck embedded in the carotid sheath with the internal jugular vein and vagus nerve. At first, it lies superficially; it then passes deep to the parotid salivary gland (Figs. 6-3 and 6-6).

The internal carotid artery leaves the neck by passing into the cranial cavity through the carotid canal in the petrous part of the temporal bone. It then passes upward and forward in the cavernous venous sinus (without communicating with it). The artery then leaves the sinus and passes upward again medial to the anterior clinoid process of the sphenoid bone. The internal carotid artery then inclines backward, lateral to the optic chiasma, and terminates by dividing into the anterior and the middle cerebral arteries.

## Relations of the Internal Carotid Artery

- **Anterolaterally: Below the digastric** lie the skin, the fascia, the anterior border of the sternocleidomastoid, and the hypoglossal nerve (Fig. 6-3). **Above the digastric** lie the stylohyoid muscle, the stylopharyngeus muscle, the glossopharyngeal nerve, the pharyngeal branch of the vagus, the parotid gland, and the external carotid artery (Figs. 6-4 and 6-6).
- **Posteriorly:** The sympathetic trunk (Fig. 6-4), the longus capitis muscle, and the transverse processes of the upper three cervical vertebrae.
- **Medially:** The pharyngeal wall and the superior laryngeal nerve.
- **Laterally:** The internal jugular vein and the vagus nerve.

## Branches of the Internal Carotid Artery

There are no branches in the neck. Many important branches, however, are given off in the skull.

### Ophthalmic Artery

The ophthalmic artery arises from the internal carotid artery as it emerges from the cavernous sinus. It passes forward into the orbital cavity through the optic canal, and it gives off the central artery of the retina, which enters the optic nerve and runs forward to enter the eye ball. The central artery is an end artery and the only blood supply to the retina.

### Posterior Communicating Artery

The posterior communicating artery runs backward to join the posterior cerebral artery (Fig. 6-7).

### Anterior Cerebral Artery

The anterior cerebral artery is a terminal branch of the internal carotid artery (Fig. 6-7). It passes forward between the cerebral hemispheres and then winds around the corpus callosum of the brain to supply the medial and the superolateral surfaces of the cerebral hemisphere. It is joined to the artery of the opposite side by the anterior communicating artery.

### Middle Cerebral Artery

The middle cerebral artery is the largest terminal branch of the internal carotid artery (Fig. 6-7), and it runs laterally in the lateral cerebral sulcus of the brain. It supplies the entire lateral surface of the cerebral hemisphere except the narrow strip along the superolateral margin (which is supplied by the anterior cerebral artery) and the occipital pole and inferolateral surface of the hemisphere (both of which are supplied by the posterior cerebral artery). The middle cerebral artery thus supplies all of the motor area of the cerebral cortex except the leg area. It also gives off central branches that supply central masses of gray matter and the internal capsule of the brain.

## Circle of Willis

The circle of Willis lies in the subarachnoid space at the base of the brain. It is formed by the anastomosis between the branches of the two internal carotid arteries and the two vertebral arteries (Fig. 6-7). The anterior communicating, posterior cerebral, and basilar (formed by the junction of the two vertebral arteries) arteries are all arteries that contribute to the circle. Cortical and central branches arise from the circle and supply the brain.

# Subclavian Arteries

## Right Subclavian Artery

The right subclavian artery arises from the brachiocephalic artery, behind the right sternoclavicular joint (Figs. 6-1 and 6-2). It arches upward and laterally over the pleura and between the scalenus anterior and medius muscles. At the outer border of the first rib, it becomes the axillary artery.

## Left Subclavian Artery

The left subclavian artery arises from the arch of the aorta in the thorax. It ascends to the root of the neck and then arches laterally in a manner similar to that of the right subclavian artery (Fig. 6-2).

The scalenus anterior muscle passes anterior to the artery on each side and divides it into three parts.

### First Part of the Subclavian Artery

The first part of the subclavian artery extends from the origin of the subclavian artery to the medial border of the scalenus anterior muscle (Fig. 6-2). This part gives off the vertebral artery, the thyrocervical trunk, and the internal thoracic artery.

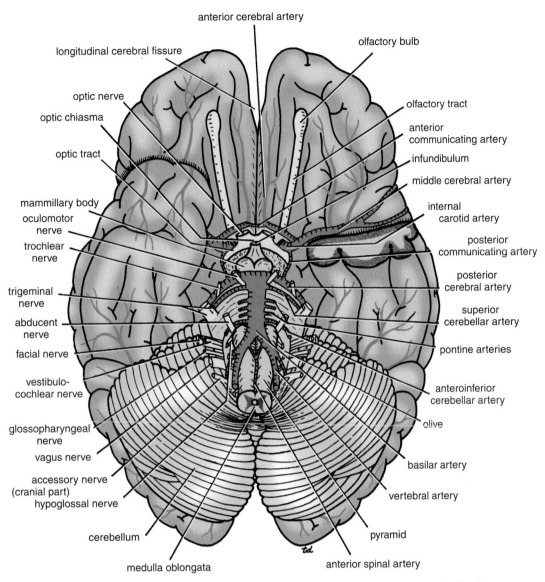

**Figure 6-7** Arteries and cranial nerves seen on the inferior surface of the brain. To show the course of the middle cerebral artery, the anterior pole of the left temporal lobe has been removed.

### Branches

The **vertebral artery** ascends in the neck through the foramina in the transverse processes of the **upper** six cervical vertebrae (Fig. 6-2). It passes medially above the posterior arch of the atlas and then ascends through the foramen magnum into the skull. On reaching the anterior surface of the medulla oblongata of the brain at the level of the lower border of the pons, it joins the vessel of the opposite side to form the basilar artery.

The **basilar artery** (Fig. 6-7) ascends in a groove on the anterior surface of the pons. It gives off branches to the pons, the cerebellum, and the internal ear. It finally divides into the two posterior cerebral arteries.

On each side, the **posterior cerebral artery** (Fig. 6-7) curves laterally and backward around the midbrain. Corti-

cal branches supply the inferolateral surfaces of the temporal lobe and the visual cortex on the lateral and the medial surfaces of the occipital lobe.

- **Branches in the neck:** Spinal and muscular arteries.
- **Branches in the skull:** Meningeal, anterior and posterior spinal, posterior inferior cerebellar, medullary arteries.

The **thyrocervical trunk** is a short trunk that gives off three terminal branches (Fig. 6-2).

- The **inferior thyroid artery** ascends to the posterior surface of the thyroid gland, where it is closely related to the recurrent laryngeal nerve. It supplies the thyroid and the inferior parathyroid glands.

- The **superficial cervical artery** is a small branch that crosses the brachial plexus (Fig. 6-2).
- The **suprascapular artery** runs laterally over the brachial plexus and follows the suprascapular nerve onto the back of the scapula (Fig. 6-2).
- The **internal thoracic artery** descends into the thorax behind the first costal cartilage and in front of the pleura (Fig. 6-2). It descends vertically one fingerbreadth lateral to the sternum; in the sixth intercostal space, it divides into the superior epigastric and the musculophrenic arteries.

### Second Part of the Subclavian Artery
The second part of the subclavian artery lies behind the scalenus anterior muscle (Fig. 6-2).

#### Branches
The **costocervical trunk** runs backward over the dome of the pleura and divides into the **superior intercostal artery**, which supplies the first and the second intercostal spaces, and the **deep cervical artery**, which supplies the deep muscles of the neck.

### Third Part of the Subclavian Artery
The third part of the subclavian artery extends from the lateral border of the scalenus anterior muscle (Fig. 6-2) across the posterior triangle of the neck to the lateral border of the first rib where it becomes the axillary artery. Here, in the root of the neck, it is closely related to the nerves of the brachial plexus.

#### Branches
The third part of the subclavian artery usually has no branches. Occasionally, however, the superficial cervical arteries, the suprascapular arteries, or both arise from this part.

# Veins of the Head and Neck

The veins of the head and neck may be divided into:

- The veins of the brain, venous sinuses, diploic veins, and emissary veins.
- The veins of the scalp, face, and neck.

## Veins of the Brain
The veins of the brain are thin walled and have no valves. They consist of the cerebral veins, the cerebellar veins, and the veins of the brain stem, all of which drain into the neighboring venous sinuses.

## Venous Sinuses
The venous sinuses are situated between the periosteal and the meningeal layer of the dura mater (Fig. 6-8). They have thick, fibrous walls, but they possess no valves. They receive tributaries from the brain, the skull bones, the orbit, and the internal ear.

## Superior Sagittal Sinus
The superior sagittal sinus lies in the upper fixed border of the falx cerebri (Fig. 6-8). It runs backward and becomes continuous with the right transverse sinus, and it communicates on each side with the **venous lacunae**. Numerous arachnoid villi and granulations project into the lacunae.

## Inferior Sagittal Sinus
The inferior sagittal sinus lies in the lower free margin of the falx cerebri (Fig. 6-8). It runs backward and joins the great cerebral vein to form the straight sinus.

## Straight Sinus
The straight sinus lies at the junction of the falx cerebri and the tentorium cerebelli (Fig. 6-8). Formed by the union of the inferior sagittal sinus with the great cerebral vein, it drains into the left transverse sinus.

## Transverse Sinuses
The right transverse sinus begins as a continuation of the superior sagittal sinus; the left transverse sinus is usually a continuation of the straight sinus (Fig. 6-8). Each sinus lies in the lateral attached margin of the tentorium cerebelli, and they end on each side by becoming the sigmoid sinus.

## Sigmoid Sinuses
The sigmoid sinuses are a direct continuation of the transverse sinuses (Fig. 6-8). Each sinus curves downward behind the mastoid antrum and then leaves the skull through the jugular foramen to become the internal jugular vein.

## Occipital Sinus
The occipital sinus lies in the attached margin of the falx cerebelli. It communicates with the vertebral veins through the foramen magnum and the transverse sinuses.

## Cavernous Sinuses and Associated Structures
Each cavernous sinus lies on the lateral side of the body of the sphenoid bone (Figs. 6-8 and 14-15). Anteriorly, the sinus receives the inferior ophthalmic vein and the central vein of the retina. The sinus drains posteriorly into the transverse sinus through the superior petrosal sinus. Intercavernous sinuses connect the two cavernous sinuses through the sella turcica.

The important structures associated with the cavernous sinuses are:

- The internal carotid artery and the sixth cranial nerve, which travel through it (Figs. 6-8 and 14-15).
- In the lateral wall, the third and the fourth cranial nerves and the ophthalmic and the maxillary divisions of the fifth cranial nerve (Fig. 14-15).

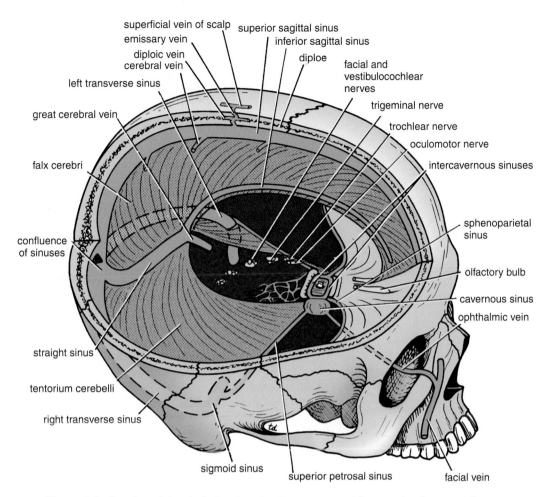

**Figure 6-8** Interior of the skull showing the dura mater and its contained venous sinuses. Note the connections of the veins of the scalp and the veins of the face with the venous sinuses.

- The pituitary gland, which lies medially in the sella turcica (Fig. 14-15).
- The veins of the face, which are connected with the cavernous sinus via the facial vein and inferior ophthalmic vein (and are an important route for the spread of infection from the face).

## Superior and Inferior Petrosal Sinuses

The petrosal sinuses run along the upper and the lower borders of the petrous part of the temporal bone (Fig. 6-8).

## Diploic Veins

The diploic veins occupy channels within the bones of the vault of the skull (Fig. 6-8).

## Emissary Veins

The emissary veins are valveless veins that pass through the skull bones (Fig. 6-8). They connect the veins of the scalp to

the venous sinuses (and are an important route for the spread of infection).

## Veins of the Face and the Neck

### Facial Vein

The facial vein is formed at the medial angle of the eye by the union of the **supraorbital** and **supratrochlear veins** (Fig. 6-9). It is connected through the ophthalmic veins with the cavernous sinus. The facial vein descends down the face with the facial artery and passes round the lateral side of the mouth. It then crosses the mandible, is joined by the anterior division of the retromandibular vein, and drains into the internal jugular vein.

### Superficial Temporal Vein

The superficial temporal vein is formed on the side of the scalp (Fig. 6-9). It follows the superficial temporal artery and the auriculotemporal nerve and then enters the parotid

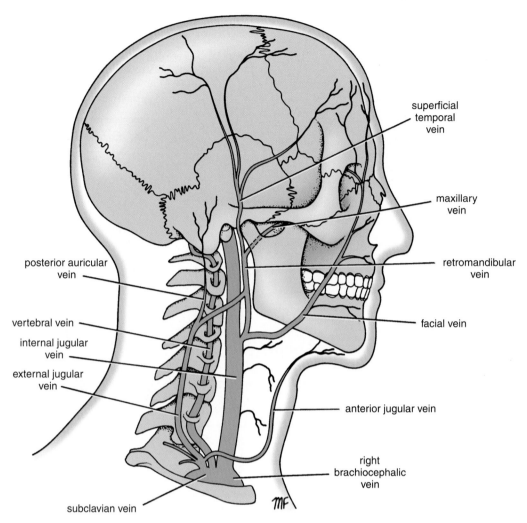

**Figure 6-9**   Main veins of the head and neck.

salivary gland; where it joins the maxillary vein to form the retromandibular vein.

## Maxillary Vein

The maxillary vein is formed in the infratemporal fossa from the pterygoid venous plexus (Fig. 6-9). The maxillary vein joins the superficial temporal vein to form the retromandibular vein.

## Retromandibular Vein

The retromandibular vein is formed by the union of the superficial temporal and the maxillary veins (Fig. 6-9). On leaving the parotid salivary gland, it divides into an anterior branch, which joins the facial vein, and a posterior branch, which joins the posterior auricular vein to form the external jugular vein.

## External Jugular Vein

The external jugular vein is formed behind the angle of the jaw by the union of the posterior auricular vein with the posterior division of the retromandibular vein (Fig. 6-9). It descends across the sternocleidomastoid muscle and beneath the platysma muscle, and it drains into the subclavian vein behind the middle of the clavicle.

### Tributaries

▪ **Posterior external jugular** vein from the back of the scalp.
▪ **Transverse cervical vein** from the skin and the fascia over the posterior triangle.
▪ **Suprascapular vein** from the back of the scapula.
▪ **Anterior jugular vein.**

## Anterior Jugular Vein

The anterior jugular vein descends in the front of the neck close to the midline (Fig. 6-9). Just above the sternum, it is joined to the opposite vein by the **jugular arch.** The anterior jugular vein joins the external jugular vein deep to the sternocleidomastoid muscle.

## Internal Jugular Vein

The internal jugular vein is a large vein that receives blood from the brain, face, and neck (Fig. 6-9). It starts as a continuation of the sigmoid sinus and leaves the skull through the jugular foramen. It then descends through the neck in the carotid sheath lateral to the vagus nerve and the internal and common carotid arteries. It ends by joining the subclavian vein behind the medial end of the clavicle to form the brachiocephalic vein (Figs. 6-2 and 6-9). Throughout its course, it is closely related to the **deep cervical lymph nodes.**

The vein has a dilatation at its upper end called the **superior bulb** and another near its termination called the **inferior bulb**. Directly above the inferior bulb is a bicuspid valve.

### Relations of the Internal Jugular Vein

■ **Anterolaterally:** The skin, the fascia, the sternocleidomastoid, and the parotid salivary gland. Its lower part is covered by the sternothyroid, sternohyoid, and omohyoid muscles, which intervene between the vein and the sternocleidomastoid (Fig. 6-3). Higher up, it is crossed by the stylohyoid, the posterior belly of the digastric, and the spinal part of the accessory nerve. The chain of deep cervical lymph nodes runs alongside the vein.
■ **Posteriorly:** The transverse processes of the cervical vertebrae, the levator scapulae, the scalenus medius, the scalenus anterior, the cervical plexus, the phrenic nerve, the thyrocervical trunk, the vertebral vein, and the first part of the subclavian artery (Fig. 6-2). On the left side, it passes in front of the thoracic duct.
■ **Medially:** Above lie the internal carotid artery and the ninth, tenth, eleventh, and twelfth cranial nerves. Below lie the common carotid artery and the vagus nerve.

### Tributaries of the Internal Jugular Vein

■ Inferior petrosal sinus (Fig. 18-16 ).
■ Facial vein (Fig. 6-9).
■ Pharyngeal veins.
■ Lingual vein (Fig. 2-8).
■ Superior thyroid vein (Fig. 6-3).
■ Middle thyroid vein (Fig. 24-9).

## Subclavian Vein

The subclavian vein is a continuation of the axillary vein at the outer border of the first rib (Fig. 6-2). It joins the internal jugular vein to form the brachiocephalic vein, and it receives the external jugular vein. In addition, it often receives the **thoracic duct** on the left side and the **right lymphatic duct** on the right side.

# Surface Anatomy of the Large Blood Vessels of the Head and Neck

## Carotid Sheath

The carotid sheath, which is a dense connective tissue sheath containing the **carotid arteries**, the **internal jugular vein**, the **vagus nerve**, and the **deep cervical lymph nodes**, can be marked out by a line joining the sternoclavicular joint to a point midway between the tip of the mastoid process and the angle of the mandible. At the level of the upper border of the thyroid cartilage, the **common carotid artery** divides into the internal and external carotid arteries (Fig. 6-10). The pulsations of these arteries can be felt at this level.

## Superficial Temporal Artery

The pulsations of the superficial temporal artery can be felt as it crosses the zygomatic arch, immediately in front of the auricle of the ear (Fig. 6-1).

## Facial Artery

The pulsations of the facial artery can be felt as it crosses the lower margin of the body of the mandible, at the anterior border of the masseter muscle (Fig. 6-10).

## Third Part of the Subclavian Artery

The course of the third part of the subclavian artery may be indicated by a curved line that passes upward from the sternoclavicular joint for about 0.5 in. (1.3 cm) and then downward to the middle of the clavicle (Fig. 6-10). It is here, where the artery lies on the upper surface of the first rib, that its pulsations can be felt easily. The **subclavian vein** lies behind the clavicle and cannot be palpated.

## External Jugular Vein

The external jugular vein lies in the superficial fascia deep to the platysma muscle. It passes downward from the region of the angle of the mandible to the middle of the clavicle (Fig. 6-10).

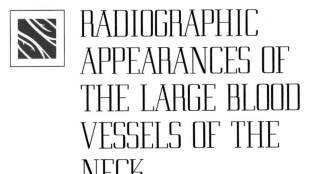

# RADIOGRAPHIC APPEARANCES OF THE LARGE BLOOD VESSELS OF THE NECK

An aortic angiogram showing the large arteries at the root of the neck is shown in Figures 5-5 and 5-6.

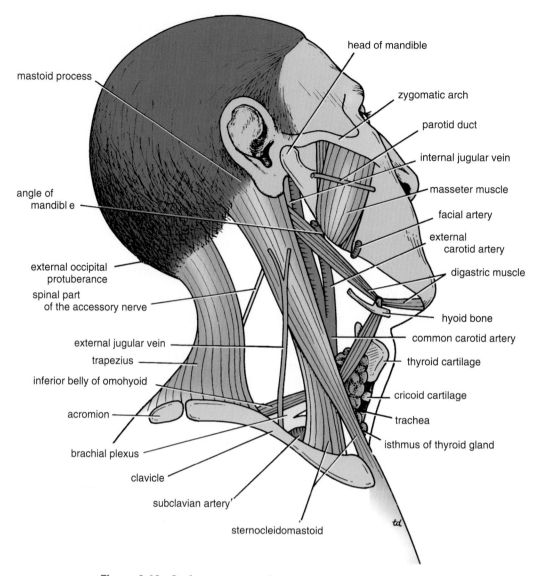

**Figure 6-10** Surface anatomy of the neck from the lateral aspect.

# Review Questions

## Completion Questions

**Each of the numbered items or incomplete statements in this section is followed by answers or by completions of the statement. Select the ONE lettered answer or completion that is BEST in each case.**

1. The common carotid artery ends above at the level of the
   A. cricoid cartilage.
   B. fifth cervical vertebra.
   C. lower border of the thyroid cartilage.
   D. hyoid bone.
   E. upper border of the thyroid cartilage.

2. The common carotid artery ends by dividing into the
   A. superficial temporal artery and the maxillary artery.
   B. vertebral artery and the ascending pharyngeal artery.
   C. internal carotid artery and the external carotid artery.
   D. superior thyroid artery and lingual arteries.
   E. occipital artery and superficial temporal arteries.

3. The carotid sinus is a
   A. dilatation of the internal jugular vein near the common carotid artery.
   B. chemoreceptor.
   C. part of the external carotid artery where the tunica media is thickened.
   D. reflex pressoreceptor in which a rise in blood pressure results in an increase in the heart rate.
   E. dilatation of the terminal part of the common carotid artery or the beginning of the internal carotid artery.

4. The three carotid arteries are nearest the skin at
   A. the anterior border of the sternocleidomastoid muscle at the level of the upper border of the thyroid cartilage.
   B. the posterior border of the sternocleidomastoid muscle at the level of the upper border of the thyroid cartilage.
   C. the lower border of the mandible.
   D. the upper border of the sternoclavicular joint.
   E. the level of the cricoid cartilage.

5. The external carotid artery ends above inside the
   A. submandibular salivary gland.
   B. thyroid gland.
   C. sublingual salivary gland.
   D. posterior cranial fossa of the skull.
   E. parotid salivary gland.

6. The terminal branches of the external carotid artery are the
   A. superficial temporal artery and the maxillary artery.
   B. facial artery and the posterior auricular artery.
   C. superior thyroid artery and the ascending pharyngeal artery.
   D. maxillary artery and the lingual artery.
   E. vertebral artery and the thyrocervical trunk.

7. The middle meningeal artery enters the skull through the
   A. foramen ovale.
   B. foramen rotundum.
   C. foramen magnum.
   D. foramen spinosum.
   E. jugular foramen.

8. The anterior branch of the middle meningeal artery lies close to
   A. sensory area of the cerebral cortex.
   B. posteroinferior angle of the parietal bone.
   C. lower part of the greater wing of the sphenoid bone.
   D. mortor area of the cerebral cortex.
   E. inferior sagittal venous sinus.

9. The internal carotid artery ascends the neck in company with the internal jugular vein and the
   A. glossopharyngeal nerve.
   B. hypoglossal nerve.
   C. vagus nerve.
   D. facial nerve.
   E. accessory nerve.

10. The internal carotid artery enters the skull through the
    A. foramen magnum.
    B. foramen ovale.
    C. jugular foramen.
    D. carotid canal.
    E. foramen rotundum.

11. The third part of the subclavian artery is closely related to the
    A. cervical plexus.
    B. lower trunk of the brachial plexus.
    C. second rib.
    D. sympathetic trunk.
    E. acromioclavicular joint.

12. The vertebral artery ascends the neck by passing through the
    A. vertebral foramina of the upper six cervical vertebrae.
    B. foramina in the transverse processes of the lower six cervical vertebrae.
    C. foramina in the transverse processes of the upper six cervical vertebrae.
    D. foramina in the transverse processes in all the cervical vertebrae.
    E. stylomastoid foramen.

13. The internal jugular vein descends through the neck enclosed within the
    A. digastric triangle.
    B. carotid sheath.
    C. prevertebral layer of deep cervical fascia.
    D. submental triangle.
    E. superficial fascia.

14. The internal jugular vein is a continuation of the
    A. transverse venous sinus.
    B. straight sinus.
    C. cavernous sinus.
    D. superior petrosal sinus.
    E. sigmoid sinus.

15. The internal jugular vein drains directly into the
    A. brachiocephalic vein.
    B. subclavian vein.
    C. superior vena cava.
    D. external jugular vein.
    E. sigmoid sinus.

16. The course of the internal jugular vein is as follows:
    A. Between the tip of the styloid process and the midpoint of the clavicle.
    B. Between the sternoclavicular joint and the point midway between the styloid process and the angle of the mandible.

C. Between the sternoclavicular joint and the point midway between the tip of the mastoid process and the angle of the mandible.

D. Follows the course of the anterior border of the sternocleidomastoid muscle.

E. Between the midpoint of the clavicle to the mastoid process.

17. The external jugular vein is formed by the junction of the
   A. posterior auricular vein and the posterior division of the retromandibular vein.
   B. transverse cervical vein and the suprascapular vein.
   C. posterior external jugular vein and the anterior jugular vein.
   D. posterior auricular vein and the anterior division of the retromandibular vein.
   E. inferior thyroid vein and the facial vein.

18. The external jugular vein descends across the neck
   A. deep to the sternocleidomastoid muscle.
   B. superficial to the platysma muscle.
   C. deep to the deep cervical fascia.
   D. to pass behind the middle of the clavicle.
   E. deep to the levator scapulae muscle.

19. The subclavian vein is a continuation of the
   A. suprascapular vein.
   B. axillary vein.
   C. brachial veins.
   D. transverse cervical vein.
   E. deep cervical vein.

20. The subclavian vein receives the
   A. thoracic duct on the right side.
   B. external jugular vein on both sides.
   C. anterior jugular vein on the left side only.
   D. inferior thyroid veins.
   E. pharyngeal veins.

# Answers and Explanations

1. **E is correct.** The common carotid artery ends above at the level of the upper border of the thyroid cartilage, i.e., the fourth cervical vertebra.

2. **C is correct.** The common carotid artery ends by dividing into the internal and external carotid arteries.

3. **E is correct.** The carotid sinus is a dilatation of the terminal part of the common carotid artery or the beginning of the internal carotid artery. It is a pressoreceptor.

4. **A is correct.** The three carotid arteries are nearest the skin at the anterior border of the sternocleidomastoid muscle at the level of the upper border of the thyroid cartilage.

5. **E is correct.** The external carotid artery ends above behind the neck of the mandible inside the parotid salivary gland (Figs. 6-3 and 6-6).

6. **A is correct.** The external carotid artery terminates by dividing into the superficial temporal artery and the maxillary artery.

7. **D is correct.** The middle meningeal artery enters the skull through the foramen spinosum in the greater wing of the sphenoid bone.

8. **D is correct.** The anterior branch of the middle meningeal artery inside the skull lies close to the motor area of the cerebral cortex.

9. **C is correct.** The internal carotid artery ascends the neck in company with the internal jugular vein and the vagus nerve within the carotid sheath (Fig. 6-5).

10. **D is correct.** The internal carotid artery enters the skull through the carotid canal in the temporal bone.

11. **B is correct.** The third part of the subclavian artery, as it passes over the upper surface of the first rib, is closely related to the lower trunk of the brachial plexus.

12. **C is correct.** There are seven cervical vertebrae and the vertebral artery ascends the neck by passing through the foramina in the transverse processes of the upper six cervical vertebrae. The foramen in the transverse process of the seventh cervical vertebra is small and allows passage of the vertebral veins.

13. **B is correct.** The internal jugular vein descends through the neck enclosed within the carotid sheath, which is formed of deep fascia. Accompanying the vein are the common and internal carotid arteries, the vagus nerve, and deep cervical lymph nodes.

14. **E is correct.** The internal jugular vein is a continuation of the sigmoid sinus.

15. **A is correct.** The internal jugular vein drains directly into the brachiocephalic vein (Fig. 6-9).

16. **C** is correct. The course of the internal jugular vein is from the point midway between the tip of the mastoid process and the angle of the jaw to the sternoclavicular joint.

17. **A** is correct. The external jugular vein is formed by the junction of the posterior auricular vein and the posterior division of the retromandibular vein.

18. **D** is correct. The external jugular vein descends across the neck to pass behind the middle of the clavicle.

19. **B** is correct. The subclavian vein is a continuation of the axillary vein at the outer border of the first rib.

20. **B** is correct. The subclavian vein receives the external jugular vein on both sides (Fig. 6-9).

# 7

# The Blood Vessels of the Upper Extremity

All clinical material relevant to this chapter can be found on the CD-ROM.

# Chapter Outline

rterial injuries in the upper extremity are common. Fortunately, the presence of a good collateral circulation ensures a good prognosis. The close proximity of many of the upper limb arteries to veins and nerves may result in multisystem injury, with the subsequent development of arteriovenous fistulas, alterations of sensation, and muscular paralysis. The use of upper limb arteries and veins as the site for vascular access and invasive monitoring can become the subject for iatrogenic injuries.

The purpose of this chapter is to familiarize the health care professional with the basic anatomy of the blood vessels of the upper extremity in order that diagnosis of vascular injury or disease can be made promptly and vascular access can be established quickly and accurately.

# BASIC ANATOMY

## Arteries of the Upper Extremity

### Axillary Artery

The axillary artery (Figs. 7-1, 7-2, 7-3, and 7-4) begins at the lateral border of the first rib as a continuation of the subclavian (Fig. 7-5) and ends at the lower border of the teres major muscle, where it continues as the brachial artery. Throughout its course, the artery is closely related to the cords of the brachial plexus and their branches and is enclosed with them in a connective tissue sheath called the **axillary sheath.**

The pectoralis minor muscle crosses in front of the axillary artery and thus divides it (for purposes of description) into three parts (Figs. 7-2, 7-3, and 7-5).

### First Part of the Axillary Artery

The first part of the axillary artery extends from the lateral border of the first rib to the upper border of the pectoralis minor (Fig. 7-5).

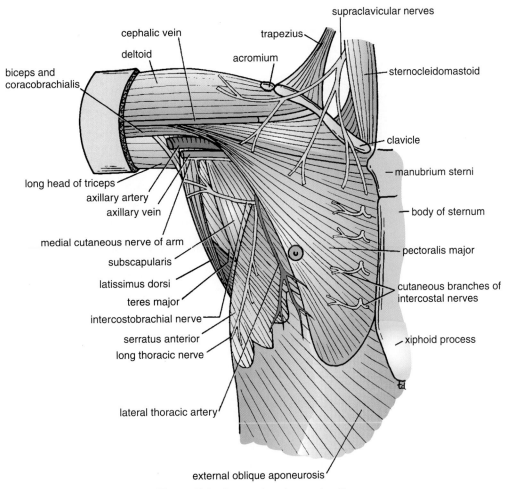

**Figure 7-1** Pectoral region and axilla.

## Relations

Relations are as follows:

- **Anteriorly:** The pectoralis major and the covering fasciae and skin. The cephalic vein crosses the artery (Figs. 7-1, 7-2, and 7-3).
- **Posteriorly:** The long thoracic nerve (nerve to the serratus anterior) (Fig. 7-3).
- **Laterally:** The three cords of the brachial plexus (Fig. 7-3 and 7-6).
- **Medially:** The axillary vein (Fig. 7-3).

## Second Part of the Axillary Artery

The second part of the axillary artery lies behind the pectoralis minor muscle (Fig. 7-3).

### Relations

Relations are as follows:

- **Anteriorly:** The pectoralis minor, the pectoralis major, and the covering fasciae and skin (Figs. 7-2 and 7-5).

- **Posteriorly:** The posterior cord of the brachial plexus, the subscapularis muscle, and the shoulder joint (Fig. 7-3).
- **Laterally:** The lateral cord of the brachial plexus (Fig. 7-3).
- **Medially:** The medial cord of the brachial plexus and the axillary vein (Figs. 7-3 and 7-6).

## Third Part of the Axillary Artery

The third part of the axillary artery extends from the lower border of the pectoralis minor to the lower border of the teres major (Fig. 7-5).

### Relations

Relations are as follows:

- **Anteriorly:** The pectoralis major for a short distance; lower down, the artery is crossed by the medial root of the median nerve (Figs. 7-2 and 7-6).

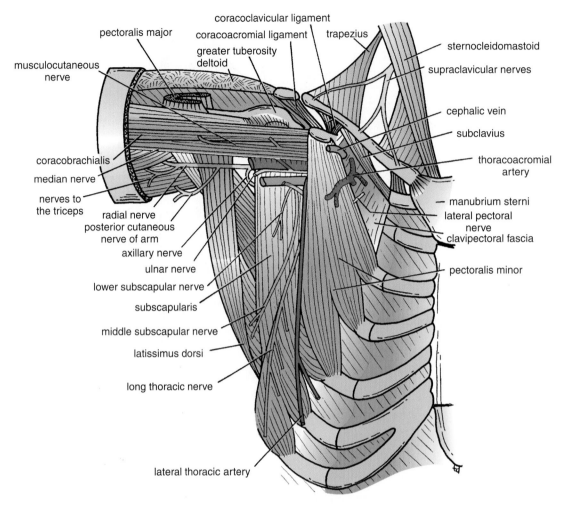

**Figure 7-2** Pectoral region and axilla; the pectoralis major muscle has been removed to display the underlying structures.

- **Posteriorly:** The subscapularis, the latissimus dorsi, and the teres major. The axillary and radial nerves also lie behind the artery (Fig. 7-3).
- **Laterally:** The coracobrachialis, the biceps, and the humerus. The lateral root of the median and the musculocutaneous nerves also lie on the lateral side (Fig. 7-2).
- **Medially:** The ulnar nerve, the axillary vein, and the medial cutaneous nerve of the arm (Fig. 7-2).

## Branches of the Axillary Artery

The branches of the axillary artery supply the thoracic wall and the shoulder region. The first part of the artery gives off one branch (the highest thoracic artery), the second part gives off two branches (the thoracoacromial artery and the lateral thoracic artery), and the third part gives off three branches (the subscapular artery, the anterior circumflex humeral artery, and the posterior circumflex humeral artery) (Fig. 7-5).

The **highest thoracic artery** is small and runs along the upper border of the pectoralis minor. The thoracoacromial artery immediately divides into terminal branches. The **lateral thoracic artery** runs along the lower border of the pectoralis minor (Fig. 7-5). The **subscapular artery** runs along the lower border of the subscapularis muscle. The **anterior and posterior circumflex humeral arteries** wind around the front and the back of the surgical neck of the humerus, respectively (Fig. 7-5).

---

**P H Y S I O L O G I C   N O T E**

### Arterial Anastomosis around the Shoulder Joint

The extreme mobility of the shoulder joint may result in kinking of the axillary artery and a temporary occlusion of its lumen. To compensate for this, an important

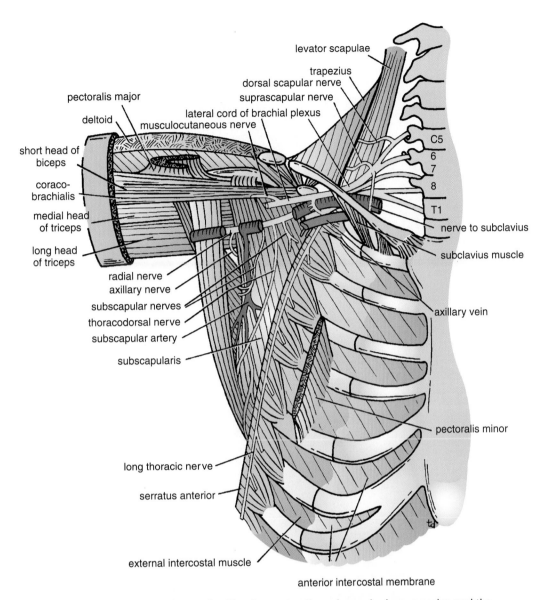

pectoralis major
deltoid
short head of biceps
coraco-brachialis
medial head of triceps
long head of triceps
radial nerve
axillary nerve
subscapular nerves
thoracodorsal nerve
subscapular artery
subscapularis

lateral cord of brachial plexus
musculocutaneous nerve

levator scapulae
trapezius
dorsal scapular nerve
suprascapular nerve

C5
6
7
8
T1

nerve to subclavius
subclavius muscle

axillary vein

pectoralis minor

long thoracic nerve
serratus anterior

external intercostal muscle

anterior intercostal membrane

**Figure 7-3** Pectoral region and axilla; the pectoralis major and minor muscles and the clavipectoral fascia have been removed to display the underlying structures.

arterial anastomosis exists between the branches of the subclavian artery and the axillary artery, thus ensuring that an adequate blood flow takes place into the upper limb irrespective of the position of the arm (Fig. 7-7).

### Branches from the Subclavian Artery

■ The **suprascapular artery**, which is distributed to the supraspinous and infraspinous fossae of the scapula.
■ The **superficial cervical artery**, which gives off a deep branch that runs down the medial border of the scapula.

### Branches from the Axillary Artery

■ The **subscapular artery** and its circumflex scapular branch supply the subscapular and infraspinous fossae of the scapula, respectively.
■ The **anterior circumflex humeral artery.**
■ The **posterior circumflex humeral artery.**

Both of the circumflex arteries form an anastomosing circle around the surgical neck of the humerus (Fig. 7-7).

## Brachial Artery

The **brachial artery** (Figs. 7-8 and 7-9) begins at the lower border of the teres major muscle as a continuation of the

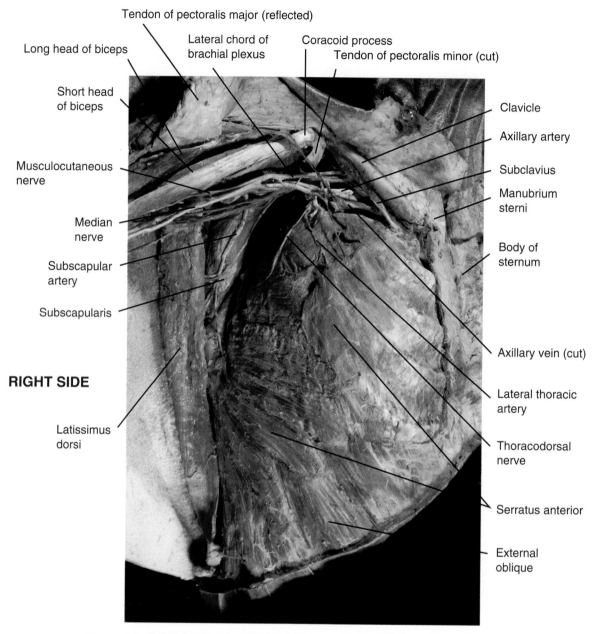

**Figure 7-4**   Dissection of the right axilla. The pectoralis major and minor muscles and the clavipectoral fascia have been removed to display the underlying structures.

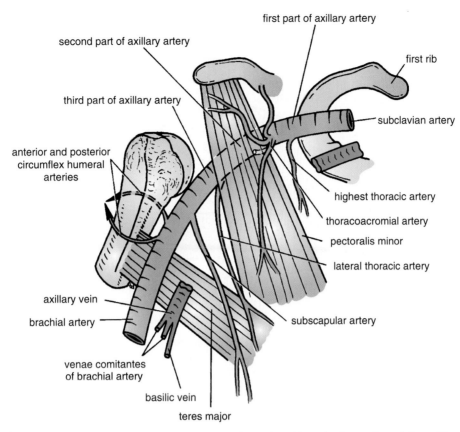

second part of axillary artery

first part of axillary artery

third part of axillary artery

first rib

subclavian artery

anterior and posterior circumflex humeral arteries

highest thoracic artery

thoracoacromial artery

pectoralis minor

lateral thoracic artery

axillary vein

brachial artery

subscapular artery

venae comitantes of brachial artery

basilic vein

teres major

**Figure 7-5** Parts of the axillary artery and its branches. Note the formation of the axillary vein at the lower border of the teres major muscle.

axillary artery. It provides the main arterial supply to the arm (Fig. 7-8). It terminates opposite the neck of the radius by dividing into the radial and ulnar arteries.

## Relations

Relations are as follows:

▦ **Anteriorly:** The vessel is superficial and is overlapped from the lateral side by the coracobrachialis and biceps. The medial cutaneous nerve of the forearm lies in front of the upper part; the median nerve crosses its middle part; and the bicipital aponeurosis crosses its lower part (Fig. 7-9).
▦ **Posteriorly:** The artery lies on the triceps, the coracobrachialis insertion, and the brachialis (Fig. 7-9).
▦ **Medially:** The ulnar nerve and the basilic vein in the upper part of the arm; in the lower part of the arm, the median nerve lies on its medial side (Fig. 7-9).
▦ **Laterally:** The median nerve and the coracobrachialis and biceps muscles above; the tendon of the biceps lies lateral to the artery in the lower part of its course (Fig. 7-9).

## Branches of the Brachial Artery

▦ **Muscular branches** to the anterior compartment of the upper arm.
▦ The **nutrient artery** to the humerus.
▦ The **profunda artery** arises near the beginning of the brachial artery and follows the radial nerve into the spiral groove of the humerus (Fig. 7-10).
▦ The **superior ulnar collateral artery** arises near the middle of the upper arm and follows the ulnar nerve (Fig. 7-10).
▦ The **inferior ulnar collateral artery** arises near the termination of the artery and takes part in the anastomosis around the elbow joint (Fig. 7-10).

> ### PHYSIOLOGIC NOTE
>
> ### Arterial Anastomosis around the Elbow Joint
>
> To compensate for temporary occlusion of the brachial artery during movements of the elbow joint, the following arteries anastomose with one another. The profunda brachii, the superior and inferior ulnar collateral arteries

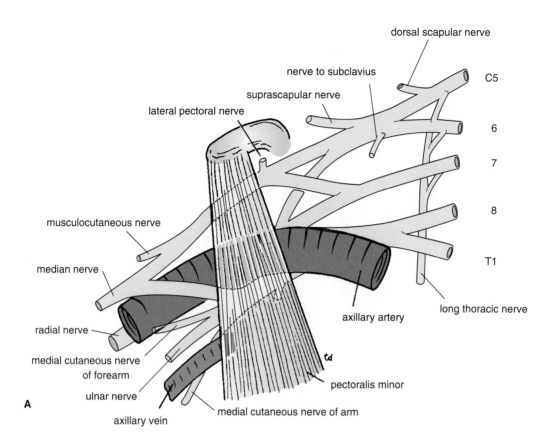

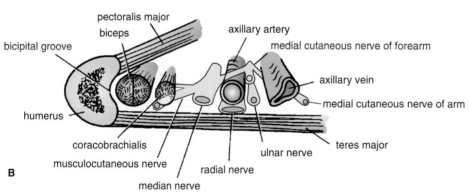

**Figure 7-6** **A.** Relations of the brachial plexus and its branches to the axillary artery and vein. **B.** Section through the axilla at the level of the teres major muscle.

from the brachial artery, anastomose with the radial and ulnar recurrent arteries and the posterior interosseous recurrent artery (branch of common interosseous artery from the ulnar artery) inferiorly (Fig. 7-10).

## Radial Artery

The radial artery is the smaller of the terminal branches of the brachial artery. It begins in the cubital fossa at the level of the neck of the radius (Figs. 7-11 and 7-12). It passes downward and laterally, beneath the brachioradialis mus-

cle, and rests on the deep muscles of the forearm. In the middle third of its course, the superficial branch of the radial nerve lies on its lateral side.

In the distal part of the forearm, the radial artery lies on the anterior surface of the radius and is covered only by skin and fascia. Here, the artery has the tendon of brachioradialis on its lateral side (Fig. 7-13) and the tendon of flexor carpi radialis on its medial side (**site for taking the radial pulse**).

At the wrist, the artery winds backward around the lateral side of the carpus to the proximal end of the space between the first and second metacarpal bones, where it passes

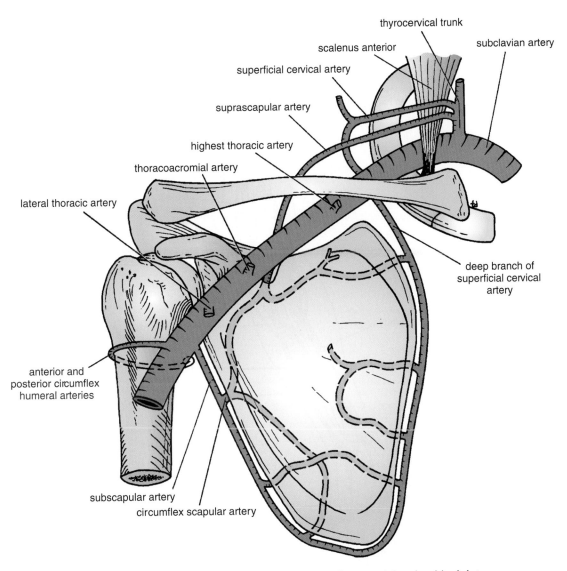

thyrocervical trunk

scalenus anterior

subclavian artery

superficial cervical artery

suprascapular artery

highest thoracic artery

thoracoacromial artery

lateral thoracic artery

deep branch of superficial cervical artery

anterior and posterior circumflex humeral arteries

subscapular artery

circumflex scapular artery

**Figure 7-7**  Arteries that take part in anastomosis around the shoulder joint.

anteriorly into the palm between the two heads of the first dorsal interosseous muscle. It now curves medially between the oblique and transverse heads of the adductor pollicis and joins the deep branch of the ulnar artery, thus forming the deep palmar arch (Fig. 7-14).

## Branches in the Forearm

- **Muscular branches** to neighboring muscles.
- **Recurrent branch**, which takes part in the arterial anastomosis around the elbow joint (Fig. 7-12).
- **Superficial palmar branch**, which arises just above the wrist (Fig. 7-12), enters the palm of the hand, and frequently joins the ulnar artery to form the superficial palmar arch.

## Branches in the Palm

Immediately on entering the palm, the radial artery gives of the arteria radialis indicis, which supplies the lateral side of the index finger, and the arterial princeps pollicis, which divides into two and supplies the lateral and medial sides of the thumb.

## The Deep Palmar Arch and Its Branches

The deep palmar arch is a direct continuation of the radial artery (Fig. 7-14). It is deeply placed in the palm and curves medially beneath the long flexor tendons and in front of the metacarpal bones and the interosseous muscles. The arch is completed on the medial side by the deep branch of the

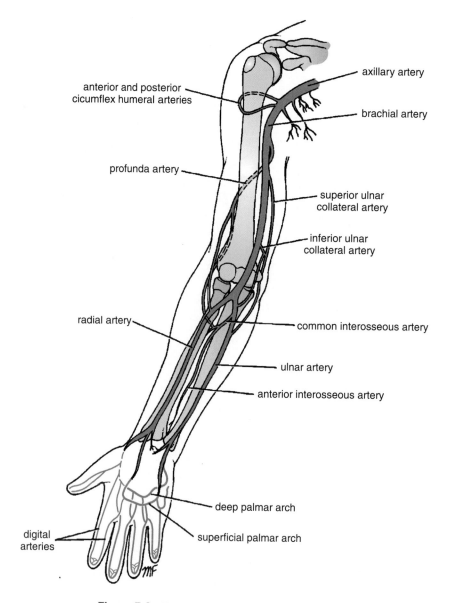

anterior and posterior cicumflex humeral arteries

axillary artery

brachial artery

profunda artery

superior ulnar collateral artery

inferior ulnar collateral artery

radial artery

common interosseous artery

ulnar artery

anterior interosseous artery

deep palmar arch

digital arteries

superficial palmar arch

**Figure 7-8** The main arteries of the upper limb.

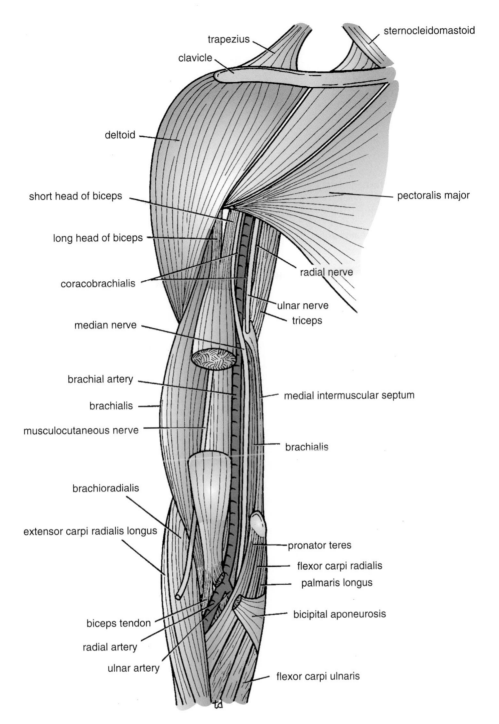

**Figure 7-9**  Anterior view of the upper arm. The middle portion of the biceps brachii has been removed to show the musculocutaneous nerve lying in front of the brachialis.

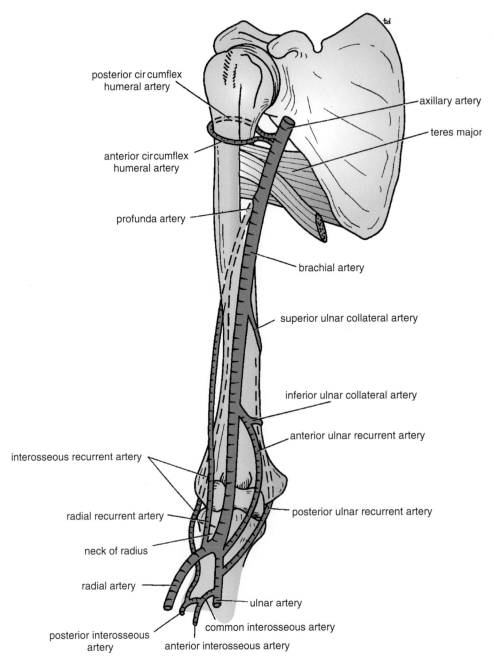

**Figure 7-10** Main arteries of the upper arm. Note the arterial anastomosis around the elbow joint.

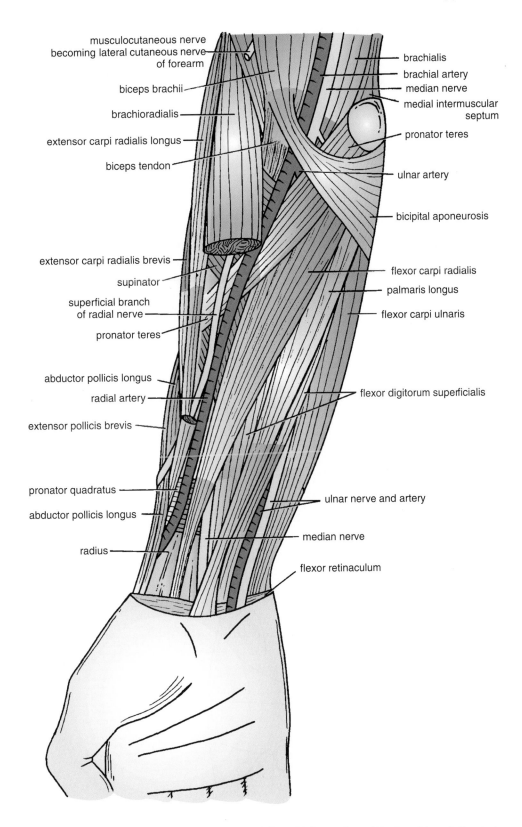

musculocutaneous nerve
becoming lateral cutaneous nerve
of forearm

biceps brachii

brachioradialis

extensor carpi radialis longus

biceps tendon

extensor carpi radialis brevis

supinator

superficial branch
of radial nerve

pronator teres

abductor pollicis longus

radial artery

extensor pollicis brevis

pronator quadratus

abductor pollicis longus

radius

brachialis

brachial artery

median nerve

medial intermuscular
septum

pronator teres

ulnar artery

bicipital aponeurosis

flexor carpi radialis

palmaris longus

flexor carpi ulnaris

flexor digitorum superficialis

ulnar nerve and artery

median nerve

flexor retinaculum

**Figure 7-11**    Anterior view of the forearm. The middle portion of the brachioradialis muscle has been removed to display the superficial branch of the radial nerve and the radial artery.

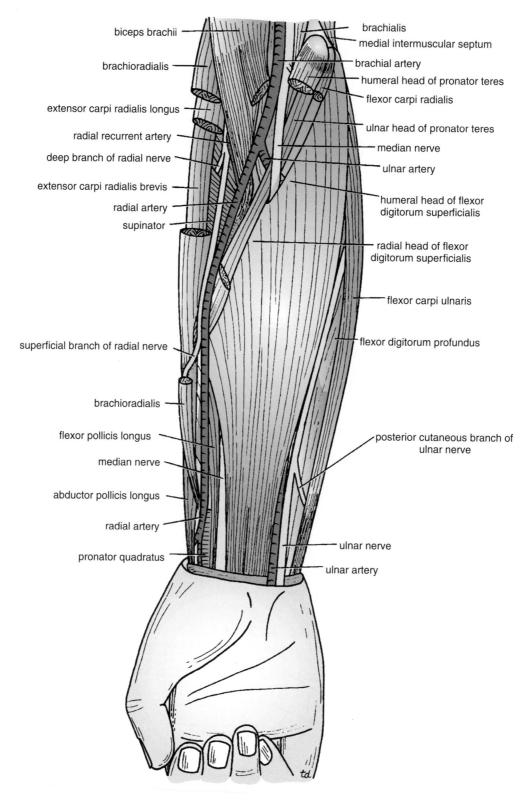

biceps brachii

brachioradialis

extensor carpi radialis longus

radial recurrent artery

deep branch of radial nerve

extensor carpi radialis brevis

radial artery

supinator

superficial branch of radial nerve

brachioradialis

flexor pollicis longus

median nerve

abductor pollicis longus

radial artery

pronator quadratus

brachialis

medial intermuscular septum

brachial artery

humeral head of pronator teres

flexor carpi radialis

ulnar head of pronator teres

median nerve

ulnar artery

humeral head of flexor digitorum superficialis

radial head of flexor digitorum superficialis

flexor carpi ulnaris

flexor digitorum profundus

posterior cutaneous branch of ulnar nerve

ulnar nerve

ulnar artery

**Figure 7-12** Anterior view of the forearm. Most of the superficial muscles have been removed to display the flexor digitorum superficialis, median nerve, superficial branch of the radial nerve, and radial artery. Note that the ulnar head of the pronator teres separates the median nerve from the ulnar artery.

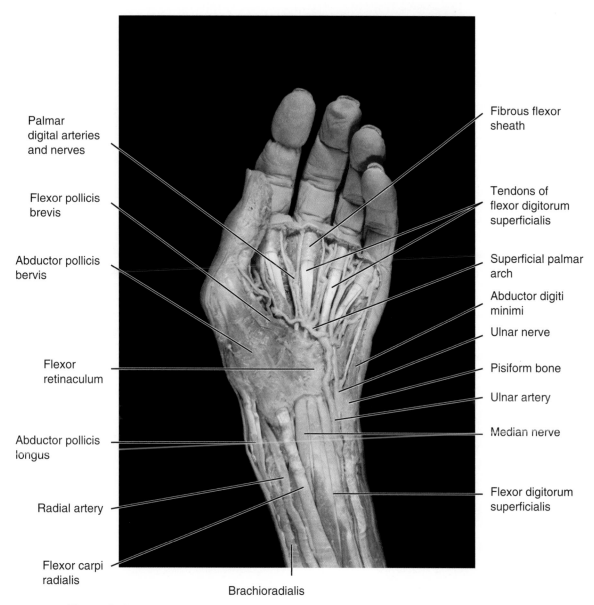

Palmar digital arteries and nerves

Flexor pollicis brevis

Abductor pollicis bervis

Flexor retinaculum

Abductor pollicis longus

Radial artery

Flexor carpi radialis

Fibrous flexor sheath

Tendons of flexor digitorum superficialis

Superficial palmar arch

Abductor digiti minimi

Ulnar nerve

Pisiform bone

Ulnar artery

Median nerve

Flexor digitorum superficialis

Brachioradialis

**Figure 7-13** Dissection of the front of the left forearm and hand showing the superficial structures.

ulnar artery. The curve of the arch lies at a level with the proximal border of the extended thumb.

The deep palmar arch sends recurrent branches superiorly, which take part in the anastomosis around the wrist joint, and inferiorly, to join the digital branches of the superficial palmar arch.

## Branches

The branches are as follows:

- Palmar
- Metacarpal
- Perforating
- Recurrent

## The Ulnar Artery and Its Branches

The ulnar artery is the larger of the two terminal branches of the brachial artery (Figs. 7-8 and 7-12). It begins in the cubital fossa at the level of the neck of the radius. It descends through the anterior compartment of the forearm and enters the palm **in front of** the flexor retinaculum in company with the ulnar nerve (Figs. 7-13 and 7-15). It ends by forming the superficial palmar arch, often anastomosing with the superficial palmar branch of the radial artery (Fig. 7-15).

In the upper part of its course, the ulnar artery lies deep to most of the flexor muscles. Below, it becomes superficial and lies between the tendons of the flexor carpi ulnaris and the tendons of the flexor digitorum superficialis. In

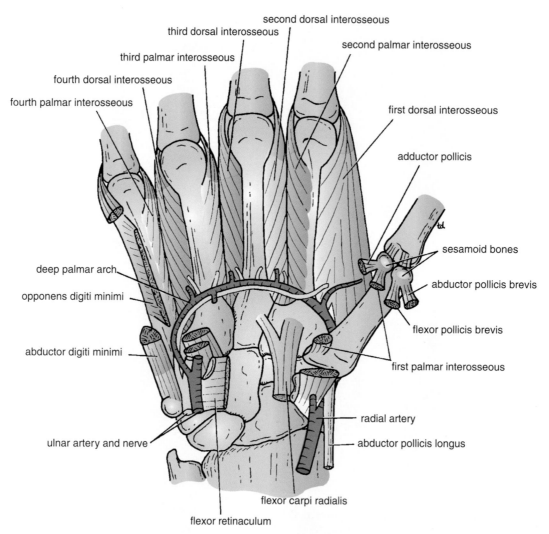

second dorsal interosseous

third dorsal interosseous

second palmar interosseous

third palmar interosseous

fourth dorsal interosseous

first dorsal interosseous

fourth palmar interosseous

adductor pollicis

sesamoid bones

deep palmar arch

abductor pollicis brevis

opponens digiti minimi

flexor pollicis brevis

abductor digiti minimi

first palmar interosseous

radial artery

ulnar artery and nerve

abductor pollicis longus

flexor carpi radialis

flexor retinaculum

**Figure 7-14** Anterior view of the palm of the hand showing the deep palmar arch and the deep terminal branch of the ulnar nerve. The interossei are also shown.

front of the flexor retinaculum, it lies just lateral to the pisiform bone and is covered only by skin and fascia (**site for taking ulnar pulse**).

## Branches

The branches are as follows:

- **Muscular branches** to neighboring muscles.
- **Recurrent branches** that take part in the arterial anastomosis around the elbow joint (Fig. 7-16).
- **Branches that take part in the arterial anastomosis around the wrist joint.**
- The **common interosseous artery,** which arises from the upper part of the ulnar artery and, after a brief course, divides into the **anterior and posterior interosseous arteries** (Fig. 7-10). The interosseous arteries are distributed to the muscles lying in front and behind the interosseous membrane; they provide nutrient arteries to the radius and ulna bone.

- The **deep palmar branch** arises in front of the flexor retinaculum, passes between the abductor digiti minimi and the flexor digiti minimi, and joins the radial artery to complete the deep palmar arch (Figs. 7-14 and 7-17).

## Superficial Palmar Arch

The superficial palmar arch is a direct continuation of the ulnar artery (Fig. 7-15). On entering the palm, it curves laterally behind the palmar aponeurosis and in front of the long flexor tendons. The arch is completed on the lateral side by one of the branches of the radial artery. The curve of the arch lies across the palm, level with the distal border of the fully extended thumb.

### Branches

**Four digital arteries** arise from the convexity of the arch and pass to the fingers (Fig. 7-15).

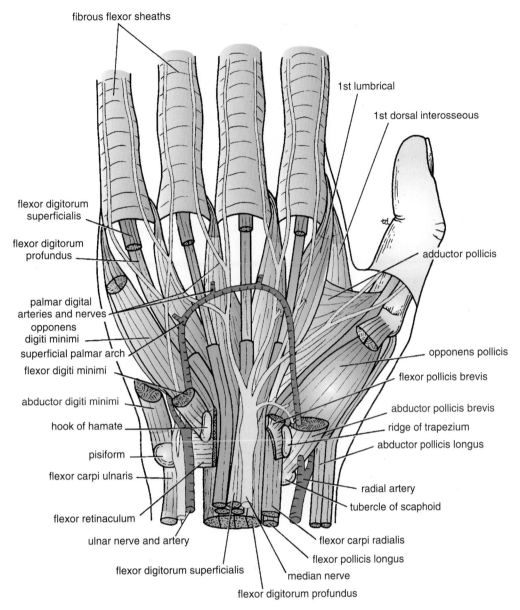

fibrous flexor sheaths

1st lumbrical

1st dorsal interosseous

flexor digitorum superficialis

flexor digitorum profundus

palmar digital arteries and nerves

opponens digiti minimi

superficial palmar arch

flexor digiti minimi

abductor digiti minimi

hook of hamate

pisiform

flexor carpi ulnaris

flexor retinaculum

ulnar nerve and artery

flexor digitorum superficialis

flexor digitorum profundus

adductor pollicis

opponens pollicis

flexor pollicis brevis

abductor pollicis brevis

ridge of trapezium

abductor pollicis longus

radial artery

tubercle of scaphoid

flexor carpi radialis

flexor pollicis longus

median nerve

**Figure 7-15** Anterior view of the palm of the hand. The palmar aponeurosis and the greater part of the flexor retinaculum have been removed to display the superficial palmar arch, the median nerve, and the long flexor tendons. Segments of the tendons of the flexor digitorum superficialis have been removed to show the underlying tendons of the flexor digitorum profundus.

# Nerve Supply of the Arteries of the Upper Extremity

The arteries of the upper limb are innervated by sympathetic nerves. The preganglionic fibers originate from cell bodies in the second to eighth thoracic segments of the spinal cord. They ascend in the sympathetic trunk and synapse in the middle cervical, inferior cervical, first thoracic, or stellate ganglia. The postganglionic fibers join the nerves that form the brachial plexus and are distributed to the arteries within the branches of the plexus. For example, the digital arteries of the fingers are supplied by postganglionic sympathetic fibers that run in the digital nerves.

## PHYSIOLOGIC NOTE

### Peripheral Resistance

The small diameter of the lumen of the arteriole, which is controlled by the smooth muscle in its wall, effectively provides resistance at the periphery of the arterial system to the passage of blood into the veins. If the arterioles

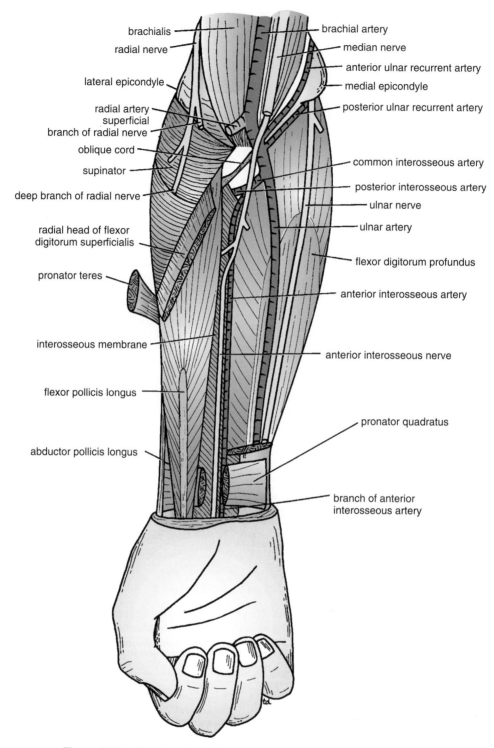

brachialis

radial nerve

lateral epicondyle

radial artery
superficial
branch of radial nerve

oblique cord

supinator

deep branch of radial nerve

radial head of flexor
digitorum superficialis

pronator teres

interosseous membrane

flexor pollicis longus

abductor pollicis longus

brachial artery

median nerve

anterior ulnar recurrent artery

medial epicondyle

posterior ulnar recurrent artery

common interosseous artery

posterior interosseous artery

ulnar nerve

ulnar artery

flexor digitorum profundus

anterior interosseous artery

anterior interosseous nerve

pronator quadratus

branch of anterior
interosseous artery

**Figure 7-16**   Anterior view of the forearm showing the deep structures.

are constricted by the activity of the sympathetic nerves, the blood is held back in the arterial system, and the systolic blood pressure rises. The elastic arterial walls are stretched further, and consequently, the diastolic blood pressure rises also. Should the arterioles become dilated, the opposite effect is produced, and the blood pressure falls.

Most blood vessels of the body are constricted by the activity of the sympathetic nerve fibers. The sympathetic fibers to skeletal muscle blood vessels and cardiac muscle vessels, however, inhibit the smooth muscle in their walls and thus produce moderate vasodilatation. The vasodilator effect is minimal on the general circulation and does not affect the blood pressure.

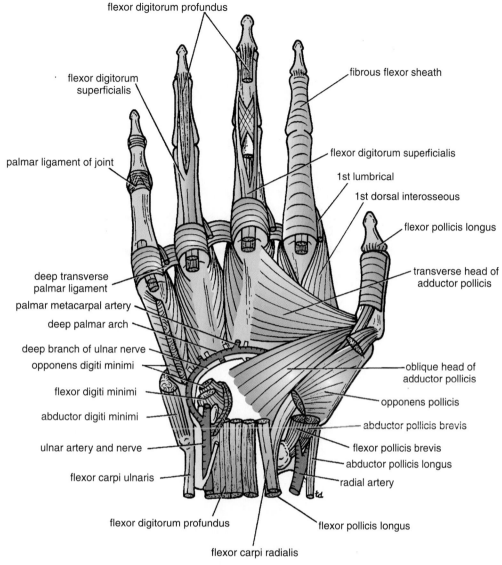

flexor digitorum profundus

flexor digitorum superficialis

palmar ligament of joint

deep transverse palmar ligament

palmar metacarpal artery

deep palmar arch

deep branch of ulnar nerve

opponens digiti minimi

flexor digiti minimi

abductor digiti minimi

ulnar artery and nerve

flexor carpi ulnaris

flexor digitorum profundus

flexor carpi radialis

fibrous flexor sheath

flexor digitorum superficialis

1st lumbrical

1st dorsal interosseous

flexor pollicis longus

transverse head of adductor pollicis

oblique head of adductor pollicis

opponens pollicis

abductor pollicis brevis

flexor pollicis brevis

abductor pollicis longus

radial artery

flexor pollicis longus

**Figure 7-17** Anterior view of the palm of the hand. The long flexor tendons have been removed from the palm, but their method of insertion into the fingers is shown.

# Veins of the Upper Extremity

The veins of the upper limb can be divided into two groups: superficial and deep. The deep veins comprise the venae comitantes, which accompany all the large arteries, usually in pairs, and the axillary vein.

## Superficial Veins

The superficial veins lie in the superficial fascia and are of great clinical importance.

### Dorsal Venous Network (arch)

The dorsal venous network lies in the subcutaneous tissue proximal to the metacarpophalangeal joints and drains on the lateral side into the cephalic vein and on the medial side into the basilic vein (Fig. 7-18). The greater part of the blood from the whole hand and the fingers drains into the network, and it freely communicates with the deep veins of the palm through the interosseous spaces.

### Veins of the Palm

Superficial and deep palmar arterial arches are accompanied by superficial and deep palmar venous arches, receiving corresponding tributaries.

### The Cephalic Vein

The cephalic vein arises from the lateral side of the dorsal venous network on the back of the hand and winds around the lateral border of the forearm; it then ascends in the superficial fascia into the cubital fossa and up the front of the arm on the lateral side of the biceps. On reaching the

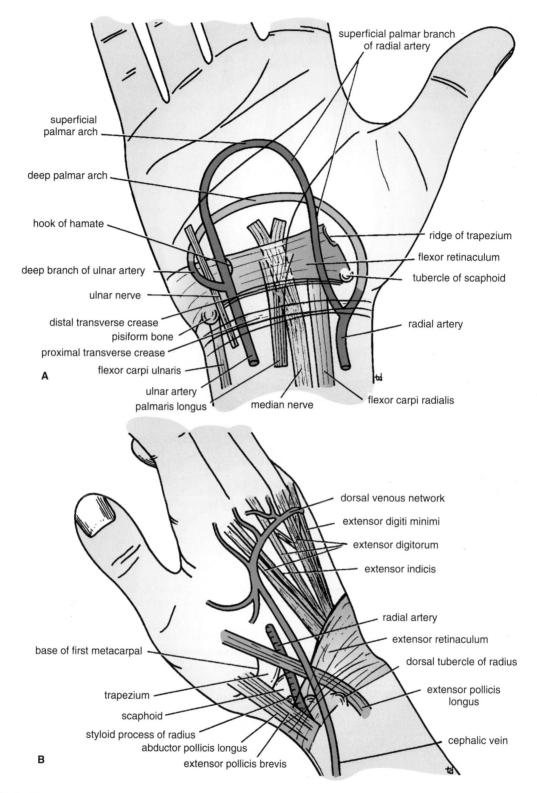

superficial palmar branch
of radial artery

superficial
palmar arch

deep palmar arch

hook of hamate

deep branch of ulnar artery

ulnar nerve

distal transverse crease

pisiform bone

proximal transverse crease

flexor carpi ulnaris

ulnar artery

palmaris longus

ridge of trapezium

flexor retinaculum

tubercle of scaphoid

radial artery

median nerve

flexor carpi radialis

A

dorsal venous network

extensor digiti minimi

extensor digitorum

extensor indicis

radial artery

extensor retinaculum

dorsal tubercle of radius

extensor pollicis
longus

cephalic vein

base of first metacarpal

trapezium

scaphoid

styloid process of radius

abductor pollicis longus

extensor pollicis brevis

B

**Figure 7-18** Surface anatomy of the wrist region.

interval between the deltoid and the pectoralis major muscles, the cephalic vein pierces the deep fascia and joins the axillary vein. As the cephalic vein passes up the upper limb, it receives a variable number of tributaries from the lateral and posterior surfaces of the limb (Fig. 7-19). The median cubital vein, a branch of the cephalic vein in the cubital fossa, runs upward and medially and joins the basilic vein. In the cubital fossa, the median cubital vein crosses in front of the brachial artery and the median nerve, but it is separated from them by the bicipital aponeurosis.

## The Basilic Vein

The basilic vein arises from the medial side of the dorsal venous network on the back of the hand and winds around the medial border of the forearm; it then ascends in the superficial fascia on the posterior surface of the forearm. Just below the elbow, it inclines forward to reach the cubital fossa (Fig. 7-19). The vein then ascends on the medial side of the biceps and pierces the deep fascia at approximately

the middle of the arm, after which it joins the venae comitantes of the brachial artery to form the axillary vein. The basilic vein receives the median cubital vein and a variable number of tributaries from the medial and posterior surfaces of the upper limb.

## The Median Vein of the Forearm

This small but important vein arises in the palm and ascends on the front of the forearm (Fig. 7-19). It drains into the basilic vein, or the median cubital vein, or divides into two branches, one of which joins the basilic vein (median basilic vein) whereas the other joins the cephalic vein (median cephalic vein).

## Deep Veins
### Venae Comitantes

The deep veins accompany the respective arteries as venae comitantes. The two venae comitantes of the brachial artery

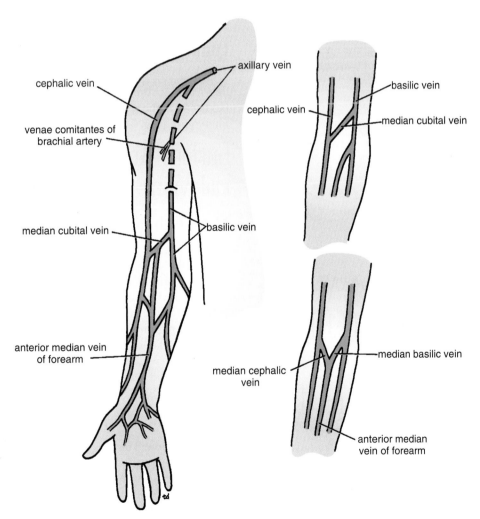

**Figure 7-19** Superficial veins of the upper limb. Note the common variations seen in the region of the elbow.

join the basilic vein at the lower border of the teres major muscle to form the axillary vein.

## The Axillary Vein

The axillary vein is formed by the union of the venae comitantes of the brachial artery with the basilic vein on the posterior wall of the axilla. It then ascends along the medial border of the axillary artery and becomes the subclavian vein at the outer border of the first rib (Figs. 7-3 and 7-4). The axillary vein possesses several valves. It receives tributaries that correspond to the branches of the axillary artery, and it also receives the cephalic vein.

## The Subclavian Vein

The subclavian vein is a vein of the root of the neck and is considered on page 185.

# Nerve Supply of the Veins of the Upper Extremity

Like the arteries, the smooth muscle in the wall of the veins is innervated by sympathetic postganglionic nerve fibers that provide vasomotor tone. The origin of these fibers is similar to those of the arteries.

---

**PHYSIOLOGIC NOTE**

### Nervous Control of Veins

Sympathetic stimulation of the smooth muscle in the walls of veins reduces the size of the lumen so that the circulation of blood continues but at a diminished rate.

---

# SURFACE ANATOMY OF THE ARTERIES AND VEINS OF THE UPPER EXTREMITY

# Surface Anatomy of the Arteries

## Subclavian Artery

The subclavian artery, as it crosses over the first rib to become the axillary artery, may be palpated in the root of the posterior triangle of the neck (see Fig. 6-10).

## Axillary Artery

The first and second parts of the axillary artery cannot be palpated because they lie deep to the pectoral muscles high up in the armpit. The third part of the axillary artery may be felt in the axilla, where it lies in front of the teres major muscle (Fig. 7-1).

## Brachial Artery

The brachial artery can be palpated in the arm as it lies on the brachialis muscle and is overlapped from the lateral side by the biceps brachii muscle (Fig. 7-9).

## Radial Artery

The radial artery lies superficially in front of the distal end of the radius, between the tendons of the brachioradialis and the flexor carpi radialis muscles (Fig. 7-13). It is here that the radial pulse can be easily felt. If the pulse cannot be detected, try feeling for the radial artery on the other wrist; occasionally, a congenitally abnormal radial artery may be difficult to feel. The radial artery may be less easily felt as it crosses the anatomical snuffbox (Fig. 7-18).

## Ulnar Artery

The ulnar artery may be palpated as it crosses anterior to the flexor retinaculum along with the ulnar nerve. The artery lies lateral to the pisiform bone and separated from it by the ulnar nerve (Fig. 7-13).

# Surface Anatomy of the Superficial Veins of the Upper Extremity

## The Cephalic Vein

**Wrist region:** The cephalic vein crosses the anatomical snuffbox (Fig. 7-18) and winds around the lateral side of the forearm to reach the anterior aspect. It is constantly found in the superficial fascia posterior to the styloid process of the radius.

**Elbow region:** The cephalic vein ascends into the arm along the lateral border of the biceps muscle (Fig. 7-19).

**Shoulder region:** The cephalic vein lies in the groove between the deltoid and pectoralis major muscles (Figs. 7-1 and 7-19).

## The Basilic Vein

**Wrist region:** The basilic vein ascends from the dorsum of the hand by slowly curving around the medial side of the forearm to reach the cubital fossa (Fig. 7-19).

**Elbow region:** The basilic vein ascends from the cubital fossa along the medial border of the biceps muscle; it pierces the deep fascia at about the middle of the arm (Fig. 7-19).

## The Median Cubital Vein

The median cubital vein links the cephalic vein to the basilic vein in the cubital fossa. It is separated from the brachial artery at this site by the bicipital aponeurosis (see p 211). In about 30% of individuals, the median cubital vein is replaced by the median cephalic and median basilic veins (Fig. 7-19).

## The Median Vein of the Forearm

The median vein of the forearm ascends on the anterior aspect of the forearm along a variable course (Fig. 7-19). It joins either the basilic vein or the median cubital vein or divides into the median cephalic and median basilic veins.

To identify these veins easily, apply firm pressure around the upper arm and repeatedly clench and relax the fist. By this means, the veins become distended with blood.

---

**PHYSIOLOGIC NOTE**

### Visualization of the Superficial Veins of the Upper Extremity

A tourniquet applied to the upper arm with sufficient pressure to occlude the superficial veins dams back the venous blood as it attempts to return to the right atrium of the heart. The clenching and relaxing of the fist merely increases the venous return and distends the superficial veins.

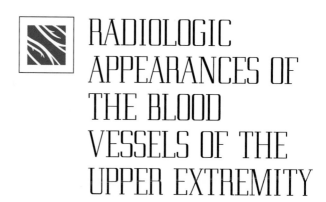

# RADIOLOGIC APPEARANCES OF THE BLOOD VESSELS OF THE UPPER EXTREMITY

Arteriograms and venograms with explanatory diagrams are shown in Figs. 7-20, 7-21, 7-22, and 7-23).

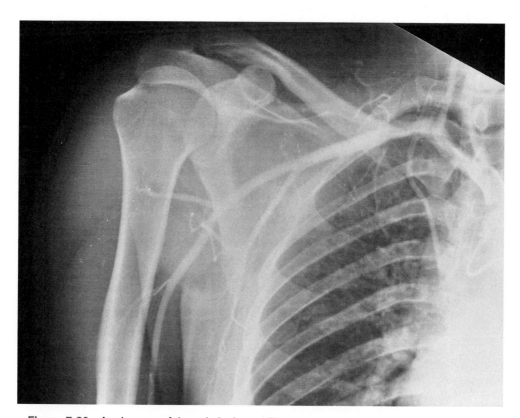

**Figure 7-20** Angiogram of the subclavian, axillary, and brachial arteries.

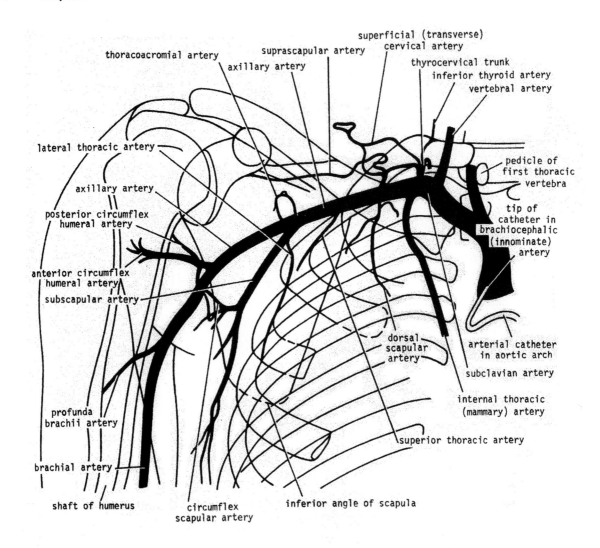

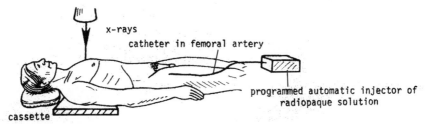

**Figure 7-21** Main features seen in the angiogram shown in Figure 7-20.

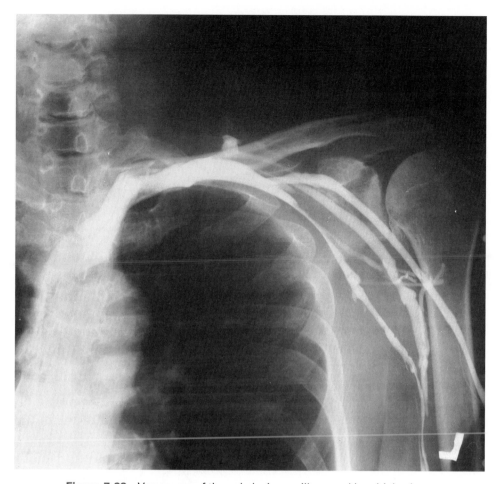

**Figure 7-22**   Venogram of the subclavian, axillary, and brachial veins.

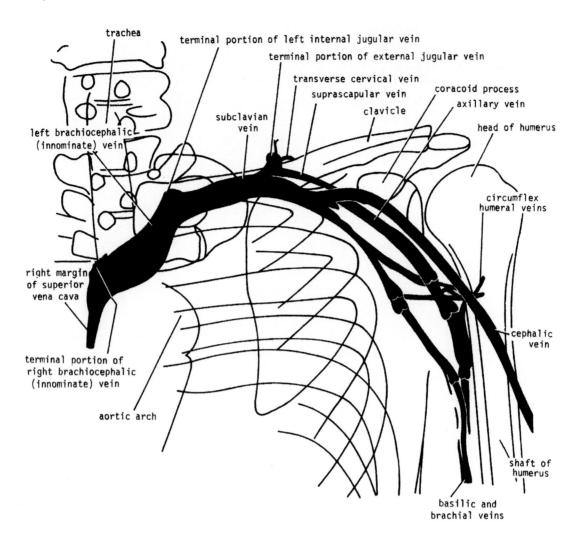

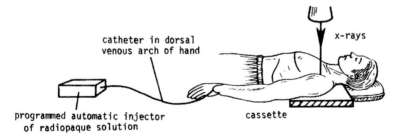

**Figure 7-23** Main features seen in the venogram shown in Figure 7-22.

# Review Questions

**Each of the numbered items or incomplete statements in this section is followed by answers or by completions of the statement. Select the one lettered answer or completion that is BEST in each case.**

1. The axillary artery begins at the _____ as a continuation of the subclavian artery.
   A. outer border of the scalenus anterior muscle
   B. outer border of the second rib
   C. outer border of the first rib
   D. upper border of the pectoralis minor muscle
   E. lower border of the teres major muscle

2. The axillary artery is most easily palpated
   A. behind the clavicle in the root of the neck.
   B. between the anterior and posterior axillary folds in the armpit.
   C. in the deltopectoral groove between the deltoid muscle and the pectoralis major muscle.
   D. through the pectoralis major muscle.
   E. through the posterior wall of the axilla.

3. The brachial artery divides into its terminal branches
   A. at the level of the neck of the radius.
   B. midway between the medial and lateral epicondyles of the humerus.
   C. at the lower border of the pronator teres muscle.
   D. behind the flexor digitorum superficialis muscle.
   E. behind the medial epicondyle of the humerus.

4. The brachial pulse is mostly easily felt
   A. on the lateral side of the tendon of the biceps brachii muscle.
   B. on the medial side of the belly of the biceps muscle.
   C. along the lateral margin of the triceps tendon.
   D. in front of the coracobrachialis muscle.
   E. on the medial side of the biceps brachii tendon in front of the elbow joint.

5. When taking the blood pressure with a sphygmomanometer, the stethoscope diaphragm is held over the brachial artery on the
   A. lateral side of the belly of the biceps brachii muscle about halfway down the arm.
   B. medial side of the belly of the biceps brachii muscle about halfway down the arm.
   C. medial side of the biceps brachii tendon in front of the elbow joint.
   D. lateral side of the biceps brachii tendon in front of the elbow joint.
   E. lateral side of the bicipital.

6. The radial artery ends below
   A. by becoming continuous with the deep branch of the ulnar artery.
   B. by becoming continuous with the main trunk of the ulnar artery.
   C. by dividing into the arteria princeps pollicis and the arteria radialis indicis muscles.
   D. by dividing into the superficial palmar and the deep palmar arches.
   E. by dividing into muscular branches.

7. The radial artery at the front of the wrist lies between the
   A. palmaris longus tendon and the brachioradialis tendon.
   B. flexor carpi radialis tendon and the brachioradialis tendon.
   C. palmaris longus tendon and the flexor carpi radialis tendon.
   D. flexor carpi radialis tendon and the flexor carpi ulnaris tendon.
   E. flexor digitorum superficialis tendon and the flexor carpi ulnaris tendon.

8. The superficial veins of the forearm originate below from the
   A. dorsal venous network.
   B. deep veins in the palm.
   C. metacarpal veins.
   D. digital veins.
   E. veins of the index finger and thumb.

9. The cephalic vein arises from the
   A. lateral side of the deep palmar arch veins.
   B. lateral side of the dorsal venous network on the dorsum of the hand.
   C. lateral side of the superficial palmar veins.
   D. venae comitantes of the radial artery.
   E. plexus of veins in front of the wrist.

10. The cephalic vein drains into the
    A. venae comitantes of the brachial artery.
    B. median vein of the forearm.
    C. axillary vein.
    D. external jugular vein.
    E. subclavian vein.

11. The median cubital vein is
    A. formed by the venae comitantes of the ulnar artery.
    B. deep to the median nerve in the cubital fossa.
    C. a tributary of the median vein of the forearm.
    D. deep to the bicipital aponeurosis.
    E. a branch of the cephalic vein.

12. The basilic vein ends by
    A. joining the median cubital vein.
    B. draining into the subclavian vein.
    C. joining the medial end of the dorsal venous network.
    D. joining the venae comitantes of the brachial artery to form the axillary vein.
    E. traveling in the deltopectoral groove.

13. The cephalic vein is constantly found in the superficial fascia posterior to the
    A. styloid process of the ulna.
    B. pisiform bone.
    C. lunate bone.
    D. styloid process of the radius.
    E. flexor retinaculum.

14. The following is true about the axillary vein.
    A. It has no valves.
    B. It receives no tributaries in the axilla.
    C. It drains into the brachiocephalic vein.
    D. It becomes continuous with the subclavian vein.
    E. It lies on the lateral side of the axillary artery.

# Answers and Explanations

1. **C is correct.** The axillary artery begins at the outer border of the first rib as a continuation of the subclavian artery (Fig. 7-5).

2. **B is correct.** The pulsating axillary artery is most easily palpated by pressing upward and laterally in the armpit between the anterior and posterior axillary folds.

3. **A is correct.** The brachial artery divides into its terminal branches in the cubital fossa at the level of the neck of the radius (Fig. 7-8).

4. **E is correct.** The pulse of the brachial artery is most easily felt on the medial side of the biceps brachii tendon in front of the elbow joint (Fig. 7-9).

5. **C is correct.** When taking the blood pressure with a sphygmomanometer, the stethoscope diaphragm is held over the brachial artery on the medial side of the biceps brachii tendon in front of the elbow joint (Fig. 7-9).

6. **A is correct.** The radial artery ends below by becoming continuous with the deep branch of the ulnar artery to complete the deep palmar arch (Fig. 7-14).

7. **B is correct.** The radial artery, just above the front of the wrist, lies between the tendons of the flexor carpi radialis and brachioradialis muscles (Figs. 7-11 and 7-13).

8. **A is correct.** The two major superficial veins of the forearm, namely the cephalic vein and the basilic vein, begin below from the dorsal venous network located on the back of the hand (Figs. 7-18 and 7-19).

9. **B is correct.** The cephalic vein arises from the lateral side of the dorsal venous network on the dorsum of the hand and proceeds to wind around the lateral side of the wrist to ascend on the lateral side of the front of the forearm.

10. **C is correct.** The cephalic vein drains into the axillary vein (Figs. 7-19, 7-22, and 7-23).

11. **E is correct.** The median cubital vein is a branch of the cephalic vein in the front of the elbow joint. It ascends medially to join the basilic vein (Fig. 7-19). Note the variations possible in this arrangement.

12. **D is correct.** The basilic vein ends by piercing the deep fascia halfway along the medial side of the biceps brachii muscle and joins the venae comitantes of the brachial artery to form the axillary vein (Fig. 7-19).

13. **D is correct.** The important cephalic vein is constantly found in the superficial fascia posterior to the styloid process of the radius. Clinically, this is a good site for introducing a catheter into the vein.

14. **D is correct.** The axillary vein becomes continuous with the subclavian vein at the outer border of the first rib (Figs. 7-3, 7-22, and 7-23).

# 8 The Blood Vessels of the Abdomen, Pelvis, and Perineum

All clinical material relevant to this chapter can be found on the CD-ROM.

## Chapter Outline

Within the abdomen and pelvis lie the aorta and its branches, the inferior vena cava and its tributaries, and the important portal vein. Because most major blood vessels of the abdomen are retroperitoneal, blunt injuries may not result in an immediate fatal intraperitoneal hemorrhage. Because bleeding may initially be confined to the retroperitoneal space, making the diagnosis may be difficult or delayed. Penetrating injuries usually give the blood access to the peritoneal cavity, and the diagnosis of intraperitoneal hemorrhage is readily made.

The purpose of this chapter is to review the major arterial and venous vessels of the abdomen, pelvis, and perineum relative to clinical practice.

# BASIC ANATOMY

## Abdominal Aorta

### Location and Description

The aorta enters the abdomen through the aortic opening of the diaphragm in front of the twelfth thoracic vertebra (Fig. 8-1). It descends behind the peritoneum on the anterior surface of the bodies of the lumbar vertebrae. At the level of the fourth lumbar vertebra, it divides into the two common iliac arteries (Fig. 8-2).

### Maintaining the Diastolic Blood Pressure

The large amount of elastic tissue present in the tunica media of the aorta (including other conducting arteries such as the pulmonary, brachiocephalic, common carotid, and iliac arteries) allows the arterial wall to become distended by the blood ejected from the heart during ventricular systole. The elastic recoil of these arteries that occurs when the ventricles are relaxing is largely responsible for maintaining the arterial blood pressure (diastolic pressure) during diastole of the heart.

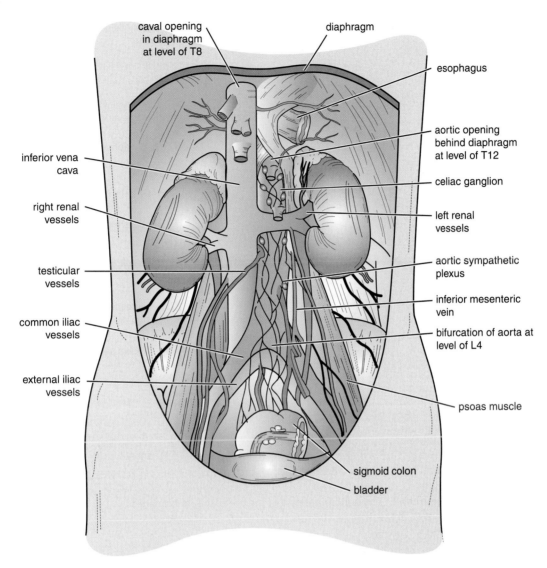

**Figure 8-1**　Posterior abdominal wall showing the aorta and inferior vena cava.

## Relations of the Abdominal Aorta

▨ On its **right side** lie the inferior vena cava, the cisterna chyli, and the beginning of the azygos vein.

▨ On its **left side** lies the left sympathetic trunk.

▨ On the **anterior surface**, the aorta is related to the stomach, celiac plexus, pancreas, splenic vein, left renal vein, third part of the duodenum, coils of small intestine, and peritoneum.

The surface markings of the aorta are shown in Figure 8-3.

## Branches of the Abdominal Aorta (Fig. 8-2)

▨ **Three anterior visceral branches**: the celiac artery, superior mesenteric artery, and inferior mesenteric artery.

▨ **Three lateral visceral branches**: the suprarenal artery, renal artery, and testicular or ovarian artery.

▨ **Five lateral abdominal wall branches**: the inferior phrenic artery and four lumbar arteries.

▨ **Three terminal branches**: the two common iliac arteries and the median sacral artery.

These branches are summarized in Diagram 8-1.

## Celiac Artery

The celiac artery or trunk is very short and arises from the commencement of the abdominal aorta at the level of the twelfth thoracic vertebra (Fig. 8-4). It is surrounded by the celiac plexus and lies behind the lesser sac of peritoneum. It has three terminal branches: the left gastric, splenic, and hepatic arteries.

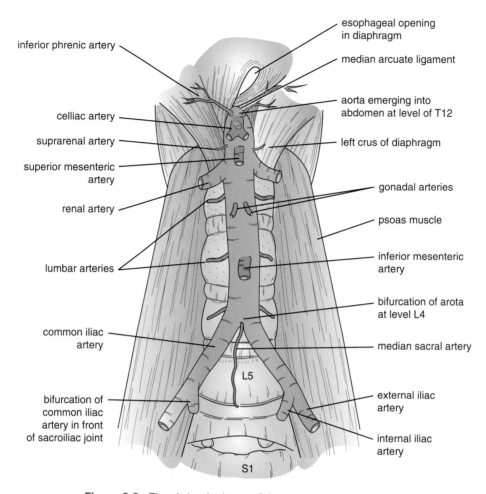

**Figure 8-2** The abdominal part of the aorta and its branches.

## Left Gastric Artery

The small left gastric artery runs to the cardiac end of the stomach, gives off a few esophageal branches, and then turns to the right along the lesser curvature of the stomach. It anastomoses with the right gastric artery (Fig. 8-4).

## Splenic Artery

The large splenic artery runs to the left in a wavy course along the upper border of the pancreas and behind the stomach (Fig. 8-5). On reaching the left kidney, the artery enters the splenicorenal ligament and runs to the hilum of the spleen (Fig. 8-6).

### Branches

■ Pancreatic branches.
■ The **left gastroepiploic artery** arises near the hilum of the spleen and reaches the greater curvature of the stomach in the gastrosplenic omentum. It passes to the right along the greater curvature of the stomach between the layers of the greater omentum. It anastomoses with the right gastroepiploic artery (Fig. 8-4).

■ The **short gastric arteries**, five or six in number, arise from the end of the splenic artery and reach the fundus of the stomach in the gastrosplenic omentum. They anastomose with the left gastric artery and the left gastroepiploic artery (Fig. 8-4).

## Hepatic Artery

The medium-size hepatic artery* runs forward and to the right and then ascends between the layers of the lesser omentum (Figs. 8-5 and 8-7). It lies in front of the opening into the lesser sac and is placed to the left of the bile duct and in front of the portal vein. At the porta hepatis, it divides into right and left branches to supply the corresponding lobes of the liver.

---

* For purposes of description, the hepatic artery is sometimes divided into the **common hepatic artery**, which extends from its origin to the gastroduodenal branch, and the **hepatic artery proper**, which is the remainder of the artery.

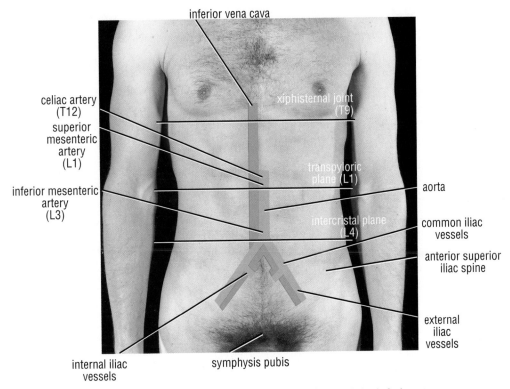

**Figure 8-3** Surface markings of the aorta and its branches and the inferior vena cava on the anterior abdominal wall.

### Branches

▦ The **right gastric artery** arises from the hepatic artery at the upper border of the pylorus and runs to the left in the lesser omentum along the lesser curvature of the stomach. It anastomoses with the left gastric artery (Fig. 8-4).

▦ The **gastroduodenal artery** is a large branch that descends behind the first part of the duodenum. It divides into the right gastroepiploic artery that runs along the greater curvature of the stomach between the layers of the greater omentum and the superior pancreaticoduodenal artery that descends between the second part of the duodenum and the head of the pancreas (Figs. 8-4 and 8-5).

▦ The **right and left hepatic arteries** enter the porta hepatis. The right hepatic artery usually gives off the cystic artery, which runs to the neck of the gallbladder (Fig. 8-8).

## Superior Mesenteric Artery

The superior mesenteric artery supplies the distal part of the duodenum, the jejunum, the ileum, the cecum, the appendix, the ascending colon, and most of the transverse colon. It arises from the front of the abdominal aorta just below the celiac artery (Fig. 8-9) and runs downward and to the right behind the neck of the pancreas and in front of the third part of the duodenum. It continues downward to the right between the layers of the mesentery of the small intestine and ends by anastomosing with the ileal branch of its own ileocolic branch.

### Branches of the Superior Mesenteric Artery

▦ The **inferior pancreaticoduodenal artery** passes to the right as a single or double branch along the upper border of the third part of the duodenum and the head of the pancreas. It supplies the pancreas and the adjoining part of the duodenum.

▦ The **middle colic artery** runs forward in the transverse mesocolon to supply the transverse colon and divides into right and left branches.

▦ The **right colic artery** is often a branch of the ileocolic artery. It passes to the right to supply the ascending colon and divides into ascending and descending branches.

▦ The **ileocolic artery** passes downward and to the right. It gives rise to a **superior branch** that anastomoses with the right colic artery and an **inferior branch** that anastomoses with the end of the superior mesenteric artery. The inferior branch gives rise to the **anterior** and **posterior cecal arteries**; the **appendicular artery** is a branch of the posterior cecal artery (Fig. 8-10).

▦ The **jejunal and ileal branches** are 12 to 15 in number and arise from the left side of the superior mesenteric

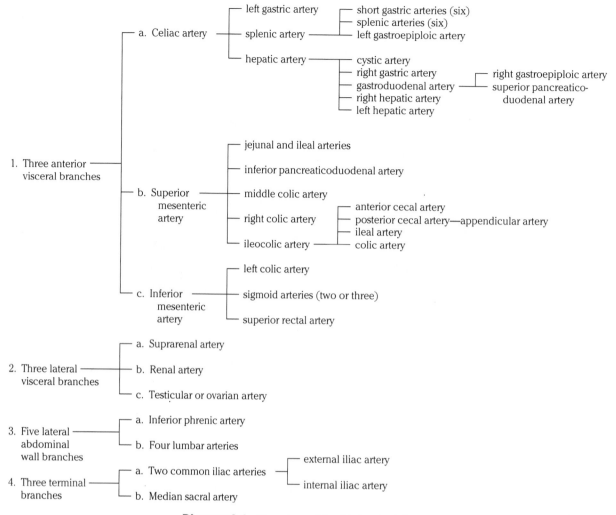

1. Three anterior visceral branches
   - a. Celiac artery
     - left gastric artery
     - splenic artery
       - short gastric arteries (six)
       - splenic arteries (six)
       - left gastroepiploic artery
     - hepatic artery
       - cystic artery
       - right gastric artery
       - gastroduodenal artery
         - right gastroepiploic artery
         - superior pancreatico-duodenal artery
       - right hepatic artery
       - left hepatic artery
   - b. Superior mesenteric artery
     - jejunal and ileal arteries
     - inferior pancreaticoduodenal artery
     - middle colic artery
     - right colic artery
     - ileocolic artery
       - anterior cecal artery
       - posterior cecal artery—appendicular artery
       - ileal artery
       - colic artery
   - c. Inferior mesenteric artery
     - left colic artery
     - sigmoid arteries (two or three)
     - superior rectal artery

2. Three lateral visceral branches
   - a. Suprarenal artery
   - b. Renal artery
   - c. Testicular or ovarian artery

3. Five lateral abdominal wall branches
   - a. Inferior phrenic artery
   - b. Four lumbar arteries

4. Three terminal branches
   - a. Two common iliac arteries
     - external iliac artery
     - internal iliac artery
   - b. Median sacral artery

**Diagram 8-1**    Branches of the Abdominal Aorta

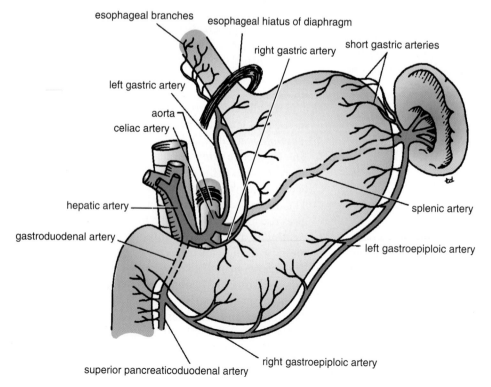

**Figure 8-4**    Arteries that supply the stomach. Note that all the arteries are derived from branches of the celiac artery.

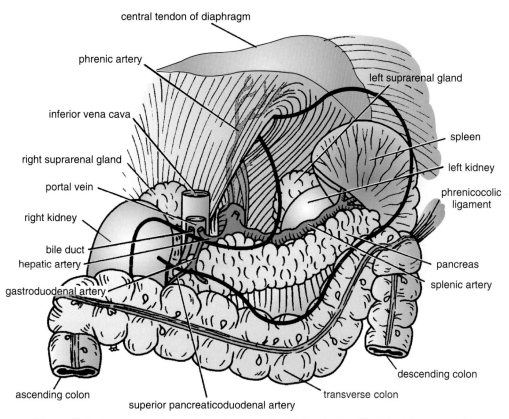

**Figure 8-5**  Structures situated on the posterior abdominal wall behind the stomach.

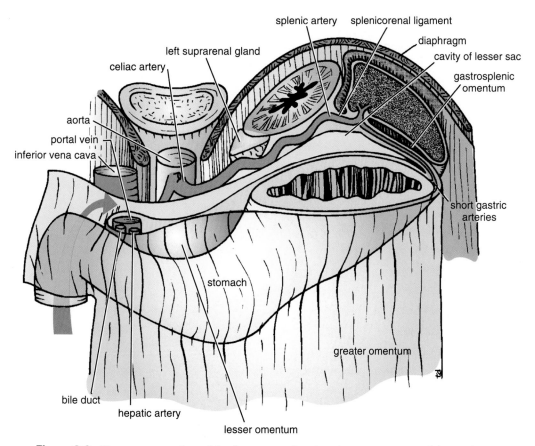

**Figure 8-6**  Transverse section of the lesser sac showing the arrangement of the peritoneum in the formation of the lesser omentum, the gastrosplenic omentum, and the splenicorenal ligament. *Arrow* indicates the position of the opening of the lesser sac.

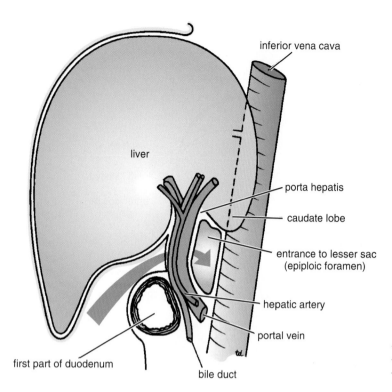

**Figure 8-7** Sagittal section through the entrance into the lesser sac showing the important structures that form boundaries to the opening. (Note the *arrow* passing from the greater sac through the epiploic foramen into the lesser sac.)

artery (Fig. 8-9). Each artery divides into two vessels, which unite with adjacent branches to form a series of arcades. Branches from the arcades divide and unite to form a second, third, and fourth series of arcades. Fewer arcades supply the jejunum than supply the ileum. From the terminal arcades, small straight vessels supply the intestine.

## Inferior Mesenteric Artery

The inferior mesenteric artery supplies the distal third of the transverse colon, the left colic flexure, the descending colon, the sigmoid colon, the rectum, and the upper half of the anal canal. It arises from the abdominal aorta about 1.5 in. (3.8 cm) above its bifurcation (Fig. 8-11). The artery runs

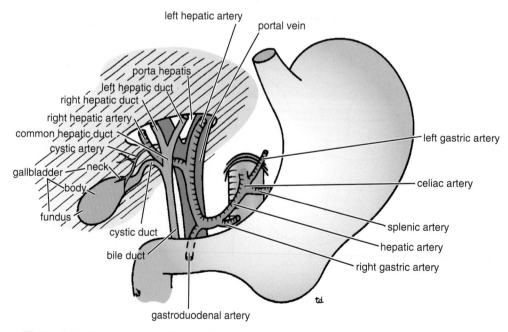

**Figure 8-8** Structures entering and leaving the porta hepatis.

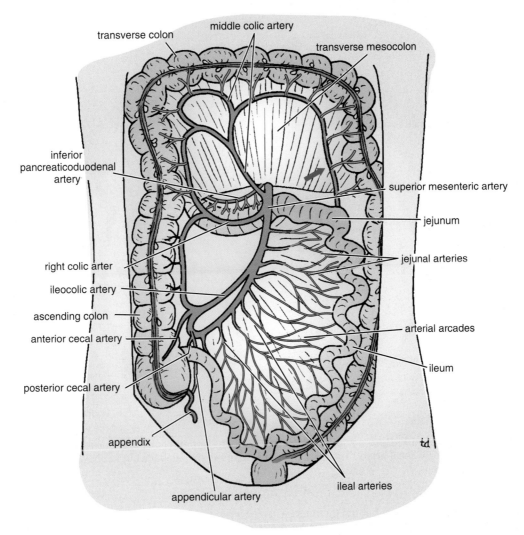

**Figure 8-9** Superior mesenteric artery and its branches. Note that this artery supplies blood to the gut from halfway down the second part of the duodenum to the distal third of the transverse colon (*arrow*).

downward and to the left and crosses the left common iliac artery. Here, it becomes the superior rectal artery.

### Branches of the Inferior Mesenteric Artery

- The **left colic artery** runs upward and to the left and supplies the distal third of the transverse colon, the left colic flexure, and the upper part of the descending colon. It divides into ascending and descending branches.
- The **sigmoid arteries** are two or three in number and supply the descending and sigmoid colon.
- The **superior rectal artery** is a continuation of the inferior mesenteric artery as it crosses the left common iliac artery. It descends into the pelvis behind the rectum. The artery supplies the rectum and upper half of the anal canal and anastomoses with the middle rectal and inferior rectal arteries.

### Marginal Artery

The anastomosis of the colic arteries around the concave margin of the large intestine forms a single arterial trunk called the marginal artery. This begins at the ileocecal junction, where it anastomoses with the ileal branches of the superior mesenteric artery, and it ends where it anastomoses less freely with the superior rectal artery (Fig. 8-11).

## Middle Suprarenal Arteries

The middle suprarenal artery arises on each side of the aorta and runs horizontal laterally to the suprarenal gland (Fig. 8-2).

## Renal Arteries

The renal artery arises on each side of the aorta just below the origin of the superior mesenteric artery (Figs. 8-1 and

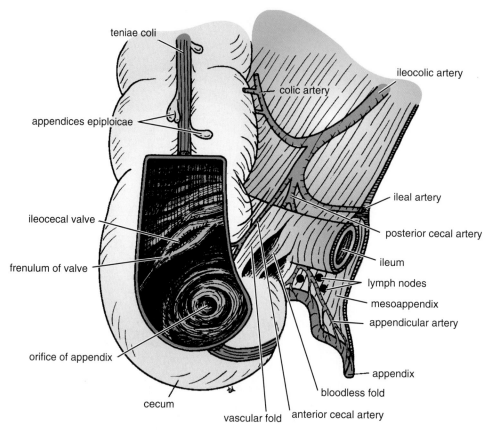

**Figure 8-10**  Cecum and appendix. Note that the appendicular artery is a branch of the posterior cecal artery. The edge of the mesoappendix has been cut to show the peritoneal layers.

8-2). The right artery is longer and passes behind the inferior vena cava. The renal artery gives off the inferior suprarenal artery.

## Testicular (or Ovarian) Arteries

The testicular arteries arise from the front of the aorta just below the origin of the renal arteries (Fig. 8-1). Each artery is long and slender and passes obliquely downward and laterally behind the peritoneum. The artery crosses the ureter and the external iliac artery to reach the deep inguinal ring, where it joins the spermatic cord (Fig. 22-3). Having passed through the inguinal canal, it enters the scrotum and supplies the testis and the epididymis.

In the female, the **ovarian artery** has a similar abdominal course. Having crossed the external iliac artery at the pelvic inlet, it enters the suspensory ligament of the ovary. It then passes into the broad ligament and enters the ovary by way of the mesovarium.

## Inferior Phrenic Arteries

The inferior phrenic arteries arise from the aorta just beneath the diaphragm. They run upward and laterally, supplying the undersurface of the diaphragm.

## Lumbar Arteries

The four pairs of lumbar arteries arise from the back of the aorta and pass round the bodies of the upper four lumbar vertebrae. They supply the muscles of the abdominal wall and back. The first lumbar artery gives off branches to the lower part of the spinal cord.

## Median Sacral Artery

The median sacral artery (Fig. 8-2) is a small branch that arises at the bifurcation of the aorta. It descends into the pelvis in front of the sacrum.

## Common Iliac Arteries

The right and left common iliac arteries are the terminal branches of the aorta. They arise at the level of the fourth lumbar vertebra and run downward and laterally along the medial border of the psoas muscle (Figs. 8-1 and 8-2). Each artery ends in front of the sacroiliac joint by dividing into the external and internal iliac arteries. At the bifurcation, the common iliac artery on each side is crossed anteriorly by the ureter (Figs. 8-1 and 8-12).

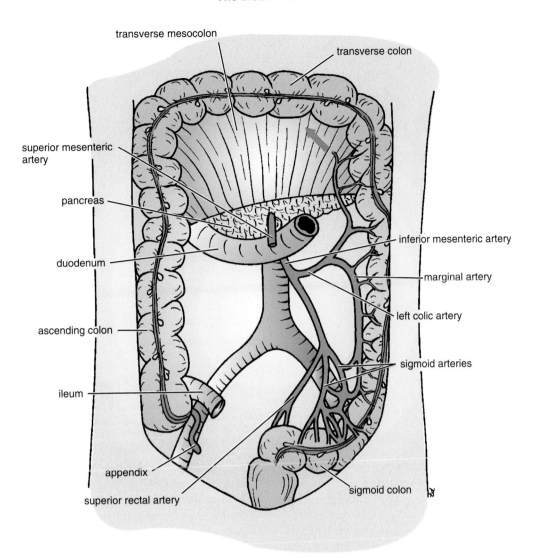

**Figure 8-11**  Inferior mesenteric artery and its branches. Note that this artery supplies the large bowel from the distal third of the transverse colon to halfway down the anal canal. It anastomoses with the middle colic branch of the superior mesenteric artery (*arrow*).

## External Iliac Artery

The external iliac artery runs along the medial border of the psoas, following the pelvic brim (Fig. 8-12). It gives off the inferior epigastric and deep circumflex iliac branches (Fig. 8-12).

The artery enters the thigh by passing under the inguinal ligament to become the femoral artery. The inferior epigastric artery arises just above the inguinal ligament. It passes upward and medially along the medial margin of the deep inguinal ring (Fig. 19-11) and enters the rectus sheath behind the rectus abdominis muscle. The deep circumflex iliac artery arises close to the inferior epigastric artery (Fig. 8-12). It ascends laterally to the anterior superior iliac spine and the iliac crest, supplying the muscles of the anterior abdominal wall.

## Internal Iliac Artery

The internal iliac artery passes down into the pelvis to the upper margin of the greater sciatic foramen, where it divides into anterior and posterior divisions (Fig. 8-12). The branches of these divisions supply the pelvic viscera, the perineum, the pelvic walls, and the buttocks. The origin of the terminal branches is subject to variation, but the usual arrangement is shown in Diagram 8-2.

### Branches of the Anterior Division

- **Umbilical artery:** From the proximal patent part of the umbilical artery arises the **superior vesical artery,** which supplies the upper portion of the bladder (Fig. 8-12).
- **Obturator artery:** This artery runs forward along the lateral wall of the pelvis with the obturator nerve and leaves the pelvis through the obturator canal.
- **Inferior vesical artery:** This artery supplies the base of the bladder and the prostate and seminal vesicles in the male; it also gives off the **artery to the vas deferens.**
- **Middle rectal artery:** Commonly, this artery arises with the inferior vesical artery (Fig. 8-12). It supplies the

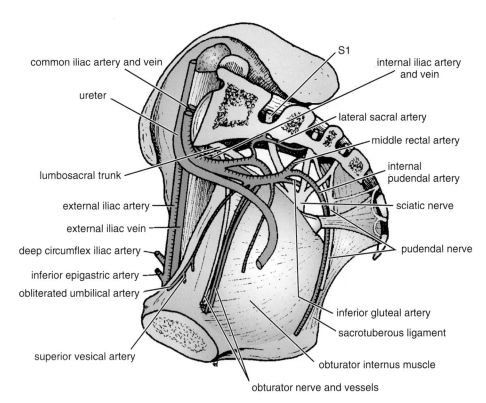

**Figure 8-12** Lateral wall of the pelvis.

muscle of the lower rectum and anastomoses with the superior rectal and inferior rectal arteries.

▦ **Internal pudendal artery:** This artery leaves the pelvis through the greater sciatic foramen and enters the gluteal region below the piriformis muscle (Fig. 8-12). It enters the perineum by passing through the lesser sciatic foramen. The artery then passes forward in the pudendal canal with the pudendal nerve and, by means of its branches, supplies the musculature of the anal canal and the skin and muscles of the perineum.

▦ **Inferior gluteal artery:** This artery leaves the pelvis through the greater sciatic foramen below the piriformis

muscle (Fig. 8-12). It passes between the first and second or second and third sacral nerves.

▦ **Uterine artery:** This artery runs medially on the floor of the pelvis and **crosses the ureter superiorly** (Fig. 8-13). It passes above the lateral fornix of the vagina to reach the uterus. Here, it ascends between the layers of the broad ligament along the lateral margin of the uterus. It ends by following the uterine tube laterally, where it anastomoses with the ovarian artery. The uterine artery gives off a vaginal branch.

▦ **Vaginal artery:** This artery usually takes the place of the inferior vesical artery present in the male. It supplies the vagina and the base of the bladder.

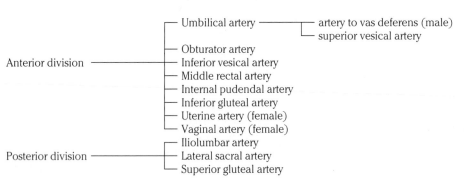

**Diagram 8-2** Branches of the Internal Iliac Artery

*Branches of the Posterior Division*

- **Iliolumbar artery:** This artery ascends across the pelvic inlet posterior to the external iliac vessels, psoas, and iliacus muscles.
- **Lateral sacral arteries:** These arteries descend in front of the sacral plexus, giving off branches to neighboring structures (Fig. 8-12).
- **Superior gluteal artery:** This artery leaves the pelvis through the greater sciatic foramen above the piriformis muscle. It supplies the gluteal region.

# SURFACE ANATOMY OF THE ABDOMINAL AORTA AND ITS MAIN BRANCHES

The surface markings of the aorta and its main branches are shown in Figure 8-3.

# RADIOLOGIC APPEARANCES OF THE ABDOMINAL AORTA AND SOME OF ITS MAIN BRANCHES

Arteriography of the superior mesenteric artery, inferior mesenteric artery, and iliac arteries is shown in Figures 8-14, 8-15, and 8-16. CT scans of the abdomen showing the aorta and the common iliac vessels are seen in Figures 8-17 and 8-18.

## Inferior Vena Cava

### Location and Description

The inferior vena cava conveys most of the blood from the body below the diaphragm to the right atrium of the heart. It

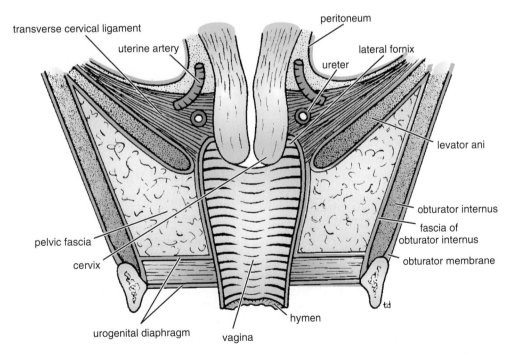

**Figure 8-13** Coronal section of the pelvis showing relation of the levatores ani muscles and transverse cervical ligaments to the uterus and vagina. Note that the transverse cervical ligaments are formed from a condensation of visceral pelvic fascia. Note also the uterine artery crossing above the ureter on each side.

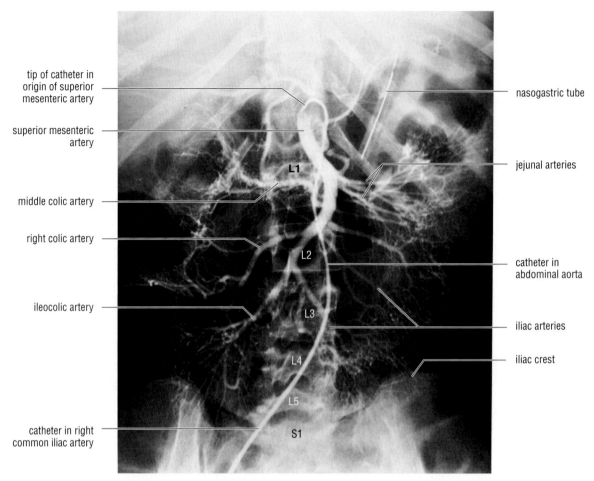

tip of catheter in origin of superior mesenteric artery

superior mesenteric artery

middle colic artery

right colic artery

ileocolic artery

catheter in right common iliac artery

nasogastric tube

jejunal arteries

catheter in abdominal aorta

iliac arteries

iliac crest

L1
L2
L3
L4
L5
S1

**Figure 8-14**    An arteriogram of the superior mesenteric artery. The catheter has been inserted into the right femoral artery and has passed up the external and common iliac arteries to ascend the aorta to the origin of the superior mesenteric artery. A nasogastric tube is also in position.

is formed by the union of the common iliac veins behind the right common iliac artery at the level of the fifth lumbar vertebra (Fig. 8-1). It ascends on the right side of the aorta, pierces the central tendon of the diaphragm at the level of the eighth thoracic vertebra, and drains into the right atrium of the heart.

PHYSIOLOGIC NOTE

### Abdominothoracic Pump

The tunica media in the wall of the vena cavae is relatively thin, consisting of sparsely arranged smooth muscle fibers mixed with connective tissue. Thus, the lumen of the vessels is subject to change brought about by changes in extravascular pressures. During inspiration, the decrease in intrathoracic pressure and the increase in intra-abdominal pressure following the descent of the diaphragm results in venous blood being sucked up into the right atrium. At the same time, the blood is forced up from below by pressure on the infe-

rior vena cava within the abdomen. This cyclical change in the intrathoracic pressure and intra-abdominal pressure is very effective and is known as the **abdominothoracic pump**.

## Relations of the Inferior Vena Cava

As the inferior vena cava passes up the posterior abdominal wall, it has the following important relations.

**Anteriorly:** Coils of small intestine, third part of the duodenum, head of the pancreas, first part of the duodenum, entrance into the lesser sac of peritoneum (which separates the inferior vena cava from the portal vein, common bile duct, and hepatic artery) (Figs. 8-6 and 8-7), and the liver.

**Laterally:** The right sympathetic trunk lies behind its right margin, and the right ureter lies 0.5 in. (1.3 cm) from its right border. The entrance into the lesser sac separates the inferior vena cava from the portal vein (see Fig. 5-7).

**Medially:** Abdominal aorta.

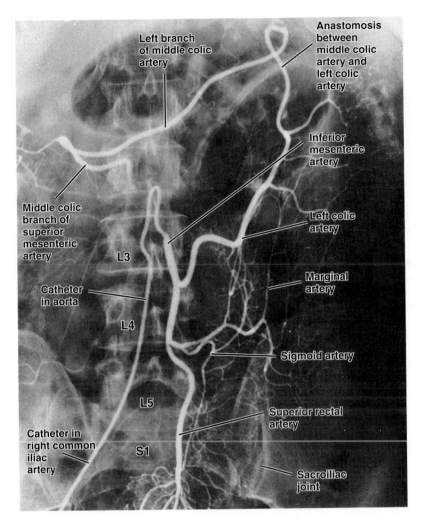

**Figure 8-15** An arteriogram of the inferior mesenteric artery. The catheter has been inserted into the right femoral artery and has passed up the external and common iliac arteries to ascend the aorta to the origin of the inferior mesenteric artery. The radiopaque dye has spread to enter the middle colic branch of the superior mesenteric artery.

## Tributaries of the Inferior Vena Cava

The inferior vena cava has the following tributaries (Fig. 8-1):

■ **Two anterior visceral tributaries:** the hepatic veins.
■ **Three lateral visceral tributaries:** the right suprarenal vein (the left vein drains into the left renal vein), renal veins, and right testicular or ovarian vein (the left vein drains into the left renal vein).
■ **Five lateral abdominal wall tributaries:** the inferior phrenic vein and four lumbar veins.
■ **Three veins of origin:** two common iliac veins and the median sacral vein.

The tributaries of the inferior vena cava are summarized in Diagram 8-3.

If one remembers that the venous blood from the abdominal portion of the gastrointestinal tract drains to the liver by means of the tributaries of the portal vein and that the left suprarenal and testicular or ovarian veins drain first into the left renal vein, then it is apparent that the tributaries of the inferior vena cava correspond rather closely to the branches of the abdominal portion of the aorta.

## Surface Anatomy of the Inferior Vena Cava

The surface markings of the inferior vena cava are shown in Figure 8-3.

# Portal Vein (Hepatic Portal Vein)

## Location and Description

This important vein (Fig. 8-19) drains blood from the abdominal part of the gastrointestinal tract from the lower third of the esophagus to halfway down the anal canal; it also drains blood from the spleen, pancreas, and gallbladder. The portal vein enters the liver and breaks up into sinusoids, from which blood passes into the hepatic veins that join the inferior vena cava. The portal vein is about 2 in. (5 cm) long

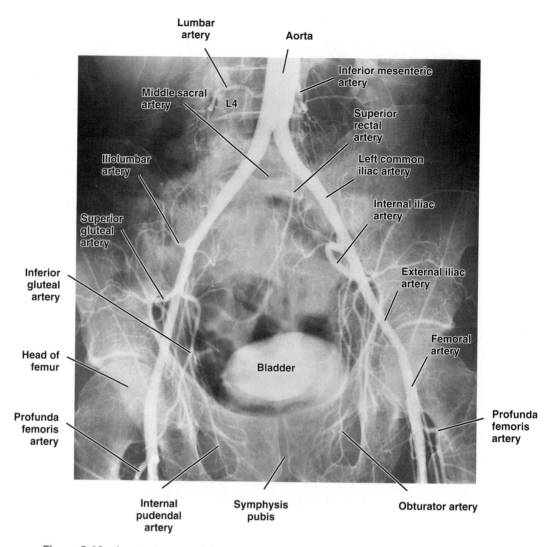

**Figure 8-16** An arteriogram of the lower part of the abdominal aorta showing the iliac arteries and their branches. The catheter (not visible) has been inserted into the left femoral artery. Some of the radiopaque material has already been excreted in the urine and shows in the bladder.

and is formed behind the neck of the pancreas by the union of the superior mesenteric and splenic veins (Fig. 8-20). It ascends to the right, behind the first part of the duodenum, and enters the lesser omentum (Figs. 8-6 and 8-7). It then runs upward in front of the opening into the lesser sac to the porta hepatis, where it divides into right and left terminal branches.

The portal circulation begins as a capillary plexus in the organs it drains and ends by emptying its blood into sinusoids within the liver.

## Tributaries of the Portal Vein

The tributaries of the portal vein are the splenic vein, superior mesenteric vein, left gastric vein, right gastric vein, and cystic veins.

■ **Splenic vein:** This vein leaves the hilum of the spleen and passes to the right in the splenicorenal ligament lying below the splenic artery. It unites with the superior mesenteric vein behind the neck of the pancreas to form the portal vein (Fig. 8-20). It receives the short gastric, left gastroepiploic, inferior mesenteric, and pancreatic veins.

■ **Inferior mesenteric vein:** This vein ascends on the posterior abdominal wall and joins the splenic vein behind the body of the pancreas (Fig. 8-20). It receives the superior rectal veins, the sigmoid veins, and the left colic vein.

■ **Superior mesenteric vein:** This vein ascends in the root of the mesentery of the small intestine on the right side of the artery. It passes in front of the third part of the duodenum and joins the splenic vein behind the neck of the pancreas (Fig. 8-20). It receives the jejunal, ileal, ileo-

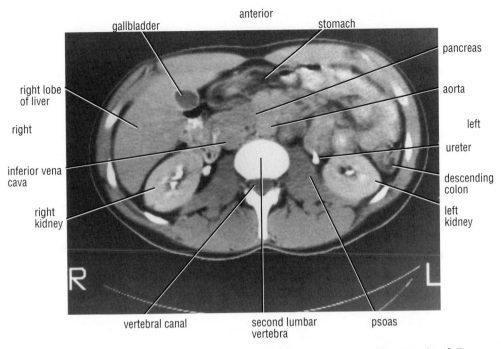

**Figure 8-17**  A CT scan of the abdomen at the level of the second lumbar vertebra following an intravenous pyelogram. The aorta and the inferior vena cava can be seen on the posterior abdominal wall.

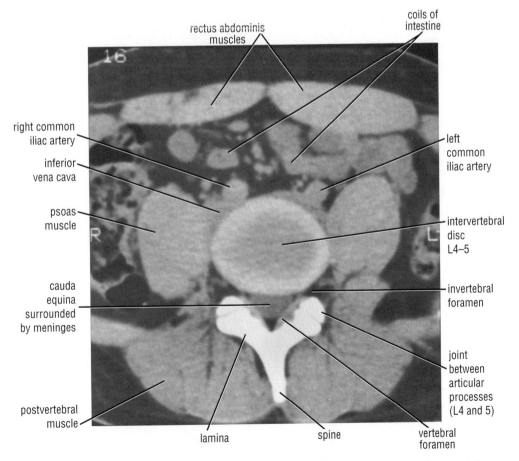

**Figure 8-18**  A CT scan of the abdomen at the level of the intervertebral disc between the fourth and fifth lumbar vertebrae. The scan shows the right and left common iliac arteries close to their origin and the beginning of the inferior vena cava.

1. Two anterior visceral tributaries—the hepatic veins

2. Three lateral visceral tributaries
   - a. Right suprarenal vein
     (the left drains into the left renal vein)
   - b. Renal veins
   - c. Right testicular or ovarian vein
     (the left drains into the left renal vein)

3. Five lateral abdominal wall tributaries
   - a. Inferior phrenic vein
   - b. Four lumbar veins

4. Three tributaries of origin
   - a. Two common iliac veins
     - external iliac vein
     - internal iliac vein
   - b. Median sacral vein

**Diagram 8-3** Tributaries of Inferior Vena Cava

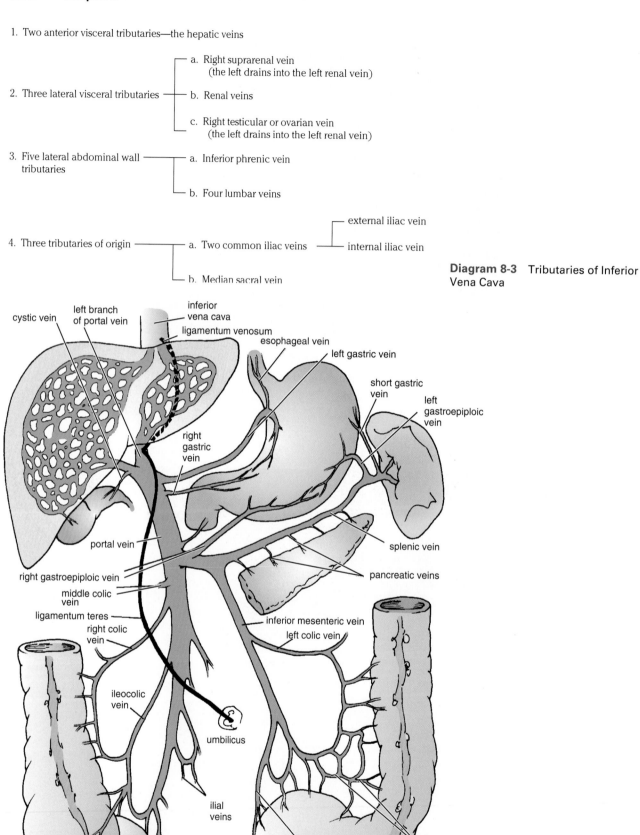

**Figure 8-19** Tributaries of the portal vein.

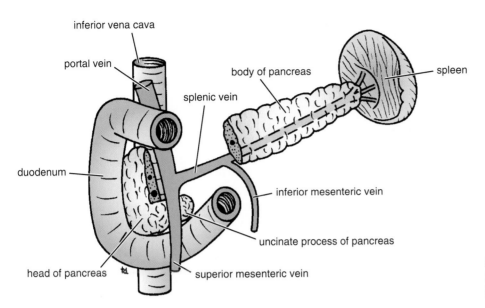

inferior vena cava

portal vein

body of pancreas

spleen

splenic vein

duodenum

inferior mesenteric vein

uncinate process of pancreas

head of pancreas

superior mesenteric vein

**Figure 8-20** Formation of the portal vein behind the neck of the pancreas.

colic, right colic, middle colic, inferior pancreaticoduodenal, and right gastroepiploic veins.

■ **Left gastric vein:** This vein drains the left portion of the lesser curvature of the stomach and the distal part of the esophagus. It opens directly into the portal vein (Fig. 8-19).

■ **Right gastric vein:** This vein drains the right portion of the lesser curvature of the stomach and drains directly into the portal vein (Fig. 8-19).

■ **Cystic veins:** These veins either drain the gallbladder directly into the liver or join the portal vein (Fig. 8-19).

# Review Questions

**Each of the numbered items or incomplete statements in this section is followed by answers or by completions of the statement. Select the one lettered answer or completion that is BEST in each case.**

1. The aorta enters the abdomen through the aortic opening in the diaphragm in front of the _____ vertebra.
   A. second lumbar
   B. twelfth thoracic
   C. eleventh thoracic
   D. first lumbar
   E. third lumbar

2. The abdominal aorta is related on its right side to the
   A. left crus of the diaphragm.
   B. inferior mesenteric vein.
   C. inferior vena cava.
   D. left sympathetic trunk.
   E. right ureter.

3. The abdominal aorta divides below in front of the _____ vertebra into the two common iliac arteries.
   A. first sacral
   B. third lumbar
   C. second lumbar
   D. fourth lumbar
   E. fifth lumbar

4. Which of the following statements concerning the celiac artery is correct?
   A. It gives direct origin to the right gastric artery.
   B. It has four terminal branches.
   C. It lies within the lesser sac of peritoneum.
   D. It lies behind the duodenum.
   E. It gives origin to the splenic artery.

5. The splenic artery gives rise to the following arteries.
   A. The left gastroepiploic artery, the short gastric arteries, and pancreatic arteries.
   B. The right gastric artery, the hepatic artery, and the cystic artery.
   C. The left gastric artery, the gastroduodenal artery, and the superior mesenteric artery.
   D. The inferior pancreaticoduodenal artery.
   E. Middle colic artery.

6. The inferior mesenteric artery supplies the
   A. fourth part of the duodenum.
   B. cecum.

C. appendix.
D. distal third of the transverse colon.
E. ascending colon.

7. The marginal artery supplies the
   A. greater curvature of the stomach.
   B. concave margin of the large intestine.
   C. appendix.
   D. gall bladder.
   E. concave margin of the duodenum.

8. The common iliac arteries divide in front of the
   _____ into the internal and external iliac arteries.
   A. psoas muscle
   B. fourth lumbar vertebra
   C. sacroiliac joint
   D. promontory of the sacrum
   E. spine of the ischium.

9. The external iliac artery becomes the femoral artery at
   the _____.
   A. sacrospinous ligament
   B. iliolumbar ligament
   C. greater sciatic foramen
   D. inguinal ligament
   E. sacrotuberous ligament

10. The inferior vena cava is formed in front of the
    A. fourth lumbar vertebra.
    B. eighth thoracic vertebra.
    C. fifth lumbar vertebra.
    D. twelfth thoracic vertebra.
    E. third lumbar vertebra.

11. As the inferior vena cava passes behind the entrance
    into the lesser peritoneal sac, it is related in front to the
    A. splenic artery.
    B. portal vein.

C. stomach.
D. pancreas.
E. first part of the duodenum.

12. The following important statement is correct concern-
    ing the portal vein.
    A. It is formed behind the body of the pancreas.
    B. It is about 5 inches long.
    C. It begins as a capillary plexus in the organs that it
       drains and ends by emptying its blood into the sinu-
       soids of the liver.
    D. It passes in front of the first part of the duodenum.
    E. It drains blood from the gastrointestinal tract from
       the duodenal jejunal junction to halfway down the
       anal canal.

13. The portal vein ascends to the liver in the _____
    and divides into right and left terminal branches.
    A. lienorenal ligament
    B. greater omentum
    C. gastrosplenic omentum
    D. lesser omentum
    E. falciform ligament

14. The portal vein is formed by the union of the splenic
    vein with the _____ vein.
    A. superior mesenteric
    B. hepatic
    C. inferior mesenteric
    D. superior rectal
    E. left gastroepiploic

# Answers and Explanations

1. **B** is correct. The aorta enters the abdomen through the aortic opening in the diaphragm in front of the twelfth thoracic vertebra (Fig. 8-1).

2. **C** is correct. The abdominal aorta is related on the right side to the inferior vena cava (Fig. 8-1).

3. **D** is correct. The abdominal aorta divides below in front of the fourth lumbar vertebra into the two common iliac arteries (Fig. 8-2).

4. **E** is correct. The celiac artery has three terminal branches: the splenic, hepatic, and left gastric arteries (see p 221).

5. **A** is correct. The splenic artery gives rise to the left gastroepiploic artery, the short gastric arteries, and the pancreatic arteries (Fig. 8-4). The pancreatic branches are given off the splenic artery as it passes to the left along the upper border of the body of the pancreas.

6. **D** is correct. The inferior mesenteric artery supplies the large bowel from the distal third of the transverse colon down to halfway down the anal canal.

7. **B** is correct. The marginal artery is formed from the anastomosis of the colic arteries around the concave margin of the large intestine. It begins at the ileocecal junction, where it anastomoses with the ileal branches of the superior mesenteric artery to where it ends by anastomosing less freely with the superior rectal artery (Fig. 8-11).

8. **C** is correct. The common iliac arteries divide in front of the sacroiliac joint into the internal and external iliac arteries (Fig. 8-1).

9. **D** is correct. The external iliac artery becomes the femoral artery as it passes posterior to the inguinal ligament to enter the thigh.

10. **C** is correct. The inferior vena cava is formed in front of the fifth lumbar vertebra by the union of the right and left common iliac veins (Fig. 8-1).

11. **B** is correct. As the inferior vena cava ascends behind the entrance into the lesser peritoneal sac, it is related anteriorly to the portal vein, hepatic artery, and the bile duct (Fig. 8-7).

12. **C** is correct. The portal vein begins as a capillary plexus in the organs that it drains and ends by emptying its blood into the sinusoids of the liver.

13. **D** is correct. The portal vein ascends to the liver in the lesser omentum and, in the porta hepatis, divides into right and left terminal branches (Fig. 8-7).

14. **A** is correct. The portal vein is formed by the union of the splenic vein with the superior mesenteric vein (Figs. 8-19 and 8-20).

# 9

# The Blood Vessels of the Lower Extremity

All clinical material relevant to this chapter can be found on the CD-ROM.

# Chapter Outline

I n the lower extremity, acute limb-threatening ischemia requires a timely diagnosis so that treatment may be quickly implemented and the limb saved. A knowledge of the collateral circulation in these circumstances is imperative. The venous system of the lower limb is the source of many emergency problems. The health care professional is aided by knowledge of venous anatomy in diagnosing phlebitis and obtaining lower extremity venous access.

The purpose of this chapter is to review the major arterial and venous vessels of the lower limbs relative to clinical situations.

# BASIC ANATOMY

## Arteries of the Lower Extremity

### Femoral Artery

The femoral artery enters the thigh by passing behind the inguinal ligament, as a continuation of the external iliac artery (Figs. 9-1, 9-2, and 9-3). Here, it lies midway between the anterior superior iliac spine and the symphysis pubis. The femoral artery is the main arterial supply to the lower limb. It descends almost vertically toward the adductor tubercle of the femur and ends at the opening in the adductor magnus muscle by entering the popliteal space as the popliteal artery (Fig. 9-4).

### Relations of the Femoral Artery

- **Anteriorly:** In the upper part of its course, it is superficial and is covered by skin and fascia. In the lower part of its course, it passes behind the sartorius muscle (Fig. 9-1).
- **Posteriorly:** The artery lies on the psoas, which separates it from the hip joint, the pectineus, and the adductor longus (Fig. 9-1). The femoral vein intervenes between the artery and the adductor longus.
- **Medially:** It is related to the femoral vein in the upper part of its course (Fig. 9-1).
- **Laterally:** The femoral nerve and its branches (Fig. 9-1).

The femoral artery is accompanied by the femoral vein, which lies on the medial side at the inguinal ligament and posterior to it at the apex of the femoral triangle. At the opening in the adductor magnus, the vein lies on the lateral

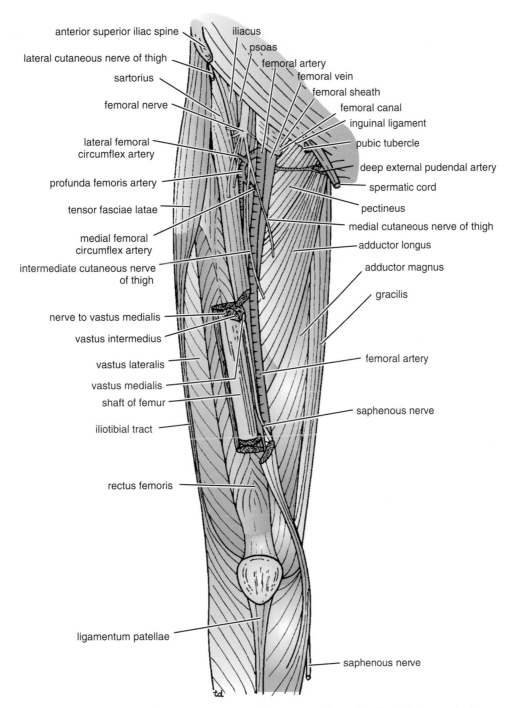

**Figure 9-1** Femoral triangle and adductor (subsartorial) canal in the right lower limb.

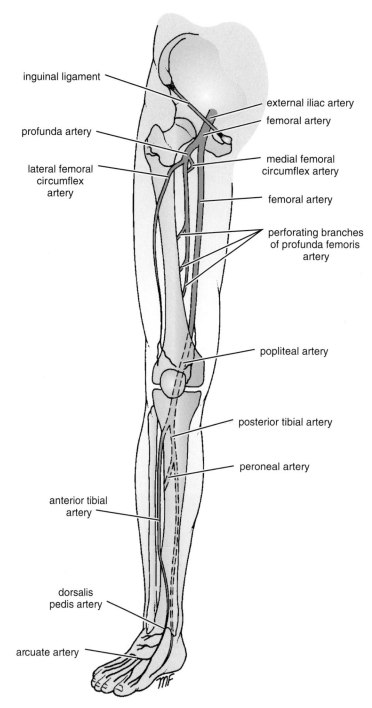

**Figure 9-2**  Major arteries of the lower limb.

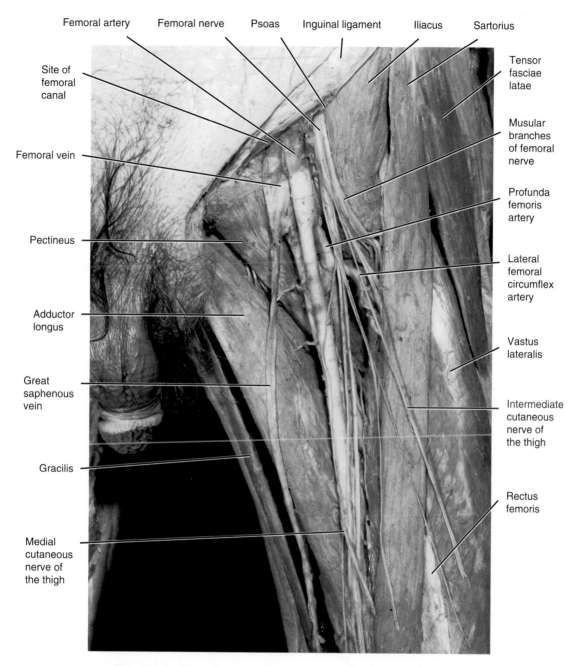

Femoral artery    Femoral nerve    Psoas    Inguinal ligament    Iliacus    Sartorius

Site of femoral canal

Femoral vein

Pectineus

Adductor longus

Great saphenous vein

Gracilis

Medial cutaneous nerve of the thigh

Tensor fasciae latae

Musular branches of femoral nerve

Profunda femoris artery

Lateral femoral circumflex artery

Vastus lateralis

Intermediate cutaneous nerve of the thigh

Rectus femoris

**Figure 9-3**    Dissection of the femoral triangle in the left lower limb.

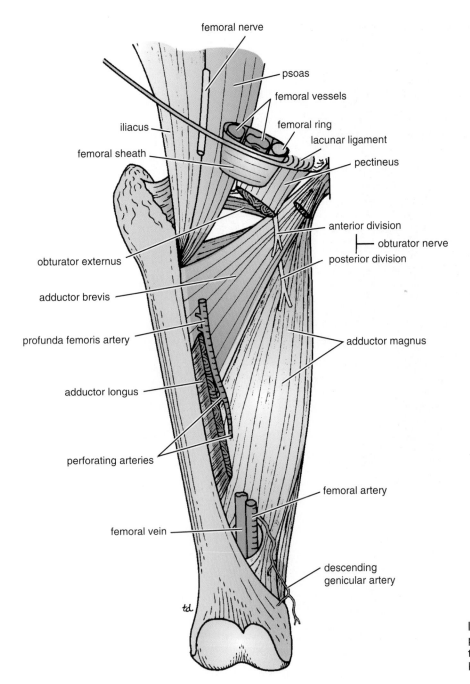

femoral nerve

psoas

femoral vessels

femoral ring

lacunar ligament

iliacus

femoral sheath

pectineus

anterior division

obturator nerve

posterior division

obturator externus

adductor brevis

adductor magnus

profunda femoris artery

adductor longus

perforating arteries

femoral artery

femoral vein

descending
genicular artery

**Figure 9-4** Relationship between the profunda artery, obturator nerve, and the adductor muscles in the right lower limb.

side of the artery, i.e., the vein changes its mediolateral relationship to the artery, moving from being medial at the groin to being lateral at the lower part of the femur.

## Branches

- The **superficial circumflex iliac artery** is a small branch that runs up to the region of the anterior superior iliac spine (Fig. 9-5).
- The **superficial epigastric artery** is a small branch that crosses the inguinal ligament and runs to the region of the umbilicus (Fig. 9-5).

- The **superficial external pudendal artery** (Fig. 9-5) is a small branch that runs medially to supply the skin of the scrotum (or labium majus).
- The **deep external pudendal artery** (Fig. 9-1) runs medially and supplies the skin of the scrotum (or labium majus).
- The **profunda femoris artery** is a large and important branch that arises from the lateral side of the femoral artery about 1.5 in. (4 cm) below the inguinal ligament (Figs. 9-1, 9-2, and 9-3). It passes medially behind the femoral vessels and enters the medial fascial compartment of the thigh (Figs. 9-4 and 9-6). It ends by becom-

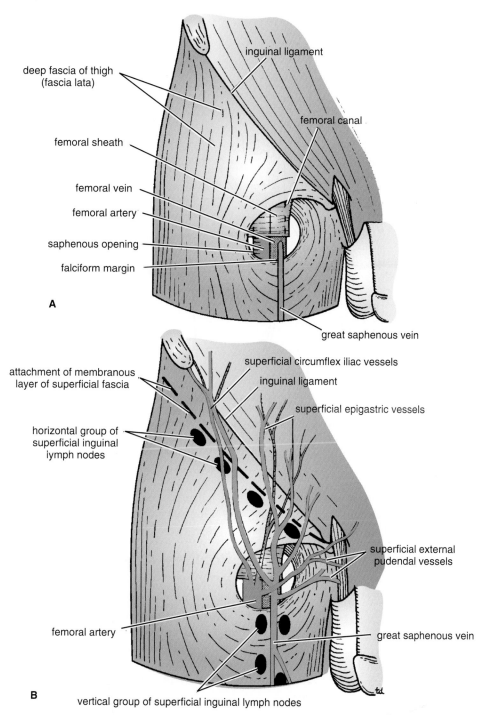

deep fascia of thigh (fascia lata)

femoral sheath

femoral vein

femoral artery

saphenous opening

falciform margin

inguinal ligament

femoral canal

**A**

great saphenous vein

attachment of membranous layer of superficial fascia

horizontal group of superficial inguinal lymph nodes

superficial circumflex iliac vessels

inguinal ligament

superficial epigastric vessels

superficial external pudendal vessels

femoral artery

great saphenous vein

**B**

vertical group of superficial inguinal lymph nodes

**Figure 9-5 A, B.** Superficial veins, arteries, and lymph nodes over the right femoral triangle. Note the saphenous opening in the deep fascia and its relationship to the femoral sheath. Note also the line of attachment of the membranous layer of superficial fascia to the deep fascia, about a finger's breadth below the inguinal ligament.

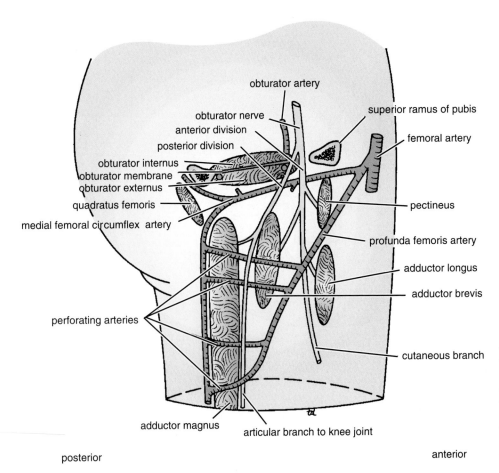

**Figure 9-6** Vertical section of the medial compartment of the thigh. Note the courses taken by the obturator nerve and its divisions and the profunda femoris artery and its branches. Note also the anastomosis between the perforating arteries and the medial femoral circumflex artery.

ing the **fourth perforating artery**. At its origin, it gives off the **medial** and **lateral femoral circumflex arteries**, and during its course, it gives off **three perforating arteries** (Fig. 9-6).

■ The **descending genicular artery** is a small branch that arises from the femoral artery near its termination (Fig. 9-4). It assists in supplying the knee joint.

## Trochanteric Anastomosis

The trochanteric anastomosis provides the main blood supply to the head of the femur (in adults) via the following arteries:

■ Superior gluteal artery.
■ Inferior gluteal artery.
■ Medial femoral circumflex artery.
■ Lateral femoral circumflex artery.

## Cruciate Anastomosis

Together with the trochanteric anastomosis, the cruciate anastomosis provides the important connection between the internal iliac and the femoral arteries. The following arteries are involved:

■ Inferior gluteal artery.
■ Medial femoral circumflex artery.

■ Lateral femoral circumflex artery.
■ First perforating artery, which is a branch of the profunda artery.

## Popliteal Artery

The popliteal artery is deeply placed and enters the popliteal fossa through the opening in the adductor magnus, as a continuation of the femoral artery (Figs. 9-7 and 9-8). It ends at the level of the lower border of the popliteus muscle by dividing into anterior and posterior tibial arteries.

### Relations of the Popliteal Artery

■ **Anteriorly:** The popliteal surface of the femur, the knee joint, and the popliteus muscle (Fig. 9-8).
■ **Posteriorly:** The popliteal vein and the tibial nerve, fascia, and skin (Fig. 9-7).

### Branches of the Popliteal Artery

The popliteal artery has the following branches:

■ **Muscular branches.**
■ **Articular branches.**
■ **Terminal branches**, the **anterior** and **posterior tibial arteries.**

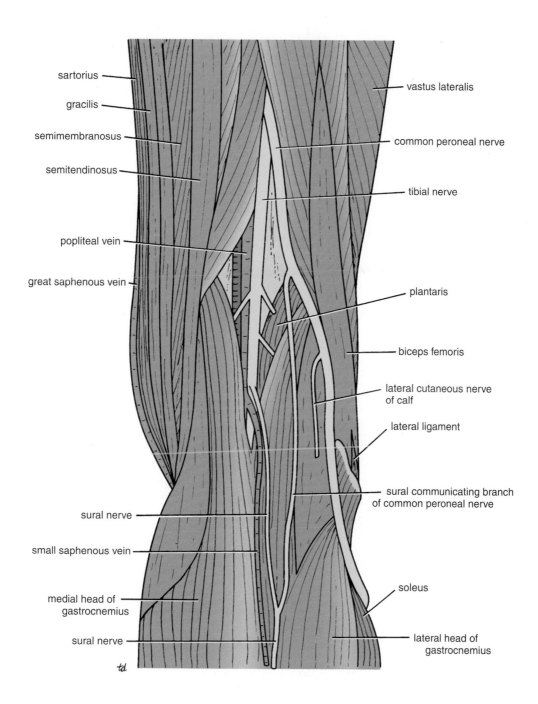

**Figure 9-7**  Boundaries and contents of the right popliteal fossa.

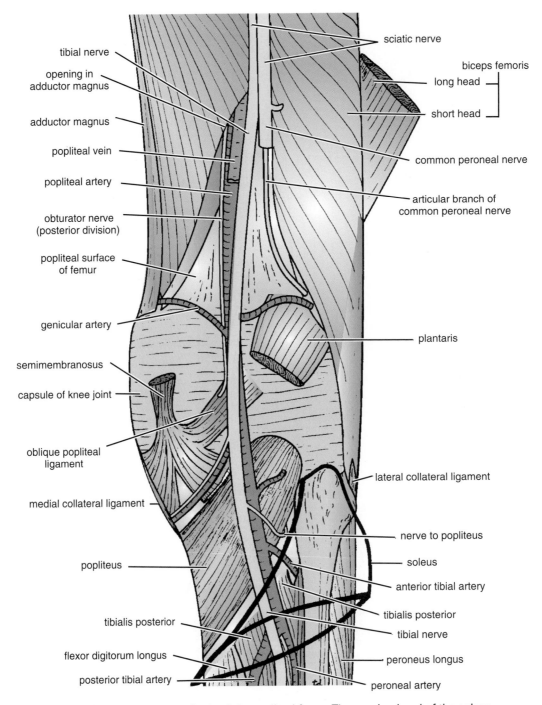

**Figure 9-8** Deep structures in the right popliteal fossa. The proximal end of the soleus muscle is shown in outline only.

## Arterial Anastomosis around the Knee Joint

To compensate for the narrowing of the popliteal artery, which occurs during extreme flexion of the knee, around the knee joint is a profuse anastomosis of small branches of the femoral artery with muscular and articular branches of the popliteal artery and with branches of the anterior and posterior tibial arteries.

## Anterior Tibial Artery

The anterior tibial artery is the smaller of the terminal branches of the popliteal artery. It arises at the level of the lower border of the popliteus muscle and passes forward into the anterior compartment of the leg through an opening in the upper part of the interosseous membrane (Fig. 9-8). It descends on the anterior surface of the interosseous membrane, accompanied by the deep peroneal nerve (Fig. 9-9).

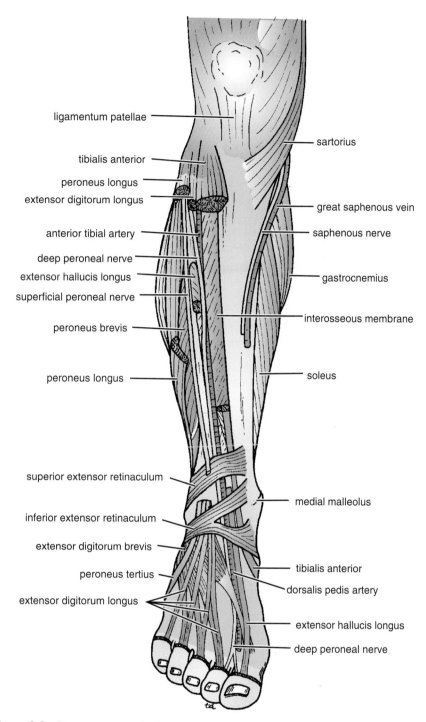

ligamentum patellae

sartorius

tibialis anterior

peroneus longus

extensor digitorum longus

great saphenous vein

anterior tibial artery

saphenous nerve

deep peroneal nerve

extensor hallucis longus

gastrocnemius

superficial peroneal nerve

peroneus brevis

interosseous membrane

peroneus longus

soleus

superior extensor retinaculum

medial malleolus

inferior extensor retinaculum

extensor digitorum brevis

tibialis anterior

peroneus tertius

dorsalis pedis artery

extensor digitorum longus

extensor hallucis longus

deep peroneal nerve

**Figure 9-9**   Deep structures in the anterior and lateral aspects of the right leg and the dorsum of the foot.

In the upper part of its course, it lies deep beneath the muscles of the compartment. In the lower part of its course, it lies superficial in front of the lower end of the tibia (Figs. 9-9 and 9-10). Having passed behind the superior extensor retinaculum, it has the tendon of the extensor hallucis longus on its medial side and the deep peroneal nerve and the tendons of extensor digitorum longus on its lateral side. It is here that its pulsations can easily be felt in the living subject. In front of the ankle joint, the artery becomes the dorsalis pedis artery.

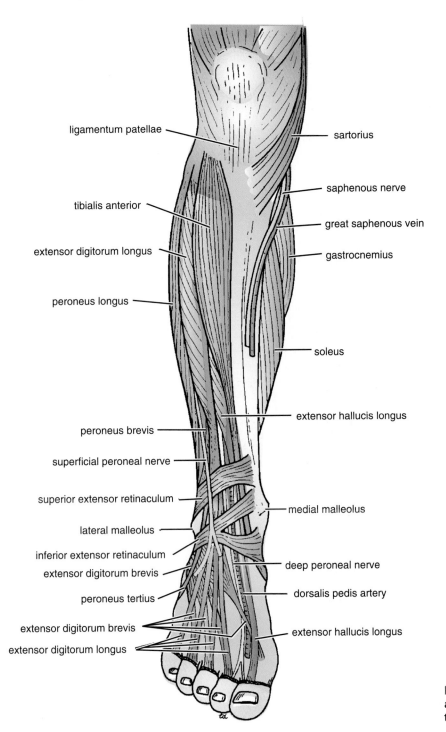

ligamentum patellae

sartorius

saphenous nerve

tibialis anterior

great saphenous vein

extensor digitorum longus

gastrocnemius

peroneus longus

soleus

extensor hallucis longus

peroneus brevis

superficial peroneal nerve

superior extensor retinaculum

medial malleolus

lateral malleolus

inferior extensor retinaculum

extensor digitorum brevis

deep peroneal nerve

peroneus tertius

dorsalis pedis artery

extensor digitorum brevis

extensor digitorum longus

extensor hallucis longus

**Figure 9-10** Structures in the anterior and lateral aspects of the right leg and the dorsum of the foot.

## Branches

The branches are as follows:

- **Muscular branches** to neighboring muscles.
- **Anastomotic branches** that anastomose with branches of other arteries around the knee and ankle joints.

## Dorsalis Pedis Artery

The dorsalis pedis artery (the dorsal artery of the foot) begins in front of the ankle joint as a continuation of the anterior tibial artery. It terminates by passing downward into the sole between the two heads of the first dorsal interosseous muscle, where it joins the lateral plantar artery

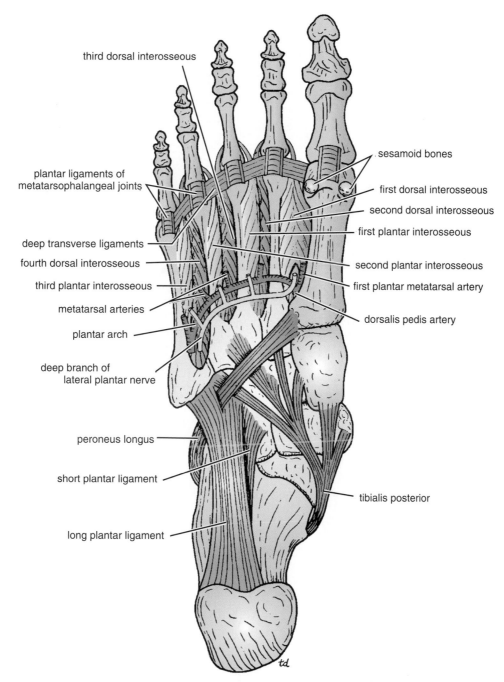

third dorsal interosseous

sesamoid bones

plantar ligaments of
metatarsophalangeal joints

first dorsal interosseous

second dorsal interosseous

first plantar interosseous

deep transverse ligaments

second plantar interosseous

fourth dorsal interosseous

first plantar metatarsal artery

third plantar interosseous

metatarsal arteries

dorsalis pedis artery

plantar arch

deep branch of
lateral plantar nerve

peroneus longus

short plantar ligament

tibialis posterior

long plantar ligament

**Figure 9-11**   Fourth layer of the plantar muscles of the right foot. The deep branch of the lateral plantar nerve and the dorsalis pedis artery and the plantar arterial arch are also shown. Note the deep transverse ligaments.

and completes the plantar arch (Fig. 9-11). It is superficial in position and is crossed by the inferior extensor retinaculum and the first tendon of extensor digitorum brevis (Fig. 9-12). On its lateral side lie the terminal part of the deep peroneal nerve and the extensor digitorum longus tendons. On the medial side lies the tendon of extensor hallucis longus (Fig. 9-12). **Its pulsations can easily be felt.**

## Branches of the Dorsalis Pedis Artery

The branches are as follows:

- **Lateral tarsal artery,** which crosses the dorsum of the foot just below the ankle joint (Fig. 9-12).
- **Arcuate artery,** which runs laterally under the extensor tendons opposite the bases of the metatarsal

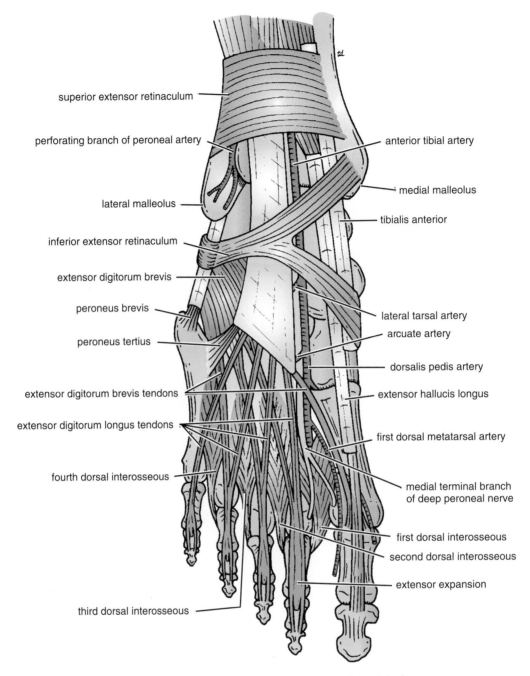

**Figure 9-12** Structures in the dorsal aspect of the right foot.

bones (Fig. 9-12). It gives off metatarsal branches to the toes.

■ **First dorsal metatarsal artery,** which supplies both sides of the big toe (Fig. 9-12).

## Posterior Tibial Artery

The posterior tibial artery is one of the terminal branches of the popliteal artery. It begins at the level of the lower border of the popliteus muscle and passes downward deep to the gastrocnemius and soleus and the deep transverse fascia of the leg (Figs. 9-8 and 9-13). It lies on the posterior surface of the tibialis posterior muscle above and on the posterior surface of the tibia below. In the lower part of the leg, the artery is covered only by skin and fascia. The artery passes behind the medial malleolus deep to the flexor retinaculum and terminates by dividing into medial and lateral plantar arteries (Fig. 9-14).

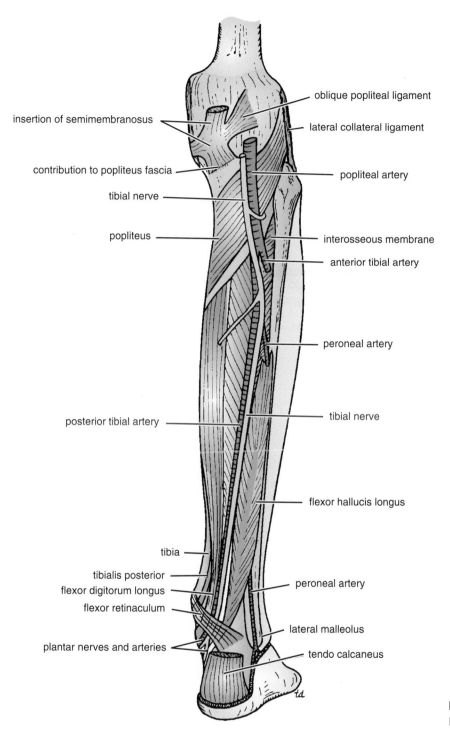

insertion of semimembranosus

oblique popliteal ligament

lateral collateral ligament

contribution to popliteus fascia

popliteal artery

tibial nerve

popliteus

interosseous membrane

anterior tibial artery

peroneal artery

tibial nerve

posterior tibial artery

flexor hallucis longus

tibia

tibialis posterior

flexor digitorum longus

peroneal artery

flexor retinaculum

plantar nerves and arteries

lateral malleolus

tendo calcaneus

**Figure 9-13**  Deep structures in the posterior aspect of the right leg.

## Branches of the Posterior Tibial Artery

The branches are as follows:

■ **Peroneal artery,** which is a large artery that arises close to the origin of the posterior tibial artery (Fig. 9-13). It descends behind the fibula, either within the substance of the flexor hallucis longus muscle or posterior to it. The peroneal artery gives off numerous **muscular branches** and a **nutrient artery to the fibula** and ends by taking part in the anastomosis around the ankle joint. A **perforating branch** pierces the interosseous membrane to reach the lower part of the front of the leg.

■ **Muscular branches** are distributed to muscles in the posterior compartment of the leg.

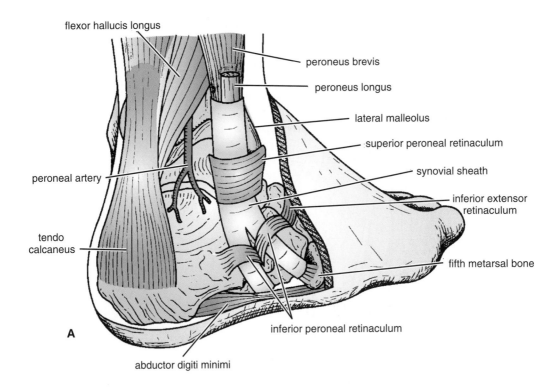

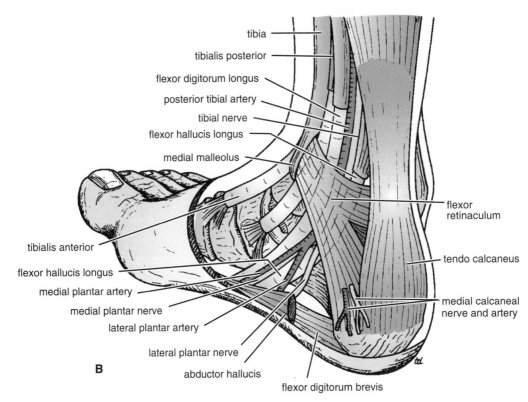

**Figure 9-14**   Structures passing behind the lateral malleolus (**A**) and the medial malleolus (**B**). Synovial sheaths of the tendons are shown in *blue*. Note the positions of the retinacula.

■ Nutrient artery to the tibia.
■ **Anastomotic branches,** which join other arteries around the ankle joint.
■ Medial and lateral plantar arteries.

## Medial Plantar Artery

The medial plantar artery is the smaller of the terminal branches of the posterior tibial artery. It arises beneath the flexor retinaculum and passes forward deep to the abductor hallucis muscle (Fig. 9-14). It ends by supplying the medial side of the big toe (Fig. 9-15). **During its course, it gives off numerous muscular, cutaneous, and articular branches.**

## Lateral Plantar Artery

The lateral plantar artery is the larger of the terminal branches of the posterior tibial artery. It arises beneath the flexor retinaculum and passes forward deep to the abductor hallucis and the flexor digitorum brevis (Figs. 9-14, 9-15, and 9-16). On reaching the base of the fifth metatarsal bone, the artery curves medially to form the **plantar arch** and, at the proximal end of the first intermetatarsal space, joins the dorsalis pedis artery (Fig. 9-11). During its course, it gives off numerous muscular, cutaneous, and articular branches. The plantar arch gives off plantar digital arteries to the toes.

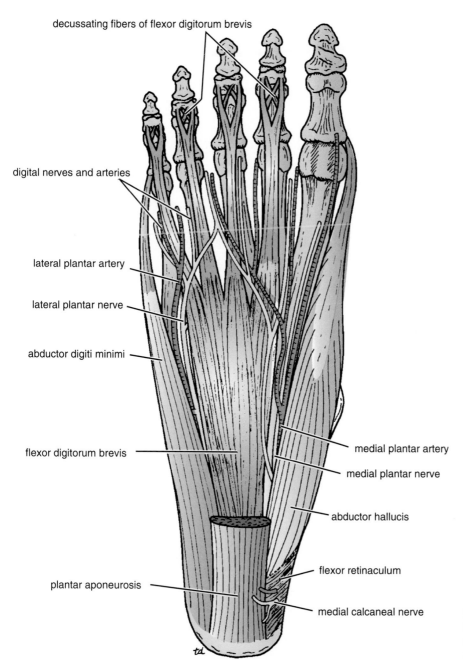

decussating fibers of flexor digitorum brevis

digital nerves and arteries

lateral plantar artery

lateral plantar nerve

abductor digiti minimi

flexor digitorum brevis

medial plantar artery

medial plantar nerve

abductor hallucis

flexor retinaculum

plantar aponeurosis

medial calcaneal nerve

**Figure 9-15** First layer of the plantar muscles of the right foot. Medial and lateral plantar arteries and nerves are also shown.

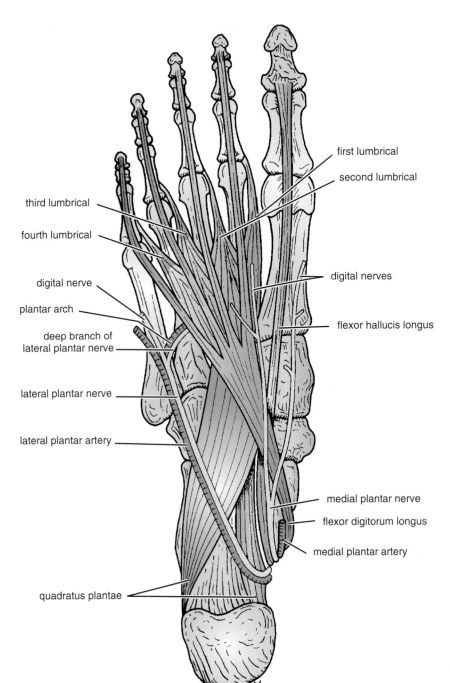

first lumbrical

second lumbrical

third lumbrical

fourth lumbrical

digital nerve

plantar arch

deep branch of
lateral plantar nerve

lateral plantar nerve

lateral plantar artery

digital nerves

flexor hallucis longus

medial plantar nerve

flexor digitorum longus

medial plantar artery

quadratus plantae

**Figure 9-16**  Second layer of the plantar muscles of the right foot. Medial and lateral plantar arteries and nerves are also shown.

# Sympathetic Innervation of the Arteries of the Lower Extremity

Sympathetic innervation of the arteries to the leg is derived from the lower three thoracic and upper two or three lumbar segments of the spinal cord. The preganglionic fibers pass to the lower thoracic and upper lumbar ganglia via white rami. The fibers synapse in the lumbar and sacral ganglia, and the postganglionic fibers reach the blood vessels via branches of the lumbar and sacral plexuses. The femoral artery receives its sympathetic fibers from the femoral and obturator nerves. The more distal arteries receive their postganglionic fibers via the common peroneal and tibial nerves.

# Veins of the Lower Extremity

The veins of the lower extremity may be divided into superficial and deep groups (Fig. 9-17). The superficial veins are of great clinical importance. They lie in the superficial fascia and have relatively thick muscle walls. The deep veins accompany the main arteries and have thin walls.

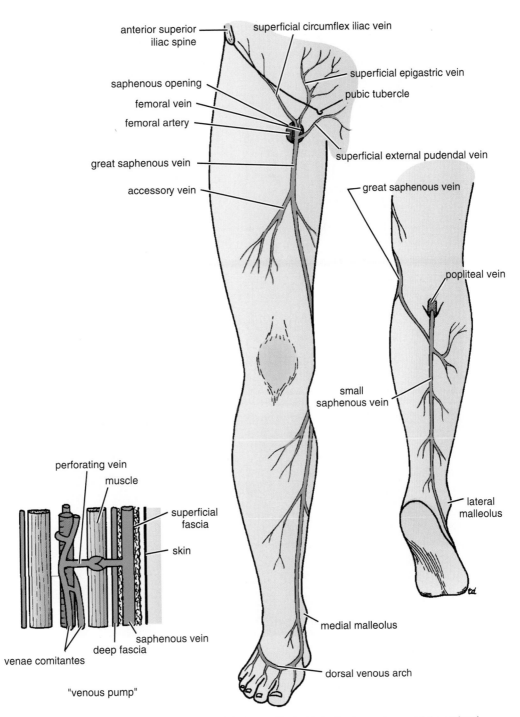

**Figure 9-17** Superficial veins of the right lower limb. Note the importance of the valved perforating veins in the "venous pump."

## Superficial Veins

### Dorsal Venous Network (Arch)

This network of veins lies on the dorsum of the foot (Fig. 9-17). The greater part of the blood from the whole foot drains into the network via digital veins and communicating veins that pass through the interosseous spaces. The dorsal venous

network is drained on the medial side by the great saphenous vein and on the lateral side by the small saphenous vein.

### Great Saphenous Vein

The great saphenous vein drains the medial end of the dorsal venous network of the foot and passes upward **directly in front of** the medial malleolus (Figs. 9-17 and 9-18). It then ascends

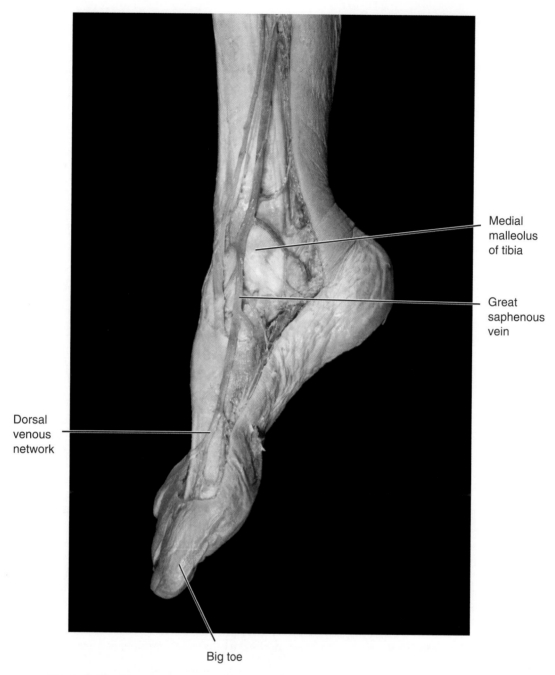

Medial
malleolus
of tibia

Great
saphenous
vein

Dorsal
venous
network

Big toe

**Figure 9-18** Dissection of the right ankle region showing the origin of the great saphenous vein from the dorsal venous arch. Note that the great saphenous vein ascends in front of the medial malleolus of the tibia.

in company with the saphenous nerve in the superficial fascia over the medial side of the leg. The vein passes behind the knee and curves forward around the medial side of the thigh. It passes through the lower part of the saphenous opening in the deep fascia and joins the femoral vein about 1.5 in. (4 cm) below and lateral to the pubic tubercle (Figs. 9-5 and 9-17).

The great saphenous vein possesses numerous valves and is connected to the small saphenous vein by one or two branches that pass behind the knee. Several **perforating veins** connect the great saphenous vein with the deep veins along the medial side of the calf (Fig. 9-17).

### Tributaries of the Great Saphenous Vein

The great saphenous vein receives a variable number of subcutaneous tributaries, and near its termination at the saphenous opening in the deep fascia, it receives three named tributaries (Figs. 9-5 and 9-17):

■ The **superficial circumflex iliac vein.**
■ The **superficial epigastric vein.**
■ The **superficial external pudendal vein.**

These veins correspond with the three branches of the femoral artery found in this region. An additional vein, known as the **accessory vein,** usually joins the main vein about the middle of the thigh or higher up at the saphenous opening.

## Small Saphenous Vein

The small saphenous vein arises from the lateral part of the dorsal venous network of the foot (Fig. 9-17). It ascends **behind** the lateral malleolus in company with the sural nerve. It follows the lateral border of the tendo calcaneus and then runs up the middle of the back of the leg. The vein pierces the deep fascia and passes between the two heads of the gastrocnemius muscle in the lower part of the popliteal fossa (Figs. 9-7 and 9-17); it ends in the popliteal vein. The small saphenous vein has numerous valves along its course.

## Tributaries of the Small Saphenous Vein

■ Numerous **small veins** from the back of the leg.
■ **Communicating veins** with the deep veins of the foot.
■ Important **anastomotic branches** that run upward and medially and join the great saphenous vein (Fig. 9-17).

The mode of termination of the small saphenous vein is subject to variation: It may join the popliteal vein; it may join the great saphenous vein; or it may split in two, one division joining the popliteal vein and the other joining the great saphenous vein.

# Deep Veins

## Venae Comitantes

The deep veins accompany the respective arteries as venae comitantes. The venae comitantes of the anterior and the posterior tibial arteries unite in the popliteal fossa to form the popliteal vein.

## Popliteal Vein

The popliteal vein is formed by the union of the venae comitantes of the anterior and the posterior tibial arteries. It ends by passing through the opening in the adductor magnus muscle to become the femoral vein (Fig. 9-4). The popliteal vein lies posterior to the popliteal artery and receives numerous tributaries, including the small saphenous vein at the lower end of the popliteal space.

## Femoral Vein

The femoral vein enters the thigh by passing through the opening in the adductor magnus as a continuation of the popliteal vein (Fig. 9-4). It ascends through the thigh, lying at first on the lateral side of the artery, then posterior to it, and finally on its medial side (Fig. 9-3). It leaves the thigh in the intermediate compartment of the femoral sheath and passes behind the inguinal ligament to become the external iliac vein.

### Tributaries of the Femoral Vein

The tributaries of the femoral vein are the **great saphenous vein** and veins that correspond to the branches of the femoral artery (Fig. 9-5). The great saphenous vein enters the femoral vein 1.5 in. (4 cm) below and lateral to the pubic tubercle (Fig. 9-17). (The superficial circumflex iliac vein, the superficial epigastric vein, and the external pudendal veins drain into the great saphenous vein.)

---

**P H Y S I O L O G I C   N O T E**

### Venous Pump of the Lower Extremity

Within the closed fascial compartments of the lower limb, the thin-walled, valved venae comitantes are subjected to intermittent pressure at rest and during exercise. The pulsations of the adjacent arteries help move the blood up the limb. However, the contractions of the large muscles within the compartments during exercise compress these deeply placed veins and force the blood up the limb.

The superficial saphenous veins, except near their termination, lie within the superficial fascia and are not subject to these compression forces. The valves in the perforating veins prevent the high-pressure venous blood from being forced outward into the low-pressure superficial veins. Moreover, as the muscles within the closed fascial compartments relax, venous blood is sucked from the superficial into the deep veins.

---

**P H Y S I O L O G I C   N O T E**

### Nervous Control of the Veins of the Lower Extremity

The smooth muscle in the walls of the superficial veins of the lower extremity is influenced by the activity of the sympathetic part of the autonomic nervous system. An increase in sympathetic activity results in an increase in the tone of the muscle and a narrowing of the lumen of the veins. The postganglionic nerve fibers reach the veins via the peripheral nerves, for example, the saphenous nerve for the great saphenous vein and the sural nerve for the short saphenous vein.

 # SURFACE ANATOMY OF THE ARTERIES AND VEINS OF THE LOWER EXTREMITY

## Arteries

The **femoral artery** enters the thigh behind the inguinal ligament (Fig. 9-1) at the midpoint of a line joining the symphysis pubis to the anterior superior iliac spine; its pulsations are easily felt (Fig. 9-19).

The **popliteal artery** can be felt by gentle palpation in the depths of the popliteal fossa, provided that the deep fascia is fully relaxed by passively flexing the knee joint.

The pulsations of the **dorsalis pedis artery** can be felt between the tendons of extensor hallucis longus and extensor digitorum longus, midway between the two malleoli on the front of the ankle (Fig. 9-20).

The pulsations of the posterior tibial artery can be felt by gentle palpation midway between the medial malleolus and the heel (Fig. 9-21). It lies here between the tendons of flexor digitorum longus and flexor hallucis longus.

## Veins

The **dorsal venous arch** or **plexus** can be seen on the dorsal surface of the foot proximal to the toes (Figs. 9-17 and 9-18). The **great saphenous vein** leaves the medial part of the plexus and passes upward **in front** of the medial malleolus (Fig. 9-18). The **small saphenous vein** drains the lateral part of the plexus and passes up **behind** the lateral malleolus (Fig. 9-17).

The **femoral vein** leaves the thigh by passing behind the inguinal ligament medial to the pulsating femoral artery (Figs. 9-3 and 9-19).

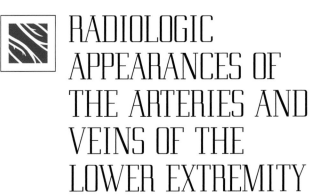 # RADIOLOGIC APPEARANCES OF THE ARTERIES AND VEINS OF THE LOWER EXTREMITY

Arteriography of the femoral and popliteal arteries is shown in Figures 9-22 and 9-23. Venography of the veins of the lower extremity is shown in Figure 9-24.

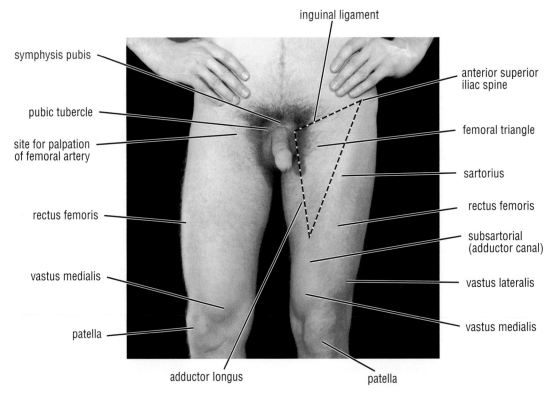

**Figure 9-19**  Anterior aspect of the thigh of a 27-year-old man. The *broken lines* indicate the boundaries of the femoral triangle. The right leg is laterally rotated at the hip joint.

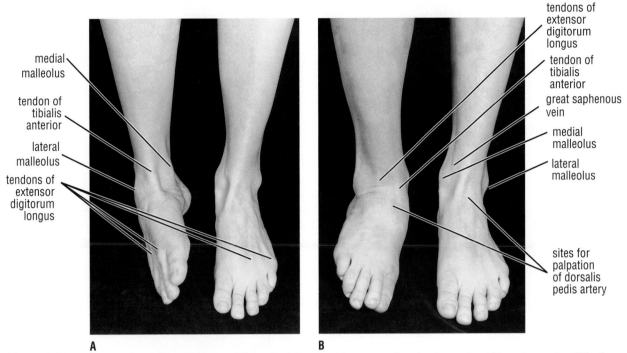

medial malleolus

tendon of tibialis anterior

lateral malleolus

tendons of extensor digitorum longus

tendons of extensor digitorum longus

tendon of tibialis anterior

great saphenous vein

medial malleolus

lateral malleolus

sites for palpation of dorsalis pedis artery

**A**    **B**

**Figure 9-20**   Anterior view of the ankles and feet of a 29-year-old woman showing inversion (**A**) and eversion (**B**) of the right foot.

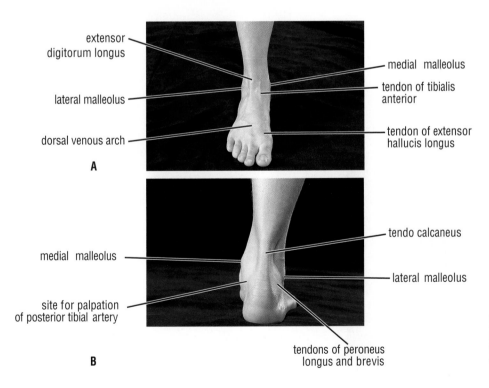

extensor digitorum longus

lateral malleolus

dorsal venous arch

**A**

medial malleolus

tendon of tibialis anterior

tendon of extensor hallucis longus

medial malleolus

site for palpation of posterior tibial artery

tendo calcaneus

lateral malleolus

tendons of peroneus longus and brevis

**B**

**Figure 9-21**   Anterior aspect (**A**) and posterior aspect (**B**) of the right foot and ankle of a 29-year-old woman.

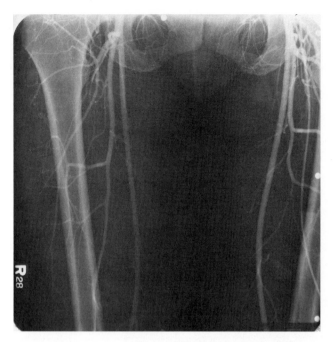

**Figure 9-22**  Arteriogram of the femoral artery and its branches.

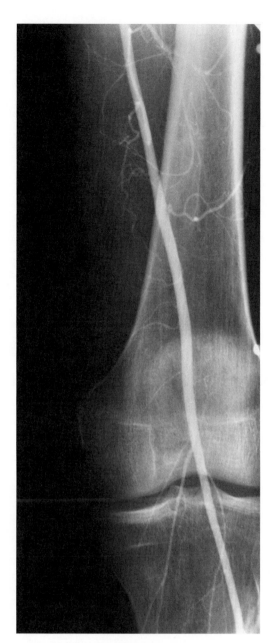

**Figure 9-23**  Arteriogram of the femoral artery and popliteal arteries and their branches.

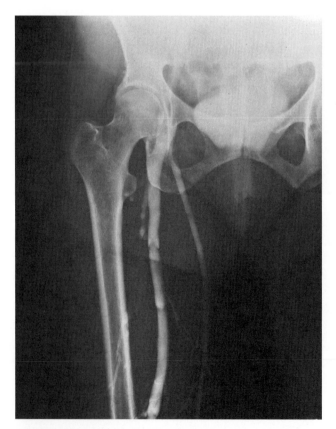

**Figure 9-24**  Venogram of the external iliac vein and the femoral vein and its major tributaries.

# Review Questions

## Completion Questions

**Select the phrase that BEST completes each statement.**

1. The femoral artery begins as a continuation of
   A. the internal iliac artery.
   B. the common iliac artery.
   C. the external iliac artery.
   D. the profunda artery.
   E. the superficial circumflex iliac artery.

2. The surface marking of the femoral artery as it emerges from beneath the inguinal ligament is
   A. midway between the anterior superior iliac spine and the pubic tubercle.
   B. midway between the anterior inferior iliac spine and the symphysis pubis.
   C. midway between the anterior inferior iliac spine and the pubic tubercle.
   D. midway between the anterior superior iliac spine and the symphysis pubis.
   E. midway between the iliac tubercle on the iliac crest and the pubic tubercle.

3. The femoral artery leaves the front of the thigh by passing through the
   A. hiatus in the adductor magnus muscle.
   B. obturator foramen.
   C. hiatus in the adductor brevis muscle.
   D. femoral canal.
   E. sartorius muscle.

4. The popliteal artery ends below at the lower border of the
   A. soleus muscle.
   B. soleal line of the tibia.
   C. popliteus muscle.
   D. capsule of the knee joint.
   E. superior tibiofibular joint.

5. The dorsalis pedis artery is a continuation of the
   A. peroneal artery.
   B. lateral plantar artery.
   C. medial plantar artery.
   D. anterior tibial artery.
   E. posterior tibial artery.

6. The pulse of the dorsalis pedis artery may be felt on the foot
   A. between the tendons of extensor hallucis longus and the most medial tendon of the extensor digitorum longus muscle.
   B. just medial to the medial malleolus of the tibia.
   C. between the tendons of extensor hallucis longus and the tibialis anterior muscle.
   D. between the tendons of extensor digitorum brevis and the extensor digitorum longus muscle.
   E. between the tendons of peroneus longus and peroneus brevis muscles.

## Multiple Choice Questions

**Select the BEST answer for each question.**

7. The following statements regarding the great saphenous vein are correct **except** which?
   A. It arises on the dorsum of the foot.
   B. It enters the leg by passing anterior to the medial malleolus.
   C. It drains into the femoral vein approximately 1.5 in. (3.8 cm) below and lateral to the pubic tubercle.
   D. It is accompanied by the saphenous nerve.
   E. It has no communication with the deep veins of the leg.

8. The following statements regarding the termination of the great saphenous vein are correct **except** which?
   A. It joins the femoral vein.
   B. It passes through the lower part of the saphenous opening in the deep fascia of the thigh.
   C. It is joined by the superficial circumflex iliac vein.
   D. At its termination, it lies 1.5 in. (4 cm) below and lateral to the pubic tubercle.
   E. It becomes reduced in size.

9. The following statements regarding the dorsal venous network (arch) are correct **except** which?
   A. It receives the greater part of the venous blood from the foot.
   B. It receives digital veins from the toes.
   C. On the medial side, it is drained by the great saphenous vein.
   D. It only rarely can be identified on the dorsum of the foot.
   E. On the lateral side, it is drained by the small saphenous vein.

10. The following statements regarding the femoral vein are correct **except** which?
    A. It is formed as a continuation of the popliteal vein.
    B. It is located in the intermediate compartment of the femoral sheath.
    C. It passes behind the inguinal ligament medial to the femoral artery.

D. The vein ends by passing through an opening in the adductor magnus muscle.

E. It receives the great saphenous vein.

## Fill in the Blank Questions

**Fill in the blank with the BEST answer.**

11. The _____ pulse can be felt midway between the medial malleolus and the heel.
    A. medial plantar artery
    B. posterior tibial artery
    C. dorsalis pedis artery

D. anterior tibial artery
E. peroneal artery

12. The dorsalis pedis artery enters the sole by passing through the _____.
    A. interval between the two heads of the fourth dorsal interosseous muscle
    B. interosseous membrane
    C. interval between the two heads of the first dorsal interosseous muscle
    D. gap between the second and third metatarsal bones
    E. plantar aponeurosis

# Answers and Explanations

1. **C is correct.** The femoral artery begins as a continuation of the external iliac artery as it passes behind the inguinal ligament (Fig. 9-2).

2. **D is correct.** The surface marking of the femoral artery as it emerges from beneath the inguinal ligament is midway between the anterior superior iliac spine and the symphysis pubis (Fig. 9-1).

3. **A is correct.** The femoral artery leaves the front of the thigh by passing through the hiatus in the adductor magnus muscle to become continuous with the popliteal artery behind the knee (Fig. 9-4).

4. **C is correct.** The popliteal artery ends below at the lower border of the popliteus muscle by dividing into the anterior and posterior tibial arteries (Fig. 9-8).

5. **D is correct.** The dorsalis pedis artery is a continuation of the anterior tibial artery in front of the ankle joint (Fig. 9-9).

6. **A is correct.** The pulse of the dorsalis pedis artery may be felt on the dorsum of the foot between the tendons of extensor hallucis longus and the most medial tendon of the extensor digitorum longus muscle (Figs. 9-9 and 9-20).

7. **E is incorrect.** The great saphenous vein has numerous communications with the deep veins of the leg through the valved perforating veins (Fig. 9-17).

8. **E is incorrect.** The great saphenous vein is largest near its termination.

9. **D is incorrect.** The dorsal venous arch can usually be identified on the dorsum of the foot as it lies in the superficial fascia (Figs. 9-17 and 9-18).

10. **D is incorrect.** The femoral vein does not end by passing through an opening in the adductor magnus muscle. The vein ascends the leg and ends by passing behind the inguinal ligament to become continuous with the external iliac vein (Fig. 9-1).

11. **B is correct.** The posterior tibial artery pulse can be felt by gentle palpation midway between the medial malleolus and the heel (Figs. 9-14 and 9-21). The plantar arteries are usually too small to feel their pulses at this location.

12. **C is correct.** The dorsalis pedis artery enters the sole by passing through the interval between the two heads of the first dorsal interosseous muscle (Figs. 9-11 and 9-12).

# The Lymphatic System

# 10

# The Lymph Vessels and Lymph Tissue

All clinical material relevant to this chapter can be found on the CD-ROM.

# Chapter Outline

The lymphatic system consists of lymphatic vessels and lymphatic tissues. The lymphatic vessels assist the capillaries and the venules of the cardiovascular system in the removal of tissue fluid from the tissues and return it to the blood. Lymphatic tissue is a type of connective tissue that contains large numbers of lymphocytes and is essential for the immunologic defenses of the body against bacteria and viruses.

The lymphatic system is of vital importance to medical personnel since it may be the channel along which infection may spread and a conduit for the spread of malignant disease. Moreover, lymphatic tissue itself may be the site of primary tumors which include lymphomas, Hodgkin's disease, and lymphatic leukemia.

The objective of this chapter is to review the main lymph vessels of the body and the main lymph organs. Especially important is the learning of the location of the regional lymph nodes that drain the different areas of the body. Only by having this knowledge will medical personnel be able to establish the site of the primary infection that has spread to a lymph node or the site of the primary tumor when the lymph node is enlarged by metastases.

# BASIC ANATOMY

## Lymph Vessels

The lymphatic capillaries begin as blind-ended tubes. They differ from blood capillaries in that they can absorb proteins and large particulate matter from the tissue spaces, whereas the fluid adsorbed by the blood capillaries is an aqueous solution of inorganic salts and sugar. **Lymph** is the name given to tissue fluid once it has entered a lymphatic vessel.

Lymph from the peripheral capillary plexuses passes into larger collecting vessels. At strategic points along the course of these vessels are small, ovoid masses of lymphatic tissue called **lymph nodes** (Fig.10-1). The direction of the flow of lymph is determined by the valves of the lymphatic vessels. Lymphatic vessels tend to run alongside blood

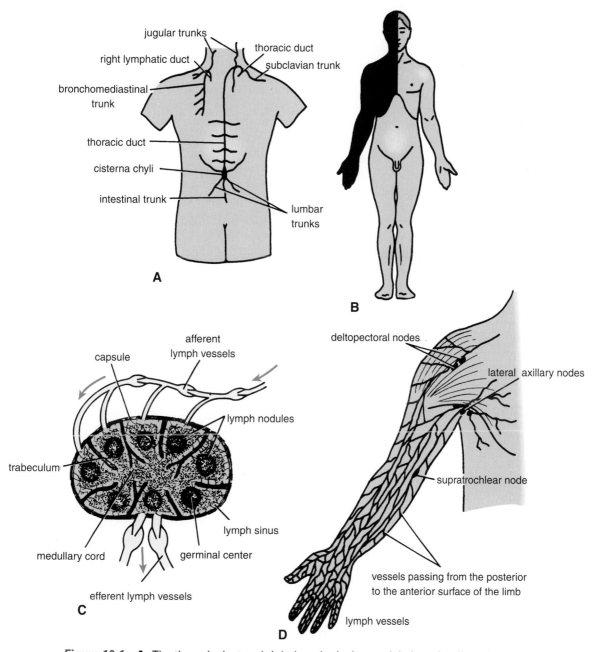

**Figure 10-1    A.** The thoracic duct and right lymphatic duct and their main tributaries.
**B.** The areas of body drained into the thoracic duct (*clear*) and right lymphatic duct (*black*).
**C.** General structure of a lymph node. **D.** Lymph vessels and nodes of the upper limb.

vessels. In the limbs, the superficial lymphatic vessels of the skin and subcutaneous tissue tend to follow the superficial veins; the deeper lymphatic vessels follow the deep arteries and veins.

Lymphatic vessels are found in most tissues and organs in the body but are absent from the central nervous system, the eyeball, the internal ear, the epidermis, the cartilage, and the bone.

The lymph from the greater part of the body reaches the blood via the thoracic duct. The lymph from the right side of the head and neck, the right upper limb, and the right side of the thorax, however, reaches the blood via the right lymphatic duct (Fig. 10-1).

## Major Lymphatic Ducts

### Thoracic Duct

The thoracic duct begins below in the abdomen at the level of the second lumbar vertebra as a dilated sac, the **cisterna chyli** (Fig.10-1). It ascends through the aortic opening in the diaphragm, on the right side of the descending aorta. It gradually crosses the median plane behind the esophagus and reaches the left border of the esophagus at the level of the fourth thoracic vertebra. It then ascends along the left border of the esophagus to the root of the neck. Here it turns laterally to the left behind the carotid sheath (containing the common carotid artery, vagus nerve, and internal jugular vein) and then turns down and crosses the subclavian artery to enter the beginning of the left brachiocephalic vein. It often enters the vein as several branches. At the termination, the thoracic duct receives the left jugular, subclavian, and mediastinal lymph trunks, although these trunks may drain independently into the neighboring large veins in this region.

**The thoracic duct thus conveys to the blood all the lymph from the lower limbs, pelvic cavity, abdominal cavity, left side of the thorax, left side of the head and neck, and left upper limb (Fig. 10-1).**

### PHYSIOLOGIC NOTE

### Factors Influencing Lymph Flow

The following factors influence the flow of lymph from the lymphatic capillaries to the bloodstream.

■ The pressure of tissue fluid.
■ The lymphatic capillary pump and the valves of lymphatic vessels.
■ The contraction of smooth muscle in the walls of lymphatic vessels.
■ The pressure on the thin-walled lymphatic vessels by surrounding skeletal muscles and the pulsations of adjacent arteries.
■ The thoracoabdominal pump during respiration.

### Right Lymphatic Duct

The right jugular, subclavian, and bronchomediastinal trunks, which drain the right side of the head and neck, the right upper limb, and the right side of the thorax, respectively, may join to form the right lymphatic duct (Fig. 10-1). This common duct, if present, is short, about ½ in. (1.3 cm) long, and opens into the beginning of the right brachiocephalic vein. Alternatively, the trunks open independently into the great veins at the root of the neck.

# Lymphatic Tissue

Lymphatic tissue has a basic network of reticular fibers and reticular cells. Lying within the spaces of the reticular network are large numbers of lymphocytes, which may or may not be associated with plasma cells. Lymphatic tissue is found in the following forms: the lymph nodes, the thymus, the spleen, and the lymph nodules.

## Lymph Nodes

Lymph nodes are widely distributed throughout the body and lie along the course of lymphatic vessels (Fig. 10-1). They are ovoid or kidney-shaped and vary in size from a few millimeters to as much as 2 cm in length. Lymph nodes are usually found in groups that are associated with the lymphatic drainage of a particular region or organ.

Each lymph node is surrounded by a tough fibrous capsule that sends into the node a number of fibrous partitions called **trabeculae**. Suspended from the trabeculae is a three-dimensional network of reticular fibers. The meshes of the network are filled with lymphocytes (Fig. 10-1). Lymph enters a lymph node by a number of valved **afferent lymphatic vessels** that pierce the **capsule** on its convex surface. The lymph moves through the **subscapular sinus** and then percolates through the meshwork until it reaches the **medulla**. It finally leaves the node by one or two **efferent lymphatic vessels** that emerge from the hilus.

### PHYSIOLOGIC NOTE

### Functions of a Lymph Node

Essentially, a lymph node serves as a filter. Any foreign particles in the lymph, whether bacterial or inert material, are trapped in the nodes as the lymph slowly diffuses through the meshwork of reticular fibers. A good example of this process is seen on examination of a bronchial lymph node. Inhaled carbon particles pass in the lymph from the aveolae and are trapped in the bronchial lymph nodes. The macrophages attached to the reticular fibers phagocytose the particles as they pass by.

Toxins in the entering lymph initiate an immune response from the lymphocytes. Specific antibodies (gamma globulins) to the antigen enter the lymph leaving the node. Eventually, the antibodies reach the blood in the neck via the thoracic duct and the right lymphatic duct, and they are widely disseminated throughout the body. T lymphocytes also respond to the antigens and develop a group of specific cytotoxic lymphocytes that are disseminated throughout the body.

The efferent lymph is therefore cleaner; it is richer in antibodies and contains more lymphocytes than the afferent lymph.

# Lymphatic Drainage of the Head and Neck

The lymph nodes of the head and neck (Fig. 10-2) are arranged as a regional collar that extends from below the chin to the back of the head and as a deep vertical terminal group that is embedded in the carotid sheath in the neck.

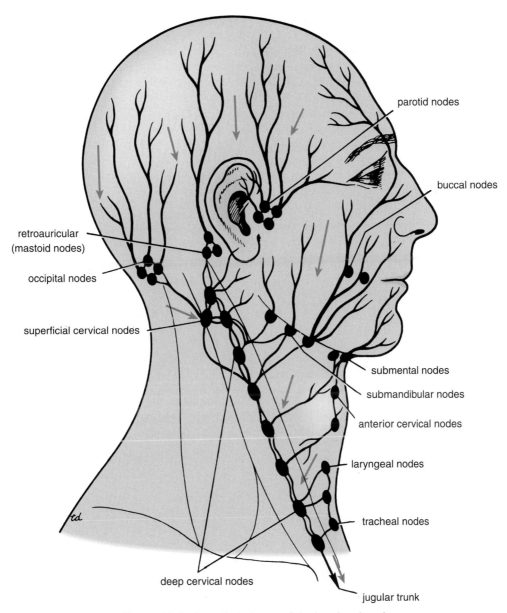

**Figure 10-2**  Lymph drainage of the head and neck.

## Regional Nodes

The regional nodes are arranged as follows:

- **Occipital nodes:** These are situated over the occipital bone on the back of the skull. They receive lymph from the back of the scalp.
- **Retroauricular (mastoid ) nodes:** These lie behind the ear over the mastoid process. They receive lymph from the scalp above the ear, the auricle, and the external auditory meatus.
- **Parotid nodes:** These are situated on or within the parotid salivary gland. They receive lymph from the scalp above the parotid gland, the eyelids, the parotid gland, the auricle, and the external auditory meatus.

- **Buccal (facial) nodes:** One or two nodes lie in the cheek over the buccinator muscle. They drain lymph that ultimately passes into the submandibular nodes.
- **Submandibular nodes:** These lie superficial to the submandibular salivary gland just below the lower margin of the jaw. They receive lymph from the front of the scalp, the nose, the cheek, the upper lip and the lower lip (except the central part), the frontal, maxillary, and ethmoid sinuses, the upper and lower teeth (except the lower incisors), the anterior two thirds of the tongue (except the tip), the floor of the mouth and vestibule, and the gums.
- **Submental nodes:** These lie in the submental triangle just below the chin. They drain lymph from the tip of the tongue, the floor of the anterior part of the mouth, the

incisor teeth, the center part of the lower lip, and the skin over the chin.

- **Anterior cervical nodes:** These lie along the course of the anterior jugular veins in the front of the neck. They receive lymph from the skin and superficial tissues of the front of the neck.
- **Superficial cervical nodes:** These lie along the course of the external jugular vein on the side of the neck. They drain lymph from the skin over the angle of the jaw, the skin over the lower part of the parotid gland, and the lobe of the ear.
- **Retropharyngeal nodes:** These lie behind the pharynx and in front of the vertebral column. They receive lymph from the nasal pharynx, the auditory tube, and the vertebral column.
- **Laryngeal nodes:** These lie in front of the larynx. They receive lymph from the larynx.
- **Tracheal (paratracheal) nodes:** These lie alongside the trachea. They receive lymph from neighboring structures, including the thyroid gland.

## Deep Cervical Nodes

The deep cervical nodes form a vertical chain along the course of the internal jugular vein within the carotid sheath (Fig. 10-2). They receive lymph from all the groups of regional nodes. The **jugulodigastric node,** which is located below and behind the angle of the jaw, is mainly concerned with drainage of the tonsil and the tongue. The **jugulo-omohyoid node,** which is situated close to the

omohyoid muscle, is mainly associated with drainage of the tongue.

The efferent lymph vessels from the deep cervical lymph nodes join to form the **jugular trunk,** which drains into the thoracic duct or the right lymphatic duct (Fig. 10-1).

# Lymphatic Drainage of the Upper Limb

The lymph vessels of the upper limb are arranged as superficial and deep sets. The superficial vessels ascend the limb in the superficial fascia and accompany the superficial veins. The deep lymph vessels lie deep to the deep fascia and follow the deep arteries and veins. All the lymph vessels of the upper limb ultimately drain into lymph nodes situated in the axilla.

## Axillary Lymph Nodes

The axillary lymph nodes (Fig. 10-3) drain lymph vessels from the entire upper limb. In addition, they drain vessels from the lateral part of the breast and the superficial lymph vessels from the thoracoabdominal walls above the level of the umbilicus.

The lymph nodes are 20 to 30 in number and are located as follows:

- **Anterior (pectoral) nodes:** These lie along the lower border of the pectoralis minor behind the pectoralis major

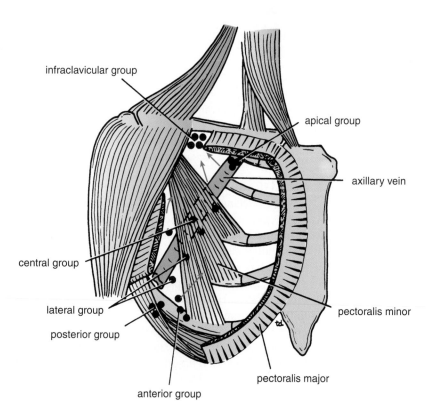

**Figure 10-3** Different groups of lymph nodes in the axilla.

muscle, i.e., behind the anterior wall of the axilla. They receive lymph from the lateral part of the breast and the superficial vessels from the thoracoabdominal wall above the level of the umbilicus.

- **Posterior (subscapular) nodes:** These lie in front of the subscapularis muscle on the posterior wall of the axilla. They receive superficial lymph vessels from the back, down as far as the level of the iliac crests.
- **Lateral nodes:** These lie along the medial side of the axillary vein. They receive most of the lymph vessels of the upper limb (except the superficial vessels draining the lateral side; see Infraclavicular nodes).
- **Central nodes:** These lie in the center of the axilla embedded in fat. They receive lymph from the above three groups.
- **Infraclavicular (deltopectoral) nodes:** These lie in the interval between the deltoid and pectoralis major muscles along the course of the cephalic vein. They receive lymph from the superficial vessels from the lateral side of the hand, the forearm, and the arm; the lymph vessels accompany the cephalic vein.

- **Apical nodes:** These lie at the apex of the axilla at the outer border of the first rib. They receive lymph from all the other axillary nodes. The apical nodes drain into the **subclavian trunk**, which on the left side, drains into the thoracic duct, but on the right side, it drains into the right lymphatic trunk.

## Supratrochlear (cubital) Lymph Node

This node lies in the superficial fascia in front of the elbow close to the trochlea of the humerus. It receives lymph from the third, fourth, and fifth fingers, the medial part of the hand, and the medial side of the forearm. The efferent lymph vessels ascend to the lateral axillary lymph nodes.

# Lymphatic Drainage of the Breast

The lymphatic drainage of the breast has been described in detail on page 92; only a brief summary will be given here.

The lateral quadrants of the breast drain into the anterior axillary nodes (Fig. 10-4). The medial quadrants

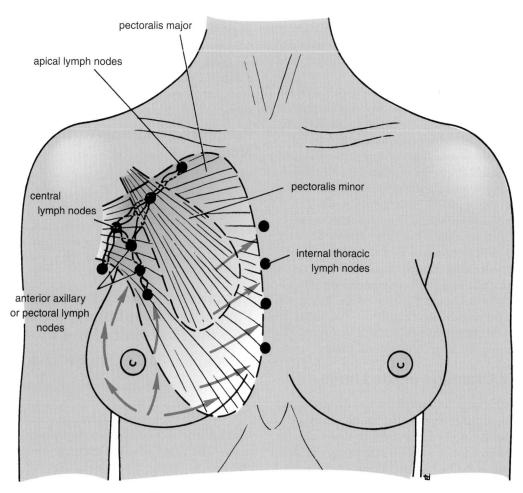

**Figure 10-4** Lymph drainage of the breast.

drain into lymph nodes lying alongside the internal thoracic artery within the thorax. Some superficial lymph vessels communicate with those of the opposite breast and those of the anterior abdominal wall. A few lymph vessels drain posteriorly into the posterior intercostal nodes.

# Lymphatic Drainage of the Thorax

The thoracic wall and the thoracic visceral are drained into the following groups of lymph nodes:

- **Axillary nodes:** The superficial lymph vessels from the skin and subcutaneous tissues of the anterior and posterior walls of the chest are drained into the axillary nodes. The anterior chest wall drains into the anterior axillary nodes, and the posterior chest wall drains into the posterior nodes.
- **Internal thoracic nodes:** These lie inside the thorax alongside the internal thoracic artery behind the costal cartilages. They receive lymph from the medial quadrants of the breast and the deep structures of the anterior thoracic and abdominal walls down as far as the umbilicus. They also receive lymph from the upper surfaces of the liver. The nodes drain into the bronchomediastinal trunk, which drains on the right side into the right lymphatic duct and on the left side into the thoracic duct.
- **Intercostal nodes:** These lie close to the heads of the ribs. They receive lymph from the intercostal spaces and some from the breast. The nodes drain into the thoracic and right lymphatic ducts.
- **Diaphragmatic nodes:** These lie on the upper surface of the diaphragm. They drain lymph from the diaphragm and the upper surface of the liver. The nodes drain into the internal thoracic and posterior mediastinal nodes.
- **Brachiocephalic nodes:** These lie with the brachiocephalic veins in the superior mediastinum. They drain lymph from the thyroid gland and the pericardium. The nodes drain into the bronchomediastinal trunks.
- **Posterior mediastinal nodes:** These lie close to the descending thoracic aorta. They drain lymph from the esophagus, the pericardium, and the diaphragmatic nodes. The nodes drain into the thoracic duct.
- **Tracheobronchial nodes:** These lie alongside the trachea, the main bronchi at the hilus of the lungs, and the bronchi within the lungs (Fig. 10-5). They drain the lymph from the lungs, the trachea, and the heart. The nodes drain into the bronchomediastinal trunk.

## Lymphatic Drainage of the Lungs

The lymph vessels begin in **superficial** and **deep plexuses** (Fig. 10-5). The superficial plexus lies beneath the visceral pleura on the surface of the lungs and then drains into the bronchopulmonary nodes at the hilus of the lung. The deep plexus accompanies the bronchi and drains via pulmonary nodes into bronchopulmonary nodes at the hilus. There is no free communication between the plexuses.

## Lymphatic Drainage of the Esophagus

The cervical part drains into the deep cervical nodes, the middle part drains into the posterior mediastinal nodes, and the lower abdominal part drains into the left gastric and then the celiac nodes.

# Lymphatic Drainage of the Abdomen and Pelvis

The lymph from most of the abdominal wall and from all the viscera (except a small part of the liver) drains into the thoracic duct. The lymph from the gastrointestinal tract, including the liver, the gallbladder, the pancreas, and the spleen, first drains into the preaortic lymph nodes (Fig. 10-6). The lymph from the remaining organs and the abdominal and pelvic walls first drains into the paraaortic (lateral aortic or lumbar ) nodes. The afferent lymph vessels to these nodes tend to accompany arteries, and there are a number of outlying small groups of nodes that the lymph passes through; these nodes take the names of the arteries along which they lie.

## Preaortic Lymph Nodes

These lie along the anterior surface of the abdominal part of the aorta (Fig. 10-6). Their efferent vessels form the **intestinal trunk**, which drains into the cisterna chyli. These nodes may be divided into the **celiac, superior mesenteric,** and **inferior mesenteric groups** that lie close to the origins of these arteries.

## Paraaortic (Lateral Aortic or Lumbar) Nodes

These are right and left groups that lie alongside the abdominal part of the aorta (Fig. 10-6). Their efferent vessels form the **right and left lumbar trunks** that drain into the cisterna chyli. Lymph from the pelvis passes first through a number of peripheral nodes that are associated with arteries. These nodes include **internal iliac, external iliac,** and **common iliac nodes** (Fig. 10-6).

## Lymphatic Drainage of the Abdominal Part of the Esophagus, the Stomach, and the First Half of the Duodenum

### Abdominal Part of the Esophagus

This part of the esophagus drains into the left gastric nodes.

### Stomach

The left half of the lesser curvature drains into left gastric nodes. The right half of the lesser curvature drains into the right gastric nodes. The fundus and the left half of the greater curvature drain into the left gastroepiploic nodes and the pancreaticosplenic nodes. The right half of the greater

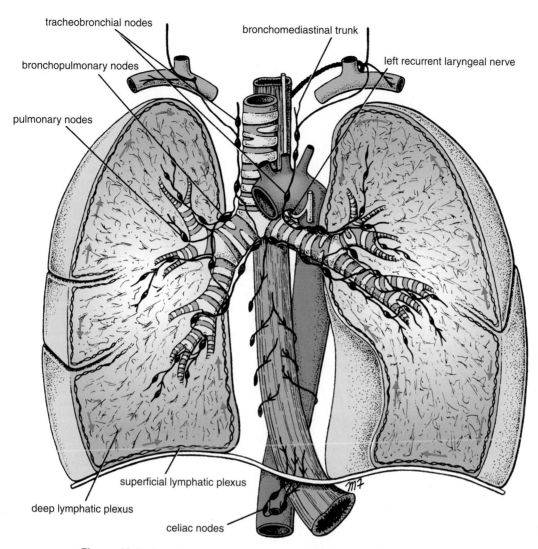

tracheobronchial nodes

bronchomediastinal trunk

bronchopulmonary nodes

left recurrent laryngeal nerve

pulmonary nodes

superficial lymphatic plexus

deep lymphatic plexus

celiac nodes

**Figure 10-5**    Lymph drainage of the lung and lower end of the esophagus.

curvature drains into the right gastroepiploic nodes and the gastroduodenal nodes.

## First Half of the Duodenum

The upper half of the duodenum drains into the pyloric nodes (superior pancreaticoduodenal nodes) and the gastroduodenal nodes. All these nodes drain into the celiac nodes.

## Lymphatic Drainage of the Lower Half of the Duodenum, the Jejunum, the Ileum, the Cecum, the Appendix, the Ascending Colon, and the Proximal Two Thirds of the Transverse Colon

The lymph passes through lymph nodes that lie along the terminal branches of the superior mesenteric artery. All these nodes drain finally into the superior mesenteric nodes.

## Lymphatic Drainage of the Distal Third of the Transverse Colon, the Descending Colon, the Sigmoid Colon, the Rectum, and the Upper Half of the Anal Canal

The lymph passes through lymph nodes that lie along the terminal branches of the inferior mesenteric artery. All these nodes drain finally into the inferior mesenteric nodes.

## Lymphatic Drainage of the Liver

The lymph passes to the **hepatic nodes** in the porta hepatis and then to the celiac nodes. The bare areas of the liver drain through the diaphragmatic nodes to the posterior mediastinal nodes.

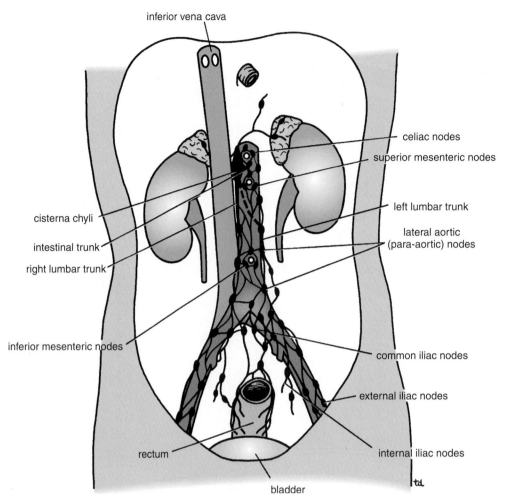

**Figure 10-6**    Lymph vessels and nodes on the posterior abdominal wall.

## Lymphatic Drainage of the Pancreas

The lymph drains to nodes that lie along the arterial supply, namely the pancreaticoduodenal nodes, the splenic nodes, and, finally, the celiac nodes.

## Lymphatic Drainage of the Spleen

The lymph drains into the pancreaticosplenic nodes.

## Lymphatic Drainage of the Suprarenal Gland and the Kidney

The lymph drains into the lateral aortic nodes.

## Lymphatic Drainage of the Bladder

The lymph drains into the internal and external iliac nodes.

## Lymphatic Drainage of the Testis and Ovary

The lymph ascends on the posterior abdominal wall in company with the gonadal blood vessels to drain into the lateral aortic nodes at the level of the first lumbar vertebra.

## Lymphatic Drainage of the Prostate

The lymph drains into the internal iliac nodes.

## Lymphatic Drainage of the Penis and Scrotum

The lymph from the superficial tissues drains into the superficial inguinal nodes. The lymph from the glans penis passes to the deep inguinal and external iliac nodes. The lymph from the erectile tissue passes to the internal iliac nodes.

## Lymphatic Drainage of the Uterus

The lymph from the body and cervix passes to the internal and external iliac nodes. Lymph vessels from the fundus accompany the ovarian vessels to the lateral aortic lymph nodes at the level of the first lumbar vertebra. A few lymph vessels pass with the round ligament of the uterus to the superficial inguinal nodes.

## Lymphatic Drainage of the Vagina

Lymph from the upper part of the vagina drains into the internal and external iliac nodes. The lymph from the orifice and from the vulva drains into the superficial inguinal nodes.

## Lymph Drainage of the Lower Half of the Anal Canal

Lymph descends to the anus and then drains into the superficial inguinal nodes.

# Lymphatic Drainage of the Lower Limb

The lymph vessels of the lower limb are arranged as superficial and deep sets (Fig. 10-7). The superficial vessels ascend the limb with the superficial veins. The deep lymph vessels lie deep to the deep fascia and follow the deep arteries and veins. All the lymph vessels from the lower limb ultimately drain into the deep inguinal group of nodes that are situated in the groin.

## Superficial Inguinal Lymph Nodes

These nodes lie in the superficial fascia just below the inguinal ligament (Fig. 10-7). They may be divided into a horizontal and a vertical group. The **horizontal group** receives lymph from the superficial lymph vessels of the anterior abdominal wall below the umbilicus, from the perineum, the external genitalia of both sexes (but not the testes), and the lower half of the anal canal. They also receive lymph from the superficial lymph vessels of the buttocks.

The **vertical group** lies along the terminal part of the great saphenous vein and receives the majority of the superficial lymph vessels of the lower limb except from the back and lateral side of the calf and lateral side of the foot.

The superficial inguinal nodes all drain into the deep inguinal nodes.

## Deep Inguinal Nodes

There are usually three nodes situated along the medial side of the femoral vein and in the femoral canal (Fig. 10-7). They receive all the lymph from the superficial inguinal nodes and from all the deep structures of the lower limb. The efferent lymph vessels ascend through the femoral canal into the abdominal cavity and drain into the external iliac nodes.

## Popliteal Nodes

These nodes lie in the popliteal fossa behind the knee. They receive superficial lymph vessels that accompany the small saphenous vein from the lateral side of the foot and the back and lateral side of the calf. They also receive lymph from the

deep structures of the leg below the knee. The efferent vessels drain upward to the deep inguinal nodes.

# Thymus

The thymus is a flat bilobed structure lying in the superior mediastinum and the anterior mediastinum of the thorax. In the newborn infant, it reaches its largest size relative to the size of the body. It continues to grow until puberty but thereafter undergoes involution. Lymphatic vessels do not drain into it, but large numbers leave it. It is one of the most important organs concerned with the defense against infection and is the site for the development of T (thymic) lymphocytes.

The blood supply is the inferior thyroid and internal thoracic arteries.

# Spleen

The spleen is the largest single mass of lymphatic tissue in the body. It lies in the abdomen just beneath the left half of the diaphragm. Unlike lymph nodes, it does not lie along the course of lymphatic vessels but along the course of the systemic circulation, having a large splenic artery and a splenic vein.

The interior of the spleen is filled with **splenic pulp.** On section, the spleen is seen to possess two types of pulp, white and red. The **white pulp** forms small grayish islands less than a millimeter in diameter that are scattered throughout the remainder of the pulp, which is the red pulp. The white pulp consists of sheaths of lymphoid tissue around small branches of the splenic artery. The **red pulp** consists of blood cells circulating through the network of reticular fibers.

The spleen is reddish in color and ovoid in shape, with a notched anterior border. It is surrounded by peritoneum that passes from the hilus to the stomach as the **gastrosplenic omentum (ligament)** and to the left kidney as the **lienorenal ligament**. The gastrosplenic omentum contains the short gastric and left gastroepiploic vessels, and the lienorenal ligament contains the splenic vessels and the tail of the pancreas.

The spleen is related **anteriorly** to the stomach, the tail of the pancreas, and the left colic flexure; the left kidney lies along its medial border. Posteriorly lie the diaphragm, the left lung, and the ninth, tenth, and eleventh ribs.

The arterial supply is the splenic artery branch of the celiac artery. The splenic vein joins the superior mesenteric vein to form the portal vein.

# Lymphatic Nodules

Lymphatic nodules are rounded collections of lymphatic tissue found in the cortex of lymph nodes (Fig. 10-1), in the spleen, and in the connective tissue of the mucous membrane of the respiratory system and intestinal system.

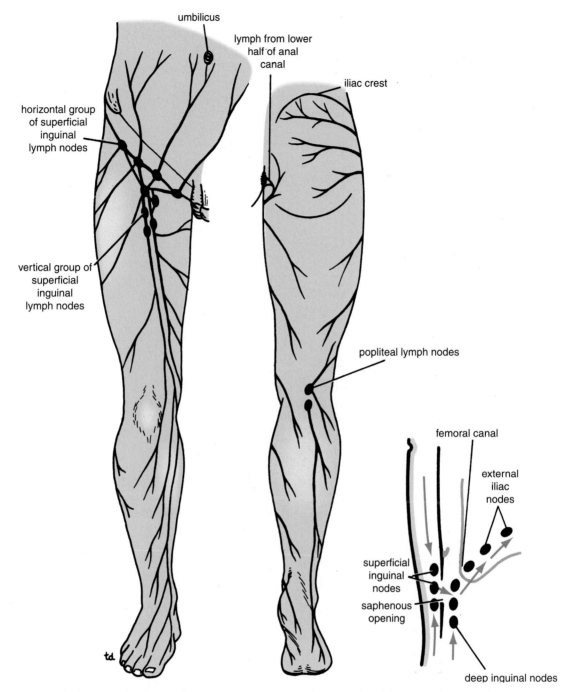

**Figure 10-7**   Lymph drainage for the superficial tissues of the right lower limb and the abdominal walls below the level of the umbilicus. Note the arrangement of the superficial and deep inguinal lymph nodes and their relationship to the saphenous opening in the deep fascia. Note also that all lymph from these nodes ultimately drains into the external iliac nodes via the femoral canal.

# Tonsils

The tonsils form a discontinuous ring of lymphatic tissue around the entrance of the mouth and nose into the pharynx. They consist of paired palatine tonsils, paired tubal tonsils, a lingual tonsil, and a nasopharyngeal tonsil; there is also some lymphatic tissue in the soft palate

(Fig.10-8). The tonsils are strategically placed at the entrance to the respiratory and digestive systems and respond immunologically to foreign antigens entering these systems.

*Additional details of the lymph drainage of organs and tissues of the body is given, where appropriate, throughout the different systems in the text.*

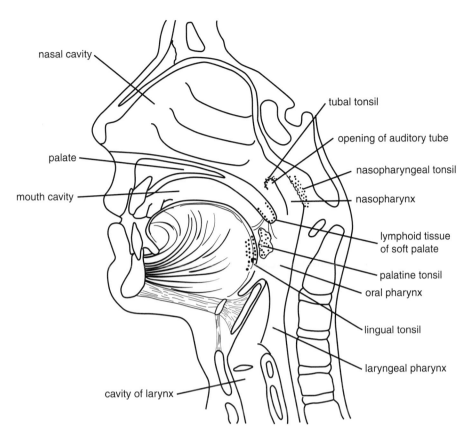

nasal cavity

tubal tonsil

opening of auditory tube

palate

nasopharyngeal tonsil

mouth cavity

nasopharynx

lymphoid tissue
of soft palate

palatine tonsil

oral pharynx

lingual tonsil

laryngeal pharynx

cavity of larynx

**Figure 10-8** Distribution of lymphatic tissue around the entrance of the mouth and nose into the pharynx.

# Review Questions

## Completion Questions

### Select the phrase that BEST completes each statement.

1. Lymphatic capillaries begin as
   A. a continuation of a venule.
   B. blind-ended tubes.
   C. a continuation of an arteriole.
   D. a branch of a blood capillary.
   E. small open-ended tubes.

2. Lymphatic capillaries differ from blood capillaries in that they can absorb
   A. water.
   B. tissue fluid containing inorganic salts only.
   C. tissue fluid containing proteins only.
   D. tissue fluid containing proteins and large particulate matter.
   E. tissue fluid containing sugar only.

3. Lymph is the name given to
   A. tissue fluid outside the wall of the lymph capillary.
   B. tissue fluid that is present only in the thoracic duct.
   C. tissue fluid once it has entered a lymphatic vessel.

   D. all body fluids outside a blood vessel.
   E. fluids in the semicircular canals of the inner ear.

4. A lymph node is found
   A. along the course of lymph vessels.
   B. along the course of blood vessels.
   C. always as a single mass of lymphoid tissue.
   D. at the commencement of a lymphatic capillary.
   E. without a fibrous capsule.

5. Lymphatic vessels are absent from the following tissues
   A. liver.
   B. kidney and pancreas.
   C. thyroid and parathyroid glands.
   D. testis and ovary.
   E. central nervous system.

6. The thoracic duct begins
   A. as a dilated sac, the cisterna chyli.
   B. in the pelvis by draining the internal iliac nodes.
   C. as a continuation of the intestinal trunk.
   D. in front of the tenth thoracic vertebra.
   E. as a continuation of the right lumbar trunk.

7. The thoracic duct ends by joining the
   A. right brachiocephalic vein.
   B. right subclavian vein.
   C. superior vena cava.
   D. left brachiocephalic vein.
   E. right jugular trunk.

## Matching Questions

**For each of the lymphatic structures listed below, select an appropriate structure or function.**

8. Lymph capillary

9. Thoracic duct

10. Right lymphatic duct

11. Lymph node

   A. Present in the central nervous system.
   B. Drains lymph directly from the tissues.
   C. Contains lymphatic tissue and has both afferent and efferent vessels.
   D. Drains lymph from the right side of the head and neck, the right upper limb, and the right side of the thorax.
   E. Drains lymph from the right side of the abdomen.

**For each of the lymph nodes listed below, select an appropriate area of the body that is drained by the nodes.**

12. Submental nodes

13. Jugulodigastric node

14. Jugulo-omohyoid node

15. Parotid nodes

   A. Lateral side of the tongue
   B. Tonsil
   C. Tip of the tongue
   D. Eyelids
   E. Back of the scalp

## Completion Questions

**Select the phrase that BEST completes each statement.**

16. The lymph from the upper lateral quadrant of the breast drains mainly into the
   A. lateral axillary nodes.
   B. internal thoracic nodes.
   C. posterior axillary nodes.
   D. anterior axillary nodes.
   E. deltopectoral group of nodes.

17. The superficial lymphatic plexus in the lungs drains into the
   A. internal thoracic nodes.
   B. intercostal nodes.
   C. brachiocephalic nodes.
   D. bronchopulmonary nodes.
   E. diaphragmatic nodes.

18. The lymphatic drainage of the abdominal part of the esophagus is into the
   A. superior mesenteric nodes.
   B. left gastric nodes.
   C. superior pancreaticoduodenal nodes.
   D. left lumbar nodes.
   E. diaphragmatic nodes.

19. The lymphatic drainage of the right kidney is into the
   A. preaortic nodes.
   B. diaphragmatic nodes.
   C. lateral aortic nodes.
   D. celiac nodes.
   E. superior mesenteric nodes.

20. The lymphatic drainage of the left testis is into the
   A. preaortic nodes at the level of the first lumbar vertebra.
   B. left superficial inguinal nodes.
   C. left deep inguinal nodes.
   D. lateral aortic nodes at the pelvic inlet.
   E. lateral aortic nodes at the level of the first lumbar vertebra.

21. The lymphatic drainage of the descending colon is into the
   A. celiac nodes.
   B. superior mesenteric nodes.
   C. inferior mesenteric nodes.
   D. left lateral aortic nodes.
   E. left common iliac nodes.

22. The lymphatic drainage of the skin on the medial side of the big toe is directly into the
   A. medial group of the horizontal group of superficial inguinal nodes.
   B. lateral group of the horizontal group of superficial inguinal nodes.
   C. vertical group of superficial inguinal nodes.
   D. deep inguinal lymph nodes.
   E. external iliac lymph nodes.

23. The thymus reaches its maximum size relative to the size of the body at
   A. birth.
   B. puberty.
   C. 40 years of age.
   D. after 65 years of age.
   E. during pregnancy.

## Multiple Choice Questions

24. The spleen has the following important characteristics **except** which?
    A. It is the largest single mass of lymphoid tissue in the body.
    B. It does not lie along the course of lymphatic vessels.
    C. It has a large splenic artery and vein.
    D. It is surrounded by peritoneum.
    E. It is composed only of red pulp.

25. Lymphatic nodules are found in the following locations **except** which?
    A. In the cortex of lymph nodes.
    B. In the spleen.
    C. In the Peyers patches of the small intestine.
    D. In the walls of the ureters.
    E. In the mucous membrane of the respiratory system.

# Answers and Explanations

1. **B** is correct. Lymphatic capillaries begin in the tissues as blind-ended tubes.

2. **D** is correct. Lymphatic capillaries differ from blood capillaries in that they can absorb tissue fluid containing proteins and large particulate matter. Blood capillaries, on the other hand, can only absorb tissue fluid containing inorganic salts and sugar.

3. **C** is correct. Lymph is the name given to tissue fluid once it has entered a lymphatic vessel and applies to any lymphatic vessel not just the thoracic duct.

4. **A** is correct. A lymph node is found along the course of lymph vessels either singly or in groups (Fig. 10-1).

5. **E** is correct. Lymphatic vessels are absent from the following tissues: the central nervous system, the eyeball, the internal ear, the epidermis, the cartilage, and the bone.

6. **A** is correct. The thoracic duct begins below in the abdomen as a dilated sac, the cisterna chyli, in front of the body of the second lumbar vertebra on the right side of the aorta (Fig. 10-6).

7. **D** is correct. The thoracic duct ends by joining the left brachiocephalic vein.

8. **B** is correct. The lymph capillary drains lymph directly from the tissues.

9. **E** is correct. The thoracic duct conveys to the blood all the lymph from the lower limbs, pelvic cavity, left side of the thorax, and left side of the head, neck, and left upper limb (Fig. 10-1).

10. **D** is correct. The right lymphatic duct drains lymph from the right side of the head and neck, the right upper limb, and the right side of the thorax (Fig. 10-1).

11. **C** is correct. The lymph node contains lymphatic tissue and has both afferent and efferent lymphatic vessels (Fig. 10-1).

12. **C** is correct. The submental nodes form part of the collar of regional lymph nodes around the neck (Fig. 10-2). They are located just beneath the chin and drain the tip of the tongue, the floor of the anterior part of the mouth, the incisor teeth, the central part of the lower lip, and the skin covering the chin.

13. **B** is correct. The jugulodigastric node is a member of the deep cervical group of lymph nodes lying in the carotid sheath alongside the internal jugular vein. It is named by the fact that the posterior belly of the digastric muscle crosses the node at this site. The jugulodigastric node drains the tonsil.

14. **A** is correct. The jugulo-omohyoid node is a member of the deep cervical group of nodes lying in the carotid sheath alongside the internal jugular vein. It is named by the fact that the omohyoid muscle crosses the node at this site. The jugulo-omohyoid node drains the side of the tongue.

15. **D** is correct. The parotid nodes, which lie on the surface and inside the parotid salivary gland (Fig. 10-2), drain lymph from the scalp above the parotid gland, the eyelids, the parotid gland itself, the auricle, and the external auditory meatus.

16. **D** is correct. The lymph from the upper lateral quadrant of the breast drains into the anterior axillary nodes (Fig. 10-4).

17. **D** is correct. The superficial lymphatic plexus in the lungs drains into the bronchopulmonary nodes at the hilus of the lung (Fig. 10-5).

18. **B** is correct. The lymphatic drainage of the abdominal part of the esophagus is into the left gastric nodes located alongside the left gastric artery.

19. **C** is correct. The lymphatic drainage of the right kidney is into the lateral aortic nodes (Fig. 10-6).

20. **E** is correct. The lymphatic drainage of the left testis is into the lateral aortic nodes at the level of the first lumbar vertebra.

21. **C** is correct. The lymphatic drainage of the descending colon is into the inferior mesenteric nodes lying around the origin of the inferior mesenteric artery.

22. **C** is correct. The lymphatic drainage of the skin on the medial side of the big toe is directly into the vertical group of superficial inguinal nodes (Fig. 10-7).

23. **A** is correct. The thymus reaches its maximum size relative to the size of the body at birth.

24. **E** is correct. The spleen has both white pulp (lymphoid tissue) and red pulp (blood) flowing through it.

25. **D** is correct. Lymphatic nodules are not found in the walls of the ureters.

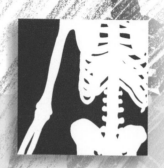

# The Musculoskeletal System

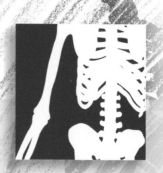

# 11

# Bones and Cartilage

All clinical material relevant to this chapter can be found on the CD-ROM.

# Chapter Outline

# Chapter Outline

The bony framework of the body, particularly the skull, vertebral column, and the pelvis, gives support and protection for some of the softer organs. The bones provide attachments for skeletal muscles and serve as levers. In certain regions, the bones are supplemented by cartilage.

Disease of bones is a common finding in clinical practice: fractures; osteoporosis; tumors, such as sarcomas; rickets; osteomalacia; and Paget's disease, to name just a few. In addition, bone marrow is the site of many marrow dysplasias. This chapter provides the basic material for bones and cartilage that a practicing medical professional needs to know in order to make an informed diagnosis and provide adequate treatment.

# BASIC ANATOMY

## Bones

Bone is a living tissue capable of changing its structure as the result of the stresses to which it is subjected. It is continuously remodeling with new bone formation and resorption. Like other connective tissues, bone consists of cells, fibers, and matrix. It is hard because of the calcification of its extracellular matrix and possesses a degree of elasticity because of the presence of organic fibers. Bone has a protective function; the skull and vertebral column, for example, protect the brain and spinal cord from injury; the sternum and ribs protect the thoracic and upper abdominal viscera (Fig. 11-1). It serves as a lever, as seen in the long bones of the limbs, and as an important storage area for calcium salts. It houses and protects within its cavities the delicate blood-forming bone marrow.

Bone exists in two forms: **compact** and **cancellous.** Compact bone appears as a solid mass; cancellous bone consists of a branching network of **trabeculae** (Fig. 11-2). The trabeculae are arranged in such a manner as to resist the stresses and strains to which the bone is exposed.

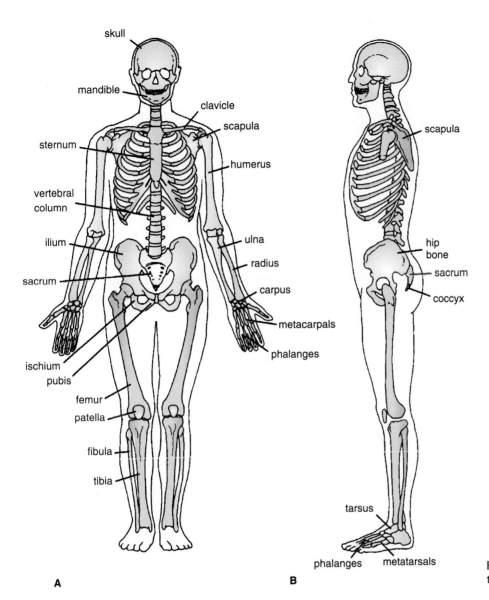

**Figure 11-1** The skeleton. **A.** Anterior view. **B.** Lateral view.

## Classification of Bones

Bones may be classified regionally or according to their general shape. The regional classification is summarized in Table 11-1. Bones are grouped as follows based on their general shape: long bones, short bones, flat bones, irregular bones, and sesamoid bones.

## Long Bones

Long bones are found in the limbs (e.g., the humerus, femur, metacarpals, metatarsals, and phalanges). Their length is greater than their breadth. They have a tubular shaft, the **diaphysis,** and usually an **epiphysis** at each end. During the growing phase, the diaphysis is separated from the epiphysis by an **epiphyseal cartilage.** The part of the diaphysis that lies adjacent to the epiphyseal cartilage is called the **metaphysis.** The shaft has a central **marrow cavity** containing **bone marrow.** The outer part of the shaft is composed of compact bone that is covered by a connective tissue sheath, the **periosteum.**

The ends of long bones are composed of cancellous bone surrounded by a thin layer of compact bone. The articular surfaces of the ends of the bones are covered by hyaline cartilage.

## Short Bones

Short bones are found in the hand and foot (e.g., the scaphoid, lunate, talus, and calcaneum). They are roughly cuboidal in shape and are composed of cancellous bone surrounded by a thin layer of compact bone. Short bones are covered with periosteum, and the articular surfaces are covered by hyaline cartilage.

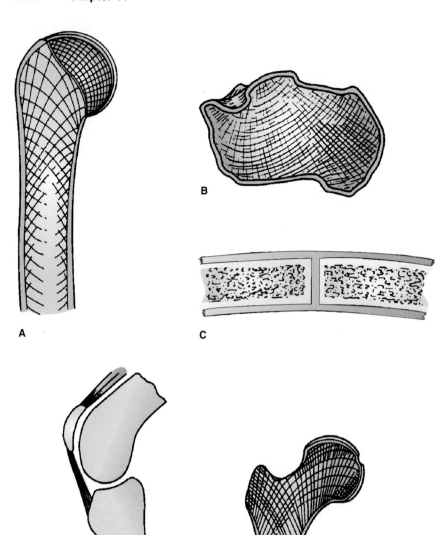

**Figure 11-2** Sections of different types of bones. **A.** Long bone (humerus). **B.** Irregular bone (calcaneum). **C.** Flat bone (two parietal bones separated by the sagittal suture). **D.** Sesamoid bone (patella). **E.** Note arrangement of trabeculae to act as struts to resist both compression and tension forces in the upper end of the femur.

## Flat Bones

Flat bones are found in the vault of the skull (e.g., the frontal and parietal bones). They are composed of thin inner and outer layers of compact bone, the **tables**, separated by a layer of cancellous bone, the **diploë**. The scapulae, although irregular, are included in this group.

## Irregular Bones

Irregular bones include those not assigned to the previous groups (e.g., the bones of the skull, the vertebrae, and the pelvic bones). They are composed of a thin shell of compact bone with an interior made up of cancellous bone.

## Sesamoid Bones

Sesamoid bones are small nodules of bone that are found in certain tendons where they rub over bony surfaces. The greater part of a sesamoid bone is buried in the tendon, and the free surface is covered with cartilage. The largest sesamoid bone is the patella, which is located in the tendon of the quadriceps femoris. Other examples are found in the tendons of the flexor pollicis brevis and flexor hallucis brevis. The function of a sesamoid bone is to reduce friction on the tendon; it can also alter the direction of pull of a tendon.

## Surface Markings of Bones

The surfaces of bones show various markings or irregularities. Where bands of fascia, ligaments, tendons, or aponeuroses are attached to bone, the surface is raised or roughened. These roughenings are not present at birth. They appear at puberty and become progressively more obvious during adult life. The pull of these fibrous structures

| Table 11-1 | Regional Classification of Bones | |
|---|---|---|
| **Region of Skeleton** | **Number of Bones** | |
| Axial skeleton | | |
| Skull | | |
|    Cranium | 8 | |
|    Face | 14 | |
|    Auditory ossicles | 6 | |
|    Hyoid | 1 | |
|    Vertebrae | 26 | |
|    (including sacrum and coccyx) | | |
|    Sternum | 1 | |
|    Ribs | 24 | |
| Appendicular skeleton | | |
|   Shoulder girdles | | |
|     Clavicle | 2 | |
|     Scapula | 2 | |
|   Upper extremities | | |
|    Humerus | 2 | |
|    Radius | 2 | |
|    Ulna | 2 | |
|    Carpals | 16 | |
|    Metacarpals | 10 | |
|    Phalanges | 28 | |
|   Pelvic girdle | | |
|    Hip bone | 2 | |
|   Lower extremities | | |
|    Femur | 2 | |
|    Patella | 2 | |
|    Fibula | 2 | |
|    Tibia | 2 | |
|    Tarsals | 14 | |
|    Metatarsals | 10 | |
|    Phalanges | 28 | |
| | 206 | |

From Snell RS: Clinical Anatomy. 7th Ed. Philadelphia: Lippincott Williams & Wilkins, 2004, p. 37.

causes the periosteum to be raised and new bone to be deposited beneath.

In certain situations, the surface markings are large and are given special names. Some of the more important markings are summarized in Table 11-2.

## Bone Marrow

Bone marrow occupies the marrow cavity in long and short bones and the interstices of the cancellous bone in flat and irregular bones.

**PHYSIOLOGIC NOTE**

### The Changing of Marrow Type with Age

At birth, the marrow of all the bones of the body is red and hematopoietic. This blood-forming activity gradually lessens with age, and the red marrow is replaced by yellow marrow. At 7 years of age, yellow marrow begins to appear in the distal bones of the limbs. This replacement of marrow gradually moves proximally, so that by the time the person becomes adult, red marrow is restricted to the bones of the skull, the vertebral column, the thoracic cage, the girdle bones, and the head of the humerus and femur.

All bone surfaces, other than the articulating surfaces, are covered by a thick layer of fibrous tissue called the **periosteum.** The periosteum has an abundant vascular supply, and the cells on its deeper surface are osteogenic. The periosteum is particularly well united to bone at sites where muscles, tendons, and ligaments are attached to bone. Bundles of collagen fibers known as Sharpey's fibers extend from the periosteum into the underlying bone. The periosteum receives a rich nerve supply and is very sensitive.

## Periosteum of Bone

**EMBRYOLOGIC NOTE**

### Development of Bone

See Chapter 1, page 24.

# Cartilage

Cartilage is a form of connective tissue in which the cells and fibers are embedded in a gel-like matrix, the latter being responsible for its firmness and resilience. Except on the exposed surfaces in joints, a fibrous membrane called the **perichondrium** covers the cartilage. There are three types of cartilage:

- **Hyaline cartilage** has a high proportion of amorphous matrix that has the same refractive index as the fibers embedded in it. Throughout childhood and adolescence, it plays an important part in the growth in length of long bones (epiphyseal plates are composed of hyaline cartilage). It has a great resistance to wear and covers the articular surfaces of nearly all synovial joints. Hyaline cartilage is incapable of repair when fractured; the defect is filled with fibrous tissue.

- **Fibrocartilage** has many collagen fibers embedded in a small amount of matrix and is found in the discs within

| Table 11-2 | Surface Markings of Bones |
|---|---|

| Bone Marking | Example |
|---|---|
| Linear elevation | |
|    Line | Superior nuchal line of the occipital bone |
|    Ridge | The medial and lateral supracondylar ridges of the humerus |
|    Crest | The iliac crest of the hip bone |
| Rounded elevation | |
|    Tubercle | Pubic tubercle |
|    Protuberance | External occipital protuberance |
|    Tuberosity | Greater and lesser tuberosities of the humerus |
|    Malleolus | Medial malleolus of the tibia, lateral malleolus of the fibula |
|    Trochanter | Greater and lesser trochanters of the femur |
| Sharp elevation | |
|    Spine or spinous process | Ischial spine, spine of vertebra |
|    Styloid process | Styloid process of temporal bone |
| Expanded ends for articulation | |
|    Head | Head of humerus, head of femur |
|    Condyle (knucklelike process) | Medial and lateral condyles of femur |
|    Epicondyle (a prominence situated just above condyle) | Medial and lateral epicondyles of femur |
| Small flat area for articulation | |
|    Facet | Facet on head of rib for articulation with vertebral body |
| Depressions | |
|    Notch | Greater sciatic notch of hip bone |
|    Groove or sulcus | Bicipital groove of humerus |
|    Fossa | Olecranon fossa of humerus, acetabular fossa of hip bone |
| Openings | |
|    Fissure | Superior orbital fissure |
|    Foramen | Infraorbital foramen of the maxilla |
|    Canal | Carotid canal of temporal bone |
|    Meatus | External acoustic meatus of temporal bone |

From Snell RS: Clinical Anatomy. 7th Ed. Philadelphia: Lippincott Williams & Wilkins, 2004, p. 38.

joints (e.g., the temporomandibular joint, sternoclavicular joint, and knee joint) and on the articular surfaces of the clavicle and mandible. Fibrocartilage, if damaged, repairs itself slowly in a manner similar to fibrous tissue elsewhere. Joint discs have a poor blood supply and, therefore, do not repair themselves when damaged.

- **Elastic cartilage** possesses large numbers of elastic fibers embedded in matrix. As would be expected, it is flexible and is found in the auricle of the ear, the external auditory meatus, the auditory tube, and the epiglottis. Elastic cartilage, if damaged, repairs itself with fibrous tissue.

Hyaline cartilage and fibrocartilage tend to calcify or even ossify in later life.

# Skeleton

The skeleton is arranged in two main divisions: the axial skeleton and the appendicular skeleton (Fig. 11-1).

The axial skeleton is made up of the bones that form the main axis of the support of the body, namely, the skull, the hyoid bone, the vertebral column, the sternum, and the ribs; and the appendicular skeleton is made up of bones of the upper and lower limbs that are attached as appendages to the axial skeleton. The appendicular skeleton includes the girdles, which connect the bones of the limbs to the axial skeleton. As you are reading the descriptions of each bone, it would be helpful to have a set of bones to study and, if possible, to have access to an articulated skeleton.

# Axial Skeleton

## Skull

### Composition

The skull is composed of several separate bones united at immobile joints called **sutures.** The connective tissue between the bones is called a **sutural ligament.** The mandible is an exception to this rule, for it is united to the skull by the mobile temporomandibular joint.

The bones of the skull can be divided into those of the **cranium** and those of the face. The **vault** is the upper part of the cranium, and the **base of the skull** is the lowest part of the cranium (Fig. 11-3).

The skull bones are made up of **external** and **internal tables** of compact bone separated by a layer of spongy bone called the **diploë** (Fig. 11-4). The internal table is thinner and more brittle than the external table. The bones are covered on the outer and inner surfaces with periosteum.

The **cranium** consists of the following bones, two of which are paired (Figs. 11-3 and 11-5):

- Frontal bone    1
- Parietal bones    2
- Occipital bone    1
- Temporal bones    2
- Sphenoid bone    1
- Ethmoid bone    1

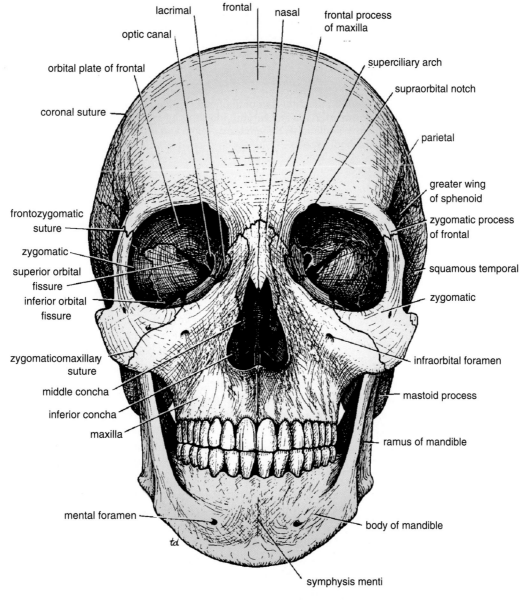

**Figure 11-3**  Bones of the anterior aspect of the skull.

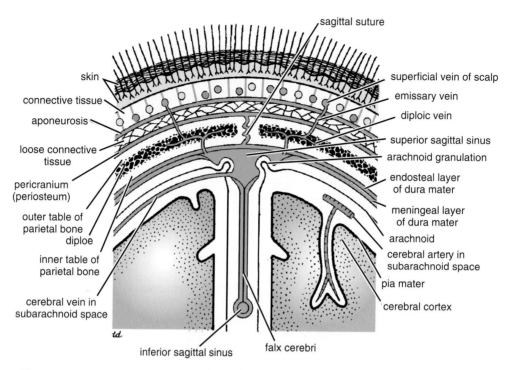

**Figure 11-4** Coronal section of the upper part of the head showing the layers of the scalp, the sagittal suture of the skull, the falx cerebri, the superior and inferior sagittal venous sinuses, the arachnoid granulations, the emissary veins, and the relation of cerebral blood vessels to the subarachnoid space.

The **facial bones** consist of the following, two of which are single:

- Zygomatic bones   2
- Maxillae   2
- Nasal bones   2
- Lacrimal bones   2
- Vomer   1
- Palatine bones   2
- Inferior conchae   2
- Mandible   1

It is unnecessary for students of medicine to know the detailed structure of each individual skull bone. However, students should be familiar with the skull as a whole and should have a dried skull available for reference as they read the following description.

*Anterior View of the Skull*

The **frontal bone**, or forehead bone, curves downward to make the upper margins of the orbits (Fig. 11-3). The **superciliary arches** can be seen on either side, and the **supraorbital notch**, or foramen, can be recognized. Medially, the frontal bone articulates with the frontal processes of the maxillae and with the nasal bones. Laterally, the frontal bone articulates with the zygomatic bone.

The **orbital margins** are bounded by the frontal bone superiorly, the zygomatic bone laterally, the maxilla inferiorly, and the processes of the maxilla and frontal bone medially.

Within the **frontal bone**, just above the orbital margins, are two hollow spaces lined with mucous membrane called the **frontal air sinuses**. These communicate with the nose and serve as voice resonators.

The two **nasal bones** form the bridge of the nose. Their lower borders, with the maxillae, make the **anterior nasal aperture**. The nasal cavity is divided into two by the bony nasal septum, which is largely formed by the **vomer**. The **superior** and **middle conchae** are shelves of bone that project into the nasal cavity from the **ethmoid** on each side; the **inferior conchae** are separate bones.

The two **maxillae** form the upper jaw, the anterior part of the hard palate, part of the lateral walls of the nasal cavities, and part of the floors of the orbital cavities. The two bones meet in the midline at the **intermaxillary suture** and form the lower margin of the nasal aperture. Below the orbit, the maxilla is perforated by the **infraorbital foramen**. The **alveolar process** projects downward and, together with the fellow of the opposite side, forms the **alveolar arch**, which carries the upper teeth. Within each maxilla is a large, pyramid-shaped cavity lined with

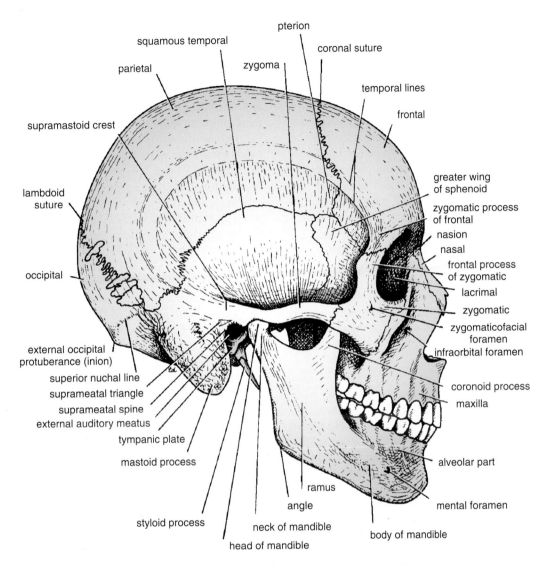

**Figure 11-5** Bones of the lateral aspect of the skull.

mucous membrane called the **maxillary sinus.** This communicates with the nasal cavity and serves as a voice resonator.

The **zygomatic bone** forms the prominence of the cheek and part of the lateral wall and floor of the orbital cavity. Medially, it articulates with the maxilla, and laterally, it articulates with the zygomatic process of the temporal bone to form the **zygomatic arch.** The zygomatic bone is perforated by two foramina for the zygomaticofacial and zygomaticotemporal nerves.

The **mandible,** or lower jaw, consists of a horizontal **body** and two vertical **rami.**

### Lateral View of the Skull
The **frontal bone** forms the anterior part of the side of the skull and articulates with the parietal bone at the **coronal suture** (Fig. 11-5).

The **parietal bones** form the sides and roof of the cranium and articulate with each other in the midline at the **sagittal suture.** They articulate with the occipital bone behind at the **lambdoid suture.**

The skull is completed at the side by the squamous part of the **occipital bone;** parts of the **temporal bone,** namely, the **squamous, tympanic, mastoid process, styloid process,** and **zygomatic process;** and the **greater wing of the sphenoid.** Note the position of the external auditory meatus. The ramus and body of the mandible lie inferiorly.

Note that the thinnest part of the lateral wall of the skull is where the anteroinferior corner of the parietal bone articulates with the greater wing of the sphenoid; this point is referred to as the **pterion.**

Clinically, the pterion is an important area because it overlies the anterior division of the **middle meningeal artery** and **vein.**

Identify the **superior** and **inferior temporal lines,** which begin as a single line from the posterior margin of the zygomatic process of the frontal bone and diverge as they arch backward. The **temporal fossa** lies below the inferior temporal line.

The **infratemporal fossa** lies below the **infratemporal crest** on the greater wing of the sphenoid. The **pterygomaxillary fissure** is a vertical fissure that lies within the fossa between the pterygoid process of the sphenoid bone and back of the maxilla. It leads medially into the **pterygopalatine fossa.**

The **inferior orbital fissure** is a horizontal fissure between the greater wing of the sphenoid bone and the maxilla. It leads forward into the orbit.

The **pterygopalatine fossa** is a small space behind and below the orbital cavity. It communicates laterally with the infratemporal fossa through the pterygomaxillary fissure, medially with the nasal cavity through the **sphenopalatine foramen,** superiorly with the skull through the **foramen rotundum,** and anteriorly with the orbit through the **inferior orbital fissure.**

### Posterior View of the Skull

The posterior parts of the two parietal bones (Fig. 11-6) with the intervening **sagittal suture** are seen above. Below, the parietal bones articulate with the squamous part of the occipital bone at the **lambdoid suture.** On each side, the occipital bone articulates with the temporal bone. In the midline of the occipital bone is a roughened elevation called the **external occipital protuberance,** which gives attachment to muscles and the ligamentum nuchae. On either side of the protuberance, the **superior nuchal lines** extend laterally toward the temporal bone.

### Superior View of the Skull

Anteriorly, the frontal bone (Fig. 11-6) articulates with the two parietal bones at the **coronal suture.** Occasionally, the two halves of the frontal bone fail to fuse, leaving a midline **metopic suture.** Behind, the two parietal bones articulate in the midline at the sagittal suture.

### Inferior View of the Skull

If the mandible is discarded, the anterior part of this aspect of the skull is seen to be formed by the **hard palate** (Fig. 11-7).

The **palatal processes of the maxillae** and the **horizontal plates of the palatine bones** can be identified. In the midline anteriorly is the **incisive fossa** and **foramen.** Posterolaterally are the **greater** and **lesser palatine foramina.**

Above the posterior edge of the hard palate are the **choanae** (posterior nasal apertures). These are separated from each other by the posterior margin of the **vomer** and are bounded laterally by the **medial pterygoid plates** of the sphenoid bone. The inferior end of the **medial pterygoid plate** is prolonged as a curved spike of bone, the **pterygoid hamulus.**

Posterolateral to the **lateral pterygoid plate,** the greater wing of the sphenoid is pierced by the large **foramen ovale** and the small **foramen spinosum.** Posterolateral to the foramen spinosum is the **spine of the sphenoid.**

Behind the spine of the sphenoid, in the interval between the greater wing of the sphenoid and the petrous part of the temporal bone, is a groove for the cartilaginous part of the **auditory tube.** The opening of the bony part of the tube can be identified.

The **mandibular fossa** of the temporal bone and the **articular tubercle** form the upper articular surfaces for the temporomandibular joint. Separating the mandibular fossa from the tympanic plate posteriorly is the **squamotympanic fissure,** through the medial end of which the chorda tympani exits from the tympanic cavity.

The **styloid process** of the temporal bone projects downward and forward from its inferior aspect. The opening of the **carotid canal** can be seen on the inferior surface of the petrous part of the temporal bone.

The medial end of the petrous part of the temporal bone is irregular and, together with the basilar part of the occipital bone and the greater wing of the sphenoid, forms the **foramen lacerum.** During life, the foramen lacerum is closed with fibrous tissue, and only a few small vessels pass through this foramen from the cavity of the skull to the exterior.

The **tympanic plate,** which forms part of the temporal bone, is C shaped on section and forms the bony part of the **external auditory meatus.** While examining this region, identify the **suprameatal crest** on the lateral surface of the squamous part of the temporal bone, the **suprameatal triangle,** and the **suprameatal spine.**

In the interval between the styloid and mastoid processes, the **stylomastoid foramen** can be seen. Medial to the styloid process, the petrous part of the temporal bone has a deep notch, which, together with a shallower notch on the occipital bone, forms the **jugular foramen.**

Behind the posterior apertures of the nose and in front of the foramen magnum are the sphenoid bone and the basilar part of the occipital bone. The **pharyngeal tubercle** is a small prominence on the undersurface of the basilar part of the occipital bone in the midline.

The **occipital condyles** should be identified; they articulate with the superior aspect of the lateral mass of the first cervical vertebra, the atlas. Superior to the occipital condyle is the **hypoglossal canal** for transmission of the hypoglossal nerve (Fig. 11-8).

Posterior to the foramen magnum in the midline is the external occipital protuberance. The superior nuchal lines should be identified as they curve laterally on each side.

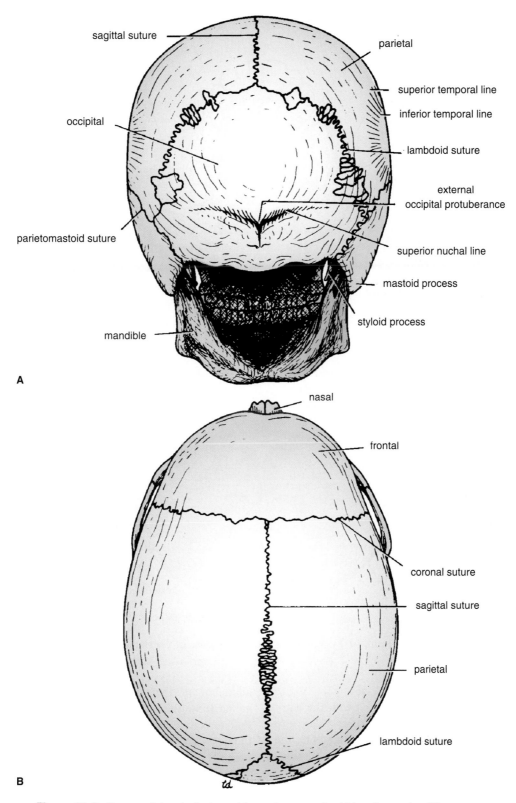

**Figure 11-6** Bones of the skull viewed from the posterior (**A**) and superior (**B**) aspects.

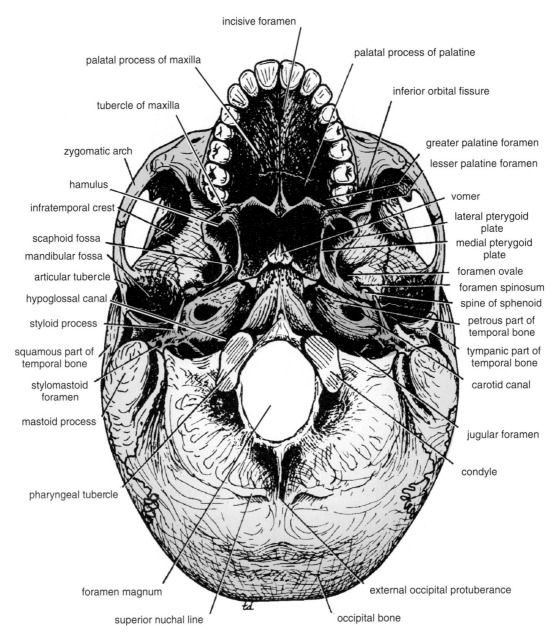

incisive foramen

palatal process of maxilla

palatal process of palatine

tubercle of maxilla

inferior orbital fissure

zygomatic arch

greater palatine foramen

lesser palatine foramen

hamulus

vomer

infratemporal crest

lateral pterygoid plate

scaphoid fossa

medial pterygoid plate

mandibular fossa

foramen ovale

articular tubercle

foramen spinosum

hypoglossal canal

spine of sphenoid

styloid process

petrous part of temporal bone

squamous part of temporal bone

tympanic part of temporal bone

stylomastoid foramen

carotid canal

mastoid process

pharyngeal tubercle

jugular foramen

condyle

foramen magnum

external occipital protuberance

superior nuchal line

occipital bone

**Figure 11-7**   Inferior surface of the base of the skull.

### Neonatal Skull

The newborn skull (Fig. 11-9), compared with the adult skull, has a disproportionately large cranium relative to the face. In childhood, the growth of the mandible, the maxillary sinuses, and the alveolar processes of the maxillae results in a great increase in length of the face.

The bones of the skull are smooth and unilaminar, there being no diploë present. Most of the skull bones are ossified at birth, but the process is incomplete, and the bones are mobile on each other, being connected by fibrous tissue or cartilage. The bones of the vault are ossified in membrane; the bones of the base are ossified in cartilage. The bones of the vault are not closely knit at sutures, as in the adult, but are separated by unossified membranous intervals called **fontanelles.** Clinically, the anterior and posterior fontanelles are most important and are easily examined in the midline of the vault.

The **anterior fontanelle** is diamond shaped and lies between the two halves of the frontal bone in front and the two parietal bones behind (Fig. 11-9). The fibrous membrane forming the floor of the anterior fontanelle is replaced by bone and is closed by 18 months of age. The **posterior fontanelle** is triangular and lies between the two parietal bones in front and the occipital bone behind. By the end of the 1st year, the fontanelle is usually closed and can no longer be palpated.

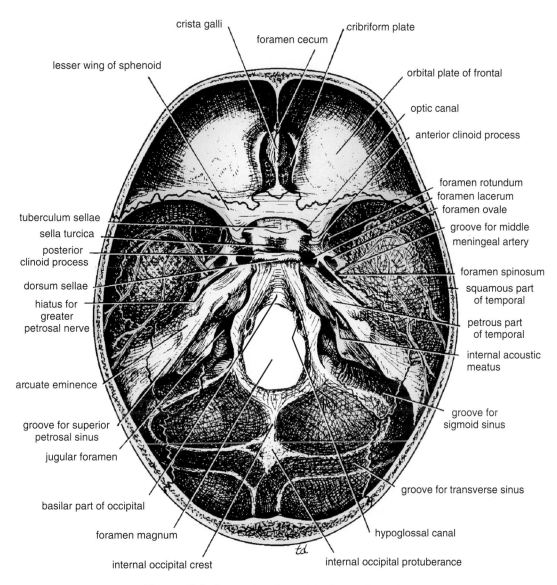

crista galli
foramen cecum
cribriform plate
lesser wing of sphenoid
orbital plate of frontal
optic canal
anterior clinoid process
foramen rotundum
foramen lacerum
foramen ovale
groove for middle
meningeal artery
tuberculum sellae
sella turcica
posterior
clinoid process
foramen spinosum
squamous part
of temporal
dorsum sellae
petrous part
of temporal
hiatus for
greater
petrosal nerve
internal acoustic
meatus
arcuate eminence
groove for superior
petrosal sinus
groove for
sigmoid sinus
jugular foramen
groove for transverse sinus
basilar part of occipital
hypoglossal canal
foramen magnum
internal occipital crest
internal occipital protuberance

**Figure 11-8**   Internal surface of the base of the skull.

The **tympanic part of the temporal bone** is merely a C-shaped ring at birth, compared with a C-shaped curved plate in the adult. This means that the external auditory meatus is almost entirely cartilaginous in the newborn, and the **tympanic membrane** is nearer the surface. Although the tympanic membrane is nearly as large as in the adult, it faces more inferiorly. During childhood, the tympanic plate grows laterally, forming the bony part of the meatus, and the tympanic membrane comes to face more directly laterally.

The **mastoid process** is not present at birth (Fig. 11-9) and develops later in response to the pull of the sternocleidomastoid muscle when the child moves his or her head.

At birth, the mastoid antrum lies about 3 mm deep to the floor of the **suprameatal triangle.** As growth of the skull continues, the lateral bony wall thickens so that,

at puberty, the antrum may lie as much as 15 mm from the surface.

The mandible has right and left halves at birth, united in the midline with fibrous tissue. The two halves fuse at the **symphysis menti** by the end of the 1st year.

The **angle of the mandible** at birth is obtuse (Fig. 11-9), the head being placed level with the upper margin of the body and the coronoid process lying at a superior level to the head. It is only after eruption of the permanent teeth that the angle of the mandible assumes the adult shape and the head and neck grow so that the head comes to lie higher than the coronoid process.

In old age, the size of the mandible is reduced when the teeth are lost. As the alveolar part of the bone becomes smaller, the ramus becomes oblique in position so that the head is bent posteriorly.

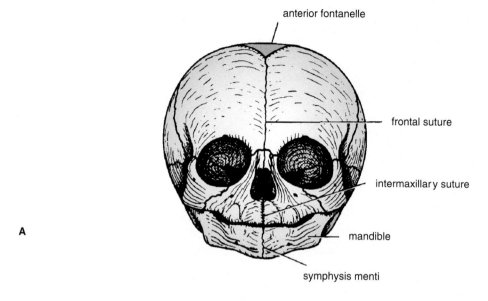

A

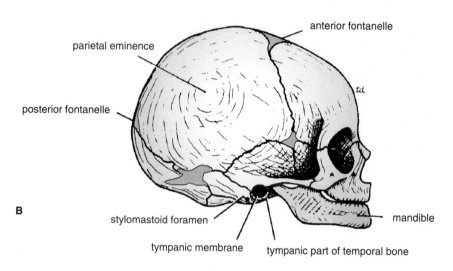

B

**Figure 11-9** Neonatal skull as seen from the anterior (**A**) and lateral (**B**) aspects.

## The Cranial Cavity

The cranial cavity contains the brain and its surrounding meninges, portions of the cranial nerves, arteries, veins, and venous sinuses.

## Vault of the Skull

The internal surface of the vault shows the coronal, sagittal, and lambdoid sutures. In the midline is a shallow sagittal groove that lodges the **superior sagittal sinus.** On each side of the groove are several small pits, called **granular pits,** which lodge the **lateral lacunae** and **arachnoid granulations.** Several narrow grooves are present for the anterior and posterior divisions of the **middle meningeal vessels** as they pass up the side of the skull to the vault.

## Base of the Skull

The interior of the base of the skull (Fig. 11-8) is divided into three cranial fossae: anterior, middle, and posterior.

The anterior cranial fossa is separated from the middle cranial fossa by the lesser wing of the sphenoid, and the middle cranial fossa is separated from the posterior cranial fossa by the petrous part of the temporal bone.

### Anterior Cranial Fossa

The anterior cranial fossa lodges the frontal lobes of the cerebral hemispheres. It is bounded anteriorly by the inner surface of the frontal bone and, in the midline, is a crest for the attachment of the **falx cerebri.** Its posterior boundary is the sharp lesser wing of the sphenoid, which articulates laterally with the frontal bone and meets the anteroinferior angle of the parietal bone, or the pterion. The medial end of the lesser wing of the sphenoid forms the **anterior clinoid process** on each side, which gives attachment to the **tentorium cerebelli.** The median part of the anterior cranial fossa is limited posteriorly by the groove for the optic chiasma.

The floor of the fossa is formed by the ridged orbital plates of the frontal bone laterally and by the **cribriform plate** of the ethmoid medially (Fig. 11-8). The **crista galli** is a sharp upward projection of the ethmoid bone in the midline for the attachment of the falx cerebri. Between the crista galli and the crest of the frontal bone is a small aperture, the **foramen cecum,** for the transmission of a small vein from the nasal mucosa to the superior sagittal sinus. Alongside the crista galli is a narrow slit in the cribriform plate for the passage of the **anterior ethmoidal nerve** into the nasal cavity. The upper surface of the cribriform plate supports the **olfactory bulbs,** and the small perforations in the cribriform plate are for the **olfactory nerves.**

### Middle Cranial Fossa

The middle cranial fossa consists of a small median part and expanded lateral parts (Fig. 11-8). The median raised part is formed by the body of the sphenoid, and the expanded lateral parts form concavities on either side, which lodge the **temporal lobes** of the **cerebral hemispheres.**

It is bounded anteriorly by the lesser wings of the sphenoid and posteriorly by the superior borders of the petrous parts of the temporal bones. Laterally lie the squamous parts of the temporal bones, the greater wings of the sphenoid, and the parietal bones.

The floor of each lateral part of the middle cranial fossa is formed by the greater wing of the sphenoid and the squamous and petrous parts of the temporal bone.

The sphenoid bone resembles a bat having a centrally placed **body** with **greater** and **lesser wings** that are outstretched on each side. The body of the sphenoid contains the **sphenoid air sinuses,** which are lined with mucous membrane and communicate with the nasal cavity; they serve as voice resonators.

Anteriorly, the **optic canal** transmits the optic nerve and the ophthalmic artery, a branch of the internal carotid artery, to the orbit. The **superior orbital fissure,** which is a slitlike opening between the lesser and greater wings of the sphenoid, transmits the lacrimal, frontal, trochlear, oculomotor, nasociliary, and abducent nerves, together with the superior ophthalmic vein. The sphenoparietal venous sinus runs medially along the posterior border of the lesser wing of the sphenoid and drains into the cavernous sinus.

The **foramen rotundum,** which is situated behind the medial end of the superior orbital fissure, perforates the greater wing of the sphenoid and transmits the maxillary nerve from the trigeminal ganglion to the pterygopalatine fossa.

The **foramen ovale** lies posterolateral to the foramen rotundum (Fig. 11-8). It perforates the greater wing of the sphenoid and transmits the large sensory root and small motor root of the mandibular nerve to the infratemporal fossa; the lesser petrosal nerve also passes through it.

The small **foramen spinosum** lies posterolateral to the foramen ovale and also perforates the greater wing of the sphenoid. The foramen transmits the middle meningeal artery from the infratemporal fossa into the cranial cavity.

The artery then runs forward and laterally in a groove on the upper surface of the squamous part of the temporal bone and the greater wing of the sphenoid (see Chapter 14, p. 541). After a short distance, the artery divides into anterior and posterior branches. The anterior branch passes forward and upward to the anteroinferior angle of the parietal bone (see Fig. 14–19). Here, the bone is deeply grooved or tunneled by the artery for a short distance before it runs backward and upward on the parietal bone. It is at this site that the artery may be damaged after a blow to the side of the head. The posterior branch passes backward and upward across the squamous part of the temporal bone to reach the parietal bone.

The large and irregularly shaped **foramen lacerum** lies between the apex of the petrous part of the temporal bone and the sphenoid bone (Fig. 11-8). The inferior opening of the foramen lacerum in life is filled by cartilage and fibrous tissue, and only small blood vessels pass through this tissue from the cranial cavity to the neck.

The **carotid canal** opens into the side of the foramen lacerum above the closed inferior opening. The internal carotid artery enters the foramen through the carotid canal and immediately turns upward to reach the side of the body of the sphenoid bone. Here, the artery turns forward in the cavernous sinus to reach the region of the anterior clinoid process. At this point, the internal carotid artery turns vertically upward, medial to the anterior clinoid process, and emerges from the cavernous sinus (see p. 180).

Lateral to the foramen lacerum is an impression on the apex of the petrous part of the temporal bone for the **trigeminal ganglion.** On the anterior surface of the petrous bone are two grooves for nerves; the largest medial groove is for the **greater petrosal nerve,** a branch of the facial nerve; the smaller lateral groove is for the **lesser petrosal nerve,** a branch of the tympanic plexus. The greater petrosal nerve enters the foramen lacerum deep to the trigeminal ganglion and joins the **deep petrosal nerve** (sympathetic fibers from around the internal carotid artery) to form the **nerve of the pterygoid canal.** The lesser petrosal nerve passes forward to the foramen ovale.

The abducent nerve bends sharply forward across the apex of the petrous bone, medial to the trigeminal ganglion. Here, it leaves the posterior cranial fossa and enters the cavernous sinus.

The **arcuate eminence** is a rounded eminence found on the anterior surface of the petrous bone and is caused by the underlying **superior semicircular canal.**

The **tegmen tympani,** a thin plate of bone, is a forward extension of the petrous part of the temporal bone and adjoins the squamous part of the bone (Fig. 11-8). From behind forward, it forms the roof of the mastoid antrum, the tympanic cavity, and the auditory tube. This thin plate of bone is the only major barrier that separates infection in the tympanic cavity from the temporal lobe of the cerebral hemisphere (see Fig. 18-16).

The median part of the middle cranial fossa is formed by the body of the sphenoid bone (Fig. 11-8). In front is the

**sulcus chiasmatis,** which is related to the optic chiasma and leads laterally to the **optic canal** on each side. Posterior to the sulcus is an elevation, the **tuberculum sellae.** Behind the elevation is a deep depression, the **sella turcica,** which lodges the **hypophysis cerebri.** The sella turcica is bounded posteriorly by a square plate of bone called the **dorsum sellae.** The superior angles of the dorsum sellae have two tubercles, called the **posterior clinoid processes,** which give attachment to the fixed margin of the tentorium cerebelli.

The cavernous sinus is directly related to the side of the body of the sphenoid (see Fig. 14-15). It carries in its lateral wall the third and fourth cranial nerves and the ophthalmic and maxillary divisions of the fifth cranial nerve (see Fig. 14-15). The internal carotid artery and the sixth cranial nerve pass forward through the sinus.

### Posterior Cranial Fossa

The posterior cranial fossa is deep and lodges the parts of the hindbrain, namely, the **cerebellum, pons,** and **medulla oblongata.** Anteriorly, the fossa is bounded by the superior border of the petrous part of the temporal bone, and posteriorly, it is bounded by the internal surface of the squamous part of the occipital bone (Fig. 11-8). The floor of the posterior fossa is formed by the basilar, condylar, and squamous parts of the occipital bone and the mastoid part of the temporal bone.

The roof of the fossa is formed by a fold of dura, the **tentorium cerebelli,** which intervenes between the cerebellum below and the occipital lobes of the cerebral hemispheres above (see Fig. 14-12).

The **foramen magnum** occupies the central area of the floor and transmits the medulla oblongata and its surrounding meninges, the ascending spinal parts of the accessory nerves, and the two vertebral arteries.

The **hypoglossal canal** is situated above the anterolateral boundary of the foramen magnum (Fig. 11-8) and transmits the **hypoglossal nerve.**

The **jugular foramen** lies between the lower border of the petrous part of the temporal bone and the condylar part of the occipital bone. It transmits the following structures from before backward: the **inferior petrosal sinus;** the **ninth, tenth,** and **eleventh cranial nerves;** and the large **sigmoid sinus.** The inferior petrosal sinus descends in the groove on the lower border of the petrous part of the temporal bone to reach the foramen. The sigmoid sinus turns down through the foramen to become the **internal jugular vein.**

The **internal acoustic meatus** pierces the posterior surface of the petrous part of the temporal bone. It transmits the vestibulocochlear nerve and the motor and sensory roots of the facial nerve.

The **internal occipital crest** runs upward in the midline posteriorly from the foramen magnum to the **internal occipital protuberance;** it is attached to the small **falx cerebelli** over the **occipital sinus.**

On each side of the internal occipital protuberance is a wide groove for the **transverse sinus** (Fig. 11-8). This groove sweeps around on either side, on the internal surface of the occipital bone, to reach the posteroinferior angle or corner of the parietal bone. The groove now passes onto the mastoid part of the temporal bone, and here, the transverse sinus becomes the **sigmoid sinus.** The **superior petrosal sinus** runs backward along the upper border of the petrous bone in a narrow groove and drains into the sigmoid sinus. As the sigmoid sinus descends to the jugular foramen, it deeply grooves the back of the petrous bone and the mastoid part of the temporal bone. Here, it lies directly posterior to the mastoid antrum.

Table 11-3 provides a summary of the more important openings in the base of the skull and the structures that pass through them.

## Mandible

The mandible or lower jaw is the largest and strongest bone of the face, and it articulates with the skull at the **temporomandibular joint.**

The mandible consists of a horseshoe-shaped **body** and a pair of **rami.** The body of the mandible meets the ramus on each side at the **angle of the mandible** (Fig. 11-10).

The **body of the mandible,** on its external surface in the midline, has a faint ridge indicating the line of fusion of the two halves during development at the **symphysis menti.** The **mental foramen** can be seen below the second premolar tooth; it transmits the terminal branches of the inferior alveolar nerve and vessels.

On the medial surface of the body of the mandible in the median plane are seen the **mental spines;** these give origin to the genioglossus muscles above and the geniohyoid muscles below (Fig. 11-10). The **mylohyoid line** can be seen as an oblique ridge that runs backward and laterally from the area of the mental spines to an area below and behind the third molar tooth. The **submandibular fossa,** for the superficial part of the submandibular salivary gland, lies below the posterior part of the mylohyoid line. The **sublingual fossa,** for the sublingual gland, lies above the anterior part of the mylohyoid line (Fig. 11-10).

The upper border of the body of the mandible is called the **alveolar** part; in the adult, it contains 16 sockets for the roots of the teeth.

The lower border of the body of the mandible is called the **base.** The **digastric fossa** is a small, roughened depression on the base, on either side of the symphysis menti (Fig. 11-10). It is in these fossae that the anterior bellies of the digastric muscles are attached.

The **ramus of the mandible** is vertically placed and has an anterior **coronoid process** and a posterior **condyloid process,** or **head;** the two processes are separated by the mandibular notch (Fig. 11-10).

On the lateral surface of the ramus are markings for the attachment of the masseter muscle. On the medial surface is

| Table 11-3 | Summary of the More Important Openings in the Base of the Skull and the Structures that Pass through Them | |
|---|---|---|
| **Opening in Skull** | **Bone of Skull** | **Structures Transmitted** |
| **Anterior Cranial Fossa** | | |
| Perforations in cribriform plate | Ethmoid | Olfactory nerves |
| **Middle Cranial Fossa** | | |
| Optic canal | Lesser wing of sphenoid | Optic nerve, ophthalmic artery |
| Superior orbital fissure | Between lesser and greater wings of sphenoid | Lacrimal, frontal, trochlear oculomotor, nasociliary, and abducent nerves; superior ophthalmic vein |
| Foramen rotundum | Greater wing of sphenoid | Maxillary division of the trigeminal nerve |
| Foramen ovale | Greater wing of sphenoid | Mandibular division of the trigeminal nerve, lesser petrosal nerve |
| Foramen spinosum | Greater wing of sphenoid | Middle meningeal artery |
| Foramen lacerum | Between petrous part of temporal and sphenoid | Internal carotid artery |
| **Posterior Cranial Fossa** | | |
| Foramen magnum | Occipital | Medulla oblongata, spinal part of accessory nerve, and right and left vertebral arteries |
| Hypoglossal canal | Occipital | Hypoglossal nerve |
| Jugular foramen | Between petrous part of temporal and condylar part of occipital | Glossopharyngeal, vagus, and accessory nerves; sigmoid sinus becomes internal jugular vein |
| Internal acoustic meatus | Petrous part of temporal | Vestibulocochlear and facial nerves |

From Snell RS: Clinical Anatomy. 7th Ed. Philadelphia: Lippincott Williams & Wilkins, 2004, p. 801.

the **mandibular foramen,** for the inferior alveolar nerve and vessels. In front of the foramen is a projection of bone, called the **lingula,** for the attachment of the **sphenomandibular ligament** (Fig. 11-10). The foramen leads into the **mandibular canal,** which opens on the lateral surface of the body of the mandible at the **mental foramen** (see p. 302). The **incisive canal** is a continuation forward of the mandibular canal beyond the mental foramen and below the incisor teeth.

The **coronoid process** receives on its medial surface the attachment of the temporalis muscle. Below the **condyloid process,** or **head,** is a short **neck** (Fig. 11-10).

The important muscles and ligaments attached to the mandible are shown in Figure 11-10.

# RADIOGRAPHIC APPEARANCES OF THE SKULL AND MANDIBLE

The radiographic appearances of the skull and mandible are shown in Figures 11-11 to 11-18. For CT scans and MRIs of the skull, see Figures 14-7 to 14-9.

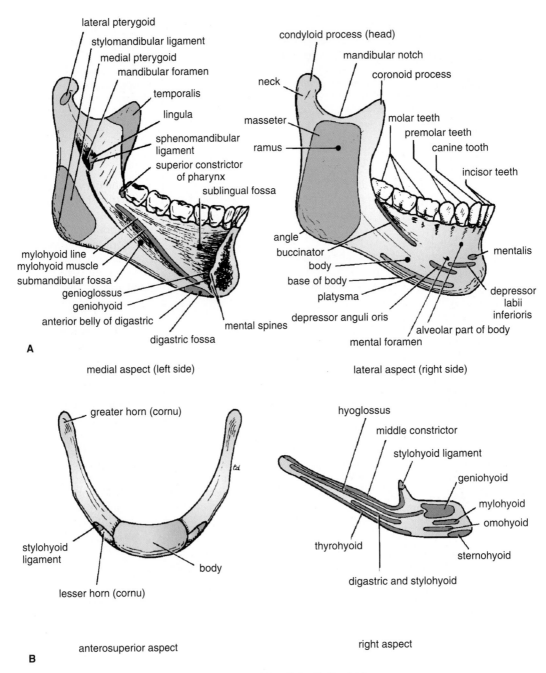

**Figure 11-10   A.** Mandible. **B.** Hyoid bone.

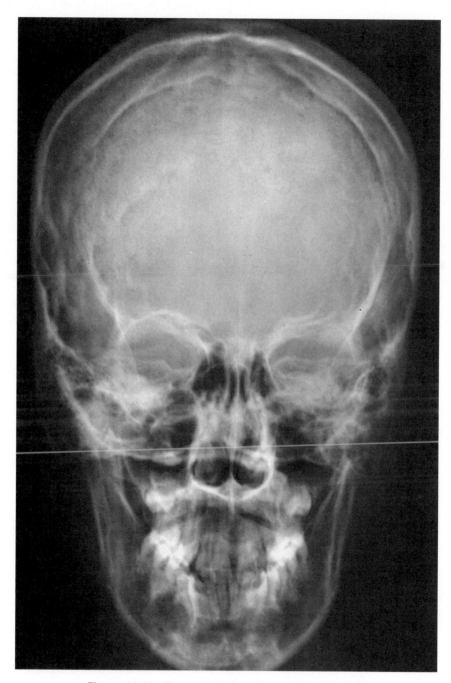

**Figure 11-11**   Posteroanterior radiograph of the skull.

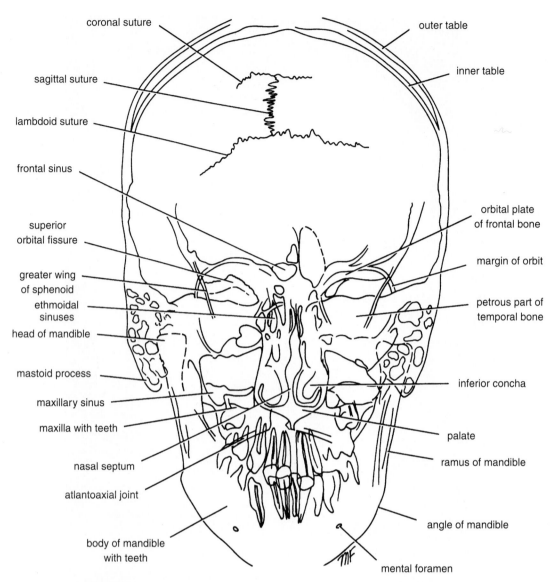

coronal suture

sagittal suture

lambdoid suture

frontal sinus

superior
orbital fissure

greater wing
of sphenoid

ethmoidal
sinuses

head of mandible

mastoid process

maxillary sinus

maxilla with teeth

nasal septum

atlantoaxial joint

body of mandible
with teeth

outer table

inner table

orbital plate
of frontal bone

margin of orbit

petrous part of
temporal bone

inferior concha

palate

ramus of mandible

angle of mandible

mental foramen

**Figure 11-12**   Main features that can be seen in the posteroanterior radiograph of the
skull in Figure 11-11.

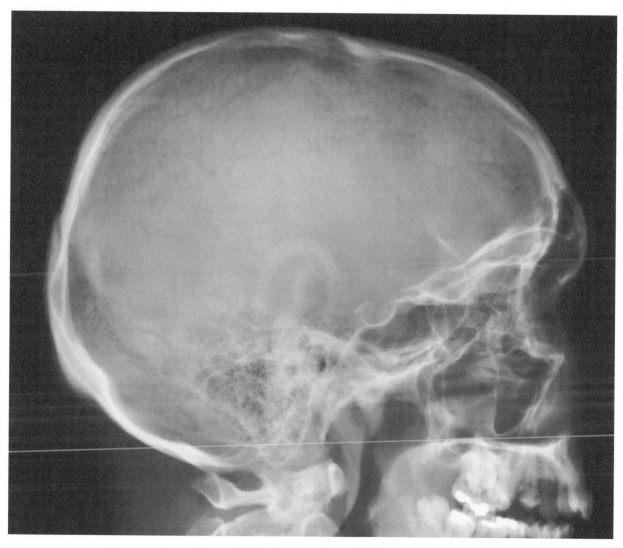

**Figure 11-13**    Lateral radiograph of the skull.

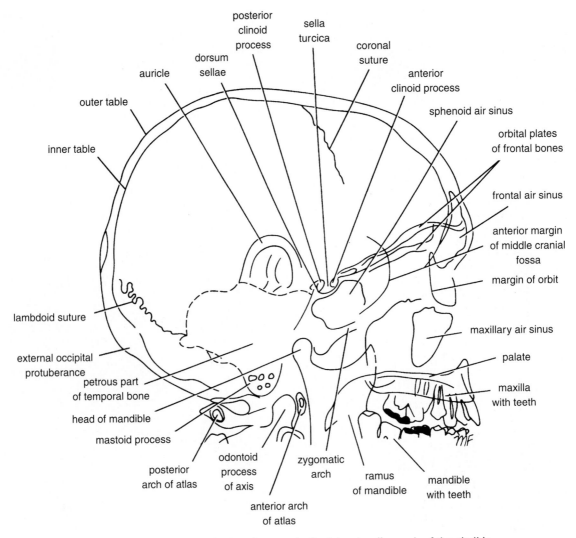

**Figure 11-14** Main features that can be seen in the lateral radiograph of the skull in Figure 11-13.

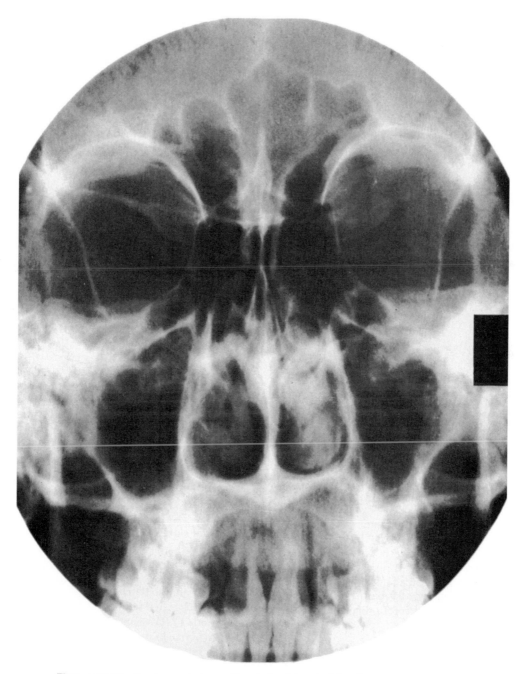

**Figure 11-15**   Posteroanterior radiograph of the skull for the paranasal sinuses.

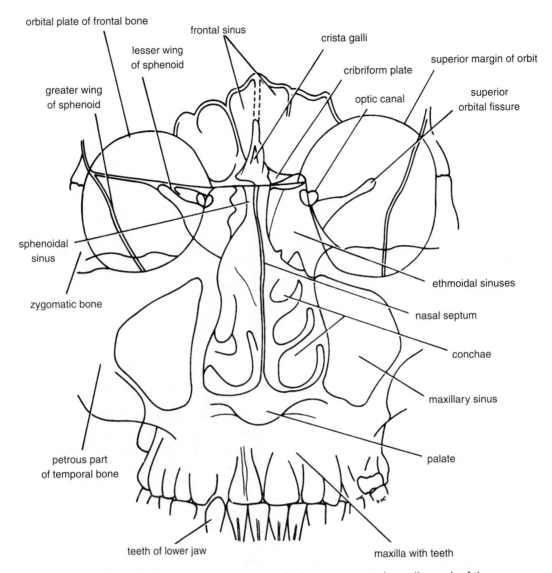

orbital plate of frontal bone

lesser wing
of sphenoid

frontal sinus

crista galli

cribriform plate

optic canal

superior margin of orbit

superior
orbital fissure

greater wing
of sphenoid

sphenoidal
sinus

zygomatic bone

ethmoidal sinuses

nasal septum

conchae

maxillary sinus

petrous part
of temporal bone

palate

teeth of lower jaw

maxilla with teeth

**Figure 11-16**   Main features that can be seen in the posteroanterior radiograph of the skull in Figure 11-15.

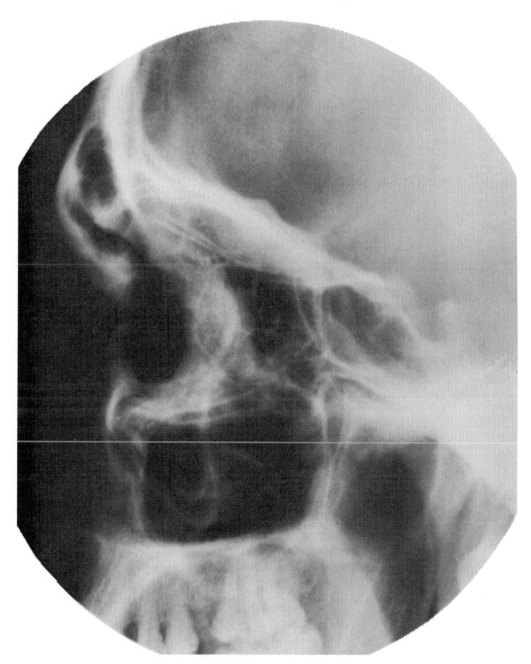

**Figure 11-17**  Lateral radiograph of the skull for the paranasal sinuses.

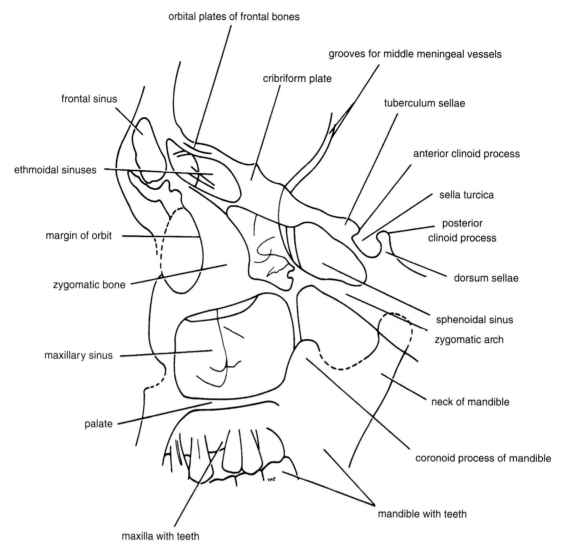

**Figure 11-18** Main features that can be seen in the lateral radiograph of the skull in Figure 11-17.

## The Hyoid Bone

The hyoid bone is a single bone found in the midline of the neck below the mandible and above the larynx. It does not articulate with any other bones. The hyoid bone is U shaped and consists of a body and two greater and two lesser cornua (Fig. 11-10). It is attached to the skull by the stylohyoid ligament and to the thyroid cartilage by the thyrohyoid membrane. It is mobile and lies in the neck just above the larynx and below the mandible. The hyoid bone forms a base for the tongue and is suspended in position by muscles that connect it to the mandible, to the styloid process of the temporal bone, to the thyroid cartilage, to the sternum, and to the scapula.

The important muscles attached to the hyoid bone are shown in Figure 11-10.

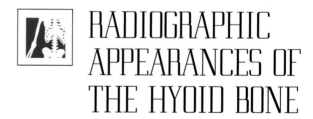

# RADIOGRAPHIC APPEARANCES OF THE HYOID BONE

The radiographic appearances of the hyoid bone are shown in Figure 11-26.

## The Vertebral Column

The vertebral column is the central bony pillar of the body. It supports the skull, pectoral girdle, upper limbs, and thoracic cage and, by way of the pelvic girdle, transmits body weight to the lower limbs. Within its cavity lie the spinal cord, the roots of the spinal nerves, and the covering meninges, to which the vertebral column gives great protection.

### Composition of the Vertebral Column

The vertebral column (Figs. 11-19 and 11-20) is composed of 33 vertebrae arranged in the following groups:

- Cervical (7)
- Thoracic (12)
- Lumbar (5)
- Sacral (5 fused to form the sacrum)
- Coccygeal (4, the lower 3 are commonly fused)

Because it is segmented and made up of vertebrae, joints, and pads of fibrocartilage called **intervertebral discs,**

it is a flexible structure. The intervertebral discs form about one fourth of the length of the column.

### Curves of the Vertebral Column

#### Curves in the Sagittal Plane

In the fetus, the vertebral column has one continuous anterior concavity. As development proceeds, the lumbosacral angle appears. After birth, when the child becomes able to raise his or her head and keep it poised on the vertebral column, the cervical part of the vertebral column becomes concave posteriorly (Fig. 11-21). Toward the end of the first year, when the child begins to stand upright, the lumbar part of the vertebral column becomes concave posteriorly. The development of these secondary curves is largely

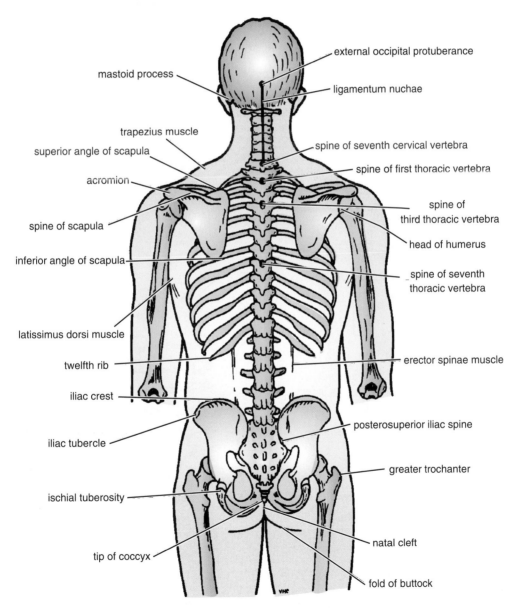

mastoid process

external occipital protuberance

ligamentum nuchae

trapezius muscle

superior angle of scapula

acromion

spine of scapula

inferior angle of scapula

latissimus dorsi muscle

twelfth rib

iliac crest

iliac tubercle

ischial tuberosity

tip of coccyx

spine of seventh cervical vertebra

spine of first thoracic vertebra

spine of third thoracic vertebra

head of humerus

spine of seventh thoracic vertebra

erector spinae muscle

posterosuperior iliac spine

greater trochanter

natal cleft

fold of buttock

**Figure 11-19**   Posterior view of the skeleton showing the surface markings on the back.

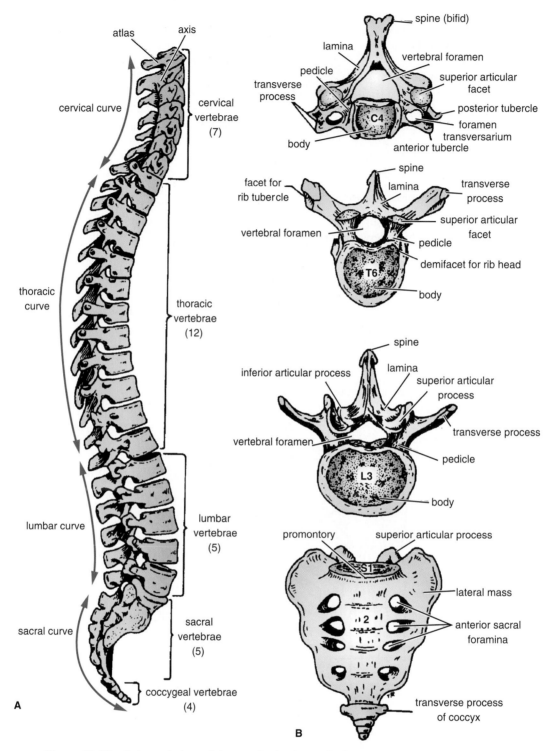

**Figure 11-20** **A.** Lateral view of the vertebral column. **B.** General features of different kinds of vertebrae.

A

newborn infant

B

baby holds head up steadily
(3–4 months)

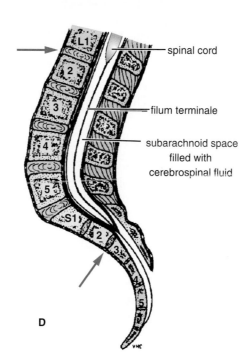

spinal cord

filum terminale

subarachnoid space
filled with
cerebrospinal fluid

D

C          adult

**Figure 11-21** **A, B, and C.**
Curves of the vertebral column at
different ages. **D.** In the adult, the
lower end of the spinal cord lies
at the level of the lower border of
the body of the first lumbar vertebra (*top arrow*), and the subarachnoid space ends at the lower
border of the body of the second
sacral vertebra (*bottom arrow*).

caused by modification in the shape of the intervertebral discs.

In the adult in the standing position (Fig. 11-21), the vertebral column exhibits in the sagittal plane the following regional curves: cervical, posterior concavity; thoracic, posterior convexity; lumbar, posterior concavity; and sacral, posterior convexity. During the later months of pregnancy, with the increase in size and weight of the fetus, women tend to increase the posterior lumbar concavity in an attempt to preserve their center of gravity. In old age, the intervertebral discs atrophy, resulting in a loss of height and a gradual return of the vertebral column to a continuous anterior concavity.

### Curves in the Coronal Plane
In late childhood, it is common to find the development of minor lateral curves in the thoracic region of the vertebral column. This is normal and is usually caused by the predominant use of one of the upper limbs. For example, right-handed persons will often have a slight right-sided thoracic convexity. Slight compensatory curves are always present above and below such a curvature.

### General Characteristics of a Vertebra
Although vertebrae show regional differences, they all possess a common pattern (Fig. 11-20).

A **typical vertebra** consists of a rounded **body** anteriorly and a **vertebral arch** posteriorly. These enclose a space called the **vertebral foramen**, through which run the spinal cord and its coverings. The vertebral arch consists of a pair of cylindrical **pedicles**, which form the sides of the arch, and a pair of flattened **laminae**, which complete the arch posteriorly.

The vertebral arch gives rise to seven processes: one spinous, two transverse, and four articular (Fig. 11-20).

The **spinous process**, or **spine**, is directed posteriorly from the junction of the two laminae. The transverse processes are directed laterally from the junction of the laminae and the pedicles. Both the spinous and transverse processes serve as levers and receive attachments of muscles and ligaments.

The **articular processes** are vertically arranged and consist of two superior and two inferior processes. They arise from the junction of the laminae and the pedicles, and their articular surfaces are covered with hyaline cartilage.

The two superior articular processes of one vertebral arch articulate with the two inferior articular processes of the arch above, forming two synovial joints.

The pedicles are notched on their upper and lower borders, forming the **superior** and **inferior vertebral notches**. On each side, the superior notch of one vertebra and the inferior notch of an adjacent vertebra together form an **intervertebral foramen**. These foramina, in an articulated skeleton, serve to transmit the spinal nerves and blood vessels. The anterior and posterior nerve roots of a spinal nerve unite within these foramina with their coverings of dura to form the segmental spinal nerves.

### Characteristics of a Typical Cervical Vertebra

A typical cervical vertebra has the following characteristics (Fig. 11-22):

- The transverse processes possess a **foramen transversarium** for the passage of the vertebral artery and veins (note that the vertebral artery passes through the transverse processes C1 to 6 and not through C7).
- The spines are small and bifid.
- The body is small and broad from side to side.
- The vertebral foramen is large and triangular.
- The superior articular processes have facets that face backward and upward; the inferior processes have facets that face downward and forward.

### Characteristics of the Atypical Cervical Vertebrae

The first, second, and seventh cervical vertebrae are atypical.

The **first cervical vertebra**, or **atlas** (Fig. 11-22), does not possess a body or a spinous process. It has an anterior and posterior arch. It has a lateral mass on each side with articular surfaces on its upper surface for articulation with the occipital condyles (**atlanto-occipital joints**) and articular surfaces on its lower surface for articulation with the axis (**atlanto-axial joints**).

The **second cervical vertebra**, or **axis** (Fig. 11-22), has a peglike **odontoid process** that projects from the superior surface of the body (representing the body of the atlas that has fused with the body of the axis).

The **seventh cervical vertebra**, or **vertebra prominens** (Fig. 11-22), is so named because it has the longest spinous process, and the process is not bifid. The transverse process is large, but the foramen transversarium is small and transmits the vertebral vein or veins.

### Characteristics of a Typical Thoracic Vertebra

A typical thoracic vertebra has the following characteristics (Fig. 11-20):

- The body is of medium size and heart shaped.
- The vertebral foramen is small and circular.
- The spines are long and inclined downward.
- Costal facets are present on the sides of the bodies for articulation with the heads of the ribs.
- Costal facets are present on the transverse processes for articulation with the tubercles of the ribs (T11 and 12 have no facets on the transverse processes).
- The superior articular processes bear facets that face backward and laterally, whereas the facets on the inferior articular processes face forward and medially. The inferior articular processes of the twelfth vertebra face laterally, as do those of the lumbar vertebrae.

### Characteristics of a Typical Lumbar Vertebra

A typical lumbar vertebra has the following characteristics (Fig. 11-20):

- The body is large and kidney shaped.
- The pedicles are strong and directed backward.
- The laminae are thick.
- The vertebral foramina are triangular.
- The transverse processes are long and slender.
- The spinous processes are short, flat, and quadrangular and project backward.
- The articular surfaces of the superior articular processes face medially, and those of the inferior articular processes face laterally.

Note that the lumbar vertebrae have no facets for articulation with ribs and no foramina in the transverse processes.

### Sacrum

The sacrum (Fig. 11-20) consists of five rudimentary vertebrae fused together to form a wedge-shaped bone, which is concave anteriorly. The upper border, or base, of the bone articulates with the fifth lumbar vertebra. The narrow inferior border articulates with the coccyx. Laterally, the sacrum articulates with the two iliac bones to form the sacroiliac joints (see Fig. 11-50). The anterior and upper margin of the first sacral vertebra bulges forward as the posterior margin of the pelvic inlet and is known as the **sacral promontory**. The

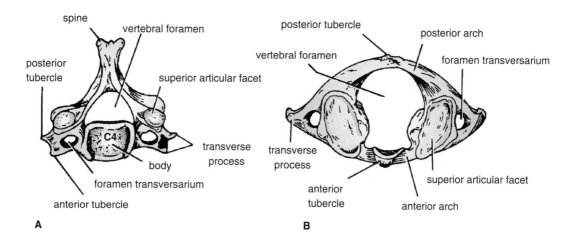

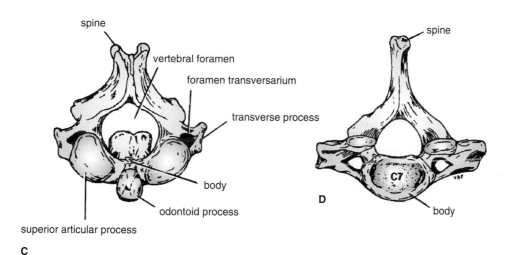

**Figure 11-22** **A.** Typical cervical vertebra, superior aspect. **B.** Atlas, or first cervical vertebra, superior aspect. **C.** Axis, or second cervical vertebra, from above and behind. **D.** Seventh cervical vertebra, superior aspect; the foramen transversarium forms a passage for the vertebral vein but not for the vertebral artery.

sacral promontory in the female is of considerable obstetric importance and is used when measuring the size of the pelvis.

The vertebral foramina are present and form the **sacral canal.** The laminae of the fifth sacral vertebra, and sometimes those of the fourth sacral vertebra as well, fail to meet in the midline, forming the **sacral hiatus** (Fig. 11-23). The sacral canal contains the anterior and posterior roots of the sacral and coccygeal spinal nerves, the filum terminale, and fibrofatty material. It also contains the lower part of the subarachnoid space down as far as the lower border of the second sacral vertebra.

The anterior and posterior surfaces of the sacrum each have four foramina on each side for the passage of the anterior and posterior rami of the upper four sacral nerves.

### Coccyx

The coccyx consists of four vertebrae fused together to form a single, small triangular bone that articulates at its base with the lower end of the sacrum (Fig. 11-20). The first coccygeal vertebra is usually not fused or is incompletely fused with the second vertebra.

Knowledge of the preceding basic anatomy of the vertebral column is important when interpreting radiographs and when noting the precise sites of bony pathologic features relative to soft-tissue injury.

### Important Variations in the Vertebrae

The number of cervical vertebrae is constant, but the seventh cervical vertebra may possess a **cervical rib.**

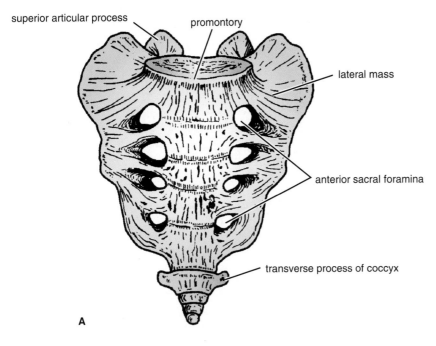

superior articular process

promontory

lateral mass

anterior sacral foramina

transverse process of coccyx

**A**

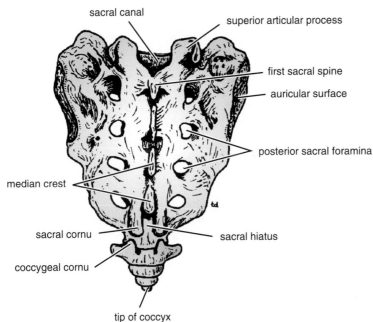

sacral canal

superior articular process

first sacral spine

auricular surface

posterior sacral foramina

median crest

sacral cornu

sacral hiatus

coccygeal cornu

tip of coccyx

**B**

**Figure 11-23**    Sacrum. **A.** Anterior view. **B.** Posterior view.

The thoracic vertebrae may be increased in number by the addition of the first lumbar vertebra, which may have a rib. The fifth lumbar vertebra may be incorporated into the sacrum; this is usually incomplete and may be limited to one side. The first sacral vertebra may remain partially or completely separate from the sacrum and resemble a sixth lumbar vertebra. A large extent of the posterior wall of the sacral canal may be absent because the laminae and spines fail to develop.

The coccyx, which usually consists of four fused vertebrae, may have three or five vertebrae. The first coccygeal vertebra may be separate. In this condition, the free vertebra usually projects downward and anteriorly from the apex of the sacrum.

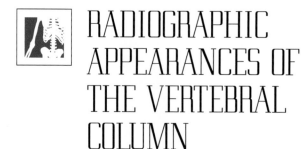

# RADIOGRAPHIC APPEARANCES OF THE VERTEBRAL COLUMN

The radiographic appearances of the vertebral column are shown in Figures 11-24 to 11-29. CT scans and MRIs of the vertebral column are shown in Figures 16-19 to 16-21.

## Thoracic Bones

The thorax or chest has in its walls a bony and cartilaginous cage formed by the sternum, the costal cartilages, the ribs, and the bodies of the thoracic vertebrae (Fig. 11-30). The thoracic cage is cone shaped, having a narrow inlet (clinically referred to as the outlet) superiorly and a broad outlet inferiorly. The cage is flattened from front to back. The function of the cage is to participate in the movements of respiration and to protect the underlying thoracic viscera, especially the heart and the lungs, and the upper abdominal viscera, namely, the liver, the spleen, and the stomach.

## Sternum

The sternum lies in the midline of the anterior chest wall. It is a flat bone that can be divided into three parts: manubrium sterni, body of the sternum, and xiphoid process.

The **manubrium** is the upper part of the sternum, and it articulates with the clavicles and the first and upper part of the second costal cartilages on each side (Fig. 11-30). It lies opposite the third and fourth thoracic vertebrae.

The **body of the sternum** articulates above with the manubrium by means of a fibrocartilaginous joint, the **manubriosternal joint**. Below, it articulates with the xiphoid process at the xiphisternal joint. On each side are notches for articulation with the lower part of the second costal cartilage and the third to the seventh costal cartilages (Fig. 11-30). The second to the seventh costal cartilages articulate with the sternum at synovial joints.

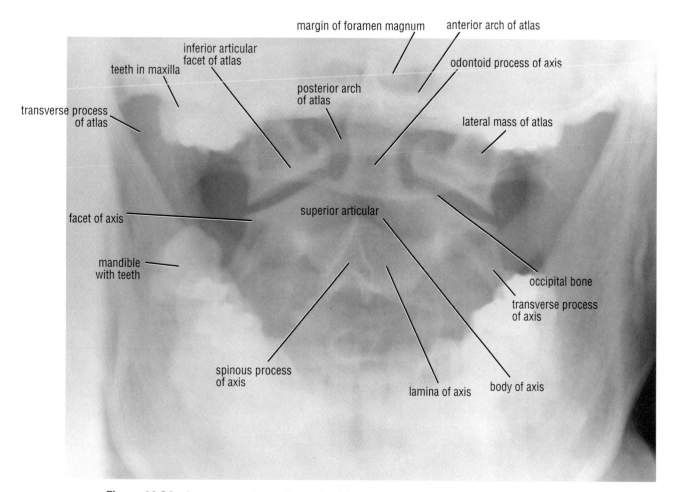

**Figure 11-24** Anteroposterior radiograph of the upper cervical region of the vertebral column with the patient's mouth open to show the odontoid process of the axis.

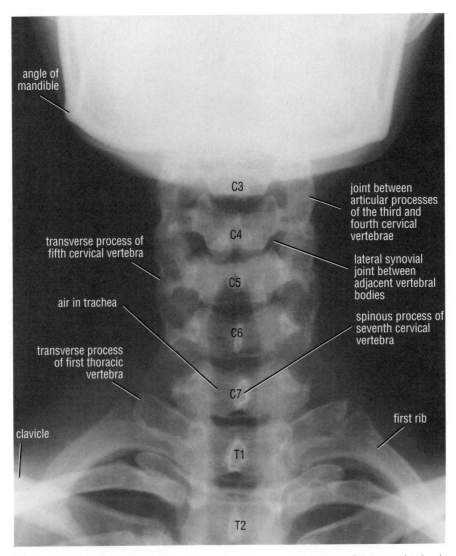

**Figure 11-25**    Anteroposterior radiograph of the cervical region of the vertebral column.

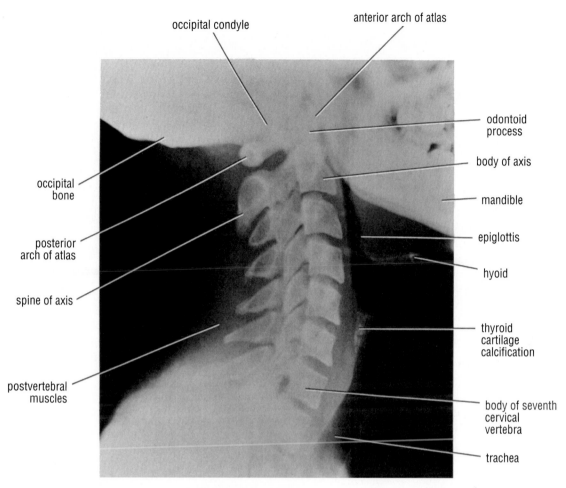

**Figure 11-26**   Lateral radiograph of the cervical region of the vertebral column.

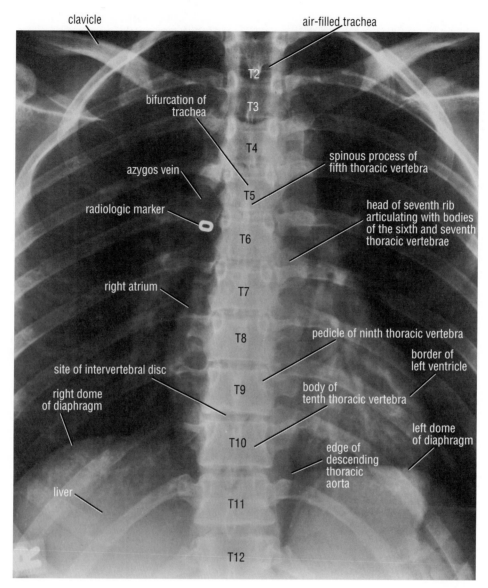

clavicle

air-filled trachea

T2

bifurcation of trachea

T3

T4

spinous process of fifth thoracic vertebra

azygos vein

T5

radiologic marker

head of seventh rib articulating with bodies of the sixth and seventh thoracic vertebrae

T6

right atrium

T7

pedicle of ninth thoracic vertebra

T8

border of left ventricle

site of intervertebral disc

T9

body of tenth thoracic vertebra

right dome of diaphragm

left dome of diaphragm

T10

edge of descending thoracic aorta

liver

T11

T12

**Figure 11-27**   Anteroposterior radiograph of the thoracic region of the vertebral column.

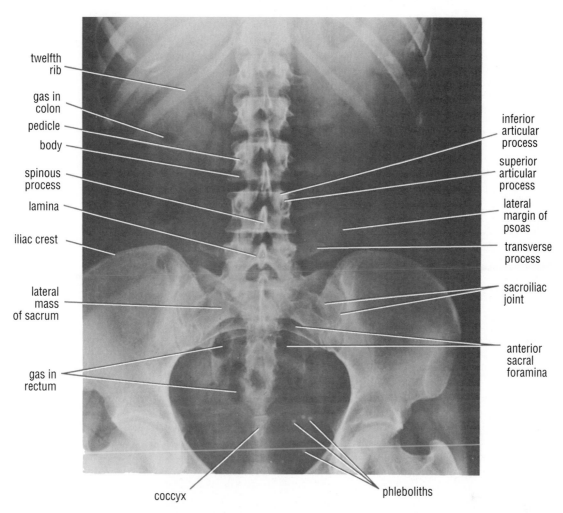

twelfth rib

gas in colon

pedicle

body

spinous process

lamina

iliac crest

lateral mass of sacrum

gas in rectum

inferior articular process

superior articular process

lateral margin of psoas

transverse process

sacroiliac joint

anterior sacral foramina

coccyx

phleboliths

**Figure 11-28**  Anteroposterior radiograph of the lower thoracic, lumbar, and sacral regions of the vertebral column.

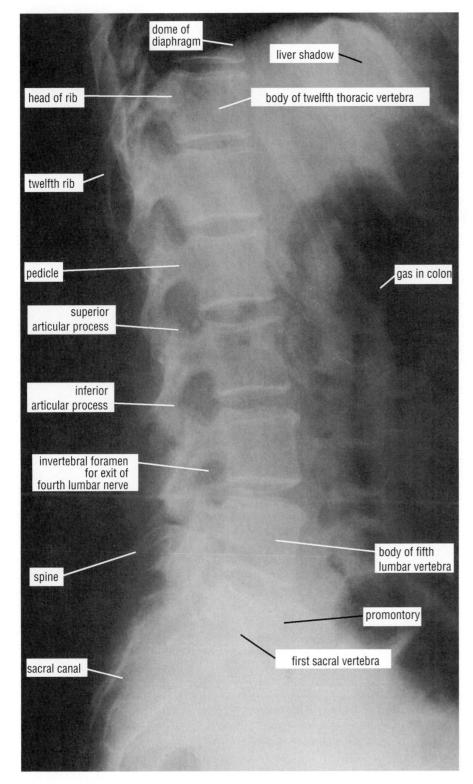

**Figure 11-29** Lateral radiograph of the lower thoracic, lumbar, and sacral regions of the vertebral column.

The **xiphoid process** (Fig. 11-30) is the lowest and smallest part of the sternum. It is a thin plate of hyaline cartilage that becomes ossified at its proximal end in adult life. No ribs or costal cartilages are attached to it.

The **sternal angle** (angle of Louis), formed by the articulation of the manubrium with the body of the sternum, can be recognized by the presence of a transverse ridge on the anterior aspect of the sternum (Fig. 11-30). The transverse ridge lies at the level of the second costal cartilage, the point from which all costal cartilages and ribs are counted. The sternal angle lies opposite the intervertebral disc between the fourth and fifth thoracic vertebrae.

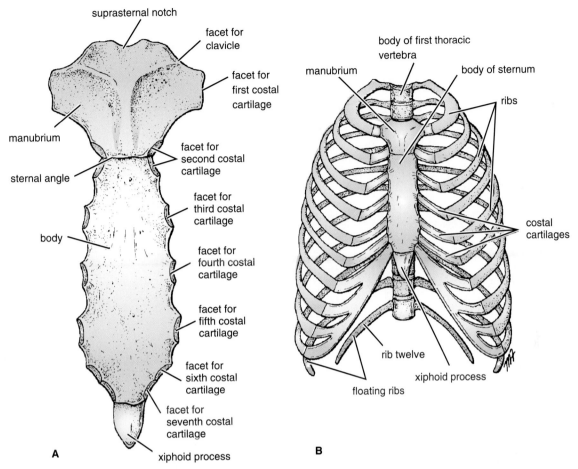

**Figure 11-30  A.** Anterior view of the sternum. **B.** Sternum, ribs, and costal cartilages forming the thoracic skeleton.

The **xiphisternal joint** lies opposite the body of the ninth thoracic vertebra.

## Costal Cartilages

Costal cartilages are bars of hyaline cartilage connecting the upper seven ribs to the lateral edge of the sternum and the eighth, ninth, and tenth ribs to the cartilage immediately above. The cartilages of the eleventh and twelfth ribs end in the abdominal musculature (Fig. 11-30).

The costal cartilages contribute significantly to the elasticity and mobility of the thoracic walls. In old age, the costal cartilages tend to lose some of their flexibility as the result of superficial calcification.

## Ribs

There are twelve pairs of ribs, all of which are attached posteriorly to the thoracic vertebrae (Figs. 11-30, 11-31, and 11-32). The upper seven pairs are attached anteriorly to the sternum by their costal cartilages. The eighth, ninth, and tenth pairs of ribs are attached anteriorly to each

other and to the seventh rib by means of their costal cartilages and small synovial joints. The eleventh and twelfth pairs have no anterior attachment and are referred to as **floating ribs.**

A **typical rib** is a long, twisted, flat bone having a rounded, smooth superior border and a sharp, thin inferior border (Figs. 11-31 and 11-32). The inferior border overhangs and forms the **costal groove,** which accommodates the intercostal vessels and nerve.

A rib has a **head, neck, tubercle, shaft,** and **angle** (Figs. 11-31 and 11-32). The **head** has two facets for articulation with the numerically corresponding vertebral body and that of the vertebra immediately above (Fig. 11-31). The **neck** is a constricted portion situated between the head and the tubercle.

The **tubercle** is a prominence on the outer surface of the rib at the junction of the neck with the shaft. It has a facet for articulation with the transverse process of the numerically corresponding vertebra (Fig. 11-31). The **shaft** or **body** is thin and flattened and twisted on its long axis. Its inferior border has the costal groove. The

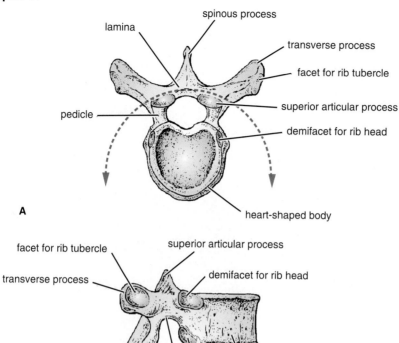

**A**

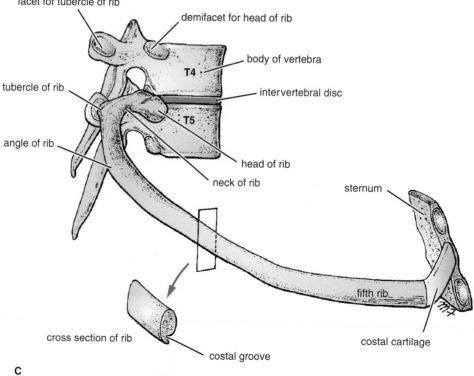

**B**

**C**

**Figure 11-31**   Fifth right rib as it articulates with the vertebral column posteriorly and the sternum anteriorly. Note that the rib head articulates with the vertebral body of its own number and that of the vertebra immediately above. Note also the presence of the costal groove along the inferior border of the rib.

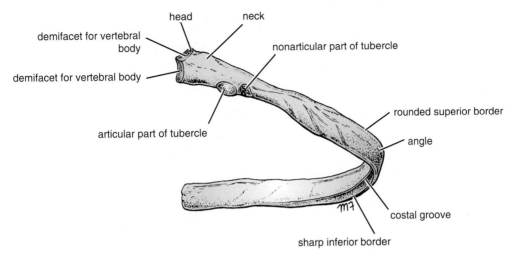

demifacet for vertebral body

head

neck

nonarticular part of tubercle

demifacet for vertebral body

articular part of tubercle

rounded superior border

angle

costal groove

sharp inferior border

**Figure 11-32** Fifth right rib, as seen from the posterior aspect.

angle is where the shaft of the rib bends sharply forward. The anterior end of each rib is attached to the corresponding costal cartilage.

The first rib is **atypical.** It is important clinically because of its close relationship to the lower nerves of the brachial plexus and the main vessels to the arm, namely, the subclavian artery and vein (Fig. 11-33). This rib is flattened from above downward. It has a tubercle on the inner border, known as the **scalene tubercle,** for the insertion of the scalenus anterior muscle. Anterior to the tubercle, the subclavian vein crosses the rib; posterior to the tubercle is the **subclavian groove,** where the subclavian artery and the lower trunk of the brachial plexus cross the rib and lie in contact with the bone.

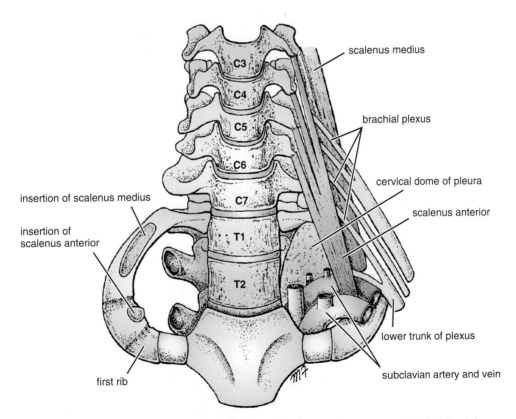

scalenus medius

brachial plexus

cervical dome of pleura

scalenus anterior

insertion of scalenus medius

insertion of scalenus anterior

lower trunk of plexus

subclavian artery and vein

first rib

C3
C4
C5
C6
C7
T1
T2

**Figure 11-33** Thoracic outlet showing the cervical dome of pleura on the left side of the body and its relationship to the inner border of the first rib. Note also the presence of brachial plexus and subclavian vessels. (Anatomists often refer to the thoracic outlet as the thoracic inlet.)

# RADIOGRAPHIC APPEARANCES OF THE THORACIC BONES

The radiographic appearances of the thoracic bones are shown in Figure 11-34.

## Appendicular Skeleton

### Bones of the Upper Limb

The bones of the upper limbs consist of the shoulder girdle, and those of the arm, the forearm, and the hand.

#### Bones of the Shoulder Girdle

The bones of the shoulder girdle consist of the clavicle and the scapula, which articulate with one another at the acromioclavicular joint (Fig. 11-1).

#### Clavicle

The clavicle is a long, slender bone that lies horizontally across the root of the neck. It articulates with the sternum and first costal cartilage medially and with the acromion process of the scapula laterally (Fig. 11-35). The clavicle acts as a strut that holds the arm away from the trunk. It also transmits forces from the upper limb to the axial skeleton and provides attachment for muscles.

The clavicle is subcutaneous throughout its length; its medial two thirds are convex forward and its lateral third is concave forward. The important muscles and ligaments attached to the clavicle are shown in Figure 11-36.

#### Scapula

The scapula is a flat triangular bone (Fig. 11-37) that lies on the posterior thoracic wall between the second and the seventh ribs. On its posterior surface, the **spine of the scapula** projects backward. The lateral end of the spine is free and forms the **acromion,** which articulates with the clavicle. The superolateral angle of the scapula forms the pear-shaped **glenoid cavity,** or **fossa,** which articulates with the head of the humerus at the shoulder joint. The **coracoid process** projects upward and forward above the glenoid cavity and provides attachment for muscles and ligaments. Medial to the base of the coracoid process is the **suprascapular notch** (Fig. 11-37).

The anterior surface of the scapula is concave and forms the shallow **subscapular fossa.** The posterior surface of the scapula is divided by the spine into the **supraspinous fossa** above and an **infraspinous fossa** below (Fig. 11-37). The **inferior angle** of the scapula can be palpated easily in the living subject and marks the level of the seventh rib and the spine of the seventh thoracic vertebra.

The important muscles and ligaments attached to the scapula are shown in Figure 11-37.

#### Bones of the Arm

##### Humerus

The humerus articulates with the scapula at the shoulder joint and with the radius and ulna at the elbow joint. The upper end of the humerus has a head (Fig. 11-38), which forms about one third of a sphere and articulates with the glenoid cavity of the scapula. Immediately below the head is the anatomic neck. Below the neck are the greater and lesser tuberosities, separated from each other by the bicipital groove. Where the upper end of the humerus joins the shaft is a narrow surgical neck. About halfway down the lateral aspect of the shaft is a roughened elevation called the deltoid tuberosity. Behind and below the tuberosity is a spiral groove, which accommodates the radial nerve (Fig. 11-38).

The lower end of the humerus possesses the **medial** and **lateral epicondyles** for the attachment of muscles and ligaments, the rounded **capitulum** for articulation with the head of the radius, and the pulley-shaped **trochlea** for articulation with the trochlear notch of the ulna (Fig. 11-38). Above the capitulum is the **radial fossa,** which receives the head of the radius when the elbow is flexed. Above the trochlea anteriorly is the **coronoid fossa,** which during the same movement receives the coronoid process of the ulna. Above the trochlea posteriorly is the **olecranon fossa,** which receives the olecranon process of the ulna when the elbow joint is extended (Fig. 11-38).

#### Bones of the Forearm

The bones of the forearm are the radius and the ulna.

##### Radius

The radius is the lateral bone of the forearm (Fig. 11-39). Its proximal end articulates with the humerus at the elbow joint and with the ulna at the proximal radioulnar joint. Its distal end articulates with the scaphoid and lunate bones of the hand at the wrist joint and with the ulna at the distal radioulnar joint.

At the proximal end of the radius is the small circular **head** (Fig. 11-39). The upper surface of the head is concave and articulates with the convex capitulum of the humerus. The circumference of the head articulates with the radial notch of the ulna. Below the head, the bone is constricted to form the **neck.** Below the neck is the **bicipital tuberosity** for the insertion of the biceps muscle.

The shaft of the radius, in contradistinction to that of the ulna, is wider below than above (Fig. 11-39). It has a

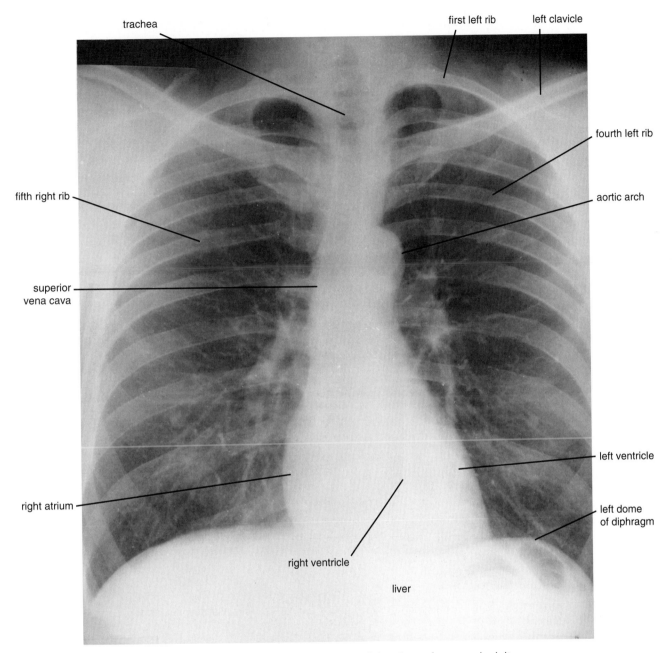

**Figure 11-34** Posteroanterior radiograph of the chest of a normal adult man.

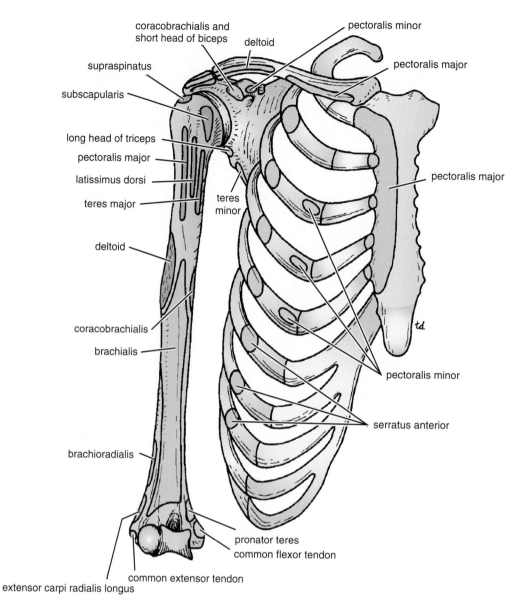

**Figure 11-35** Muscle attachments to the bones of the thorax, clavicle, scapula, and humerus.

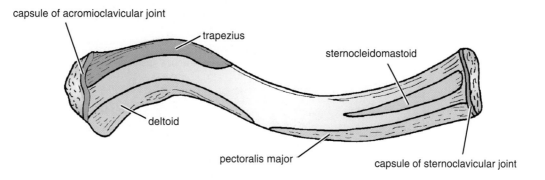

capsule of acromioclavicular joint

trapezius

sternocleidomastoid

deltoid

pectoralis major

capsule of sternoclavicular joint

superior surface

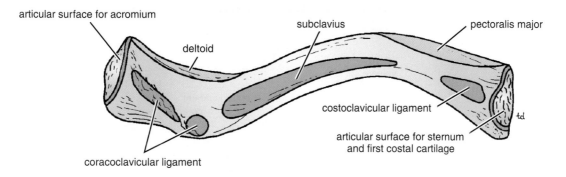

articular surface for acromium

subclavius

pectoralis major

deltoid

coracoclavicular ligament

costoclavicular ligament

articular surface for sternum and first costal cartilage

inferior surface

**Figure 11-36**   Important muscular and ligamentous attachments to the right clavicle.

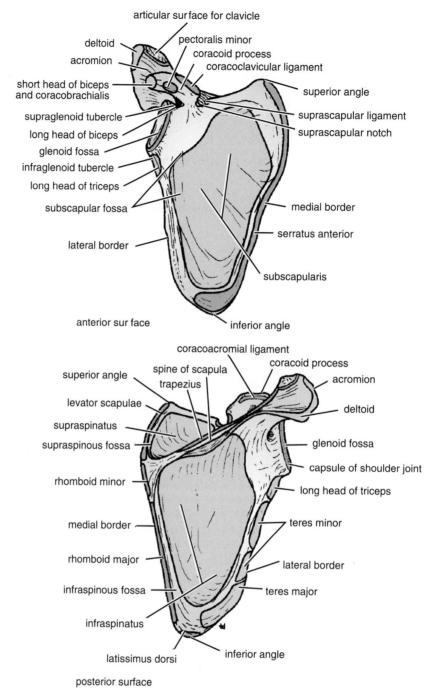

**Figure 11-37** Important muscular and ligamentous attachments to the right scapula.

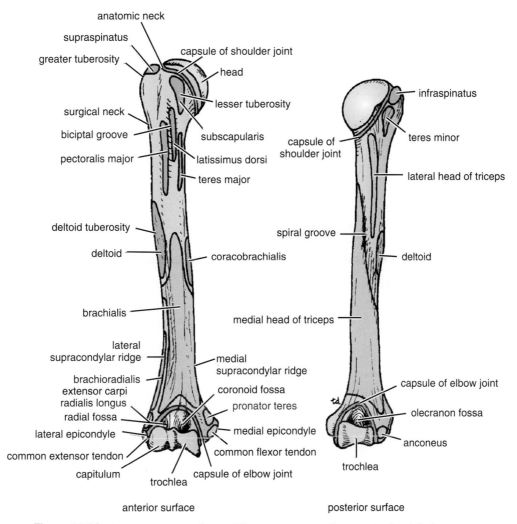

**Figure 11-38** Important muscular and ligamentous attachments to the right humerus.

sharp **interosseous border** medially for the attachment of the interosseous membrane that binds the radius and ulna together. The **pronator tubercle,** for the insertion of the pronator teres muscle, lies halfway down on its lateral side.

At the distal end of the radius is the **styloid process;** this projects distally from its lateral margin (Fig. 11-39). On the medial surface is the **ulnar notch,** which articulates with the round head of the ulna. The inferior articular surface articulates with the scaphoid and lunate bones. On the posterior aspect of the distal end is a small tubercle, the **dorsal tubercle,** which is grooved on its medial side by the tendon of the extensor pollicis longus (Fig. 11-39).

The important muscles and ligaments attached to the radius are shown in Figure 11-39.

## Ulna

The ulna is the medial bone of the forearm (Fig. 11-39). Its proximal end articulates with the humerus at the elbow joint and with the head of the radius at the proximal ra-

dioulnar joint. Its distal end articulates with the radius at the distal radioulnar joint, but it is excluded from the wrist joint by the articular disc.

The proximal end of the ulna is large and is known as the **olecranon process** (Fig. 11-39); this forms the prominence of the elbow. It has a notch on its anterior surface, the **trochlear notch,** which articulates with the trochlea of the humerus. Below the trochlear notch is the triangular **coronoid process,** which has on its lateral surface the **radial notch** for articulation with the head of the radius.

The **shaft** of the ulna tapers from above down (Fig. 11-39). It has a sharp **interosseous border** laterally for the attachment of the interosseous membrane. The posterior border is rounded and subcutaneous and can be easily palpated throughout its length. Below the radial notch is a depression, the **supinator fossa,** which gives clearance for the movement of the bicipital tuberosity of the radius. The posterior border of the fossa is sharp and is known as the **supinator crest;** it gives origin to the supinator muscle.

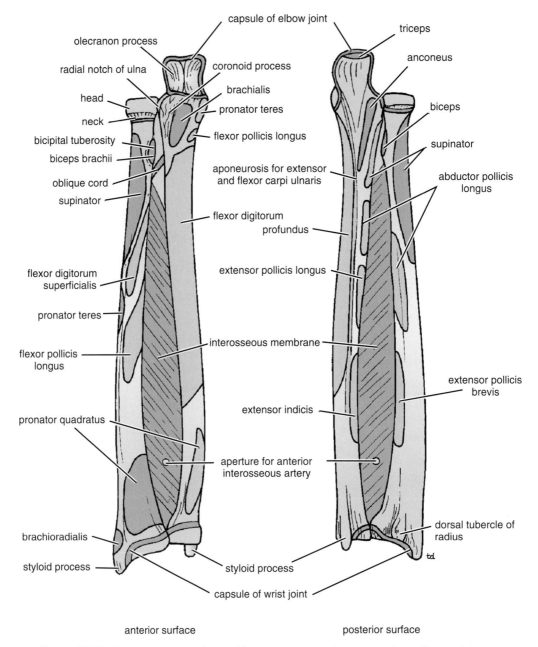

capsule of elbow joint

olecranon process

radial notch of ulna

head

neck

bicipital tuberosity

biceps brachii

oblique cord

supinator

flexor digitorum superficialis

pronator teres

flexor pollicis longus

pronator quadratus

brachioradialis

styloid process

coronoid process

brachialis

pronator teres

flexor pollicis longus

aponeurosis for extensor and flexor carpi ulnaris

flexor digitorum profundus

extensor pollicis longus

interosseous membrane

extensor indicis

aperture for anterior interosseous artery

styloid process

capsule of wrist joint

triceps

anconeus

biceps

supinator

abductor pollicis longus

extensor pollicis brevis

dorsal tubercle of radius

anterior surface

posterior surface

**Figure 11-39**   Important muscular and ligamentous attachments to the radius and the ulna.

At the distal end of the ulna is the small rounded **head,** which has projecting from its medial aspect the **styloid process** (Fig. 11-39).

The important muscles and ligaments attached to the ulna are shown in Figure 11-39.

### Bones of the Hand

There are eight carpal bones, made up of two rows of four (Figs. 11-40 and 11-41). The **proximal row** consists of (from lateral to medial) the **scaphoid, lunate, triquetral,** and **pisiform** bones. The **distal row** consists of (from lateral to medial) the **trapezium, trapezoid, capitate,** and **hamate**

bones. Together, the bones of the carpus present on their anterior surface a concavity, to the lateral and medial edges of which is attached a strong membranous band called the **flexor retinaculum.** In this manner, an osteofascial tunnel, **the carpal tunnel,** is formed for the passage of the median nerve and the flexor tendons of the fingers.

The bones of the hand are cartilaginous at birth. The capitate begins to ossify during the 1st year, and the other bones begin to ossify at intervals thereafter until the 12th year, when all the bones are ossified.

Although detailed knowledge of the bones of the hand is unnecessary for a medical student, the position, shape,

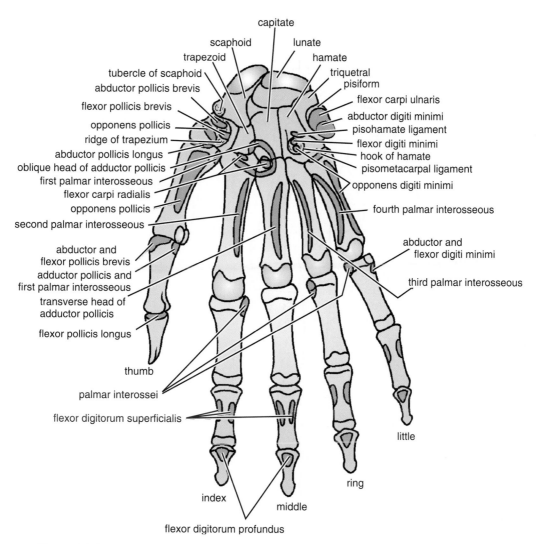

**Figure 11-40** Important muscular attachments to the anterior surfaces of the bones of the hand.

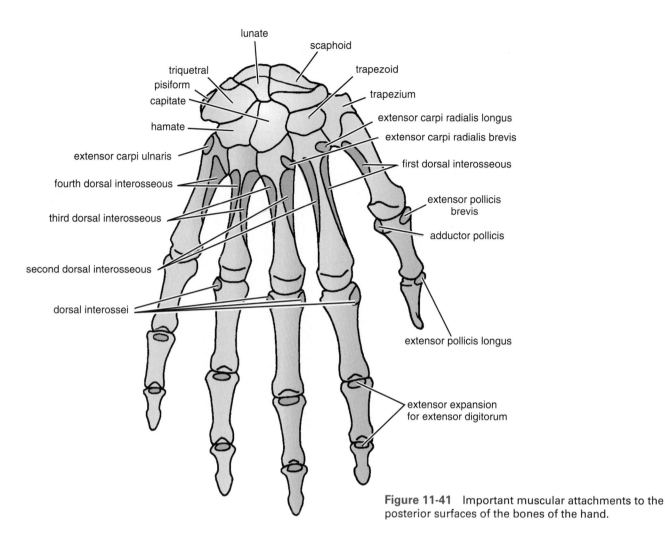

lunate

scaphoid

triquetral

trapezoid

pisiform

trapezium

capitate

extensor carpi radialis longus

hamate

extensor carpi radialis brevis

extensor carpi ulnaris

first dorsal interosseous

fourth dorsal interosseous

extensor pollicis brevis

third dorsal interosseous

adductor pollicis

second dorsal interosseous

dorsal interossei

extensor pollicis longus

extensor expansion for extensor digitorum

**Figure 11-41**  Important muscular attachments to the posterior surfaces of the bones of the hand.

and size of the scaphoid bone should be studied because it is commonly fractured. The ridge of the trapezium and the hook of the hamate should be examined.

### The Metacarpals and Phalanges

There are five metacarpal bones, each of which has a **base**, a **shaft**, and a **head** (Figs. 11-40 and 11-41).

The first metacarpal bone of the thumb is the shortest and most mobile. It does not lie in the same plane as the others but occupies a more anterior position. It is also rotated medially through a right angle so that its extensor surface is directed laterally and not backward.

The bases of the metacarpal bones articulate with the distal row of the carpal bones; the heads, which form the knuckles, articulate with the proximal phalanges (Figs. 11-40 and 11-41). The shaft of each metacarpal bone is slightly concave forward and is triangular in transverse section. Its surfaces are posterior, lateral, and medial.

There are three phalanges for each of the fingers but only two for the thumb.

The important muscles attached to the bones of the hand and fingers are shown in Figures 11-40 and 11-41.

# RADIOGRAPHIC APPEARANCES OF THE BONES OF THE UPPER LIMB

The radiographic appearances of the bones of the upper limb are shown in Figures 11-42 to 11-49.

## Bones of the Lower Limb

The bones of the lower limbs consist of the pelvic girdle, and those of the thigh, the leg, and the foot.

### Bones of the Pelvic Girdle

The pelvic girdle is composed of four bones: the two **hip bones**, the **sacrum**, and the **coccyx** (Figs. 11-1 and 11-50). The pelvic girdle provides a strong and stable connection

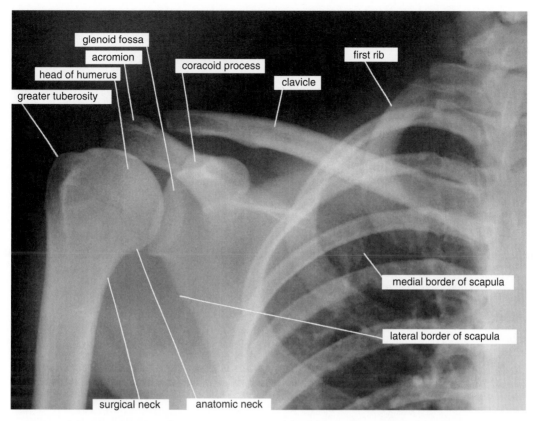

Figure 11-42   Anteroposterior radiograph of the shoulder region in the adult.

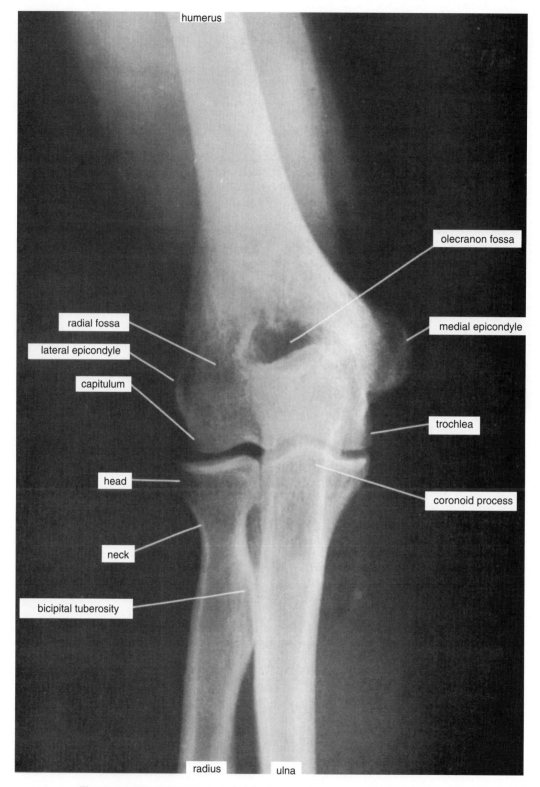

**Figure 11-43**    Anteroposterior radiograph of the elbow region in the adult.

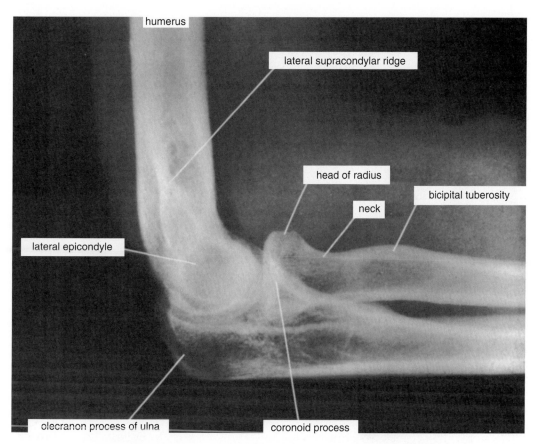

**Figure 11-44** Lateral radiograph of the elbow region in the adult.

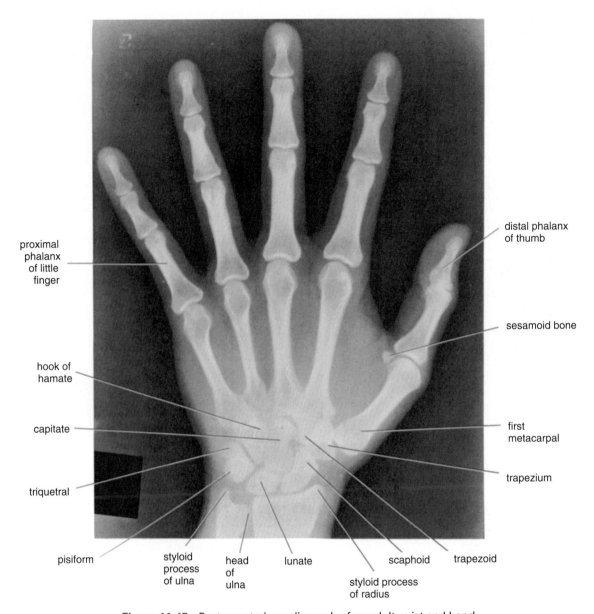

**Figure 11-45**   Posteroanterior radiograph of an adult wrist and hand.

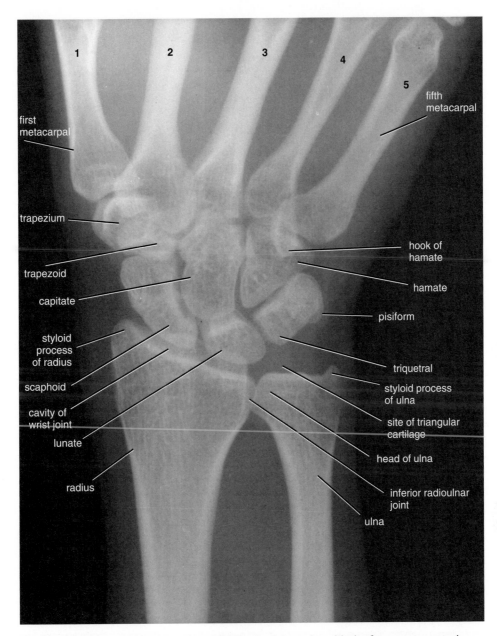

first
metacarpal

trapezium

trapezoid

capitate

styloid
process
of radius

scaphoid

cavity of
wrist joint

lunate

radius

fifth
metacarpal

hook of
hamate

hamate

pisiform

triquetral

styloid process
of ulna

site of triangular
cartilage

head of ulna

inferior radioulnar
joint

ulna

**Figure 11-46**   Posteroanterior radiograph of the wrist with the forearm pronated.

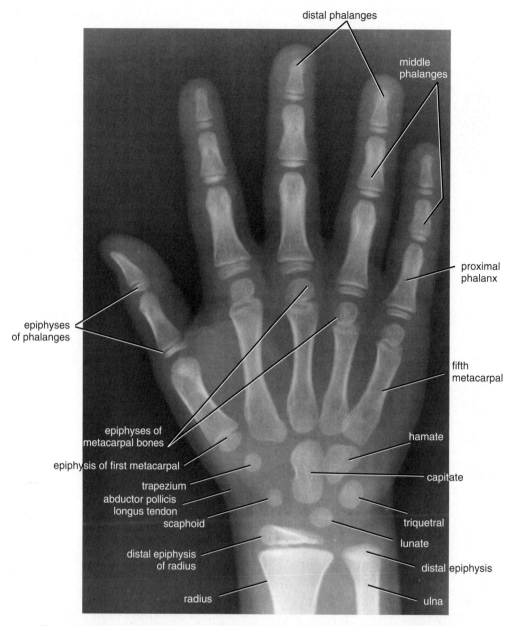

**Figure 11-47** Posteroanterior radiograph of the wrist and hand of an 8-year-old boy.

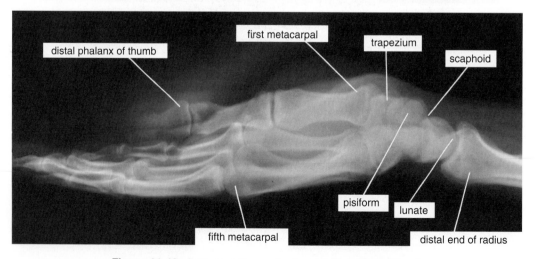

**Figure 11-48** Lateral radiograph of an adult wrist and hand.

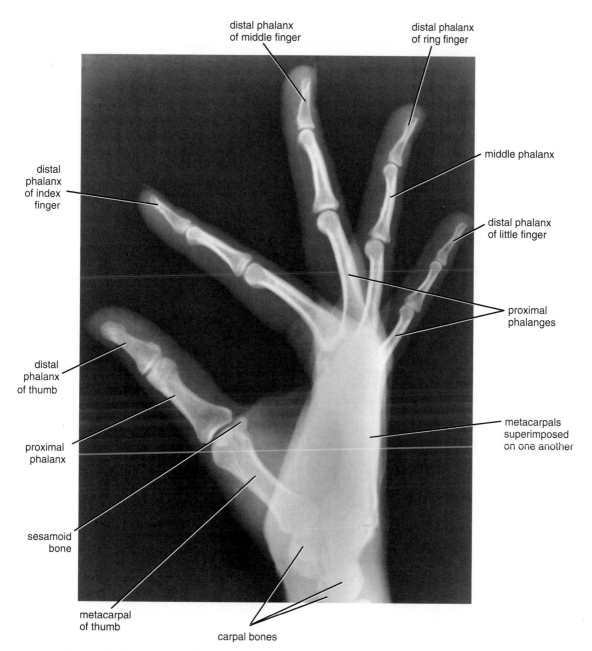

distal phalanx
of middle finger

distal phalanx
of ring finger

middle phalanx

distal
phalanx
of index
finger

distal phalanx
of little finger

proximal
phalanges

distal
phalanx
of thumb

metacarpals
superimposed
on one another

proximal
phalanx

sesamoid
bone

metacarpal
of thumb

carpal bones

**Figure 11-49**  Lateral radiograph of an adult wrist and hand with the fingers at different degrees of flexion.

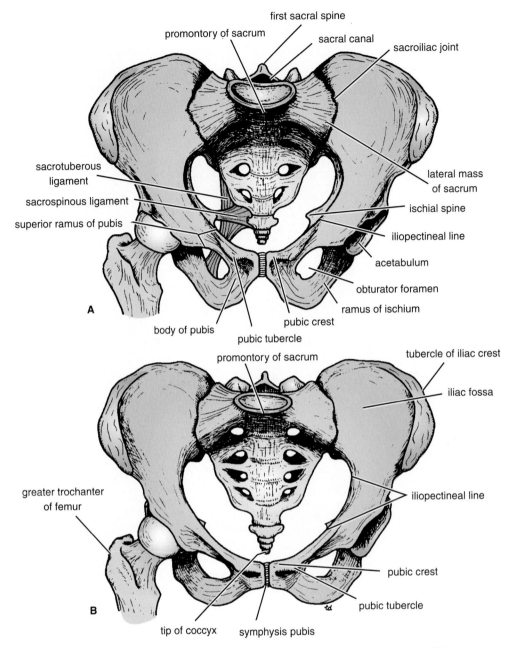

first sacral spine

promontory of sacrum

sacral canal

sacroiliac joint

sacrotuberous ligament

sacrospinous ligament

superior ramus of pubis

lateral mass of sacrum

ischial spine

iliopectineal line

acetabulum

obturator foramen

ramus of ischium

**A**

body of pubis

pubic crest

pubic tubercle

promontory of sacrum

tubercle of iliac crest

iliac fossa

greater trochanter of femur

iliopectineal line

pubic crest

pubic tubercle

**B**

tip of coccyx

symphysis pubis

**Figure 11-50** Anterior view of the male pelvis (**A**) and female pelvis (**B**).

between the trunk and the lower limbs; it has to be strong because of the large weight it carries. It is interesting to compare the pelvic girdle with the shoulder girdle because the latter is relatively unstable but possesses a great deal more mobility than the pelvic girdle.

The two hip bones articulate with each other anteriorly at the **symphysis pubis** and posteriorly with the sacrum at the **sacroiliac joints** (Fig. 11-50). The pelvic girdle with its joints forms a strong basin-shaped structure called the **pelvis.**

The pelvis is divided into two parts by the **pelvic brim,** which is formed by the **sacral promontory** behind (Fig. 11-

51), the **iliopectineal lines** laterally, and the **symphysis pubis** anteriorly. Above the brim is the **false pelvis,** which forms part of the abdominal cavity. Below the brim is the **true pelvis.**

### Orientation of the Pelvis

It is important for the student, at the outset, to understand the correct orientation of the bony pelvis relative to the trunk, with the individual standing in the anatomic position. The front of the symphysis pubis and the anterior superior iliac spines should lie in the same vertical plane (Fig. 11-51). This means that the pelvic surface of the symphysis pubis faces upward and backward and the an-

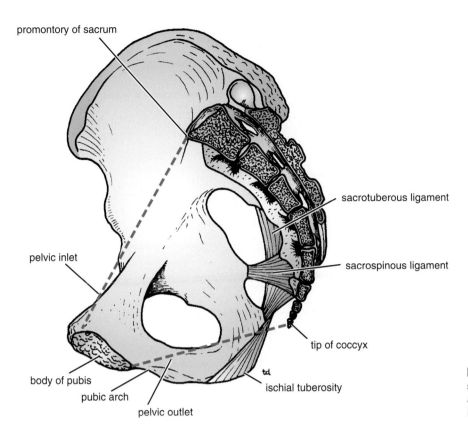

promontory of sacrum

sacrotuberous ligament

sacrospinous ligament

pelvic inlet

tip of coccyx

body of pubis

pubic arch

ischial tuberosity

pelvic outlet

**Figure 11-51** Right half of the pelvis showing the pelvic inlet, pelvic outlet, and sacrotuberous and sacrospinous ligaments.

terior surface of the sacrum is directed forward and downward.

### False Pelvis

The false pelvis is of little clinical importance. It is bounded behind by the lumbar vertebrae, laterally by the iliac fossae and the iliacus muscles, and in front by the lower part of the anterior abdominal wall. The false pelvis flares out at its upper end and should be considered as part of the abdominal cavity. It supports the abdominal contents and, after the third month of pregnancy, helps support the gravid uterus. During the early stages of labor, it helps guide the fetus into the true pelvis.

### True Pelvis

Knowledge of the shape and dimensions of the female pelvis is of great importance for obstetrics because it is the bony canal through which the child passes during birth.

The true pelvis has an inlet, an outlet, and a cavity. The **pelvic inlet,** or **pelvic brim** (Fig. 11-51), is bounded posteriorly by the sacral promontory, laterally by the iliopectineal lines, and anteriorly by the symphysis pubis (Fig. 11-50).

The **pelvic outlet** (Fig. 11-51) is bounded posteriorly by the coccyx, laterally by the ischial tuberosities, and anteriorly by the **pubic arch** (Figs. 11-51 and 11-52). The pelvic outlet does not present a smooth outline but has

three wide notches. Anteriorly, the pubic arch is between the ischiopubic rami, and laterally, are the sciatic notches. The sciatic notches are divided by the **sacrotuberous** and **sacrospinous ligaments** (Figs. 11-50 and 11-51) into the **greater** and **lesser sciatic foramina.** From an obstetric standpoint, because the sacrotuberous ligaments are strong and relatively inflexible, they should be considered to form part of the perimeter of the pelvic outlet. Thus, the outlet is diamond shaped, with the ischiopubic rami and the symphysis pubis forming the boundaries in front and the sacrotuberous ligaments and the coccyx forming the boundaries behind.

The **pelvic cavity** lies between the inlet and the outlet. It is a short, curved canal, with a shallow anterior wall and a much deeper posterior wall (Fig. 11-51).

### Hip Bone

In children, each hip bone consists of the ilium, which lies superiorly; the ischium, which lies posteriorly and inferiorly; and the pubis, which lies anteriorly and inferiorly (Fig. 11-52). The three separate bones are joined by cartilage at the **acetabulum.** At puberty, these three bones fuse together to form one large, irregular bone. The hip bones articulate with the sacrum at the sacroiliac joints and form the anterolateral walls of the pelvis; they also articulate with one another anteriorly at the symphysis pubis.

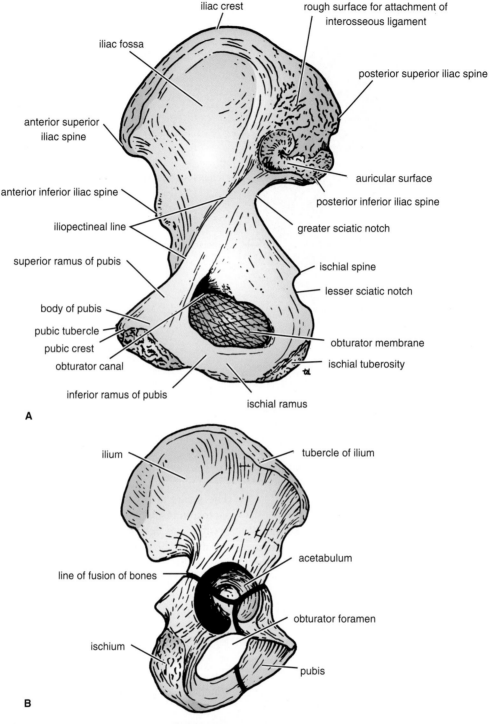

**Figure 11-52**  Right hip bone. **A.** Medial surface. **B.** Lateral surface. Note the lines of fusion between the three bones—the ilium, the ischium, and the pubis.

On the outer surface of the hip bone is a deep depression, the **acetabulum**, which articulates with the hemispherical head of the femur (Figs. 11-50, 11-52, 11-53, and 11-54). Behind the acetabulum is a large notch, the **greater sciatic notch**, which is separated from the **lesser sciatic notch** by the **spine of the ischium**. The sciatic notches are converted into the **greater** and **lesser sciatic foramina** by the presence of the **sacrotuberous** and **sacrospinous ligaments** (Fig. 11-51).

The **ilium**, which is the upper flattened part of the hip bone, possesses the **iliac crest** (Fig. 11-52). The iliac crest runs between the **anterior** and **posterior superior iliac**

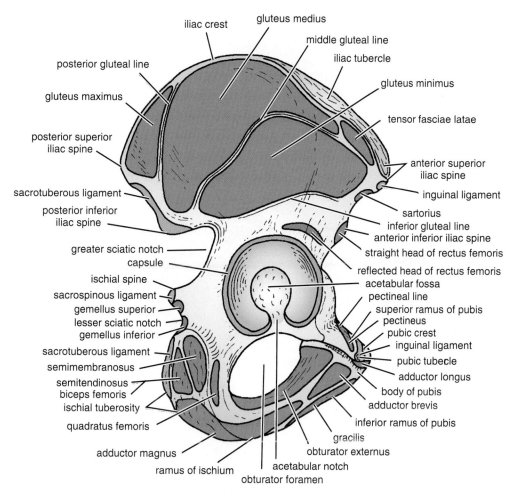

**Figure 11-53** Muscles and ligaments attached to the external surface of the right hip bone.

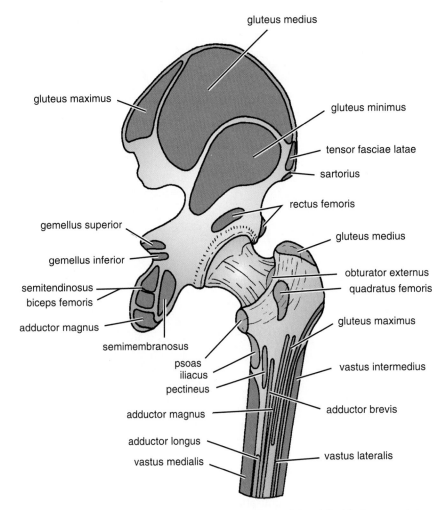

**Figure 11-54** Muscles attached to the external surface of the right hip bone and the posterior surface of the femur.

spines. Below these spines are the corresponding anterior and posterior inferior iliac spines. On the inner surface of the ilium is the large **auricular surface** for articulation with the sacrum. The **iliopectineal line** runs downward and forward around the inner surface of the ilium and serves to divide the false from the true pelvis.

The **ischium** is the inferior and posterior part of the hip bone and possesses an **ischial spine** and an **ischial tuberosity** (Fig. 11-52).

The **pubis** is the anterior part of the hip bone and has a **body** and **superior** and **inferior pubic rami.** The body of the pubis bears the **pubic crest** and the **pubic tubercle** and articulates with the pubic bone of the opposite side at the **symphysis pubis** (Fig. 11-50). In the lower part of the hip bone is a large opening, the **obturator foramen,** which is bounded by the parts of the ischium and pubis. The obturator foramen is filled in by the obturator membrane (Fig. 11-52).

### Sex Differences of the Pelvis

The sex differences of the bony pelvis are easily recognized. The more obvious differences result from the adaptation of the female pelvis for childbearing. The stronger muscles in the male are responsible for the thicker bones and more prominent bony markings (Figs. 11-50 and 11-55).

- The false pelvis is shallow in the female and deep in the male.
- The pelvic inlet is transversely oval in the female but heart shaped in the male because of the indentation produced by the promontory of the sacrum in the male.
- The pelvic cavity is roomier in the female than in the male, and the distance between the inlet and the outlet is much shorter.
- The pelvic outlet is larger in the female than in the male. The ischial tuberosities are everted in the female and turned in in the male.
- The sacrum is shorter, wider, and flatter in the female than in the male.
- The subpubic angle, or pubic arch, is more rounded and wider in the female than in the male.

### Bones of the Thigh

The bones of the thigh consist of the femur and the patella or kneecap.

### Femur

The femur articulates above with the acetabulum to form the hip joint and below with the tibia and the patella to form the knee joint.

The upper end of the femur has a head, a neck, and greater and lesser trochanters (Figs. 11-56 and 11-57). The **head** forms about two thirds of a sphere and articulates with the acetabulum of the hip bone to form the hip joint (Fig. 11-54). In the center of the head is a small depression, called the **fovea capitis,** for the attachment of the ligament of the head. Part of the blood supply to the head of the femur from

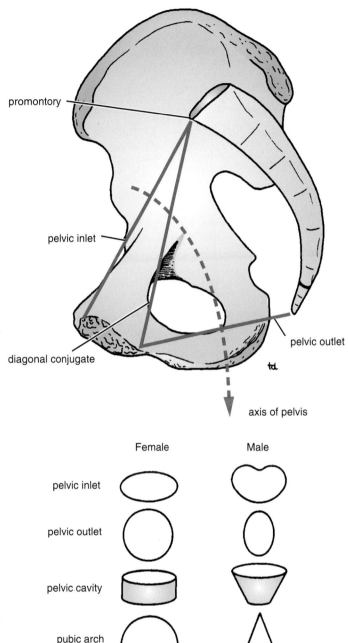

Figure 11-55 Pelvic inlet, pelvic outlet, diagonal conjugate, and axis of the pelvis. Some of the main differences between the female and the male pelvis are also shown.

the obturator artery is conveyed along this ligament and enters the bone at the fovea.

The **neck,** which connects the head to the shaft, passes downward, backward, and laterally and makes an angle of about 125° (slightly less in the female) with the long axis of the shaft. The size of this angle can be altered by disease.

The **greater** and **lesser trochanters** are large eminences situated at the junction of the neck and the shaft (Figs. 11-56 and 11-57). Connecting the two trochanters are the **intertrochanteric line** anteriorly, where the

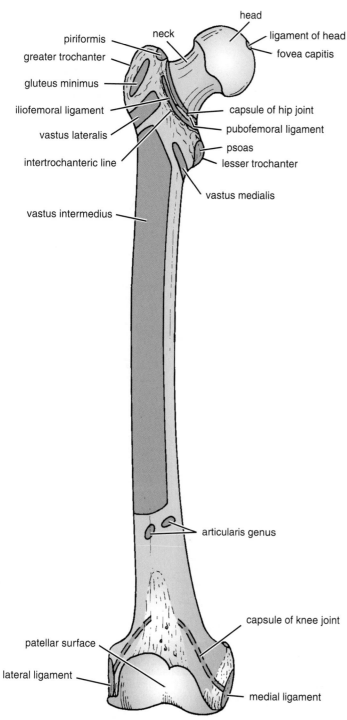

**Figure 11-56** Muscles and ligaments attached to the anterior surface of the right femur.

iliofemoral ligament is attached, and a prominent **intertrochanteric crest** posteriorly, on which is the **quadrate tubercle** (Fig. 11-57).

The **shaft** of the femur is smooth and rounded on its anterior surface but posteriorly has a ridge, the **linea aspera** (Fig. 11-57), to which are attached muscles and intermuscular septa. The margins of the linea aspera diverge above

and below. The medial margin continues below as the **medial supracondylar ridge** to the **adductor tubercle** on the medial condyle (Fig. 11-57). The lateral margin becomes continuous below with the **lateral supracondylar ridge**. On the posterior surface of the shaft below the greater trochanter is the **gluteal tuberosity** for the attachment of the gluteus maximus muscle. The shaft becomes broader toward its distal end and forms a flat, triangular area on its posterior surface called the **popliteal surface** (Fig. 11-57).

The lower end of the femur has **lateral** and **medial** condyles, separated posteriorly by the **intercondylar notch**. The anterior surfaces of the condyles are joined by an articular surface for the patella. The two condyles take part in the formation of the knee joint. Above the condyles are the **medial** and **lateral epicondyles** (Fig. 11-57). The adductor tubercle is continuous with the medial epicondyle.

The important muscles and ligaments attached to the femur are shown in Figures 11-56 and 11-57.

### Patella

The patella (Fig. 11-58) is the largest sesamoid bone (i.e., a bone that develops within the tendon of the quadriceps femoris muscle in front of the knee joint). It is triangular, and its apex lies inferiorly; the apex is connected to the tuberosity of the tibia by the **ligamentum patellae**. The posterior surface articulates with the condyles of the femur. The patella is situated in an exposed position in front of the knee joint and can easily be palpated through the skin. It is separated from the skin by an important subcutaneous bursa (see Fig. 12-28).

The upper, lateral, and medial margins give attachment to the different parts of the quadriceps femoris muscle. It is prevented from being displaced laterally during the action of the quadriceps muscle by the lower horizontal fibers of the vastus medialis and by the large size of the lateral condyle of the femur.

### Bones of the Leg

### Tibia

The tibia is the large weight-bearing medial bone of the leg (Figs. 11-58 and 11-59). It articulates with the condyles of the femur and the head of the fibula above and with the talus and the distal end of the fibula below. It has an expanded upper end, a smaller lower end, and a shaft.

At the upper end are the **lateral** and **medial condyles** (sometimes called lateral and medial **tibial plateaus**), which articulate with the lateral and medial condyles of the femur, with the **lateral** and **medial menisci** intervening. Separating the upper articular surfaces of the tibial condyles are **anterior** and **posterior intercondylar areas**; lying between these areas is the **intercondylar eminence** (Fig. 11-58).

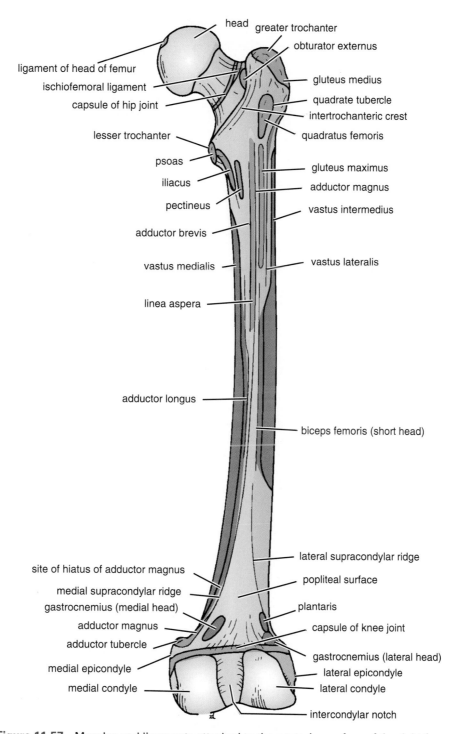

**Figure 11-57** Muscles and ligaments attached to the posterior surface of the right femur.

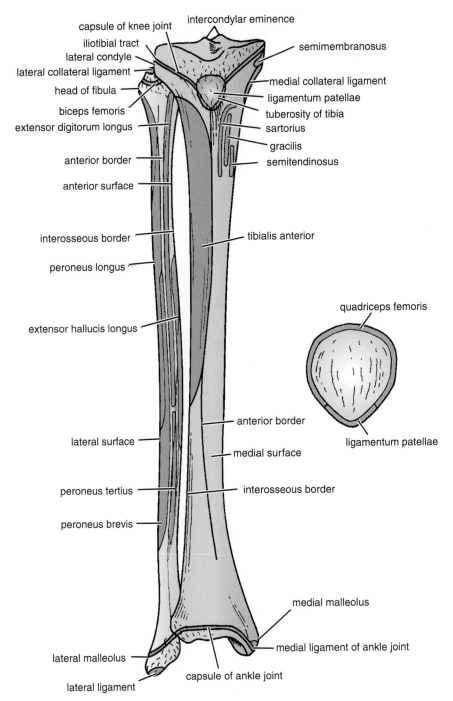

**Figure 11-58** Muscles and ligaments attached to the anterior surfaces of the right tibia and fibula. Attachments to the patella are also shown.

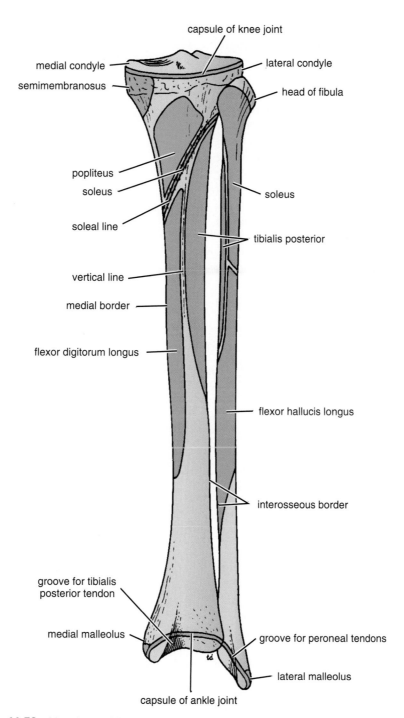

capsule of knee joint

medial condyle

semimembranosus

lateral condyle

head of fibula

popliteus

soleus

soleal line

soleus

tibialis posterior

vertical line

medial border

flexor digitorum longus

flexor hallucis longus

interosseous border

groove for tibialis
posterior tendon

medial malleolus

groove for peroneal tendons

lateral malleolus

capsule of ankle joint

**Figure 11-59** Muscles and ligaments attached to the posterior surfaces of the right tibia
and the fibula.

The lateral condyle possesses on its lateral aspect a small **circular articular facet for the head of the fibula.** The medial condyle has on its posterior aspect the insertion of the semimembranosus muscle (Fig. 11-59).

The **shaft of the tibia** is triangular in cross section, presenting three borders and three surfaces. Its anterior and medial borders, with the medial surface between them, are subcutaneous. The anterior border is prominent and forms the shin. At the junction of the anterior border with the upper end of the tibia is the **tuberosity,** which receives the attachment of the ligamentum patellae. The anterior border becomes rounded below, where it becomes continuous with the medial malleolus. The lateral or interosseous border gives attachment to the interosseous membrane.

The posterior surface of the shaft shows an oblique line, the **soleal line** (Fig. 11-59), for the attachment of the soleus muscle.

The lower end of the tibia is slightly expanded and, on its inferior aspect, shows a saddle-shaped articular surface for the talus. The lower end is prolonged downward medially to form the **medial malleolus.** The lateral surface of the medial malleolus articulates with the talus. The lower end of the tibia shows a wide, rough depression on its lateral surface for articulation with the fibula.

The important muscles and ligaments attached to the tibia are shown in Figures 11-58 and 11-59.

## Fibula

The fibula is the slender lateral bone of the leg (Figs. 11-58 and 11-59). It takes no part in the articulation at the knee joint, but below it forms the lateral malleolus of the ankle joint. It takes no part in the transmission of body weight, but it provides attachment for muscles. The fibula has an expanded upper end, a shaft, and a lower end.

The **upper end,** or **head,** is surmounted by a **styloid process.** It possesses an **articular surface** for articulation with the lateral condyle of the tibia.

The **shaft of the fibula** is long and slender. Typically, it has four borders and four surfaces. The medial or interosseous border gives attachment to the interosseous membrane.

The **lower end of the fibula** forms the triangular lateral malleolus, which is subcutaneous. On the medial surface of the lateral malleolus is a triangular **articular facet** for articulation with the lateral aspect of the talus. Below and behind the articular facet is a depression called the **malleolar fossa.**

The important muscles and ligaments attached to the fibula are shown in Figures 11-58 and 11-59.

## Bones of the Foot

The bones of the foot are the **tarsal bones,** the **metatarsals,** and the **phalanges.**

## Tarsal Bones

The tarsal bones are the calcaneum, the talus, the navicular, the cuboid, and the three cuneiform bones. Only the talus articulates with the tibia and the fibula at the ankle joint.

## Calcaneum

The calcaneum is the largest bone of the foot and forms the prominence of the heel (Figs. 11-60, 11-61, and 11-62). It articulates above with the talus and in front with the cuboid. It has six surfaces.

- The **anterior surface** is small and forms the articular facet that articulates with the cuboid bone.
- The **posterior surface** forms the prominence of the heel and gives attachment to the tendo calcaneus (Achilles tendon).
- The **superior surface** is dominated by two articular facets for the talus, separated by a roughened groove, the **sulcus calcanei.**
- The **inferior surface** has an **anterior tubercle** in the midline and a large **medial** and a smaller **lateral** tubercle at the junction of the inferior and posterior surfaces.
- The **medial surface** possesses a large, shelflike process, termed the **sustentaculum tali,** which assists in the support of the talus.
- The **lateral surface** is almost flat. On its anterior part is a small elevation called the **peroneal tubercle,** which separates the tendons of the peroneus longus and brevis muscles.

The important muscles and ligaments attached to the calcaneum are shown in Figures 11-61 and 11-62.

## Talus

The talus articulates above at the ankle joint with the tibia and fibula, below with the calcaneum, and in front with the navicular bone. It possesses a head, a neck, and a body (Figs. 11-60 and 11-61).

The **head** of the talus is directed distally and has an oval convex articular surface for articulation with the navicular bone. This articular surface is continued on its inferior surface, where it rests on the sustentaculum tali behind and the calcaneonavicular ligament in front.

The **neck** of the talus lies posterior to the head and is slightly narrowed. Its upper surface is roughened and gives attachment to ligaments, and its lower surface shows a deep groove, the **sulcus tali.** The sulcus tali and the sulcus calcanei in the articulated foot form a tunnel, the **sinus tarsi,** which is occupied by the strong **interosseous talocalcaneal ligament.**

The **body** of the talus is cuboidal. Its superior surface articulates with the distal end of the tibia; it is convex from before backward and slightly concave from side to side. Its lateral surface presents a triangular **articular facet** for articulation with the lateral malleolus of the fibula. Its medial surface has a small, comma-shaped **articular facet** for articulation with the medial malleolus of the tibia. The posterior surface is marked by two small **tubercles,** separated by a groove for the flexor hallucis longus tendon.

Numerous important ligaments are attached to the talus, but no muscles are attached to this bone.

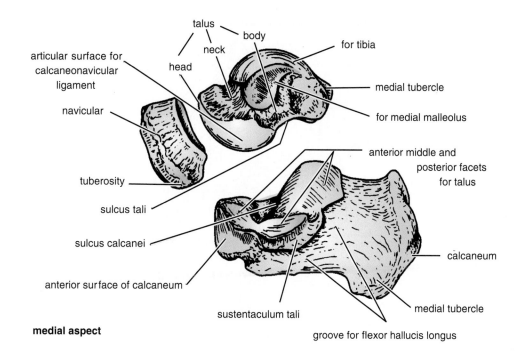

talus
body
neck
head
for tibia
articular surface for calcaneonavicular ligament
navicular
medial tubercle
for medial malleolus
tuberosity
anterior middle and posterior facets for talus
sulcus tali
sulcus calcanei
calcaneum
anterior surface of calcaneum
medial tubercle
sustentaculum tali
groove for flexor hallucis longus

**medial aspect**

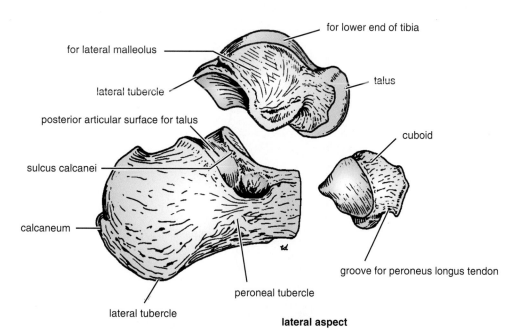

for lower end of tibia
for lateral malleolus
lateral tubercle
talus
posterior articular surface for talus
cuboid
sulcus calcanei
calcaneum
groove for peroneus longus tendon
peroneal tubercle
lateral tubercle

**lateral aspect**

**Figure 11-60** Calcaneum, talus, navicular, and cuboid bones.

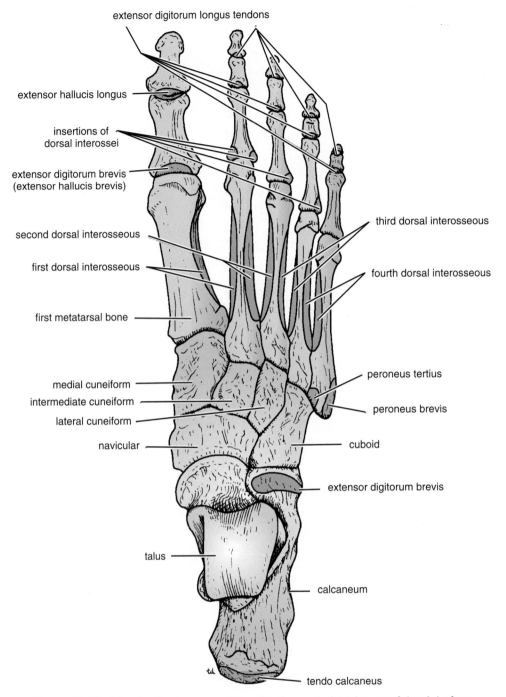

extensor digitorum longus tendons

extensor hallucis longus

insertions of dorsal interossei

extensor digitorum brevis (extensor hallucis brevis)

second dorsal interosseous

first dorsal interosseous

first metatarsal bone

medial cuneiform

intermediate cuneiform

lateral cuneiform

navicular

talus

third dorsal interosseous

fourth dorsal interosseous

peroneus tertius

peroneus brevis

cuboid

extensor digitorum brevis

calcaneum

tendo calcaneus

**Figure 11-61** Muscle attachments on the dorsal aspect of the bones of the right foot.

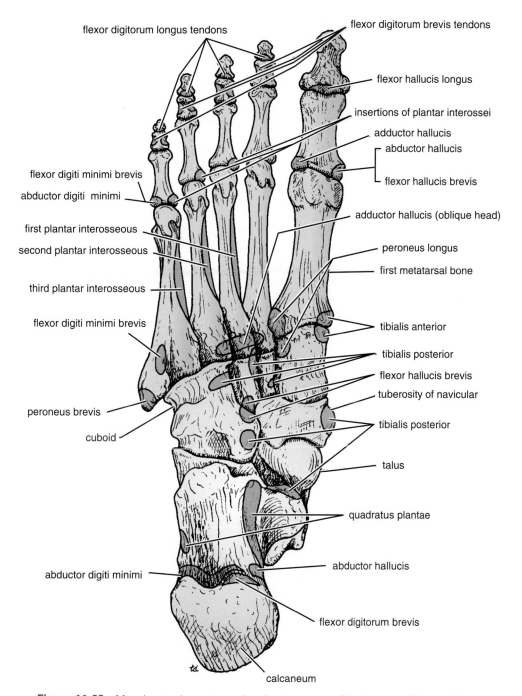

flexor digitorum longus tendons

flexor digitorum brevis tendons

flexor hallucis longus

insertions of plantar interossei

adductor hallucis

abductor hallucis

flexor hallucis brevis

flexor digiti minimi brevis

abductor digiti minimi

first plantar interosseous

second plantar interosseous

adductor hallucis (oblique head)

peroneus longus

first metatarsal bone

third plantar interosseous

flexor digiti minimi brevis

tibialis anterior

tibialis posterior

flexor hallucis brevis

tuberosity of navicular

peroneus brevis

tibialis posterior

cuboid

talus

quadratus plantae

abductor hallucis

abductor digiti minimi

flexor digitorum brevis

calcaneum

**Figure 11-62** Muscle attachments on the plantar aspect of the bones of the right foot.

The remaining tarsal bones should be identified and the following important features noted.

### Navicular Bone

The tuberosity of the navicular bone (Figs. 11-60, 11-61, and 11-62) can be seen and felt on the medial border of the foot 1 in. (2.5 cm) in front of and below the medial malleolus; it gives attachment to the main part of the tibialis posterior tendon.

### Cuboid Bone

A deep groove on the inferior aspect of the cuboid bone (Figs. 11-60, 11-61, and 11-62) lodges the tendon of the peroneus longus muscle.

### Cuneiform Bones

The three small, wedge-shaped cuneiform bones (Figs. 11-61 and 11-62) articulate proximally with the navicular bone and distally with the first three metatarsal bones. Their wedge shape contributes greatly to the formation and maintenance of the transverse arch of the foot

The tarsal bones, unlike those of the carpus, start to ossify before birth. Centers of ossification for the calcaneum and the talus, and often for the cuboid, are present at birth. By the fifth year, ossification is taking place in all the tarsal bones.

### Metatarsals and Phalanges

The metatarsal bones and phalanges (Figs. 11-61 and 11-62) resemble the metacarpals and phalanges of the hand, and each possesses a **head** distally, a **shaft**, and a **base** proximally. The five metatarsals are numbered from the medial to the lateral side.

The **first metatarsal bone** is large and strong and plays an important role in supporting the weight of the body. The head is grooved on its inferior aspect by the medial and lateral **sesamoid bones** in the tendons of the flexor hallucis brevis.

The **fifth metatarsal** has a prominent tubercle on its base that can be easily palpated along the lateral border of the foot. The tubercle gives attachment to the peroneus brevis tendon.

Each toe has three phalanges except the big toe, which possesses only two.

# RADIOGRAPHIC APPEARANCES OF THE BONES OF THE LOWER LIMB

The radiographic appearances of the bones of the lower limb are shown in Figures 11-63 to 11-74.

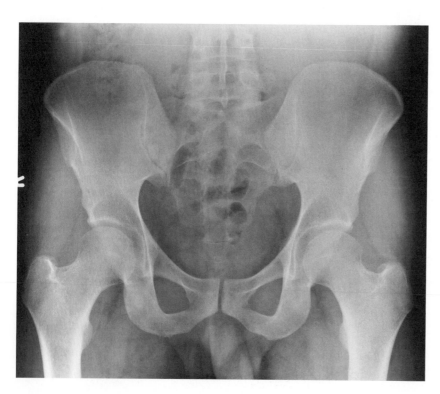

**Figure 11-63**  Anteroposterior radiograph of the male pelvis.

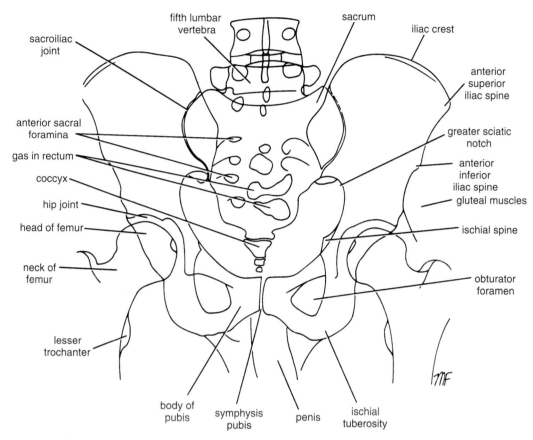

**Figure 11-64** Representation of the radiograph of the pelvis seen in Figure 11-63.

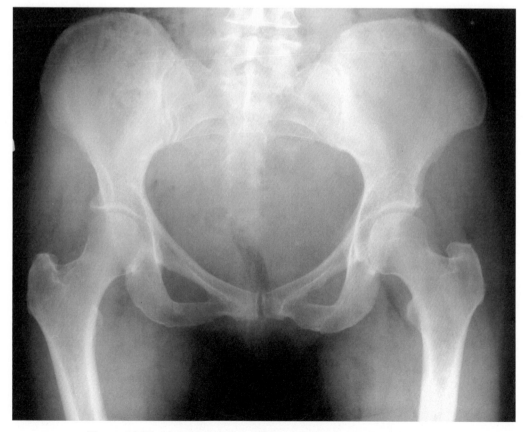

**Figure 11-65** Anteroposterior radiograph of the adult female pelvis.

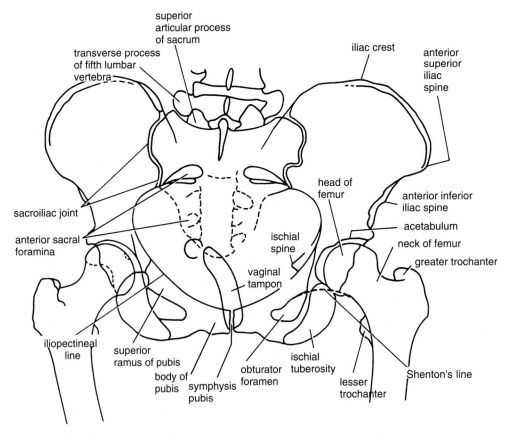

**Figure 11-66**    Representation of the radiograph of the pelvis seen in Figure 11-65.

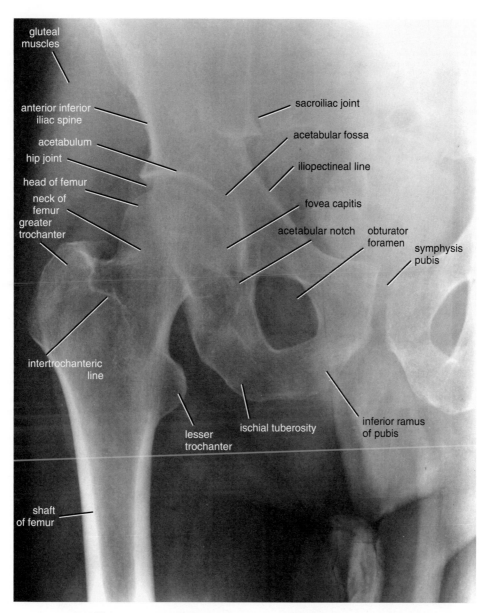

**Figure 11-67**    Anteroposterior radiograph of the hip joint.

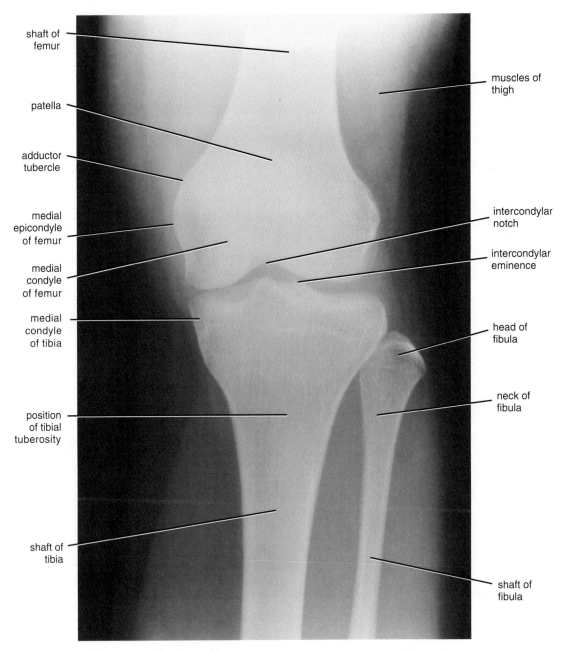

shaft of femur

muscles of thigh

patella

adductor tubercle

medial epicondyle of femur

intercondylar notch

intercondylar eminence

medial condyle of femur

medial condyle of tibia

head of fibula

neck of fibula

position of tibial tuberosity

shaft of tibia

shaft of fibula

**Figure 11-68**  Anteroposterior radiograph of the adult knee.

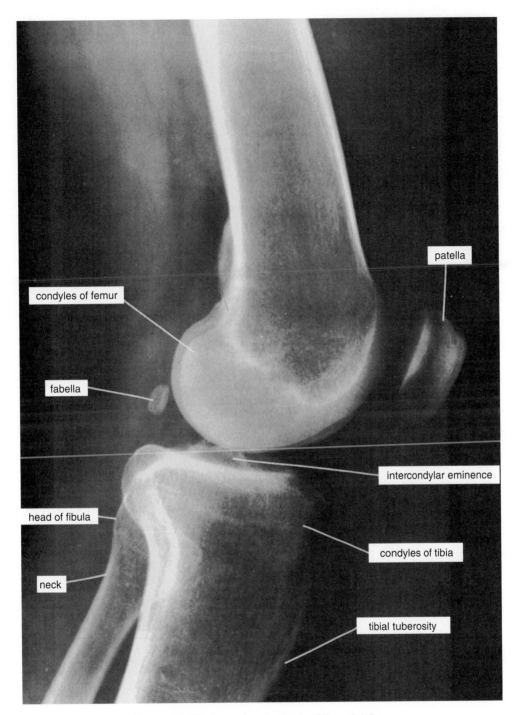

condyles of femur

patella

fabella

intercondylar eminence

head of fibula

condyles of tibia

neck

tibial tuberosity

**Figure 11-69**  Lateral radiograph of the adult knee.

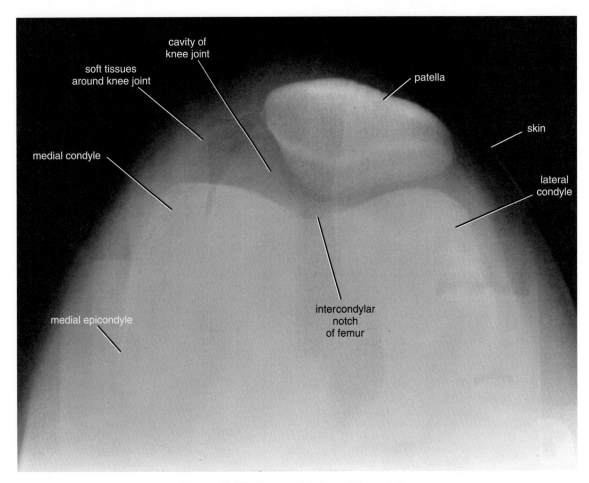

Figure 11-70   Tangential view of the patella.

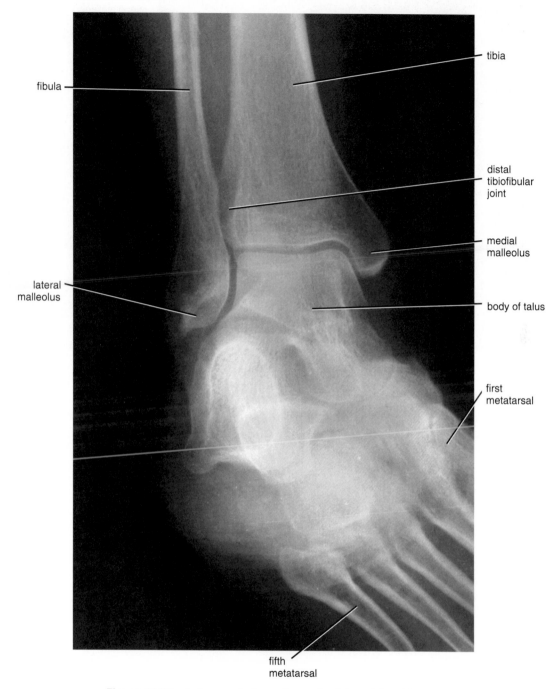

fibula

tibia

distal
tibiofibular
joint

medial
malleolus

lateral
malleolus

body of talus

first
metatarsal

fifth
metatarsal

**Figure 11-71**  Anteroposterior radiograph of the adult ankle.

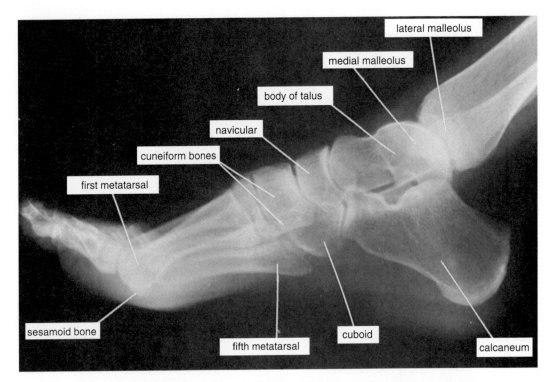

**Figure 11-72**   Lateral radiograph of the adult ankle.

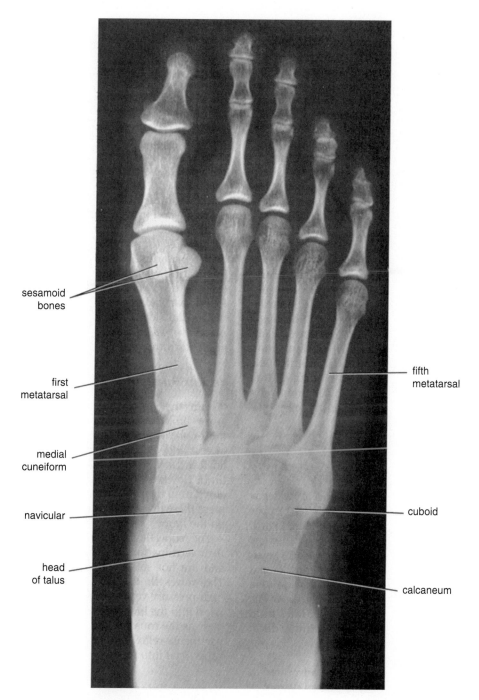

sesamoid
bones

first
metatarsal

medial
cuneiform

navicular

head
of talus

fifth
metatarsal

cuboid

calcaneum

**Figure 11-73**   Anteroposterior radiograph of the adult foot.

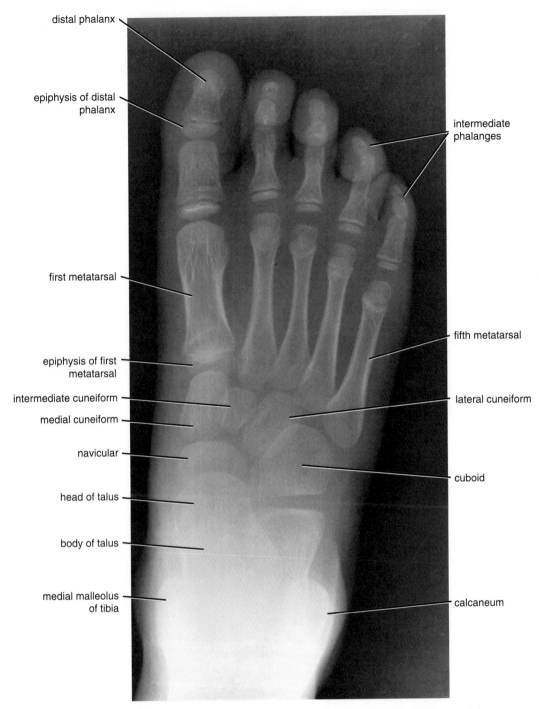

distal phalanx

epiphysis of distal
phalanx

intermediate
phalanges

first metatarsal

fifth metatarsal

epiphysis of first
metatarsal

intermediate cuneiform

lateral cuneiform

medial cuneiform

navicular

cuboid

head of talus

body of talus

medial malleolus
of tibia

calcaneum

Figure 11-74   Anteroposterior radiograph of the foot showing the epiphyses of the phalanges and metatarsal bones (10-year-old boy).

# Review Questions

## Skull

**Select the phrase that BEST completes each statement.**

1. The orbital margins are bounded by the frontal bone superiorly, the _____ bone laterally, the maxilla inferiorly, and the processes of the maxilla and frontal bone medially.
   A. greater wing of the sphenoid
   B. maxilla
   C. parietal
   D. lesser wing of the sphenoid
   E. zygomatic

2. The two parietal bones articulate with each other in the midline at the
   A. lambdoid suture.
   B. sagittal suture.
   C. coronal suture.
   D. squamotympanic suture.
   E. pterion.

3. The styloid process projects downward and forward from the
   A. occipital bone.
   B. sphenoid bone.
   C. temporal bone.
   D. palatine bone.
   E. maxilla.

4. The optic canal is an opening in the
   A. lesser wing of the sphenoid bone.
   B. occipital bone.
   C. petrous part of the temporal bone.
   D. frontal bone.
   E. squamous part of the temporal bone.

5. The carotid canal is located in the
   A. frontal bone.
   B. occipital bone.
   C. petrous part of the temporal bone.
   D. greater wing of the sphenoid bone.
   E. parietal bone.

6. The foramen spinosum is located in the
   A. sphenoid bone.
   B. occipital bone.
   C. frontal bone.
   D. petrous part of the temporal bone.
   E. squamous part of the temporal bone.

7. The hypoglossal canal is located in the
   A. squamous part of the temporal bone.
   B. occipital bone.
   C. frontal bone.
   D. sphenoid bone.
   E. parietal bone.

8. The foramen rotundum is located in the
   A. lesser wing of the sphenoid bone.
   B. frontal bone.
   C. petrous part of the temporal bone.
   D. occipital bone.
   E. greater wing of the sphenoid bone.

9. The foramen magnum is located in the
   A. sphenoid bone.
   B. temporal bone.
   C. parietal bone.
   D. frontal bone.
   E. occipital bone.

10. The mandibular division of the trigeminal nerve leaves the skull through the
    A. superior orbital fissure.
    B. foramen rotundum.
    C. foramen ovale.
    D. jugular foramen.
    E. foramen magnum.

11. The vagus nerve leaves the skull through the
    A. jugular foramen.
    B. occipital foramen.
    C. inferior orbital fissure.
    D. foramen rotundum.
    E. foramen spinosum.

12. The ophthalmic division of the trigeminal nerve leaves the skull through the
    A. inferior orbital fissure.
    B. foramen ovale.
    C. foramen rotundum.
    D. superior orbital fissure.
    E. pterygopalatine foramen.

13. The maxillary division of the trigeminal nerve leaves the skull through the
    A. foramen spinosum.
    B. foramen rotundum.
    C. superior orbital fissure.
    D. foramen ovale.
    E. jugular foramen.

## Vertebral Column

**Select the phrase that BEST completes each statement.**

14. The characteristic feature of the second cervical vertebra is its
    A. absent body.
    B. odontoid process.
    C. heart-shaped body.
    D. massive body.
    E. trifid spinous process.

15. The seventh cervical vertebra is characterized by having
    A. the longest spinous process.
    B. a large foramen transversarium.
    C. a heart-shaped body.
    D. a massive body.
    E. an odontoid process.

16. The sixth thoracic vertebra is characterized by
    A. its heart-shaped body.
    B. its bifid spinous process.
    C. its massive body.
    D. having the superior articular processes face medially and those of the inferior articular process face laterally.
    E. its thick lamina.

17. The characteristic feature of the first cervical vertebra is its
    A. odontoid process.
    B. massive body.
    C. absent body.
    D. long spinous process.
    E. absent foramen transversarium.

18. The characteristic feature of the fifth lumbar vertebra is its
    A. heart-shaped body.
    B. rounded vertebral foramen.
    C. small pedicles.
    D. massive body.
    E. short and thick transverse process.

## Thoracic Bones

**Select the phrase that BEST completes each statement.**

19. Clinicians define the thoracic outlet as
    A. the lower opening in the thoracic cage.
    B. the gap between the crurae of the diaphragm.
    C. the esophageal opening in the diaphragm.
    D. the upper opening in the thoracic cage.
    E. the gap between the sternal and costal origins of the diaphragm.

20. The costal margin is formed by the
    A. sixth, eighth, and tenth ribs.
    B. inner margins of the first ribs.
    C. edge of the xiphoid process.
    D. costal cartilages of the seventh, eighth, ninth, and tenth ribs.
    E. costal cartilages of the seventh to the tenth ribs and the ends of the cartilages of the eleventh and twelfth ribs.

## Shoulder Girdle

**Select the phrase that BEST completes each statement.**

21. The clavicle articulates with the _____ laterally.
    A. coracoid process
    B. superior angle of the scapula
    C. acromion
    D. base of the spine of the scapula
    E. glenoid fossa

22. The lateral end of the spine of the scapula forms the
    A. Acromion.
    B. glenoid fossa.
    C. coracoid process.
    D. suprascapular notch.
    E. superior angle of the scapula.

23. The scapula is a flat triangular bone that lies on the posterior chest wall between the
    A. first and sixth ribs.
    B. second and seventh ribs.
    C. third and eighth ribs.
    D. fourth and ninth ribs.
    E. fifth and tenth ribs.

## Bone of the Arm

**Select the phrase that BEST completes the statement.**

24. The anatomical neck of the humerus lies
    A. where the expanded upper end joins the shaft.
    B. between the greater and lesser tuberosities.
    C. at the deltoid tuberosity.
    D. immediately below the head.
    E. just above the bicipital groove.

25. Above the trochlea of the humerus anteriorly is the
    A. coronoid fossa.
    B. radial fossa.
    C. capitulum.
    D. olecranon fossa.
    E. deltoid tuberosity.

## Bones of the Forearm

**Select the phrase that BEST completes each statement.**

26. The dorsal tubercle of the radius is situated
    A. just below the head.
    B. on the posterior surface of the distal end.
    C. on the lateral margin of the distal end.
    D. on the interosseous border.
    E. halfway down on the lateral side of the shaft.

27. The olecranon process of the ulna lies
    A. at the distal end of the bone.
    B. halfway down the lateral border.
    C. halfway down the medial border.
    D. at the proximal end.
    E. below the radial notch.

## Bones of the Hand

**Select the BEST answer for each question.**

28. The following bones form the proximal row of carpal bones **except** which?
    A. Lunate
    B. Pisiform
    C. Scaphoid
    D. Triquetral
    E. Trapezium

29. The bones of the carpus form part of the carpal tunnel, which is completed by a strong fibrous band called the
    A. extensor retinaculum.
    B. bicipital aponeurosis.
    C. pisohamate ligament.
    D. flexor retinaculum.
    E. palmar aponeurosis.

## Pelvic Girdle

**Select the BEST answer for each question.**

30. The following statements concerning the pelvis are correct **except** which?
    A. The ilium, ischium, and pubis are three separate bones that fuse together to form the hip bone in the 25th year of life.
    B. The platypelloid type of pelvis occurs in about 2% of women.
    C. External pelvic measurements have little practical importance in determining whether a disproportion between the size of the fetal head and the size of the pelvic inlet is likely.
    D. The pelvic outlet is formed by the symphysis pubis anteriorly, the ischial tuberosities laterally, the sacrotuberous ligaments laterally, and the coccyx posteriorly.

E. The sacrum is shorter, wider, and flatter in the female than in the male.

31. The following statements concerning the bony pelvis are correct **except** which?
    A. When the patient is in the standing position, the anterior superior iliac spines lie vertically above the anterior surface of the symphysis pubis.
    B. Very little movement is possible at the sacrococcygeal joint.
    C. The false pelvis helps guide the fetus into the true pelvis during labor.
    D. The female sex hormones cause a relaxation of the ligaments of the pelvis during pregnancy.
    E. Obliteration of the cavity of the sacroiliac joint often occurs in both sexes after middle age.

## Leg Bones

## Bones of the Thigh

**Select the BEST answer for each question.**

32. In the adult, the neck of the femur makes an angle of about _____ degrees with the long axis of the shaft.
    A. 155
    B. 165
    C. 125
    D. 115
    E. 145

33. In the adult, the greater part of the blood supply to the head of the femur is from the
    A. obturator artery.
    B. external pudendal artery.
    C. perforating branches of the profunda femoris artery.
    D. medial and lateral circumflex femoral arteries.
    E. superficial circumflex iliac artery.

34. The posterior surface of the tibia shows an oblique line for the attachment of the soleus muscle. The line is referred to as the
    A. interosseous line.
    B. popliteal line.
    C. marginal line.
    D. soleal line.
    E. malleolar line.

35. At the junction of the anterior border of the tibia with the upper end is the _____ for the attachment of the ligamentum patellae.
    A. tibial plateau
    B. medial condyle
    C. intercondylar eminence
    D. medial malleolus
    E. tibial tuberosity

36. The lower end of the fibula forms the triangular
    A. lateral malleolus.
    B. medial malleolus.
    C. styloid process.
    D. interosseous border.
    E. malleolar fossa.

## Bones of the Foot

**Select the BEST answer for each question.**

37. The medial surface of the calcaneum has a large shelflike process called the
    A. peroneal tubercle.
    B. anterior tubercle.
    C. sustentaculum tali.
    D. medial tubercle.
    E. lateral tubercle.

38. The head of the talus has an oval convex articular surface for articulation with the
    A. cuboid bone.
    B. medial cuneiform bone.
    C. intermediate cuneiform bone.
    D. navicular bone.
    E. calcaneum.

39. The fifth metatarsal bone has a prominent tubercle on its base for the attachment of the
    A. peroneus brevis muscle.
    B. flexor digitorum longus muscle.
    C. flexor digiti minimi brevis.
    D. peroneus longus muscle.
    E. peroneus tertius muscle.

# Answers and Explanations

1. **E is correct.** The orbital margins are bounded by the frontal bone superiorly, the zygomatic bone laterally, the maxilla inferiorly, and the processes of the maxilla and frontal bone medially (Fig. 11-3).

2. **B is correct.** The two parietal bones articulate with each other in the midline at the sagittal suture (Fig. 11-6).

3. **C is correct.** The styloid process projects downward and forward from the undersurface of the temporal bone (Fig. 11-5).

4. **A is correct.** The optic canal is an opening in the lesser wing of the sphenoid bone (Figs. 11-3 and 11-8).

5. **C is correct.** The carotid canal is located in the petrous part of the temporal bone (Fig. 11-7).

6. **A is correct.** The foramen spinosum is located in the greater wing of the sphenoid bone; the middle meningeal artery passes through this foramen into the middle cranial fossa from the infratemporal fossa (Fig. 11-8).

7. **B is correct.** The hypoglossal canal is situated above the anterolateral boundary of the foramen magnum (Fig. 11-8).

8. **E is correct.** The foramen rotundum is located in the greater wing of the sphenoid bone (Fig. 11-8).

9. **E is correct.** The foramen magnum is located in the occipital bone. It transmits the medulla oblongata, the spinal part of the accessory nerve, and the right and left vertebral arteries (Fig. 11-8).

10. **C is correct.** The mandibular division of the trigeminal nerve leaves the skull through the foramen ovale in the greater wing of the sphenoid bone (Fig. 11-8).

11. **A is correct.** The vagus nerve leaves the skull through the jugular foramen in company with the glossopharyngeal nerve and accessory nerve; the sigmoid venous sinus also passes through the posterior part of the same foramen to become the internal jugular vein.

12. **D is correct.** The ophthalmic division of the trigeminal nerve leaves the skull through the superior orbital fissure as its three terminal branches—namely, the lacrimal, frontal, and nasociliary nerves.

13. **B is correct.** The maxillary division of the trigeminal nerve leaves the skull through the foramen rotundum in the greater wing of the sphenoid bone (Fig. 11-8).

14. **B is correct.** The characteristic feature of the second cervical vertebra is its odontoid process (Figs. 11-22 and 11-24).

15. **A is correct.** The seventh cervical vertebra is characterized by having the longest spinous process (Fig. 11-22).

16. **A is correct.** The sixth thoracic vertebra is characterized by its heart-shaped body.

17. **C is correct.** The characteristic feature of the first cervical vertebra or atlas is its absent body. During development, the centrum of the atlas fuses with that of the axis to form the odontoid process of the axis.

18. **D** is correct. The characteristic of the fifth lumbar vertebral is its massive body; the pedicles are also large, and the transverse processes are long and slender; the vertebral foramen is triangular.

19. **D** is correct. Clinicians define the thoracic outlet as the upper opening in the thoracic cage.

20. **E** is correct. The costal margin is formed by the costal cartilages of the seventh to the tenth ribs and the ends of the cartilages of the eleventh and twelfth ribs (Fig. 11-1).

21. **C** is correct. The clavicle articulates with the acromion laterally.

22. **A** is correct. The lateral end of the spine of the scapula forms the acromion (Fig. 11-37).

23. **B** is correct. The scapula is a flat triangular bone that lies on the posterior chest wall between the second and seventh ribs (Fig. 11-19).

24. **D** is correct. The anatomical neck of the humerus lies immediately below the head (Fig. 11-38).

25. **A** is correct. Above the trochlea of the humerus is the coronoid fossa for the reception of the coronoid process of the ulna during the movement of flexion (Fig. 11-38).

26. **B** is correct. The dorsal tubercle of the radius is situated on the posterior surface of the distal end. It serves as a pulley around which the tendon of extensor pollicis longus changes direction and extends onto the posterior surface of the hand (Fig. 11-39).

27. **D** is correct. The olecranon process of the ulna is situated at the proximal end and forms the point of the elbow (Fig. 11-39).

28. **E** is incorrect. The trapezium is in the distal row of carpal bones (Fig. 11-40).

29. **D** is correct. The bones of the carpus together with the flexor retinaculum form the carpal tunnel through which pass the median nerve and the long flexor tendons of the fingers with their surrounding synovial sheaths.

30. **A** is incorrect. At puberty, three separate bones, the ilium, ischium, and pubis, fuse together to form one large irregular bone, the hip bone (Fig. 11-52).

31. **B** is incorrect. The sacrococcygeal joint is a cartilaginous joint and can perform a great deal of movement.

32. **C** is correct. The neck of the femur makes an angle of about 125 degrees with the shaft.

33. **D** is correct. In the adult, the greater part of the blood supply to the head of the femur is from the medial and lateral circumflex femoral arteries (Fig. 12-25).

34. **D** is correct. The posterior surface of the tibia shows an oblique line for the attachment of the soleus muscle and is referred to as the soleal line (Fig. 11-59).

35. **E** is correct. At the junction of the anterior border of the tibia with the upper end is the tibial tuberosity for the attachment of the ligamentum patellae (tendon of insertion of the quadriceps femoris muscle) (Fig. 11-58).

36. **A** is correct. The lower end of the fibula forms the triangular lateral malleolus (Fig. 11-58).

37. **C** is correct. The medial surface of the calcaneum has a large shelflike process called the sustentaculum tali for the support of the overlying talus (Fig. 11-60).

38. **D** is correct. The head of the talus has an oval convex articular surface for articulation with the navicular bone (Fig. 11-60).

39. **A** is correct. The fifth metatarsal bone has a prominent tubercle on its base for the attachment of the tendon of the peroneus brevis muscle (Fig. 11-62).

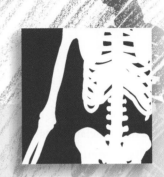

# 12

# Joints

All clinical material relevant to this chapter can be found on the CD-ROM.

# Chapter Outline

J oint disease is a very common problem seen by medical personnel. This chapter is primarily concerned with the presentation of the basic anatomy of joints to assist the medical professional in making the diagnosis and initiating prompt treatment.

 # BASIC ANATOMY

The term joint or articulation is used to describe the site where two or more bones of the skeleton come together. At most joints, the bones are held together by flexible connective tissues that permit muscles to act on the bones and, therefore, bring about movements of the different parts of the body.

# Classification of Joints

Joints may be classified according to their structure, namely fibrous, cartilaginous, and synovial.

# Types of Synovial Joints

▦ **Plane joints:** In plane joints, the apposed articular surfaces are flat or almost flat, and this permits the bones to slide on one another. Examples of these joints are the sternoclavicular and acromioclavicular joints (Fig. 12-1).

▦ **Hinge joints:** Hinge joints resemble the hinge on a door, so that flexion and extension movements are possible.

Examples of these joints are the elbow, knee, and ankle joints (Fig. 12-1).

▦ **Pivot joints:** In pivot joints, a central bony pivot is surrounded by a bony-ligamentous ring (Fig. 12-1), and rotation is the only movement possible. The atlantoaxial and superior radioulnar joints are good examples.

▦ **Condyloid joints:** Condyloid joints have two distinct convex surfaces that articulate with two concave surfaces.

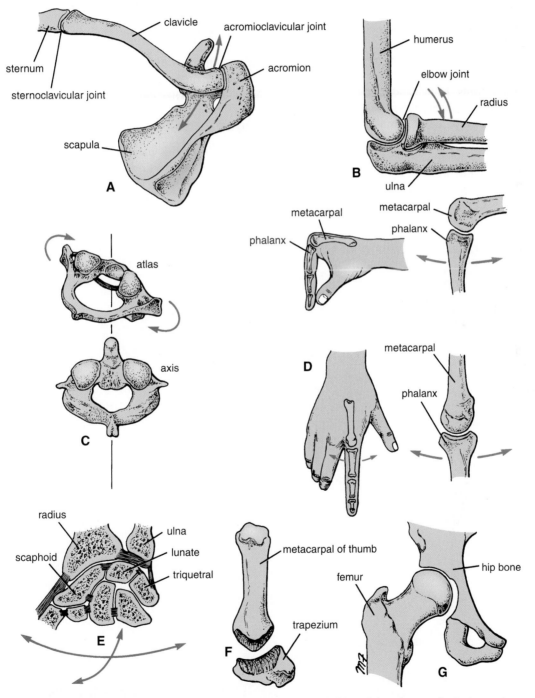

**Figure 12-1**  Examples of different types of synovial joints. **A.** Plane joints (sternoclavicular and acromioclavicular joints). **B.** Hinge joint (elbow joint). **C.** Pivot joint (atlantoaxial joint). **D.** Condyloid joint (metacarpophalangeal joint). **E.** Ellipsoid joint (wrist joint). **F.** Saddle joint (carpometacarpal joint of the thumb). **G.** Ball-and-socket joint (hip joint).

The movements of flexion, extension, abduction, and adduction are possible together with a small amount of rotation. The metacarpophalangeal joints or knuckle joints are good examples (Fig. 12-1).

- **Ellipsoid joints:** In ellipsoid joints, an elliptical convex articular surface fits into an elliptical concave articular surface. The movements of flexion, extension, abduction, and adduction can take place, but rotation is impossible. The wrist joint is a good example (Fig. 12-1).
- **Saddle joints:** In saddle joints, the articular surfaces are reciprocally concavoconvex and resemble a saddle on a horse's back. These joints permit flexion, extension, abduction, adduction, and rotation. The best example of this type of joint is the carpometacarpal joint of the thumb (Fig. 12-1).
- **Ball-and-socket joints:** In ball-and-socket joints, a ball-shaped head of one bone fits into a socketlike concavity of another. This arrangement lateral permits free movements, including flexion, extension, abduction, adduction, medial rotation, lateral rotation, and circumduction. The shoulder and hip joints are good examples of this type of joint (Fig. 12-1).

# Stability of Joints

The stability of a joint depends on three main factors: the shape, size, and arrangement of the articular surfaces; the ligaments; and the tone of the muscles around the joint.

## Articular Surfaces

The ball-and-socket arrangement of the hip joint (Fig. 12-2) and the mortise arrangement of the ankle joint are good examples of how bone shape plays an important role in joint stability. Other examples of joints, however, in which the shape of the bones contributes little or nothing to the stability include the acromioclavicular joint, the calcaneocuboid joint, and the knee joint.

## Ligaments

**Fibrous ligaments** prevent excessive movement in a joint (Fig. 12-2), but if the stress is continued for an excessively long period, then fibrous ligaments stretch. For example, the

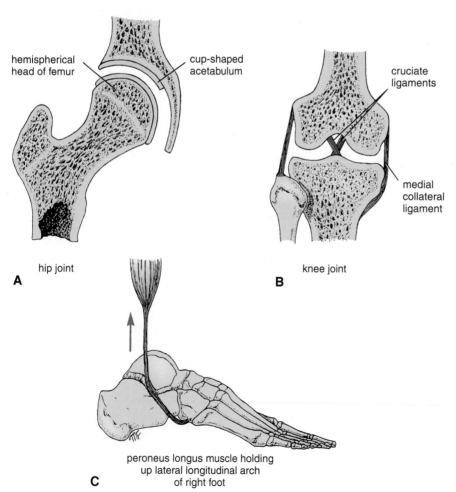

hemispherical head of femur

cup-shaped acetabulum

cruciate ligaments

medial collateral ligament

hip joint

knee joint

**A**

**B**

peroneus longus muscle holding up lateral longitudinal arch of right foot

**C**

arch of foot

**Figure 12-2** The three main factors responsible for stabilizing a joint. **A.** Shape of articular surfaces. **B.** Ligaments. **C.** Muscle tone.

ligaments of the joints between the bones forming the arches of the feet will not by themselves support the weight of the body. Should the tone of the muscles that normally support the arches become impaired by fatigue, then the ligaments will stretch and the arches will collapse, producing flat feet.

**Elastic ligaments,** conversely, return to their original length after stretching. The elastic ligaments of the auditory ossicles play an active part in supporting the joints and assisting in the return of the bones to their original position after movement.

## Muscle Tone

In most joints, muscle tone is the major factor controlling stability. For example, the muscle tone of the short muscles around the shoulder joint keeps the hemispherical head of the humerus in the shallow glenoid cavity of the scapula. Without the action of these muscles, very little force would be required to dislocate this joint. The knee joint is very unstable without the tonic activity of the quadriceps femoris muscle. The joints between the small bones forming the arches of the feet are largely supported by the tone of the muscles of the leg, whose tendons are inserted into the bones of the feet (Fig. 12-2).

# Nerve Supply of Joints

The capsule and ligaments receive an abundant sensory nerve supply. A sensory nerve supplying a joint also supplies the muscles moving the joint and the skin overlying the insertions of these muscles, a fact that has been codified as **Hilton's law.** The blood vessels in a joint receive autonomic sympathetic fibers. The cartilage covering the articular surfaces possesses only a few nerve endings near its edges. Overstretching of the capsule and ligaments produces reflex contraction of muscles around the joint; excessive stretching produces pain. The stretch receptors in the capsule and ligaments are continually sending proprioceptive information up to the central nervous system, keeping it informed of the position of the joints. This supplements the information passing to the nervous system from the muscle and tendon spindles, helps maintain postural tone, and coordinates voluntary movements.

# Joints of the Skull

In the vault of the skull, the flat bones are joined together by fibrous joints called **sutures** (Fig.12-3). The periosteum covering the outer surface of the bones becomes continuous with the endosteum covering the inner surface at the suture, forming the **sutural ligament.** These dense ligaments do not permit any movement between the bones. Examples are the **sagittal suture,** the **coronal suture,** and the **lambdoid suture** (Fig. 12-3).

# Temporomandibular Joint

## Articulation

Articulation occurs between the articular tubercle and the anterior portion of the mandibular fossa of the temporal bone above and the head (condyloid process) of the mandible below (Figs. 12-4 and 12-5). The articular surfaces are covered with fibrocartilage.

## Type of Joint

The temporomandibular joint is synovial. The articular disc divides the joint into upper and lower cavities (Fig. 12-6).

## Capsule

The capsule encloses the joint.

## Ligaments

The **lateral temporomandibular ligament** is attached above to the tubercle on the root of the zygomatic arch and below to the neck of the mandible (Fig. 12-4). The fibers extend downwards and backwards. This ligament limits the posterior movement of the mandible, thus protecting the external auditory meatus.

The **sphenomandibular ligament** lies on the medial side of the joint (Fig. 12-4). It is attached above to the spine of the sphenoid bone and below to the lingula of the mandibular foramen. It represents the remains of the first pharyngeal arch, and its function is unknown.

The **stylomandibular ligament** is attached to the apex of the styloid process above and to the angle of the mandible below (Fig. 12-4). Its function is unknown.

The **articular disc** is an oval disc of fibrocartilage (Fig. 12-6) that divides the joint into upper and lower cavities. It is attached circumferentially to the capsule. It is also attached in front to the tendon of the lateral pterygoid muscle and by fibrous bands to the head of the mandible. These bands ensure that the disc moves forward and backward with the head of the mandible during protraction and retraction of the mandible. The upper surface of the disc is concavo-convex from before backward to fit the shape of the articular tubercle and the mandibular fossa; the lower surface is concave to fit the head of the mandible. The function of the disc is to permit gliding movement in the upper part of the joint and hinge movement in the lower part of the joint.

## Synovial Membrane

This lines the capsule in the upper and lower cavities of the joint (Fig. 12-6).

## Nerve Supply

Auriculotemporal and masseteric branches of the mandibular division of the trigeminal nerve.

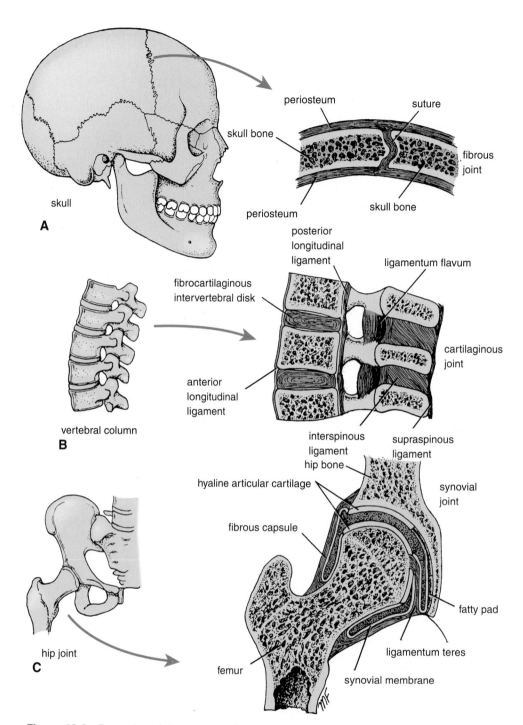

**Figure 12-3** Examples of three types of joints. **A.** Fibrous joint (coronal suture of skull). **B.** Cartilaginous joint (joint between two lumbar vertebral bodies). **C.** Synovial joint (hip joint).

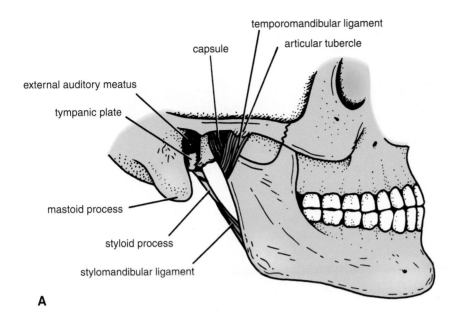

external auditory meatus

tympanic plate

capsule

temporomandibular ligament

articular tubercle

mastoid process

styloid process

stylomandibular ligament

**A**

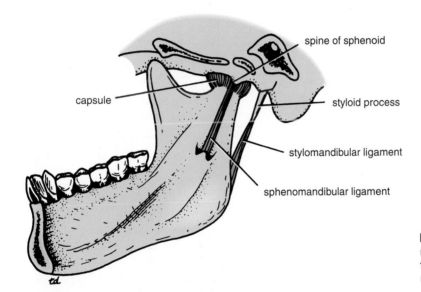

capsule

spine of sphenoid

styloid process

stylomandibular ligament

sphenomandibular ligament

**B**

**Figure 12-4**   Temporomandibular joint as seen from the lateral (**A**) and medial (**B**) aspects.

## Movements and the Muscles that Produce Movement

The mandible can be depressed or elevated, protruded or retracted. Rotation can also occur, as in chewing. In the position of rest, the teeth of the upper and lower jaws are slightly apart. On closure of the jaws, the teeth come into contact.

### Depression of the Mandible

As the mouth is opened, the head of the mandible rotates on the undersurface of the articular disc around a horizontal axis. To prevent the angle of the jaw impinging unnecessarily on the parotid gland and the sternocleidomastoid muscle, the mandible is pulled forward. This is accomplished by the contraction of the lateral pterygoid muscle, which pulls forward the neck of

the mandible and the articular disc so that the latter moves onto the articular tubercle (Fig. 12-6). The forward movement of the disc is limited by the tension of the fibroelastic tissue, which tethers the disc to the temporal bone posteriorly.

Depression of the mandible is brought about by contraction of the digastrics, the geniohyoids, and the mylohyoids; the lateral pterygoids play an important role by pulling the mandible forward.

### Elevation of the Mandible

The movements in depression of the mandible are reversed. First, the head of the mandible and the disc move backward, and then the head rotates on the lower surface of the disc.

Elevation of the mandible is brought about by contraction of the temporalis, the masseter, and the medial

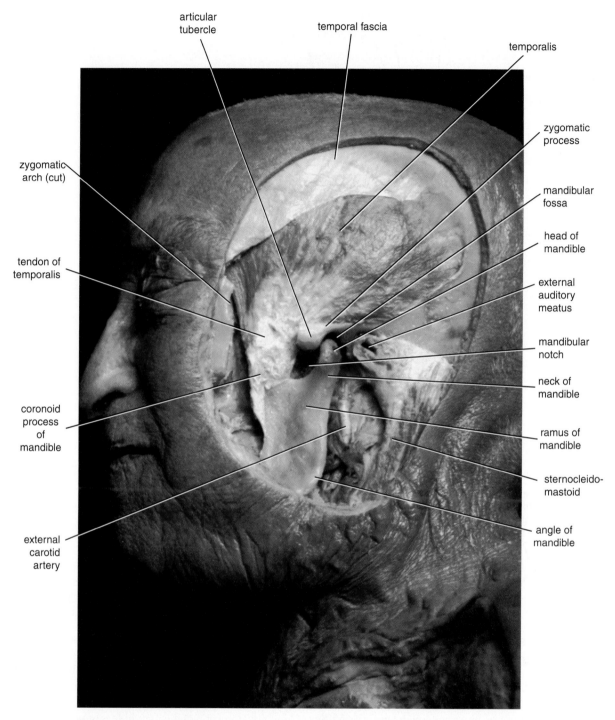

articular tubercle

temporal fascia

temporalis

zygomatic process

zygomatic arch (cut)

mandibular fossa

head of mandible

tendon of temporalis

external auditory meatus

mandibular notch

neck of mandible

coronoid process of mandible

ramus of mandible

sternocleido-mastoid

external carotid artery

angle of mandible

**Figure 12-5** A dissection of the left temporomandibular joint. The capsule and lateral temporomandibular ligament have been removed to reveal the interior of the joint. Note the articular tubercle and mandibular fossa of the temporal bone and the head of the mandible. The articular disc is present within the joint cavity on the upper surface of the head of the mandible.

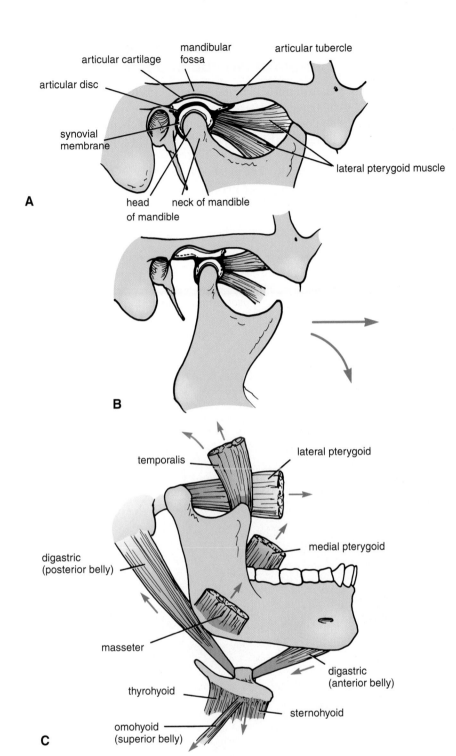

Figure 12-6   Temporomandibular joint with mouth closed (**A**) and with the mouth open (**B**). Note the position of the head of the mandible and articular disc in relation to the articular tubercle in each case. **C.** The attachment of the muscles of mastication to the mandible. The *arrows* indicate the direction of their actions.

pterygoids. The head of the mandible is pulled backward by the posterior fibers of the temporalis. The articular disc is pulled backward by the fibroelastic tissue, which tethers the disc to the temporal bone posteriorly.

**Protrusion of the Mandible**
The articular disc is pulled forward onto the anterior tubercle, carrying the head of the mandible with it. All

movement thus takes place in the upper cavity of the joint. In protrusion, the lower teeth are drawn forward over the upper teeth, which is brought about by contraction of the lateral pterygoid muscles of both sides, assisted by both medial pterygoids.

**Retraction of the Mandible**
The articular disc and the head of the mandible are pulled backward into the mandibular fossa. Retraction is

brought about by contraction of the posterior fibers of the temporalis.

- **Lateral Chewing Movements**
These are accomplished by alternately protruding and retracting the mandible on each side. For this to take place, a certain amount of rotation occurs, and the muscles responsible on both sides work alternately and not in unison.

The muscles of mastication are summarized in Tables 13-2 and 13-3. See also Figure 12-6.

## Important Relations of the Temporomandibular Joint

- **Anteriorly:** The mandibular notch and the masseteric nerve and artery (Fig. 12-7).
- **Posteriorly:** The tympanic plate of the external auditory meatus (Fig. 12-4) and the glenoid process of the parotid gland.
- **Laterally:** The parotid gland, fascia, and skin (Fig. 12-8).
- **Medially:** The maxillary artery and vein and the auriculotemporal nerve.

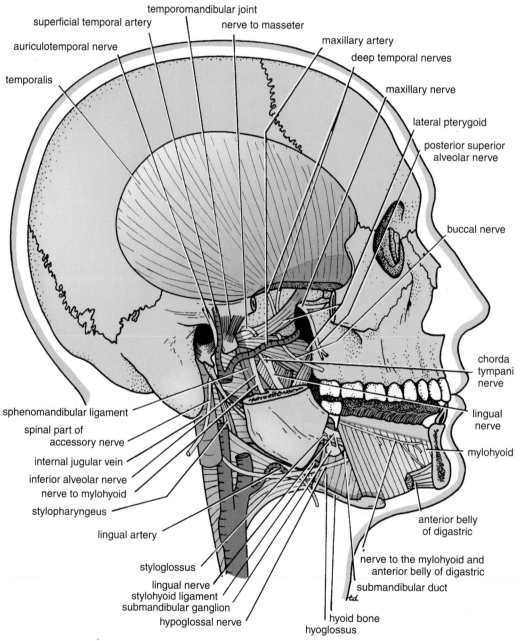

**Figure 12-7** Infratemporal and submandibular regions. Parts of the zygomatic arch, the ramus, and the body of the mandible have been removed to display deeper structures.

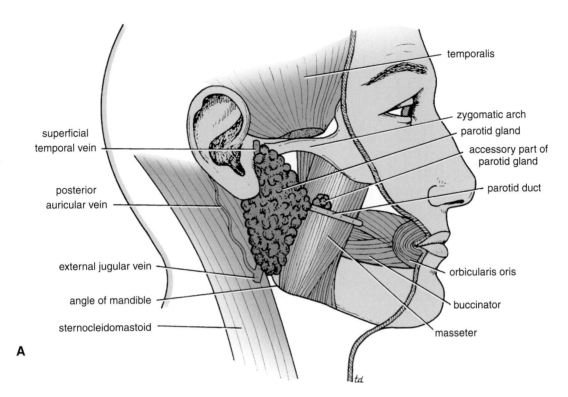

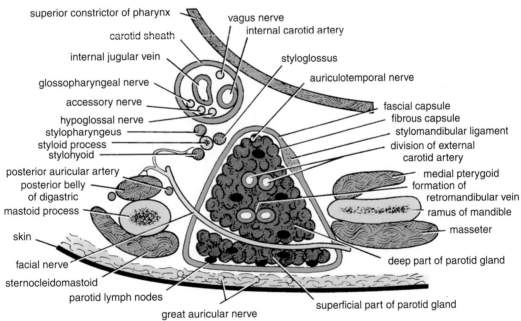

**Figure 12-8** Parotid gland and its relations. **A.** Lateral surface of the gland and the course of the parotid duct. **B.** Horizontal section of the parotid gland.

# RADIOLOGIC APPEARANCES OF THE JOINTS OF THE SKULL AND THE TEMPORO-MANDIBULAR JOINT

The radiologic appearances of the joints of the skull and the temporomandibular joint are shown in Figures 11-11 to 11-14.

# Joints of the Vertebral Column

## Atlanto-Occipital Joints

The atlanto-occipital joints are synovial joints that are formed between the occipital condyles, which are found on either side of the foramen magnum above and the facets on the superior surfaces of the lateral masses of the atlas below (Fig. 12-9). They are enclosed by a capsule.

## Ligaments

- **Anterior atlanto-occipital membrane:** This is a continuation of the anterior longitudinal ligament, which runs as a band down the anterior surface of the vertebral column. The membrane connects the anterior arch of the atlas to the anterior margin of the foramen magnum.
- **Posterior atlanto-occipital membrane:** This membrane is similar to the ligamentum flavum and connects the posterior arch of the atlas to the posterior margin of the foramen magnum.

## Movements

Flexion, extension, and lateral flexion. Rotation is not possible.

## Atlanto-Axial Joints

The atlanto-axial joints are three synovial joints; one is between the odontoid process and the anterior arch of the atlas, and the other two are between the lateral masses of the bones (Fig. 12-9). The joints are enclosed by capsules.

## Ligaments

- **Apical ligament:** This median-placed structure connects the apex of the odontoid process to the anterior margin of the foramen magnum (Fig. 12-9).
- **Alar ligaments:** These lie one on each side of the apical ligament and connect the odontoid process to the medial sides of the occipital condyles (Fig. 12-9).
- **Cruciate ligament:** This ligament consists of a transverse part and a vertical part (Fig. 12-9). The transverse part is attached on each side to the inner aspect of the lateral mass of the atlas and binds the odontoid process to the anterior arch of the atlas. The vertical part runs from the posterior surface of the body of the axis to the anterior margin of the foramen magnum (Fig. 12-9).
- **Membrana tectoria:** This is an upward continuation of the posterior longitudinal ligament (Fig. 12-9). It is attached above to the occipital bone just within the foramen magnum. It covers the posterior surface of the odontoid process and the apical, alar, and cruciate ligaments.

## Movements

There can be extensive rotation of the atlas and thus of the head on the axis.

# Joints of the Vertebral Column Below the Axis

With the exception of the first two cervical vertebrae, the remainder of the mobile vertebrae articulate with each other by means of cartilaginous joints between their bodies and by synovial joints between their articular processes (Fig. 12-10).

## Joints between Two Vertebral Bodies

The upper and lower surfaces of the bodies of adjacent vertebrae are covered by thin plates of hyaline cartilage. Sandwiched between the plates of hyaline cartilage is an intervertebral disc of fibrocartilage (Fig. 12-10). The collagen fibers of the disc strongly unite the bodies of the two vertebrae.

In the lower cervical region, small synovial joints are present at the sides of the intervertebral disc between the upper and lower surfaces of the bodies of the vertebrae.

## Intervertebral Discs

The intervertebral discs are responsible for one fourth of the length of the vertebral column (Fig. 12-10). They are thickest in the cervical and lumbar regions, where the movements of the vertebral column are greatest. They may be regarded as semielastic discs, which lie between the rigid bodies of adjacent vertebrae (Fig. 12-10). Their physical characteristics permit them to serve as shock absorbers when

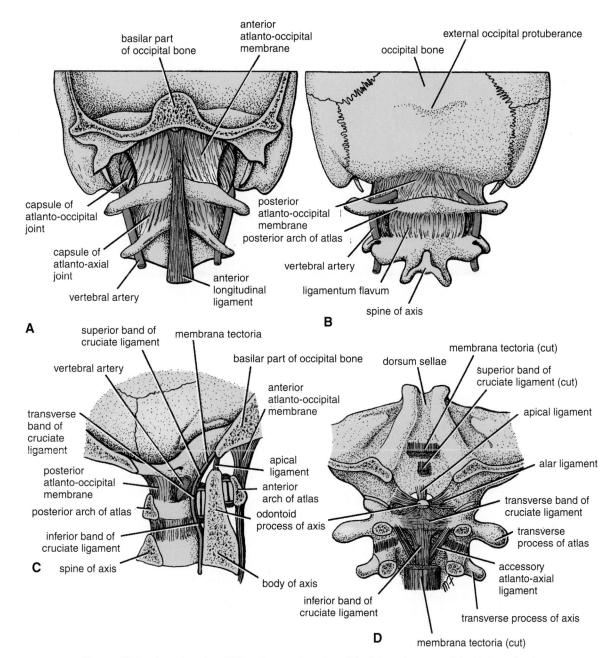

**Figure 12-9** Anterior view (**A**) and posterior view (**B**) of the atlanto-occipital joints. Sagittal section (**C**) and posterior view (**D**) of the atlanto-axial joints. Note that the posterior arch of the atlas and the laminae and spine of the axis have been removed.

the load on the vertebral column is suddenly increased, as when one is jumping from a height. Their elasticity allows the rigid vertebrae to move one on the other. Unfortunately, their resilience is gradually lost with advancing age.

Each disc consists of a peripheral part, the anulus fibrosus, and a central part, the nucleus pulposus (Fig. 12-10).

The **anulus fibrosus** is composed of fibrocartilage, in which the collagen fibers are arranged in concentric layers or sheets. The collagen bundles pass obliquely between adjacent vertebral bodies, and their inclination is reversed in alternate sheets. The more peripheral fibers are strongly at-

tached to the anterior and posterior longitudinal ligaments of the vertebral column.

The **nucleus pulposus** in children and adolescents is an ovoid mass of gelatinous material containing a large amount of water, a small number of collagen fibers, and a few cartilage cells. It is normally under pressure and situated slightly nearer to the posterior than to the anterior margin of the disc.

The upper and lower surfaces of the bodies of adjacent vertebrae that abut onto the disc are covered with thin plates of hyaline cartilage. No discs are found between the first two cervical vertebrae or in the sacrum or coccyx.

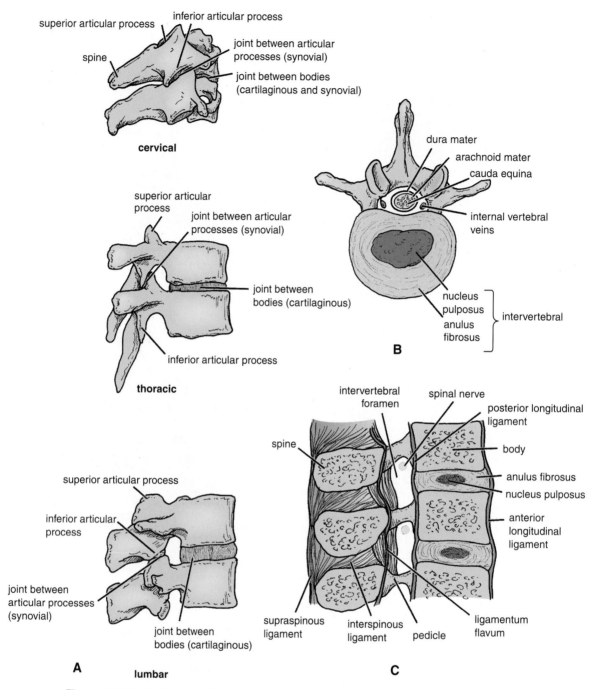

**Figure 12-10   A.** Joints in the cervical, thoracic, and lumbar regions of the vertebral column. **B.** Third lumbar vertebra seen from above showing the relationship between intervertebral disc and cauda equina. **C.** Sagittal section through three lumbar vertebrae showing ligaments and intervertebral discs. Note the relationship between the emerging spinal nerve in an intervertebral foramen and the intervertebral disc.

### Function of the Intervertebral Disc

The semifluid nature of the nucleus pulposus allows it to change shape and permits one vertebra to rock forward or backward on another, as in flexion and extension of the vertebral column.

A sudden increase in the compression load on the vertebral column causes the semifluid nucleus pulposus to become flattened. The outward thrust of the nucleus is accommodated by the resilience of the surrounding anulus fibrosus. Sometimes, the outward thrust is too great for the anulus fibrosus and it ruptures, allowing the nucleus pulposus to herniate and protrude into the vertebral canal, where it may press on the spinal nerve roots, the spinal nerve, or even the spinal cord.

With advancing age, the water content of the nucleus pulposus diminishes and is replaced by fibrocartilage. The collagen fibers of the anulus degenerate and, as a result, the anulus cannot always contain the nucleus pulposus under stress. In old age, the discs are thin and less elastic, and it is no longer possible to distinguish the nucleus from the anulus.

### Ligaments

The **anterior** and **posterior longitudinal ligaments** run as continuous bands down the anterior and posterior surfaces of the vertebral column from the skull to the sacrum (Fig. 12-10 and CD Fig 12-XX). The anterior ligament is wide and is strongly attached to the front and sides of the vertebral bodies and to the intervertebral discs. The posterior ligament is weak and narrow and is attached to the posterior borders of the discs. These ligaments hold the vertebrae firmly together but, at the same time, permit a small amount of movement to take place between them.

## Joints between Two Vertebral Arches

The joints between two vertebral arches consist of synovial joints between the superior and inferior articular processes of adjacent vertebrae (Fig. 12-10). The articular facets are covered with hyaline cartilage, and the joints are surrounded by a capsular ligament.

## Ligaments

- **Supraspinous ligament** (Fig. 12-10): This runs between the tips of adjacent spines.
- **Interspinous ligament** (Fig. 12-10): This connects adjacent spines.
- **Intertransverse ligaments**: These run between adjacent transverse processes.
- **Ligamentum flavum** (Fig. 12-10): This connects the laminae of adjacent vertebrae.

In the cervical region, the supraspinous and interspinous ligaments are greatly thickened to form the strong **ligamentum nuchae**. The latter extends from the spine of the seventh cervical vertebra to the external occipital protuberance of the skull, with its anterior border being strongly attached to the cervical spines in between.

## Nerve Supply of Vertebral Joints

The joints between the vertebral bodies are innervated by the small meningeal branches of each spinal nerve (Fig. 12-11). The nerve arises from the spinal nerve as it exits from

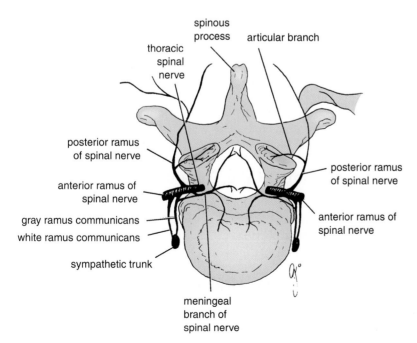

**Figure 12-11** The innervation of vertebral joints. At any particular vertebral level, the joints receive nerve fibers from two adjacent spinal nerves.

the intervertebral foramen. It then reenters the vertebral canal through the intervertebral foramen and supplies the meninges, ligaments, and intervertebral discs. The joints between the articular processes are innervated by branches from the posterior rami of the spinal nerves (Fig. 12-11). It should be noted that the joints of any particular level receive nerve fibers from two adjacent spinal nerves.

## Curves of the Vertebral Column

The curves of the vertebral column are fully described on page 310.

## Movements of the Vertebral Column

As has been seen in the previous sections, the vertebral column consists of several separate vertebrae accurately positioned one on the other and separated by intervertebral discs. The vertebrae are held in position relative to one another by strong ligaments that severely limit the degree of movement possible between adjacent vertebrae. Nevertheless, the summation of all these movements gives the vertebral column as a whole a remarkable degree of mobility.

The following movements are possible: flexion, extension, lateral flexion, rotation, and circumduction.

- **Flexion** is a forward movement, and **extension** is a backward movement. Both are extensive in the cervical and lumbar regions but restricted in the thoracic region.
- **Lateral flexion** is the bending of the body to one or the other side. It is extensive in the cervical and lumbar regions but restricted in the thoracic region.
- **Rotation** is a twisting of the vertebral column. This is least extensive in the lumbar region.
- **Circumduction** is a combination of all these movements.

The type and range of movements possible in each region of the column largely depend on the thickness of the intervertebral discs and the shape and direction of the articular processes. In the thoracic region, the ribs, the costal cartilages, and the sternum severely restrict the range of movement.

The **atlanto-occipital joints** permit extensive flexion and extension of the head. The **atlanto-axial joints** allow a wide range of rotation of the atlas and thus of the head on the axis.

The vertebral column is moved by numerous muscles, many of which are attached directly to the vertebrae, whereas others, such as the sternocleidomastoid and the abdominal wall muscles, are attached to the skull or to the ribs or fasciae.

In the **cervical region,** flexion is produced by the longus cervicis, scalenus anterior, and sternocleidomastoid muscles. Extension is produced by the postvertebral muscles. Lateral flexion is produced by the scalenus anterior and medius and the trapezius and sternocleidomastoid muscles. Rotation is produced by the sternocleidomastoid on one side and the splenius on the other side.

In the **thoracic region**, rotation is produced by the semispinalis and rotatores muscles, assisted by the oblique muscles of the anterolateral abdominal wall.

In the **lumbar region,** flexion is produced by the rectus abdominis and the psoas muscles. Extension is produced by the postvertebral muscles. Lateral flexion is produced by the postvertebral muscles, the quadratus lumborum, and the oblique muscles of the anterolateral abdominal wall. The psoas may also play a part in this movement. Rotation is produced by the rotatores muscles and the oblique muscles of the anterolateral abdominal wall.

# RADIOGRAPHIC APPEARANCES OF THE JOINTS OF THE VERTEBRAL COLUMN

The radiographic appearances of the joints of the vertebral column are shown in Figures 11-24 to 11-29.

## Joints of the Ribs

A typical rib articulates by its head and by its tubercle with the vertebral column posteriorly. Anteriorly, the first seven ribs with their costal cartilages articulate with the sternum (Fig. 12-12); the eighth, ninth, and tenth ribs and their costal cartilages articulate with the costal cartilages above. The eleventh and twelfth ribs are free-floating ribs anteriorly.

### Joints of the Heads of the Ribs

The first rib and the three lowest ribs have a single synovial joint with their corresponding vertebral body. For the second to the ninth ribs, the head articulates by means of a synovial joint on the body of the corresponding vertebra and on that of the vertebra above (Fig. 12-12).

### Joints of the Tubercles of the Ribs

The tubercle articulates by means of a synovial joint with the transverse process of the vertebra to which it corresponds numerically (Fig. 12-12). (It is absent on the eleventh and twelfth ribs.)

### Costochondral Joints

Costochondral joints are cartilaginous joints, and no movement is possible.

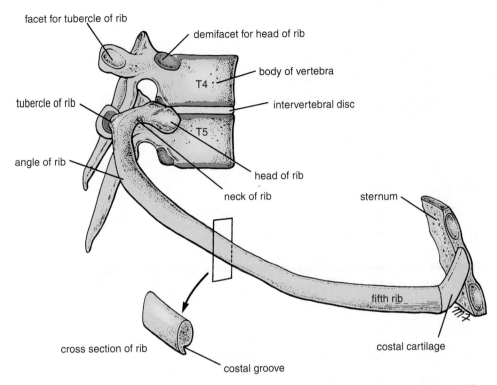

**Figure 12-12** Fifth right rib as it articulates with the vertebral column posteriorly and the sternum anteriorly. Note that the rib head articulates with the vertebral body of its own number and that of the vertebra immediately above.

# Joints of the Costal Cartilages with the Sternum

The first costal cartilages are attached to the manubrium, and no movement is possible (Fig. 12-13). The second costal cartilages articulate with the manubrium and body of the sternum by a mobile synovial joint. The third to the seventh costal cartilages articulate with the lateral border of the sternum by synovial joints. (The sixth, seventh, eighth, ninth, and tenth costal cartilages articulate with each other along their borders by small synovial joints. The cartilages of the eleventh and twelfth ribs are embedded in the abdominal musculature.)

## Movements of the Ribs and Costal Cartilages

The first ribs and their costal cartilages are fixed to the manubrium and are immobile. The raising and lowering of the ribs during respiration are accompanied by movements in both the joints of the head and the tubercle, permitting the neck of each rib to rotate around its own axis.

# Joints of the Sternum

The **manubriosternal joint** is a cartilaginous joint between the manubrium and the body of the sternum (Fig. 12-13). A small amount of angular movement is possible during respiration. The **xiphisternal joint** is a cartilaginous joint between the xiphoid process (cartilage) and the body of the sternum (Fig. 12-13). The xiphoid process usually fuses with the body of the sternum during middle age.

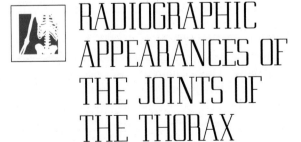

# RADIOGRAPHIC APPEARANCES OF THE JOINTS OF THE THORAX

The radiographic appearances of the joints of the thorax are shown in Figures 3-37 to 3-42.

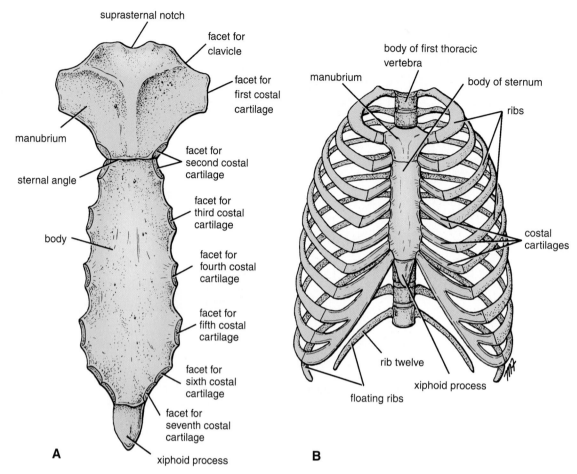

**Figure 12-13** **A.** Anterior view of the sternum. **B.** Sternum, ribs, and costal cartilages forming the thoracic skeleton.

# Joints of the Upper Extremity

## Sternoclavicular Joint

- **Articulation:** This occurs between the sternal end of the clavicle, the manubrium sterni, and the first costal cartilage (Fig. 12-14).
- **Type:** Synovial double-plane joint.
- **Capsule:** This surrounds the joint and is attached to the margins of the articular surfaces.
- **Ligaments:** The capsule is reinforced in front of and behind the joint by the strong **sternoclavicular ligaments.**
- **Articular disc:** This flat fibrocartilaginous disc lies within the joint and divides the joint's interior into two compartments (Fig. 12-14). Its circumference is attached to the interior of the capsule, but it is also strongly attached to the superior margin of the articular surface of the clavicle above and to the first costal cartilage below.
- **Accessory ligament:** The **costoclavicular ligament** is a strong ligament that runs from the junction of the first rib

with the first costal cartilage to the inferior surface of the sternal end of the clavicle (Fig. 12-14).
- **Synovial membrane:** This lines the capsule and is attached to the margins of the cartilage covering the articular surfaces.
- **Nerve supply:** The supraclavicular nerve and the nerve to the subclavius muscle.

## Movements

Forward and backward movement of the clavicle takes place in the medial compartment. Elevation and depression of the clavicle take place in the lateral compartment.

## Muscles Producing Movement

The forward movement of the clavicle is produced by the serratus anterior muscle. The backward movement is produced by the trapezius and rhomboid muscles. Elevation of the clavicle is produced by the trapezius, sternocleidomastoid, levator scapulae, and rhomboid muscles. Depression of the clavicle is produced by the pectoralis minor and the subclavius muscles (Fig. 12-15).

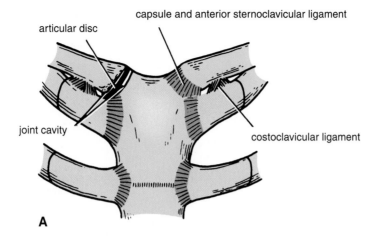

articular disc

capsule and anterior sternoclavicular ligament

joint cavity

costoclavicular ligament

**A**

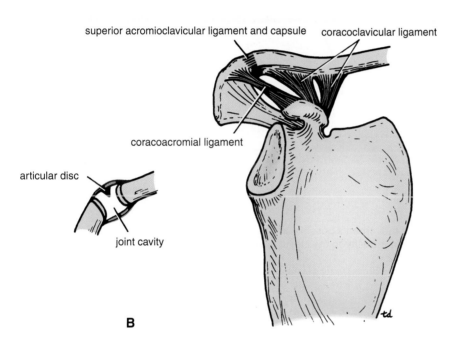

superior acromioclavicular ligament and capsule    coracoclavicular ligament

coracoacromial ligament

articular disc

joint cavity

**B**

**Figure 12-14   A.** Sternoclavicular joint. **B.** Acromioclavicular joint.

## Important Relations

▦ **Anteriorly:** The skin and some fibers of the sternocleidomastoid and pectoralis major muscles.

▦ **Posteriorly:** The sternohyoid muscle; on the right, the brachiocephalic artery; on the left, the left brachiocephalic vein and the left common carotid artery.

## Acromioclavicular Joint

▦ **Articulation:** This occurs between the acromion of the scapula and the lateral end of the clavicle (Fig. 12-14).

▦ **Type:** Synovial plane joint.

▦ **Capsule:** This surrounds the joint and is attached to the margins of the articular surfaces.

▦ **Ligaments: Superior** and **inferior acromioclavicular ligaments** reinforce the capsule; from the capsule, a

wedge-shaped **fibrocartilaginous disc** projects into the joint cavity from above (Fig. 12-14).

▦ **Accessory ligament:** The very strong **coracoclavicular ligament** extends from the coracoid process to the undersurface of the clavicle (Fig. 12-14). It is largely responsible for suspending the weight of the scapula and the upper limb from the clavicle.

▦ **Synovial membrane:** This lines the capsule and is attached to the margins of the cartilage covering the articular surfaces.

▦ **Nerve supply:** The suprascapular nerve.

## Movements

A gliding movement takes place when the scapula rotates or when the clavicle is elevated or depressed (Fig. 12-15).

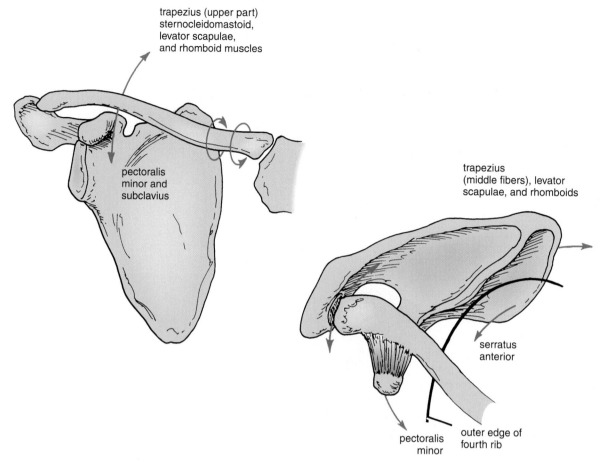

trapezius (upper part)
sternocleidomastoid,
levator scapulae,
and rhomboid muscles

pectoralis
minor and
subclavius

trapezius
(middle fibers), levator
scapulae, and rhomboids

serratus
anterior

pectoralis
minor

outer edge of
fourth rib

**Figure 12-15** The wide range of movements possible at the sternoclavicular and acromioclavicular joints gives great mobility to the clavicle and the upper limb.

## Important Relations

- **Anteriorly:** The deltoid muscle.
- **Posteriorly:** The trapezius muscle.
- **Superiorly:** The skin.

## Shoulder Joint

- **Articulation:** This occurs between the rounded head of the humerus and the shallow, pear-shaped glenoid cavity of the scapula. The articular surfaces are covered by hyaline articular cartilage, and the glenoid cavity is deepened by the presence of a fibrocartilaginous rim called the **glenoid labrum** (Figs. 12-16 and 12-17).
- **Type:** Synovial ball-and-socket joint.
- **Capsule:** This surrounds the joint and is attached medially to the margin of the glenoid cavity outside the labrum; laterally, it is attached to the anatomic neck of the humerus (Fig. 12-17). The capsule is thin and lax, allowing a wide range of movement. It is strengthened by fibrous slips from the tendons of the subscapularis, supraspinatus, infraspinatus, and teres minor muscles (the rotator cuff muscles).
- **Ligaments:** The **glenohumeral ligaments** are three weak bands of fibrous tissue that strengthen the front of the capsule. The **transverse humeral ligament** strengthens the capsule and bridges the gap between the two tuberosities (Fig. 12-16). The **coracohumeral ligament** strengthens the capsule above and stretches from the root of the coracoid process to the greater tuberosity of the humerus (Fig. 12-16).
- **Accessory ligaments:** The **coracoacromial ligament** extends between the coracoid process and the acromion. Its function is to protect the superior aspect of the joint (Fig. 12-16).
- **Synovial membrane:** This lines the capsule and is attached to the margins of the cartilage covering the articular surfaces (Figs. 12-16 and 12-17). It forms a tubular sheath around the tendon of the long head of the biceps brachii. It extends through the anterior wall of the capsule to form the **subscapularis bursa** beneath the subscapularis muscle (Fig. 12-16).
- **Nerve supply:** The axillary and suprascapular nerves.

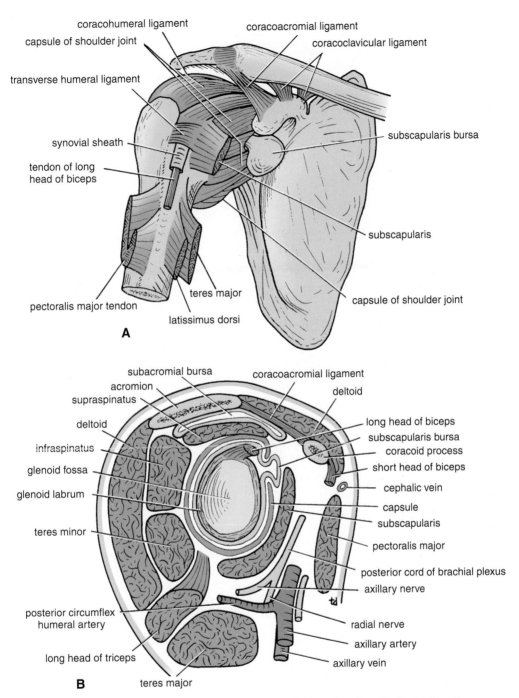

**Figure 12-16** Shoulder joint and its relations. **A.** Anterior view. **B.** Sagittal section.

## Movements and the Muscles Producing the Movements

The shoulder joint has a wide range of movement, and the stability of the joint has been sacrificed to permit this (compare with the hip joint, which is stable but limited in its movements). The strength of the joint depends on the tone of the short rotator cuff muscles that cross in front, above, and behind the joint—namely, the subscapularis, supraspinatus, infraspinatus, and teres minor. When the joint is abducted, the lower surface of the head of the humerus is supported by the long head of the triceps, which bows downward because of its length and gives little actual support to the humerus. In addition, the inferior part of the capsule is the weakest area.

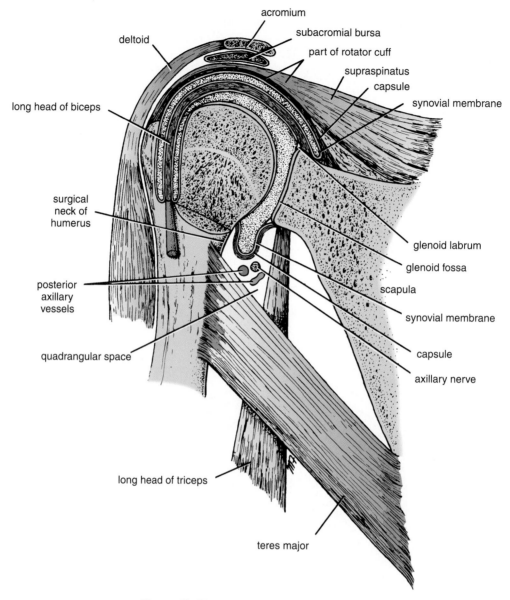

**Figure 12-17**  Interior of the shoulder joint.

The following movements are possible (Fig. 12-18):

- **Flexion:** Normal flexion is about 90° and is performed by the anterior fibers of the deltoid, pectoralis major, biceps, and coracobrachialis muscles.
- **Extension:** Normal extension is about 45° and is performed by the posterior fibers of the deltoid, latissimus dorsi, and teres major muscles.
- **Abduction:** Abduction of the upper limb occurs both at the shoulder joint and between the scapula and the thoracic wall. The middle fibers of the deltoid, assisted by the supraspinatus, are involved. The supraspinatus muscle initiates the movement of abduction and holds the head of the humerus against the glenoid fossa of the scapula; this latter function allows the deltoid muscle to contract and abduct the humerus at the shoulder joint.

- **Adduction:** Normally, the upper limb can be swung 45° across the front of the chest. This is performed by the pectoralis major, latissimus dorsi, teres major, and teres minor muscles.
- **Lateral rotation:** Normal lateral rotation is 40° to 45°. This is performed by the infraspinatus, the teres minor, and the posterior fibers of the deltoid muscle.
- **Medial rotation:** Normal medial rotation is about 55°. This is performed by the subscapularis, the latissimus dorsi, the teres major, and the anterior fibers of the deltoid muscle.
- **Circumduction:** This is a combination of the above movements.

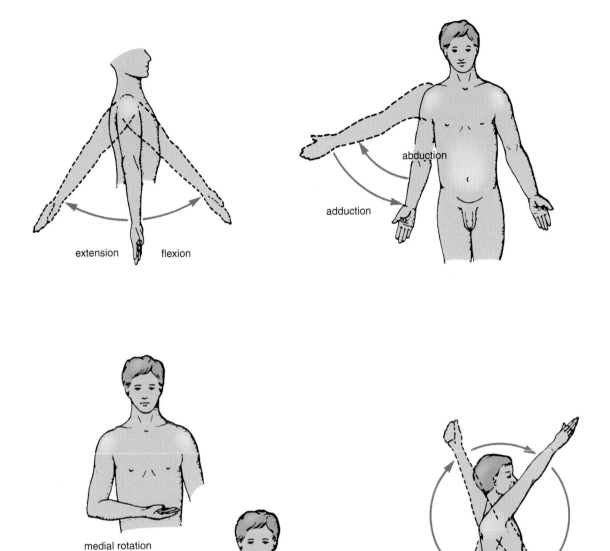

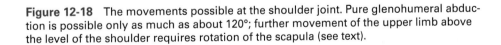

**Figure 12-18** The movements possible at the shoulder joint. Pure glenohumeral abduction is possible only as much as about 120°; further movement of the upper limb above the level of the shoulder requires rotation of the scapula (see text).

## Important Relations

- **Anteriorly:** The subscapularis muscle and the axillary vessels and brachial plexus.
- **Posteriorly:** The infraspinatus and teres minor muscles.
- **Superiorly:** The supraspinatus muscle, subacromial bursa, coracoacromial ligament, and deltoid muscle.

- **Inferiorly:** The long head of the triceps muscle, the axillary nerve, and the posterior circumflex humeral vessels.

The tendon of the long head of the biceps muscle passes through the joint and emerges beneath the transverse ligament.

### The Scapular–Humeral Mechanism

The scapula and upper limb are suspended from the clavicle by the strong coracoclavicular ligament assisted by the tone of muscles. When the scapula rotates on the chest wall so that the position of the glenoid fossa is altered, the axis of rotation may be considered to pass through the coracoclavicular ligament.

Abduction of the arm involves rotation of the scapula as well as movement at the shoulder joint. For every 3° of abduction of the arm, a 2° abduction occurs in the shoulder joint and a 1° abduction occurs by rotation of the scapula. At about 120° of abduction of the arm, the greater tuberosity of the humerus comes into contact with the lateral edge of the acromion. Further elevation of the arm above the head is accomplished by rotating the scapula. Figure 12-19 summarizes the movements of abduction of the arm and shows the direction of pull of the muscles responsible for these movements.

## Elbow Joint

- **Articulation:** This occurs between the trochlea and capitulum of the humerus and the trochlear notch of the ulna and the head of the radius (Fig. 12-20). The articular surfaces are covered with hyaline cartilage.
- **Type:** Synovial hinge joint.
- **Capsule: Anteriorly,** it is attached above to the humerus along the upper margins of the coronoid and radial fossae and to the front of the medial and lateral epicondyles and below to the margin of the coronoid process of the ulna and to the anular ligament, which surrounds the head of the radius. **Posteriorly,** it is attached above to the margins of the olecranon fossa of the humerus and below to the upper margin and sides of the olecranon process of the ulna and to the anular ligament.
- **Ligaments:** The **lateral ligament** (Fig. 12-20) is triangular and is attached by its apex to the lateral epicondyle of the humerus and by its base to the upper margin of the anular ligament. The **medial ligament** is also triangular and consists principally of three strong bands: the anterior band, which passes from the medial epicondyle of the humerus to the medial margin of the coronoid process; the posterior band, which passes from the medial epicondyle of the humerus to the medial side of the olecranon; and the transverse band, which passes between the ulnar attachments of the two preceding bands.
- **Synovial membrane:** This lines the capsule and covers fatty pads in the floors of the coronoid, radial, and olecranon fossae; it is continuous below with the synovial membrane of the proximal radioulnar joint.
- **Nerve supply:** Branches from the median, ulnar, musculocutaneous, and radial nerves.

## Movements and the Muscles Producing the Movements

The elbow joint is capable of flexion and extension. Flexion is limited by the anterior surfaces of the forearm and arm coming into contact. Extension is checked by the tension of the anterior ligament and the brachialis muscle. **Flexion** is performed by the brachialis, biceps brachii, brachioradialis, and pronator teres muscles. **Extension** is performed by the triceps and anconeus muscles.

It should be noted that the long axis of the extended forearm lies at an angle to the long axis of the arm. This angle, which opens laterally, is called the **carrying angle** and is about 170° in the male and 167° in the female. The angle disappears when the elbow joint is fully flexed.

## Important Relations

- **Anteriorly:** The brachialis, the tendon of the biceps, the median nerve, and the brachial artery.
- **Posteriorly:** The triceps muscle, a small bursa intervening.
- **Medially:** The ulnar nerve passes behind the medial epicondyle and crosses the medial ligament of the joint.
- **Laterally:** The common extensor tendon and the supinator.

## Proximal Radioulnar Joint

- **Articulation:** Between the circumference of the head of the radius and the anular ligament and the radial notch on the ulna (Figs. 12-20 and 12-21).
- **Type:** Synovial pivot joint.
- **Capsule:** The capsule encloses the joint and is continuous with that of the elbow joint.
- **Ligament:** The **anular ligament** is attached to the anterior and posterior margins of the radial notch on the ulna and forms a collar around the head of the radius (Fig. 12-21). It is continuous above with the capsule of the elbow joint. It is not attached to the radius.
- **Synovial membrane:** This is continuous above with that of the elbow joint. Below, it is attached to the inferior margin of the articular surface of the radius and the lower margin of the radial notch of the ulna.
- **Nerve supply:** Branches of the median, ulnar, musculocutaneous, and radial nerves.

## Movements

Pronation and supination of the forearm (see p. 400).

## Important Relations

- **Anteriorly:** Supinator muscle and the radial nerve.
- **Posteriorly:** Supinator muscle and the common extensor tendon.

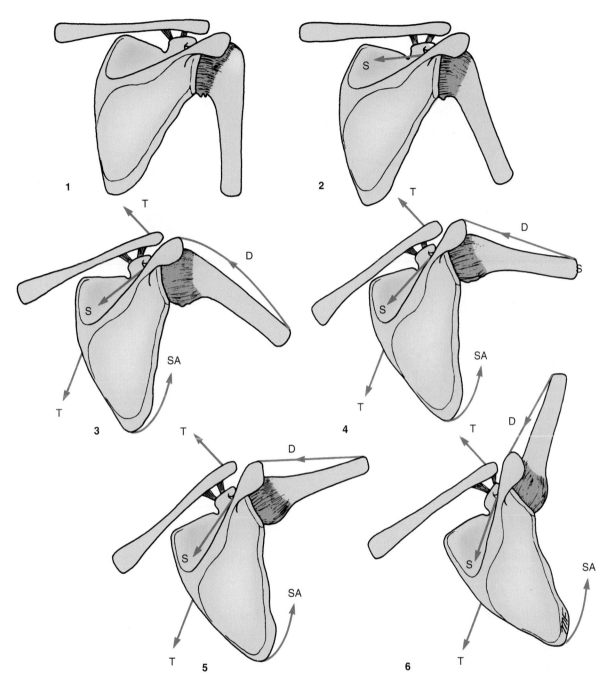

**Figure 12-19** Movements of abduction of the shoulder joint and rotation of the scapula and the muscles producing these movements. Note that for every 3° of abduction of the arm, a 2° abduction occurs in the shoulder joint, and 1° occurs by rotation of the scapula. At about 120° of abduction, the greater tuberosity of the humerus hits the lateral edge of the acromion. Elevation of the arm above the head is accomplished by rotating the scapula. S, supraspinatus; D, deltoid; T, trapezius; and SA, serratus anterior.

## Distal Radioulnar Joint

- **Articulation:** Between the rounded head of the ulna and the ulnar notch on the radius (Fig. 12-21).
- **Type:** Synovial pivot joint.
- **Capsule:** The capsule encloses the joint but is deficient superiorly.

- **Ligaments:** Weak **anterior** and **posterior** ligaments strengthen the capsule.
- **Articular disc:** This is triangular and composed of fibrocartilage. It is attached by its apex to the lateral side of the base of the styloid process of the ulna and by its base to the lower border of the ulnar notch of the radius (Fig. 12-21). It shuts off the distal

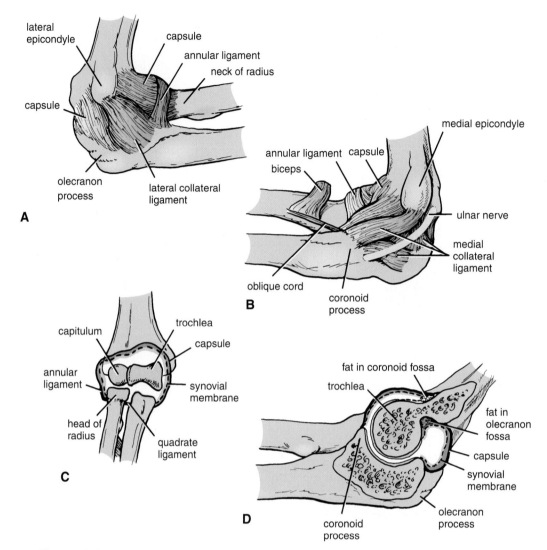

**Figure 12-20** Right elbow joint. **A.** Lateral view. **B.** Medial view. **C.** Anterior view of the interior of the joint. **D.** Sagittal section.

radioulnar joint from the wrist and strongly unites the radius to the ulna.

- **Synovial membrane:** This lines the capsule passing from the edge of one articular surface to that of the other.
- **Nerve supply:** Anterior interosseous nerve and the deep branch of the radial nerve.

## Movements

The movements of pronation and supination of the forearm involve a rotary movement around a vertical axis at the proximal and distal radioulnar joints. The axis passes through the head of the radius above and the attachment of the apex of the triangular articular disc below.

In the movement of pronation, the head of the radius rotates within the anular ligament, whereas the distal end of the radius with the hand moves bodily forward, the ulnar notch of the radius moving around the circumference of the

head of the ulna (see Fig. 12-23). In addition, the distal end of the ulna moves laterally so that the hand remains in line with the upper limb and is not displaced medially. This movement of the ulna is important when using an instrument such as a screwdriver because it prevents side-to-side movement of the hand during the repetitive movements of supination and pronation.

The movement of pronation results in the hand's rotating medially in such a manner that the palm comes to face posteriorly and the thumb lies on the medial side. The movement of supination is a reversal of this process, so that the hand returns to the anatomic position and the palm faces anteriorly.

**Pronation** is performed by the pronator teres and the pronator quadratus.

**Supination** is performed by the biceps brachii and the supinator. Supination is the more powerful of the two movements because of the strength of the biceps muscle. Because

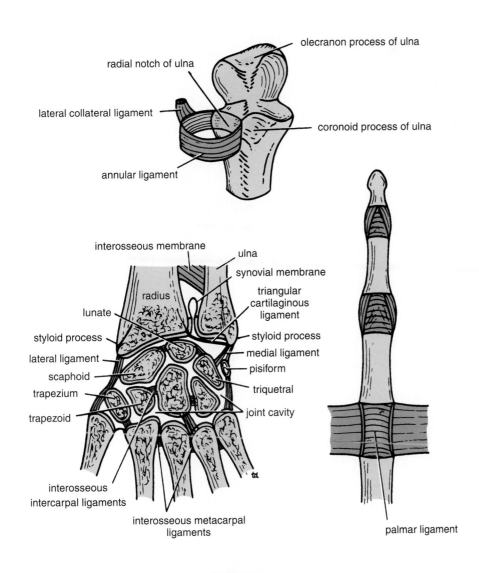

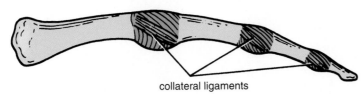

**Figure 12-21** Ligaments of the proximal and distal radioulnar joints, wrist joint, carpal joints, and joints of the fingers.

supination is the more powerful movement, screw threads and the spiral of corkscrews are made so that the screw and corkscrews are driven inward by the movement of supination in right-handed people.

## Important Relations

▧ **Anteriorly:** The tendons of flexor digitorum profundus.
▧ **Posteriorly:** The tendon of extensor digiti minimi.

## Wrist Joint (Radiocarpal Joint)

▧ **Articulation:** Between the distal end of the radius and the articular disc above and the scaphoid, lunate, and triquetral bones below (Figs. 12-21 and 12-22). The proximal articular surface forms an ellipsoid concave surface, which is adapted to the distal ellipsoid convex surface.
▧ **Type:** Synovial ellipsoid joint.
▧ **Capsule:** The capsule encloses the joint and is attached above to the distal ends of the radius and ulna and below to the proximal row of carpal bones.
▧ **Ligaments:** **Anterior** and **posterior ligaments** strengthen the capsule. The **medial ligament** is attached to the styloid process of the ulna and to the triquetral bone (Fig. 12-21). The **lateral ligament** is attached to the styloid process of the radius and to the scaphoid bone (Fig. 12-21).

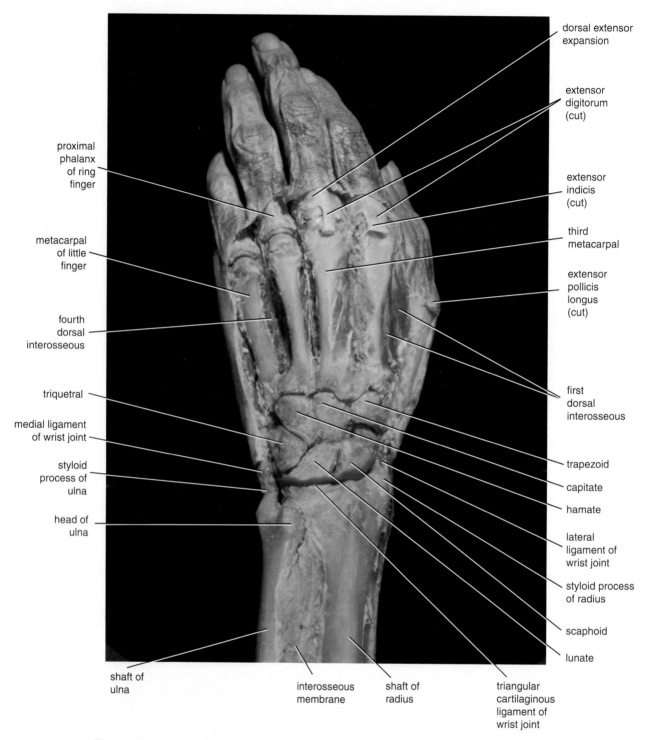

**Figure 12-22**   Dissection of the dorsal surface of the left hand and distal end of the fore-arm. Note the carpal bones and the intercarpal joints; note also the wrist joint.

■ **Synovial membrane:** This lines the capsule and is attached to the margins of the articular surfaces. The joint cavity does not communicate with that of the distal radioulnar joint or with the joint cavities of the intercarpal joints.

■ **Nerve supply:** Anterior interosseous nerve and the deep branch of the radial nerve.

## Movements and the Muscles Producing the Movements

The following movements are possible: flexion, extension, abduction, adduction, and circumduction. Rotation is **not** possible because the articular surfaces are ellipsoid shaped.

The lack of rotation is compensated for by the movements of pronation and supination of the forearm.

**Flexion** is performed by the flexor carpi radialis, the flexor carpi ulnaris, and the palmaris longus. These muscles are assisted by the flexor digitorum superficialis, the flexor digitorum profundus, and the flexor pollicis longus.

**Extension** is performed by the extensor carpi radialis longus, the extensor carpi radialis brevis, and the extensor carpi ulnaris. These muscles are assisted by the extensor digitorum, the extensor indicis, the extensor digiti minimi, and the extensor pollicis longus.

**Abduction** is performed by the flexor carpi radialis and the extensor carpi radialis longus and brevis. These muscles are assisted by the abductor pollicis longus and extensor pollicis longus and brevis.

**Adduction** is performed by the flexor and extensor carpi ulnaris.

## Important Relations

- **Anteriorly:** The tendons of the flexor digitorum profundus and superficialis, the flexor pollicis longus, the flexor carpi radialis, the flexor carpi ulnaris, and the median and ulnar nerves.
- **Posteriorly:** The tendons of the extensor carpi ulnaris, the extensor digiti minimi, the extensor digitorum, the extensor indicis, the extensor carpi radialis longus and brevis, the extensor pollicis longus and brevis, and the abductor pollicis longus.
- **Medially:** The posterior cutaneous branch of the ulnar nerve.
- **Laterally:** The radial artery.

# Joints of the Hand and Fingers

## Intercarpal Joints

- **Articulation:** Between the individual bones of the proximal row of the carpus; between the individual bones of the distal row of the carpus; and finally, the **midcarpal joint,** between the proximal and distal rows of carpal bones (Figs. 12-21 and 12-22).
- **Type:** Synovial plane joints.
- **Capsule:** The capsule surrounds each joint.
- **Ligaments:** The bones are united by strong **anterior, posterior,** and **interosseous ligaments.**
- **Synovial membrane:** This lines the capsule and is attached to the margins of the articular surfaces. The joint cavity of the midcarpal joint extends not only between the two rows of carpal bones but also upward between the individual bones forming the proximal row and downward between the bones of the distal row.
- **Nerve supply:** Anterior interosseous nerve, deep branch of the radial nerve, and deep branch of the ulnar nerve.

## Movements

A small amount of gliding movement is possible.

## Carpometacarpal and Intermetacarpal Joints

The carpometacarpal and intermetacarpal joints are synovial plane joints possessing anterior, posterior, and interosseous ligaments. They have a common joint cavity. A small amount of gliding movement is possible (Fig. 12-21).

## Carpometacarpal Joint of the Thumb

- **Articulation:** Between the trapezium and the saddle-shaped base of the first metacarpal bone (Fig. 12-21).
- **Type:** Synovial saddle-shaped joint.
- **Capsule:** The capsule surrounds the joint.
- **Synovial membrane:** This lines the capsule and forms a separate joint cavity.

## Movements

The following movements are possible:

- **Flexion:** Flexor pollicis brevis and opponens pollicis.
- **Extension:** Extensor pollicis longus and brevis.
- **Abduction:** Abductor pollicis longus and brevis.
- **Adduction:** Adductor pollicis.
- **Rotation (opposition):** The thumb is rotated medially by the opponens pollicis.

## Metacarpophalangeal Joints

- **Articulations:** Between the heads of the metacarpal bones and the bases of the proximal phalanges (Fig. 12-21).
- **Type:** Synovial condyloid joints.
- **Capsule:** The capsule surrounds the joint.
- **Ligaments:** The **palmar ligaments** are strong and contain some fibrocartilage. They are firmly attached to the phalanx but less so to the metacarpal bone (Fig. 12-21). The palmar ligaments of the second, third, fourth, and fifth joints are united by the **deep transverse metacarpal ligaments,** which hold the heads of the metacarpal bones together. The **collateral ligaments** are cordlike bands present on each side of the joints (Fig. 12-21). Each passes downward and forward from the head of the metacarpal bone to the base of the phalanx. The collateral ligaments are taut when the joint is in flexion and lax when the joint is in extension.
- **Synovial membrane:** This lines the capsule and is attached to the margins of the articular surfaces.

## Movements

The following movements are possible:

- **Flexion:** The lumbricals and the interossei, assisted by the flexor digitorum superficialis and profundus.
- **Extension:** Extensor digitorum, extensor indicis, and extensor digiti minimi.

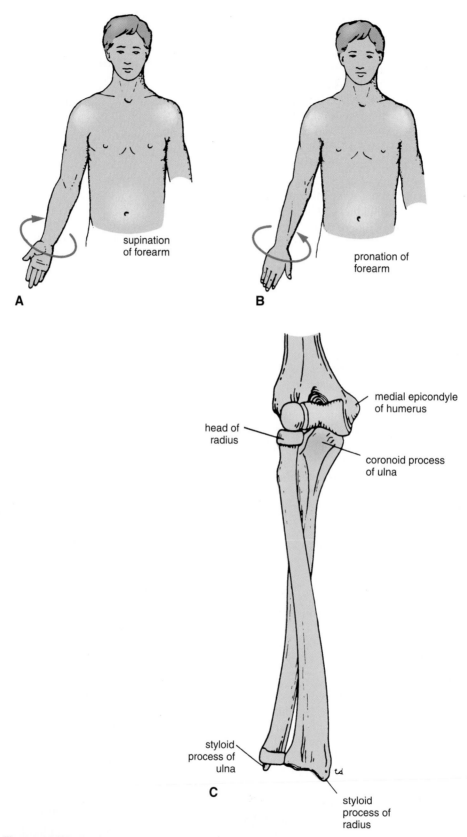

**Figure 12-23**  Movements of supination (**A**) and pronation (**B**) of the forearm that take place at the proximal and distal radioulnar joints. **C.** Relative positions of the radius and ulna when the forearm is fully pronated.

- **Abduction:** Movement away from the midline of the third finger is performed by the dorsal interossei.
- **Adduction:** Movement toward the midline of the third finger is performed by the palmar interossei. In the case of the metacarpophalangeal joint of the thumb, **flexion** is performed by the flexor pollicis longus and brevis, and **extension** is performed by the extensor pollicis longus and brevis. The movements of abduction and adduction are performed at the carpometacarpal joint.

## Interphalangeal Joints

Interphalangeal joints are synovial hinge joints that have a structure similar to that of the metacarpophalangeal joints (Fig. 12-21).

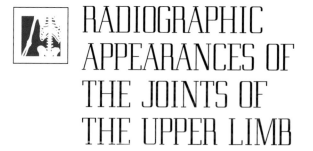

# RADIOGRAPHIC APPEARANCES OF THE JOINTS OF THE UPPER LIMB

The radiographic appearances of the joints of the upper limb are shown in Figures 11-42 to 11-49.

# Joints of the Pelvis

## Sacroiliac Joints

The sacroiliac joints are strong synovial joints and are formed between the auricular surfaces of the sacrum and the iliac bones (Fig. 12-24). The sacrum carries the weight of the trunk, and, apart from the interlocking of the irregular articular surfaces, the shape of the bones contributes little to the stability of the joints. The strong **posterior** and **interosseous sacroiliac ligaments** suspend the sacrum between the two iliac bones. The **anterior sacroiliac ligament** is thin and situated on the anterior aspect of the joint.

The weight of the trunk tends to thrust the upper end of the sacrum downward and rotate the lower end of the bone upward. This rotory movement is prevented by the strong **sacrotuberous** and **sacrospinous ligaments** described previously (Fig. 12-24). The **iliolumbar ligament** connects the tip of the fifth lumbar transverse process to the iliac crest. A small but limited amount of movement is possible at these joints. In older people, the synovial cavity disappears, and the joint becomes fibrosed. Their primary function is to transmit the weight of the body from the vertebral column to the bony pelvis.

## Nerve Supply

The sacroiliac joint is supplied by branches of the sacral spinal nerves.

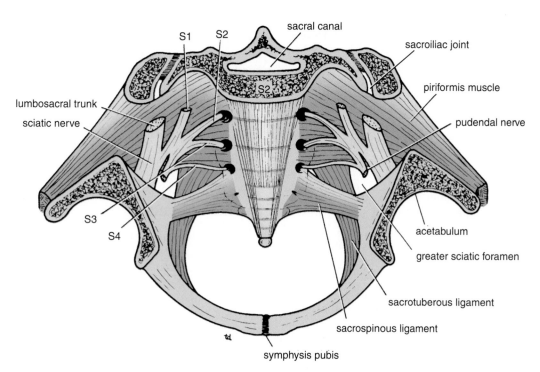

**Figure 12-24** Horizontal section through the pelvis showing the sacroiliac joints and the symphysis publis.

## Symphysis Pubis

The symphysis pubis is a cartilaginous joint between the two pubic bones (Fig. 12-24). The articular surfaces are covered by a layer of hyaline cartilage and are connected together by a fibrocartilaginous disc. The joint is surrounded by ligaments that extend from one pubic bone to the other. Almost no movement is possible at this joint.

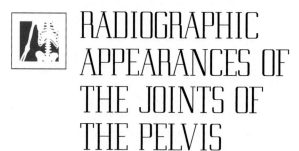 # RADIOGRAPHIC APPEARANCES OF THE JOINTS OF THE PELVIS

The radiographic appearances of the joints of the pelvis are shown in Figures 11-63 to 11-66.

# Joints of the Lower Extremity

## Hip Joint

### Articulation

The hip joint is the articulation between the hemispherical head of the femur and the cup-shaped acetabulum of the hip bone (Fig. 12-25). The articular surface of the acetabulum is horseshoe shaped and is deficient inferiorly at the **acetabular notch.** The cavity of the acetabulum is deepened by the presence of a fibrocartilaginous rim called the **acetabular labrum.** The labrum bridges across the acetabular notch and is here called the **transverse acetabular ligament** (Fig. 12-25).

The articular surfaces are covered with hyaline cartilage.

### Type

The hip joint is a synovial ball-and-socket joint.

### Capsule

The capsule encloses the joint and is attached to the acetabular labrum medially (Fig. 12-25). Laterally, it is attached to the intertrochanteric line of the femur in front and halfway along the posterior aspect of the neck of the bone behind. At its attachment to the intertrochanteric line in front, some of its fibers, accompanied by blood vessels, are reflected upward along the neck as bands called **retinacula.** These blood vessels supply the head and neck of the femur.

### Ligaments

The **iliofemoral ligament** is a strong, inverted Y-shaped ligament (Fig. 12-26). Its base is attached to the anterior inferior iliac spine above; below, the two limbs of the Y are attached to the upper and lower parts of the intertrochanteric line of the femur. This strong ligament prevents overextension during standing.

The **pubofemoral ligament** is triangular (Fig. 12-26). The base of the ligament is attached to the superior ramus of the pubis, and the apex is attached below to the lower part of the intertrochanteric line. This ligament limits extension and abduction.

The **ischiofemoral ligament** is spiral shaped and is attached to the body of the ischium near the acetabular margin (Fig. 12-26). The fibers pass upward and laterally and are attached to the greater trochanter. This ligament limits extension.

The **transverse acetabular ligament** is formed by the acetabular labrum as it bridges the acetabular notch (Fig. 12-25). The ligament converts the notch into a tunnel through which the blood vessels and nerves enter the joint.

The **ligament of the head of the femur** is flat and triangular (Fig. 12-25). It is attached by its apex to the pit on the head of the femur (fovea capitis) and by its base to the transverse ligament and the margins of the acetabular notch. It lies within the joint and is ensheathed by synovial membrane (Fig. 12-25).

## Synovial Membrane

The synovial membrane lines the capsule and is attached to the margins of the articular surfaces (Fig. 12-25). It covers the portion of the neck of the femur that lies within the joint capsule. It ensheathes the ligament of the head of the femur and covers the pad of fat contained in the acetabular fossa. A pouch of synovial membrane frequently protrudes through a gap in the anterior wall of the capsule, between the pubofemoral and iliofemoral ligaments, and forms the **psoas bursa** beneath the psoas tendon (Figs. 12-26 and 12-27).

## Nerve Supply

Femoral, obturator, and sciatic nerves and the nerve to the quadratus femoris supply the joint.

## Movements and the Muscles Producing the Movements

The hip joint has a wide range of movement, but less so than the shoulder joint. Some of the movement has been sacrificed to provide strength and stability. The strength of the joint depends largely on the shape of the bones taking part in the articulation and on the strong ligaments. When the knee is flexed, flexion is limited by the anterior surface of the thigh coming into contact with the anterior abdominal wall. When the knee is extended, flexion is limited by the tension of the hamstring group of muscles. Extension, which is the movement of the flexed thigh backward to the anatomic

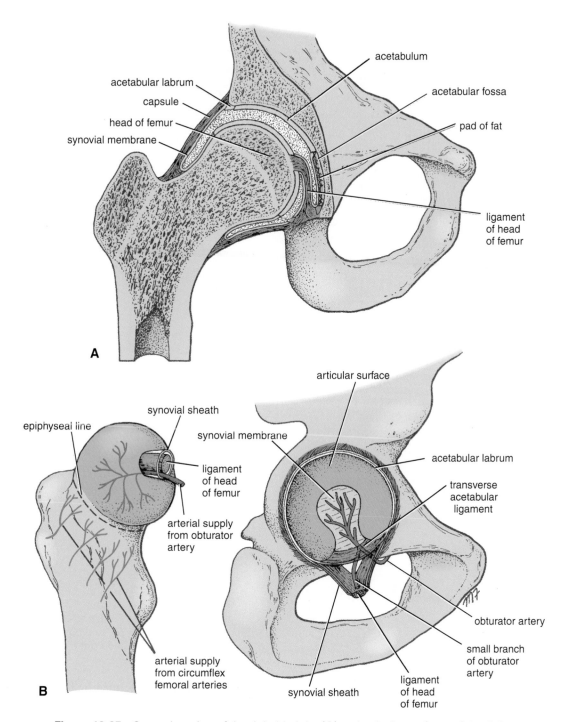

**Figure 12-25**  Coronal section of the right hip joint (**A**) and articular surfaces of the right hip joint and arterial supply of the head of the femur (**B**).

position, is limited by the tension of the iliofemoral, pub-ofemoral, and ischiofemoral ligaments. Abduction is limited by the tension of the pubofemoral ligament, and adduction is limited by contact with the opposite limb and by the tension in the ligament of the head of the femur. Lateral rotation is limited by the tension in the iliofemoral and pubofemoral ligaments, and medial rotation is limited by the ischiofemoral ligament. The following movements take place:

■ **Flexion** is performed by the iliopsoas, rectus femoris, and sartorius and also by the adductor muscles.

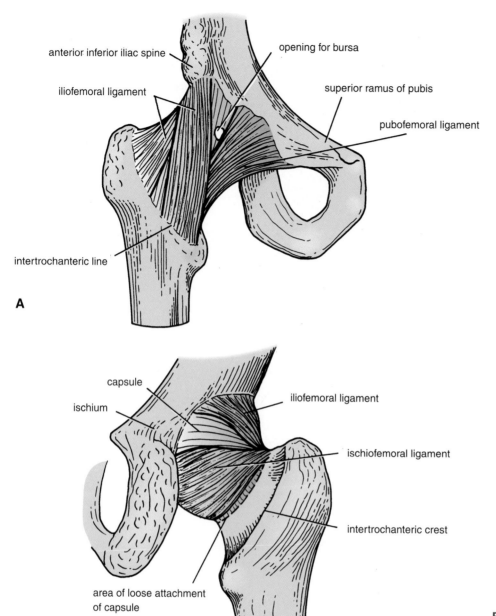

anterior inferior iliac spine

iliofemoral ligament

intertrochanteric line

**A**

opening for bursa

superior ramus of pubis

pubofemoral ligament

capsule

ischium

area of loose attachment
of capsule

**B**

iliofemoral ligament

ischiofemoral ligament

intertrochanteric crest

**Figure 12-26**   Anterior aspect
(**A**) and posterior aspect (**B**) of
the right hip joint.

- **Extension** (a backward movement of the flexed thigh) is performed by the gluteus maximus and the hamstring muscles.
- **Abduction** is performed by the gluteus medius and minimus, assisted by the sartorius, tensor fasciae latae, and piriformis.
- **Adduction** is performed by the adductor longus and brevis and the adductor fibers of the adductor magnus. These muscles are assisted by the pectineus and the gracilis.
- **Lateral rotation** is performed by the piriformis, obturator internus and externus, superior and inferior gemelli, and quadratus femoris, assisted by the gluteus maximus.

- **Medial rotation** is performed by the anterior fibers of gluteus medius and gluteus minimus and the tensor fasciae latae.
- **Circumduction** is a combination of the previous movements.

The extensor group of muscles is more powerful than the flexor group, and the lateral rotators are more powerful than the medial rotators.

## Important Relations

- **Anteriorly:** Iliopsoas, pectineus, and rectus femoris muscles. The iliopsoas and pectineus separate the femoral vessels and nerve from the joint (Fig. 12-27).

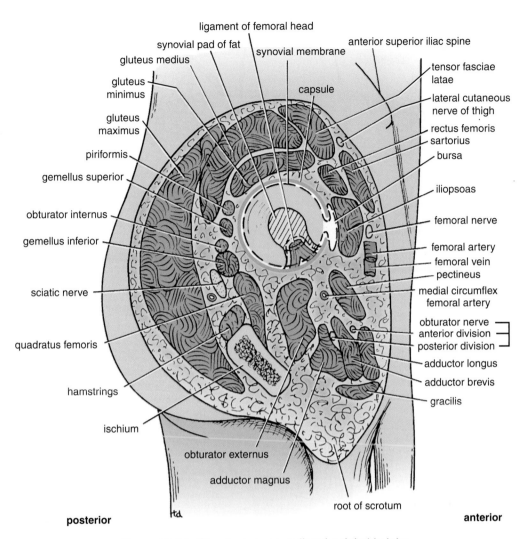

ligament of femoral head
synovial pad of fat
gluteus medius
gluteus minimus
gluteus maximus
piriformis
gemellus superior
obturator internus
gemellus inferior
sciatic nerve
quadratus femoris
hamstrings
ischium
obturator externus
adductor magnus
root of scrotum

synovial membrane
capsule

anterior superior iliac spine
tensor fasciae latae
lateral cutaneous nerve of thigh
rectus femoris
sartorius
bursa
iliopsoas
femoral nerve
femoral artery
femoral vein
pectineus
medial circumflex femoral artery
obturator nerve
anterior division
posterior division
adductor longus
adductor brevis
gracilis

posterior                anterior

**Figure 12-27**   Structures surrounding the right hip joint.

■ **Posteriorly:** The obturator internus, the gemelli, and the quadratus femoris muscles separate the joint from the sciatic nerve (Fig. 12-27).

■ **Superiorly:** Piriformis and gluteus minimus (Fig. 12-27).

■ **Inferiorly:** Obturator externus tendon (Fig. 12-27).

## The Surface Anatomy of the Region of the Hip Joint

The surface anatomy of the region of the hip joint is shown in Figs. 12-28 and 12-29.

## Knee Joint

The knee joint is the largest and most complicated joint in the body. Basically, it consists of two condylar joints between the medial and lateral condyles of the femur and the corresponding condyles of the tibia, and a gliding joint, between the patella and the patellar surface of the

femur. Note that the fibula is not directly involved in the joint.

## Articulation

Above are the rounded condyles of the femur; below are the condyles of the tibia and their cartilaginous menisci (Fig. 12-30); in front is the articulation between the lower end of the femur and the patella.

The articular surfaces of the femur, tibia, and patella are covered with hyaline cartilage. Note that the articular surfaces of the medial and lateral condyles of the tibia are often referred to clinically as the medial and lateral **tibial plateaus.**

## Type

The joint between the femur and tibia is a synovial joint of the hinge variety, but some degree of rotatory movement is possible. The joint between the patella and femur is a synovial joint of the plane gliding variety.

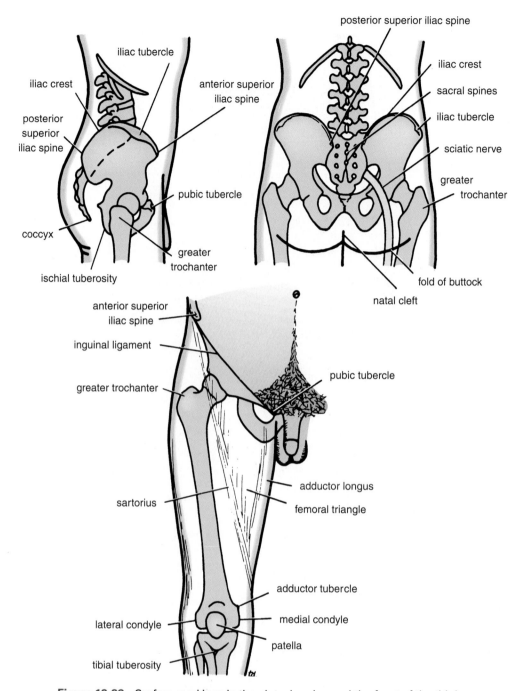

**Figure 12-28** Surface markings in the gluteal region and the front of the thigh.

## Capsule

The capsule is attached to the margins of the articular surfaces and surrounds the sides and posterior aspect of the joint. On the front of the joint, the capsule is absent, permitting the synovial membrane to pouch upward beneath the quadriceps tendon, forming the **suprapatellar bursa** (Fig. 12-30). On each side of the patella, the capsule is strengthened by expansions from the tendons of vastus lateralis and medialis. Behind the joint, the capsule is strengthened by an

expansion of the semimembranous muscle called the **oblique popliteal ligament** (Fig. 12-30). An opening in the capsule behind the lateral tibial condyle permits the tendon of the popliteus to emerge (Fig. 12-30).

## Ligaments

The ligaments may be divided into those that lie outside the capsule and those that lie within the capsule.

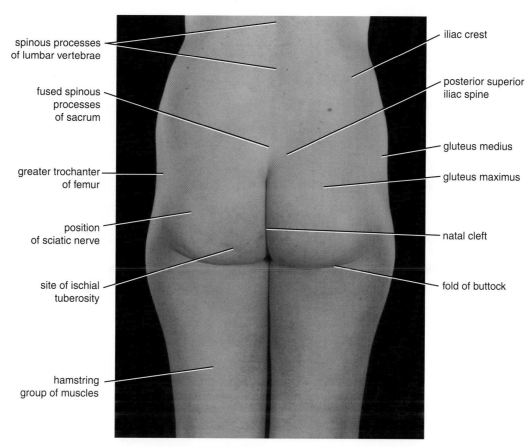

spinous processes
of lumbar vertebrae

fused spinous
processes
of sacrum

greater trochanter
of femur

position
of sciatic nerve

site of ischial
tuberosity

hamstring
group of muscles

iliac crest

posterior superior
iliac spine

gluteus medius

gluteus maximus

natal cleft

fold of buttock

**Figure 12-29** The gluteal region and the posterior aspect of the thigh of a 25-year-old woman.

### Extracapsular Ligaments

The **ligamentum patellae** is attached above to the lower border of the patella and below to the tuberosity of the tibia (Fig. 12-30). It is, in fact, a continuation of the central portion of the common tendon of the quadriceps femoris muscle.

The **lateral collateral ligament** is cordlike and is attached above to the lateral condyle of the femur and below to the head of the fibula (Fig. 12-30). The tendon of the popliteus muscle intervenes between the ligament and the lateral meniscus (Fig. 12-31).

The **medial collateral ligament** is a flat band and is attached above to the medial condyle of the femur and below to the medial surface of the shaft of the tibia (Fig. 12-30). **It is firmly attached to the edge of the medial meniscus** (Fig. 12-31).

The **oblique popliteal ligament** is a tendinous expansion derived from the semimembranosus muscle. It strengthens the posterior aspect of the capsule (Fig. 12-30).

### Intracapsular Ligaments

The **cruciate ligaments** are two strong intracapsular ligaments that cross each other within the joint cavity (Fig. 12-30). They are named anterior and posterior, according to their tibial attachments (Fig. 12-31). These important ligaments are the main bond between the femur and the tibia throughout the joint's range of movement.

**Anterior cruciate ligament:** The anterior cruciate ligament is attached to the anterior intercondylar area of the tibia and passes upward, backward, and laterally to be attached to the posterior part of the medial surface of the lateral femoral condyle (Figs. 12-30 and 12-31). The anterior cruciate ligament prevents posterior displacement of the femur on the tibia. With the knee joint flexed, the anterior cruciate ligament prevents the tibia from being pulled anteriorly.

**Posterior cruciate ligament:** The posterior cruciate ligament is attached to the posterior intercondylar area of the tibia and passes upward, forward, and medially to be attached to the anterior part of the lateral surface of the medial femoral condyle (Figs. 12-30 and 12-31). The posterior cruciate ligament prevents anterior displacement of the femur on the tibia. With the knee joint flexed, the posterior cruciate ligament prevents the tibia from being pulled posteriorly.

**Menisci:** The menisci are C-shaped sheets of fibrocarti-

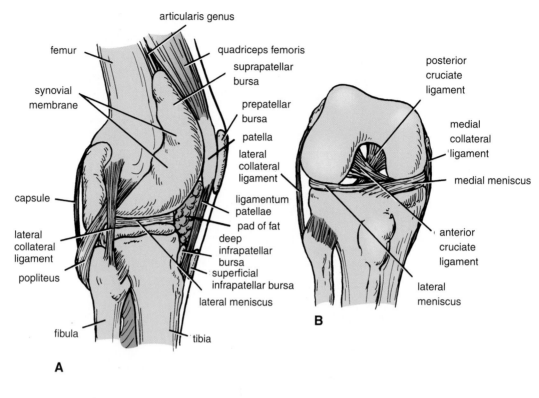

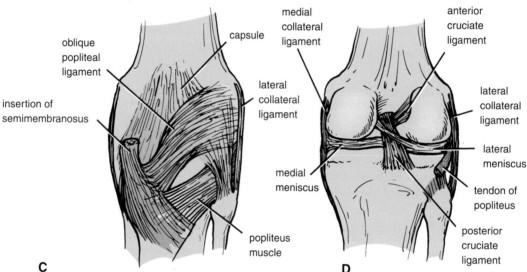

**Figure 12-30**    **A.** The right knee joint as seen from the lateral aspect. **B.** The anterior aspect, with the joint flexed. **C and D.** The posterior aspect.

lage. The peripheral border is thick and attached to the capsule, and the inner border is thin and concave and forms a free edge (Figs. 12-30 and 12-31). The upper surfaces are in contact with the femoral condyles. The lower surfaces are in contact with the tibial condyles. Their function is to deepen the articular surfaces of the tibial condyles to receive the convex femoral condyles; they also serve as cushions between the two bones. Each meniscus is attached to the upper surface of the tibia by

anterior and posterior horns. Because the medial meniscus is also attached to the medial collateral ligament, it is relatively immobile.

## Synovial Membrane

The synovial membrane lines the capsule and is attached to the margins of the articular surfaces (Figs. 12-30 and 12-31). On the front and above the joint, it forms a pouch, which ex-

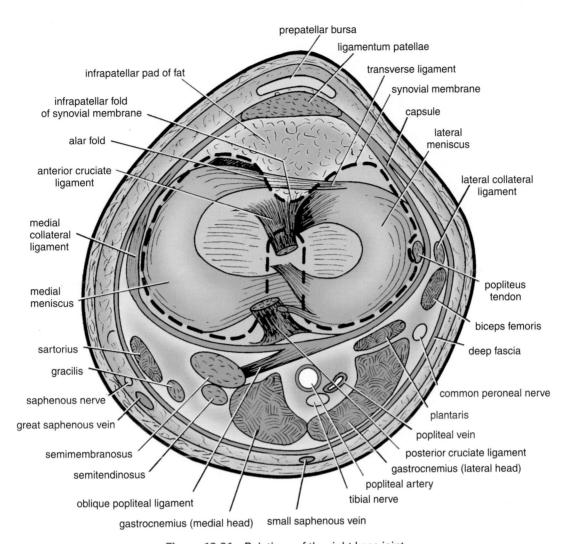

**Figure 12-31** Relations of the right knee joint.

tends up beneath the quadriceps femoris muscle for three fingerbreadths above the patella, forming the **suprapatellar bursa.** This is held in position by the attachment of a small portion of the vastus intermedius muscle, called the **articularis genus** muscle (Fig. 12-30).

At the back of the joint, the synovial membrane is prolonged downward on the deep surface of the tendon of the popliteus, forming the **popliteal bursa.** A bursa is interposed between the medial head of the gastrocnemius and the medial femoral condyle and the semimembranosus tendon; this is termed the **semimembranosus bursa,** and it frequently communicates with the synovial cavity of the joint.

The synovial membrane is reflected forward from the posterior part of the capsule around the front of the cruciate ligaments (Fig. 12-31). As a result, the cruciate ligaments lie behind the synovial cavity and are not bathed in synovial fluid.

In the anterior part of the joint, the synovial membrane is reflected backward from the posterior surface of the ligamentum patellae to form the **infrapatellar fold;** the free borders of the fold are termed the **alar folds** (Fig. 12-31).

## Bursae Related to the Knee Joint

Numerous bursae are related to the knee joint. They are found wherever skin, muscle, or tendon rubs against bone. Four are situated in front of the joint and six are found behind the joint. The suprapatellar bursa and the popliteal bursa always communicate with the joint, and the semimembranosus bursa may communicate with the joint.

### Anterior Bursae

■ The **suprapatellar bursa** lies beneath the quadriceps muscle and communicates with the joint cavity (Fig. 12-30). It is described earlier.

■ The **prepatellar bursa** lies in the subcutaneous tissue between the skin and the front of the lower half of the patella and the upper part of the ligamentum patellae (Figs. 12-30 and 12-31).

■ The **superficial infrapatellar bursa** lies in the subcutaneous tissue between the skin and the front of the lower part of the ligamentum patellae (Fig. 12-30).

▓ The **deep infrapatellar bursa** lies between the ligamentum patellae and the tibia (Fig. 12-30).

### Posterior Bursae
▓ The **popliteal bursa** is found in association with the tendon of the popliteus and communicates with the joint cavity. It was described previously.
▓ The **semimembranosus bursa** is found related to the insertion of the semimembranosus muscle and may communicate with the joint cavity. It was described previously.

The remaining four bursae are found related to the tendon of insertion of the biceps femoris; related to the tendons of the sartorius, gracilis, and semitendinosus muscles as they pass to their insertion on the tibia; beneath the lateral head of origin of the gastrocnemius muscle; and beneath the medial head of origin of the gastrocnemius muscle.

## Nerve Supply

The femoral, obturator, common peroneal, and tibial nerves supply the knee joint.

## Movements and the Muscles Producing the Movements

The knee joint can flex, extend, and rotate. As the knee joint assumes the position of full extension,* medial rotation of the femur results in a twisting and tightening of all the major ligaments of the joint, and the knee becomes a mechanically rigid structure; the cartilaginous menisci are compressed like rubber cushions between the femoral and tibial condyles. The extended knee is said to be in the locked position.

Before flexion of the knee joint can occur, it is essential that the major ligaments be untwisted and slackened to permit movements between the joint surfaces. This unlocking or untwisting process is accomplished by the popliteus muscle, which laterally rotates the femur on the tibia. Once again, the menisci have to adapt their shape to the changing contour of the femoral condyles. The attachment of the popliteus to the lateral meniscus results in that structure being pulled backward also.

When the knee joint is flexed to a right angle, a considerable range of rotation is possible. In the flexed position, the tibia can also be moved passively forward and backward on the femur. This is possible because the major ligaments, especially the cruciate ligaments, are slack in this position. The following muscles produce movements of the knee joint.

▓ **Flexion:** The biceps femoris, semitendinosus, and semimembranosus muscles, assisted by the gracilis, sartorius, and popliteus muscles produce flexion. Flexion is limited by the contact of the back of the leg with the thigh.

▓ **Extension:** The quadriceps femoris produces extension. Extension is limited by the tension of all the major ligaments of the joint.
▓ **Medial rotation:** The sartorius, gracilis, and semitendinosus produce medial rotation.
▓ **Lateral rotation:** The biceps femoris produces lateral rotation.

The stability of the knee joint depends on the tone of the strong muscles acting on the joint and the strength of the ligaments. Of these factors, the tone of the muscles is the most important, and it is the job of the physiotherapist to build up the strength of these muscles, especially the quadriceps femoris, after injury to the knee joint.

## Important Relations

▓ **Anteriorly:** The prepatellar bursa (Fig. 12-31).
▓ **Posteriorly:** The popliteal vessels; tibial and common peroneal nerves; lymph nodes; and the muscles that form the boundaries of the popliteal fossa, namely, the semimembranosus, the semitendinosus, the biceps femoris, the two heads of the gastrocnemius, and the plantaris (Fig. 12-31).
▓ **Medially:** Sartorius, gracilis, and semitendinosus muscles (Fig. 12-31).
▓ **Laterally:** Biceps femoris and common peroneal nerve (Fig. 12-31).

## The Surface Anatomy of the Region of the Knee.

The surface anatomy of the region of the knee is shown in Figure 12-32.

## Proximal Tibiofibular Joint
### Articulation

Articulation is between the lateral condyle of the tibia and the head of the fibula (Fig. 12-30). The articular surfaces are flattened and covered by hyaline cartilage.

### Type

This is a synovial, plane, gliding joint.

### Capsule

The capsule surrounds the joint and is attached to the margins of the articular surfaces.

### Ligaments

**Anterior** and **posterior ligaments** strengthen the capsule. The **interosseous membrane,** which connects the shafts of the tibia and fibula together, also greatly strengthens the joint.

---

*Note that when the foot is firmly planted on the ground when a person is standing, the femur is medially rotated on the tibia to lock and stabilize the knee joint. However, if the foot is raised off the ground, the tibia may be laterally rotated on the femur to lock the knee joint.

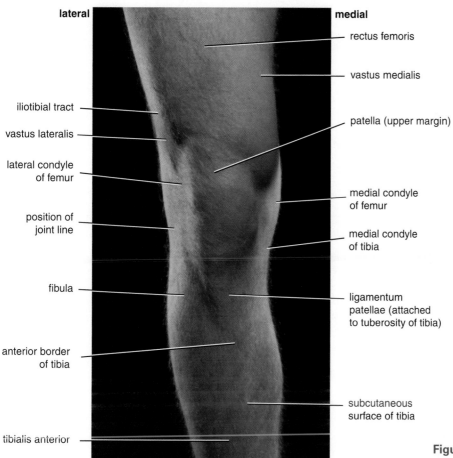

lateral

medial

rectus femoris

vastus medialis

iliotibial tract

vastus lateralis

patella (upper margin)

lateral condyle
of femur

position of
joint line

medial condyle
of femur

medial condyle
of tibia

fibula

ligamentum
patellae (attached
to tuberosity of tibia)

anterior border
of tibia

subcutaneous
surface of tibia

tibialis anterior

**Figure 12-32**    Anterior aspect of the
right knee of a 27-year-old man.

## Synovial Membrane

The synovial membrane lines the capsule and is attached to the margins of the articular surfaces.

## Nerve Supply

The common peroneal nerve supplies the joint.

## Movements

A small amount of gliding movement takes place during movements at the ankle joint.

# Distal Tibiofibular Joint

## Articulation

Articulation is between the fibular notch at the lower end of the tibia and the lower end of the fibula (Figs. 12-33 and 12-34). The opposed bony surfaces are roughened.

## Type

The distal tibiofibular joint is a fibrous joint.

## Capsule

There is no capsule.

## Ligaments

The **interosseous ligament** is a strong, thick band of fibrous tissue that binds the two bones together. The **interosseous membrane,** which connects the shafts of the tibia and fibula together, also greatly strengthens the joint.

The **anterior** and **posterior ligaments** are flat bands of fibrous tissue connecting the two bones together in front and behind the interosseous ligament.

The **inferior transverse ligament** runs from the medial surface of the upper part of the lateral malleolus to the posterior border of the lower end of the tibia.

## Nerve Supply

Deep peroneal and tibial nerves supply the joint.

## Movements

A small amount of movement takes place during movements at the ankle joint.

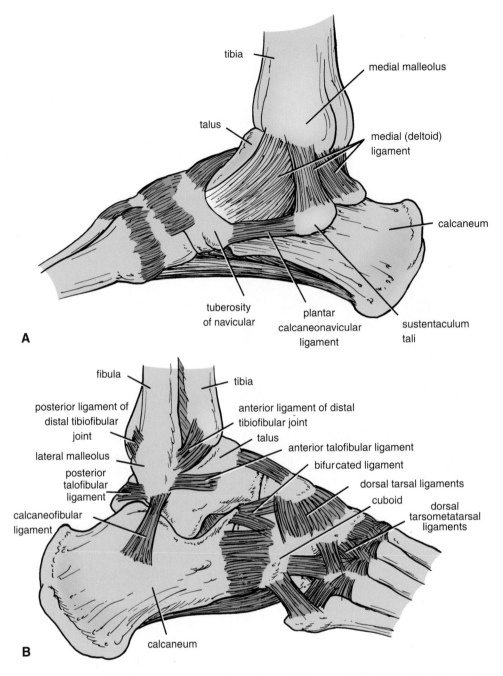

**Figure 12-33** The right ankle joint as seen from the medial aspect (**A**) and the lateral aspect (**B**).

## Ankle Joint

The ankle joint consists of a deep socket formed by the lower ends of the tibia and fibula, into which is fitted the upper part of the body of the talus. The talus is able to move on a transverse axis in a hingelike manner. The shape of the bones and the strength of the ligaments and the surrounding tendons make this joint strong and stable.

## Articulation

Articulation is between the lower end of the tibia, the two malleoli, and the body of the talus (Figs. 12-33 and 12-34). The inferior transverse tibiofibular ligament, which runs between the lateral malleolus and the posterior border of the lower end of the tibia, deepens the socket into which the body of the talus fits snugly. The articular surfaces are covered with hyaline cartilage.

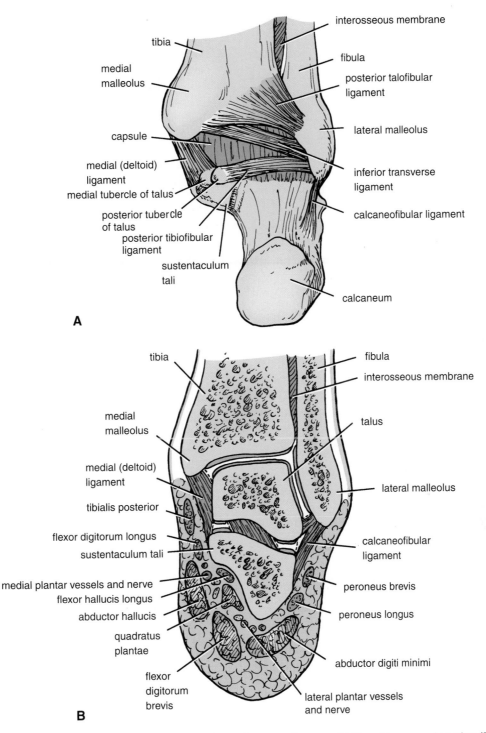

**Figure 12-34** The right ankle joint as seen from the posterior aspect (**A**) and in coronal section (**B**).

## Type

The ankle is a synovial hinge joint.

## Capsule

The capsule encloses the joint and is attached to the bones near their articular margins.

## Ligaments

The **medial**, or **deltoid**, **ligament** is strong and is attached by its apex to the tip of the medial malleolus (Fig. 12-34). Below, the deep fibers are attached to the nonarticular area on the medial surface of the body of the talus; the superficial fibers are attached to the medial side of the talus, the sustentaculum tali, the plantar calcaneonavicular ligament, and the tuberosity of the navicular bone.

The **lateral ligament** is weaker than the medial ligament and consists of three bands.

- The **anterior talofibular ligament** (Fig. 12-33) runs from the lateral malleolus to the lateral surface of the talus.
- The **calcaneofibular ligament** (Fig. 12-33) runs from the tip of the lateral malleolus downward and backward to the lateral surface of the calcaneum.
- The **posterior talofibular ligament** (Fig. 12-33) runs from the lateral malleolus to the posterior tubercle of the talus.

## Synovial Membrane

The synovial membrane lines the capsule.

## Nerve Supply

Deep peroneal and tibial nerves supply the ankle joint.

## Movements and the Muscles Producing the Movements

Dorsiflexion (toes pointing upward) and plantar flexion (toes pointing downward) are possible. The movements of inversion and eversion take place at the tarsal joints and **not at the ankle joint.**

**Dorsiflexion** is performed by the tibialis anterior, extensor hallucis longus, extensor digitorum longus, and peroneus tertius. It is limited by the tension of the tendo calcaneus, the posterior fibers of the medial ligament, and the calcaneofibular ligament.

**Plantar flexion** is performed by the gastrocnemius, soleus, plantaris, peroneus longus, peroneus brevis, tibialis posterior, flexor digitorum longus, and flexor hallucis longus. It is limited by the tension of the opposing muscles, the anterior fibers of the medial ligament, and the anterior talofibular ligament.

Note that during dorsiflexion of the ankle joint, the wider anterior part of the articular surface of the talus is forced between the medial and lateral malleoli, causing them to separate slightly and tighten the ligaments of the distal tibiofibular joint. This arrangement greatly increases the stability of the ankle joint when the foot is in the initial position for major thrusting movements in walking, running, and jumping.

Note also that, when the ankle joint is fully plantar flexed, the ligaments of the distal tibiofibular joint are less taut, and small amounts of rotation, abduction, and adduction are possible.

## Important Relations

- **Anteriorly:** The tibialis anterior, the extensor hallucis longus, the anterior tibial vessels, the deep peroneal nerve, the extensor digitorum longus, and the peroneus tertius (Fig. 12-35).

- **Posteriorly:** The tendo calcaneus and plantaris (Fig. 12-35).
- **Posterolaterally (behind the lateral malleolus):** The peroneus longus and brevis (Fig. 12-35).
- **Posteromedially (behind the medial malleolus):** The tibialis posterior, the flexor digitorum longus, the posterior tibial vessels, the tibial nerve, and the flexor hallucis longus (Fig. 12-35).

## The Surface Anatomy of the Region of the Ankle Joint

The surface anatomy of the region of the ankle joint is shown in Figures 12-36 and 12-37.

## Joints of the Foot and Toes

### Tarsal Joints

#### Subtalar Joint
The subtalar joint is the posterior joint between the talus and the calcaneum.

*Articulation*
Articulation is between the inferior surface of the body of the talus and the facet on the middle of the upper surface of the calcaneum (Fig. 12-38). The articular surfaces are covered with hyaline cartilage.

*Type*
These joints are synovial, of the plane variety.

*Capsule*
The capsule encloses the joint and is attached to the margins of the articular areas of the two bones.

*Ligaments*
**Medial** and **lateral (talocalcaneal) ligaments** strengthen the capsule. The **interosseous (talocalcaneal) ligament** (Fig. 12-34) is strong and is the main bond of union between the two bones. It is attached above to the sulcus tali and below to the sulcus calcanei.

*Synovial Membrane*
The synovial membrane lines the capsule.

*Movements*
Gliding and rotatory movements are possible.

#### Talocalcaneonavicular Joint
The talocalcaneonavicular joint is the anterior joint between the talus and the calcaneum and also involves the navicular bone (Fig. 12-38).

*Articulation*
Articulation is between the rounded head of the talus, the upper surface of the sustentaculum tali, and the posterior

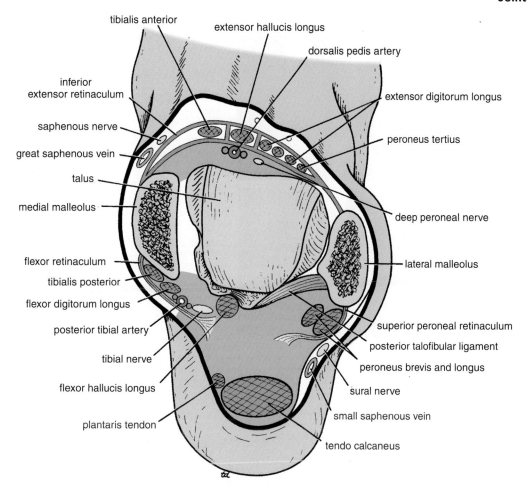

**Figure 12-35**   Relations of the right ankle joint.

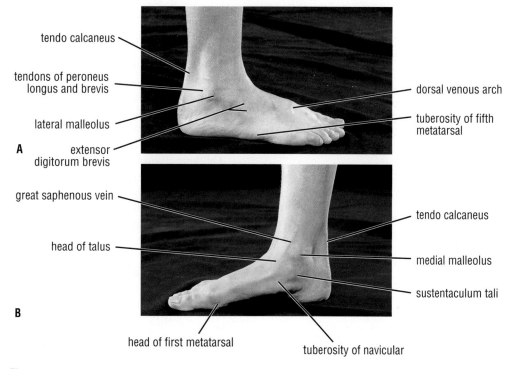

**Figure 12-36**   Lateral aspect (**A**) and medial aspect (**B**) of the right ankle of a 29-year-old woman.

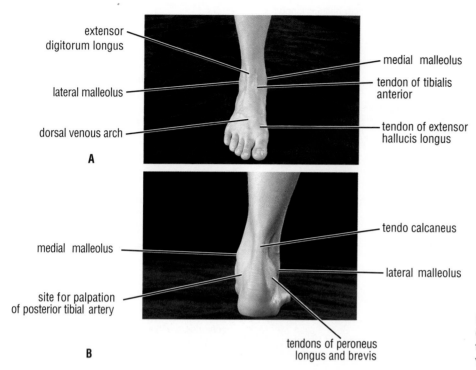

extensor digitorum longus

lateral malleolus

dorsal venous arch

**A**

medial malleolus

tendon of tibialis anterior

tendon of extensor hallucis longus

medial malleolus

site for palpation of posterior tibial artery

**B**

tendo calcaneus

lateral malleolus

tendons of peroneus longus and brevis

**Figure 12-37** Anterior aspect (**A**) and posterior aspect (**B**) of the right foot and ankle of a 29-year-old woman.

concave surface of the navicular bone. The articular surfaces are covered with hyaline cartilage.

### Type
The joint is a synovial joint.

### Capsule
The capsule incompletely encloses the joint.

### Ligaments
The **plantar calcaneonavicular ligament** is strong and runs from the anterior margin of the sustentaculum tali to the inferior surface and tuberosity of the navicular bone. The superior surface of the ligament is covered with fibrocartilage and supports the head of the talus.

### Synovial Membrane
The synovial membrane lines the capsule.

### Movements
Gliding and rotatory movements are possible.

### Calcaneocuboid Joint

### Articulation
Articulation is between the anterior end of the calcaneum and the posterior surface of the cuboid (Fig. 12-38). The articular surfaces are covered with hyaline cartilage.

### Type
The calcaneocuboid joint is synovial, of the plane variety.

### Capsule
The capsule encloses the joint.

### Ligaments
The **bifurcated ligament** is a strong ligament on the upper surface of the joint (Fig. 12-33). It is Y shaped, and the stem is attached to the upper surface of the anterior part of the calcaneum. The lateral limb is attached to the upper surface of the cuboid, and the medial limb is attached to the upper surface of the navicular bone.

The **long plantar ligament** is a strong ligament on the lower surface of the joint (Figs. 12-39 and 12-40). It is attached to the undersurface of the calcaneum behind and to the undersurface of the cuboid and the bases of the third, fourth, and fifth metatarsal bones in front. It bridges over the groove for the peroneus longus tendon, converting it into a tunnel.

The **short plantar ligament** is a wide, strong ligament that is attached to the anterior tubercle on the undersurface of the calcaneum and to the adjoining part of the cuboid bone (Fig. 12-40).

### Synovial Membrane
The synovial membrane lines the capsule.

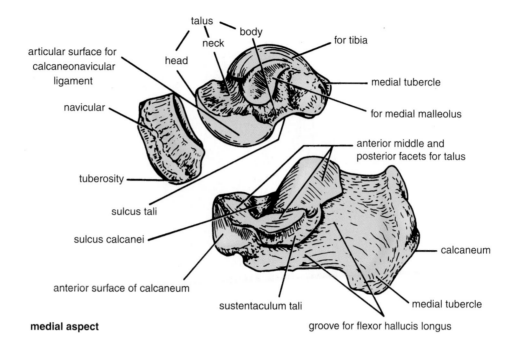

**medial aspect**

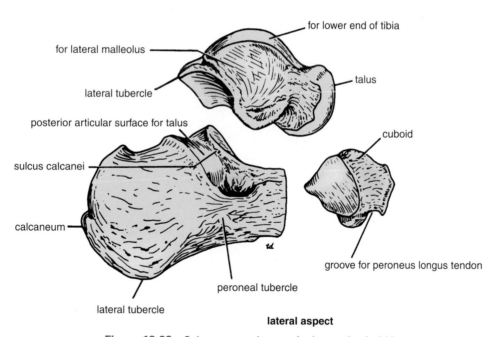

**lateral aspect**

**Figure 12-38**   Calcaneum, talus, navicular, and cuboid bones.

### Movements in the Subtalar, Talocalcaneonavicular, and Calcaneocuboid Joints

The talocalcaneonavicular and the calcaneocuboid joints are together referred to as the **midtarsal** or **transverse tarsal joints.**

The important movements of inversion and eversion of the foot take place at the subtalar and transverse tarsal joints. **Inversion** is the movement of the foot so that the sole faces medially. **Eversion** is the opposite movement of the foot so

that the sole faces in the lateral direction. The movement of inversion is more extensive than eversion.

Inversion is performed by the tibialis anterior, the extensor hallucis longus, and the medial tendons of extensor digitorum longus; the tibialis posterior also assists.

Eversion is performed by the peroneus longus, peroneus brevis, and peroneus tertius; the lateral tendons of the extensor digitorum longus also assist.

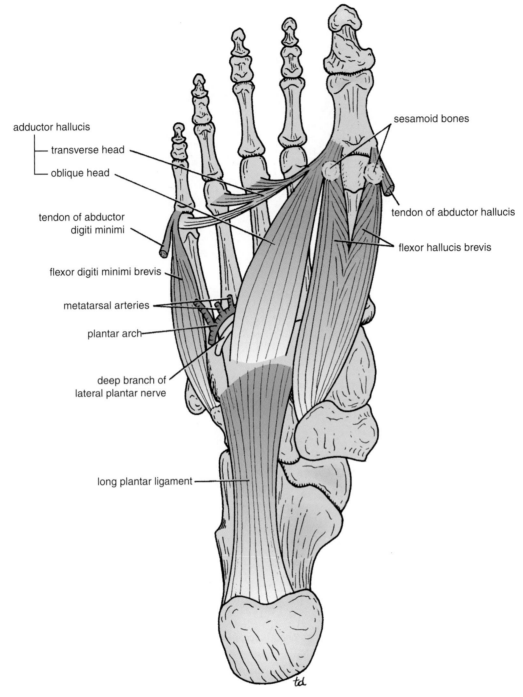

adductor hallucis
transverse head
oblique head

tendon of abductor
digiti minimi

flexor digiti minimi brevis

metatarsal arteries

plantar arch

deep branch of
lateral plantar nerve

long plantar ligament

sesamoid bones

tendon of abductor hallucis

flexor hallucis brevis

**Figure 12-39**  Third layer of the plantar muscles of the right foot. The deep branch of the lateral plantar nerve and the plantar arterial arch are also shown.

### Cuneonavicular Joint

The cuneonavicular joint is the **articulation** between the navicular bone and the three cuneiform bones. It is a synovial joint of the gliding variety. The **capsule** is strengthened by dorsal and plantar ligaments. The **joint cavity** is continuous with those of the intercuneiform and cuneocuboid joints and also with the cuneometatarsal and intermetatarsal joints, between the bases of the second and third and the third and fourth metatarsal bones.

### Cuboideonavicular Joint

The cuboideonavicular joint is usually a fibrous joint, with the two bones connected by dorsal, plantar, and interosseous ligaments.

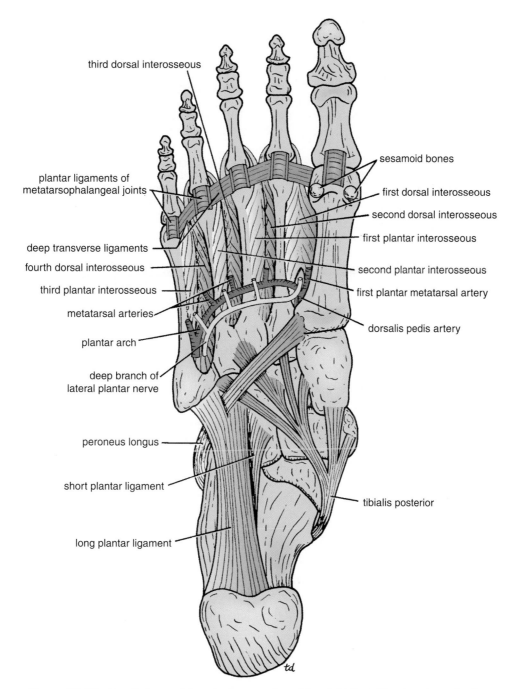

third dorsal interosseous

plantar ligaments of
metatarsophalangeal joints

deep transverse ligaments

fourth dorsal interosseous

third plantar interosseous

metatarsal arteries

plantar arch

deep branch of
lateral plantar nerve

peroneus longus

short plantar ligament

long plantar ligament

sesamoid bones

first dorsal interosseous

second dorsal interosseous

first plantar interosseous

second plantar interosseous

first plantar metatarsal artery

dorsalis pedis artery

tibialis posterior

**Figure 12-40** Fourth layer of the plantar muscles of the right foot. The deep branch of the lateral plantar nerve and the plantar arterial arch are also shown. Note the deep transverse ligaments.

### Intercuneiform and Cuneocuboid Joints

The intercuneiform and cuneocuboid joints are synovial joints of the plane variety. Their joint cavities are continuous with that of the cuneonavicular joint. The bones are connected by dorsal, plantar, and interosseous ligaments.

### Tarsometatarsal and Intermetatarsal Joints

The tarsometatarsal and intermetatarsal joints are synovial joints of the plane variety. The bones are connected by dorsal, plantar, and interosseous ligaments. The tarsometatarsal joint of the big toe has a separate joint cavity.

**Metatarsophalangeal and Interphalangeal Joints**

The metatarsophalangeal and interphalangeal joints closely resemble those of the hand. The deep transverse ligaments connect the joints of the five toes.

The movements of abduction and adduction of the toes, performed by the interossei muscles, are minimal and take place from the midline of the second digit and not the third, as in the hand.

---

### P H Y S I O L O G I C   N O T E

#### The Foot as a Functional Unit

##### The Foot as a Weight-Bearer and a Lever

The foot has two important functions: to support the body weight and to serve as a lever to propel the body forward in walking and running. If the foot possessed a single strong bone, instead of a series of small bones, it could sustain the body weight and serve well as a rigid lever for forward propulsion (Fig. 12-41). However, with such an arrangement, the foot could not adapt itself to uneven surfaces, and the forward propulsive action would depend entirely on the activities of the gastrocnemius and soleus muscles. Because the lever is segmented with multiple joints, the foot is pliable and can adapt itself to uneven surfaces. Moreover, the long flexor muscles and the small muscles of the foot can exert their action on the bones of the forepart of the foot and toes (i.e., the takeoff point of the foot) and greatly assist the forward propulsive action of the gastrocnemius and soleus muscles (Fig. 12-41).

# The Arches of the Foot

A segmented structure can hold up weight only if it is built in the form of an arch. The foot has three such arches, which are present at birth: the **medial longitudinal, lateral longitudinal,** and **transverse arches** (Fig. 12-42). In the young child, the foot appears to be flat because of the presence of a large amount of subcutaneous fat on the sole of the foot.

## The Bones of the Arches

An examination of an articulated foot or a lateral x-ray of the foot shows the bones that form the arches.

- **Medial longitudinal arch:** This consists of the calcaneum, the talus, the navicular bone, the three cuneiform bones, and the first three metatarsal bones (Fig. 12-42).
- **Lateral longitudinal arch:** This consists of the calcaneum, the cuboid, and the fourth and fifth metatarsal bones (Fig. 12-42).

- **Transverse arch:** This consists of the bases of the metatarsal bones and the cuboid and the three cuneiform bones (Fig. 12-42).

## Mechanisms of Arch Support

Apart from the shape of the individual bones, the following muscles and ligaments support the arches.

### Medial Longitudinal Arch

**Muscular support:** Medial part of flexor digitorum brevis, abductor hallucis, flexor hallucis longus, medial part of flexor digitorum longus, flexor hallucis brevis, tibialis anterior, tendinous extensions of the insertion of the tibialis posterior.

**Ligamentous support:** Plantar and dorsal ligaments including the important plantar calcaneonavicular ligament; the medial ligament of the ankle joint, and the plantar aponeurosis.

### Lateral Longitudinal Arch

**Muscular support:** Abductor digiti minimi, lateral part of the flexor digitorum longus and brevis, and peroneus longus and brevis.

**Ligamentous support:** Long and short plantar ligaments and the plantar aponeurosis.

### Transverse Arch

**Muscular support:** Dorsal interossei, transverse head of adductor hallucis, peroneus longus, and peroneus brevis.

**Ligamentous support:** Deep transverse ligaments, very strong plantar ligaments.

The marked wedge shape of the cuneiform bones and the bases of the metatarsal bones play a large part in the support of the transverse arch.

---

### P H Y S I O L O G I C   N O T E

#### The Propulsive Action of the Foot

##### Standing Immobile

The body weight is distributed via the heel behind and the heads of the metatarsal bones in front (including the two sesamoid bones under the head of the first metatarsal).

##### Walking

As the body weight is thrown forward, the weight is born successively on the lateral margin of the foot and

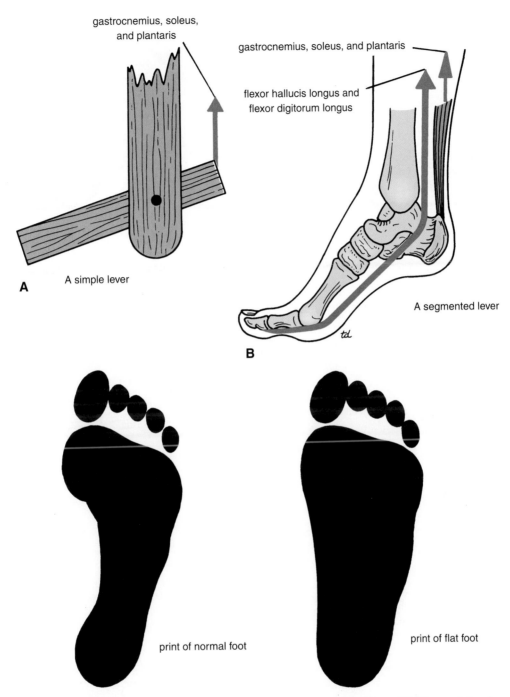

**Figure 12-41** The foot as a simple lever (**A**) and as a segmented lever (**B**). Floor prints of a normal foot and a flat foot are also shown.

the heads of the metatarsal bones. As the heel rises, the toes are extended at the metatarsophalangeal joints, and the plantar aponeurosis is pulled on, thus shortening the tie beams and heightening the longitudinal arches. The "slack" in the long flexor tendons is taken up, thereby increasing their efficiency. The body is then thrown forward by the actions of the gastrocnemius and soleus (and plantaris) on the ankle joint, using the foot as a lever, and by the toes being strongly flexed by the long and short flexors of the

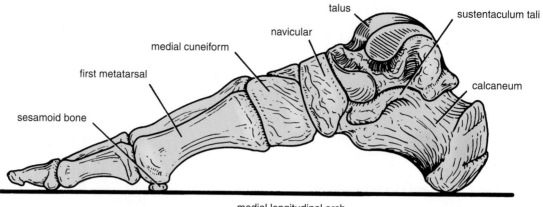

medial longitudinal arch

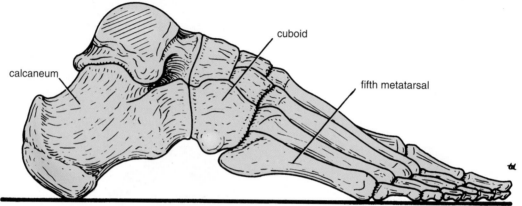

lateral longitudinal arch

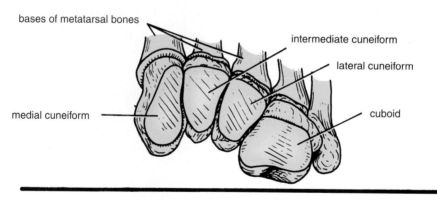

transverse arch

**Figure 12-42** Bones forming the medial longitudinal, lateral longitudinal, and transverse arches of the right foot.

foot, providing the final thrust forward. The lumbricals and interossei contract and keep the toes extended so that they do not fold under because of the strong action of the flexor digitorum longus. In this action, the long flexor tendons also assist in plantar flexing the ankle joint.

### Running

When a person runs, the weight is borne on the forepart of the foot, and the heel does not touch the ground. The forward thrust to the body is provided by the mechanisms described for walking.

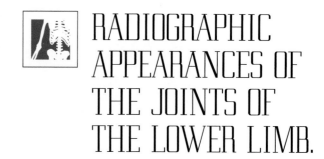

# RADIOGRAPHIC APPEARANCES OF THE JOINTS OF THE LOWER LIMB.

The radiographic appearances of the joints of the lower limb are shown in Figures 11-63 to 11-74 and 12-43 and 12-44.

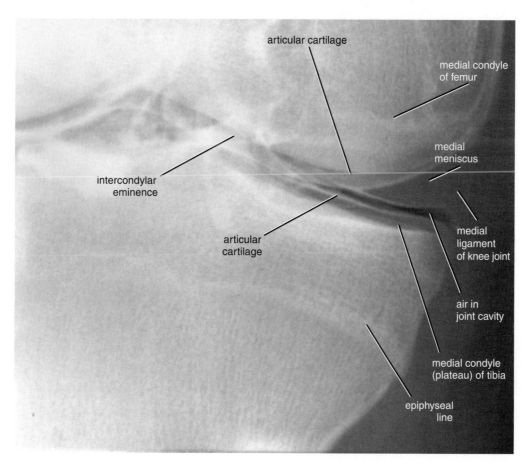

**Figure 12-43**    Pneumoarthrography of the knee.

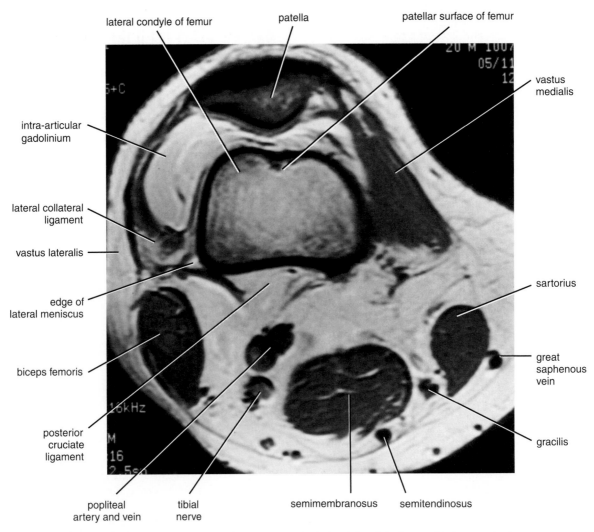

**Figure 12-44** Transverse (axial) proton density MRI of the right knee with intra-articular gadolinium–saline solution (as seen from below).

# Review Questions

## General Joint Questions

**Select the BEST answer for each question.**

## Completion Questions

1. A patient is performing the movement of flexion of the hip joint when she
   A. moves the lower limb away from the midline in the coronal plane.
   B. moves the lower limb posteriorly in the paramedian plane.
   C. moves the lower limb anteriorly in the paramedian plane.
   D. rotates the lower limb so that the anterior surface faces medially.
   E. moves the lower limb toward the median sagittal plane.

2. Inversion of the foot is the movement so that the sole faces
   A. downward and posteriorly.
   B. medially.
   C. laterally.
   D. downward.
   E. downward and laterally.

## Matching Questions

**For each joint listed below, indicate with which type of movement it is associated.**

3. Sternoclavicular joint

4. Superior radioulnar joint

5. Ankle joint
   A. Flexion
   B. Gliding
   C. Both A and B
   D. Neither A nor B

**For each joint listed below, give the most appropriate classification.**

6. Joints between vertebral bodies

7. Inferior tibiofibular joint

8. Sutures between bones of vault of skull

9. Wrist joint
   A. Synovial joint
   B. Cartilaginous
   C. Fibrous
   D. None of the choices

**For each type of synovial joint listed below, give an appropriate example from the list of joints.**

10. Hinge joint

11. Condyloid joint

12. Ball-and-socket joint

13. Saddle joint
    A. Metacarpophalangeal joint of index finger
    B. Shoulder joint
    C. Wrist joint
    D. Carpometacarpal joint of the thumb
    E. None of the choices

## Joints of the Head and Neck

## Multiple Choice Questions

**Select the BEST answer for each question.**

14. The following facts concerning the temporomandibular joint are correct **except** which?
    A. The superior articular surfaces are formed by the articular tubercle and the anterior portion of the mandibular fossa of the temporal bone.
    B. The lateral temporomandibular ligament protects the external auditor meatus from excessive backward displacement of the head of the mandible.
    C. The articular disc is composed of fibrocartilage.
    D. The upper and lower cavities of the joint normally communicate with one another.

E. The nerve supply of the joint is from the auriculotemporal nerve and the masseteric branches of the mandibular division of the trigeminal nerve.

15. The following facts concerning the movements of the temporomandibular joint are correct **except** which?
    A. Depression of the mandible is brought about by the contraction of the genioglossus muscle.
    B. Elevation of the mandible is brought about by the contraction of the temporalis, masseter, and medial pterygoid muscles.
    C. As the mouth begins to open, the head of the mandible rotates on the undersurface of the articular disc.
    D. The movement of protrusion of the mandible takes place in the upper cavity of the joint.
    E. Protrusion of the mandible is brought about by the contraction of the lateral pterygoid muscles of both sides, assisted by both medial pterygoids.

16. Which of the following muscles partially inserts on the articular disc of the temporomandibular joint?
    A. Medial pterygoid
    B. Anterior fibers of the temporalis
    C. Masseter
    D. Posterior fibers of the temporalis
    E. Lateral pterygoid

## Joints of the Vertebral Column

## Multiple Choice Questions

**Select the BEST answer for each question.**

17. The following statements concerning an intervertebral disc are correct **except** which?
    A. The nucleus pulposus is most likely to herniate in an anterolateral direction.
    B. The discs are the thickest in the lumbar region.
    C. The atlanto-axial joint possesses no disc.
    D. The discs play a major role in the development of the curvatures of the vertebral column.
    E. During aging, the fluid within the nucleus pulposus is replaced by fibrocartilage.

18. The following facts concerning the atlanto-axial joints are correct **except** which?
    A. The apical ligament connects the apex of the odontoid process to the anterior arch of the atlas.
    B. The transverse part of the cruciate ligament is attached on each side to the inner aspect of the lateral mass of the atlas and binds the odontoid process to the anterior arch of the atlas.
    C. The alar ligaments connect the odontoid process to the medial side of the occipital condyles of the occipital bone of the skull.
    D. The atlanto-axial joints are synovial joints.
    E. Extensive rotation of the atlas and thus the head can take place at this joint.

19. The following facts concerning the joints between two vertebral arches are correct **except** which?
    A. The joints are synovial joints.
    B. The joints are between the superior and inferior articular processes of adjacent vertebrae.
    C. The articular surfaces are devoid of hyaline cartilage.
    D. The joints are surrounded by a capsular ligament.
    E. In the cervical region, the supraspinous and interspinous ligaments are greatly thickened to form the ligamentum nuchae.

## Joints of the Thorax

## Multiple Choice Questions

**Select the BEST answer for each question.**

20. The sixth thoracic vertebra articulates by means of synovial joints with all the following structures **except** which?
    A. The head of the sixth rib
    B. The body of the fifth thoracic vertebra
    C. The tubercle of the sixth rib
    D. The inferior articular process of the fifth thoracic vertebra
    E. The superior articular process of the seventh thoracic vertebra

21. Which of the following costal cartilages do **not** directly articulate with the body of the sternum?
    A. Second
    B. Fourth
    C. Fifth
    D. Eighth
    E. Third

## Joints of the Upper Extremity

## Multiple Choice Questions

**Select the BEST answer for each question.**

22. Which of the following nerves is related to the inferior aspect of the shoulder joint and may be injured in dislocations of the shoulder joint?
    A. Radial
    B. Ulnar
    C. Axillary
    D. Median
    E. Musculocutaneous

23. The following facts concern the abduction of the arm at the shoulder joint **except** which?
    A. Abduction of the arm involves rotation of the scapula as well as movement at the shoulder joint.
    B. For every 3° of abduction of the arm, a 2° abduction occurs in the shoulder joint and a 1° abduction occurs by rotation of the scapula.

C. At about 120° of abduction of the arm, the greater tuberosity of the humerus contacts the lateral edge of the acromium.
D. After 120° of abduction of the arm, further abduction is accomplished by rotation of the scapula.
E. The trapezius and the serratus anterior muscles are responsible for abduction of the shoulder joint.

24. The following muscles are responsible for flexion of the elbow joint **except** which?
    A. The biceps brachii
    B. The brachioradialis
    C. The pronator teres
    D. The anconeus
    E. The brachialis

25. The medial ligament of the elbow joint is closely related to the following structure:
    A. Brachial artery
    B. Radial nerve
    C. Ulnar artery
    D. Basilic vein
    E. Ulnar nerve

26. Select the structure that is most important in strengthening the wrist joint.
    A. The capsule
    B. The tone of the flexor and extensor muscles of the wrist joint
    C. The anterior and posterior ligaments
    D. The synovial membrane
    E. The medial and lateral ligaments

27. The following facts concerning the movements of the metacarpophalangeal joints are correct **except** which?
    A. The lumbricals and the interossei, assisted by the flexor digitorum superficialis and profundus muscles, produce flexion.
    B. The movement away from the midline of the third finger is performed by the palmar interossei muscles.
    C. Extension is performed by the extensor digitorum, extensor indicis, and extensor digiti minimi.
    D. Adduction movement toward the midline of the third finger is performed by the palmar interossei muscles.
    E. No rotation is possible at these joints.

## Joints of the Pelvis

## Multiple Choice Questions

**Select the BEST answer for each question.**

28. The following statements concerning the joints of the bony pelvis are correct **except** which?
    A. Very little movement is possible at the sacroiliac joint.
    B. The female sex hormones cause a relaxation of the ligaments of the pelvis during pregnancy.

C. Obliteration of the cavity of the sacroiliac joint often occurs in both sexes after middle age.

D. The sacroiliac joint is supplied by the lumbar plexus of nerves.

E. The symphysis pubis has a fibrocartilaginous disc.

## Joints of the Lower Extremity

## Multiple Choice Questions

**Select the BEST answer for each question.**

29. Which of the following nerves innervates at least one muscle that acts on both the hip and knee joints?
    A. Ilioinguinal nerve
    B. Saphenous nerve
    C. Femoral nerve
    D. Common peroneal nerve
    E. Superficial peroneal nerve

30. Which of the following muscles dorsiflexes the foot at the ankle joint?
    A. Peroneus longus
    B. Extensor digitorum brevis
    C. Tibialis posterior
    D. Extensor hallucis brevis
    E. Tibialis anterior

## Completion Questions

31. Unlocking of the knee joint to permit flexion is caused by the action of the
    A. vastus medialis muscle.
    B. articularis genu muscle.
    C. gastrocnemius muscle.
    D. biceps femoris muscle.
    E. popliteus muscle.

32. Hyperextension of the hip joint is prevented by
    A. obturator internus tendon.
    B. ischiofemoral ligament.
    C. tensor fascia latae muscle.
    D. iliotibial tract.
    E. ligamentum teres.

## Fill in the Blank Questions

**Fill in the blank with the BEST answer.**

33. The _____ prevents dislocation of the femur backward at the knee joint.
    A. anterior cruciate ligament
    B. posterior cruciate ligament
    C. medial collateral ligament
    D. lateral collateral ligament
    E. tendon of the popliteus muscle

34. The _____ prevents abduction of the tibia at the knee joint.
    A. posterior cruciate ligament
    B. anterior cruciate ligament
    C. lateral collateral ligament
    D. lateral meniscus
    E. medial collateral ligament

35. The _____ is attached to the head of the fibula.
    A. lateral meniscus
    B. lateral collateral ligament
    C. anterior cruciate ligament
    D. posterior cruciate ligament
    E. medial meniscus

## Multiple Choice Questions

**Select the BEST answer for each question.**

36. The calcaneum participates in the formation of which arch(es) of the foot?
    A. Medial longitudinal arch only
    B. Transverse arch only
    C. Medial and lateral longitudinal arches
    D. Medial longitudinal and transverse arches
    E. Lateral longitudinal and transverse arches

37. The talus participates in the formation of which arch(es) of the foot?
    A. Transverse arch only
    B. Lateral longitudinal arch only
    C. Medial longitudinal arch only
    D. Medial and lateral longitudinal arches
    E. Transverse and medial longitudinal arches

38. The cuboid participates in the formation of which arch(es) of the foot?
    A. Medial longitudinal arch only
    B. Lateral longitudinal arch only
    C. Transverse arch only
    D. Medial longitudinal and transverse arches
    E. Lateral longitudinal and transverse arches

39. The following statements concerning the ankle joint are correct **except** which?
    A. It is strengthened by the deltoid (medial collateral) ligament.
    B. It is a hinge joint.
    C. It is formed by the articulation of the talus and the distal ends of the tibia and the fibula.
    D. It is most stable in the fully plantar-flexed position.
    E. It is a synovial joint.

40. The foot is inverted by the following muscles **except** which?
    A. The tibialis anterior
    B. The extensor hallucis longus
    C. The extensor digitorum longus
    D. The peroneus tertius
    E. The tibialis posterior

# Answers and Explanations

1. **C** is correct. Flexion of the hip joint moves the lower limb anteriorly in the paramedian plane.

2. **B** is correct. Inversion of the foot is the movement so that the sole faces medially (see Fig. 1-3).

3. **B** is correct. The movement associated with the sternoclavicular joint is gliding.

4. **D** is correct. The superior radioulnar joint is associated with rotary movement.

5. **A** is correct. Dorsiflexion (toes pointing upward) and plantar flexion (toes pointing downward) take place at the ankle joint.

6. **B** is correct. The joints between vertebral bodies are cartilaginous.

7. **C** is correct. The inferior tibiofibular joint is fibrous.

8. **C** is correct. Sutures between bones of the vault of the skull are fibrous (Fig.12-3).

9. **A** is correct. The wrist joint is a synovial joint.

10. **E** is correct. The elbow joint is a good example of a hinge joint.

11. **A** is correct. The metacarpophalangeal joints of the fingers are examples of condyloid joints (Fig. 12-1).

12. **B** is correct. The shoulder joint is a ball-and-socket joint.

13. **D** is correct. The carpometacarpal joint of the thumb is a saddle joint (Fig. 12-1).

14. **D** is incorrect. The upper and lower cavities of the temporomandibular joint normally do not communicate with one another. However, it is not uncommon in later life, as a result of wear and tear, to find such a communication.

15. **A** is incorrect. The depression of the mandible is brought about by the contraction of the digastrics, the geniohyoids, and the mylohyoids; the lateral pterygoid muscles play an important role by pulling the mandible forward (Fig.12-6).

16. **E** is correct. The lateral pterygoid muscle partially inserts on the articular disc of the temporomandibular joint and is responsible for pulling the disc and the mandible forward during the depression of the mandible.

17. **A** is incorrect. The nucleus pulposus is most likely to herniate in a posterolateral direction.

18. **A** is incorrect. In the atlanto-axial joints, the apical ligament connects the apex of the odontoid process to the anterior margin of the foramen magnum of the skull (Fig. 12-9).

19. **C** is incorrect. In the synovial joints between two vertebral arches, the articular surfaces of the superior and inferior articular processes are covered with hyaline cartilage.

20. **B** is incorrect. The sixth thoracic vertebra articulates by means of a cartilaginous joint with the body of the fifth thoracic vertebra.

21. **D** is correct. The eighth costal cartilage articulates with the seventh costal cartilage.

22. **C** is correct. The axillary nerve passes through the quadrangular space just below the shoulder joint to supply the deltoid muscle and the teres minor muscle; it also supplies the skin over the lower half of the deltoid muscle (Fig. 12-17).

23. **E** is incorrect. The abduction of the arm at the **shoulder joint** involves the contraction of the supraspinatus and deltoid muscles. The trapezius and serratus anterior muscles are involved in the rotation of the scapula, which also takes place during the abduction of the arm (Fig. 12-19).

24. **D** is incorrect. The anconeus muscle is an extensor of the elbow joint.

25. **E** is correct. The ulnar nerve is closely related to the medial ligament of the elbow joint (Fig. 12-20).

26. **E** is correct. The medial and lateral ligaments of the wrist joint are strong.

27. **B** is incorrect. The movement of the metacarpophalangeal joint of the index finger away from the midline of the middle finger is performed by the dorsal interossei muscles.

28. **D** is incorrect. The sacroiliac joint receives its nerve supply from the sacral plexus.

29. **C** is correct. The femoral nerve supplies the rectus femoris muscle, which flexes the thigh at the hip joint and extends the leg at the knee joint.

30. **E** is correct. The tibialis anterior dorsiflexes the foot at the ankle joint.

31. **A** is correct. The popliteus muscle unlocks the knee joint by rotating the tibia on the femur (or the femur on the tibia), thus untwisting and slackening the ligaments of the knee.

32. **B** is correct. The ischiofemoral ligament together with the iliofemoral ligament and the pubofemoral ligament limit extension of the hip joint.

33. **A** is correct. The anterior cruciate ligament prevents dislocation of the femur backward at the knee joint.

34. **E** is correct. The medial collateral ligament prevents abduction of the tibia at the knee joint.

35. **B** is correct. The lateral collateral ligament is attached to the head of the fibula (Fig. 12-30).

36. **C** is correct. The calcaneum participates in the formation of the medial and lateral longitudinal arches of the foot (Fig. 12-42).

37. **C** is correct. The talus participates in the formation of the medial longitudinal arch of the foot only (Fig. 12-42).

38. **E** is correct. The cuboid participates in the formation of the lateral and transverse arches of the foot (Fig. 12-42).

39. **D** is incorrect. The ankle joint is most stable in the fully dorsiflexed position.

40. **D** is incorrect. The peroneus tertius muscle is an evertor of the foot at the subtalar and transverse tarsal joints.

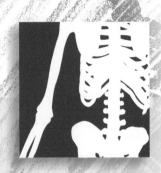

# 13 Skeletal Muscles

All clinical material relevant to this chapter can be found on the CD-ROM.

# Chapter Outline

In this chapter, the most important muscles of the body and their origin, insertion, nerve supply, and action are described briefly in Tables. If you have access to an articulated skeleton, it is helpful to imagine that you are contracting a particular muscle, and you should then move the appropriate bone or bones on the skeleton. Remember that you yourself have these muscles and skeleton and joints, so do not hesitate to reinforce your learning by performing the movement on yourself. This is particularly important in the case of the limbs. You may find it helpful to perform the muscle action in front of a mirror.

# BASIC ANATOMY

## Internal Structure of Skeletal Muscle

The muscle fibers are bound together with delicate areolar tissue, which is condensed on the surface to form a fibrous envelope, the **epimysium.** The individual fibers of a muscle are arranged either parallel or oblique to the long axis of the muscle (Fig. 13-1). Because a muscle shortens by one third to one half its resting length when it contracts, it follows that muscles whose fibers run parallel to the line of pull will bring about a greater degree of movement compared with those whose fibers run obliquely. Examples of muscles with parallel fiber arrangements (Fig. 13-1) are the sternocleidomastoid, the rectus abdominis, and the sartorius.

Muscles whose fibers run obliquely to the line of pull are referred to as **pennate muscles** (they resemble a feather) (Fig. 13-1). A **unipennate muscle** is one in which the tendon lies along one side of the muscle, and the muscle fibers pass obliquely to it (e.g., extensor digitorum longus). A **bipennate muscle** is one in which the tendon lies in the center of the muscle and the muscle fibers pass to it from two sides (e.g., rectus femoris). A **multipennate muscle** may be arranged as a series of bipennate muscles lying alongside one another (e.g., acromial fibers of the deltoid) or may have the tendon lying within its center and the muscle fibers passing to it from all sides, converging as they go (e.g., tibialis anterior).

For a given volume of muscle substance, pennate muscles have many more fibers compared to muscles with parallel fiber arrangements and are, therefore, more powerful;

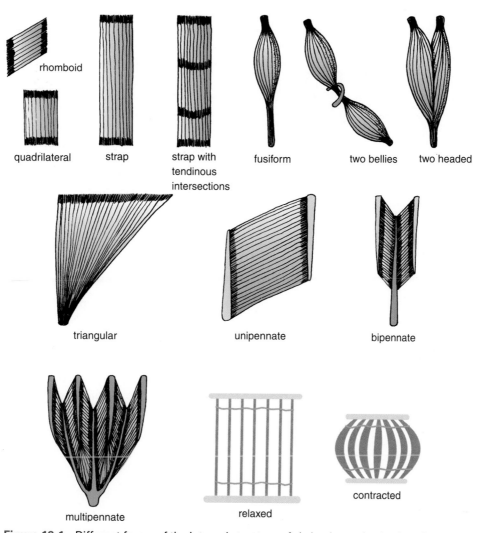

**Figure 13-1**    Different forms of the internal structure of skeletal muscle. A relaxed and a contracted muscle are also shown; note how the muscle fibers, on contraction, shorten by one third to one half of their resting length. Note also how the muscle swells.

in other words, range of movement has been sacrificed for strength.

**PHYSIOLOGIC NOTE**

### Skeletal Muscle Tone and Action

A **motor unit** consists of a motor neuron in the anterior gray horn or column of the spinal cord and all the muscle fibers it supplies (Fig. 13-2). In a large buttock muscle, such as the gluteus maximus, where fine control is unnecessary, a given motor neuron may supply as many as 200 muscle fibers. In contrast, in the small muscles of the hand or the extrinsic muscles of the eyeball, where fine control is required, one nerve fiber supplies only a few muscle fibers.

While resting, every skeletal muscle is in a partial state of contraction. This condition is referred to as **muscle tone.** Because muscle fibers are either fully contracted or fully relaxed, with no intermediate stage, it follows that a few muscle fibers within a muscle are fully contracted all the time. To bring about this state and to avoid fatigue, different groups of motor units and, thus, different groups of muscle fibers are brought into action at different times. This is accomplished by the asynchronous discharge of nervous impulses in the motor neurons in the anterior gray horn of the spinal cord.

Basically, muscle tone depends on the integrity of a simple monosynaptic reflex arc composed of two neurons in the nervous system (Fig. 13-3). The degree of tension in a muscle is detected by sensitive sensory endings called **muscle spindles** and **tendon spindles** (Fig. 13-3). The nervous impulses travel in the afferent neurons that enter the spinal cord. There, they synapse with the motor neurons in the anterior gray horn, which, in turn, send impulses down their axons to the muscle fibers (Fig. 13-3).

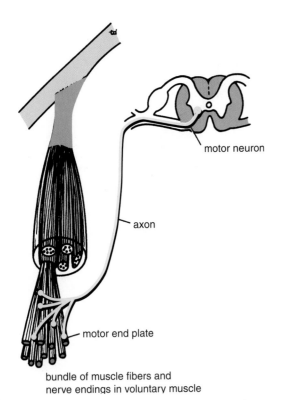

motor neuron

axon

motor end plate

bundle of muscle fibers and
nerve endings in voluntary muscle

**Figure 13-2** Components of a motor unit.

If the afferent or efferent pathways of this simple reflex arc were cut, the muscle would immediately lose its tone and become flaccid. A flaccid muscle on palpation feels like a mass of dough and has completely lost its resilience. It quickly atrophies, becoming reduced in volume. The degree of activity of the motor anterior horn cells and, therefore, the degree of muscle tone depends on the summation of the nerve impulses received by these cells from other neurons of the nervous system.

Muscle movement is accomplished by bringing into action increasing numbers of motor units and at the same time reducing the activity of the motor units of muscles that will oppose or antagonize the movement. When the maximum effort is required, all the motor units of a muscle are thrown into action.

All movements are the result of the coordinated action of many muscles. However, to understand a muscle's action it is necessary to study it individually.

A muscle may work in the following four ways:

- **Prime mover:** A muscle is a prime mover when it is the chief muscle or member of a chief group of muscles responsible for a particular movement. For example, the quadriceps femoris is a prime mover in the movement of extending the knee joint (Fig. 13-4).

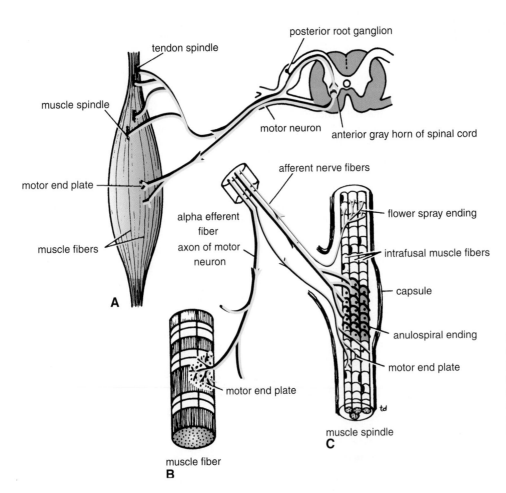

posterior root ganglion

tendon spindle

muscle spindle

motor neuron

anterior gray horn of spinal cord

motor end plate

afferent nerve fibers

flower spray ending

muscle fibers

alpha efferent fiber

axon of motor neuron

intrafusal muscle fibers

capsule

anulospiral ending

motor end plate

A

motor end plate

muscle fiber
B

muscle spindle
C

**Figure 13-3** **A.** Simple reflex arc consisting of an afferent neuron arising from muscle spindles and tendon spindles and an efferent neuron whose cell body lies in the anterior gray horn of the spinal cord. **B.** Axon from a motor neuron ending on muscle fiber at the motor end plate. **C.** Structure of a muscle spindle.

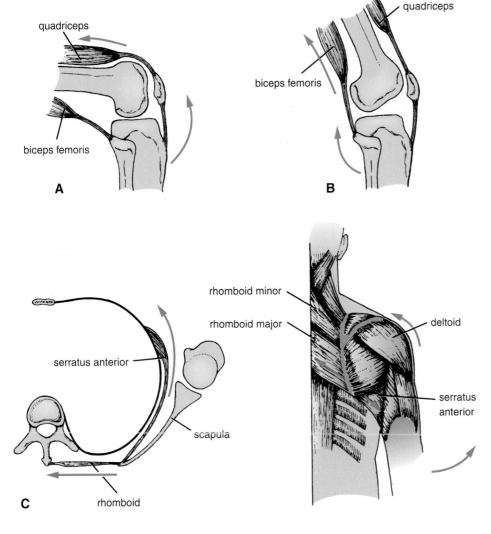

**Figure 13-4** Different types of muscle action. **A.** Quadriceps femoris extending the knee as a prime mover, and biceps femoris acting as an antagonist. **B.** Biceps femoris flexing the knee as a prime mover, and quadriceps acting as an antagonist. **C.** Muscles around shoulder girdle fixing the scapula so that movement of abduction can take place at the shoulder joint. **D.** Flexor and extensor muscles of the carpus acting as synergists and stabilizing the carpus so that long flexor and extensor tendons can flex and extend the fingers.

- **Antagonist:** Any muscle that opposes the action of the prime mover is an antagonist. For example, the biceps femoris opposes the action of the quadriceps femoris when the knee joint is extended (Fig. 13-4). Before a prime mover can contract, the antagonist muscle must be equally relaxed; this is brought about by nervous reflex inhibition.
- **Fixator:** A fixator contracts isometrically (i.e., contraction increases the tone but does not in itself produce movement) to stabilize the origin of the prime mover so that it can act efficiently. For example, the muscles attaching the shoulder girdle to the trunk con-

tract as fixators to allow the deltoid to act on the shoulder joint (Fig. 13-4).
- **Synergist:** In many locations in the body, the prime mover muscle crosses several joints before it reaches the joint at which its main action takes place. To prevent unwanted movements in an intermediate joint, groups of muscles called synergists contract and stabilize the intermediate joints. For example, the flexor and extensor muscles of the hand contract to fix the wrist joint, and this allows the long flexor and extensor muscles of the fingers to work efficiently (Fig. 13-4).

| Table 13-1 | Naming of Skeletal Muscles[a] | | | | | | |
|---|---|---|---|---|---|---|---|
| **Name** | **Shape** | **Size** | **Number of Heads or Bellies** | **Position** | **Depth** | **Attachments** | **Actions** |
| Deltoid<br>Teres<br>Rectus<br>Major<br>Latissimus<br>Longissimus<br>Biceps<br>Quadriceps<br>Digastric<br>Pectoralis<br>Supraspinatus<br><br>Brachii<br>Profundus<br>Superficialis<br>Externus<br>Sternocleido-<br>mastoid<br><br><br><br>Coraco-<br>brachialis<br><br><br>Extensor<br>Flexor<br>Constrictor | Triangular<br>Round<br>Straight | Large<br>Broadest<br>Longest | Two heads<br>Four heads<br>Two bellies | Of the chest<br>Above spine<br>of scapula<br>Of the arm | Deep<br>Superficial<br>External | From<br>sternum<br>and clavicle<br>to mastoid<br>process<br>From<br>coracoid<br>process to<br>arm | Extend<br>Flex<br>Constrict |

[a]These names are commonly used in combination; for example, flexor pollicis longus (long flexor of the thumb).
From Snell RS: Clinical Anatomy. 7th Ed. Philadelphia: Lippincott Williams & Wilkins, 2004, p. 14–15.

These terms are applied to the action of a particular muscle during a particular movement; many muscles can act as a prime mover, an antagonist, a fixator, or a synergist, depending on the movement to be accomplished.

Muscles can even contract paradoxically; for example, when the biceps brachii, a flexor of the elbow joint, contracts and controls the rate of extension of the elbow when the triceps brachii contracts.

# Nerve Supply of Skeletal Muscle

The nerve trunk to a muscle is a mixed nerve; about 60% is motor and 40% is sensory, and it also contains some sympathetic autonomic fibers. The nerve enters the muscle at about the midpoint on its deep surface, often near the margin; the place of entrance is known as the **motor point.** This arrangement allows the muscle to move with minimum interference with the nerve trunk.

The **motor fibers** are of two types: the larger **alpha fibers** derived from large cells in the anterior gray horn and the smaller **gamma fibers** derived from smaller cells in the spinal cord. Each fiber is myelinated and ends by dividing into many branches, each of which ends on a muscle fiber at the **motor end plate** (Fig. 13-3). Each muscle fiber has at least one motor end plate; longer fibers possess more.

The **sensory fibers** are myelinated and arise from specialized sensory endings lying within the muscle or tendons called **muscle spindles** and **tendon spindles,** respectively. These endings are stimulated by tension in the muscle, which may occur during active contraction or by passive stretching. The function of these sensory fibers is to convey to the central nervous system information regarding the degree of tension of the muscles. This is essential for the maintenance of muscle tone and body posture and for carrying out coordinated voluntary movements.

The **sympathetic fibers** are nonmyelinated and pass to the smooth muscle in the walls of the blood vessels supplying the muscle. Their function is to regulate blood flow to the muscles.

# Naming of Skeletal Muscles

Individual muscles are named according to their shape, size, number of heads or bellies, position, depth, attachments, or actions. Some examples of muscle names are shown in Table 13-1.

# Muscles of the Head

The muscles of the scalp, external ear, and face are derived from the second pharyngeal arch. Therefore, they are all supplied by the facial nerve (seventh cranial nerve).

## Muscles of the Scalp

The scalp consists of five layers (Fig. 13-5), the first three of which are intimately bound together and move as a whole on the skull. These layers are:

- Skin
- Connective tissue of the superficial fascia
- Aponeurosis of the occipitofrontalis muscle
- Loose connective tissue, which permits the first three layers to move on the fifth layer
- Periosteum of the skull bones

Note that the first letter of each layer combines to spell the acronym **SCALP.** The origin, insertion, nerve supply, and action of the scalp muscles are summarized in Table 13-2.

## Muscles of the External Ear

These are anterior, superior, and posterior and are vestigial in the human subject (Fig. 13-6). Only a small amount of movement of the auricle occurs in some individuals.

## Muscles of Facial Expression

Situated in the superficial fascia, the muscles of facial expression arise from the skull and are inserted into the skin (Fig. 13-6). These muscles serve as sphincters and dilators to the orbit, nose, and mouth. They also modify the expression of the face.

The origin, insertion, nerve supply, and action of the muscles of facial expression are summarized in Table 13-2. (Try using the various muscles while looking in the mirror.) It is unnecessary to know the precise attachments of these muscles.

## Muscles of Mastication

The muscles of mastication are four powerful muscles acting on the mandible (Fig. 13-7). They are developed from the first pharyngeal arch and are, therefore, all supplied by

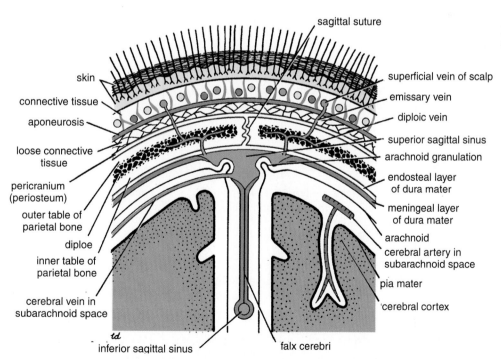

**Figure 13-5**  Coronal section of the upper part of the head showing the layers of the scalp, the sagittal suture of the skull, the falx cerebri, the superior and inferior sagittal venous sinuses, the arachnoid granulations, the emissary veins, and the relation of cerebral blood vessels to the subarachnoid space.

| Table 13-2 | Muscles of the Head | | | |
|---|---|---|---|---|
| **Muscle** | **Origin** | **Insertion** | **Nerve Supply** | **Action** |
| **Muscle of Scalp** Occipitofrontalis | | | | |
|   Occipital belly | Highest nuchal line of occipital bone | Epicranial aponeurosis | Facial nerve | Moves scalp on skull and raises eyebrows |
|   Frontal belly | Skin and superficial fascia of eyebrows | | | |
| **Muscles of Facial Expression** Orbicularis oculi | | | | |
|   Palpebral part | Medial palpebral ligament | Lateral palpebral raphe | Facial nerve | Closes eyelids and dilates lacrimal sac |
|   Orbital part | Medial palpebral ligament and adjoining bone | Loops return to origin | Facial nerve | Throws skin around orbit into folds to protect eyeball |
| Corrugator supercilii | Superciliary arch | Skin of eyebrow | Facial nerve | Vertical wrinkles of forehead, as in frowning |
| Compressor nasi | Frontal process of maxilla | Aponeurosis of bridge of nose | Facial nerve | Compresses mobile nasal cartilages |
| Dilator naris | Maxilla | Ala of nose | Facial nerve | Widens nasal aperture |
| Procerus | Nasal bone | Skin between eyebrows | Facial nerve | Wrinkles skin of nose |
| Orbicularis oris | Maxilla, mandible, and skin | Encircles oral orifice | Facial nerve | Compresses lips together |
| **Dilator Muscles of Lips** Levator labii superioris alaeque nasi Levator labii superioris Zygomaticus minor Zygomaticus major Levator anguli oris Risorius Depressor labii inferioris Depressor anguli oris Mentalis | Arise from bones and fascia around oral aperture and insert into substance of lips | | Facial nerve | Separate lips |
| Buccinator | Outer surface of alveolar margins of maxilla and mandible and pterygo-mandibular ligament | | Facial nerve | Compresses cheeks and lips against teeth |
| Platysma | See Table 13-3 | | | |
| **Muscles of Mastication** Masseter | Zygomatic arch | Lateral surface ramus of mandible | Mandibular division of trigeminal nerve | Elevates mandible to occlude teeth |
| Temporalis | Floor of temporal fossa | Coronoid process of mandible | Mandibular division of trigeminal nerve | Anterior and superior fibers elevate mandible; posterior fibers retract mandible |

**Table 13-2**     *(Continued)*

| Muscle | Origin | Insertion | Nerve Supply | Action |
|---|---|---|---|---|
| Lateral pterygoid (two heads) | Greater wing of sphenoid and lateral pterygoid plate | Neck of mandible and articular disc | Mandibular division of trigeminal nerve | Pulls neck of mandible forward |
| Medial pterygoid (two heads) | Tuberosity of maxilla and lateral pterygoid plate | Medial surface of angle of mandible | Mandibular division of trigeminal nerve | Elevates mandible |

[a]From Snell RS: Clinical Anatomy. 7th Ed. Philadelphia: Lippincott Williams & Wilkins, 2004, p. 772.

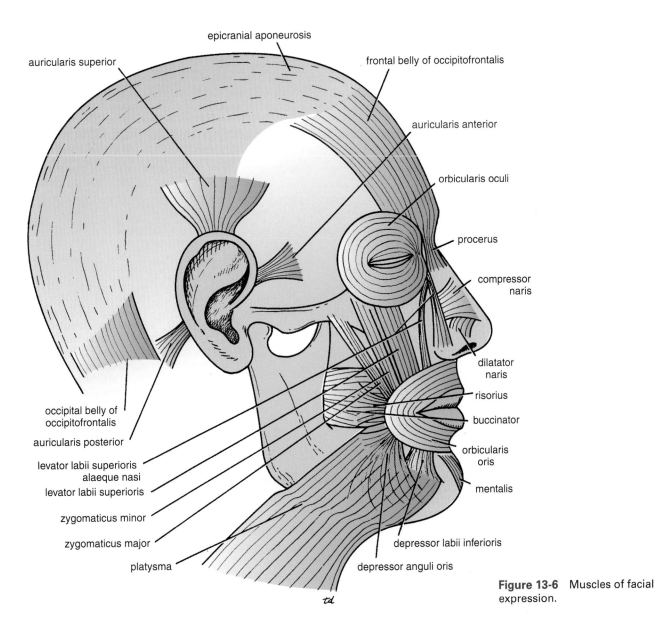

**Figure 13-6**   Muscles of facial expression.

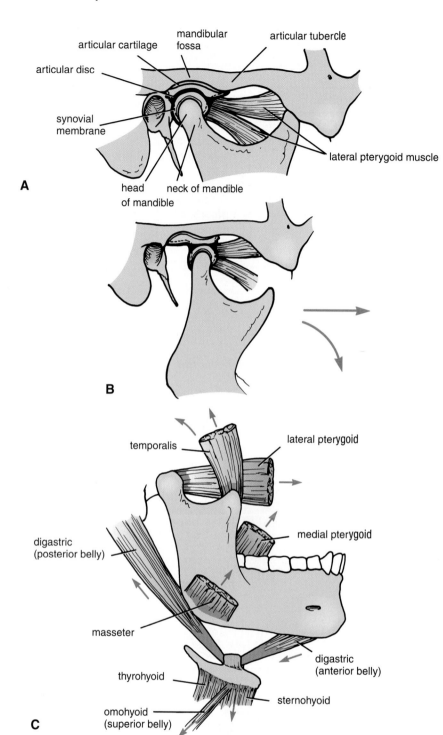

**Figure 13-7** Temporomandibular joint with mouth closed (**A**) and with the mouth open (**B**). Note the position of the head of the mandible and articular disc in relation to the articular tubercle in each case. **C.** The attachment of the muscles of mastication to the mandible. The *arrows* indicate the direction of their actions.

the mandibular division of the trigeminal nerve (fifth cranial nerve). See Table 13-2.

# Muscles of the Neck

Table 13-3 describes the superficial muscles of the side of the neck (Figs. 13-6 and 13-8), the suprahyoid and infrahyoid muscles, and the anterior and lateral vertebral muscles.

## Key Neck Muscles

### Sternocleidomastoid Muscle

When the sternocleidomastoid muscle (Figs. 13-8 and 13-9) contracts, it appears as an oblique band crossing the side of the neck from the sternoclavicular joint to the mastoid process of the skull. It divides the neck into anterior and posterior triangles. The anterior border covers the carotid arter-

| Table 13-3 | **Muscles of the Neck** | | | |
|---|---|---|---|---|
| **Muscle** | **Origin** | **Insertion** | **Nerve Supply** | **Action** |
| Platysma | Deep fascia over pectoralis major and deltoid | Body of mandible and angle of mouth | Facial nerve, cervical branch | Depresses mandible and angle of mouth |
| Sternocleido-mastoid | Manubrium sterni and medial third of clavicle | Mastoid process of temporal bone and occipital bone | Spinal part of accessory nerve and C2 and 3 | Two muscles acting together extend head and flex neck; one muscle rotates head to opposite side |
| Digastric  Posterior belly | Mastoid process of temporal bone | Intermediate tendon is held to hyoid by fascial sling | Facial nerve | Depresses mandible or elevates hyoid bone |
| Anterior belly | Body of mandible | | Nerve to mylohyoid | |
| Stylohyoid | Styloid process | Body of hyoid bone | Facial nerve | Elevates hyoid bone |
| Mylohyoid | Mylohyoid line of body of mandible | Body of hyoid bone and fibrous raphe | Inferior alveolar nerve | Elevates floor of mouth and hyoid bone or depresses mandible |
| Geniohyoid | Inferior mental spine of mandible | Body of hyoid bone | First cervical nerve | Elevates hyoid bone or depresses mandible |
| Sternohyoid | Manubrium sterni and clavicle | Body of hyoid bone | Ansa cervicalis; C1, 2, and 3 | Depresses hyoid bone |
| Sternothyroid | Manubrium sterni | Oblique line on lamina of thyroid cartilage | Ansa cervicalis; C1, 2, and 3 | Depresses larynx |
| Thyrohyoid | Oblique line on lamina of thyroid cartilage | Lower border of body of hyoid bone | First cervical nerve | Depresses hyoid bone or elevates larynx |
| Omohyoid  Inferior belly | Upper margin of scapula and supra-scapular ligament | Intermediate tendon is held to clavicle and first rib by fascial sling | Ansa cervicalis; C1, 2, and 3 | Depresses hyoid bone |
| Superior belly | Lower border of body of hyoid bone | | | |
| Scalenus anterior | Transverse processes of third, fourth, fifth, and sixth cervical vertebrae | First rib | C4, 5, and 6 | Elevates first rib; laterally flexes and rotates cervical part of vertebral column |
| Scalenus medius | Transverse processes of upper six cervical vertebrae | First rib | Anterior rami of cervical nerves | Elevates first rib; laterally flexes and rotates cervical part of vertebral column |
| Scalenus posterior | Transverse processes of lower cervical vertebrae | Second rib | Anterior rami of cervical nerves | Elevates second rib; laterally flexes and rotates cervical part of vertebral column |

(From Snell RS. Clinical Anatomy. 7th Ed. Philadelphia: Lippincott Williams & Wilkins, 2004, p. 754).

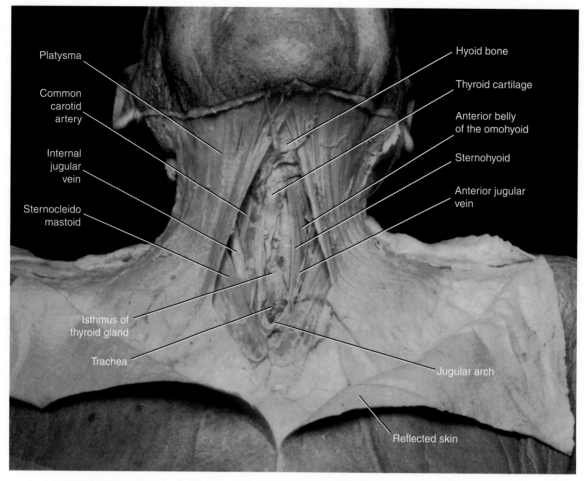

**Figure 13-8** Dissection of the anterior aspect of the neck showing the platysma muscles and the lower ends of the sternocleidomastoid muscles on both sides. The skin has been reflected downwards.

ies, the internal jugular vein, and the deep cervical lymph nodes; it also overlaps the thyroid gland. The muscle is covered superficially by skin, fascia, the platysma muscle, and the external jugular vein. The deep surface of the posterior border is related to the cervical plexus of nerves, the phrenic nerve, and the upper part of the brachial plexus. The origin, insertion, nerve supply, and action of the sternocleidomastoid muscle are summarized in Table 13-3.

## Scalenus Anterior Muscle

The scalenus anterior muscle is a key muscle in understanding the root of the neck (Fig. 13-10). It is deeply placed, and it descends almost vertically from the vertebral column to the first rib.

### Important Relations

■ **Anteriorly:** Related to the carotid arteries, the vagus nerve, the internal jugular vein, and the deep cervical lymph nodes. The transverse cervical and suprascapular arteries and the prevertebral layer of deep cervical fascia bind the phrenic nerve to the muscle.

■ **Posteriorly:** Related to the pleura, the origin of the brachial plexus, and the second part of the subclavian artery. The scalenus medius muscle lies behind the scalenus anterior muscle.

■ **Medially:** Related to the vertebral artery and vein and the sympathetic trunk. On the left side, the medial border is related to the thoracic duct.

■ **Laterally:** Related to the emerging branches of the cervical plexus, the roots of the brachial plexus, and the third part of the subclavian artery.

The origin, insertion, nerve supply, and action of the scalenus anterior muscle are summarized in Table 13-3.

## Cervical Fascia

### Superficial Cervical Fascia

The superficial cervical fascia is a thin layer of connective tissue that encloses the platysma muscle (Fig. 13-11). Also embedded within it are cutaneous nerves, superficial veins, and superficial lymph nodes.

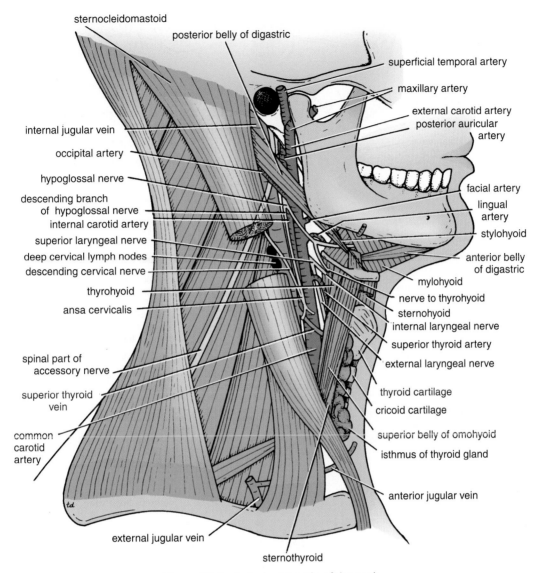

**Figure 13-9** Anterior triangle of the neck.

## Deep Cervical Fascia

The deep cervical fascia supports the muscles, the vessels, and the viscera of the neck (Fig. 13-11). In certain areas, it is condensed to form well-defined, fibrous sheets called the **investing layer**, the **pretracheal layer**, and the **prevertebral layer**. It is also condensed to form the **carotid sheath** (Fig. 13-11).

### Investing Layer

The investing layer is a thick layer that encircles the neck. It splits to enclose the trapezius and the sternocleidomastoid muscles (Fig. 13-11).

### Pretracheal Layer

The pretracheal layer is a thin layer that is attached above to the laryngeal cartilages (Fig. 13-11). It surrounds the thyroid and the parathyroid glands, forming a sheath for them, and encloses the infrahyoid muscles.

### Prevertebral Layer

The prevertebral layer is a thick layer that passes like a septum across the neck behind the pharynx and the esophagus and in front of the prevertebral muscles and the vertebral column (Fig. 13-11). It forms the fascial floor of the posterior triangle, and it extends laterally over the first rib into the axilla to form the important **axillary sheath**.

### Carotid Sheath

The carotid sheath is a local condensation of the prevertebral, the pretracheal, and the investing layers of the deep fascia that surround the **common** and **internal carotid arteries**, the **internal jugular vein**, the **vagus nerve**, and the **deep cervical lymph nodes** (Fig. 13-11).

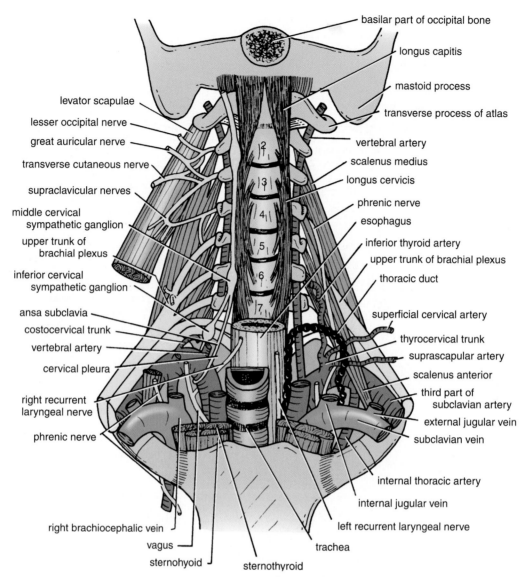

**Figure 13-10**  Prevertebral region and the root of the neck.

## Cervical Ligaments

- **Stylohyoid ligament:** Connects the styloid process to the lesser cornu of the hyoid bone.
- **Stylomandibular ligament:** Connects the styloid process to the angle of the mandible.
- **Sphenomandibular ligament:** Connects the spine of the sphenoid bone to the lingula of the mandible.
- **Pterygomandibular ligament:** Connects the hamular process of the medial pterygoid plate to the posterior end of the mylohyoid line of the mandible. It gives attachment to the superior constrictor and the buccinator muscles.

The superficial muscles of the side of the neck are shown in Table 13-3.

## Muscular Triangles of the Neck

The sternocleidomastoid muscle divides the neck into the anterior and the posterior triangles (Fig. 13-12).

### Anterior Triangle

The anterior triangle is bounded above by the body of the mandible, posteriorly by the sternocleidomastoid muscle, and anteriorly by the midline (Fig. 13-12). It is further subdivided into the **carotid triangle**, the **digastric triangle**, the **submental triangle**, and the **muscular triangle** (Fig. 13-12).

### Posterior Triangle

The posterior triangle is bounded posteriorly by the trapezius muscle, anteriorly by the sternocleidomastoid muscle,

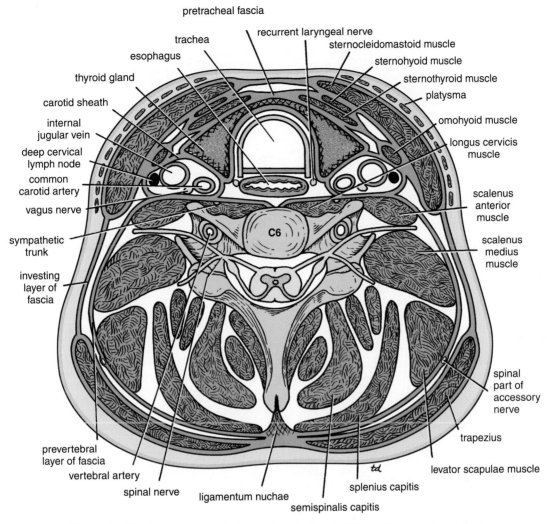

**Figure 13-11** Cross section of the neck at the level of the sixth cervical vertebra.

and inferiorly by the clavicle (Fig. 13-12). The posterior triangle of the neck is further subdivided by the inferior belly of the omohyoid muscle into a large occipital triangle and a small supraclavicular triangle (Fig. 13-12).

The suprahyoid and infrahyoid muscles and the anterior and lateral vertebral muscles are shown in Table 13-3.

# Muscles of the Back

The muscles of the back may be divided into three groups:

▪ The **superficial muscles** connected with the shoulder girdle.
▪ The **intermediate muscles** involved with movements of the thoracic cage.
▪ The **deep** or **postvertebral muscles** belonging to the vertebral column.

## Postvertebral Muscles

The postvertebral muscles are very well developed in humans and form a broad, thick column of muscle tissue that occupies the hollow on each side of the spinous processes of the vertebral column (Fig. 13-13). The spines and the transverse processes of the vertebrae serve as levers that assist with actions of the muscles. The muscles of longest length lie superficially and run from the sacrum to the rib angles, the transverse processes, and the upper vertebral spines. The muscles of intermediate length run obliquely from the transverse processes to the spines. The shortest and deepest muscle fibers run between the spines and between the transverse processes of adjacent vertebrae.

The postvertebral muscles may be classified as follows:

▪ **Superficial vertically running muscles.**
▪ **Erector spinae:** iliocostalis, longissimus, and spinalis.

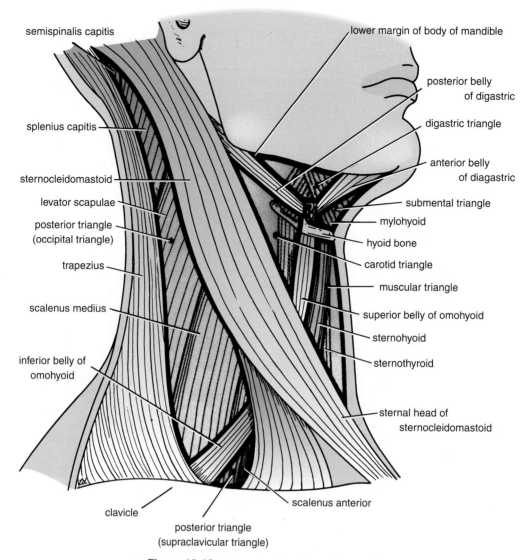

**Figure 13-12** Muscular triangles of the neck.

- **Intermediate oblique running muscles.**
- **Transversospinalis:** Semispinalis, multifidus, and rotatores.
- **Deepest muscles:** interspinales and intertransversarii.

Students are not required to learn the detailed attachments of these muscles. Figure 13-13 shows the arrangement of the deep muscles of the back.

## Muscular Triangles of the Back

### Auscultatory Triangle

The auscultatory triangle is the site on the back where breath sounds may be most easily heard with a stethoscope.

The boundaries are the latissimus dorsi, the trapezius, and the medial border of the scapula.

### Lumbar Triangle

The lumbar triangle is the site where pus may emerge from the abdominal wall. The boundaries are the latissimus dorsi, the posterior border of the external oblique muscle of the abdomen, and the iliac crest.

# Muscles of the Thoracic Wall

The muscles of the thoracic wall, including the diaphragm, are summarized in Table 13-4. The intercostal muscles and the important muscle, the diaphragm, are fully discussed in Chapter 3.

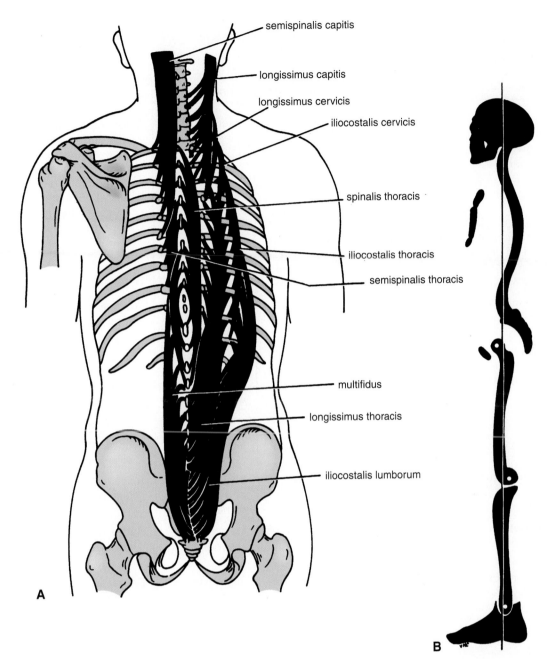

semispinalis capitis

longissimus capitis

longissimus cervicis

iliocostalis cervicis

spinalis thoracis

iliocostalis thoracis

semispinalis thoracis

multifidus

longissimus thoracis

iliocostalis lumborum

A

B

**Figure 13-13** **A.** Arrangement of the deep muscles of the back. **B.** Lateral view of the skeleton showing the line of gravity. Because the greater part of the body weight lies anterior to the vertebral column, the deep muscles of the back are important in maintaining the normal postural curves of the vertebral column in the standing position.

# Muscles of the Anterior Abdominal Wall

The muscles of the anterior abdominal wall consist mainly of three broad, thin sheets that are aponeurotic in front. From exterior to interior, these sheets are the **external oblique**, the **internal oblique**, and the **transversus** (Fig. 13-14). In addition, on either side of the midline anteriorly, there is a wide, vertical muscle called the **rectus abdominis** (Fig. 13-15). As the aponeuroses of the three sheets pass forward, they enclose the rectus abdominis to form the **rectus sheath**.

In the lower part of the rectus sheath, there may be a small muscle called the **pyramidalis**.

| Table 13-4 | Muscles of the Thorax | | | |
|---|---|---|---|---|
| **Name of Muscle** | **Origin** | **Insertion** | **Nerve Supply** | **Action** |
| External intercostal muscle (11) (fibers pass downward and forward) | Inferior border of rib | Superior border of rib below | Intercostal nerves | With first rib fixed, they raise ribs during inspiration and thus increase antero-posterior and transverse diameters of thorax. With last rib fixed by abdominal muscles, they lower ribs during expiration |
| Internal intercostal muscle (11) (fibers pass downward and backward) | Inferior border of rib | Superior border of rib below | Intercostal nerves | |
| Innermost intercostal muscle (incomplete layer) | Adjacent ribs | Adjacent ribs | Intercostal nerves | Assist external and internal intercostal muscles |
| Diaphragm (most important muscle of respiration) | Xiphoid process; lower six costal cartilages, first three lumbar vertebrae | Central tendon | Phrenic nerve | Very important muscle of inspiration; increases vertical diameter of thorax by pulling central tendon downward, assists in raising lower ribs. Also used in abdominal straining and weight lifting |
| Levatores costarum (12) | Tip of transverse process of C7 and T1–11 vertebrae | Rib below | Posterior rami of thoracic spinal | Raise ribs and therefore inspiratory muscles |
| Serratus posterior superior | Lower cervical and upper thoracic spines | Upper ribs | Intercostal nerves | Raises ribs and therefore inspiratory muscles |
| Serratus posterior inferior | Upper lumbar and lower thoracic spines | Lower ribs | Intercostal nerves | Depresses ribs and therefore expiratory muscles |

From Snell RS: Clinical Anatomy. 7th Ed. Philadelphia: Lippincott Williams & Wilkins, 2004, p. 68.

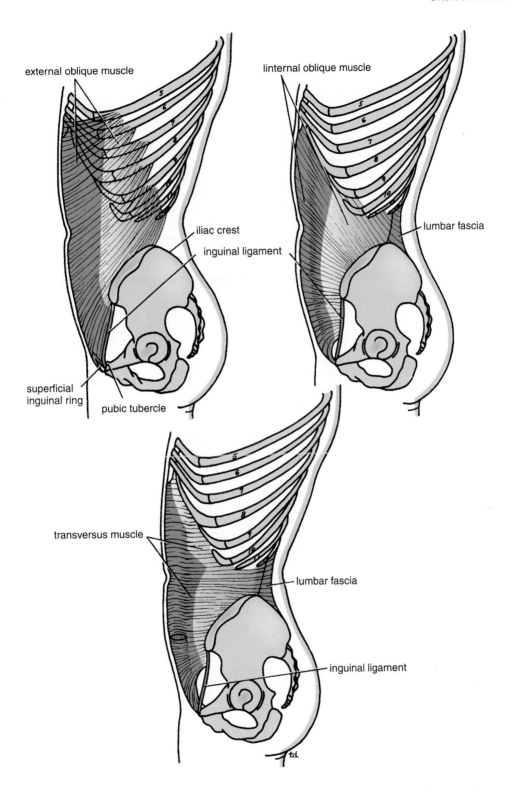

external oblique muscle

internal oblique muscle

iliac crest

inguinal ligament

lumbar fascia

superficial inguinal ring

pubic tubercle

transversus muscle

lumbar fascia

inguinal ligament

**Figure 13-14**   External oblique, internal oblique, and transversus muscles of the anterior abdominal wall.

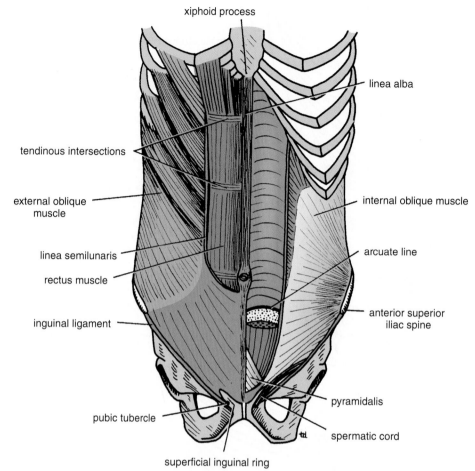

xiphoid process

linea alba

tendinous intersections

external oblique muscle

internal oblique muscle

linea semilunaris

arcuate line

rectus muscle

anterior superior iliac spine

inguinal ligament

pyramidalis

pubic tubercle

spermatic cord

superficial inguinal ring

**Figure 13-15** Anterior view of the rectus abdominis muscle and the rectus sheath. *Left:* The anterior wall of the sheath has been partly removed, revealing the rectus muscle with its tendinous intersections. *Right:* The posterior wall of the rectus sheath is shown. The edge of the arcuate line is shown at the level of the anterior superior iliac spine.

The **cremaster muscle** is derived from the lower fibers of the internal oblique; it passes inferiorly as a covering of the spermatic cord and enters the scrotum.

The muscles of the anterior abdominal wall are summarized in Table 13-5.

## Rectus Sheath

The rectus sheath (Figs. 13-15 and 13-16) is a long fibrous sheath that encloses the rectus abdominis muscle and pyramidalis muscle (if present) and contains the anterior rami of the lower six thoracic nerves and the superior and inferior epigastric vessels and lymph vessels. It is formed by the aponeuroses of the three lateral abdominal muscles. The internal oblique aponeurosis splits at the lateral edge of the rectus abdominis to form two laminae; one passes anteriorly and one passes posteriorly to the rectus. The aponeurosis of the external oblique fuses with the anterior

lamina, and the transversus aponeurosis fuses with the posterior lamina. At the level of the anterior superior iliac spines, all three aponeuroses pass anteriorly to the rectus muscle, leaving the sheath deficient posteriorly below this level. The lower, crescent-shaped edge of the posterior wall of the sheath is called the **arcuate line.** All three aponeuroses fuse with each other and with their fellows of the opposite side in the midline between the right and the left recti muscles to form a fibrous band called the **linea alba,** which extends from the xiphoid process above to the pubic symphysis below.

The posterior wall of the sheath has no attachment to the rectus abdominis muscle. The transverse **tendinous intersections,** which divide the rectus abdominis muscle into segments, are usually three in number: one at the level of the xiphoid process, one at the level of the umbilicus, and one between these two. The anterior wall of the rectus sheath is firmly attached to the tendinous intersections.

| Table 13-5 | Muscles of the Anterior Abdominal Wall | | | |
| --- | --- | --- | --- | --- |
| **Name of Muscle** | **Origin** | **Insertion** | **Nerve Supply** | **Action** |
| External oblique | Lower eight ribs | Xiphoid process, linea alba, pubic crest, pubic tubercle, iliac crest | Lower six thoracic nerves and iliohypogastric and ilioinguinal nerves (L1) | Supports abdominal contents; compresses abdominal contents; assists in flexing and rotation of trunk; assists in forced expiration, micturition, defecation, parturition, and vomiting |
| Internal oblique | Lumbar fascia, iliac crest, lateral two thirds of inguinal ligament | Lower three ribs and costal cartilages, xiphoid process, linea alba, symphysis pubis | Lower six thoracic nerves and iliohypogastric and ilioinguinal nerves (L1) | As above |
| Transversus | Lower six costal cartilages, lumbar fascia, iliac crest, lateral third of inguinal ligament | Xiphoid process linea alba, symphysis pubis | Lower six thoracic nerves and iliohypogastric and ilioinguinal nerves (L1) | Compresses abdominal contents |
| Rectus abdominis | Symphysis pubis and pubic crest | Fifth, sixth, and seventh costal cartilages and xiphoid process | Lower six thoracic nerves | Compresses abdominal contents and flexes vertebral column; accessory muscle of expiration |
| Pyramidalis (if present) | Anterior surface of pubis | Linea alba | Twelfth thoracic nerve | Tenses the linea alba |

From Snell RS: Clinical Anatomy. 7th Ed. Philadelphia: Lippincott Williams & Wilkins, 2004, p. 167.

## Linea Semilunaris

The linea semilunaris is the lateral edge of the rectus abdominis muscle. It crosses the costal margin at the tip of the ninth costal cartilage.

## Conjoint Tendon

The internal oblique muscle (Fig. 13-17) has a lower, free border that arches over the spermatic cord (or the round ligament of the uterus) and then descends behind and attaches to the pubic crest and the pectineal line. Near their insertion, the lowest tendinous fibers are joined by similar fibers from the transversus abdominis to form the **conjoint ten-** don, which strengthens the medial half of the posterior wall of the inguinal canal.

## Inguinal Ligament

The inguinal ligament (Fig. 13-17) connects the anterior superior iliac spine with the pubic tubercle. This ligament is formed by the lower border of the aponeurosis of the external oblique, which is folded back on itself (Fig. 13-14). From the medial end of the ligament, the **lacunar ligament** (Fig. 13-17) extends backward and upward to the pectineal line of the superior ramus of the pubis, where it becomes continuous with the **pectineal ligament** (a thickening of the periosteum, Fig. 13-17). The lower border of the inguinal

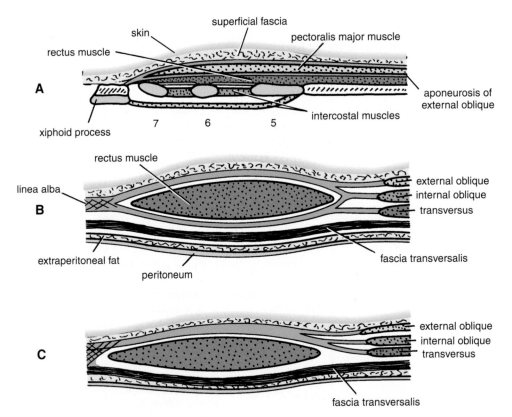

**Figure 13-16** Transverse sections of the rectus sheath seen at three levels. **A.** Above the costal margin. **B.** Between the costal margin and the level of the anterior superior iliac spine. **C.** Below the level of the anterior superior iliac spine and above the pubis.

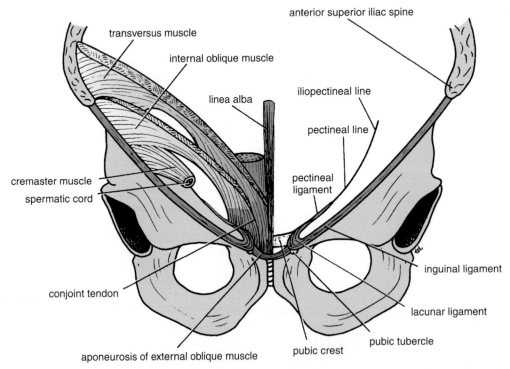

**Figure 13-17** Anterior view of the pelvis showing the attachment of the conjoint tendon to the pubic crest and the adjoining part of the pectineal line.

ligament is attached to the deep fascia of the thigh (the **fascia lata**).

## Fascia Transversalis

The fascia transversalis is a thin layer of fascia that lines the transversus muscle and is continuous with a similar layer lining the diaphragm and the iliacus muscle. The **femoral sheath** of the femoral vessels is formed from the fascia transversalis and the fascia iliaca.

# Muscles of the Posterior Abdominal Wall

The muscles of the posterior abdominal wall include the **psoas major**, the **quadratus lumborum**, and the **iliacus muscles** (Fig. 17-25).

The muscles of the posterior abdominal wall are summarized in Table 13-6.

# Muscles of the Pelvis

The **piriformis muscle** lines the posterior wall of the pelvis and lies anterior to the sacrum (Fig. 13-18). It leaves the pelvis through the greater sciatic foramen to enter the gluteal region and act on the femur at the hip joint. The **obturator internus** lines the lateral wall of the pelvis (Fig. 13-19) and lies medial to the obturator membrane. It leaves the pelvis through the lesser sciatic foramen to enter the gluteal region and act on the femur at the hip joint. The **levator ani muscles** and the **coccygeus muscles** of the two sides form, with their covering fascia, the **pelvic diaphragm** (Fig. 13-20). The pelvic diaphragm is incomplete anteriorly, to allow for the passage of the urethra and, in the female, the vagina also.

The attachments, nerve supplies, and actions of the muscles of the pelvis are given in Table 13-7.

## Pelvic Fasciae

The **parietal pelvic fascia** lines the pelvic walls and is regionally named according to the muscle it overlies. Above the pelvic inlet, it is continuous with the fascia lining the abdominal walls.

The **visceral pelvic fascia** covers all the pelvic viscera. Around the cervix in females, this fascia is called the **parametrium.**

# Muscles of the Perineum

The perineum, when seen from below, is diamond shaped and is bounded anteriorly by the **symphysis pubis**, posteri-

| Table 13-6 | Muscles of the Posterior Abdominal Wall | | | |
|---|---|---|---|---|
| **Name of Muscle** | **Origin** | **Insertion** | **Nerve Supply** | **Action** |
| Psoas | Transverse processes, bodies, and intervertebral discs of twelfth thoracic and five lumbar vertebrae | With iliacus into lesser trochanter of femur | Lumbar plexus | Flexes thigh on trunk; if thigh is fixed, it flexes trunk on thigh, as in sitting up from lying position |
| Quadratus lumborum | Iliolumbar ligament, iliac crest, tips of transverse processes of lower lumbar vertebrae | Twelfth rib | Lumbar plexus | Fixes twelfth rib during inspiration; depresses twelfth rib during forced expiration; laterally flexes vertebral column same side |
| Iliacus | Iliac fossa | With psoas into lesser trochanter of femur | Femoral nerve | Flexes thigh on trunk; if thigh is fixed, it flexes the trunk on the thigh, as in sitting up from lying position |

From Snell RS: Clinical Anatomy. 7th Ed. Philadelphia: Lippincott Williams & Wilkins, 2004, p. 188.

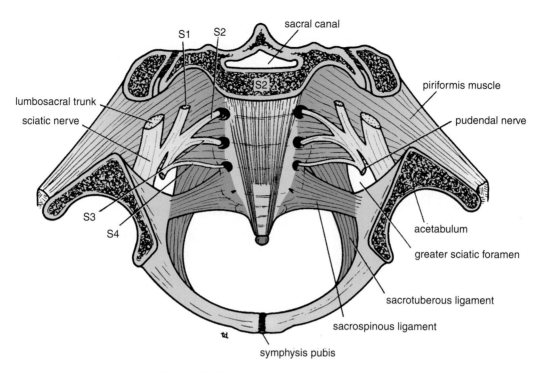

**Figure 13-18**    Posterior wall of the pelvis..

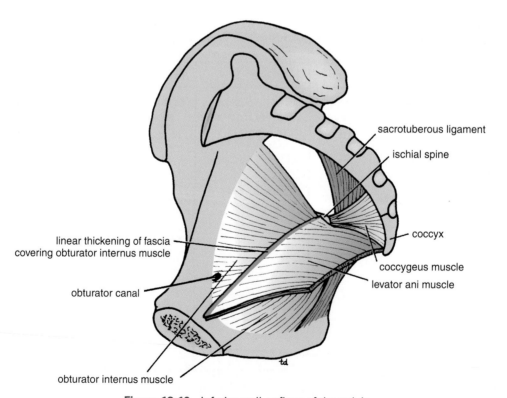

**Figure 13-19**    Inferior wall or floor of the pelvis.

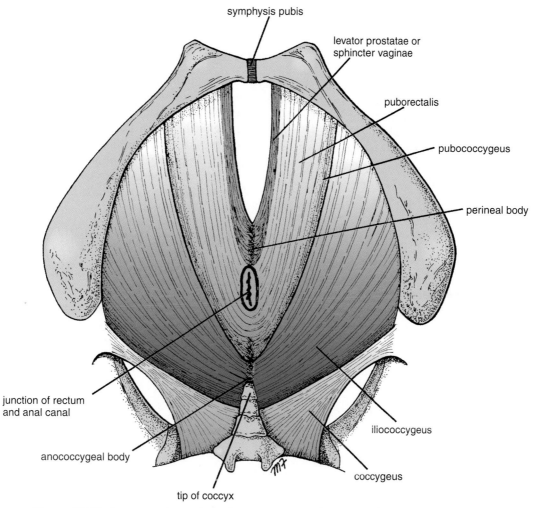

symphysis pubis

levator prostatae or
sphincter vaginae

puborectalis

pubococcygeus

perineal body

junction of rectum
and anal canal

iliococcygeus

anococcygeal body

coccygeus

tip of coccyx

**Figure 13-20** Levator ani muscle and coccygeus muscle seen on their inferior aspects.
Note that the levator ani is made up of several different muscle groups. The levator ani
and coccygeus muscles with their fascial coverings form a continuous muscular floor to
the pelvis, known as the pelvic diaphragm.

| Table 13-7 | **Muscles of the Pelvic Walls and Floor** | | | |
|---|---|---|---|---|
| **Name of Muscle** | **Origin** | **Insertion** | **Nerve Supply** | **Action** |
| Piriformis | Front of sacrum | Greater trochanter of femur | Sacral plexus | Lateral rotator of femur at hip joint |
| Obturator internus | Obturator membrane and adjoining part of hip bone | Greater trochanter of femur | Nerve to obturator internus from sacral plexus | Lateral rotator of femur at hip joint |
| Levator ani | Body of pubis, fascia of obturator internus, spine of ischium | Perineal body, anococcygeal body, walls of prostate, vagina, rectum, and anal canal | Fourth sacral nerve, pudendal nerve | Supports pelvic viscera; sphincter to anorectal junction and vagina |
| Coccygeus | Spine of ischium | Lower end of sacrum; coccyx | Fourth and fifth sacral nerve | Assists levator ani to support pelvic viscera; flexes coccyx |

From Snell RS: Clinical Anatomy. 7th Ed. Philadelphia: Lippincott Williams & Wilkins, 2004, p. 347.

orly by the **tip of the coccyx,** and laterally by the **ischial tuberosities.**

The perineum may be divided into two triangles by joining the **ischial tuberosities** by an imaginary line. The posterior triangle, which contains the anus, is called the **anal triangle;** the anterior triangle, which contains the urogenital orifices, is called the **urogenital triangle.**

## Anococcygeal Body

This is a mass of fibrous tissue that lies between the anal canal and the coccyx (Fig. 13-21).

## Perineal Body

The perineal body is a small mass of fibrous tissue that is attached to the center of the posterior margin of the urogenital diaphragm (Fig. 13-21). It is a larger structure in the female as compared to the male and serves to support the posterior vaginal wall. In both sexes, it provides a point of attachment for muscles in the perineum.

## Urogenital Diaphragm

This is a musculofascial diaphragm that fills in the gap of the pubic arch (Fig. 13-22). It is formed by the **sphincter urethrae** and the deep transverse perineal muscles, which are enclosed between a superior and an inferior layer of fascia of the urogenital diaphragm. The inferior layer of fascia is called the **perineal membrane.**

The attachments, nerve supply, and actions of the muscles of the perineum are summarized in Table 13-8.

# Muscles of the Upper Limb

## Shoulder Region

### Axilla

The axilla, or armpit, is a pyramid-shaped space between the upper part of the arm and the side of the chest. The upper end, or **apex,** is directed into the root of the neck and is

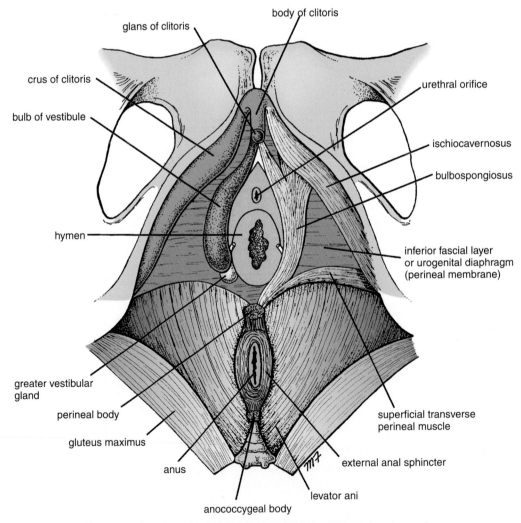

**Figure 13-21**   Root and body of the clitoris and the perineal muscles.

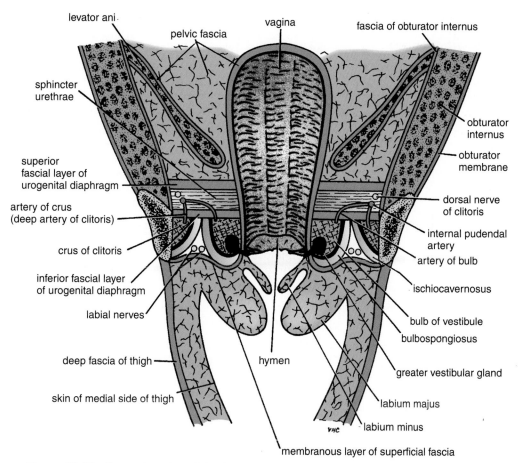

**Figure 13-22** Coronal section of the female pelvis showing the vagina, the urogenital diaphragm, and the contents of the superficial perineal pouch.

| Table 13-8 | Muscles of Perineum | | | |
|---|---|---|---|---|
| **Name of Muscle** | **Origin** | **Insertion** | **Nerve Supply** | **Action** |
| External anal sphincter | | | | |
| Subcutaneous part | Encircles anal canal, no bony attachments | | Inferior rectal nerve and perinea branch of fourth sacral nerve | Together with puborectalis muscle forms voluntary sphincter of anal canal |
| Superficial part | Perineal body | Coccyx | | |
| Deep part | Encircles anal canal, no bony attachments | | | |

## Table 13-8 (Continued)

| Name of Muscle | Origin | Insertion | Nerve Supply | Action |
|---|---|---|---|---|
| Puborectalis (part of levator ani) | Pubic bones | Sling around junction of rectum and anal canal | Perineal branch of fourth sacral nerve and from perineal branch of pudendal nerve | Together with external anal sphincter forms voluntary sphincter for anal canal |
| **Male Urogenital Muscles** | | | | |
| Bulbospongiosus | Perineal body | Fascia of bulb of penis and corpus spongiosum and cavernosum | Perineal branch of pudendal nerve | Compresses urethra and assists in erection of penis |
| Ischiocavernosus | Ischial tuberosity | Fascia covering corpus cavernosum | Perineal branch of pudendal nerve | Assists in erection of penis |
| Sphincter urethrae | Pubic arch | Surrounds urethra | Perineal branch of pudendal nerve | Voluntary sphincter of urethra |
| Superficial transverse perineal muscle | Ischial tuberosity | Perineal body | Perineal branch of pudendal nerve | Fixes perineal body |
| Deep transverse perineal muscle | Ischial ramus | Perineal body | Perineal branch of pudendal nerve | Fixes perineal body |
| **Female Urogenital Muscles** | | | | |
| Bulbospongiosus | Perineal body | Fascia of corpus cavernosum | Perineal branch of pudendal nerve | Sphincter of vagina and assists in erection of clitoris |
| Ischiocavernosus | Ischial tuberosity | Fascia covering corpus cavernosum | Perineal branch of pudendal nerve | Causes erection of clitoris |
| Sphincter urethrae | Same as in male | | | |
| Superficial transverse perineal muscle | Same as in male | | | |
| Deep transverse perineal muscle | Same as in male | | | |

From Snell RS: Clinical Anatomy. 7th Ed. Philadelphia: Lippincott Williams & Wilkins, 2004, p. 439.

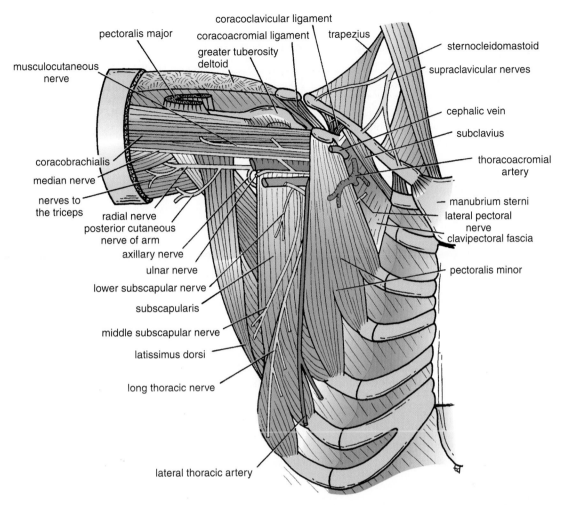

**Figure 13-23**    Pectoral region and axilla; the pectoralis major muscle has been removed to display the underlying structures.

bounded in front by the clavicle, behind by the upper border of the scapula, and medially by the outer border of the first rib. The lower end, or **base**, is bounded in front by the anterior axillary fold (formed by the lower border of the pectoralis major muscle), behind by the posterior axillary fold (formed by the tendon of latissimus dorsi and the teres major muscle), and medially by the chest wall.

The axilla contains the principal vessels and nerves to the upper limb (Fig. 13-23) and many lymph nodes.

The muscles connecting the upper limb to the thoracic wall are shown in Table 13-9, the muscles connecting the upper limb to the vertebral column are shown in Table 13-10, and the muscles connecting the scapula to the humerus are shown in Table 13-11.

### Rotator Cuff

The rotator cuff is the name given to the tendons of the subscapularis, supraspinatus, infraspinatus, and the teres minor muscles, which are fused to the underlying capsule of the shoulder joint (Fig. 13-24). The cuff plays a very important role in stabilizing the shoulder joint.

### Quadrangular Space

The quadrangular space, located immediately below the shoulder joint, is bounded above by the subscapularis muscle and below by the teres major muscle. It is bounded laterally by the surgical neck of the humerus and medially by the long head of the triceps. The axillary nerve and posterior circumflex humeral arteries pass through the space (Fig. 13-24).

## Upper Arm

The muscles of the upper arm (Figs 13-25, 13-26, and 13-27) are summarized in Table 13-12.

| Table 13-9 | | Muscles Connecting the Upper Limb to the Thoracic Wall | | | |
|---|---|---|---|---|---|
| **Muscle** | **Origin** | **Insertion** | **Nerve Supply** | **Nerve Roots***a* | **Action** |
| Pectoralis major | Clavicle, sternum and upper six costal cartilages | Lateral lip of bicipital groove of humerus | Medial and lateral pectoral nerves from brachial plexus | C5, **6, 7, 8;** T1 | Adducts arm and rotates it medially; clavicular fibers also flex arm |
| Pectoralis minor | Third, fourth, and fifth ribs | Coracoid process of scapula | Medial pectoral nerve from brachial plexus | C6, **7**, 8 | Depresses point of shoulder; if the scapula is fixed, it elevates the ribs of origin |
| Subclavius | First costal cartilage | Clavicle | Nerve to subclavius from upper trunk of brachial plexus | **C5,** 6 | Depresses the clavicle and steadies this bone during movements of the shoulder girdle |
| Serratus anterior | Upper eight ribs | Medial border and inferior angle of scapula | Long thoracic nerve | C5, **6, 7** | Draws the scapula forward around the thoracic wall; rotates scapula |

*a*The predominant nerve root supply is indicated by boldface type.
From Snell RS: Clinical Anatomy. 7th Ed. Philadelphia: Lippincott Williams & Wilkins, 2004, p. 499.

## Fascial Compartments of the Upper Arm

The upper arm is enclosed in a sheath of deep fascia (Fig. 13-28). Two fascial septa (one on the medial side and one on the lateral side) extend from this sheath and are attached to the medial and lateral borders of the humerus, respectively. By this means, the upper arm is divided into an anterior and a posterior fascial compartment, with each compartment having its own muscles, nerves, and arteries.

## Cubital Fossa

The cubital fossa is a skin depression that lies in front of the elbow and is triangular in shape (Fig. 13-27). It is bounded **laterally** by the brachioradialis muscle and **medially** by the pronator teres muscle. The **base** of the triangle is formed by an imaginary line drawn between the two epicondyles of the humerus.

From medial to lateral, the cubital fossa contains the median nerve, the bifurcation of the brachial artery into the ulnar and the radial arteries, the tendon of the biceps muscle, and the radial nerve and its deep branch.

Lying in the superficial fascia covering the cubital fossa are the **cephalic** and the **basilic veins** as well as their tributaries.

## Forearm

The muscles of the anterior fascial compartment of the forearm (Fig. 13-29) are summarized in Table 13-13. The muscles of the lateral fascial compartment (Fig. 13-29) are summarized in Table 13-14, and the muscles of the posterior fascial compartment (Fig. 13-30) are summarized in Table 13-15.

## Fascial Compartments of the Forearm

The forearm is enclosed in a sheath of deep fascia, which is attached to the periosteum of the posterior subcutaneous border of the ulna (Fig. 13-31). Together with the in-

## Table 13-10   Muscles Connecting the Upper Limb to the Vertebral Column

| Muscle | Origin | Insertion | Nerve Supply | Nerve Roots[a] | Action |
|---|---|---|---|---|---|
| Trapezius | Occipital bone, ligamentum nuchae, spine of seventh cervical vertebra, spines of all thoracic vertebrae | Upper fibers into lateral third of clavicle; middle and lower fibers into acromion and spine of scapula | Spinal part of accessory nerve (motor) and C3 and 4 (sensory) | XI **cranial nerve** (spinal part) | Upper fibers elevate the scapula; middle fibers pull scapula medially; lower fibers pull medial border of scapula downward |
| Latissimus dorsi | Iliac crest, lumbar fascia, spines of lower six thoracic vertebrae, lower three or four ribs, and inferior angle of scapula | Floor of bicipital groove of humerus | Thoracodorsal nerve | C6, **7**, 8, | Extends, adducts, and medially rotates the arm |
| Levator scapulae | Transverse processes of first four cervical vertebrae | Medial border of scapula | C3 and 4 and dorsal scapular nerve | C3, 4, 5 | Raises medial border of scapula |
| Rhomboid minor | Ligamentum nuchae and spines of seventh cervical and first thoracic vertebrae | Medial border of scapula | Dorsal scapular nerve | **C4**, 5 | Raises medial border of scapula upward and medially |
| Rhomboid major | Second to fifth thoracic spines | Medial border of scapula | Dorsal scapular nerve | **C4**, 5 | Raises medial border of scapula upward and medially |

[a]The predominant nerve root supply is indicated by boldface type.
From Snell RS: Clinical Anatomy. 7th Ed. Philadelphia: Lippincott Williams & Wilkins, 2004, p. 499.

## Table 13-11   Muscles Connecting the Scapula to the Humerus

| Muscle | Origin | Insertion | Nerve Supply | Nerve Roots[a] | Action |
|---|---|---|---|---|---|
| Deltoid | Lateral third of clavicle, acromion, spine of scapula | Middle of lateral surface of shaft of humerus | Axillary nerve | **C5**, 6 | Abducts arm; anterior fibers flex and medially rotate arm; posterior fibers extend and laterally rotate arm |
| Supraspinatus | Supraspinous fossa of scapula | Greater tuberosity of humerus; capsule of shoulder joint | Suprascapular nerve | C4, **5**, 6 | Abducts arm and stabilizes shoulder joint |

**Table 13-11** *(Continued)*

| Muscle | Origin | Insertion | Nerve Supply | Nerve Roots[a] | Action |
|---|---|---|---|---|---|
| Infraspinatus | Infraspinous fossa of scapula | Greater tuberosity of humerus; capsule of shoulder joint | Suprascapular nerve | (C4), **5**, 6 | Laterally rotates arm and stabilizes shoulder joint |
| Teres major | Lower third of lateral border of scapula | Medial lip of bicipital groove of humerus | Lower subscapular nerve | **C6**, 7 | Medially rotates and adducts arm and stabilizes shoulder joint |
| Teres minor | Upper two thirds of lateral border of scapula | Greater tuberosity of humerus; capsule of shoulder joint | Axillary nerve | (C4), **C5**, 6 | Laterally rotates arm and stabilizes shoulder joint |
| Subscapularis | Subscapular fossa | Lesser tuberosity of humerus | Upper and lower subscapular nerves | C5, **6**, 7 | Medially rotates arm and stabilizes shoulder joint |

[a]The predominant nerve root supply is indicated by boldface type.
From Snell RS: Clinical Anatomy. 7th Ed. Philadelphia: Lippincott Williams & Wilkins, 2004, p. 500.

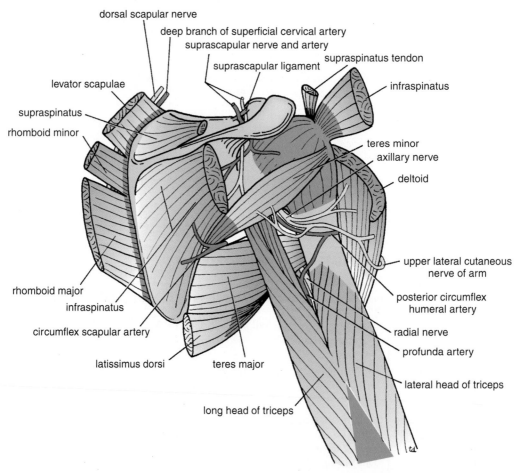

**Figure 13-24** Muscles, nerves, and blood vessels of the scapular region. Note the close relation of the axillary nerve to the shoulder joint.

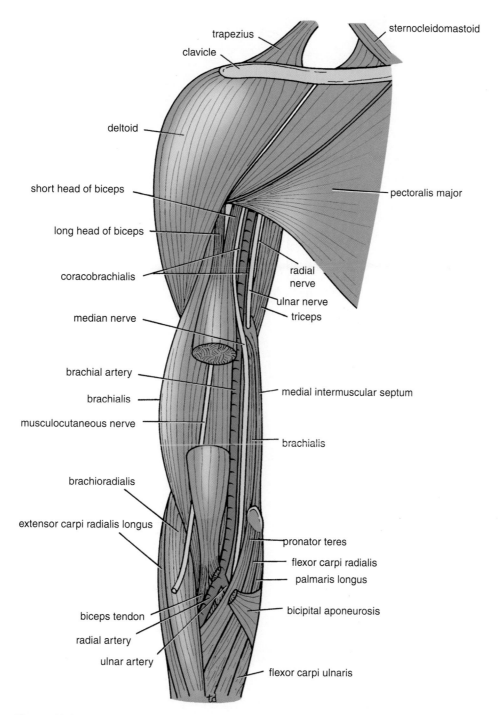

**Figure 13-25**  Anterior view of the upper arm. The middle portion of the biceps brachii has been removed to show the musculocutaneous nerve lying in front of the brachialis.

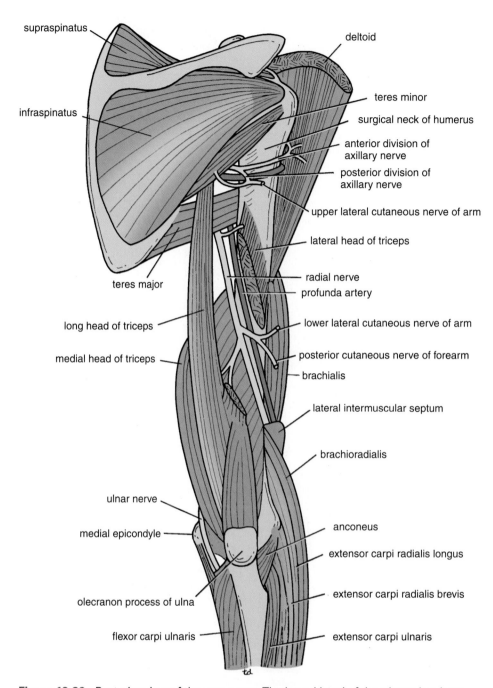

**Figure 13-26**  Posterior view of the upper arm. The lateral head of the triceps has been divided to display the radial nerve and the profunda artery in the spiral groove of the humerus.

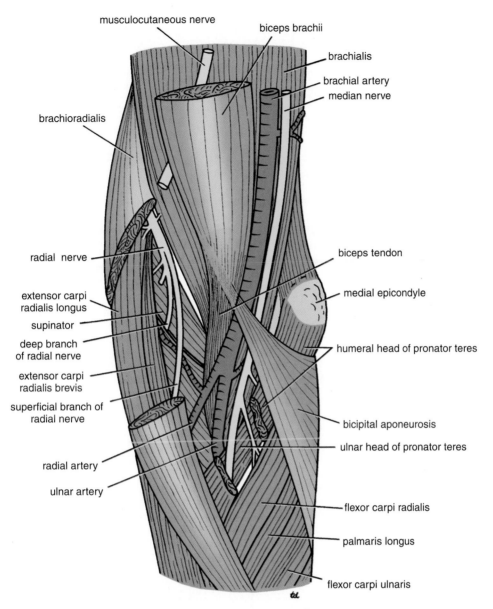

musculocutaneous nerve

biceps brachii

brachialis

brachial artery

median nerve

brachioradialis

radial nerve

biceps tendon

extensor carpi radialis longus

medial epicondyle

supinator

deep branch of radial nerve

humeral head of pronator teres

extensor carpi radialis brevis

superficial branch of radial nerve

bicipital aponeurosis

ulnar head of pronator teres

radial artery

ulnar artery

flexor carpi radialis

palmaris longus

flexor carpi ulnaris

**Figure 13-27**   Right cubital fossa.

| Table 13-12 | Muscles of the Arm | | | | |
| --- | --- | --- | --- | --- | --- |
| **Muscle** | **Origin** | **Insertion** | **Nerve Supply** | **Nerve Roots**[a] | **Action** |
| **Anterior Compartment** Biceps brachii | | | | | |
| Long head | Supraglenoid tubercle of scapula | Tuberosity of radius and bicipital aponeurosis into deep fascia of forearm | Musculocutaneous nerve | C5, **6** | Supinator of forearm and flexor of elbow joint; weak flexor of shoulder joint |
| Short head | Coracoid process of scapula | | | | |
| Coraco-brachialis | Coracoid process of scapula | Medial aspect of shaft of humerus | Musculocutaneous nerve | C5, **6**, 7 | Flexes arm and also weak adductor |
| Brachialis | Front of lower half of humerus | Coronoid process of ulna | Musculocutaneous nerve | C5, **6** | Flexor of elbow joint |
| **Posterior Compartment** Triceps | | | | | |
| Long head | Infraglenoid tubercle of scapula | | | | |
| Lateral head | Upper half of posterior surface of shaft of humerus | Olecranon process of ulna | Radial nerve | C6, 7, **8** | Extensor of elbow joint |
| Medial head | Lower half of posterior surface of shaft of humerus | | | | |

[a]The predominant nerve root supply is indicated by boldface type.
From Snell RS: Clinical Anatomy. 7th Ed. Philadelphia: Lippincott Williams & Wilkins, 2004, p. 513.

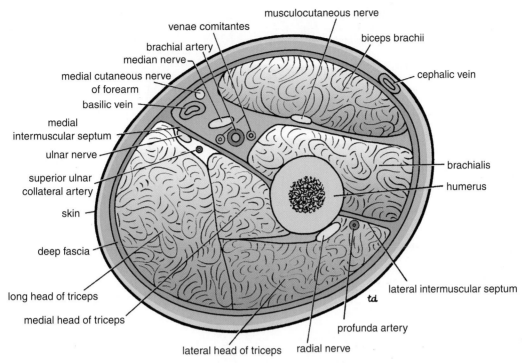

**Figure 13-28** Cross section of the upper arm just below the level of insertion of the deltoid muscle. Note the division of the arm by the humerus and the medial and lateral intermuscular septa into anterior and posterior compartments.

terosseous membrane and fibrous intermuscular septa, this fascial sheath divides the forearm into several compartments, with each compartment having its own muscles, nerves, and blood supply.

## Interosseous Membrane

The interosseous membrane is a strong membrane that unites the shafts of the radius and the ulna (Fig. 13-31). Because its fibers are taut, the forearm is most stable in the midprone position (position of function). The interosseous membrane provides attachment for the neighboring muscles.

## Wrist

### Flexor and Extensor Retinacula

The retinacula are bands of deep fascia that hold the long flexor and extensor tendons in position at the wrist (Fig. 13-32). The flexor retinaculum is attached medially to the pisiform bone and the hook of the hamate and laterally to the tubercle of the scaphoid and the trapezium. The extensor retinaculum is attached medially to the pisiform bone and the hook of the hamate and laterally to the distal end of the radius.

## Carpal Tunnel

The bones of the hand and the flexor retinaculum form the carpal tunnel (Fig. 13-32).

## Hand

The small muscles of the hand (Figs. 13-33, 13-34, and 13-35) are summarized in Table 13-16.

## Fibrous Flexor Sheaths

The anterior surface of each finger from the metacarpal head to the base of the distal phalanx is provided with a strong, fibrous sheath that is attached to the sides of the phalanges (Fig. 13-33). The sheath and the bones form a blind tunnel in which the long flexor tendons of the finger lie.

## Synovial Flexor Sheaths

In the hand, the tendons of the flexor digitorum superficialis and profundus muscles invaginate a common synovial sheath from the lateral side (Fig. 13-36). The medial part of this common sheath extends distally without interruption on the tendons of the little fingers. The lateral part of the sheath stops abruptly on the middle of the palm, and the dis-

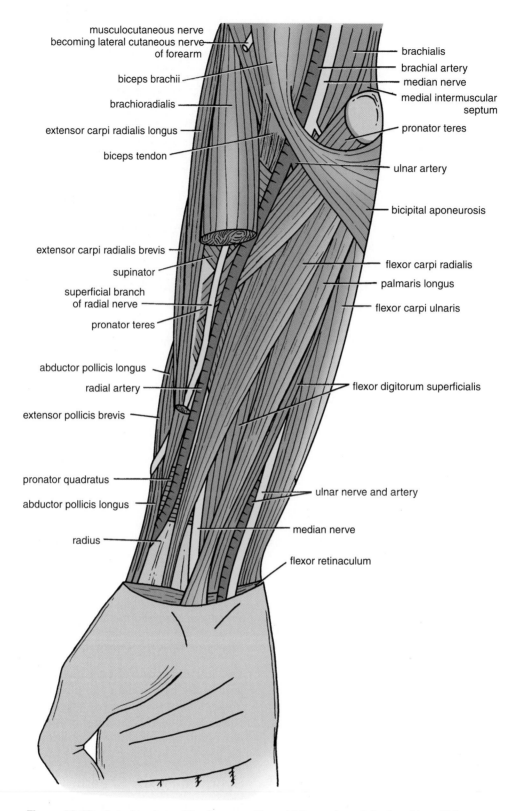

musculocutaneous nerve
becoming lateral cutaneous nerve
of forearm

biceps brachii

brachioradialis

extensor carpi radialis longus

biceps tendon

extensor carpi radialis brevis

supinator

superficial branch
of radial nerve

pronator teres

abductor pollicis longus

radial artery

extensor pollicis brevis

pronator quadratus

abductor pollicis longus

radius

brachialis

brachial artery

median nerve

medial intermuscular
septum

pronator teres

ulnar artery

bicipital aponeurosis

flexor carpi radialis

palmaris longus

flexor carpi ulnaris

flexor digitorum superficialis

ulnar nerve and artery

median nerve

flexor retinaculum

**Figure 13-29** Anterior view of the forearm. The middle portion of the brachioradialis muscle has been removed to display the superficial branch of the radial nerve and the radial artery.

| Table 13-13 | Muscles of the Anterior Fascial Compartment of the Forearm | | | | |
|---|---|---|---|---|---|
| **Muscle** | **Origin** | **Insertion** | **Nerve Supply** | **Nerve Roots**[a] | **Action** |
| Pronator teres<br>  Humeral head | Medial epicondyle<br>  of humerus | Lateral aspect of<br>  shaft of radius | Median<br>  nerve | C6, **7** | Pronation and<br>  flexion of<br>  forearm |
|   Ulnar head | Medial border of<br>  coronoid process<br>  of ulna | | | | |
| Flexor carpi<br>  radialis | Medial epicondyle<br>  of humerus | Bases of second<br>  and third<br>  metacarpal<br>  bones | Median<br>  nerve | C6, **7** | Flexes and abducts<br>  hand at wrist<br>  joint |
| Palmaris longus | Medial epicondyle<br>  of humerus | Flexor retinacu-<br>  lum and palmar<br>  aponeurosis | Median<br>  nerve | C7, 8 | Flexes hand |
| Flexor carpi<br>  ulnaris<br>  Humeral head | Medial epicondyle<br>  of humerus | Pisiform bone,<br>  hook of the<br>  hamate, base of<br>  fifth metacarpal<br>  bone | Ulnar nerve | C8; T1 | Flexes and<br>  adducts hand<br>  at wrist joint |
|   Ulnar head | Medial aspect of<br>  olecranon process<br>  and posterior<br>  border of ulna | | | | |
| Flexor digitorum<br>  superficialis<br>  Humeroulnar<br>  head | Medial epicondyle<br>  of humerus;<br>  medial border of<br>  coronoid process<br>  of ulna | Middle phalanx<br>  of medial four<br>  fingers | Median<br>  nerve | C7, **8**; T1 | Flexes middle<br>  phalanx of<br>  fingers and<br>  assists in flexing<br>  proximal pha-<br>  lanx and hand |
|   Radial head | Oblique line on<br>  anterior surface<br>  of shaft of radius | | | | |
| Flexor policis<br>  longus | Anterior surface<br>  of shaft of radius | Distal phalanx of<br>  thumb | Anterior in-<br>  terosseous<br>  branch of<br>  median<br>  nerve | **C8**; T1 | Flexes distal<br>  phalanx of<br>  thumb |

| Table 13-13 | (Continued) | | | | |
|---|---|---|---|---|---|
| **Muscle** | **Origin** | **Insertion** | **Nerve Supply** | **Nerve Roots**[a] | **Action** |
| Flexor digitorum profundus | Anteromedial surface of shaft of ulna | Distal phalanges of medial four fingers | Ulnar (medial half) and median (lateral half) nerves | **C8**; T1 | Flexes distal phalanx of fingers; then assists in flexion of middle and proximal phalanges and wrist |
| Pronator quadratus | Anterior surface of shaft of ulna | Anterior surface of shaft of radius | Anterior interosseous branch of median nerve | **C8**; T1 | Pronates forearm |

[a]The predominant nerve root supply is indicated by boldface type.
From Snell RS: Clinical Anatomy. 7th Ed. Philadelphia: Lippincott Williams & Wilkins, 2004, p. 535.

| Table 13-14 | Muscles of the Lateral Fascial Compartment of the Forearm | | | | |
|---|---|---|---|---|---|
| **Muscle** | **Origin** | **Insertion** | **Nerve Supply** | **Nerve Roots**[a] | **Action** |
| Brachioradialis | Lateral supracondylar ridge of humerus | Base of styloid process of radius | Radial nerve | C5, **6**, 7 | Flexes forearm at elbow joint; rotates forearm to the midprone position |
| Extensor carpi radialis longus | Lateral supracondylar ridge of humerus | Posterior surface of base of second metacarpal bone | Radial nerve | C6, 7 | Extends and abducts hand at wrist joint |

[a]The predominant nerve root supply is indicated by boldface type.
From Snell RS: Clinical Anatomy. 7th Ed. Philadelphia: Lippincott Williams & Wilkins, 2004, p. 536.

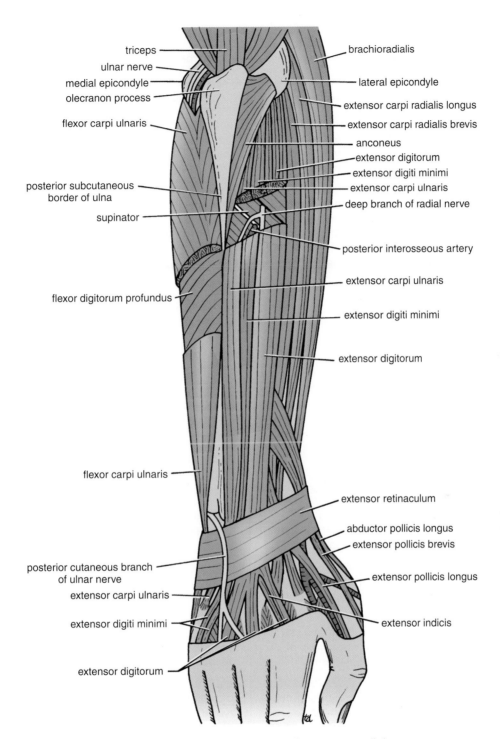

triceps
ulnar nerve
medial epicondyle
olecranon process
flexor carpi ulnaris

posterior subcutaneous
border of ulna
supinator

flexor digitorum profundus

flexor carpi ulnaris

posterior cutaneous branch
of ulnar nerve
extensor carpi ulnaris

extensor digiti minimi

extensor digitorum

brachioradialis

lateral epicondyle
extensor carpi radialis longus
extensor carpi radialis brevis
anconeus
extensor digitorum
extensor digiti minimi
extensor carpi ulnaris
deep branch of radial nerve

posterior interosseous artery

extensor carpi ulnaris

extensor digiti minimi

extensor digitorum

extensor retinaculum
abductor pollicis longus
extensor pollicis brevis
extensor pollicis longus

extensor indicis

**Figure 13-30**   Posterior view of the forearm. Parts of the extensor digitorum, extensor digiti minimi, and extensor carpi ulnaris have been removed to show the deep branch of the radial nerve and the posterior interosseous artery.

| Table 13-15 | | Muscles of the Posterior Fascial Compartment of the Forearm | | | |
|---|---|---|---|---|---|
| Muscle | Origin | Insertion | Nerve Supply | Nerve Roots[a] | Action |
| Extensor carpi radialis brevis | Lateral epicondyle of humerus | Posterior surface of base of third metacarpal bone | Deep branch of radial nerve | **C7**, 8 | Extends and abducts hand at wrist joint |
| Extensor digitorum | Lateral epicondyle of humerus | Middle and distal phalanges of medial four fingers | Deep branch of radial nerve | **C7**, 8 | Extends fingers and hand (see text for details) |
| Extensor digiti minimi | Lateral epicondyle of humerus | Extensor expansion of little finger | Deep branch of radial nerve | **C7**, 8 | Extends metacarpal phalangeal joint of little finger |
| Extensor carpi ulnaris | Lateral epicondyle of humerus | Base of fifth metacarpal bone | Deep branch of radial nerve | C7, **8** | Extends and adducts hand at wrist joint |
| Anconeus | Lateral epicondyle of humerus | Lateral surface of olecranon process of ulna | Radial nerve | C7, 8; T1 | Extends elbow joint |
| Supinator | Lateral epicondyle of humerus, anular ligament of proximal radioulnar joint, and ulna | Neck and shaft of radius | Deep branch of radial nerve | C5, 6 | Supination of forearm |
| Abductor pollicis longus | Posterior surface of shafts of radius and ulna | Base of first metacarpal bone | Deep branch of radial nerve | C7, **8** | Abducts and extends thumb |
| Extensor pollicis brevis | Posterior surface of shaft of radius | Base of proximal phalanx of thumb | Deep branch of radial nerve | C7, **8** | Extends metacarpophalangeal joints of thumb |
| Extensor pollicis longus | Posterior surface of shaft of ulna | Base of distal phalanx of thumb | Deep branch of radial nerve | C7, **8** | Extends distal phalanx of thumb |
| Extensor indicis | Posterior surface of shaft of ulna | Extensor expansion of index finger | Deep branch of radial nerve | C7, **8** | Extends metacarpophalangeal joint of index finger |

[a]The predominant nerve root supply is indicated by boldface type.
From Snell RS: Clinical Anatomy. 7th Ed. Philadelphia: Lippincott Williams & Wilkins, 2004, p. 536.

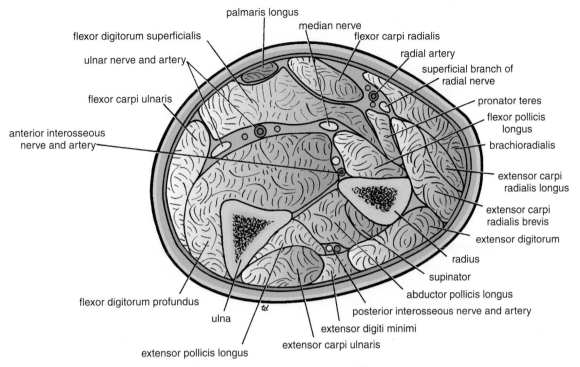

**Figure 13-31** Cross section of the forearm at the level of insertion of the pronator teres muscle.

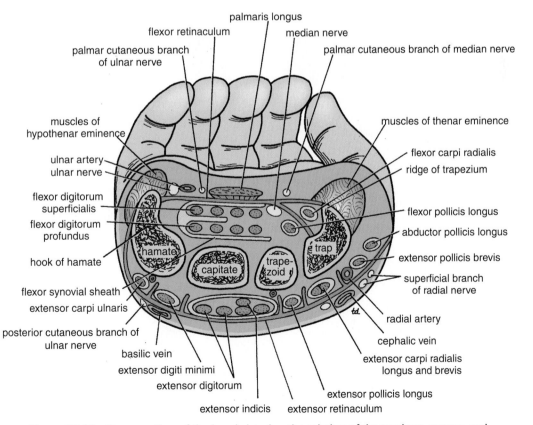

**Figure 13-32** Cross section of the hand showing the relation of the tendons, nerves, and arteries to the flexor and extensor retinacula.

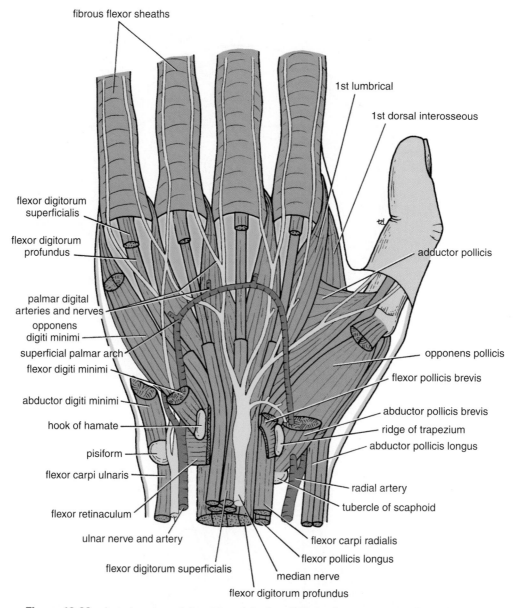

fibrous flexor sheaths

1st lumbrical

1st dorsal interosseous

flexor digitorum
superficialis

flexor digitorum
profundus

adductor pollicis

palmar digital
arteries and nerves

opponens
digiti minimi

superficial palmar arch

opponens pollicis

flexor digiti minimi

flexor pollicis brevis

abductor digiti minimi

abductor pollicis brevis

hook of hamate

ridge of trapezium

abductor pollicis longus

pisiform

flexor carpi ulnaris

radial artery

tubercle of scaphoid

flexor retinaculum

ulnar nerve and artery

flexor carpi radialis

flexor pollicis longus

flexor digitorum superficialis

median nerve

flexor digitorum profundus

**Figure 13-33**   Anterior view of the palm of the hand. The palmar aponeurosis and the
greater part of the flexor retinaculum have been removed to display the superficial palmar
arch, the median nerve, and the long flexor tendons. Segments of the tendons of the
flexor digitorum superficialis have been removed to show the underlying tendons of the
flexor digitorum profundus.

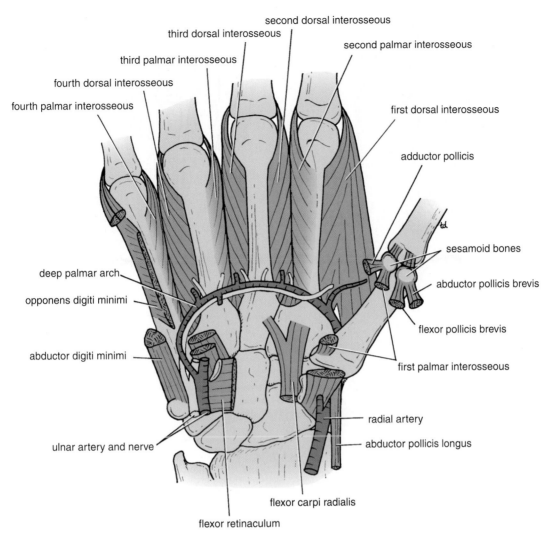

**Figure 13-34** Anterior view of the palm of the hand showing the deep palmar arch and the deep terminal branch of the ulnar nerve. The interossei are also shown.

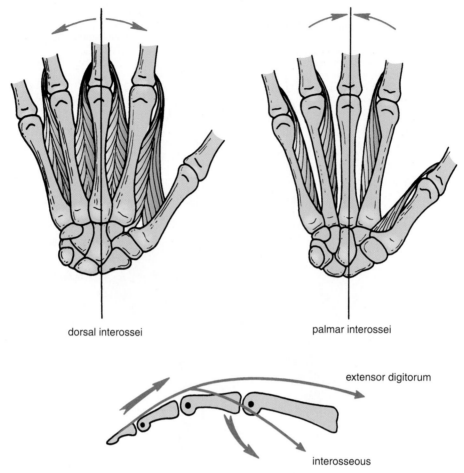

dorsal interossei

palmar interossei

extensor digitorum

interosseous

**Figure 13-35** Origins and insertion of the palmar and the dorsal interossei muscles. The actions of these muscles are also shown.

tal ends of the long flexor tendons of the index, the middle, and the ring fingers acquire **digital synovial sheaths** as they enter the fingers. The flexor pollicis longus tendon has its own synovial sheath that passes into the thumb. These sheaths allow the long tendons to move smoothly, with a minimum of friction, beneath the flexor retinaculum and the fibrous flexor sheaths.

## Insertion of the Long Flexor Tendons

Each tendon of the flexor digitorum superficialis is divided into two halves that pass around the profundus tendon and meet on its posterior surface, where partial decussation of the fibers occurs (Fig. 13-37). The superficialis tendon, having united again, then divides into two further slips, which are attached to the borders of the middle phalanx. Each tendon of the flexor digitorum profundus, having passed through the superficialis tendon, is inserted into the base of the distal phalanx.

## Insertion of the Long Extensor Tendons

The four tendons of the extensor digitorum fan out over the dorsum of the hand. The tendon to the index finger is joined on its **medial side** (Figs. 13-38 and 13-39) by the tendon of the extensor indicis. The tendon of the little finger is joined on its **medial side** by the two tendons of the extensor digiti minimi.

On the posterior surface of each finger, the extensor tendon widens to form the **extensor expansion.** Near the proximal interphalangeal joint, the extensor expansion then splits into three parts: a **central part**, which is inserted into the base of the middle phalanx, and **two lateral parts,** which converge to be inserted into the base of the distal phalanx.

In addition, the extensor expansion receives the tendon of insertion of the corresponding interosseous muscle on each side. Farther distally, the extensor expansion also receives the tendon of the lumbrical muscle on the lateral side.

| Table 13-16 | Small Muscles of the Hand | | | | |
|---|---|---|---|---|---|
| **Muscle** | **Origin** | **Insertion** | **Nerve Supply** | **Nerve Roots**[a] | **Action** |
| Palmaris brevis | Flexor retinaculum, palmar aponeurosis | Skin of palm | Superficial branch of ulnar nerve | C8; **T1** | Corrugates skin to improve grip of palm |
| Lumbricals (4) | Tendons of flexor digitorum profundus | Extensor expansion of medial four fingers | First and second, i.e., lateral two, median nerve; third and fourth deep branch of ulnar nerve | C8; **T1** | Flex metacarpophalangeal joints and extend interphalangeal joints of fingers except thumb |
| Interossei (8) Palmar (4) | First arises from base of first metacarpal; remaining three from anterior surface of shafts of second, fourth, and fifth metacarpals | Proximal phalanges of thumb, index, ring, and little fingers and dorsal extensor expansion of each finger | Deep branch of ulnar nerve | C8; **T1** | Palmar interossei adduct fingers toward center of third finger |
| Dorsal (4) | Contiguous sides of shafts of metacarpal bones | Proximal phalanges of index, middle and ring fingers and dorsal extensor expansion (Fig. 13-35) | Deep branch of ulnar nerve | C8; **T1** | Dorsal interossei abduct fingers from center of third finger; both palmar and dorsal flex metacarpophalangeal joints and extend interphalangeal joints |
| **Short Muscles of Thumb** | | | | | |
| Abductor pollicis brevis | Scaphoid, trapezium, flexor retinaculum | Base of proximal phalanx of thumb | Median nerve | **C8**; T1 | Abduction of thumb |
| Flexor pollicis brevis | Flexor retinaculum | Base of proximal phalanx of thumb | Median nerve | **C8**; T1 | Flexes metacarpophalangeal joint of thumb |
| Opponens pollicis | Flexor retinaculum | Shaft of metacarpal bone of thumb | Median nerve | **C8**; T1 | Pulls thumb medially and forward across palm |

**Table 13-16** *(Continued)*

| Muscle | Origin | Insertion | Nerve Supply | Nerve Roots[a] | Action |
|--------|--------|-----------|--------------|------------------|--------|
| Adductor pollicis | Oblique head; second and third metacarpal bones; transverse head; third metacarpal bone | Base of proximal phalanx of thumb | Deep branch of ulnar nerve | C8; **T1** | Adduction of thumb |
| **Short Muscles of Little Finger** | | | | | |
| Abductor digiti minimi | Pisiform bone | Base of proximal phalanx of little finger | Deep branch of ulnar nerve | C8; **T1** | Abducts little finger |
| Flexor digiti minimi | Flexor retinaculum | Base of proximal phalanx of little finger | Deep branch of ulnar nerve | C8; **T1** | Flexes little finger |
| Opponens digiti minimi | Flexor retinaculum | Medial border fifth metacarpal bone | Deep branch of ulnar nerve | C8; **T1** | Pulls fifth metacarpal forward as in cupping the palm |

[a] The predominant nerve root supply is indicated by boldface type.
From Snell RS: Clinical Anatomy. 7th Ed. Philadelphia: Lippincott Williams & Wilkins, 2004, p. 546.

## Palmar Aponeurosis

In the palm, the deep fascia is greatly thickened to protect the underlying tendons, nerves, and blood vessels and is called the **palmar aponeurosis.** It is continuous proximally with the palmaris longus tendon, and it is attached to the flexor retinaculum. The distal edge of the aponeurosis divides at the bases of the fingers into four slips that pass into the fingers. It is also continuous with the fasciae covering the thenar and the hypothenar eminences.

# Muscles of the Lower Limb

## Gluteal Region

The gluteal region is bounded superiorly by the iliac crest and inferiorly by the fold of the buttock. This region consists largely of the gluteal muscles and a thick layer of superficial fascia.

The muscles of the gluteal region (Figs. 13-40 and 13-41) are described in Table 13-17.

## Fascia

### Superficial Fascia
The superficial fascia is thick (especially in women) and is impregnated with large quantities of fat.

### Deep Fascia
The deep fascia is continuous below with the fascia lata of the thigh, and it splits to enclose the gluteus maximus muscle.

### Important Ligaments
The sacrotuberous and the sacrospinous ligaments stabilize the sacrum and prevent its rotation by the weight of the vertebral column.

### Sacrotuberous Ligament
The sacrotuberous ligament connects the posteroinferior iliac spine, the lateral part of the sacrum, and the coccyx to the ischial tuberosity (Fig. 13-41).

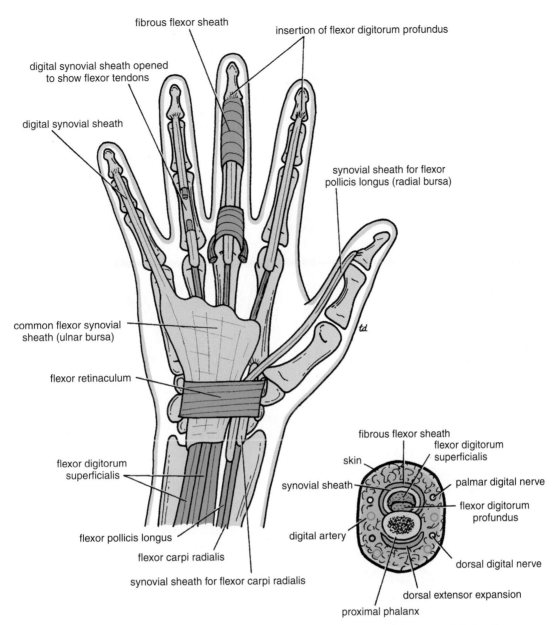

fibrous flexor sheath

insertion of flexor digitorum profundus

digital synovial sheath opened
to show flexor tendons

digital synovial sheath

synovial sheath for flexor
pollicis longus (radial bursa)

common flexor synovial
sheath (ulnar bursa)

flexor retinaculum

flexor digitorum
superficialis

flexor pollicis longus

flexor carpi radialis

synovial sheath for flexor carpi radialis

fibrous flexor sheath

skin

flexor digitorum
superficialis

synovial sheath

palmar digital nerve

digital artery

flexor digitorum
profundus

dorsal digital nerve

dorsal extensor expansion

proximal phalanx

**Figure 13-36**   Anterior view of the palm of the hand showing the flexor synovial sheaths.
Cross section of a finger is also shown.

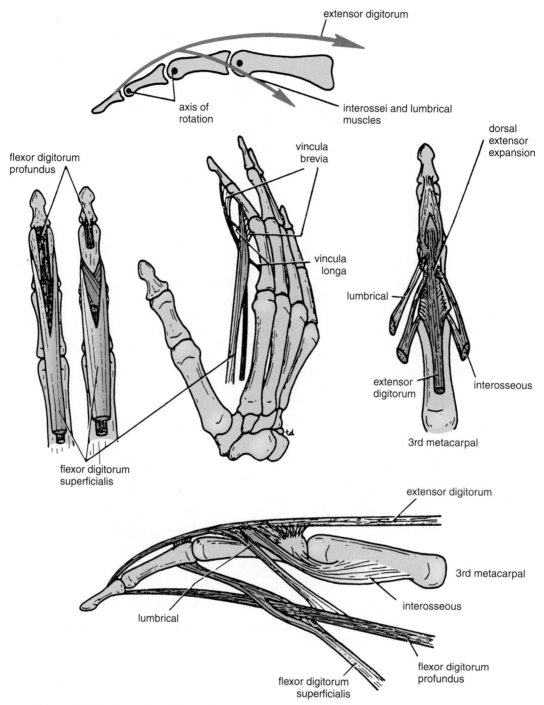

**Figure 13-37**  Insertions of long flexor and extensor tendons in the fingers. Insertions of the lumbrical and interossei muscles are also shown. The uppermost figure illustrates the action of the lumbrical and interossei muscles in flexing the metacarpophalangeal joints and extending the interphalangeal joints.

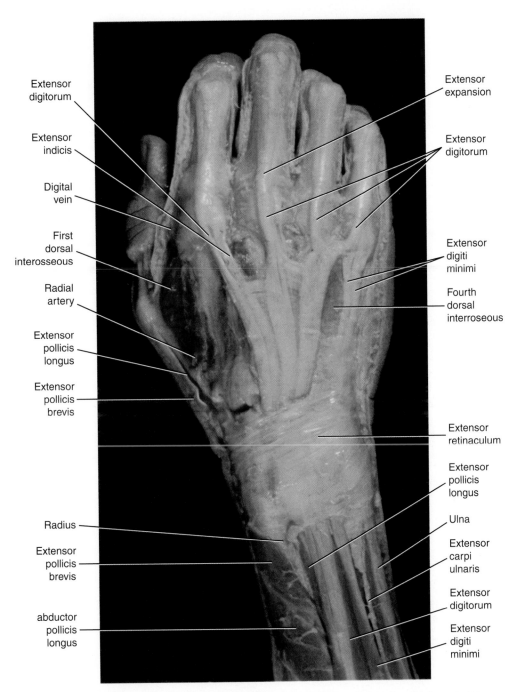

Extensor
digitorum

Extensor
indicis

Digital
vein

First
dorsal
interosseous

Radial
artery

Extensor
pollicis
longus

Extensor
pollicis
brevis

Radius

Extensor
pollicis
brevis

abductor
pollicis
longus

Extensor
expansion

Extensor
digitorum

Extensor
digiti
minimi

Fourth
dorsal
interroseous

Extensor
retinaculum

Extensor
pollicis
longus

Ulna

Extensor
carpi
ulnaris

Extensor
digitorum

Extensor
digiti
minimi

**Figure 13-38** Dissection of the dorsal surface of the right hand showing the long extensor tendons and the extensor retinaculum.

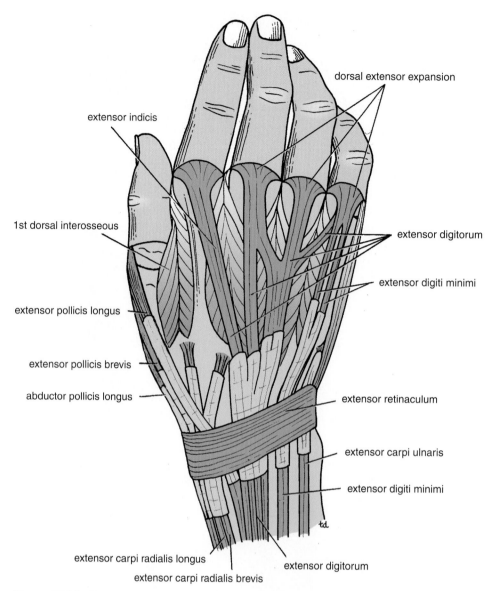

**Figure 13-39**   Dorsal surface of the hand showing the long extensor tendons and their synovial sheaths.

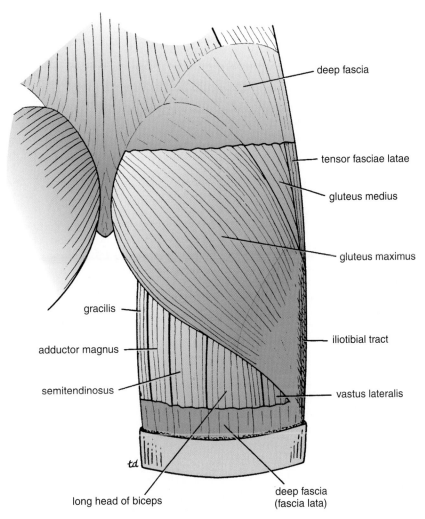

deep fascia

tensor fasciae latae

gluteus medius

gluteus maximus

gracilis

iliotibial tract

adductor magnus

semitendinosus

vastus lateralis

*td*

long head of biceps

deep fascia
(fascia lata)

**Figure 13-40**   Right gluteus maximus muscle.

*Sacrospinous Ligament*

The sacrospinous ligament connects the lateral part of the sacrum and the coccyx to the spine of the ischium (Fig. 13-41).

**Important Foramina**
*Greater Sciatic Foramen*

The greater sciatic foramen is formed by the conversion of the greater sciatic notch of the hip bone into a foramen by the presence of the sacrotuberous and the sacrospinous ligaments.

The following structures pass through the foramen:

■ Piriformis muscle
■ Sciatic nerve
■ Posterior cutaneous nerve of the thigh
■ Superior and inferior gluteal nerves
■ Nerves to obturator internus and quadratus femoris muscles

■ Pudendal nerve
■ Superior and inferior gluteal arteries and veins
■ Internal pudendal artery and vein

*Lesser Sciatic Foramen*

The lesser sciatic foramen is formed by the conversion of the lesser sciatic notch of the hip bone into a foramen by the presence of the sacrotuberous and the sacrospinous ligaments.

The following structures pass through the foramen:

■ Tendon of the obturator internus muscle
■ Nerve to the obturator internus muscle
■ Pudendal nerve
■ Internal pudendal artery and vein

## Thigh

The muscles of the anterior fascial compartment (Fig. 13-42) are described in Table 13-18. The muscles of the medial

## Table 13-17    Muscles of the Gluteal Region

| Muscle | Origin | Insertion | Nerve Supply | Nerve Root[a] | Action |
|---|---|---|---|---|---|
| Gluteus maximus | Outer surface of ilium, sacrum, coccyx, sacrotuberous ligament | Iliotibial tract and gluteal tuberosity of femur | Inferior gluteal nerve | L5; **S1, 2** | Extends and laterally rotates hip joint; through iliotibial tract, it extends knee joint |
| Gluteus medius | Outer surface of ilium | Lateral surface of greater trochanter of femur | Superior gluteal nerve | **L5**; S1 | Abducts thigh at hip joint; tilts pelvis when walking to permit opposite leg to clear ground |
| Gluteus minimus | Outer surface of ilium | Anterior surface of greater trochanter of femur | Superior gluteal nerve | **L5**; S1 | Abducts thigh at hip joint; tilts pelvis when walking to permit opposite leg to clear ground |
| Tensor fasciae latae | Iliac crest | Iliotibial tract | Superior gluteal nerve | L4; 5 | Assists gluteus maximus in extending the knee joint |
| Piriformis | Anterior surface of sacrum | Upper border of greater trochanter of femur | First and second sacral nerves | L5; **S1**, 2 | Lateral rotator of thigh at hip joint |
| Obturator internus | Inner surface of obturator membrane | Upper border of greater trochanter of femur | Sacral plexus | L5; **S1** | Lateral rotator of thigh at hip joint |
| Gemellus superior | Spine of ischium | Upper border of greater trochanter of femur | Sacral plexus | L5; S1 | Lateral rotator of thigh at hip joint |
| Gemellus inferior | Ischial tuberosity | Upper border of greater trochanter of femur | Sacral plexus | L5; S1 | Lateral rotator of thigh at hip joint |
| Quadratus femoris | Lateral border of ischial tuberosity | Quadrate tubercle of femur | Sacral plexus | L5; S1 | Lateral rotator of thigh at hip joint |

[a] The predominant nerve root supply is indicated by bold face type.
From Snell RS: Clinical Anatomy. 7th Ed. Philadelphia: Lippincott Williams & Wilkins, 2004, p. 611.

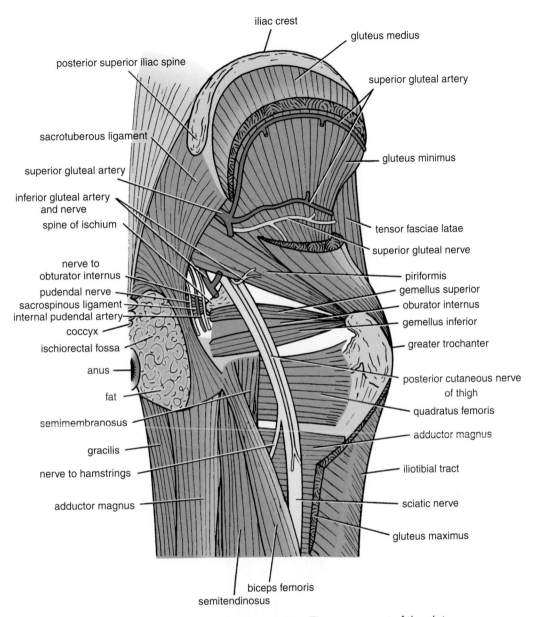

**Figure 13-41** Structures in the right gluteal region. The greater part of the gluteus maximus and part of the gluteus medius have been removed.

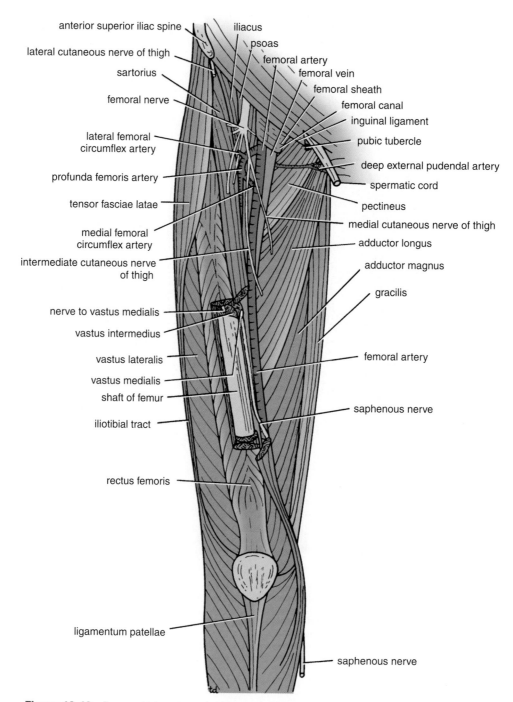

**Figure 13-42** Femoral triangle and adductor (subsartorial) canal in the right lower limb.

| Table 13-18 | Muscles of the Anterior Fascial Compartment of the Thigh | | | | |
|---|---|---|---|---|---|
| **Muscle** | **Origin** | **Insertion** | **Nerve Supply** | **Nerve Root**[a] | **Action** |
| Sartorius | Anterior superior iliac spine | Upper medial surface of shaft of tibia | Femoral nerve | L2, 3 | Flexes, abducts, laterally rotates thigh at hip joint; flexes and medially rotates leg at knee joint |
| Iliacus | Iliac fossa of hip bone | With psoas into lesser trochanter of femur | Femoral nerve | **L2,** 3 | Flexes thigh on trunk; if thigh is fixed, it flexes the trunk on the thigh as in sitting up from lying down |
| Psoas | Transverse processes, bodies, and intervertebral discs of the twelfth thoracic and five lumbar vertebrae | With iliacus into lesser trochanter of femur | Lumbar plexus | **L1, 2,** 3 | Flexes thigh on trunk; if thigh is fixed, it flexes the trunk on thigh as in sitting up from lying down |
| Pectineus | Superior ramus of pubis | Upper end of linea aspera of shaft of femur | Femoral nerve | **L2,** 3 | Flexes and adducts thigh at hip joint |
| Quadriceps femoris Rectus femoris | Straight head: anterior inferior iliac spine  Reflected head: ilium above acetabulum | Quadriceps tendon into patella, then via ligamentum patellae into tubercle of tibia | Femoral nerve | L2, **3, 4** | Extension of leg at knee joint; flexes thigh at hip joint |
| Vastus lateralis | Upper end and shaft of femur | Quadriceps tendon into patella, then via ligamentum patellae into tubercle of tibia | Femoral nerve | L2, **3, 4** | Extension of leg at knee joint |
| Vastus medialis | Upper end and shaft of femur | Quadriceps tendon into patella, then via ligamentum patellae into tubercle of tibia | Femoral nerve | L2, **3, 4** | Extension of leg at knee joint; stabilizes patella |
| Vastus inter- medius | Anterior and lateral surfaces of shaft of femur | Quadriceps tendon into patella, then via ligamentum patellae into tubercle of tibia | Femoral nerve | L2, **3, 4** | Extension of leg at knee joint; articularis genus retracts synovial membrane |

[a] The predominant nerve root supply is indicated by boldface type.
From Snell RS: Clinical Anatomy. 7th Ed. Philadelphia: Lippincott Williams & Wilkins, 2004, p. 625.

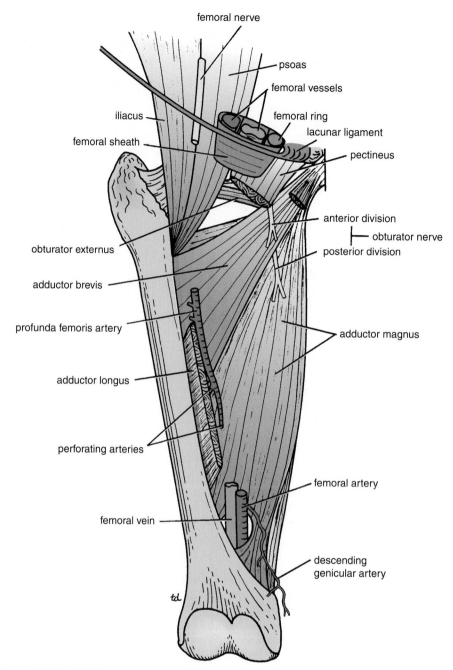

**Figure 13-43** Relationship between the obturator nerve and the adductor muscles in the right lower limb.

**Table 13-19** Muscles of the Medial Fascial Compartment of the Thigh

| Muscle | Origin | Insertion | Nerve Supply | Nerve Root[a] | Action |
|---|---|---|---|---|---|
| Gracilis | Inferior ramus of pubis, ramus of ischium | Upper part of shaft of tibia on medial surface | Obturator nerve | **L2**, 3 | Adducts thigh at hip joint; flexes leg at knee joint |
| Adductor longus | Body of pubis, medial to pubic tubercle | Posterior surface of shaft of femur (linea aspera) | Obturator nerve | L2, **3, 4** | Adducts thigh at hip joint and assists in lateral rotation |
| Adductor brevis | Inferior ramus of pubis | Posterior surface of shaft of femur (linea aspera) | Obturator nerve | L2, **3, 4** | Adducts thigh at hip joint and assists in lateral rotation |
| Adductor magnus | Inferior ramus of pubis, ramus of ischium, ischial tuberosity | Posterior surface of shaft of femur, adductor tubercle of femur | Adductor portion: obturator nerve<br><br>Hamstring portion: sciatic nerve | L2, **3, 4** | Adducts thigh at hip joint and assists in lateral rotation; hamstring portion extends thigh at hip joint |
| Obturator externus | Outer surface of obturator membrane and pubic and ischial rami | Medial surface of greater trochanter | Obturator nerve | L3, **4** | Laterally rotates thigh at hip joint |

[a] The predominant nerve root supply is indicated by boldface type.
From Snell RS: Clinical Anatomy. 7th Ed. Philadelphia: Lippincott Williams & Wilkins, 2004, p. 633.

fascial compartment (Fig. 13-43) are described in Table 13-19, and the muscles of the posterior fascial compartment (Fig. 13-44) are described in Table 13-20.

## Deep Fascia of the Thigh (Fascia Lata)

The deep fascia encloses the thigh as a trouser leg would (Fig. 13-45). The upper end is attached to the pelvis and its associated ligaments.

### Iliotibial Tract
The iliotibial tract is a thickening of the fascia lata on its lateral side (Fig. 13-45). It is attached above to the iliac tubercle and below to the lateral condyle of the tibia. It receives the insertion of the greater part of the gluteus maximus and the tensor fasciae latae muscles.

### Saphenous Opening
The saphenous opening is a gap in the deep fascia in the front of the thigh and just below the inguinal ligament (Fig. 13-46). It allows passage of the great saphenous vein, some small branches of the femoral artery, and lymph vessels. The opening is filled with loose connective tissue called the **cribriform fascia**.

## Fascial Compartments of the Thigh

Three fascial septa pass from the inner aspect of the deep facial sheath of the thigh to the linea aspera of the femur (Fig. 13-45). By this means, the thigh is divided into three compartments, with each having muscles, nerves, and arteries. The compartments are as follows:

- Anterior with the femoral nerve
- Medial (adductor) with the obturator nerve
- Posterior with the sciatic nerve

## Femoral Triangle

The femoral triangle is situated in the upper part of the front of the thigh (Fig. 13-42). Its boundaries are as follows:

- **Superiorly:** The inguinal ligament
- **Laterally:** The sartorius muscle
- **Medially:** The adductor longus muscle

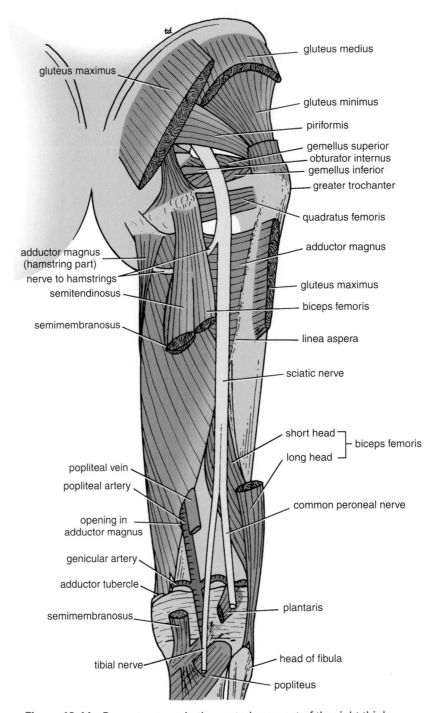

**Figure 13-44**   Deep structures in the posterior aspect of the right thigh.

**Table 13-20** **Muscles of the Posterior Fascial Compartment of the Thigh**

| Muscle | Origin | Insertion | Nerve Supply | Nerve Root[a] | Action |
|---|---|---|---|---|---|
| Biceps femoris | Long head: ischial tuberosity<br><br>Short head: linea aspera, lateral supracondylar ridge of shaft of femur | Head of fibula | Long head: tibial portion of sciatic nerve<br>Short head: common peroneal portion of sciatic nerve | L5; **S1**, 2 | Flexes and laterally rotates leg at knee joint; long head also extends thigh at hip joint |
| Semi-tendinosus | Ischial tuberosity | Upper part of medial surface of shaft of tibia | Tibial portion of sciatic nerve | **L5; S1**, 2 | Flexes and medially rotates leg at knee joint; extends thigh at hip joint |
| Semimem-branosus | Ischial tuberosity | Medial condyle of tibia | Tibial portion of sciatic nerve | **L5; S1**, 2 | Flexes and medially rotates leg at knee joint; extends thigh at hip joint |
| Adductor magnus (hamstring portion) | Ischial tuberosity | Adductor tubercle of femur | Tibial portion of sciatic nerve | L2, **3, 4** | Extends thigh at hip joint |

[a] The predominant nerve root supply is indicated by boldface type.
From Snell RS: Clinical Anatomy. 7th Ed. Philadelphia: Lippincott Williams & Wilkins, 2004, p. 636.

The femoral triangle contains the terminal part of the femoral nerve and its branches, the femoral sheath, the femoral artery and its branches, the femoral vein and its tributaries, and the inguinal lymph nodes.

## Femoral Sheath

The femoral sheath is a downward protrusion from the abdomen into the thigh of the fascia transversalis and the fascia iliaca. The sheath surrounds the femoral blood vessels and lymph vessels for approximately 1 in. (2.5 cm) below the inguinal ligament (Fig. 13-46). As the **femoral artery** enters the thigh beneath the inguinal ligament, it occupies the **lateral compartment** of the sheath. The **femoral vein** occupies the **intermediate compartment**, and the **lymph vessels** (and usually one lymph node) occupy the most **medial compartment.**

### Femoral Canal
The femoral canal (Fig. 13-46) is the small, medial compartment of the femoral sheath occupied by the lymphatics. It is approximately 0.5 in. (1.3 cm) in length. It is also a potentially weak area in the wall of the abdomen; a protrusion

of peritoneum could be forced down the femoral canal to form a femoral hernia.

### Femoral Ring
The femoral ring is the upper opening of the femoral. It is filled by a plug of extra peritoneal fat called the **femoral septum.**

### Important Relations

- **Anteriorly:** Inguinal ligament
- **Posteriorly:** Superior ramus of the pubis and the pectineal ligament
- **Laterally:** Femoral vein
- **Medially:** Lacunar ligament (an extension of the inguinal ligament; see Fig. 13-17).

## Adductor (Subsartorial) Canal

The adductor canal is an intermuscular cleft on the medial aspect of the middle third of the thigh beneath the sartorius muscle (Fig. 13-42). The posterior wall is formed by

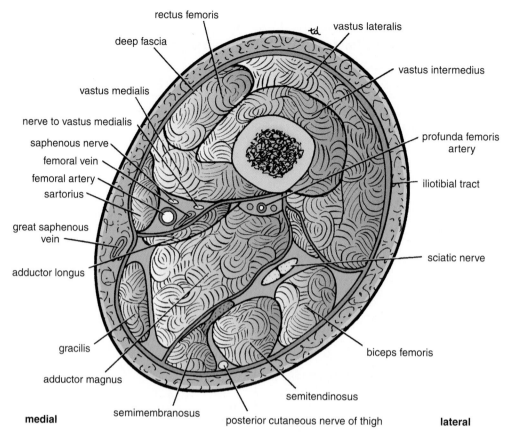

**Figure 13-45** Transverse section through the middle of the right thigh as seen from above.

the adductor magnus muscle, the lateral wall by the vastus medialis muscle, and the anteromedial wall by the sartorius muscle and fascia. The canal contains the femoral artery and vein, the deep lymph vessels, the saphenous nerve, and the nerve to the vastus medialis muscle.

## Knee Region

### Popliteal Fossa

The popliteal fossa is a diamond-shaped, intermuscular space at the back of the knee (Fig. 13-47). It contains the popliteal vessels, the small saphenous vein, the common peroneal and tibial nerves, the posterior cutaneous nerve of the thigh, connective tissue, and lymph nodes.

#### Boundaries

- **Laterally:** The biceps femoris muscle above and the lateral head of the gastrocnemius and the plantaris muscles below.
- **Medially:** The semimembranosus and semitendinosus muscles above and the medial head of the gastrocnemius muscle below.

## Leg

The muscles of the anterior fascial compartment (Fig. 13-48) are described in Table 13-21. The muscles of the lateral fascial compartment (Fig. 13-49) are described in Table 13-22, and the muscles of the posterior fascial compartment (Fig. 13-50) are described in Table 13-23. The muscle on the dorsum of the foot (Fig. 13-51) is described in Table 13-21.

### Fascial Compartments of the Leg

The deep fascia surrounds the leg and is continuous above with the deep fascia of the thigh (Fig. 13-52). It is attached to the anterior and the medial borders of the tibia, and two intermuscular septa pass from its deep aspect to be attached to the fibula. Together with the interosseous membrane, the septa divide up the leg into three compartments, with each having its own muscles, blood supply, and nerve supply. The compartments are as follows:

- Anterior with the deep peroneal nerve
- Lateral (peroneal) with the superficial peroneal nerve
- Posterior with the tibial nerve

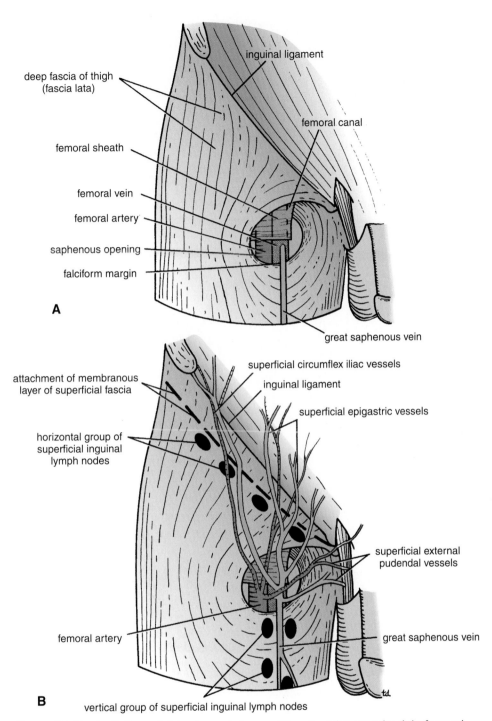

inguinal ligament

deep fascia of thigh
(fascia lata)

femoral canal

femoral sheath

femoral vein

femoral artery

saphenous opening

falciform margin

**A**

great saphenous vein

attachment of membranous
layer of superficial fascia

superficial circumflex iliac vessels

inguinal ligament

superficial epigastric vessels

horizontal group of
superficial inguinal
lymph nodes

superficial external
pudendal vessels

femoral artery

great saphenous vein

**B**

vertical group of superficial inguinal lymph nodes

**Figure 13-46  A, B.** Superficial veins, arteries, and lymph nodes over the right femoral triangle. Note the saphenous opening in the deep fascia and its relationship to the femoral sheath. Note also the line of attachment of the membranous layer of superficial fascia to the deep fascia, about a finger's breadth below the inguinal ligament.

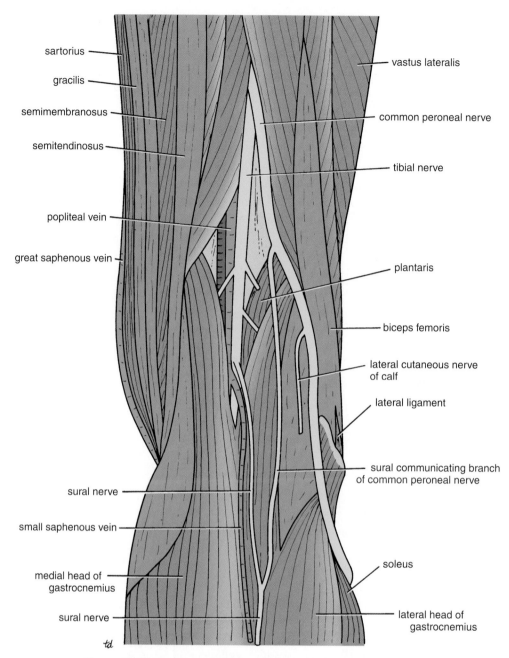

sartorius

gracilis

semimembranosus

semitendinosus

popliteal vein

great saphenous vein

sural nerve

small saphenous vein

medial head of
gastrocnemius

sural nerve

vastus lateralis

common peroneal nerve

tibial nerve

plantaris

biceps femoris

lateral cutaneous nerve
of calf

lateral ligament

sural communicating branch
of common peroneal nerve

soleus

lateral head of
gastrocnemius

**Figure 13-47** Boundaries and contents of the right popliteal fossa.

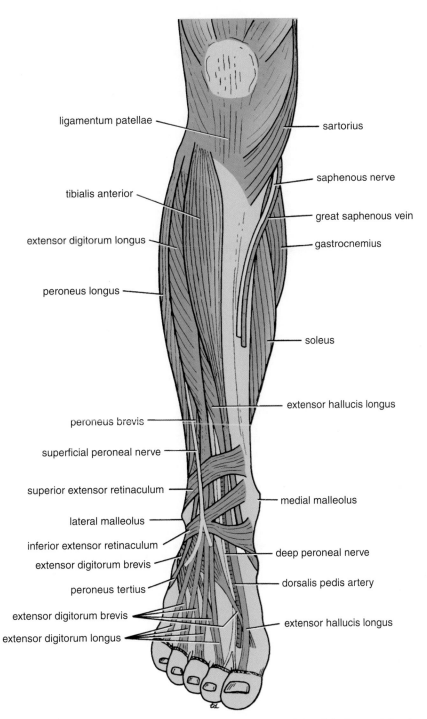

ligamentum patellae

tibialis anterior

extensor digitorum longus

peroneus longus

peroneus brevis

superficial peroneal nerve

superior extensor retinaculum

lateral malleolus

inferior extensor retinaculum

extensor digitorum brevis

peroneus tertius

extensor digitorum brevis

extensor digitorum longus

sartorius

saphenous nerve

great saphenous vein

gastrocnemius

soleus

extensor hallucis longus

medial malleolus

deep peroneal nerve

dorsalis pedis artery

extensor hallucis longus

**Figure 13-48** Structures in the anterior and lateral aspects of the right leg and the dorsum of the foot.

| Table 13-21 | Muscles of the Anterior Fascial Compartment of the Leg |
|---|---|

| Muscle | Origin | Insertion | Nerve Supply | Nerve Root[a] | Action |
|---|---|---|---|---|---|
| Tibialis anterior | Lateral surface of shaft of tibia and interosseous membrane | Medial cuneiform and base of first metatarsal bone | Deep peroneal nerve | **L4,** 5 | Extends[b] foot at ankle joint; inverts foot at subtalar and transverse tarsal joints; holds up medial longitudinal arch of foot |
| Extensor digitorum-longus | Anterior surface of shaft of fibula | Extensor expansion of lateral four toes | Deep peroneal nerve | L5; S1 | Extends toes; extends foot at ankle joint |
| Peroneus tertius | Anterior surface of shaft of fibula | Base of fifth metatarsal bone | Deep peroneal nerve | L5; S1 | Extends foot at ankle joint; everts foot at subtalar and transverse tarsal joints |
| Extensor hallucis longus | Anterior surface of shaft of fibula | Base of distal phalanx of great toe | Deep peroneal nerve | L5; S1 | Extends big toe; extends foot at ankle joint; inverts foot at subtalar and transverse tarsal joints |
| Extensor digitorum brevis | Calcaneum | By four tendons into the proximal phalanx of big toe and long extensor tendons to second, third, and fourth toes | Deep peroneal nerve | S1, 2 | Extends toes |

[a] The predominant nerve root supply is indicated by boldface type.
[b] Extension, or dorsiflexion, of the ankle is the movement of the foot away from the ground.
From Snell RS: Clinical Anatomy. 7th Ed. Philadelphia: Lippincott Williams & Wilkins, 2004, p. 660.

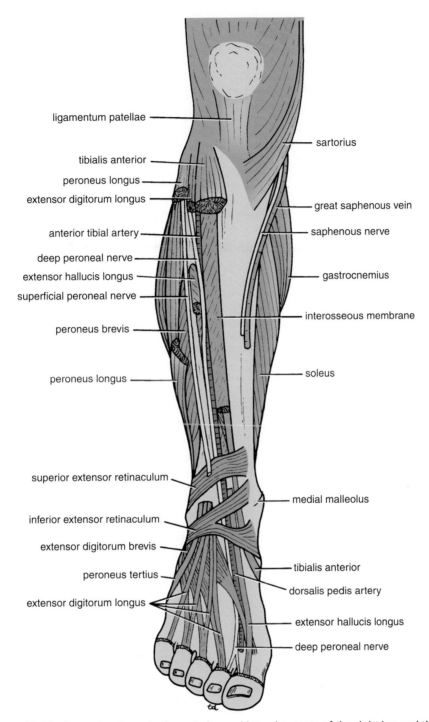

ligamentum patellae

tibialis anterior

peroneus longus

extensor digitorum longus

anterior tibial artery

deep peroneal nerve

extensor hallucis longus

superficial peroneal nerve

peroneus brevis

peroneus longus

superior extensor retinaculum

inferior extensor retinaculum

extensor digitorum brevis

peroneus tertius

extensor digitorum longus

sartorius

great saphenous vein

saphenous nerve

gastrocnemius

interosseous membrane

soleus

medial malleolus

tibialis anterior

dorsalis pedis artery

extensor hallucis longus

deep peroneal nerve

**Figure 13-49** Deep structures in the anterior and lateral aspects of the right leg and the dorsum of the foot.

| Table 13-22 | Muscles of the Lateral Fascial Compartment of the Leg | | | | |
|---|---|---|---|---|---|
| **Muscle** | **Origin** | **Insertion** | **Nerve Supply** | **Nerve Root**[a] | **Action** |
| Peroneus longus | Lateral surface of shaft of fibula | Base of first metatarsal and the medial cuneiform | Superficial peroneal nerve | **L5; S1**, 2 | Plantar flexes foot at ankle joint; everts foot at subtalar and transverse tarsal joints; supports lateral longitudinal and transverse arches of foot |
| Peroneus brevis | Lateral surface of shaft of fibula | Base of fifth metatarsal bone | Superficial peroneal nerve | **L5; S1**, 2 | Plantar flexes foot at ankle joint; everts foot at subtalar and transverse tarsal joint; supports lateral longitudinal arch of foot |

[a] The predominant nerve root supply is indicated by boldface type.
From Snell RS: Clinical Anatomy. 7th Ed. Philadelphia: Lippincott Williams & Wilkins, 2004, p. 661.

## Interosseous Membrane

The interosseous membrane is thick and strong and binds the tibia and the fibula together and provides attachment for the muscles (Fig. 13-52).

# Ankle

## Retinacula

The retinacula are thickenings of the deep fascia that keep the long tendons around the ankle joint in position and act as pulleys (Figs. 13-48 and 13-53).

### Superior Extensor Retinaculum

The superior extensor retinaculum is attached to the distal ends of the anterior borders of the fibula and the tibia (Fig. 13-48).

### Inferior Extensor Retinaculum

The inferior extensor retinaculum is a Y-shaped band located in front of the ankle joint (Figs. 13-48, 13-53, and 13-54).

### Flexor Retinaculum

The flexor retinaculum extends from the medial malleolus to the medial surface of the calcaneum (Figs. 13-54 and 13-55). It binds the deep muscles of the back of the leg to the back of the medial malleolus as they pass forward to enter the sole.

### Superior Peroneal Retinaculum

The superior peroneal retinaculum connects the lateral malleolus to the lateral surface of the calcaneum (Figs. 13-54 and 13-55). It binds the tendons of the peroneus longus and brevis muscles to the back of the lateral malleolus.

### Inferior Peroneal Retinaculum

The inferior peroneal retinaculum binds the tendons of the peroneus longus and brevis muscles to the lateral side of the calcaneum (Figs. 13-53 and 13-55).

# Sole of the Foot

The muscles of the sole (Figs. 13-56, 13-57, 13-58, 13-59, and 13-60) are usually described in four layers (from inferior to superior). These muscles are described in Table 13-24.

## Deep Fascia

### Plantar Aponeurosis

The plantar aponeurosis is a triangular thickening of the deep fascia that protects the underlying nerves, blood vessels, and muscles. Its apex is attached to the medial and lateral tubercles of the calcaneum. The base of the aponeurosis divides into five slips that pass into the toes.

## Arches of the Foot

The arches of the foot and their muscular support are fully described on page 424.

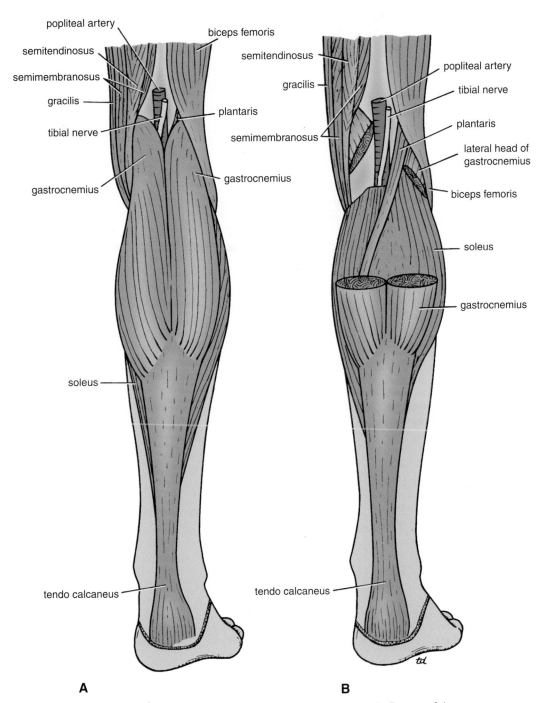

**A**

**B**

**Figure 13-50** Structures in the posterior aspect of the right leg. In **B,** part of the gastrocnemius has been removed.

| Table 13-23 | Muscles of the Posterior Fascial Compartment of the Leg | | | | |
|---|---|---|---|---|---|
| Muscle | Origin | Insertion | Nerve Supply | Nerve Root[a] | Action |
| **Superficial Group** | | | | | |
| Gastrocnemius | Lateral head from lateral condyle of femur and medial head from above medial condyle | Via tendo calcaneus into posterior surface of calcaneum | Tibial nerve | **S1**, 2 | Plantar flexes foot at ankle joint; flexes knee joint |
| Plantaris | Lateral supra-condylar ridge of femur | Posterior surface of calcaneum | Tibial nerve | **S1**, 2 | Plantar flexes foot at ankle joint; flexes knee joint |
| Soleus | Shafts of tibia and fibula | Via tendo calcaneus into posterior surface of calcaneum | Tibial nerve | S1, **2** | Together with gastrocnemius and plantaris is the powerful plantar flexor of the ankle joint; provides main propulsive force in walking and running |
| **Deep Group** | | | | | |
| Popliteus | Lateral surface of lateral condyle of femur | Posterior surface of shaft of tibia above soleal line | Tibial nerve | L4, 5; S1 | Flexes leg at knee joint; unlocks knee joint by lateral rotation of femur on tibia and slackens ligaments of joint |
| Flexor digitorum longus | Posterior surface of shaft of tibia | Bases of distal phalanges of lateral four toes | Tibial nerve | **S2**, 3 | Flexes distal phalanges of lateral four toes; plantar flexes foot at ankle joint; supports medial and lateral longitudinal arches of foot |
| Flexor hallucis longus | Posterior surface of shaft of fibula | Base of distal phalanx of big toe | Tibial nerve | **S2**, 3 | Flexes distal phalanx of big toe; plantar flexes foot at ankle joint; supports medial longitudinal arch of foot |
| Tibialis posterior | Posterior surface of shafts of tibia and fibula and interosseous membrane | Tuberosity of navicular bone and other neighbouring bones | Tibial nerve | L4, 5 | Plantar flexes foot at ankle joint; inverts foot at subtalar and transverse tarsal joints; supports medial longitudinal arch of foot |

[a] The predominant nerve root supply is indicated by boldface type.
From Snell RS: Clinical Anatomy. 7th Ed. Philadelphia: Lippincott Williams & Wilkins, 2004, p. 665.

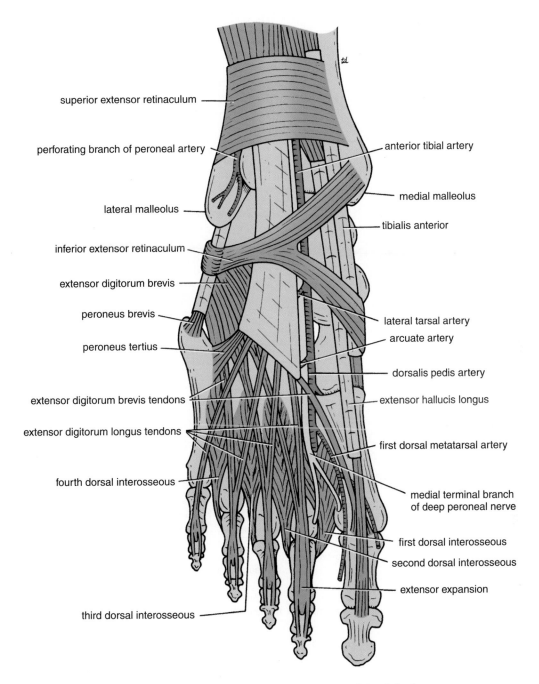

superior extensor retinaculum

perforating branch of peroneal artery

lateral malleolus

inferior extensor retinaculum

extensor digitorum brevis

peroneus brevis

peroneus tertius

extensor digitorum brevis tendons

extensor digitorum longus tendons

fourth dorsal interosseous

third dorsal interosseous

anterior tibial artery

medial malleolus

tibialis anterior

lateral tarsal artery

arcuate artery

dorsalis pedis artery

extensor hallucis longus

first dorsal metatarsal artery

medial terminal branch of deep peroneal nerve

first dorsal interosseous

second dorsal interosseous

extensor expansion

**Figure 13-51**　Structures in the dorsal aspect of the right foot.

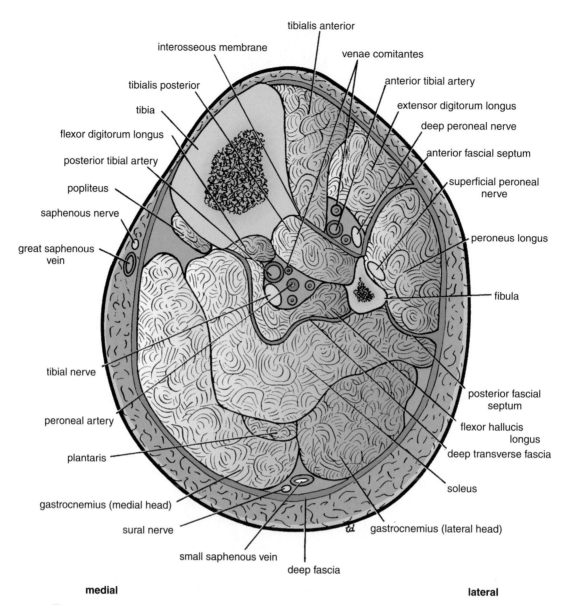

**Figure 13-52** Transverse section through the middle of the right leg as seen from above.

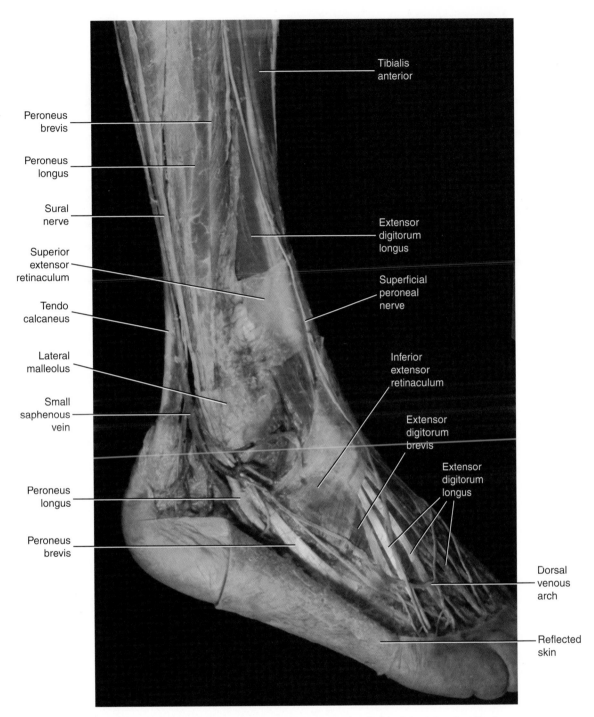

Peroneus
brevis

Peroneus
longus

Sural
nerve

Superior
extensor
retinaculum

Tendo
calcaneus

Lateral
malleolus

Small
saphenous
vein

Peroneus
longus

Peroneus
brevis

Tibialis
anterior

Extensor
digitorum
longus

Superficial
peroneal
nerve

Inferior
extensor
retinaculum

Extensor
digitorum
brevis

Extensor
digitorum
longus

Dorsal
venous
arch

Reflected
skin

**Figure 13-53** Dissection showing the structures passing behind the lateral malleolus. Note the position of the retinacula.

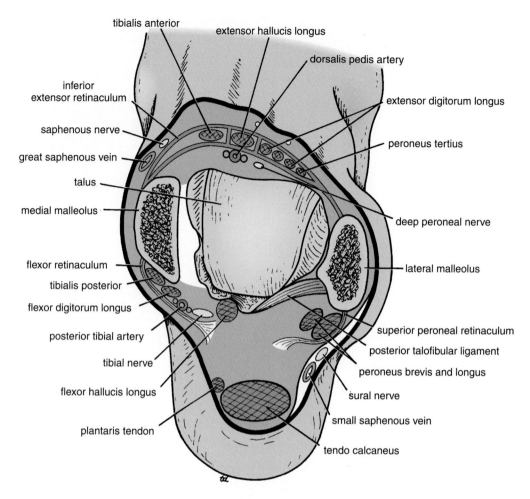

**Figure 13-54**  Relations of the right ankle joint.

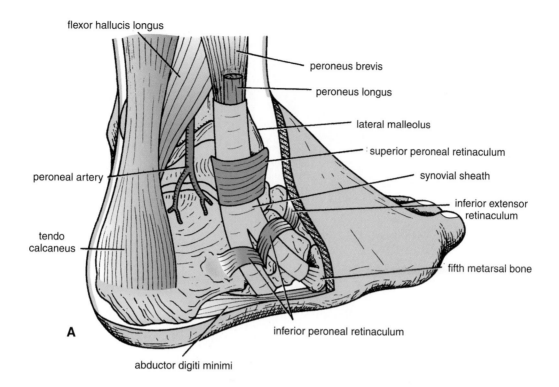

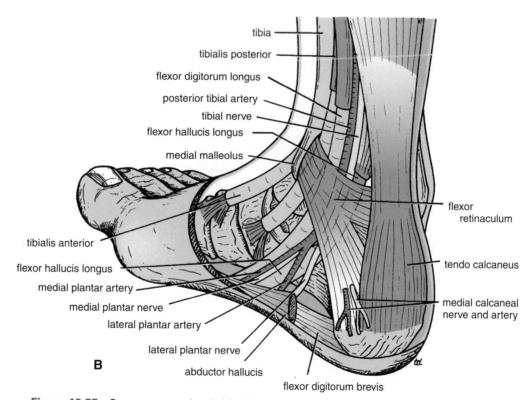

**Figure 13-55**   Structures passing behind the lateral malleolus (**A**) and the medial malleolus (**B**). Synovial sheaths of the tendons are shown in blue. Note the positions of the retinacula.

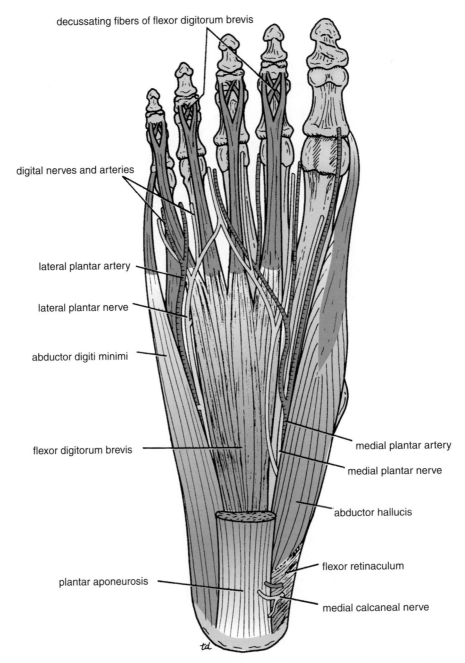

decussating fibers of flexor digitorum brevis

digital nerves and arteries

lateral plantar artery

lateral plantar nerve

abductor digiti minimi

flexor digitorum brevis

medial plantar artery

medial plantar nerve

abductor hallucis

flexor retinaculum

plantar aponeurosis

medial calcaneal nerve

**Figure 13-56**   First layer of the plantar muscles of the right foot. Medial and lateral plantar arteries and nerves are also shown.

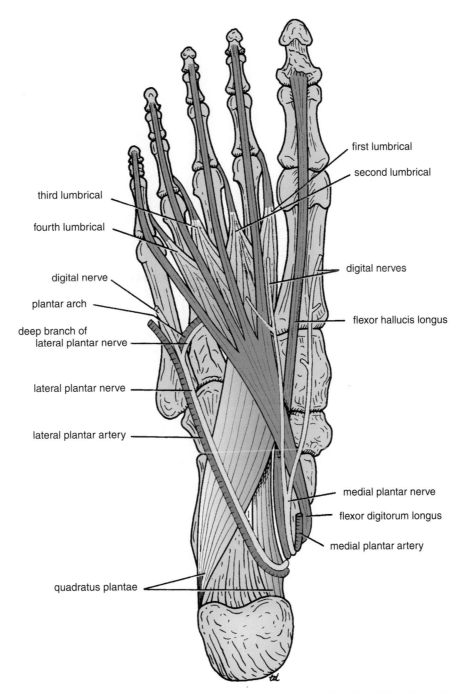

first lumbrical

second lumbrical

third lumbrical

fourth lumbrical

digital nerves

digital nerve

plantar arch

flexor hallucis longus

deep branch of
lateral plantar nerve

lateral plantar nerve

medial plantar nerve

lateral plantar artery

flexor digitorum longus

medial plantar artery

quadratus plantae

**Figure 13-57**   Second layer of the plantar muscles of the right foot. Medial and lateral plantar arteries and nerves are also shown.

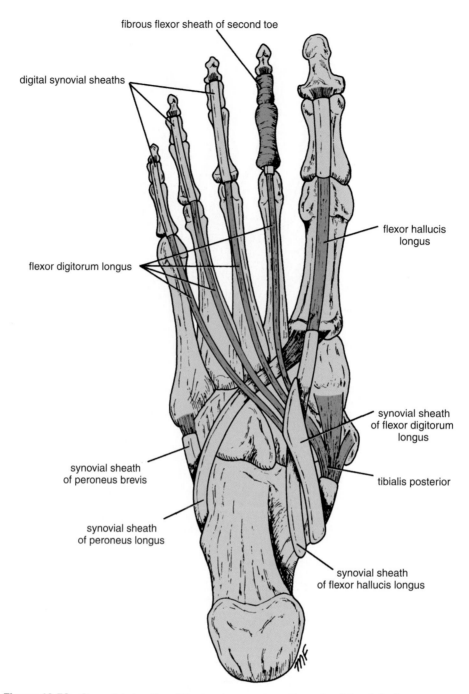

**Figure 13-58** Synovial sheaths of the tendons seen on the sole of the right foot.

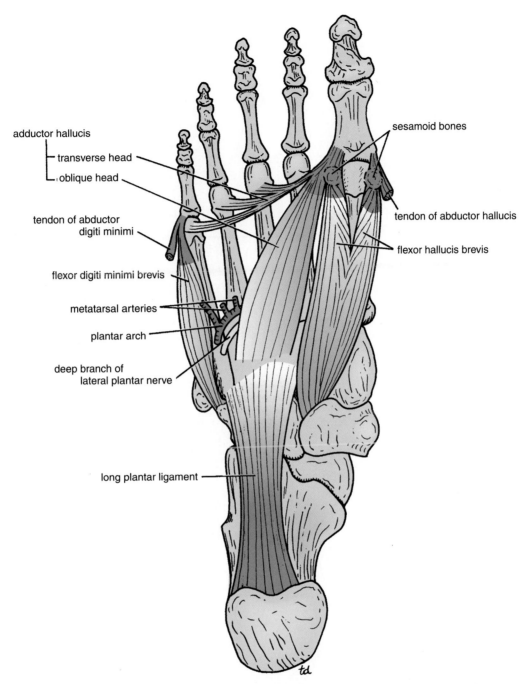

adductor hallucis
— transverse head
— oblique head

tendon of abductor
digiti minimi

flexor digiti minimi brevis

metatarsal arteries

plantar arch

deep branch of
lateral plantar nerve

long plantar ligament

sesamoid bones

tendon of abductor hallucis

flexor hallucis brevis

**Figure 13-59**  Third layer of the plantar muscles of the right foot. The deep branch of the lateral plantar nerve and the plantar arterial arch are also shown.

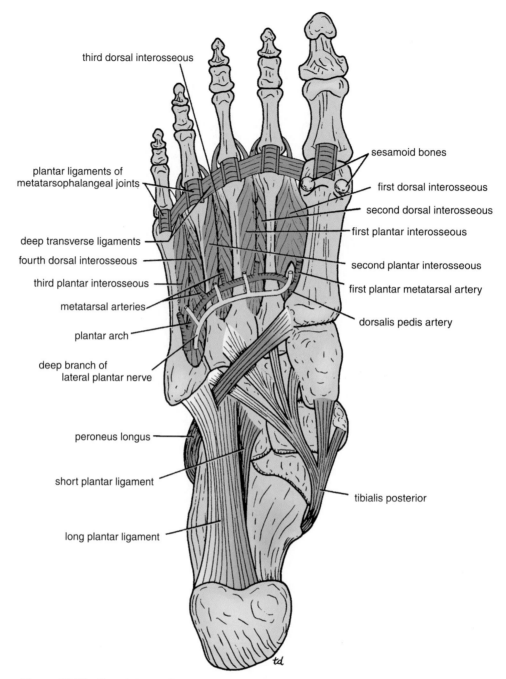

third dorsal interosseous

plantar ligaments of
metatarsophalangeal joints

deep transverse ligaments

fourth dorsal interosseous

third plantar interosseous

metatarsal arteries

plantar arch

deep branch of
lateral plantar nerve

peroneus longus

short plantar ligament

long plantar ligament

sesamoid bones

first dorsal interosseous

second dorsal interosseous

first plantar interosseous

second plantar interosseous

first plantar metatarsal artery

dorsalis pedis artery

tibialis posterior

**Figure 13-60**  Fourth layer of the plantar muscles of the right foot. The deep branch of the lateral plantar nerve and the plantar arterial arch are also shown. Note the deep transverse ligaments.

## Table 13-24    Muscles of the Sole of the Foot

| Muscle | Origin | Insertion | Nerve Supply | Nerve Root[a] | Action |
|---|---|---|---|---|---|
| **First Layer** | | | | | |
| Abductor hallucis | Medial tuberosity of calcaneum and flexor retinaculum | Base of proximal phalanx of big toe | Medial plantar nerve | S2, **3** | Flexes and abducts big toe; braces medial longitudinal arch |
| Flexor digitorum brevis | Medial tubercle of calcaneum | Four tendons to four lateral toes—inserted into borders of middle phalanx; tendons perforated by those of flexor digitorum longus | Medial plantar nerve | S2, **3** | Flexes lateral four toes; braces medial and lateral longitudinal arches |
| Abductor digiti minimi | Medial and lateral tubercles of calcaneum | Base of proximal phalanx of fifth toe | Lateral plantar nerve | S2, **3** | Flexes and abducts fifth toe; braces lateral longitudinal arch |
| **Second Layer** | | | | | |
| Quadratus plantae | Medial and lateral sides of calcaneum | Tendon of flexor digitorum longus | Lateral plantar nerve | S2, **3** | Assists flexor digitorum longus in flexing lateral four toes |
| Lumbricals (4) | Tendons of flexor digitorum longus | Dorsal extensor expansion; bases of proximal phalanges of lateral four toes | First lumbrical: medial plantar nerve; remainder: lateral plantar nerve | S2, **3** | Extends toes at interphalangeal joints |
| Flexor digitorum longus tendon | See Table 13-23 | | | | |
| Flexor hallucis longus tendon | See Table 13-23 | | | | |
| **Third Layer** | | | | | |
| Flexor hallucis brevis | Cuboid, lateral cuneiform, tibialis posterior insertion | Medial tendon into medial side of base of proximal phalanx of big toe; lateral tendon into lateral side of base of proximal phalanx of big toe | Medial plantar nerve | S2, **3** | Flexes metatarsophalangeal joint of big toe; supports medial longitudinal arch |

| Table 13-24 | (continued) | | | | |
|---|---|---|---|---|---|
| **Muscle** | **Origin** | **Insertion** | **Nerve Supply** | **Nerve Root**[a] | **Action** |
| Adductor hallucis | Oblique head bases of second, third, and fourth metatarsal bones; transverse head from plantar ligaments | Lateral side of base of proximal phalanx of big toe | Deep branch lateral plantar nerve | S2, **3** | Flexes metatarsophalangeal joint of big toe; holds together metatarsal bones |
| Flexor digiti minimi brevis | Base of fifth metatarsal bone | Lateral side of base of proximal phalanx of little toe | Lateral plantar nerve | S2, **3** | Flexes metatarsophalangeal joint of little toe |
| **Fourth Layer** | | | | | |
| Interossei Dorsal (4) | Adjacent sides of metatarsal bones | Bases of proximal phalanges—first: medial side of second toe; remainder: lateral sides of second, third, and fourth toes—also dorsal extensor expansion | Lateral plantar nerve | S2, **3** | Abduction of toes; flexes metarsophalangeal joints and extends interphalangeal joints |
| Plantar (3) | Inferior surfaces of third, fourth, and fifth metatarsal bones | Medial side of bases of proximal phalanges of lateral three toes | Lateral plantar nerve | S2, **3** | Adduction of toes; flexes metarsophalangeal joints and extends interphalangeal joints |
| Peroneus longus tendon | See Table 13-22 | | | | |
| Tibialis posterior tendon | See Table 13-23 | | | | |

[a] The predominant nerve root supply is indicated by boldface type.
From Snell RS: Clinical Anatomy. 7th Ed. Philadelphia: Lippincott Williams & Wilkins, 2004, p. 669.

# Review Questions

## General Muscle Information

### Matching Questions

**Match each structure listed below with a structure or occurrence with which it is most closely associated. Each lettered answer may be used more than once.**

1. Superficial fascia

2. Deep fascia

3. Skeletal muscle
   A. Divides up interior of limbs into compartments
   B. Adipose tissue
   C. Tendon spindles
   D. None of the above

**For each type of muscle action listed below, select the most appropriate definition.**

4. Prime mover

5. Fixator

6. Synergist

7. Antagonist
   A. A muscle that contracts isometrically to stabilize the origin of another muscle
   B. A muscle that opposes the action of a flexor muscle
   C. A muscle that is chiefly responsible for a particular movement
   D. A muscle that prevents unwanted movements in an intermediate joint so that another muscle can cross that joint and act primarily on a distal joint
   E. A muscle that opposes the action of a prime mover

## Head and Neck Muscles

### Completion Questions

**Select the phrase that BEST completes each statement.**

8. The occipitofrontalis muscle
   A. raises the skin on the back of the neck.
   B. raises the eyebrows.
   C. raises the upper eyelid.
   D. moves the ears upward.
   E. moves the scalp laterally.

9. The sternocleidomastoid muscle
   A. extends the head.
   B. extends the neck.
   C. flexes the neck.
   D. flexes the head.
   E. (the two muscles acting together) extend the head and flex the neck.

10. The scalenus anterior muscle is attached below to the
    A. first rib.
    B. third rib.
    C. transverse process of the seventh cervical vertebra.
    D. second rib.
    E. manubrium sterni.

11. The genioglossus muscle _____ the tongue.
    A. retracts
    B. depresses
    C. elevates
    D. protrudes
    E. changes the shape of

12. The hyoglossus muscle
    A. changes the shape of the tongue.
    B. elevates the tongue.
    C. depresses the tongue.
    D. protrudes the tongue.
    E. retracts the tongue upward and backward.

13. The styloglossus muscle
    A. protrudes the tongue.
    B. depresses the tongue.
    C. retracts the tongue upward and backward.
    D. changes the shape of the tongue.
    E. elevates the tongue.

14. The palatoglossus muscle
    A. retracts the tongue upward and backward.
    B. elevates the tongue.
    C. changes the shape of the tongue.
    D. depresses the tongue.
    E. protrudes the tongue.

### Multiple Choice Questions

**Select the BEST answer for each question.**

15. Which of the following muscles elevates the soft palate during swallowing?
    A. Tensor veli palatini
    B. Palatoglossus
    C. Palatopharyngeus
    D. Levator veli palatini
    E. Salpingopharyngeus

16. Which of the following muscles partially inserts on the articular disc of the temporomandibular joint?
    A. Medial pterygoid
    B. Anterior fibers of the temporalis

C. Masseter

D. Posterior fibers of the temporalis

E. Lateral pterygoid

## Muscles of the Back

## Multiple Choice Question

**Select the BEST answer for the question.**

17. The deep muscles of the back (postvertebral muscles) have the following characteristics **except** which?
    A. They form a broad thick column of muscle tissue situated in the hollow on each side of the spinous processes of the vertebrae.
    B. All the muscles are supplied by the anterior rami of the spinal nerves.
    C. They are composed of many separate muscles of different length.
    D. They are well developed in humans since they are largely responsible for maintaining the vertical posture.
    E. Their postural tone is the major factor for the maintenance of the normal curves of the vertebral column.

## Muscles of the Thoracic Wall

## Multiple Choice Question

**Select the BEST answer for the question.**

18. The following statements concerning the diaphragm are correct **except** which?
    A. The right crus provides a muscular sling around the esophagus and possibly prevents regurgitation of stomach contents into the esophagus.
    B. On contraction, the diaphragm raises the intra-abdominal pressure and assists in the return of the venous blood to the right atrium of the heart.
    C. The level of the diaphragm is higher in the recumbent position than in the standing position.
    D. On contraction, the central tendon descends, reducing the intrathoracic pressure.
    E. The diaphragm receives its motor nerve supply from the lower six intercostal nerves.

## Abdominal Muscles

## Multiple Choice Questions

**Select the BEST answer for each question.**

19. The following statements concerning the conjoint tendon are correct **except** which?
    A. It is continuous with the inguinal ligament.
    B. It is formed by the fusion of the aponeuroses of the transversus abdominis and internal oblique muscles.
    C. It is attached medially to the linea alba.
    D. It is attached to the pubic crest and the pectineal line.
    E. It may bulge forward in a direct inguinal hernia.

20. The following statements are correct concerning the muscles forming the posterior abdominal wall **except** which?
    A. The psoas major muscle has a fascial sheath that extends down into the thigh as far as the lesser trochanter of the femur.
    B. The quadratus lumborum is covered anteriorly by fascia that forms the lateral arcuate ligament.
    C. The iliacus muscle is innervated by the femoral nerve.
    D. The transversus abdominis muscle does not form part of the posterior abdominal wall.
    E. The diaphragm does contribute to the musculature on the posterior abdominal wall.

21. The rectus abdominis muscle has the following characteristics **except** which?
    A. It is a long strap muscle.
    B. It is separated from the muscle of the opposite side by the linea alba.
    C. It is inserted into the eighth, ninth, and tenth ribs.
    D. Its lateral margin forms a curved ridge termed the linea semilunaris.
    E. The muscle is divided into distinct segments by tendinous intersections.

22. The external oblique muscle of the anterior abdominal wall has the following characteristics **except** which?
    A. It is inserted into the xiphoid process, the linea alba, the pubic crest, the pubic tubercle, and the anterior half of the iliac crest.
    B. A triangular defect in the aponeurosis is called the superficial inguinal ring.
    C. The lower border of the aponeurosis forms the inguinal ligament.
    D. It contributes to the anterior wall of the rectus sheath.
    E. It assists the diaphragm during expiration by relaxing as the diaphragm descends.

## Pelvic Muscles

## Multiple Choice Questions

**Select the BEST answer for each question.**

23. The following statements concerning the muscles and fascia in the pelvis are correct **except** which?
    A. The pelvic diaphragm is strong and has no openings.
    B. In the pelvis, the fascia is divided into parietal and visceral layers.
    C. The iliococcygeus muscle arises from a thickening of the obturator internus fascia.
    D. The levator ani muscle is innervated by the perineal branch of the fourth sacral nerve and from the perineal branch of the pudendal nerve.
    E. The visceral layer of pelvic fascia forms important ligaments that help support the uterus.

24. The following statements concerning the motor nerve supply of the muscles of the pelvic walls are correct **except** which?
    A. The sacral nerves, or plexus, supply the obturator internus muscle.
    B. The obturator nerve supplies the piriformis muscle.
    C. The sacral nerves, or plexus, supply the iliococcygeus muscle.
    D. The sacral nerves, or plexus, supply the coccygeus muscle.
    E. The perineal branch of the fourth sacral nerve and the perineal branch of the pudendal nerve supply the levator ani muscle.

## Perineal Muscles

## Multiple Choice Question

### Select the BEST answer for the question.

25. The urogenital diaphragm is formed by the following structures **except** which?
    A. Deep transverse perineal muscle
    B. Perineal membrane
    C. Sphincter urethrae
    D. Colles' fascia (membranous layer of superficial fascia)
    E. Parietal pelvic fascia covering the upper surface of the sphincter urethrae muscle

## Muscles of the Upper Limb

## Multiple Choice Questions

### Select the BEST answer for each question.

26. The following tendons are inserted into the base of the proximal phalanx of the thumb **except** which?
    A. Extensor pollicis brevis
    B. Abductor pollicis longus
    C. Oblique head of adductor pollicis
    D. Flexor pollicis brevis
    E. First palmar interosseous

27. The following muscles abduct the hand at the wrist joint **except** which?
    A. Flexor carpi radialis
    B. Abductor pollicis longus
    C. Extensor carpi radialis longus
    D. Extensor digiti minimi
    E. Extensor pollicis longus

28. The tendons of the following muscles form the rotator cuff **except** which?
    A. Teres major
    B. Supraspinatus
    C. Subscapularis
    D. Teres minor
    E. Infraspinatus

29. The quadrangular space is bounded by the following structures **except** which?
    A. Surgical neck of the humerus
    B. Long head of triceps
    C. Deltoid
    D. Teres major
    E. Teres minor

30. The following structures are attached to the greater tuberosity of the humerus **except** which?
    A. Supraspinatus muscle
    B. Coracohumeral ligament
    C. Teres minor muscle
    D. Infraspinatus muscle
    E. Subscapularis muscle

## Muscles of the Lower Limb

## Multiple Choice Questions

### Select the BEST answer for each question.

31. In walking, the hip bone of the suspended leg is raised by which of the following muscles acting on the supported side of the body?
    A. Gluteus maximus
    B. Obturator internus
    C. Gluteus medius
    D. Obturator externus
    E. Quadratus femoris

32. Which of the following muscles is a flexor of the thigh?
    A. Superior gemellus
    B. Adductor longus
    C. Gracilis
    D. Psoas
    E. Obturator internus

33. Which of the following muscles dorsiflexes the foot at the ankle joint?
    A. Peroneus longus
    B. Extensor digitorum brevis
    C. Tibialis posterior
    D. Extensor hallucis brevis
    E. Tibialis anterior

34. Unlocking of the knee joint to permit flexion is caused by the action of the
    A. vastus medialis muscle.
    B. popliteus muscle.
    C. gastrocnemius muscle.
    D. biceps femoris muscle.
    E. articularis genu muscle.

35. The following structures contribute to the boundaries of the popliteal fossa **except** which?
    A. The semimembranosus muscle
    B. The plantaris
    C. The biceps femoris muscle

D. The medial head of the gastrocnemius muscle

E. The soleus

36. The following structures pass through the greater sciatic foramen **except** which?
    A. The superior gluteal artery
    B. The sciatic nerve
    C. The obturator internus tendon
    D. The pudendal nerve
    E. The inferior gluteal vein

37. The floor of the femoral triangle is formed by the following muscles **except** which?
    A. The pectineus

B. The adductor brevis

C. The iliacus

D. The psoas

E. The adductor longus

38. The foot is inverted by the following muscles **except** which?
    A. The tibialis anterior
    B. The extensor hallucis longus
    C. The extensor digitorum longus
    D. The peroneus tertius
    E. The tibialis posterior

# Answers and Explanations

1. **B** is correct. The superficial fascia contains adipose tissue, which is more plentiful in the female.

2. **A** is correct. The deep fascia forms fibrous septa that divides up the interior of limbs into compartments (Figs. 13-28, 13-31, 13-45, and 13-52).

3. **C** is correct. Skeletal muscle contains sensory endings, the tendon spindles, and muscle spindles, whose function is to detect the degree of tension in a muscle (Fig. 13-3).

4. **C** is correct. A prime mover is a muscle that is chiefly responsible for a particular movement.

5. **A** is correct. A fixator is a muscle that contracts isometrically to stabilize the origin of another muscle.

6. **D** is correct. A synergist is a muscle that prevents unwanted movements in an intermediate joint so that another muscle can cross that joint and act primarily on a distal joint.

7. **E** is correct. An antagonist is a muscle that opposes the action of a prime mover.

8. **B** is correct. The occipitofrontalis muscle raises the eyebrows (Fig. 13-6).

9. **E** is correct. The sternocleidomastoid muscles, when acting together, extend the head at the atlanto-occipital joint and flex the neck at the cervical intervertebral joints.

10. **A** is correct. The scalenus anterior muscle is inserted below onto the first rib (Fig. 13-10).

11. **D** is correct. The genioglossus muscle protrudes the tongue. Remember that contraction of the right genioglossus muscle (for example) points the tip of the tongue to the patient's left.

12. **C** is correct. The hyoglossus muscle depresses the tongue.

13. **C** is correct. The styloglossus muscle retracts the tongue upward and backward.

14. **A** is correct. The palatoglossus muscle retracts the tongue upward and backward.

15. **D** is correct. The levator veli palatini muscle elevates the soft palate during swallowing.

16. **E** is correct. The lateral pterygoid muscle partially inserts on the articular disc of the temporomandibular joint (Fig. 13-7).

17. **B** is incorrect. The postvertebral muscles are innervated by the posterior primary rami of the spinal nerves.

18. **E** is incorrect. The diaphragm receives its motor nerve supply from the phrenic nerves.

19. **A** is incorrect. The conjoint tendon is not continuous with the inguinal ligament (Fig. 13-17).

20. **D** is incorrect. The transversus abdominis muscle does form part of the lateral area of the posterior abdominal wall (Fig. 17-25).

21. **C** is incorrect. The rectus abdominis muscle is inserted into the fifth, sixth, and seventh ribs (Fig. 13-15).

22. **E** is incorrect. The external oblique muscle of the anterior abdominal wall assists the diaphragm during inspiration by relaxing as the diaphragm descends so that the abdominal viscera can be accommodated.

23. **A** is incorrect. The pelvic diaphragm is a gutter-shaped sheet of muscle formed by the levatores ani and coccygeus muscles and their covering fasciae. Its function is to support the pelvic viscera. The pelvic diaphragm is

incomplete anteriorly, forming an opening to allow passage of the urethra in males and the urethra and the vagina in females.

24. **B** is incorrect. The piriformis muscle receives its motor nerve supply from the sacral plexus.

25. **D** is incorrect. Colles' fascia (membranous layer of superficial fascia) takes no part in the formation of the urogenital diaphragm; it is too superficial and lies just beneath the skin (Fig. 13-22).

26. **B** is incorrect. The abductor pollicis longus is inserted into the base of the first metacarpal bone.

27. **D** is incorrect. The extensor digiti minimi extends the metacarpophalangeal joint of the little finger and adducts the hand at the wrist joint.

28. **A** is incorrect. The teres major tendon is inserted into the medial lip of the bicipital groove of the humerus (Figs. 13-24 and 11–38).

29. **C** is incorrect. The deltoid muscle is inserted into the deltoid tuberosity halfway down the lateral side of the shaft of the humerus (Fig. 13-26).

30. **E** is incorrect. The subscapularis muscle is inserted into the lesser tuberosity of the humerus.

31. **C** is correct. The gluteus medius muscle acts with the gluteus minimus muscle to raise the pelvis on the opposite side. This permits the leg on the opposite side to clear the ground when walking.

32. **D** is correct. The psoas muscle is a flexor of the thigh at the hip joint.

33. **E** is correct. The tibialis anterior muscle dorsiflexes the foot at the ankle joint.

34. **B** is correct. The rotatory action of the popliteus muscle slackens the ligaments of the extended knee joint, thus permitting flexion to take place.

35. **E** is incorrect. The soleus muscle does not contribute to the boundaries of the popliteal fossa (Fig. 13-47).

36. **C** is incorrect. The obturator internus tendon passes through the lesser sciatic foramen to be inserted into the upper border of the greater trochanter of the femur (Fig. 13-41).

37. **B** is incorrect. The adductor brevis muscle is situated beneath the floor of the femoral triangle (Fig. 13-43).

38. **D** is incorrect. The peroneus tertius muscle extends or dorsiflexes the ankle joint and everts the foot at the subtalar and transverse tarsal joints.

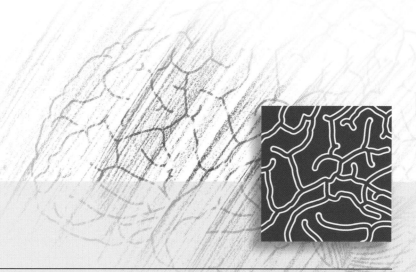

# The Nervous System

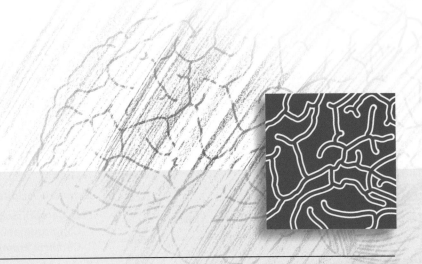

# 14

# The Skull, the Brain, the Meninges, and the Blood Supply of the Brain Relative to Trauma and Intracranial Hemorrhage

All clinical material relevant to this chapter can be found on the CD-ROM.

# Chapter Outline

H ead injuries from blunt trauma and penetrating missiles are associated with a
high mortality and disabling morbidity.
The cerebrovascular accident (stroke) still remains as the third leading cause of morbidity and
death in the United States.

The purpose of this chapter is to review briefly
the anatomy of the skull and its contents and to
highlight those areas that are important to understanding the pathophysiology of head injuries and
intracranial hemorrhage.

# BASIC ANATOMY

## The Skull

The bones forming the anterior wall, lateral wall, and base
of the skull have been fully discussed in Chapter 11 and are
shown in Figures 14-1 through 14-3.

## Tables of the Skull

The skull bones are made up of **external** and **internal tables**
of compact bone, separated by a layer of spongy bone called
the **diploë**. The bones are covered on the outer and inner
surfaces with periosteum (see Fig. 14-11).

## Sutures of the Skull

The bones of the skull are united at immobile joints called
**sutures.** The **coronal suture** lies between the frontal bones

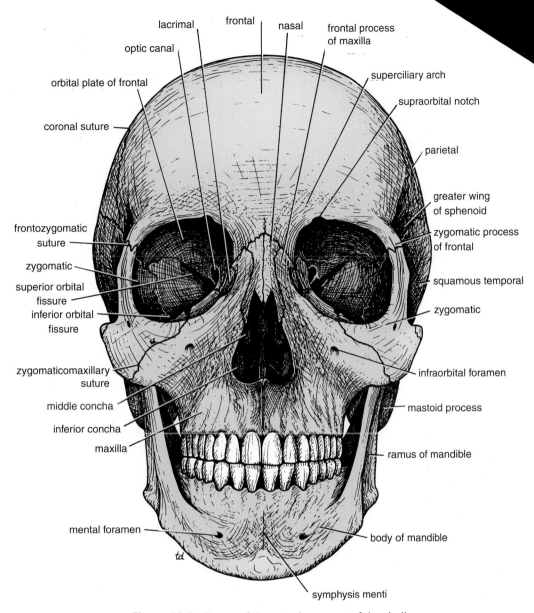

lacrimal
frontal
nasal
frontal process
of maxilla

optic canal

superciliary arch

orbital plate of frontal

supraorbital notch

coronal suture

parietal

greater wing
of sphenoid

frontozygomatic
suture

zygomatic process
of frontal

zygomatic

squamous temporal

superior orbital
fissure

zygomatic

inferior orbital
fissure

zygomaticomaxillary
suture

infraorbital foramen

middle concha

mastoid process

inferior concha

maxilla

ramus of mandible

mental foramen

body of mandible

*td*

symphysis menti

**Figure 14-1**   Bones of the anterior aspect of the skull.

and the parietal bones, the **lambdoid suture** lies between the parietal bones and the occipital bone, and the **sagittal suture** lies between the two parietal bones (Figs. 14-1 and 14-2).

## Fontanelles

At birth, areas of membrane still remain between the bones; these soft areas are known as fontanelles. The anterior fontanelle and posterior fontanelle are described on page 298.

## Base of the Skull

The interior of the base of the skull is conveniently divided up into three cranial fossae: anterior, middle, and posterior (Fig. 14-3). The anterior cranial fossa is separated from the middle cranial fossa by the lesser wing of the sphenoid, and the middle cranial fossa is separated from the posterior cranial fossa by the petrous part of the temporal bone.

The **anterior cranial fossa** lodges the frontal lobes of the cerebral hemispheres, the lateral parts of the **middle cranial fossa** lodge the temporal lobes of the cerebral hemispheres, and the very deep **posterior cranial fossa** lodges parts of the cerebellum, the pons, and the medulla oblongata.

The **sphenoid bone** occupies the central position in the cranial floor. It has a centrally placed **body** with **greater** and **lesser wings** that are outstretched on each side. The sphenoid bone stabilizes the center of the skull by being attached by sutures to the frontal, parietal, occipital, and

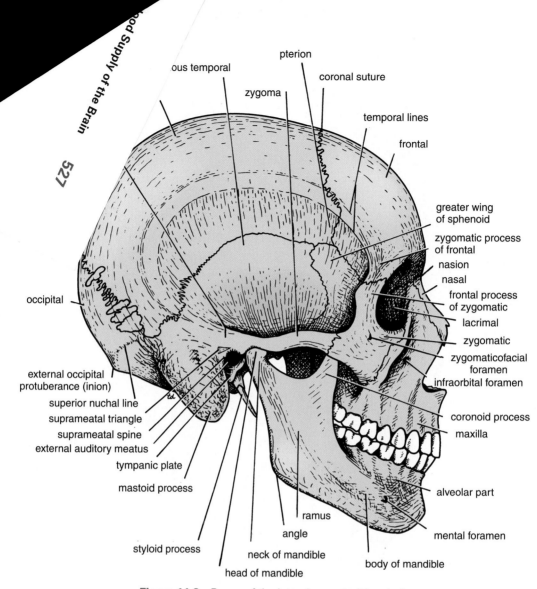

pterion

coronal suture

ous temporal

zygoma

temporal lines

frontal

greater wing
of sphenoid

zygomatic process
of frontal

nasion

nasal

frontal process
of zygomatic

lacrimal

zygomatic

zygomaticofacial
foramen

infraorbital foramen

coronoid process

maxilla

occipital

external occipital
protuberance (inion)

superior nuchal line

suprameatal triangle

suprameatal spine

external auditory meatus

tympanic plate

mastoid process

styloid process

neck of mandible

head of mandible

ramus

angle

body of mandible

alveolar part

mental foramen

**Figure 14-2**   Bones of the lateral aspect of the skull.

ethmoid bones. The body of the sphenoid contains the **sphenoid air sinuses.**

The following important foramina can be identified in Figure 14-3.

In the anterior cranial fossa, the perforations of the **cribriform plate of the ethmoid** can be seen; these transmit the olfactory nerves. In the middle cranial fossa, the **optic canal** is present in the lesser wing of the sphenoid; this transmits the optic nerve and the ophthalmic artery. The slitlike **superior orbital fissure** that exists between the lesser and greater wings of the sphenoid transmits the oculomotor, trochlear, branches of the ophthalmic division of the trigeminal, and the abducent nerves. The **foramen rotundum** in the greater wing of the sphenoid transmits the maxillary division of the trigeminal nerve. The **foramen ovale** perforates the greater wing and transmits the mandibular division of the trigeminal nerve. The small **foramen spinosum,** which is also in the greater wing,

transmits the middle meningeal artery. The larger irregular **foramen lacerum** lies between the greater wing of the sphenoid and the petrous part of the temporal bone and allows the passage of the internal carotid artery from the carotid canal into the cranial cavity.

In the posterior cranial fossa, the large **foramen magnum** in the occipital bone transmits the medulla oblongata. Here, the medulla becomes continuous with the spinal cord. The foramen also allows the passage of the spinal roots of the accessory nerves and the two vertebral arteries.

The **hypoglossal canal** transmits the hypoglossal nerve, and the **jugular foramen** transmits the glossopharyngeal, vagus, and accessory nerves. Here, the sigmoid venous sinus leaves the skull to become the internal jugular vein.

The **internal acoustic meatus** pierces the posterior surface of the petrous part of the temporal bone and transmits the vestibulocochlear nerve and the facial nerve.

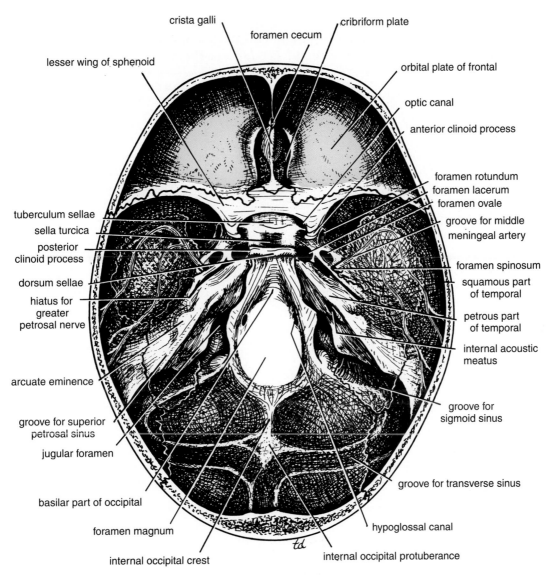

**Figure 14-3**   Internal surface of the base of the skull.

A summary of the more important openings in the base of the skull and the structures that pass through them is shown in Table 14-1.

# SECTIONS OF THE HEAD AND NECK

Before studying the radiographic appearances of the skull, the reader is encouraged to examine photographs of sections of the head and neck (Figs. 14-4, 14-5, and 14-6).

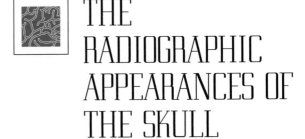

# THE RADIOGRAPHIC APPEARANCES OF THE SKULL

The radiographic appearances of the skull can be seen in Figures 11-11 to 11-18. CT scans and MRIs of the head are shown in Figures 14-7, 14-8, and 14-9.

| Table 14-1 | Summary of the More Important Openings in the Base of the Skull and the Structures that Pass through Them | |
|---|---|---|

| Opening in Skull | Bone of Skull | Structures Transmitted |
|---|---|---|
| **Anterior Cranial Fossa** | | |
| Perforations in cribriform plate | Ethmoid | Olfactory nerves |
| **Middle Cranial Fossa** | | |
| Optic canal | Lesser wing of sphenoid | Optic nerve, ophthalmic artery |
| Superior orbital fissure | Between lesser and greater wings of sphenoid | Lacrimal, frontal, trochlear oculomotor, nasociliary, and abducent nerves; superior ophthalmic vein |
| Foramen rotundum | Greater wing of sphenoid | Maxillary division of the trigeminal nerve |
| Foramen ovale | Greater wing of sphenoid | Mandibular division of the trigeminal nerve, lesser petrosal nerve |
| Foramen spinosum | Greater wing of sphenoid | Middle meningeal artery |
| Foramen lacerum | Between petrous part of temporal and sphenoid | Internal carotid artery |
| **Posterior Cranial Fossa** | | |
| Foramen magnum | Occipital | Medulla oblongata, spinal part of accessory nerve, and right and left vertebral arteries |
| Hypoglossal canal | Occipital | Hypoglossal nerve |
| Jugular foramen | Between petrous part of temporal and condylar part of occipital | Glossopharyngeal, vagus, and accessory nerves; sigmoid sinus becomes internal jugular vein |
| Internal acoustic meatus | Petrous part of temporal | Vestibulocochlear and facial nerves |

From Snell RS: Clinical Anatomy. 7th Ed. Philadelphia: Lippincott Williams & Wilkins, 2004, p. 801.

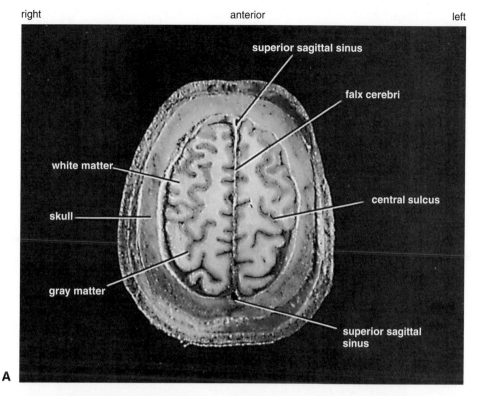

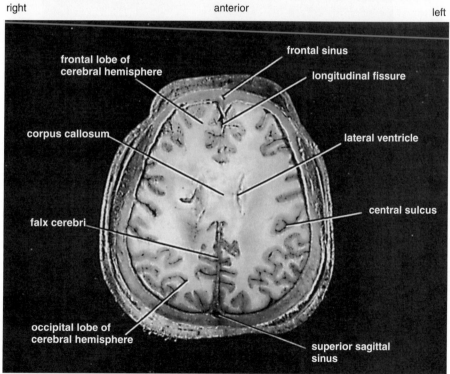

Figure 14-4   **A.** Cross section of the head a short distance beneath the vault of the skull viewed from below. **B.** Cross section of the head at the level of the corpus callosum viewed from below.

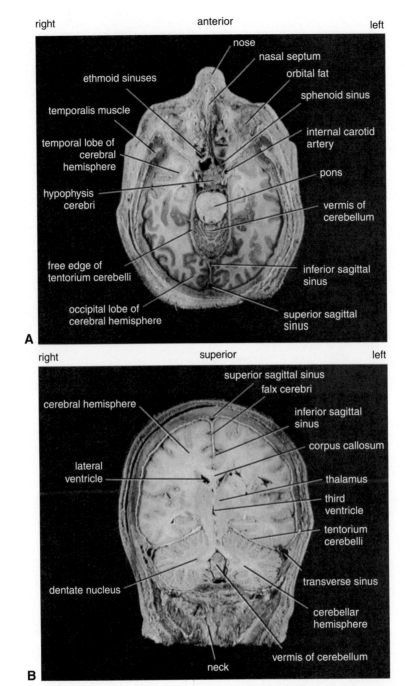

**Figure 14-5    A.** Cross section of the head viewed from below. **B.** Coronal section of the head and the upper part of the neck.

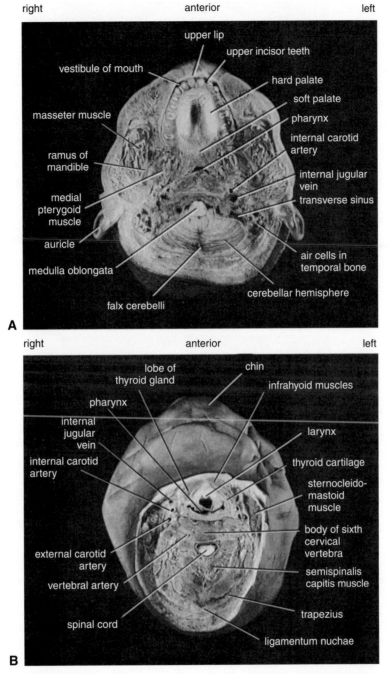

**Figure 14-6** **A.** Cross section of the head just below the level of the hard palate viewed from below. **B.** Cross section of the neck at the level of the sixth cervical vertebra viewed from below.

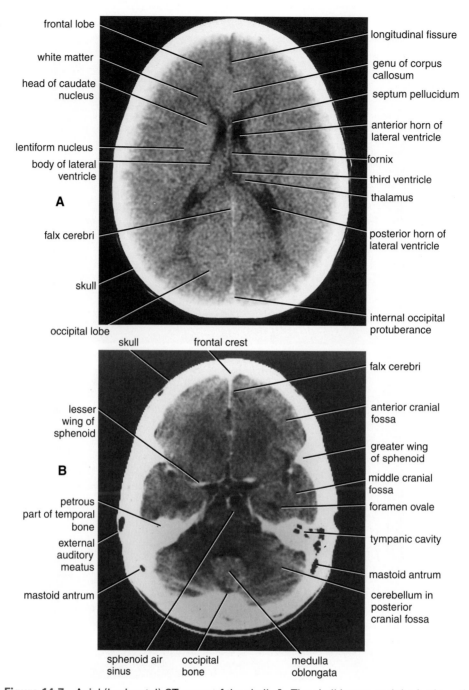

**Figure 14-7**  Axial (horizontal) CT scans of the skull. **A.** The skull bones and the brain and the different parts of the lateral ventricles. **B.** A scan made at a lower level showing the three cranial fossae.

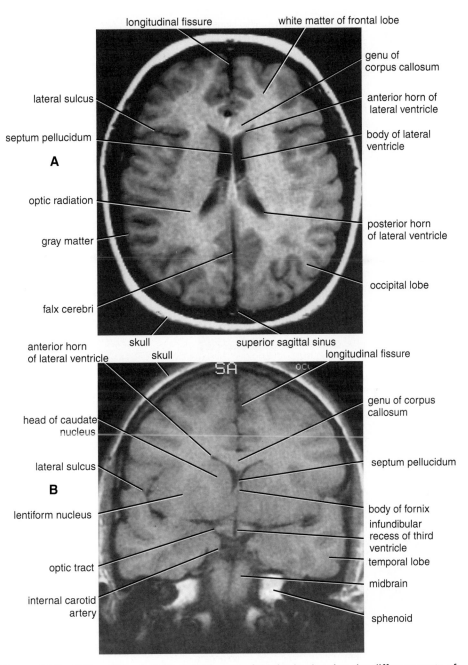

longitudinal fissure

white matter of frontal lobe

genu of corpus callosum

anterior horn of lateral ventricle

lateral sulcus

body of lateral ventricle

septum pellucidum

**A**

optic radiation

posterior horn of lateral ventricle

gray matter

occipital lobe

falx cerebri

anterior horn of lateral ventricle

skull

skull

superior sagittal sinus

longitudinal fissure

genu of corpus callosum

head of caudate nucleus

septum pellucidum

lateral sulcus

**B**

lentiform nucleus

body of fornix

infundibular recess of third ventricle

temporal lobe

optic tract

midbrain

internal carotid artery

sphenoid

**Figure 14-8**   MRIs of the skull. **A.** Axial image of the brain showing the different parts of the lateral ventricle and the lateral sulcus of the cerebral hemisphere. **B.** Coronal image through the frontal lobe of the brain showing the anterior horn of the lateral ventricle. Note the improved contrast between the gray and white matter compared with the CT scans seen in Figure 14-7.

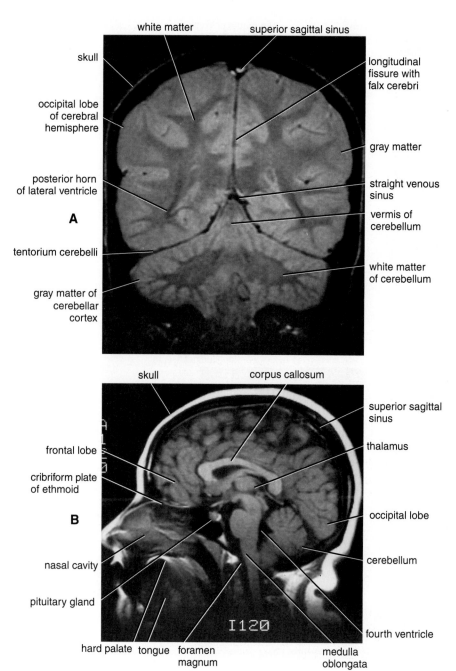

**Figure 14-9** MRIs of the skull. **A.** Coronal image through the occipital lobes of the brain showing the posterior horn of the lateral ventricle and the cerebellum. **B.** Sagittal image showing the different parts of the brain and the nasal and mouth cavities.

# Parts of the Brain

For a detailed description of the gross structure of the brain, a textbook of neuroanatomy should be consulted. In the following account, only the main parts of the brain are described.

The brain is that part of the central nervous system that lies inside the cranial cavity. It is continuous with the spinal cord through the foramen magnum.

## Cerebrum

The **cerebrum** is the largest part of the brain and consists of two **cerebral hemispheres** connected by a mass of white matter called the **corpus callosum** (Fig. 14-10). Each hemisphere extends from the frontal to the occipital bones; above the anterior and middle cranial fossae; and, posteriorly, above the tentorium cerebelli. The hemispheres are separated by a deep cleft, the **longitudinal fissure,** into which projects the **falx cerebri** (Fig. 14-10). The surface layer of

| Major Parts of the Brain | | Cavities of the Brain |
|---|---|---|
| Forebrain | | |
| | Cerebrum | Right and left lateral ventricles |
| | Diencephalon | Third ventricle |
| Midbrain | | Cerebral aqueduct |
| Hindbrain | Pons | Fourth ventricle |
| | Medulla oblongata | and central |
| | Cerebellum | canal |

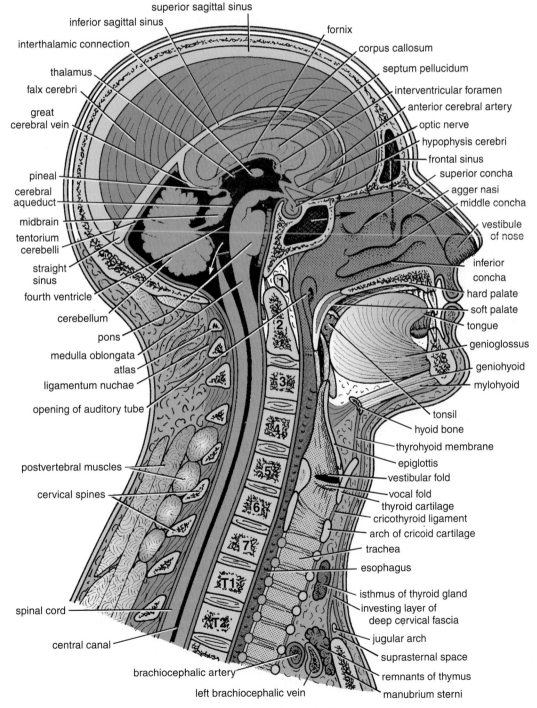

**Figure 14-10**  Sagittal section of the head and neck.

each hemisphere is called the **cortex** and is composed of **gray matter**. The cerebral cortex is thrown into folds, or **gyri**, separated by fissures, or **sulci**. By this means, the surface area of the cortex is greatly increased. Several of the large sulci conveniently subdivide the surface of each hemisphere into **lobes**. The lobes are named for the bones of the cranium under which they lie (see Fig. 14-17). The **frontal lobe** is situated in front of the **central sulcus** (see Fig. 14-17) and above the **lateral sulcus**. The **parietal lobe** is situated behind the central sulcus and above the lateral sulcus. The **occipital lobe** lies below the **parieto-occipital sulcus**. Below the lateral sulcus is situated the **temporal lobe**.

The **precentral gyrus** lies immediately anterior to the central sulcus and is known as the **motor area** (see Fig. 14-17). The large motor nerve cells in this area control voluntary movements on the opposite side of the body. Most nerve fibers cross over to the opposite side in the medulla oblongata as they descend to the spinal cord.

In the motor area, the body is represented in an inverted position, with the nerve cells controlling the movements of the feet located in the upper part and the nerve cells controlling the movements of the face and hands in the lower part (see Fig. 14-17). The **postcentral gyrus** lies immediately posterior to the central sulcus and is known as the **sensory area** (see Fig. 14-17). The small nerve cells in this area receive and interpret sensations of pain, temperature, touch, and pressure from the opposite side of the body.

The **superior temporal gyrus** lies immediately below the lateral sulcus (see Fig. 14-17). The middle of this gyrus is concerned with the reception and interpretation of sound and is known as the **auditory area.**

**Broca's area**, or the **motor speech area,** lies just above the lateral sulcus (see Fig. 14-17). It controls the movements employed in speech. It is dominant in the left hemisphere in right-handed persons and is dominant in the right hemisphere in left-handed persons.

The **visual area** is situated on the posterior pole and medial aspect of the cerebral hemisphere in the region of the **calcarine sulcus** (see Fig. 14-17). It is the receiving area for visual impressions.

The cavity present within each cerebral hemisphere is called the **lateral ventricle** (see Fig. 14-18). The lateral ventricles communicate with the third ventricle through the **interventricular foramina** (see Figs.14-10 and 14-18).

## Diencephalon

The diencephalon is almost completely hidden from the surface of the brain. It consists of a dorsal **thalamus** (Fig. 14-10) and a ventral **hypothalamus**. The thalamus is a large mass of gray matter that lies on either side of the third ventricle. It is the great relay station on the afferent sensory pathway to the cerebral cortex.

The hypothalamus forms the lower part of the lateral wall and floor of the third ventricle. The following structures are found in the floor of the third ventricle from before backward: the **optic chiasma** (see Fig. 14-16), the **tuber cinereum** and the **infundibulum,** the **mammillary bodies,** and the **posterior perforated substance.**

## Midbrain

The midbrain is the narrow part of the brain that passes through the tentorial notch and connects the forebrain to the hindbrain (Fig. 14-10). The midbrain comprises two lateral halves called the **cerebral peduncles;** each of these is divided into an anterior part, the **crus cerebri,** and a posterior part, the **tegmentum,** by a pigmented band of gray matter, the **substantia nigra** (see Fig. 14-15). The narrow cavity of the midbrain is the **cerebral aqueduct** (Figs. 14-10 and 14-18), which connects the third and fourth ventricles. The **tectum** is the part of the midbrain posterior to the cerebral aqueduct; it has four small surface swellings, namely, the **two superior** and **two inferior colliculi.** The colliculi are deeply placed between the cerebellum and the cerebral hemispheres.

The **pineal body** is a small glandular structure that lies between the superior colliculi (Fig. 14-10). It is attached by a stalk to the region of the posterior wall of the third ventricle. The pineal commonly calcifies in middle age, and thus, it can be visualized on radiographs.

## Hindbrain

The **pons** is situated on the anterior surface of the cerebellum below the midbrain and above the medulla oblongata (Fig. 14-10). It is composed mainly of nerve fibers, which connect the two halves of the cerebellum. It also contains ascending and descending fibers connecting the forebrain, the midbrain, and the spinal cord. Some of the nerve cells within the pons serve as relay stations, whereas others form cranial nerve nuclei.

The **medulla oblongata** is conical in shape and connects the pons above to the spinal cord below (Fig. 14-10). A **median fissure** is present on the anterior surface of the medulla, and on each side of this is a swelling called the **pyramid** (see Fig. 14-16). The pyramids are composed of bundles of nerve fibers that originate in large nerve cells in the precentral gyrus of the cerebral cortex. The pyramids taper below, and here most of the descending fibers cross over to the opposite side, forming the **decussation of the pyramids.**

Posterior to the pyramids are the **olives,** which are oval elevations produced by the underlying **olivary nuclei** (see Fig. 14-16). Behind the olives are the **inferior cerebellar peduncles,** which connect the medulla to the cerebellum.

On the posterior surface of the inferior part of the medulla oblongata are the **gracile** and **cuneate tubercles,** produced by the medially placed underlying **nucleus gracilis** and the laterally placed underlying **nucleus cuneatus.**

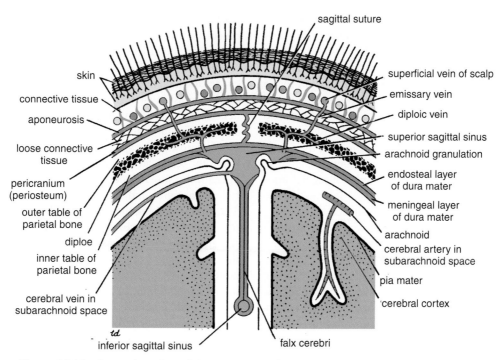

**Figure 14-11**    Coronal section of the upper part of the head showing the layers of the scalp, the sagittal suture of the skull, the falx cerebri, the superior and inferior sagittal venous sinuses, the arachnoid granulations, the emissary veins, and the relation of cerebral blood vessels to the subarachnoid space.

The **cerebellum** lies within the posterior cranial fossa beneath the tentorium cerebelli (Fig. 14-10). It is situated posterior to the pons and the medulla oblongata. It consists of two hemispheres connected by a median portion, the **vermis.** The cerebellum is connected to the midbrain by the **superior cerebellar peduncles,** to the pons by the **middle cerebellar peduncles,** and to the medulla by the **inferior cerebellar peduncles.**

The surface layer of each cerebellar hemisphere, called the **cortex,** is composed of gray matter. The cerebellar cortex is thrown into folds, or **folia,** separated by closely set transverse fissures. Certain masses of gray matter are found in the interior of the cerebellum, embedded in the white matter; the largest of these is known as the **dentate nucleus.**

The cerebellum plays an important role in the control of muscle tone and the coordination of muscle movement on the same side of the body.

The cavity of the hindbrain is the fourth ventricle (see Figs. 14-10 and 14-18). This is bounded in front by the pons and the medulla oblongata and behind by the **superior** and **inferior medullary vela** and the cerebellum. The fourth ventricle is connected above to the third ventricle by the cerebral aqueduct, and below it is continuous with the central canal of the spinal cord. It communicates with the subarachnoid space through three openings in the lower part of the roof: a median and two lateral openings.

# The Meninges

The brain and spinal cord are surrounded by three membranes, or meninges: the dura mater, the arachnoid mater, and the pia mater.

## Dura Mater of the Brain

The dura mater is conventionally described as two layers: the endosteal layer and the meningeal layer (Fig. 14-11). These are closely united except along certain lines, where they separate to form venous sinuses.

The **endosteal layer** is nothing more than the ordinary periosteum covering the inner surface of the skull bones. **It does not extend** through the foramen magnum to become continuous with the dura mater of the spinal cord. Around the margins of all the foramina in the skull, it becomes continuous with the periosteum on the outside of the skull bones. At the sutures, it is continuous with the sutural ligaments. It is most strongly adherent to the bones over the base of the skull.

The **meningeal layer** is the dura mater proper. It is a dense, strong, fibrous membrane covering the brain and is continuous through the foramen magnum with the dura mater of the spinal cord. It provides tubular sheaths for the cranial nerves as the latter pass through the foramina in the skull. Outside the skull, the sheaths fuse with the epineurium of the nerves.

The meningeal layer sends inward four septa that divide the cranial cavity into freely communicating spaces lodging the subdivisions of the brain. The function of these septa is to restrict the rotatory displacement of the brain.

The **falx cerebri** is a sickle-shaped fold of dura mater that lies in the midline between the two cerebral hemispheres (Figs. 14-10 and 14-12). Its narrow end in front is attached to the internal frontal crest and the crista galli. Its broad posterior part blends in the midline with the upper surface of the tentorium cerebelli. The superior sagittal sinus runs in its upper fixed margin, the inferior sagittal sinus runs in its lower concave free margin, and the straight sinus runs along its attachment to the tentorium cerebelli.

The **tentorium cerebelli** is a crescent-shaped fold of dura mater that roofs over the posterior cranial fossa (Figs. 14-12, 14-13, and 14-14). It covers the upper surface of the cerebellum and supports the occipital lobes of the cerebral hemispheres. In front is a gap, the **tentorial notch,** for the passage of the midbrain (Figs. 14-14 and 14-15), thus producing an inner free border and an outer attached or fixed border. The fixed border is attached to the posterior clinoid processes, the superior borders of the petrous bones, and the margins of the grooves for the transverse sinuses on the occipital bone. The free border runs forward at its two ends, crosses the attached border, and is affixed to the anterior clinoid process on each side. At the point where the two borders cross, the third and fourth cranial nerves pass forward to enter the lateral wall of the cavernous sinus (Figs. 14-14 and 14-15). Close to the apex of the petrous part of the temporal bone, the lower layer of the tentorium is pouched forward beneath the superior petrosal sinus to form a recess for the trigeminal nerve and the trigeminal ganglion (Fig. 14-14). The falx cerebri and the falx cerebelli are attached to the upper and lower surfaces of the tentorium, respectively. The straight sinus runs along its attachment to the falx cerebri, the superior petrosal sinus along its attachment to the petrous bone, and the transverse sinus along its attachment to the occipital bone (Fig. 14-13).

The **falx cerebelli** is a small, sickle-shaped fold of dura mater that is attached to the internal occipital crest and

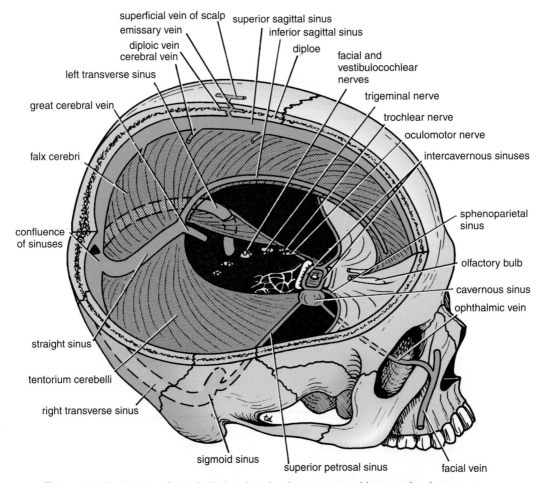

**Figure 14-12** Interior of the skull showing the dura mater and its contained venous sinuses. Note the connections of the veins of the scalp and the veins of the face with the venous sinuses.

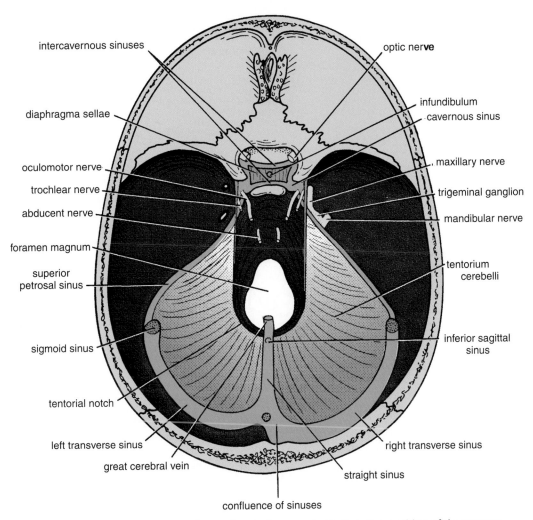

**Figure 14-13**   Diaphragma sellae and tentorium cerebelli. Note the position of the venous sinuses.

projects forward between the two cerebellar hemispheres. Its posterior fixed margin contains the occipital sinus.

The **diaphragma sellae** is a small circular fold of dura mater that forms the roof for the sella turcica (Fig. 14-3). A small opening in its center allows passage of the stalk of the hypophysis cerebri (Figs. 14-13 and 14-15).

## Dural Nerve Supply

Branches of the trigeminal, vagus, and first three cervical nerves and branches from the sympathetic system pass to the dura.

Numerous sensory endings are in the dura. The dura is sensitive to stretching, which produces the sensation of headache. Stimulation of the sensory endings of the trigeminal nerve above the level of the tentorium cerebelli produces referred pain to an area of skin on the same side of the head. Stimulation of the dural endings below the level of the tentorium produces referred pain to the back of the neck and back of the scalp along the distribution of the greater occipital nerve.

## Dural Arterial Supply

Numerous arteries supply the dura mater from the internal carotid, maxillary, ascending pharyngeal, occipital, and vertebral arteries. From a clinical standpoint, the most important is the middle meningeal artery, which is commonly damaged in head injuries.

The **middle meningeal artery** arises from the maxillary artery in the infratemporal fossa. It enters the cranial cavity and runs forward and laterally in a groove on the upper surface of the squamous part of the temporal bone (see Fig. 18-3). To enter the cranial cavity, it passes through the foramen spinosum to **lie between the meningeal and endosteal** layers of dura. The anterior (frontal) branch

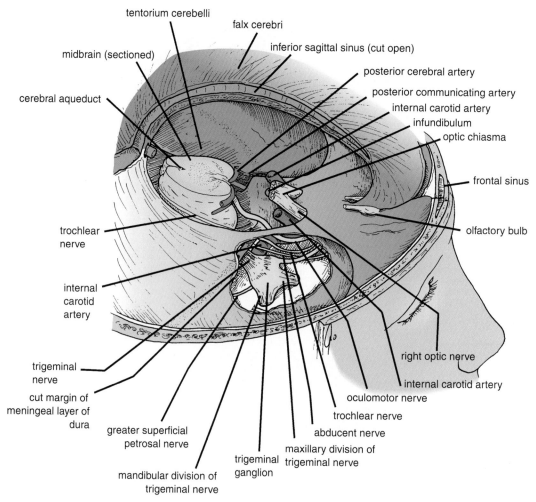

**Figure 14-14** Lateral view of the skull showing the falx cerebri, tentorium cerebelli, brainstem, and trigeminal ganglion.

deeply grooves or tunnels the anteroinferior angle of the parietal bone, and its course corresponds roughly to the line of the underlying precentral gyrus of the brain. The posterior (parietal) branch curves backward and supplies the posterior part of the dura mater.

## Dural Venous Drainage

The **meningeal veins** lie in the endosteal layer of dura. The middle meningeal vein follows the branches of the middle meningeal artery and drains into the pterygoid venous plexus or the sphenoparietal sinus. The veins lie lateral to the arteries.

## Arachnoid Mater of the Brain

The arachnoid mater is a delicate, impermeable membrane covering the brain and lying between the pia mater internally and the dura mater externally (Fig. 14-11). It is separated from the dura by a potential space, the **subdural**

**space,** and from the pia by the **subarachnoid space,** which is filled with **cerebrospinal fluid.**

The arachnoid bridges over the sulci on the surface of the brain, and in certain situations, the arachnoid and pia are widely separated to form the **subarachnoid cisternae.**

In certain areas, the arachnoid projects into the venous sinuses to form **arachnoid villi.** The arachnoid villi are most numerous along the superior sagittal sinus. Aggregations of arachnoid villi are referred to as **arachnoid granulations** (see Figs. 14-11 and 14-18). Arachnoid villi serve as sites where the cerebrospinal fluid diffuses into the bloodstream.

It is important to remember that structures passing to and from the brain to the skull or its foramina must pass through the subarachnoid space. All the cerebral arteries and veins lie in the space, as do the cranial nerves (Fig. 14-11). The arachnoid fuses with the epineurium of the nerves at their point of exit from the skull. In the case of the optic nerve, the arachnoid forms a sheath for the nerve that ex-

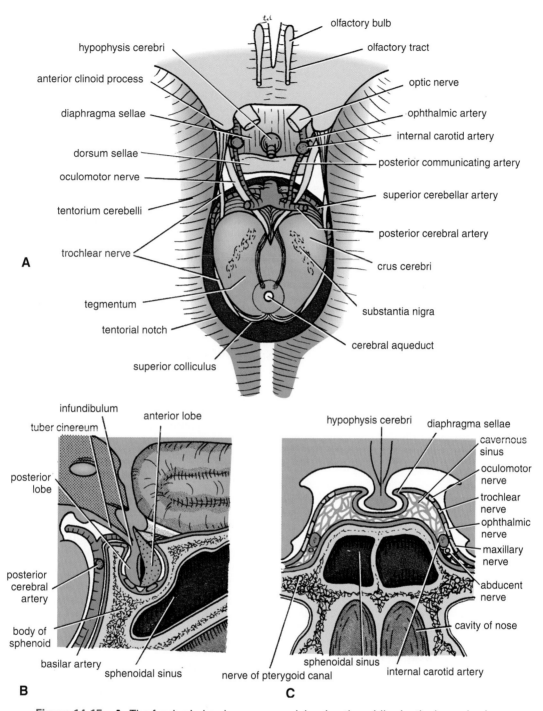

**Figure 14-15** **A.** The forebrain has been removed, leaving the midbrain, the hypophysis cerebri, and the internal carotid and basilar arteries in position. **B.** Sagittal section through the sella turcica showing the hypophysis cerebri. **C.** Coronal section through the body of the sphenoid showing the hypophysis cerebri and the cavernous sinuses. Note the position of the cranial nerves.

tends into the orbital cavity through the optic canal and fuses with the sclera of the eyeball (see Fig. 18-9). Thus, the subarachnoid space extends around the optic nerve as far as the eyeball hemispheres and downward around the spinal cord. The spinal subarachnoid space extends down as far as the **second sacral vertebra** (see Fig. 14-18).

## Pia Mater of the Brain

The pia mater is a vascular membrane that closely invests the brain, covering the gyri and descending into the deepest sulci (Fig. 14-11). It extends over the cranial nerves and fuses with their epineurium. The cerebral arteries entering the substance of the brain carry a sheath of pia with them.

# The Venous Blood Sinuses

The venous blood sinuses are fully described in Chapter 6, p. 182.

# Blood Supply of the Brain

## Arteries of the Brain

The brain is supplied by the two internal carotid and the two vertebral arteries. The four arteries anastomose on the inferior surface of the brain and form the **circle of Willis** (circulus arteriosus).

## Internal Carotid Artery

The internal carotid artery emerges from the cavernous sinus on the medial side of the anterior clinoid process. It then turns backward to the region of the lateral cerebral sulcus. Here, it divides into the anterior and middle cerebral arteries (Fig. 14-16).

### Branches of the Cerebral Portion of the Internal Carotid Artery

- The **ophthalmic artery** arises as the internal carotid artery emerges from the cavernous sinus (Fig. 18-3). It enters the orbit through the optic canal, below and lateral to the optic nerve. It supplies the eye and other orbital structures, and its terminal branches supply the frontal area of the scalp, the ethmoid and frontal sinuses, and the dorsum of the nose.
- The **posterior communicating artery** is a small vessel that runs backward to join the posterior cerebral artery (Fig. 14-16).
- The **choroidal artery**, a small branch, passes backward, enters the inferior horn of the lateral ventricle, and ends in the choroid plexus.
- The **anterior cerebral artery** runs forward and medially and enters the longitudinal fissure of the cerebrum (Fig. 14-16). It is joined to the artery of the opposite side by the **anterior communicating artery.** It curves backward over the corpus callosum, and its **cortical branches** supply all the medial surface of the cerebral cortex as far back as the parieto-occipital sulcus (Fig. 14-17). The branches also supply a strip of cortex about 1 in (2.5 cm) wide on the adjoining lateral surface. The anterior cerebral artery thus supplies the "leg area" of the precentral gyrus. Several **central branches** pierce the brain substance and supply the deep masses of gray matter within the cerebral hemisphere.
- The **middle cerebral artery**, the largest branch of the internal carotid, runs laterally in the lateral cerebral sulcus (Fig. 14-16). **Cortical branches** supply the entire lateral surface of the hemisphere, except for the narrow strip supplied by the anterior cerebral artery, the occipital pole, and the inferolateral surface of the hemisphere, which are supplied by the posterior cerebral artery. This

artery thus supplies all the motor area except the leg area. **Central branches** enter the anterior perforated substance and supply the deep masses of gray matter within the cerebral hemisphere.

## Vertebral Artery

The vertebral artery, a branch of the first part of the subclavian artery (Fig. 13-10), ascends the neck through the foramina in the transverse processes of the upper six cervical vertebrae. It enters the skull through the foramen magnum and passes upward, forward, and medially on the medulla oblongata (Fig. 14-16). At the lower border of the pons, it joins the vessel of the opposite side to form the **basilar artery.**

### Cranial Branches

- Meningeal arteries.
- Anterior and posterior spinal arteries.
- Posteroinferior cerebellar artery.
- Medullary arteries.

## Basilar Artery

The basilar artery, formed by the union of the two vertebral arteries, ascends in a groove on the anterior surface of the pons (Fig. 14-16). At the upper border of the pons, it divides into the two posterior cerebral arteries.

### Branches

- It gives off branches to the pons, cerebellum, and internal ear.
- The posterior cerebral arteries.

The **posterior cerebral artery** on each side curves laterally and backward around the midbrain (Fig. 14-16). **Cortical branches** supply the inferolateral surface of the temporal lobe and the lateral and medial surfaces of the occipital lobe (Fig. 14-17). It thus supplies the visual cortex. **Central branches** pierce the brain substance and supply the deep masses of gray matter within the cerebral hemisphere and the midbrain.

## Circle of Willis

The **circle of Willis** lies in the interpeduncular fossa at the base of the brain. It is formed by the anastomosis between the two internal carotid arteries and the two vertebral arteries (Fig. 14-16). The anterior communicating, the anterior cerebral, the internal carotid, the posterior communicating, the posterior cerebral, and the basilar arteries all contribute to the circle. The circle of Willis allows blood that enters by either internal carotid or vertebral arteries to be distributed to any part of both cerebral hemispheres. Cortical and central branches arise from the circle and supply the brain substance.

## Veins of the Brain

The veins of the brain have no muscular tissue in their thin walls, and they possess no valves. They emerge from the brain and drain into the cranial venous sinuses (Fig. 14-12).

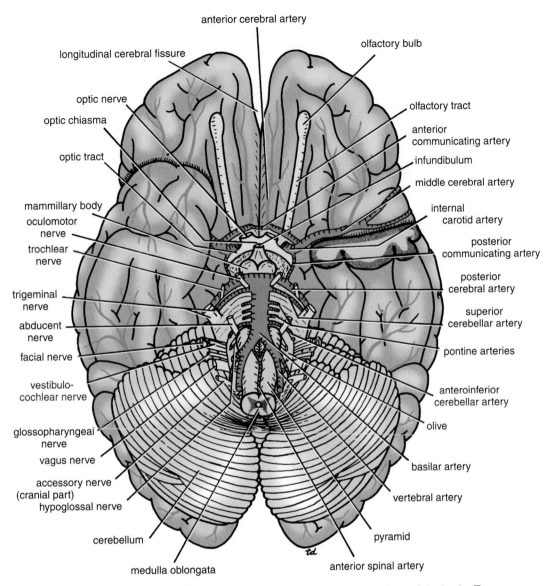

**Figure 14-16** Arteries and cranial nerves seen on the inferior surface of the brain. To show the course of the middle cerebral artery, the anterior pole of the left temporal lobe has been removed.

Cerebral and cerebellar veins and veins of the brainstem are present. The **great cerebral vein** is formed by the union of the two **internal cerebral veins** and drains into the straight sinus (Fig. 14-12).

# The Ventricular System of the Brain

The ventricles of the brain consist of the two lateral ventricles, the third ventricle, and the fourth ventricle. The two **lateral ventricles** communicate with the **third ventricle** through the **interventricular foramina** (Fig. 14-10); the third ventricle communicates with the fourth ventricle by the **cerebral aqueduct.** The fourth ventricle, in turn, is continuous with the narrow **central canal** of the spinal cord and, through the three foramina in its roof, with the subarachnoid space (Fig. 14-10).

The ventricles are filled with **cerebrospinal fluid**. The size and shape of the cerebral ventricles may be visualized clinically using CT scans and MRIs (Figs. 14-7, 14-8, and 14-9).

---

**P H Y S I O L O G I C   N O T E**

### The Cerebrospinal Fluid

Cerebrospinal fluid is a clear, colorless fluid. It possesses, in solution, inorganic salts similar to those in the blood plasma. The glucose content is about half that of blood, and there is only a trace of protein. In the lateral recumbent position, cerebrospinal fluid pressure, as measured by **spinal tap** (lumbar puncture), is about 60 to 150 mm of water. This pressure may be easily raised by straining

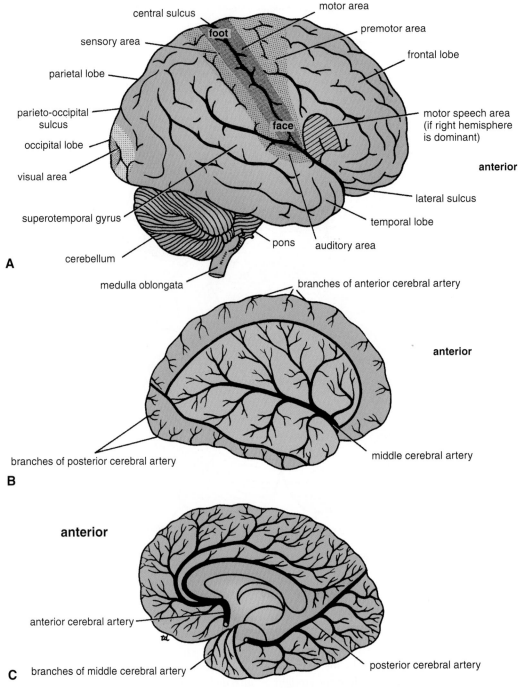

**Figure 14-17   A.** Right side of the brain showing some important localized areas of cerebral function. Note that the motor speech area is most commonly located in the left rather than the right cerebral hemisphere. **B.** Lateral surface of the cerebral hemisphere showing areas supplied by the cerebral arteries. In this and the next figure, areas colored *blue* are supplied by the anterior cerebral artery; those colored *red* are supplied by the middle cerebral artery; and those colored *green* are supplied by the posterior cerebral artery. **C.** Medial surface of the cerebral hemisphere showing the areas supplied by the cerebral arteries.

or coughing or compressing the internal jugular veins in the neck.

### Functions of the Cerebrospinal Fluid

Cerebrospinal fluid, which bathes the external and internal surfaces of the brain and spinal cord, serves as a cushion between the central nervous system and the surrounding bones, thus protecting the brain and spinal cord against mechanical trauma. The close relationship of the fluid to the nervous tissue and the blood enables the fluid to serve as reservoir and assist in the regulation of the contents of the skull. If the brain volume or the blood volume increases, the cerebrospinal fluid volume decreases. Since the cerebrospinal fluid is an ideal physiologic substrate, it plays an active part in the nourishment of the nervous tissue; it also assists in the removal of products of neuronal metabolism. It is possible that the secretions of the pineal gland influence the activities of the pituitary gland by circulating through the cerebrospinal fluid in the third ventricle.

### Formation of Cerebrospinal Fluid

Cerebrospinal fluid is formed mainly in the **choroid plexuses** of the lateral, third, and fourth ventricles; some may originate as tissue fluid formed in the brain substance. The cuboidal epithelium covering the surface of the choroid plexuses actively secretes cerebrospinal fluid.

### Circulation of Cerebrospinal Fluid

The circulation begins with its secretion from the choroid plexuses in the ventricles and its production from the brain surface. The fluid passes from the lateral ventricles into the third ventricle through the interventricular foramina (Fig. 14-18). It then passes into the fourth ventricle through the cerebral aqueduct. The circulation is aided by the arterial pulsations of the choroid plexuses.

From the fourth ventricle, the fluid passes through the median apertures and the lateral foramina in the roof of the fourth ventricle and enters the subarachnoid space. The fluid flows superiorly through the notch in the tentorium cerebelli to reach the inferior surface of the cerebrum (Fig. 14-18). It now moves superiorly over the lateral aspect of each cerebral hemisphere. Some of the cerebrospinal fluid moves inferiorly in the subarachnoid space around the spinal cord and cauda equina. The pulsations of the cerebral and spinal arteries and the movements of the vertebral column facilitate this gradual flow of fluid.

### Absorption of Cerebrospinal Fluid

The main sites for the absorption of the cerebrospinal fluid are the **arachnoid villi** that project into the dural venous sinuses, especially the superior sagittal sinus (Fig. 14-18). The arachnoid villi tend to be grouped together to form **arachnoid granulations.** The arachnoid granulations increase in number and size with age and tend to become calcified with advanced age.

Absorption of cerebrospinal fluid into the venous sinuses occurs when the pressure of the fluid exceeds that of the blood in the sinus. Some of the cerebrospinal fluid is absorbed directly into the veins in the subarachnoid space, and some possibly escapes through perineural lymphatic vessels of the cranial and spinal nerves.

### Turnover of Cerebrospinal Fluid

Cerebrospinal fluid is produced continuously at a rate of about 0.5 mL/min and with a total volume of about 130 mL; this corresponds to a turnover time of about 5 hours.

# SURFACE ANATOMY OF THE SKULL AND BRAIN

The important surface landmarks of the skull are as follows.

# Nasion

This is the depression in the midline of the root of the nose (Fig. 14-19).

# External Occipital Protuberance

This is a bony prominence in the middle of the squamous part of the occipital bone (Fig. 14-19). It lies in the midline at the junction of the head and neck and gives attachment to the ligamentum nuchae.

# Falx Cerebri, Superior Sagittal Sinus, and the Longitudinal Cerebral Fissure between the Cerebral Hemispheres

The position of these structures can be indicated by passing a line over the vertex of the skull in the sagittal plane that joins the nasion to the external occipital protuberance.

# Parietal Eminence

This is a raised area on the lateral surface of the parietal bone and can be felt about 2 in. (5 cm) above the auricle. It lies close to the lower end of the **central cerebral sulcus of the brain** (Fig. 14-19).

# Pterion

This the point where the greater wing of the sphenoid meets the anteroinferior angle of the parietal bone. Lying 1½ in. (4 cm) above the midpoint of the zygomatic arch (Fig. 14-19), it is not marked by an eminence or a depression, but it is important since the **anterior branches of the middle meningeal artery and vein lie beneath it.**

# Mastoid Process of the Temporal Bone

The mastoid process projects downward and forward from behind the ear (Fig. 14-19). It is undeveloped in the newborn and

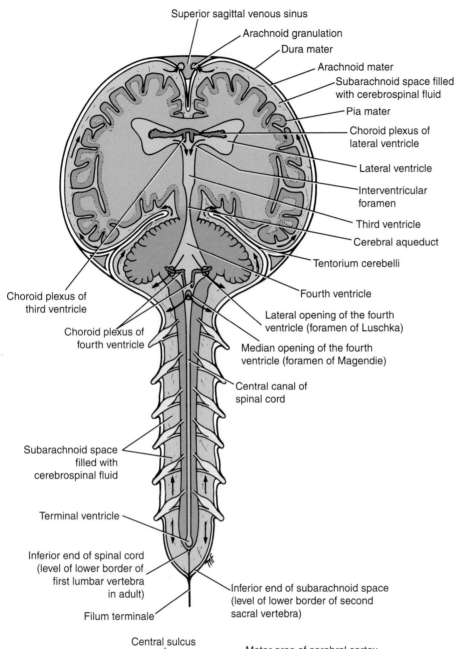

Superior sagittal venous sinus

Arachnoid granulation

Dura mater

Arachnoid mater

Subarachnoid space filled with cerebrospinal fluid

Pia mater

Choroid plexus of lateral ventricle

Lateral ventricle

Interventricular foramen

Third ventricle

Cerebral aqueduct

Tentorium cerebelli

Fourth ventricle

Lateral opening of the fourth ventricle (foramen of Luschka)

Median opening of the fourth ventricle (foramen of Magendie)

Central canal of spinal cord

Choroid plexus of third ventricle

Choroid plexus of fourth ventricle

Subarachnoid space filled with cerebrospinal fluid

Terminal ventricle

Inferior end of spinal cord (level of lower border of first lumbar vertebra in adult)

Filum terminale

Inferior end of subarachnoid space (level of lower border of second sacral vertebra)

**Figure 14-18** Origin and circulation of the cerebrospinal fluid.

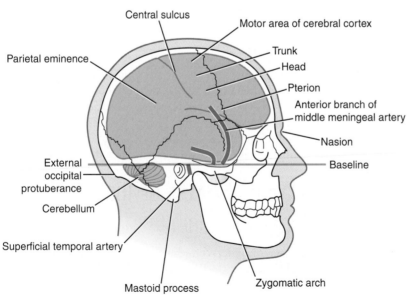

Central sulcus

Motor area of cerebral cortex

Trunk

Head

Pterion

Anterior branch of middle meningeal artery

Nasion

Baseline

Parietal eminence

External occipital protuberance

Cerebellum

Superficial temporal artery

Mastoid process

Zygomatic arch

**Figure 14-19** Surface landmarks on the right side of the head. The relations of the middle meningeal artery and brain to the surface of the skull are shown.

grows only as the result of the pull of the sternocleidomastoid muscle, as the child moves his head. It may be recognized as a bony projection at the end of the second year of life.

# Zygomatic Arch

The zygomatic arch extends forward in front of the ear and ends in front in the zygomatic bone (Fig. 14-19). Above the zygomatic arch is the **temporal fossa**, which is filled with the **temporalis muscle.** Attached to the lower margin of the arch is the masseter muscle. Contraction of both the tempo-

ralis and masseter muscles (testing for the integrity of the motor part of the mandibular division of the trigeminal nerve) may be felt by clenching the teeth.

# Anatomical Baseline of the Skull

This baseline extends from the lower margin of the orbit backward through the upper margin of the external auditory meatus. The **cerebrum** lies entirely above the line, and the **cerebellum** lies in the posterior cranial fossa below the posterior third of the line (Fig. 14-19).

# Review Questions

## Completion Questions

**Based on the lateral radiograph of the skull, select the phrase that BEST completes each statement.**

1. Structure 1 is the
   A. maxillary sinus.
   B. frontal sinus.
   C. anterior arch of the atlas.
   D. sella turcica.
   E. temporomandibular joint.

2. Structure 2 is the
   A. orbital cavity.
   B. maxillary sinus.
   C. sphenoid sinus.

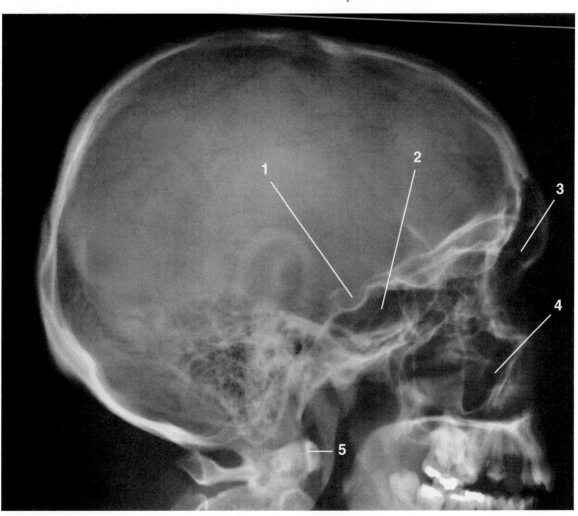

D. tympanic cavity.
E. frontal sinus.

3. Structure 3 is the
   A. orbital cavity.
   B. nasal cavity.
   C. frontal sinus.
   D. maxillary sinus.
   E. sphenoid sinus.

4. Structure 4 is the
   A. mouth cavity.
   B. frontal sinus.
   C. nasal cavity.
   D. sphenoid sinus.
   E. maxillary sinus.

5. Structure 5 is the
   A. posterior arch of the atlas.
   B. body of the axis.
   C. anterior arch of the atlas.
   D. odontoid process of axis.
   E. anterior longitudinal ligament.

**Based on the axial (horizontal) CT scan of the skull, select the phrase that BEST completes each statement.**

6. Structure 1 is the
   A. crista galli.
   B. nasal septum.

C. superior sagittal sinus.
D. falx cerebri.
E. frontal bone.

7. Structure 2 is the
   A. internal carotid artery.
   B. middle cerebral artery.
   C. foramen rotundum.
   D. optic canal.
   E. sphenoid sinus.

8. Structure 3 is the
   A. greater wing of the sphenoid.
   B. petrous part of the temporal bone.
   C. mastoid process.
   D. malleus.
   E. head of the mandible.

9. Structure 4 is the
   A. mastoid antrum.
   B. internal carotid artery.
   C. tympanic cavity.
   D. pharyngotympanic tube.
   E. external auditory meatus.

10. Structure 5 is the
    A. sella turcica.
    B. sphenoid air sinus.
    C. third ventricle of the brain.
    D. pineal gland.
    E. foramen spinosum.

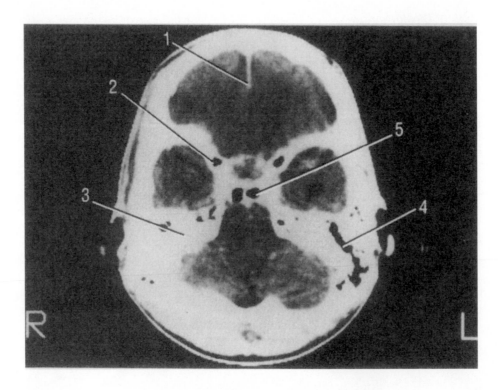

**Based on the axial (horizontal) MRI study of the skull, select the phrase that BEST completes each statement.**

11. Structure 1 is the
    A. white matter of the occipital lobe.
    B. falx cerebri.
    C. septum pellucidum.
    D. genu of the corpus callosum.
    E. optic chiasma.

12. Structure 2 is the
    A. posterior edge of the lesser wing of the sphenoid bone.
    B. lateral sulcus of the cerebral hemisphere.
    C. middle cerebral artery.
    D. middle meningeal artery.
    E. anterior inferior end of the parietal bone.

13. Structure 3 is the
    A. falx cerebri.
    B. septum pellucidum of the brain.
    C. basilar artery.
    D. foramen magnum.
    E. fourth ventricle of the brain.

14. Structure 4 is the
    A. gray matter of the brain.
    B. white matter of the occipital lobe of the brain.
    C. genu of the corpus callosum of the brain.
    D. tentorium cerebelli.
    E. cerebellum.

15. Structure 5 is the
    A. third ventricle of the brain.
    B. fourth ventricle of the brain.
    C. body of the lateral ventricle of the brain.
    D. inferior horn of the lateral ventricle of the brain.
    E. thalamus of the brain.

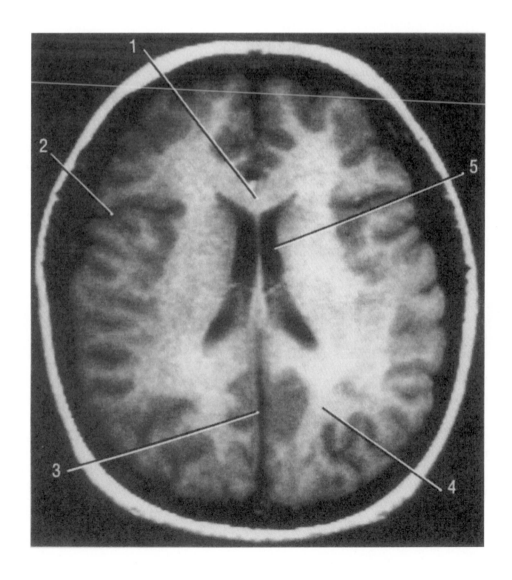

## Multiple Choice Questions

**Select the one BEST answer for each question.**

16. The following statements concerning the central nervous system are correct **except** which?
    A. A CT brain scan can distinguish between white matter and gray matter.
    B. The lateral ventricles of the brain are in direct communication with the fourth ventricle.
    C. An MRI of the brain uses the magnetic properties of the hydrogen nucleus excited by radiofrequency radiation transmitted by a coil surrounding the patient's head.
    D. Following trauma and sudden movement of the brain within the skull, the large arteries at the base of the brain are rarely torn.
    E. The movement of the brain at the time of head injuries may damage the small fourth cranial nerve.

17. The following statements concerning the cerebrospinal fluid are correct **except** which?
    A. The cerebrospinal fluid in the central canal of the spinal cord is unable to enter the fourth ventricle.
    B. With the patient in the recumbent position, the normal pressure is about 60 to 150 mm of water.
    C. It protects the brain and spinal cord from traumatic injury.
    D. Compression of the internal jugular veins in the neck raises the cerebrospinal fluid pressure.
    E. The subarachnoid space is filled with cerebrospinal fluid.

18. The following statements concerning the blood supply to the dura mater within the skull are correct **except** **which?**
    A. The arteries include branches of the internal carotid, maxillary, and vertebral arteries.
    B. The middle meningeal artery arises from the maxillary artery.
    C. The middle meningeal artery enters the skull through the foramen spinosum.
    D. The middle meningeal artery runs between the bone and the endosteal layer of dura.
    E. The anterior branch of the middle meningeal artery grooves the anterior inferior angle of the parietal bone, and it is here that it is commonly injured.

19. The following statements concerning the arterial supply to the brain are correct **except** which?
    A. The main arteries that supply the brain lie within the subarachnoid space.
    B. The basilar artery is formed by the union of the two vertebral arteries.
    C. The cerebral arteries anastomose on the surface of the brain.
    D. The gray matter situated in the interior of the brain receives its nourishment by diffusion of tissue fluid from the blood vessels situated on the surface of the brain.
    E. The sympathetic nerve fibers have very little control over the diameter of the cerebral arteries.

20. The following statements concerning the circle of Willis are correct **except** which?
    A. It lies in the interpeduncular fossa at the base of the brain.
    B. It lies in the subarachnoid space.
    C. It permits the arterial blood to flow forward or backward should the internal carotid or vertebral artery be occluded.
    D. It permits the arterial blood to flow across the midline to the opposite side of the brain.
    E. It communicates with branches of the external carotid artery.

# Answers and Explanations

1. **D** is correct. 1 is the sella turcica that lies on the superior surface of the body of the sphenoid bone.

2. **C** is correct. 2 is the sphenoid air sinus, which lies within the body of the sphenoid bone.

3. **C** is correct. 3 is the frontal air sinus that lies within the frontal bone.

4. **E** is correct. 4 is the maxillary air sinus that lies within the body of the maxillary bone.

5. **C** is correct. 5 is the anterior arch of the atlas, which lies anterior to the odontoid process of the axis in the radiograph.

6. **D** is correct. 1 is the falx cerebri (formed of dura), which is attached anteriorly to the internal surface of the frontal bone.

7. **D** is correct. 2 is the interior of the left optic canal.

8. **B** is correct. 3 is the petrous part of the left temporal bone.

9. **C** is correct. 4 is the right tympanic cavity of the middle ear.

10. **B** is correct. 5 is the cavity of the right sphenoid sinus.

11. **D** is correct. 1 is the genu of the corpus callosum, a major commissure connecting the two cerebral hemispheres.

12. **B** is correct. 2 is the lateral sulcus of the left cerebral cortex.

13. **A** is correct. 3 is the posterior part of the falx cerebri lying in the longitudinal fissure between the cerebral hemispheres.

14. **B** is correct. 4 is the white matter of the occipital lobe of the right cerebral hemisphere.

15. **C** is correct. 5 is the body of the lateral ventricle of the right cerebral hemisphere.

16. **B** is incorrect. The lateral ventricles of the brain communicate first through the interventricular foramen with the third ventricle, which in turn communicates through the cerebral aqueduct of the midbrain with the fourth ventricle of the hind brain (Figs. 14-10 and 14-18).

17. **A** is incorrect. The cerebrospinal fluid in the central canal of the spinal cord can flow freely into the fourth ventricle.

18. **D** is incorrect. The middle meningeal artery runs between the periosteal layer of dura and the meningeal layer of dura once it has entered the skull through the foramen spinosum.

19. **D** is incorrect. The masses of gray matter that lie within the substance of the cerebral hemispheres (e.g., the thalamus, lentiform nucleus, and caudate nucleus) receive their nourishment from the central branches of the cerebral arteries.

20. **E** is incorrect. The branches of the external carotid artery do not communicate with the circle of Willis.

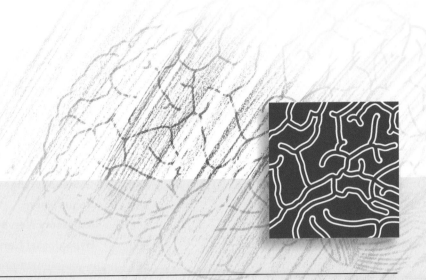

# 15 The Cranial Nerves and Trigeminal Nerve Blocks

All clinical material relevant to this chapter can be found on the CD-ROM.

# Chapter Outline

There are 12 pairs of cranial nerves, which leave the brain and pass through foramina and fissures in the skull. All the nerves are distributed in the head and neck, except the tenth, which also supplies structures in the thorax and abdomen. The cranial nerves are commonly involved in disease, whether it be disease of the nerve itself, such as herpes zoster, or involvement in hemorrhage or tumor formation in the central nervous system. The smaller, slender nerves are commonly damaged in patients with trauma to the head. Furthermore, the cranial nerves in the neck may be damaged in penetrating injuries or involved in disease processes such as metastatic carcinomas.

For these reasons, a systematic examination of the 12 cranial nerves is an important part of the physical examination of every neurologic patient. It may reveal a lesion of a cranial nerve nucleus or its central connections, or it may show an interruption of the lower motor neuron.

# BASIC ANATOMY

## Organization of the Cranial Nerves

The cranial nerves are named as follows:

- I. Olfactory
- II. Optic
- III. Oculomotor
- IV. Trochlear
- V. Trigeminal
- VI. Abducent
- VII. Facial
- VIII. Vestibulocochlear
- IX. Glossopharyngeal
- X. Vagus
- XI. Accessory
- XII. Hypoglossal

The olfactory, optic, and vestibulocochlear nerves are entirely sensory; the oculomotor, trochlear, abducent, accessory, and hypoglossal nerves are entirely motor; and the remaining nerves are mixed. The letter symbols commonly used to indicate the functional components of each cranial nerve are shown in Table 15-1. The different components of the cranial nerves, their functions, and the openings in the skull through which the nerves leave the cranial cavity are summarized in Table 15-2.

| Table 15-1 | The Letter Symbols Commonly Used to Indicate the Functional Components of Each Cranial Nerve | |
| --- | --- | --- |
| Component | Function | Letter Symbols |
| **Afferent Fibers** | Sensory | |
| General somatic afferent | General sensations | GSA |
| Special somatic afferent | Hearing, balance, vision | SSA |
| General visceral afferent | Viscera | GVA |
| Special visceral afferent | Smell, taste | SVA |
| **Efferent Fibers** | | |
| General somatic efferent | Somatic striated muscles | GSE |
| General visceral efferent | Glands and smooth muscles (parasympathetic innervation) | GVE |
| Special visceral efferent | Branchial arch striated muscles | SVE |

From Snell RS. Clinical Neuroanatmy. 6th Ed. Philadelphia: Lippincott Williams & Wilkins, 2006. p. 327

| Table 15-2 | Cranial Nerves | | | |
| --- | --- | --- | --- | --- |
| Number | Name | Components[a] | Function | Opening to Skull |
| I | Olfactory | Sensory (SVA) | Smell | Openings in cribriform plate of ethmoid |
| II | Optic | Sensory (SSA) | Vision | Optic canal |
| III | Oculomotor | Motor (GSE, GVE) | Raises upper eyelid, turns eyeball upward, downward, and medially; constricts pupil; accommodates eye | Superior orbital fissure |
| IV | Trochlear | Motor (GSE) | Assists in turning eyeball downward and laterally | Superior orbital fissure |
| V | Trigeminal[b] | | | |
| | Ophthalmic division | Sensory (GSA) | Cornea, skin of forehead, scalp, eyelids, and nose; also mucous membrane of paranasal sinuses and nasal cavity | Superior orbital fissure |
| | Maxillary division | Sensory (GSA) | Skin of face over maxilla; teeth of upper jaw; mucous membrane of nose, the maxillary sinus, and palate | Foramen rotundum |
| | Mandibular division | Motor (SVE) | Muscles of mastication, mylohyoid, anterior belly of digastric, tensor veli palatine, and tensor tympani | Foramen ovale |
| | | Sensory (GSA) | Skin of cheek, skin over mandible and side of head, teeth of lower jaw and temporomandibular joint; mucous membrane of mouth and anterior part of tongue | |
| VI | Abducent | Motor (GSE) | Lateral rectus muscle turns eyeball laterally | Superior orbital fissure |

| **Table 15-2** | (continued) | | | |
|---|---|---|---|---|
| **Number** | **Name** | **Components**[a] | **Function** | **Opening to Skull** |
| VII | Facial | Motor (SVE) | Muscles of face and scalp, stapedius muscle, posterior belly of digastric and stylohyoid muscles | Internal acoustic meatus, facial canal, stylomastoid foramen |
| | | Sensory (SVA) | Taste from anterior two thirds of tongue, from floor of mouth and palate | |
| | | Secretomotor (GVE) parasympathetic | Submandibular and sublingual salivary glands, the lacrimal gland, and glands of nose and palate | |
| VIII | Vestibulo-cochlear | | | |
| | Vestibular | Sensory (SSA) | From utricle and saccule and semicircular canals—position and movement of head | Internal acoustic meatus |
| | Cochlear | Sensory (SSA) | Organ or Corti—hearing | |
| IX | Glossopharyngeal | Motor (SVE) | Stylopharyngeus muscle—assists swallowing | Jugular foramen |
| | | Secretomotor (GVE) parasympathetic | Parotid salivary gland | |
| | | Sensory (GVA, SVA, GSA) | General sensation and taste from posterior one third of tongue and pharynx; carotid sinus (baroreceptor); and carotid body (chemoreceptor) | |
| X | Vagus | Motor (GVE, SVE) Sensory (GVA, SVA, GSA) | Heart and great thoracic blood vessels; larynx, trachea, bronchi, and lungs; alimentary tract from pharynx to splenic flexure of colon; liver, kidneys, and pancreas | Jugular foramen |
| XI | Accessory | | | |
| | Cranial root | Motor (SVE) | Muscles of soft palate (except tensor veli palatini), pharynx (except stylopharyngeus), and larynx (except cricothyroid) in branches of vagus | Jugular foramen |
| | Spinal root | Motor (SVE) | Sternocleidomastoid and trapezius muscles | |
| XII | Hypoglossal | Motor (GVE) | Muscles of tongue (except palatoglossus controlling its shape and movement | Hypoglossal canal |

[a] The letter symbols are explained in Table 15-1.
[b] The trigeminal nerve also carries proprioceptive impulses from the muscles of mastication and the facial and extraocular muscles.
From Snell RS: Clinical Neuroanatomy. 6th Ed. Philadelphia: Lippincott Williams & Wilkins, 2006, p. 328.

# Olfactory Nerves

The olfactory nerves arise from **olfactory receptor nerve cells** in the olfactory mucous membrane. The olfactory mucous membrane is situated in the upper part of the nasal cavity above the level of the superior concha (Fig. 15-1). Bundles of these olfactory nerve fibers pass through the openings of the cribriform plate of the ethmoid bone to enter the **olfactory bulb** in the cranial cavity. The olfactory bulb is connected to the olfactory area of the cerebral cortex by the **olfactory tract.**

# Optic Nerve

The optic nerve is composed of the axons of the cells of the **ganglionic layer** of the retina. The optic nerve emerges from the back of the eyeball and leaves the orbital cavity through the optic canal to enter the cranial cavity (Fig. 15-3). The optic nerve then unites with the optic nerve of the opposite side to form the optic chiasma (Fig. 15-3).

In the chiasma, the fibers from the medial half of each retina cross the midline and enter the **optic tract** of the

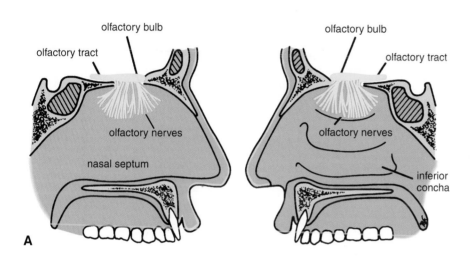

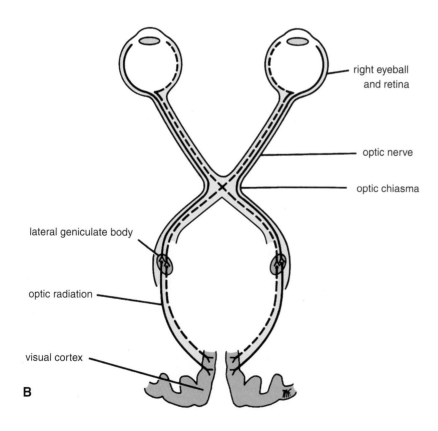

**Figure 15-1  A.** Distribution of the olfactory nerves on the nasal septum and the lateral wall of the nose. **B.** The optic nerve and its connections.

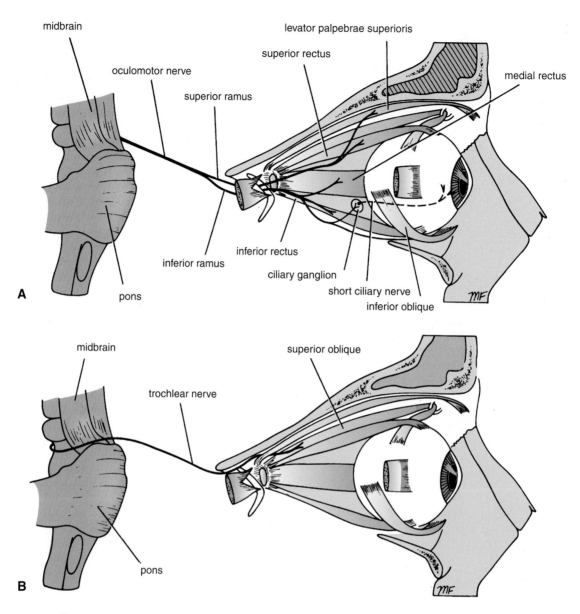

**Figure 15-2** **A.** Origin and distribution of the oculomotor nerve. **B.** Origin and distribution of the trochlear nerve.

opposite side, whereas the fibers from the lateral half of each retina pass posteriorly in the optic tract of the same side. Most of the fibers of the optic tract terminate by synapsing with nerve cells in the **lateral geniculate body** (Fig. 15-1). A few fibers pass to the pretectal nucleus and the superior colliculus and are concerned with light reflexes.

The axons of the nerve cells of the lateral geniculate body pass posteriorly as the **optic radiation** and terminate in the **visual cortex** of the cerebral hemisphere (Fig. 15-1).

# Oculomotor Nerve

The oculomotor nerve emerges on the anterior surface of the midbrain (Fig. 15-2). It passes forward between the pos-

terior cerebral and superior cerebellar arteries (Fig. 15-3). It then continues into the middle cranial fossa in the lateral wall of the cavernous sinus. Here, it divides into a **superior** and an **inferior ramus,** which enter the orbital cavity through the superior orbital fissure (see Fig. 18-2).

The oculomotor nerve supplies the following:

- The **extrinsic muscles of the eye:** the levator palpebrae superioris, superior rectus, medial rectus, inferior rectus, and inferior oblique (Fig. 15-2).
- The **intrinsic muscles of the eye:** the constrictor pupillae of the iris and the ciliary muscles are supplied by the parasympathetic component of the oculomotor nerve. These fibers synapse in the ciliary ganglion and reach the eyeball in the short ciliary nerves.

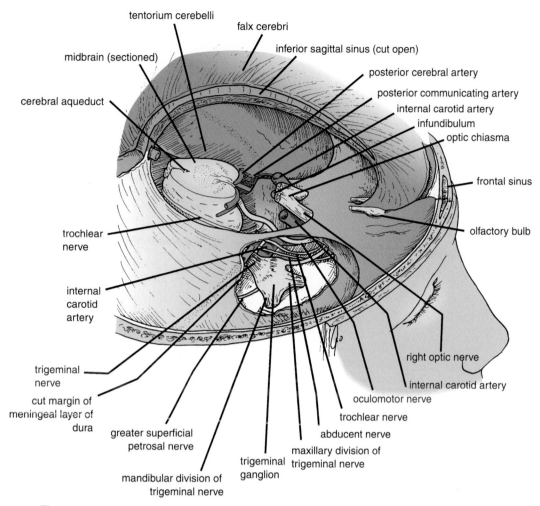

**Figure 15-3** Lateral view of the skull showing the falx cerebri, tentorium cerebelli, brainstem, and trigeminal ganglion.

The oculomotor nerve, therefore, is entirely motor. It is responsible for lifting the upper eyelid; turning the eye upward, downward, and medially; constricting the pupil; and accommodation of the eye.

# Trochlear Nerve

The trochlear nerve is the most slender of the cranial nerves. Having crossed the nerve of the opposite side, it leaves the posterior surface of the midbrain (Fig. 15-2). It then passes forward through the middle cranial fossa in the lateral wall of the cavernous sinus and enters the orbit through the superior orbital fissure (Fig. 15-3).

The trochlear nerve supplies the superior oblique muscle of the eyeball (extrinsic muscle). The trochlear nerve is entirely motor and assists in turning the eye downward and laterally.

# Trigeminal Nerve

The trigeminal nerve is the largest cranial nerve (Fig. 15-4). It leaves the anterior aspect of the pons as a small **motor root** and a large **sensory root**, and it passes forward, out of the posterior cranial fossa, to reach the apex of the petrous part of the temporal bone in the middle cranial fossa. Here, the large sensory root expands to form the **trigeminal ganglion** (Figs. 15-3 and 15-4). The trigeminal ganglion lies within a pouch of dura mater called the **trigeminal cave.** The motor root of the trigeminal nerve is situated below the sensory ganglion and is completely separate from it. The ophthalmic (V1), maxillary (V2), and mandibular (V3) nerves arise from the anterior border of the ganglion (Figs. 15-3 and 15-4).

## Ophthalmic Nerve

The ophthalmic nerve is purely sensory (Figs. 15-4 and 15-5). It runs forward in the lateral wall of the cavernous sinus in the middle cranial fossa and divides into three branches, the lacrimal, frontal, and nasociliary nerves, which enter the orbital cavity through the superior orbital fissure.

### Branches

The **lacrimal nerve** runs forward on the upper border of the lateral rectus muscle. It is joined by the zygomaticotemporal

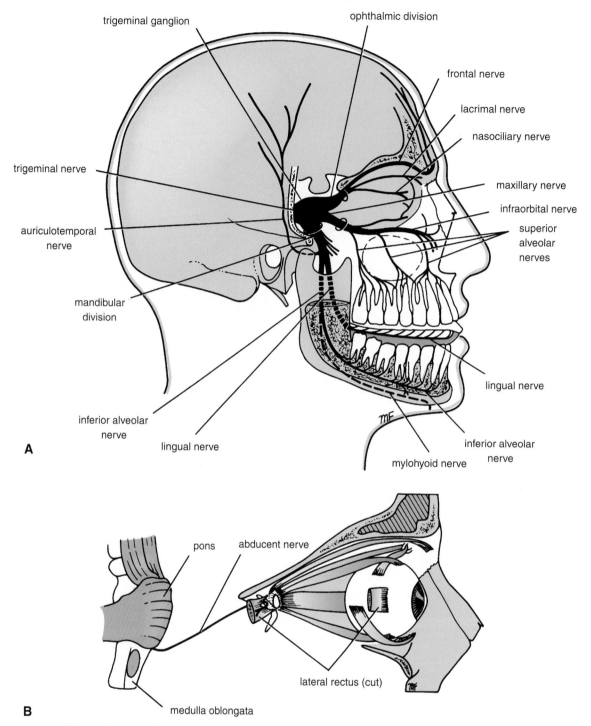

**Figure 15-4  A.** Distribution of the trigeminal nerve. **B.** Origin and distribution of the abducent nerve.

branch of the maxillary nerve, which contains the parasympathetic secretomotor fibers to the lacrimal gland. The lacrimal nerve then enters the lacrimal gland and gives branches to the conjunctiva and the skin of the upper eyelid.

The **frontal nerve** runs forward on the upper surface of the levator palpebrae superioris muscle and divides into the **supraorbital** and **supratrochlear nerves** (see Fig. 18-3).

These nerves leave the orbital cavity and supply the frontal air sinus and the skin of the forehead and the scalp.

The **nasociliary nerve** crosses the optic nerve, runs forward on the upper border of the medial rectus muscle (see Fig. 18-3), and continues as the **anterior ethmoid nerve** through the anterior ethmoidal foramen to enter the cranial cavity. It then descends through a slit at the side of the

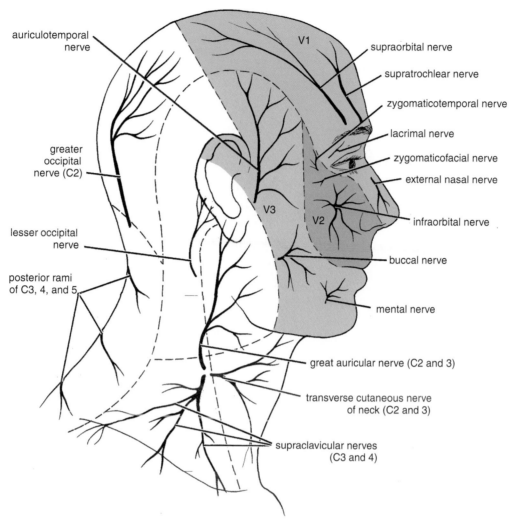

**Figure 15-5** The facial cutaneous distribution of the ophthalmic (V1) (*red*), maxillary (V2) (*blue*), and mandibular (V3) (*green*) divisions of the trigeminal nerve. Note that the skin over the angle of the jaw is supplied by the great auricular nerve (C2 and 3 segments of the spinal cord).

crista galli to enter the nasal cavity. It gives off two **internal nasal branches,** and it then supplies the skin of the tip of the nose with the **external nasal nerve.** Its branches include the following:

- **Sensory fibers** to the ciliary ganglion (Fig. 15-6).
- **Long ciliary nerves** that contain sympathetic fibers to the dilator pupillae muscle and sensory fibers to the cornea.
- **Infratrochlear nerve** that supplies the skin of the eyelids.
- **Posterior ethmoidal nerve** that is sensory to the ethmoid and sphenoid sinuses.

## Maxillary Nerve

The maxillary nerve is purely sensory (Fig. 15-4). It leaves the skull through the foramen rotundum (Fig. 15-3) and crosses the pterygopalatine fossa to enter the orbit through the inferior orbital fissure (Fig. 15-6). It then continues as

the **infraorbital nerve** in the infraorbital groove, and it emerges on the face through the infraorbital foramen. It gives sensory fibers to the skin of the face and the side of the nose.

## Branches

- **Meningeal branches.**
- **Zygomatic branch** (Fig. 15-6), which divides into the **zygomaticotemporal** and the **zygomaticofacial nerves** that supply the skin of the face. The zygomaticotemporal branch gives parasympathetic secretomotor fibers to the lacrimal gland via the lacrimal nerve.
- **Ganglionic branches,** which are two short nerves that suspend the pterygopalatine ganglion in the pterygopalatine fossa (Fig. 15-6). They contain sensory fibers that have passed through the ganglion from the nose, the palate, and the pharynx. They also contain postganglionic parasympathetic fibers that are going to the lacrimal gland.

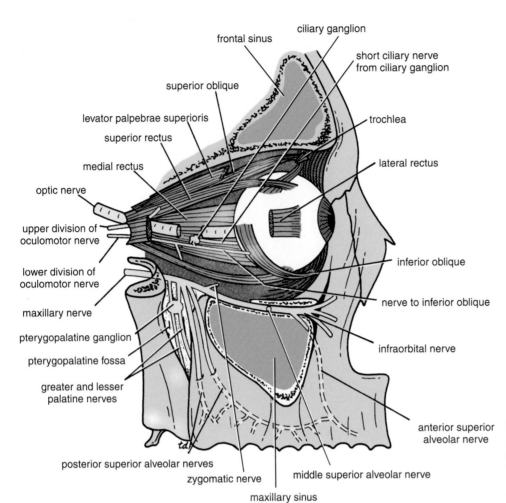

frontal sinus
ciliary ganglion
short ciliary nerve from ciliary ganglion
superior oblique
trochlea
levator palpebrae superioris
superior rectus
lateral rectus
medial rectus
optic nerve
upper division of oculomotor nerve
inferior oblique
lower division of oculomotor nerve
nerve to inferior oblique
maxillary nerve
infraorbital nerve
pterygopalatine ganglion
pterygopalatine fossa
greater and lesser palatine nerves
anterior superior alveolar nerve
posterior superior alveolar nerves
zygomatic nerve
middle superior alveolar nerve
maxillary sinus

**Figure 15-6** Muscles, nerves, and ciliary ganglion of the right orbit viewed from the lateral side. The maxillary nerve and the pterygopalatine ganglion are also shown.

- **Posterior superior alveolar nerve** (Fig. 15-6), which supplies the maxillary sinus as well as the upper molar teeth and adjoining parts of the gum and the cheek.
- **Middle superior alveolar nerve** (Fig. 15-6), which supplies the maxillary sinus as well as the upper premolar teeth, the gums, and the cheek.
- **Anterior superior alveolar nerve** (Fig. 15-6), which supplies the maxillary sinus as well as the upper canine and the incisor teeth.

## Pterygopalatine Ganglion

The pterygopalatine ganglion is a parasympathetic ganglion that is suspended from the maxillary nerve in the pterygopalatine fossa (Fig. 15-6). It is secretomotor to the lacrimal and nasal glands.

### Branches

- **Orbital branches**, which enter the orbit through the inferior orbital fissure.
- **Greater and lesser palatine nerves** (Fig. 15-6), which supply the palate, the tonsil, and the nasal cavity.
- **Pharyngeal branch**, which supplies the roof of the nasopharynx.

## Mandibular Nerve

The mandibular nerve is both motor and sensory (Figs. 15-3 and 15-4). The sensory root leaves the trigeminal ganglion and passes out of the skull through the foramen ovale to enter the infratemporal fossa. The motor root of the trigeminal nerve also leaves the skull through the foramen ovale and joins the sensory root to form the trunk of the mandibular nerve and then divides into a small anterior and a large posterior division (see Fig. 15-8).

### Branches from the Main Trunk of the Mandibular Nerve

- **Meningeal branch.**
- **Nerve to the medial pterygoid muscle**, which supplies not only the medial pterygoid muscle but the tensor veli palatini muscle as well.

### Branches from the Anterior Division of the Mandibular Nerve

- **Masseteric nerve** to the masseter muscle (Fig. 15-7).
- **Deep temporal nerves** to the temporalis muscle (Fig. 15-7).

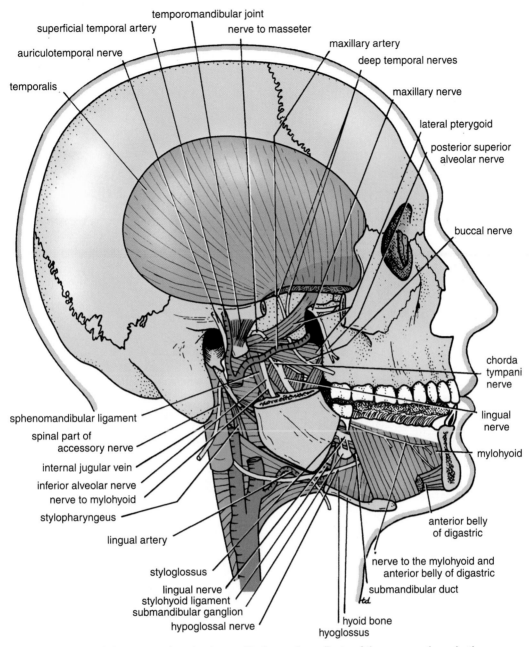

**Figure 15-7**   Infratemporal and submandibular regions. Parts of the zygomatic arch, the ramus, and the body of the mandible have been removed to display deeper structures.

■ **Nerve to the lateral pterygoid muscle.**

■ **Buccal nerve** to the skin and the mucous membranes of the cheek (Fig. 15-7). The buccal nerve **does not supply the buccinator muscle** (which is supplied by the facial nerve), and it is the **only sensory** branch of the anterior division of the mandibular nerve.

## Branches from the Posterior Division of the Mandibular Nerve

■ **Auriculotemporal nerve,** which supplies the skin of the auricle (Fig. 15-8), the external auditory meatus, the

temporomandibular joint, and the scalp. This nerve also conveys postganglionic parasympathetic secretomotor fibers from the otic ganglion to the parotid salivary gland.

■ **Lingual nerve,** which descends in front of the inferior alveolar nerve and enters the mouth (Figs. 15-7 and 15-8). It then runs forward on the side of the tongue and crosses the submandibular duct. In its course, it is joined by the **chorda tympani nerve** (Figs. 15-7 and 15-8), and it supplies the mucous membrane of the anterior two thirds of the tongue and the floor of the mouth. It also gives off **preganglionic parasympathetic secretomotor fibers** to the submandibular ganglion.

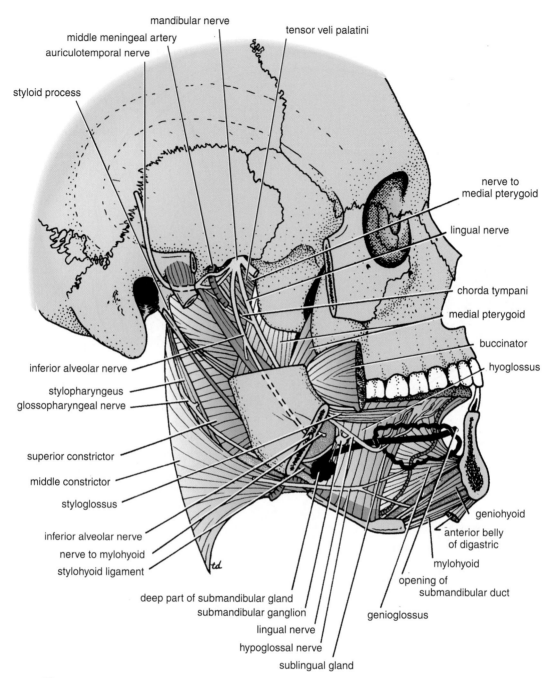

styloid process

auriculotemporal nerve
middle meningeal artery
mandibular nerve
tensor veli palatini

nerve to
medial pterygoid

lingual nerve

chorda tympani

medial pterygoid

buccinator

hyoglossus

inferior alveolar nerve

stylopharyngeus
glossopharyngeal nerve

superior constrictor

middle constrictor

styloglossus

inferior alveolar nerve
nerve to mylohyoid
stylohyoid ligament

geniohyoid
anterior belly
of digastric

mylohyoid
opening of
submandibular duct

deep part of submandibular gland
submandibular ganglion
lingual nerve
hypoglossal nerve
sublingual gland

genioglossus

**Figure 15-8** Infratemporal and submandibular regions. Parts of the zygomatic arch, the ramus, and the body of the mandible have been removed. Mylohyoid and lateral pterygoid muscles have also been removed to display deeper structures. The outline of the sublingual gland is shown as a *solid black wavy line.*

■ **Inferior alveolar nerve** (Figs. 15-7 and 15-8), which enters the mandibular canal to supply the teeth of the lower jaw and emerges through the mental foramen (mental nerve) to supply the skin of the chin (Fig. 15-5). Before entering the canal, it gives off the **mylohyoid nerve** (Fig. 15-7), which supplies the mylohyoid muscle and the anterior belly of the digastric muscle.

■ **Communicating branch,** which frequently runs from the inferior alveolar nerve to the lingual nerve.

The branches of the posterior division of the mandibular nerve are sensory (except the nerve to the mylohyoid muscle).

## Otic Ganglion

The otic ganglion is a parasympathetic ganglion that is located medial to the mandibular nerve just below the skull, and it is adherent to the nerve to the medial pterygoid muscle. The preganglionic fibers originate in the glossopharyngeal nerve, and they reach the ganglion via the lesser petrosal nerve. The postganglionic secretomotor fibers reach the parotid salivary gland via the auriculotemporal nerve.

## Submandibular Ganglion

The submandibular ganglion is a parasympathetic ganglion that lies deep to the submandibular salivary gland and is attached to the lingual nerve by small nerves (Figs. 15-7 and 15-8). Preganglionic parasympathetic fibers reach the ganglion from the facial nerve via the chorda tympani and the lingual nerves. Postganglionic secretomotor fibers pass to the submandibular and the sublingual salivary glands.

The trigeminal nerve is thus the main sensory nerve of the head and innervates the muscles of mastication. It also tenses the soft palate and the tympanic membrane.

# Abducent Nerve

This small nerve emerges from the anterior surface of the hindbrain between the pons and the medulla oblongata (Figs. 15-3 and 15-4). It passes forward with the internal carotid artery through the cavernous sinus in the middle cranial fossa and enters the orbit through the superior orbital fissure (Fig. 15-4). The abducent nerve supplies the lateral rectus muscle (see Fig.18-2A) and is therefore responsible for turning the eye laterally.

# Facial Nerve

The facial nerve emerges as a motor root and a sensory root (**nervus intermedius**) (Fig. 15-9). The nerve emerges on the anterior surface of the hindbrain between the pons and the medulla oblongata. The roots pass laterally in the posterior cranial fossa with the vestibulocochlear nerve and enter the internal acoustic meatus in the petrous part of the temporal bone (see Figs. 14-16 and 18-14A). At the bottom of the meatus, the nerve enters the facial canal that runs laterally through the inner ear. On reaching the medial wall of the middle ear (tympanic cavity), the nerve swells to form the sensory **geniculate ganglion** (Fig. 15-9; and see Figs. 18-15 and 18-16). The nerve then bends sharply backward above the promontory and, at the posterior wall of the middle ear, bends down on the medial side of the aditus of the mastoid. The nerve descends behind the pyramid, and it emerges from the temporal bone through the stylomastoid foramen. The facial nerve now passes forward through the parotid gland to its distribution (Figs. 15-9 and 15-10).

## Important Branches of the Facial Nerve

- **Greater petrosal nerve** arises from the nerve at the geniculate ganglion (Fig. 15-9). It contains preganglionic parasympathetic fibers that synapse in the pterygopalatine ganglion. The postganglionic fibers are secretomotor to the lacrimal gland and the glands of the nose and the palate. The greater petrosal nerve also contains taste fibers from the palate.
- **Nerve to stapedius** supplies the stapedius muscle in the middle ear (Fig. 15-9).
- **Chorda tympani** arises from the facial nerve in the facial canal in the posterior wall of the middle ear (Fig. 15-9). It runs forward over the medial surface of the upper part of the tympanic membrane (see Fig.18-15A) and leaves the middle ear through the **petrotympanic fissure**, thus entering the infratemporal fossa and joining the lingual nerve. The chorda tympani contains preganglionic parasympathetic secretomotor fibers to the submandibular and the sublingual salivary glands. It also contains taste fibers from the anterior two thirds of the tongue and floor of the mouth.
- **Posterior auricular, the posterior belly of the digastric, and the stylohyoid nerves** (Fig. 15-9) are muscular branches given off by the facial nerve as it emerges from the stylomastoid foramen.
- **Five terminal branches to the muscles of facial expression.** These are the **temporal**, the **zygomatic**, the **buccal**, the **mandibular**, and the **cervical branches** (Fig. 15-9).

The facial nerve lies within the parotid salivary gland (Fig. 15-10) after leaving the stylomastoid foramen, and it is located between the superficial and the deep parts of the gland. Here, it gives off the terminal branches that emerge from the anterior border of the gland and pass to the muscles of the face and the scalp. **The buccal branch supplies the buccinator muscle, and the cervical branch supplies the platysma and the depressor anguli oris muscles.**

The facial nerve thus controls facial expression, salivation, and lacrimation and is a pathway for taste sensation from the anterior part of the tongue and floor of the mouth and from the palate.

# Vestibulocochlear Nerve

The vestibulocochlear nerve is a sensory nerve that consists of two sets of fibers: **vestibular** and **cochlear**. They leave the anterior surface of the brain between the pons and the medulla oblongata (Fig. 15-11). They cross the posterior cranial fossa and enter the internal acoustic meatus with the facial nerve (see Fig. 18-14A).

## Vestibular Fibers

The vestibular fibers are the central processes of the nerve cells of the vestibular ganglion situated in the internal acoustic meatus (Fig. 15-11). The vestibular fibers originate from the vestibule and the semicircular canals; therefore, they are concerned with the sense of position and with movement of the head.

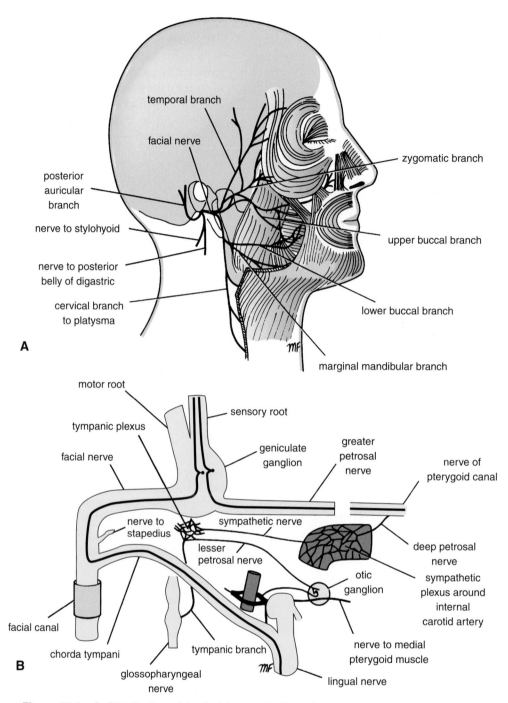

**Figure 15-9    A.** Distribution of the facial nerve. **B.** Branches of the facial nerve within the petrous part of the temporal bone; the taste fibers are shown in *black*. The glossopharyngeal nerve is also shown.

## Cochlear Fibers

The cochlear fibers are the central processes of the nerve cells of the **spiral ganglion of the cochlea** (Fig. 15-11). The cochlear fibers originate in the **spiral organ of Corti** and are therefore concerned with hearing.

# Glossopharyngeal Nerve

The glossopharyngeal nerve is a motor and sensory nerve (Fig. 15-11). It emerges from the anterior surface of the medulla oblongata between the olive and the inferior cerebellar peduncle. It passes laterally in the posterior cranial

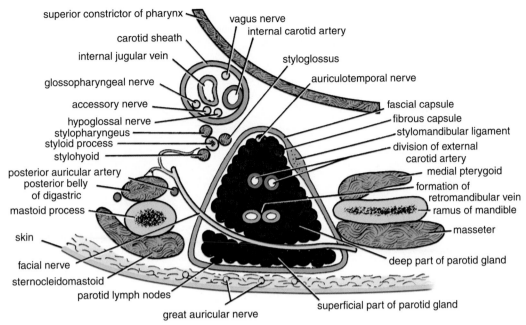

**Figure 15-10**   Horizontal section of the parotid gland showing the course of the facial nerve through the gland.

fossa and leaves the skull by passing through the jugular foramen. The **superior and inferior sensory ganglia** are located on the nerve as it passes through the foramen. The glossopharyngeal nerve then descends through the upper part of the neck to the back of the tongue (Fig. 15-11).

## Important Branches of the Glossopharyngeal Nerve

- **Tympanic branch** passes to the tympanic plexus in the middle ear (Fig. 15-11). Preganglionic parasympathetic fibers for the parotid salivary gland now leave the plexus as the **lesser petrosal nerve**, and they synapse in the otic ganglion.
- **Carotid branch** contains sensory fibers from the carotid sinus (pressoreceptor mechanism for the regulation of blood pressure) and the carotid body (chemoreceptor mechanism for the regulation of heart rate and respiration) (Fig. 15-11).
- **Nerve to the stylopharyngeus muscle.**
- **Pharyngeal branches** (Fig. 15-11) run to the **pharyngeal plexus**, which also receives branches from the vagus nerve and the sympathetic trunk.
- **Lingual branch** (Fig. 15-11) passes to the mucous membrane of the posterior third of the tongue (including the vallate papillae).

The glossopharyngeal nerve thus assists swallowing and promotes salivation. It also conducts sensation from the pharynx and the back of the tongue and carries impulses, which influence the arterial blood pressure and respiration, from the carotid sinus and carotid body.

# Vagus Nerve

The vagus nerve is composed of motor and sensory fibers (Fig. 15-12). It emerges from the anterior surface of the medulla oblongata between the olive and the inferior cerebellar peduncle. The nerve passes laterally through the posterior cranial fossa and leaves the skull through the jugular foramen. The vagus nerve has both **superior and inferior sensory ganglia**. Below the inferior ganglion, the **cranial root of the accessory nerve** joins the vagus nerve and is distributed mainly in its pharyngeal and recurrent laryngeal branches.

The vagus nerve descends through the neck alongside the carotid arteries and internal jugular vein within the carotid sheath (Fig. 15-10). It passes through the mediastinum of the thorax (Fig. 15-12), passing behind the root of the lung, and enters the abdomen through the esophageal opening in the diaphragm.

## Important Branches of the Vagus Nerve in the Neck

- **Meningeal and auricular branches.**
- **Pharyngeal branch** contains nerve fibers from the cranial part of the accessory nerve. This branch joins the pharyngeal plexus and supplies all the muscles of the pharynx (except the stylopharyngeus) and of the soft palate (except the tensor veli palatini).
- **Superior laryngeal nerve** (Fig. 15-12) divides into the internal and the external laryngeal nerves. The **internal laryngeal nerve** is sensory to the mucous membrane of the piriform fossa and the larynx down as far as the vocal

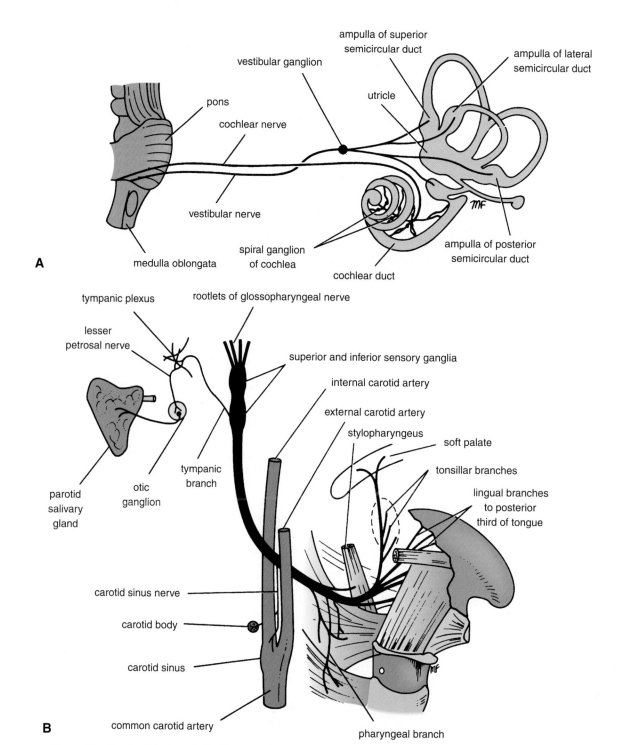

**Figure 15-11 A.** Origin and distribution of the vestibulocochlear nerve. **B.** Distribution of the glossopharyngeal nerve.

cords. The **external laryngeal nerve** is motor and is located close to the superior thyroid artery; it supplies the cricothyroid muscle.

- **Recurrent laryngeal nerve** (Fig. 15-12). On the right side, the nerve hooks around the **first part of the subclavian artery** and then ascends in the groove between the trachea and the esophagus. On the left side, the nerve hooks around the **arch of the aorta** and then ascends into the neck between the trachea and the esophagus. The nerve is closely related to the inferior thyroid artery, and it supplies all the muscles of the larynx, except the cricothyroid muscle, the mucous membrane of the larynx below the vocal cords, and the mucous membrane of the upper part of the trachea.

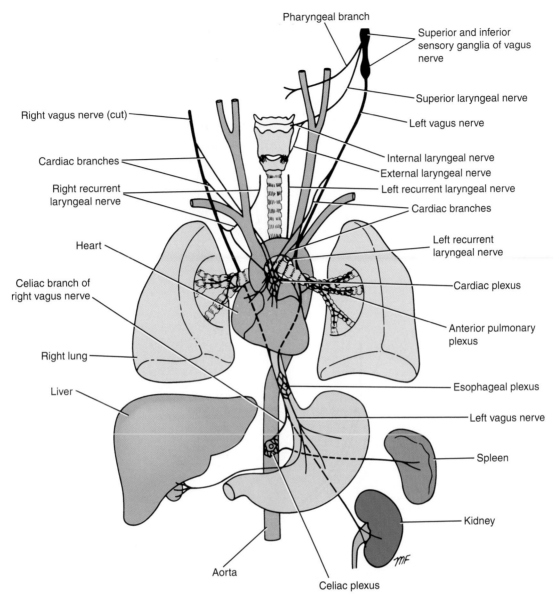

Pharyngeal branch

Superior and inferior sensory ganglia of vagus nerve

Superior laryngeal nerve

Left vagus nerve

Right vagus nerve (cut)

Internal laryngeal nerve

External laryngeal nerve

Cardiac branches

Left recurrent laryngeal nerve

Right recurrent laryngeal nerve

Cardiac branches

Heart

Left recurrent laryngeal nerve

Cardiac plexus

Celiac branch of right vagus nerve

Anterior pulmonary plexus

Right lung

Esophageal plexus

Liver

Left vagus nerve

Spleen

Kidney

Aorta

Celiac plexus

**Figure 15-12** Distribution of the vagus nerve.

■ **Cardiac branches (2 or 3)** arise in the neck, descend into the thorax, and end in the cardiac plexus (Fig. 15-12).

The vagus nerve thus innervates the heart and great vessels within the thorax; the larynx, trachea, bronchi, and lungs; and much of the alimentary tract from the pharynx to the splenic flexure of the colon. It also supplies glands associated with the alimentary tract, such as the liver and pancreas.

The vagus nerve has the most extensive distribution of all the cranial nerves and supplies the aforementioned structures with afferent and efferent fibers.

# Accessory Nerve

The accessory nerve is a motor nerve. It consists of a cranial root (part) and a spinal root (part) (Fig. 15-13).

## Cranial Root

The cranial root emerges from the anterior surface of the medulla oblongata between the olive and the inferior cerebellar peduncle (Fig. 15-13). The nerve runs laterally in the posterior cranial fossa and joins the spinal root.

## Spinal Root

The spinal root arises from nerve cells in the anterior gray column (horn) of the upper five segments of the cervical part of the spinal cord (Fig. 15-13). The nerve ascends alongside the spinal cord and enters the skull through the foramen magnum. It then turns laterally to join the cranial root.

The two roots unite and leave the skull through the jugular foramen. The roots then separate: the cranial root joins

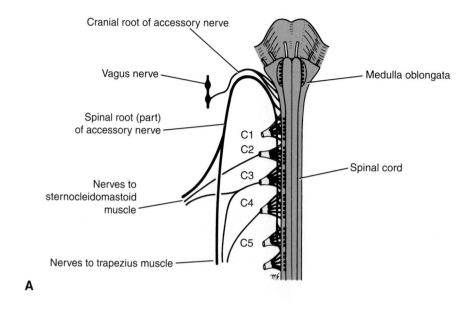

A

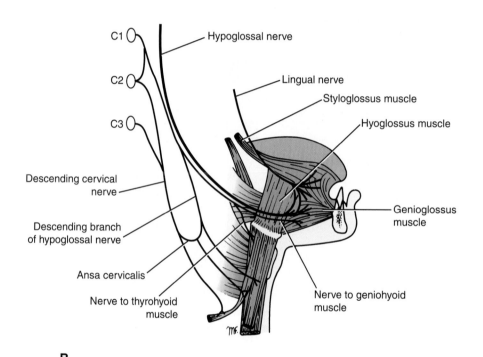

B

**Figure 15-13**   **A.** Origin and distribution of the accessory nerve. **B.** Distribution of the hypoglossal nerve.

the vagus nerves and is distributed in its branches to the muscles of the soft palate and pharynx (via the pharyngeal plexus) and to the muscles of the larynx (except the cricothyroid muscle). The spinal root runs downward and laterally and enters the deep surface of the sternocleidomastoid muscle, which it supplies, and then crosses the posterior triangle of the neck to supply the trapezius muscle (Fig. 13-9).

The accessory nerve thus brings about movements of the soft palate, pharynx, and larynx and controls the movements

of the sternocleidomastoid and trapezius muscles, two large muscles in the neck.

# Hypoglossal Nerve

The hypoglossal nerve is a motor nerve. It emerges on the anterior surface of the medulla oblongata between the pyramid and the olive, crosses the posterior cranial fossa, and leaves the skull through the hypoglossal canal. The nerve then passes downward and forward in the neck and crosses the internal and external carotid arteries to reach the tongue (Fig. 15-13). In the upper part of its course, it is joined by C1 fibers from the cervical plexus.

# Important Branches of the Hypoglossal Nerve

- Meningeal branch.
- Descending branch (C1 fibers) passes downward and joins the descending cervical nerve (C2 and 3) to form the ansa cervicalis. Branches from this loop supply the omohyoid, the sternohyoid, and the sternothyroid muscles.
- Nerve to the thyrohyoid muscle (C1).
- Muscular branches to all the muscles of the tongue except the palatoglossus (pharyngeal plexus).
- Nerve to the geniohyoid muscle (C1).

The hypoglossal nerve thus innervates the muscles of the tongue (except the palatoglossus) and therefore controls the shape and movements of the tongue.

# Review Questions

## Completion Questions

### Select the phrase that BEST completes each statement.

1. The mandibular division of the trigeminal nerve leaves the skull through the
   A. superior orbital fissure.
   B. foramen rotundum.
   C. foramen ovale.
   D. jugular foramen.
   E. foramen magnum.

2. The vagus nerve leaves the skull through the
   A. jugular foramen.
   B. occipital foramen.
   C. inferior orbital fissure.
   D. foramen rotundum.
   E. foramen spinosum.

3. The abducent nerve leaves the skull through the
   A. foramen rotundum.
   B. jugular foramen.
   C. inferior orbital fissure.
   D. superior orbital fissure.
   E. foramen ovale.

4. The ophthalmic division of the trigeminal nerve leaves the skull through the
   A. inferior orbital fissure.
   B. foramen ovale.
   C. foramen rotundum.
   D. superior orbital fissure.
   E. pterygopalatine foramen.

5. The maxillary division of the trigeminal nerve leaves the skull through the
   A. foramen spinosum.
   B. foramen rotundum.
   C. superior orbital fissure.
   D. foramen ovale.
   E. jugular foramen.

6. The oculomotor nerve leaves the skull through the
   A. inferior orbital fissure.
   B. foramen rotundum.
   C. superior orbital fissure.
   D. foramen magnum.
   E. foramen ovale.

7. The facial nerve canal is located in the
   A. temporal bone.
   B. greater wing of the sphenoid bone.
   C. occipital bone.
   D. mastoid process.
   E. lacrimal bone.

8. The vagus nerve supplies the gastrointestinal tract down as far as the
   A. pylorus of the stomach.
   B. duodenojejunal junction.
   C. ileocolic junction.
   D. hepatic flexure of the colon.
   E. splenic flexure of the colon.

9. The facial nerve supplies all the muscles of face including the
   A. masseter.
   B. buccinator.
   C. lateral pterygoid.
   D. temporalis.
   E. medial pterygoid.

## Multiple Choice Questions

### Select the BEST answer for each question.

10. The following statements concerning the chorda tympani are correct **except** which?
    A. It contains parasympathetic postganglionic fibers.
    B. It contains special sensory (taste) fibers.
    C. It joins the lingual nerve in the infratemporal fossa.
    D. It is a branch of the facial nerve in the temporal bone.
    E. It carries secretomotor fibers to the submandibular and sublingual salivary glands.

11. Assuming that the patient's eyesight is normal, in which cranial nerve is there likely to be a lesion when the direct and consensual light reflexes are absent?
    A. Trochlear nerve
    B. Optic nerve
    C. Abducent nerve
    D. Oculomotor nerve
    E. Trigeminal nerve

12. A patient is unable to taste a piece of sugar placed on the anterior part of the tongue. Which cranial nerve is likely to have a lesion?
    A. Hypoglossal
    B. Vagus
    C. Glossopharyngeal
    D. Facial
    E. Maxillary division of the trigeminal

13. On asking a patient to say "ah," the uvula is seen to be drawn upward to the right. Which cranial nerve is likely to be damaged?
    A. Left glossopharyngeal
    B. Right hypoglossal
    C. Left accessory (cranial part)
    D. Right vagus
    E. Right trigeminal

14. The following statements concerning the optic nerve are correct **except** which?
    A. The axons arise from the ganglionic layer of the retina.
    B. At the optic chiasma, the fibers from the lateral half of each retina cross the midline and enter the optic tract of the opposite side.
    C. The optic nerve leaves the orbital cavity through the optic canal.
    D. The optic nerve is surrounded by the three meninges and an extension of the subarachnoid space into the orbital cavity.
    E. The optic nerve is made up of myelinated axons.

15. When testing the sensory innervation of the face, it is important to remember that the skin of the tip of the nose is supplied by one of the following nerves.
    A. Zygomatic branch of the facial nerve
    B. Maxillary division of the trigeminal nerve
    C. Ophthalmic division of the trigeminal nerve
    D. External nasal branch of the facial nerve
    E. Buccal branch of the mandibular division of the trigeminal nerve

16. Compression of the facial nerve in the facial canal in the posterior wall of the middle ear could result in all of the following **except** which?
    A. Cessation of lacrimal secretion.
    B. Paralysis of the posterior belly of the digastric muscle.
    C. Inability to whistle.
    D. Decreased saliva in the mouth.
    E. Loss of taste sensation in the anterior two thirds of the tongue.

17. The following statements concerning the function of the oculomotor nerve are correct **except** which?
    A. It accommodates the eye.
    B. It raises the upper eyelid.
    C. It innervates the lateral rectus muscle and thus turns the eye laterally.
    D. It turns the eye downward.
    E. It constricts the pupil.

18. The following statements concerning a lesion of the trigeminal nerve are correct **except** which?
    A. The masseter muscle cannot be felt to contract.
    B. There is loss of skin sensation over the angle of the jaw.
    C. The cornea and conjunctiva are insensitive to touch.
    D. The temporalis muscle cannot be felt to contract.
    E. There is loss of skin sensation over the cheek.

19. The following statements concerning a lesion of the spinal part of the accessory nerve are correct **except** which?
    A. It arises from the first three cervical segments of the spinal cord.
    B. It enters the skull through the foramen magnum.
    C. It joins the cranial part of the accessory nerve in the posterior cranial fossa.
    D. It supplies the sternocleidomastoid and trapezius muscles.
    E. It can be injured in the posterior triangle of the neck.

20. The following statements concerning the hypoglossal nerve are correct **except** which?
    A. It supplies all the intrinsic muscles of the tongue.
    B. In lesions of the nerve, the tip of the tongue deviates to the same side when the tongue is protruded from the mouth.
    C. It supplies the palatoglossus muscle.
    D. It leaves the anterior surface of the brain between the pyramid and the olive.
    E. It supplies the styloglossus and the hyoglossus muscles.

# Answers and Explanations

1. **C is correct.** Both the motor and sensory divisions of the mandibular division of the trigeminal nerve leave the skull together through the foramen ovale and quickly unite.

2. **A is correct.** The glossopharyngeal, vagus, and accessory nerves, leave the skull through the jugular foramen; the sigmoid sinus passes through the posterior part of the same foramen to become the internal jugular vein.

3. **D is correct.** The abducent nerve leaves the skull through the superior orbital fissure.

4. **D is correct.** The ophthalmic division of the trigeminal nerve leaves the skull through the superior orbital fissure as its three terminal branches—namely, the lacrimal, frontal, and nasociliary nerve.

5. **B is correct.** The ophthalmic division of the trigeminal nerve leaves the skull through the foramen rotundum to enter the pterygopalatine fossa (Fig. 15-6).

6. **C is correct.** The oculomotor nerve passes through the superior orbital fissure as upper and lower divisions.

7. **A is correct.** The facial canal is located in the temporal bone (Fig. 18-14).

8. **E is correct.** The vagus nerve supplies the gastrointestinal tract down as far as the splenic flexure.

9. **B is correct.** The facial nerve supplies all the muscles of the face and the buccinator muscle.

10. **A is incorrect.** The chorda tympani contains parasympathetic preganglionic fibers.

11. **E is correct.** If the patient's eyesight is normal but the direct and consensual light reflexes are absent, a lesion of the oculomotor nerve is likely.

12. **D is correct.** Loss of the sensation of taste on the anterior two thirds of the tongue is caused by a lesion of the facial nerve. The sensation of taste travels in the chorda tympani nerve fibers in the lingual nerve, and they ultimately join the facial nerve in the facial nerve canal (Figs. 15-8 and 15-9).

13. **C is correct.** The left accessory nerve (cranial part) supplies the left levator veli palatini muscle via the pharyngeal plexus; the left veli palatini muscle was paralyzed in this patient.

14. **B is incorrect.** At the optic chiasma, the fibers from the medial half of each retina cross the midline and enter the optic tract of the opposite side (Fig. 15-1).

15. **C is correct.** The external nasal nerve is a continuation of the anterior ethmoidal branch of the nasociliary branch of the ophthalmic division of the trigeminal nerve.

16. **A is incorrect.** Lacrimal secretion is controlled by the lacrimal nucleus of the facial nerve. The fibers leave the facial nerve as the greater petrosal nerve on the medial wall of the middle ear **before** the facial nerve reaches the posterior wall of the middle ear.

17. **C is incorrect.** The lateral rectus muscle is innervated by the abducent nerve and not the oculomotor nerve.

18. **B is incorrect.** The skin over the angle of the jaw is supplied by the great auricular nerve (C2 and C3) and not by the trigeminal nerve.

19. **A is incorrect.** The spinal part of the accessory nerve arises from the first five cervical segments of the spinal cord (Fig. 15-13).

20. **C is incorrect.** The palatoglossus muscle is supplied by the cranial part of the accessory nerve through the pharyngeal branches of the vagus nerve.

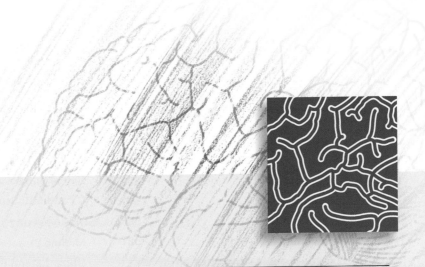

# 16

# The Vertebral Column,
# the Spinal Cord,
# and the Meninges

All clinical material relevant to this chapter can be found on the
CD-ROM.

# Chapter Outline

I njuries to the vertebral column commonly occur in automobile and motorcycle accidents, falls, sports injuries, and gunshot wounds. Spinal cord and spinal nerve damage may be associated with vertebral fractures and herniated intervertebral discs. Back injuries range from a simple acute back strain to a catastrophic injury of the spinal cord or cauda equina.

Because unprotected movement of a damaged vertebral column during initial medical care can result in injury to the delicate spinal cord, emergency personnel must be familiar with the overall anatomy of this region. The assessment of neurologic damage requires not only an understanding of the main nervous pathways within the spinal cord but an ability to correlate radiologic evidence of bone injury with segmental levels of the spinal cord and subtle neurologic deficits.

The purpose of this chapter is to review the basic anatomy of the vertebral column and related soft nervous tissue structures.

# BASIC ANATOMY

## Vertebral Column

The vertebral column is the central bony pillar of the body. It supports the skull, pectoral girdle, upper limbs, and thoracic cage and, by way of the pelvic girdle, transmits body weight to the lower limbs. Within its cavity lie the spinal cord, the roots of the spinal nerves, and the covering meninges, to which the vertebral column gives great protection.

### Composition of the Vertebral Column

The vertebral column (Figs. 16-1 and 16-2) is composed of 33 vertebrae—7 cervical, 12 thoracic, 5 lumbar, 5 sacral

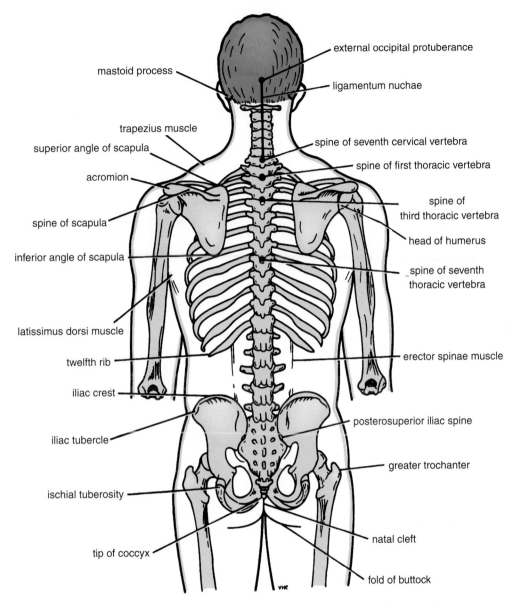

**Figure 16-1**   Posterior view of the skeleton showing the surface markings on the back.

(fused to form the sacrum), and 4 coccygeal (the lower 3 are commonly fused). Because it is segmented and made up of vertebrae, joints, and pads of fibrocartilage called **intervertebral discs**, it is a flexible structure. The intervertebral discs form about one fourth the length of the column.

## General Characteristics of a Vertebra

The general characteristics of the vertebrae in the different regions of the vertebral column should be reviewed in Chapter 11. Although vertebrae show regional differences, they all possess a common pattern (Fig. 16-2).

A **typical vertebra** consists of a rounded **body** anteriorly and a **vertebral arch** posteriorly. These enclose a space called the **vertebral foramen,** through which run the spinal cord and its coverings. The vertebral arch consists of a pair of cylindrical **pedicles,** which form the sides of the arch, and a pair of flattened **laminae,** which complete the arch posteriorly.

The vertebral arch gives rise to seven processes: one spinous, two transverse, and four articular (Fig. 16-2).

The **spinous process,** or **spine,** is directed posteriorly from the junction of the two laminae. The transverse processes are directed laterally from the junction of the laminae and the pedicles. Both the spinous and transverse processes serve as levers and receive attachments of muscles and ligaments.

The **articular processes** are vertically arranged and consist of two superior and two inferior processes. They arise

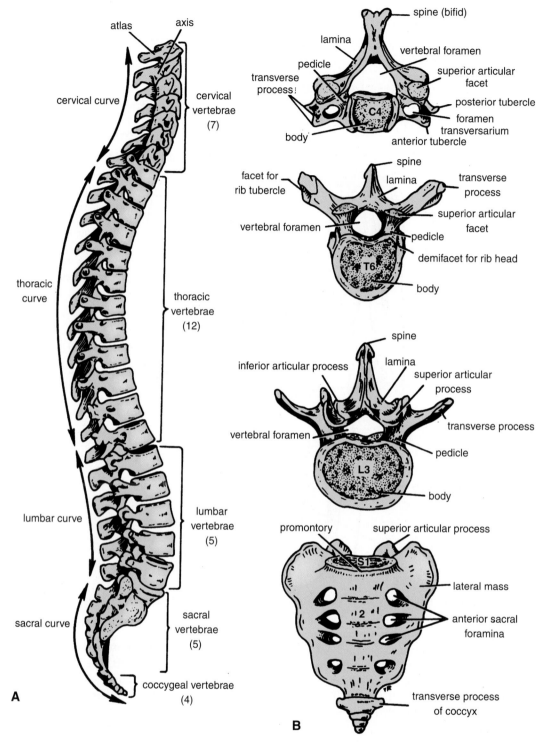

**Figure 16-2** **A.** Lateral view of the vertebral column. **B.** General features of different kinds of vertebrae.

from the junction of the laminae and the pedicles, and their articular surfaces are covered with hyaline cartilage.

The two superior articular processes of one vertebral arch articulate with the two inferior articular processes of the arch above, forming two synovial joints.

The pedicles are notched on their upper and lower borders, forming the **superior** and **inferior vertebral notches.** On each side, the superior notch of one vertebra and the inferior notch of an adjacent vertebra together form an **intervertebral foramen.** These foramina, in an articulated skeleton, serve to transmit the spinal nerves and blood vessels. The anterior and posterior nerve roots of a spinal nerve unite within these foramina with their coverings of dura to form the segmental spinal nerves.

## Joints of the Vertebral Column

The atlanto-occipital joints, the atlanto-axial joints, and the joints of the vertebral column below the axis should be reviewed in Chapter 12.

Below the axis, the vertebrae articulate with each other by means of cartilaginous joints between their bodies and by synovial joints between their articular processes. A brief review will be given here.

### Joints between Two Vertebral Bodies

The upper and lower surfaces of the bodies of adjacent vertebrae are covered by thin plates of hyaline cartilage. Sandwiched between the plates of hyaline cartilage is an intervertebral disc of fibrocartilage (Fig. 16-3).

#### Intervertebral Discs

The intervertebral discs (Fig. 16-3) are thickest in the cervical and lumbar regions, where the movements of the vertebral column are greatest. They serve as shock absorbers when the load on the vertebral column is suddenly increased. Unfortunately, their resilience is gradually lost with advancing age.

Each disc consists of a peripheral part, the anulus fibrosus, and a central part, the nucleus pulposus (Fig. 16-3).

The **anulus fibrosus** is composed of fibrocartilage, which is strongly attached to the vertebral bodies and the anterior and posterior longitudinal ligaments of the vertebral column.

The **nucleus pulposus** in the young is an ovoid mass of gelatinous material. It is normally under pressure and situated slightly nearer to the posterior than to the anterior margin of the disc. The upper and lower surfaces of the bodies of adjacent vertebrae that abut onto the disc are covered with thin plates of hyaline cartilage.

The semifluid nature of the nucleus pulposus allows it to change shape and permits one vertebra to rock forward or backward on another. A sudden increase in the compression load on the vertebral column causes the nucleus pulposus to become flattened and this is accommodated by the resilience of the surrounding anulus fibrosus. Sometimes, the outward thrust is too great for the anulus fibrosus, and it ruptures, allowing the nucleus pulposus to herniate and protrude into the vertebral canal, where it may press on the spinal nerve roots, the spinal nerve, or even the spinal cord.

With advancing age, the nucleus pulposus becomes smaller and is replaced by fibrocartilage. The collagen fibers of the anulus degenerate and, as a result, the anulus cannot always contain the nucleus pulposus under stress. In old age, the discs are thin and less elastic, and it is no longer possible to distinguish the nucleus from the anulus.

#### Ligaments

The **anterior** and **posterior longitudinal ligaments** run as continuous bands down the anterior and posterior surfaces of the vertebral column from the skull to the sacrum (Fig. 16-3 and CD Fig. 16-3). The anterior ligament is wide and is strongly attached to the front and sides of the vertebral bodies and to the intervertebral discs. The posterior ligament is weak and narrow and is attached to the posterior borders of the discs.

### Joints between Two Vertebral Arches

The joints between two vertebral arches consist of synovial joints between the superior and inferior articular processes of adjacent vertebrae (Fig. 16-3).

#### Ligaments

- **Supraspinous ligament** (Fig. 16-3): This runs between the tips of adjacent spines.
- **Interspinous ligament** (Fig. 16-3): This connects adjacent spines.
- **Intertransverse ligaments**: These run between adjacent transverse processes.
- **Ligamentum flavum** (Fig. 16-3): This connects the laminae of adjacent vertebrae.

In the cervical region, the supraspinous and interspinous ligaments are greatly thickened to form the strong **ligamentum nuchae.**

### Nerve Supply of Vertebral Joints

The joints between the vertebral bodies are innervated by the small meningeal branches of each spinal nerve (Fig. 16-4). The joints between the articular processes are innervated by branches from the posterior rami of the spinal nerves (Fig. 16-4); the joints of any particular level receive nerve fibers from two adjacent spinal nerves.

# Spinal Cord

The spinal cord is a cylindrical, grayish white structure that begins above at the foramen magnum, where it is continuous with the medulla oblongata of the brain. It terminates below in the adult at the level of the lower border of the first lumbar vertebra (Fig. 16-5). In the young child, it is relatively longer and ends at the upper border of the third lumbar vertebra. The spinal cord in the cervical region, where it gives origin to the brachial plexus, and in the lower thoracic and lumbar regions, where it gives origin to the lumbosacral plexus,

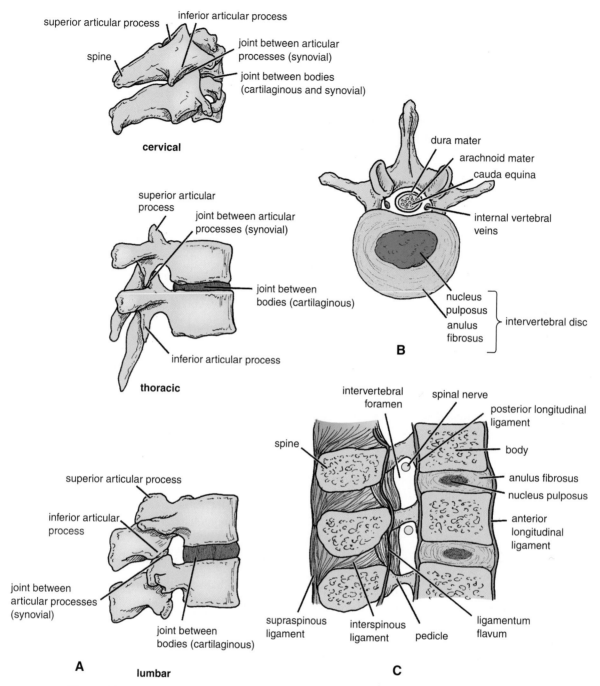

**Figure 16-3** **A.** Joints in the cervical, thoracic, and lumbar regions of the vertebral column. **B.** Third lumbar vertebra seen from above showing the relationship between intervertebral disc and cauda equina. **C.** Sagittal section through three lumbar vertebrae showing ligaments and intervertebral discs. Note the relationship between the emerging spinal nerve in an intervertebral foramen and the intervertebral disc.

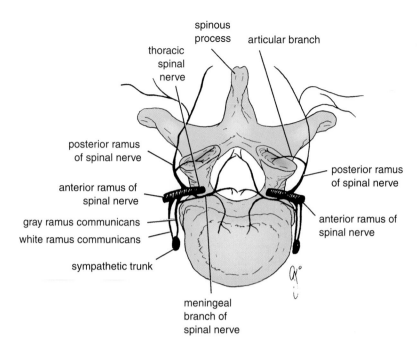

spinous
process

articular branch

thoracic
spinal
nerve

posterior ramus
of spinal nerve

anterior ramus of
spinal nerve

gray ramus communicans

white ramus communicans

sympathetic trunk

posterior ramus
of spinal nerve

anterior ramus of
spinal nerve

meningeal
branch of
spinal nerve

**Figure 16-4**   The innervation of vertebral joints. At any particular vertebral level, the joints receive nerve fibers from two adjacent spinal nerves.

has fusiform enlargements called **cervical** and **lumbar enlargements.**

Inferiorly, the spinal cord tapers off into the **conus medullaris,** from the apex of which a prolongation of the pia mater, the **filum terminale,** descends to be attached to the back of the coccyx (Figs. 16-5 and 16-6). The cord possesses in the midline anteriorly a deep longitudinal fissure, the **anterior median fissure,** and on the posterior surface a shallow furrow, the **posterior median sulcus.**

## Roots of the Spinal Nerves

Along the whole length of the spinal cord are attached 31 pairs of spinal nerves by the **anterior,** or **motor, roots** and the **posterior,** or **sensory, roots** (Fig. 16-6). Each root is attached to the cord by a series of rootlets, which extend the whole length of the corresponding segment of the cord. Each posterior nerve root possesses a posterior root ganglion, the cells of which give rise to peripheral and central nerve fibers.

The spinal nerve roots pass laterally from each spinal cord segment to the level of their respective intervertebral foramina, where they unite to form a **spinal nerve.** Here, the motor and sensory fibers become mixed so that a spinal nerve is made up of a mixture of motor and sensory fibers. Because of the disproportionate growth in length of the vertebral column during development, compared to that of the spinal cord, the length of the roots increases progressively from above downward (Fig. 16-7). In the upper cervical region, the spinal nerve roots are short and run almost horizontally, but the roots of the lumbar and sacral nerves below the level of the termination of

the cord (lower border of the first lumbar vertebra in the adult) form a vertical leash of nerves around the filum terminale. The lower nerve roots together are called the **cauda equina** (Fig. 16-6).

After emergence from the intervertebral foramen, each spinal nerve immediately divides into a large **anterior ramus** and a smaller **posterior ramus,** which contain both motor and sensory fibers.

## Blood Supply of the Spinal Cord

The spinal cord receives its arterial supply from three small, longitudinally running arteries: the two posterior spinal arteries and one anterior spinal artery. The **posterior spinal arteries,** which arise either directly or indirectly from the vertebral arteries, run down the side of the spinal cord, close to the attachments of the posterior spinal nerve roots. The **anterior spinal arteries,** which arise from the vertebral arteries, unite to form a single artery, which runs down within the anterior median fissure.

The posterior and anterior spinal arteries are reinforced by **radicular arteries,** which enter the vertebral canal through the intervertebral foramina.

The **veins** of the spinal cord drain into the internal vertebral venous plexus.

# Meninges of the Spinal Cord

The spinal cord, like the brain, is surrounded by three meninges: the dura mater, the arachnoid mater, and the pia mater (Fig. 16-6).

A     newborn infant

B     baby holds head up steadily
(3–4 months)

C     adult

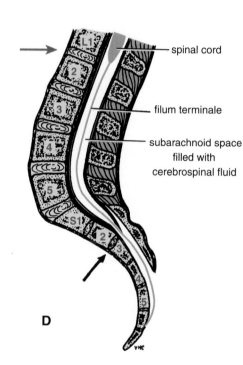

- spinal cord

- filum terminale

subarachnoid space
filled with
cerebrospinal fluid

D

**Figure 16-5   A, B,** and **C.** Curves of the vertebral column at different ages. **D.** In the adult, the lower end of the spinal cord lies at the level of the lower border of the body of the first lumbar vertebra (*top arrow*), and the subarachnoid space ends at the lower border of the body of the second sacral vertebra (*bottom arrow*).

# Dura Mater

The dura mater is the most external membrane and is a dense, strong, fibrous sheet that encloses the spinal cord and cauda equina (Figs. 16-6 and 16-8). It is continuous above through the foramen magnum with the meningeal layer of dura covering the brain. Inferiorly, it ends on the filum terminale at the level of the lower border of the second sacral vertebra (Fig. 16-5). The dural sheath lies loosely in the vertebral canal and is separated from the walls of the canal by the **extradural space** (epidural space). This contains loose areolar tissue and the internal vertebral venous plexus. The dura mater extends along each nerve root and becomes continuous with connective tissue surrounding each spinal nerve (**epineurium**) at the intervertebral foramen. The inner surface of the dura mater is separated from the arachnoid mater by the potential **subdural space.**

# Arachnoid Mater

The arachnoid mater is a delicate impermeable membrane covering the spinal cord and lying between the pia mater internally and the dura mater externally (Figs. 16-6 and 16-8). It is separated from the dura by the subdural space that contains a thin film of tissue fluid. The

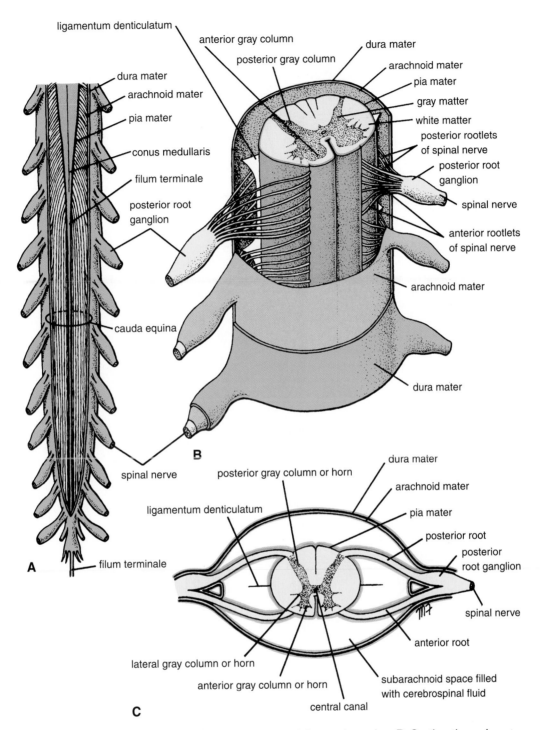

**Figure 16-6** **A.** Lower end of the spinal cord and the cauda equina. **B.** Section through the thoracic part of the spinal cord showing the anterior and posterior roots of the spinal nerves and meninges. **C.** Transverse section through the spinal cord showing the meninges and the position of the cerebrospinal fluid.

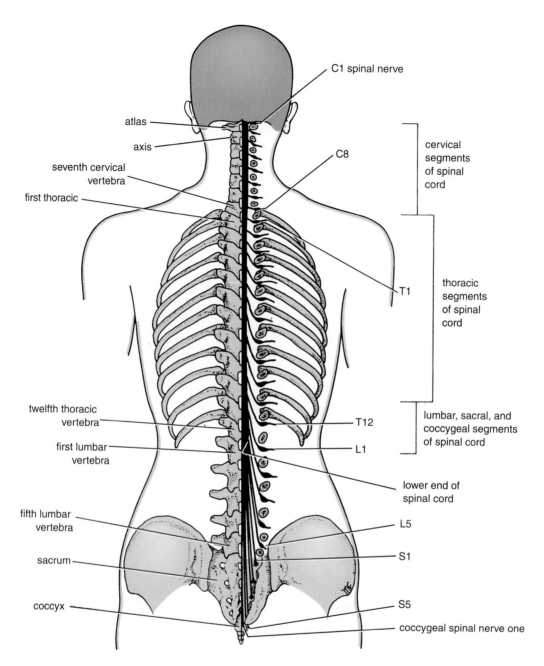

**Figure 16-7** Posterior view of the spinal cord showing the origins of the roots of the spinal nerves and their relationship to the different vertebrae. On the right, the laminae have been removed to expose the right half of the spinal cord and the nerve roots.

arachnoid is separated from the pia mater by a wide space, the **subarachnoid space,** which is filled with **cerebrospinal fluid** (Fig. 16-6). The arachnoid is continuous above through the foramen magnum with the arachnoid covering the brain. Inferiorly, it ends on the filum terminale at the level of the lower border of the second sacral vertebra (Fig. 16-5). Between the levels of the conus medullaris and the lower end of the subarachnoid space lie the nerve roots of the cauda equina bathed in cerebrospinal fluid (Fig. 16-6). The arachnoid mater is continued along the spinal nerve roots, forming small lateral extensions of the subarachnoid space.

## Pia Mater

The pia mater is a vascular membrane that closely covers the spinal cord (Figs. 16-6 and 16-8). It is continuous above through the foramen magnum with the pia covering the brain; below, it fuses with the filum terminale. The pia mater is thickened on either side between the nerve roots to form the **ligamentum denticulatum,** which passes laterally to be attached to the dura. It is by this means that the spinal cord is suspended in the middle of the dural sheath. The pia mater extends along each nerve root and becomes continuous with the connective tissue surrounding each spinal nerve (Fig. 16-6).

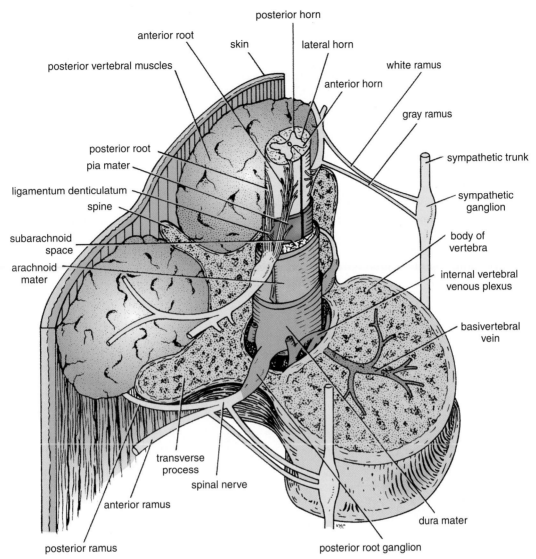

**Figure 16-8** Oblique section through the first lumbar vertebra showing the spinal cord and its covering membranes. Note the relationship between the spinal nerve and sympathetic trunk on each side. Note also the important internal vertebral venous plexus.

# Cerebrospinal Fluid

The cerebrospinal fluid is a clear, colorless fluid formed mainly by the **choroid plexuses,** within the lateral, third, and fourth ventricles of the brain. The fluid circulates through the ventricular system and enters the subarachnoid space through the three foramina in the roof of the fourth ventricle. It circulates both upward over the surface of the cerebral hemispheres and downward around the spinal cord. The spinal part of the subarachnoid space extends down as far as the lower border of the second sacral vertebra, where the arachnoid fuses with the filum terminale (Fig. 16-5). Eventually, the fluid enters the bloodstream by passing through the **arachnoid villi** into the dural venous sinuses, in particular the **superior sagittal venous sinus.**

In addition to removing waste products associated with neuronal activity, the cerebrospinal fluid provides a fluid medium that surrounds the spinal cord. This fluid, together with the bony and ligamentous walls of the vertebral canal, effectively protects the spinal cord from trauma.

---

### E M B R Y O L O G I C    N O T E

## Development of the Vertebral Column

Early in development, the embryonic mesoderm becomes differentiated into three distinct regions: **paraxial mesoderm, intermediate mesoderm,** and **lateral mesoderm.** The paraxial mesoderm is a column of tissue situated on either side of the midline of the embryo, and at about the 4th week, it becomes divided into blocks of tissue called **somites.** Each somite becomes differentiated into a ventromedial part (the **sclerotome**) and a dorsolateral part (the **dermatomyotome**). The dermatomyotome now further differentiates into the **myotome** and the **dermatome** (Fig. 16-9).

The mesenchymal cells of the sclerotome rapidly divide and migrate medially during the 4th week of development and surround the **notochord** (Fig. 16-9). The caudal half of each sclerotome now fuses with the cephalic half of the immediately succeeding sclerotome to form the mesenchymal **vertebral body** (Figs. 16-9 and 16-10). Each vertebral body is thus an intersegmental structure. The notochord degenerates completely in the region of the vertebral body; but in the intervertebral region, it enlarges to from the nucleus pulposus of the **intervertebral discs** (Fig. 16-10). The surrounding fibrocartilage, the **anulus fibrosus,** of the intervertebral disc is derived from sclerotomic mesenchyme situated between adjacent vertebral bodies (Fig. 16-10).

Meanwhile, the mesenchymal vertebral body gives rise to dorsal and lateral outgrowths on each side. The dorsal outgrowths grow around the neural tube between the segmental nerves to fuse with their fellows of the opposite side and form the mesenchymal **neural arch** (Fig. 16-9). The lateral outgrowths pass between the myotomes to form the mesenchymal **costal processes,** or primordia of the **ribs.**

Two centers of chondrification appear in the middle of each mesenchymal vertebral body. These quickly fuse to form a cartilaginous **centrum** (Fig. 16-9). A chondrification center forms in each half of the mesenchymal neural arch and spreads dorsally to fuse behind the neural

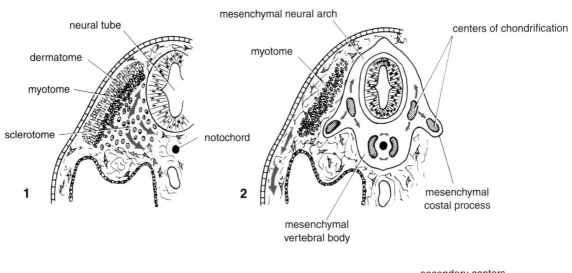

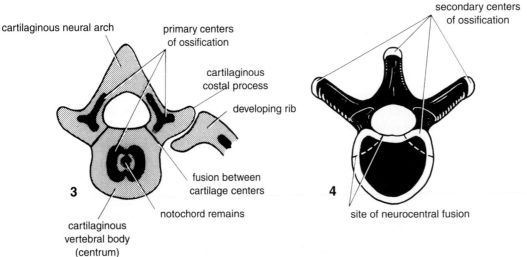

**Figure 16-9**   The stages in the formation of a thoracic vertebra.

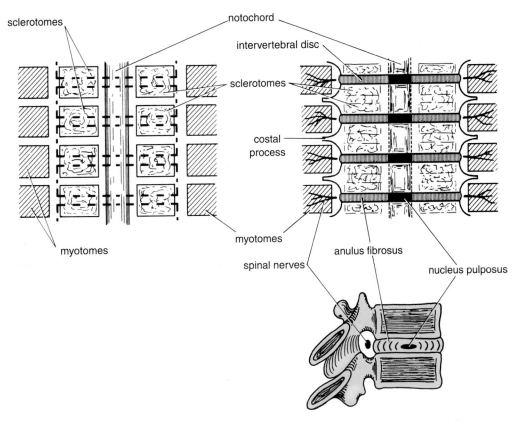

**Figure 16-10**   The formation of each mesenchymal vertebral body by the fusion of the caudal half of each sclerotome with the cephalic half of the immediately succeeding sclerotome. Each vertebral body is thus an intersegmental structure. The costal processes grow out between adjacent myotomes. Also shown is the close relationship that exists between each spinal nerve and each intervertebral disc.

tube with its fellow of the opposite side. These centers also extend anteriorly to fuse with the cartilaginous centrum and laterally into the costal processes. The condensed mesenchymal or membranous vertebra has thus been converted into a **cartilaginous vertebra.**

In the thoracic region, each costal process forms a **cartilaginous rib.** The costal processes in the cervical region remain short and form the lateral and anterior boundaries of the **foramen transversarium** of each vertebra. In the lumbar region, the costal process forms part of the transverse process; and in the sacral region, the costal processes fuse together to form the **lateral mass** of the **sacrum.**

At about the 9th week of development, primary ossification centers appear: two for each centum and one for each half of the neural arch (Fig. 16-9). The two centers for the centrum usually unite quickly, but the complete union of all the primary centers does not occur until several years after birth.

During adolescence, secondary centers appear in the cartilage covering the superior and inferior ends of the vertebral body, and the **epiphyseal plates** are formed. A secondary center also appears at the tip of each trans-

verse process and at the tip of the spinous process. By the 25th year, all the secondary centers have fused with the rest of the vertebra.

The **atlas** and **axis** develop somewhat differently. The centrum of the atlas fuses with that of the axis and becomes the part of the axis vertebra known as the **odontoid process.** This leaves only the neural arch for the atlas, which grows anteriorly and finally fuses in the midline to form the characteristic ring shape of the atlas vertebra.

In the **sacral region,** the bodies of the individual vertebrae are separated from each other in early life by intervertebral discs. At about the 18th year, the bodies start to become united by bone; this process starts caudally. Usually by the 13th year, all the sacral vertebrae are united. In the **coccygeal region,** segmental fusion also takes place, and in later life, the coccyx often fuses with the sacrum.

### Development of the Curves of the Vertebral Column

The embryonic vertebral column shows one continuous anterior (ventral) concavity. Later, the sacrovertebral angle develops. At birth, the cervical, thoracic, and lumbar regions show one continuous anterior (ventral) concavity. When the child begins to raise his or her head, the cervical curve, which is convex anteriorly, develops. Toward the end of the 1st year, when the child stands up, the lumbar curve, which is convex anteriorly, develops (Fig. 16-5).

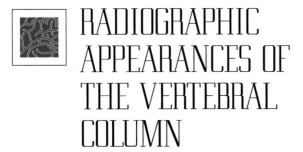

# RADIOGRAPHIC APPEARANCES OF THE VERTEBRAL COLUMN

## Cervical Region

The views commonly used for the cervical region are the anteroposterior and the lateral.

The **anteroposterior view** is taken with the patient in the supine position. The film cassette is placed behind the head and the neck, and the x-ray tube is centered over the front of the thyroid cartilage. The atlanto-axial articulation may be demonstrated by asking the patient to keep the mandible in motion while the film is being exposed or by directing the x-ray tube through the open mouth (Fig. 16-11). By using the latter method, the entire length of the odontoid process can be visualized lying between the lateral masses of the atlas.

Below the level of the third cervical vertebra, the bodies of the vertebrae are well shown and the spines are clearly seen (Fig. 16-12). The laminae can be identified. The transverse processes, the foramina transversaria, and the articular processes overlap one another and are difficult to distinguish separately. The lumen of the trachea can be seen as a tubular transradiancy that narrows at the upper end, where it becomes continuous with the cavity of the larynx.

The **lateral view** is taken with the patient sitting up and the shoulders dropped so that the seventh cervical vertebra

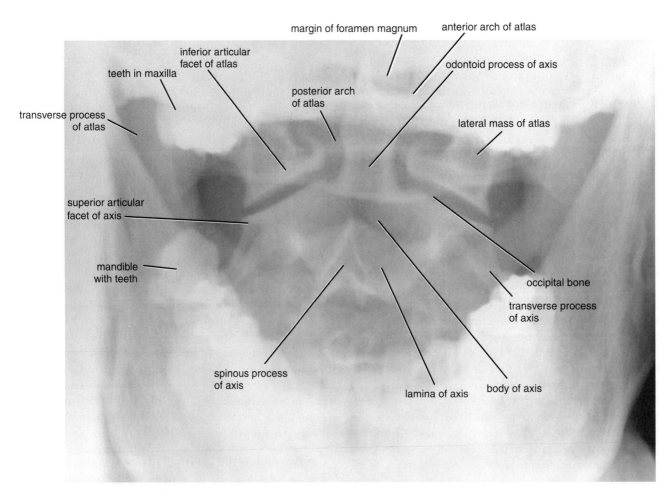

**Figure 16-11** Anteroposterior radiograph of the upper cervical region of the vertebral column with the patient's mouth open to show the odontoid process of the axis.

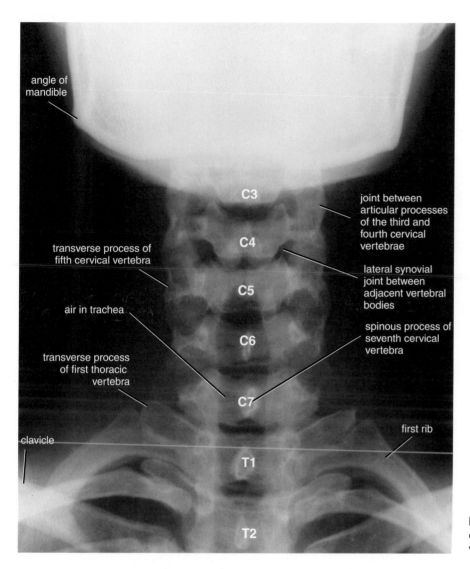

angle of mandible

transverse process of fifth cervical vertebra

air in trachea

transverse process of first thoracic vertebra

clavicle

C3

C4

C5

C6

C7

T1

T2

joint between articular processes of the third and fourth cervical vertebrae

lateral synovial joint between adjacent vertebral bodies

spinous process of seventh cervical vertebra

first rib

**Figure 16-12** Anteroposterior radiograph of the cervical region of the vertebral column.

can be demonstrated. The film cassette is placed at the side of the neck in the parasagittal plane. The x-ray tube is directed at the side of the neck at right angles to the long axis of the vertebral column and the film.

The atlanto-occipital joint is difficult to make out. The anterior and posterior arches of the atlas are well shown (Fig. 16-13), and the body of the axis is easily identified. The odontoid process of the axis extends upward, close to the posterior margin of the anterior arch of the atlas. The articular processes are well shown, and the spinous processes can be clearly seen. The transverse processes are difficult to make out because they are superimposed on the vertebral bodies. The intervertebral disc spaces between the bodies of adjacent vertebrae are easily defined and are of equal height.

The anterior and posterior surfaces of the vertebral bodies and the posterior wall of the vertebral

canal form smooth curved lines that are roughly parallel (Fig. 16-13).

# Thoracic Region

The views commonly used for the thoracic region are the anteroposterior and the lateral.

The **anteroposterior view** is taken with the patient in the supine position. The film cassette is placed behind the thorax, and the x-ray tube is centered over the front of the sternum.

Because of the curvature of the thoracic part of the vertebral column, the upper and lower margins of the bodies of adjacent vertebrae overlap. The spinous processes and laminae are superimposed on the bodies (Fig. 16-14). The transverse processes can be identified, but they are obscured by the heads and necks of the ribs. Note that the first rib and the tenth, eleventh, and twelfth ribs on each side articulate

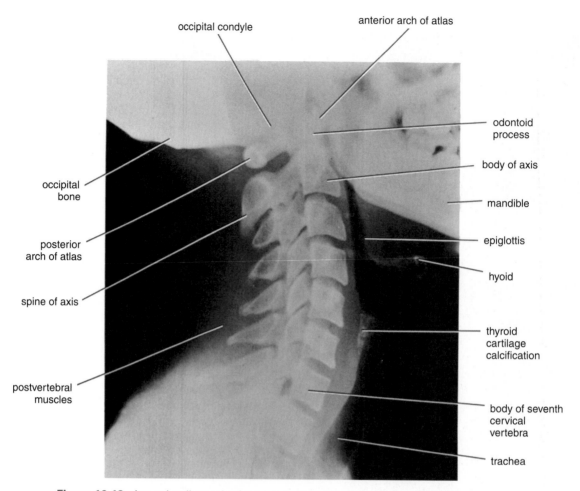

occipital condyle

anterior arch of atlas

odontoid process

body of axis

mandible

epiglottis

hyoid

thyroid cartilage calcification

body of seventh cervical vertebra

trachea

occipital bone

posterior arch of atlas

spine of axis

postvertebral muscles

**Figure 16-13**    Lateral radiograph of the cervical region of the vertebral column.

only with the bodies of the first, tenth, eleventh, and twelfth thoracic vertebrae, respectively; all the other ribs articulate with two vertebrae.

The pedicles are clearly seen as ovoid structures that are superimposed on the lateral parts of the bodies.

The transradiant trachea and the heart shadow are superimposed on the thoracic vertebrae.

The **lateral view** is taken with the patient lying on the side, with the arms stretched above the head. If it is desirable to demonstrate the postural curves, the patient assumes the standing position. The film cassette is placed against the side of the thorax, and the x-ray tube is directed laterally through the vertebral column at right angles to the film.

The rectangular vertebral bodies and the intervertebral disc spaces are clearly seen, even though the ribs and lungs are superimposed on them. The upper four vertebrae are obscured by the shadows of the shoulder girdle.

The pedicles and intervertebral foramina are well demonstrated. However, the spinous processes, the laminae, the transverse processes, and the ribs are superimposed on

one another, and their detail is obscured. The vertebral canal is well shown.

# Lumbosacral Region

The views commonly used for the lumbosacral region are the anteroposterior and the lateral.

The **anteroposterior view** is taken with the patient in the supine position. The film cassette is placed behind the lumbar region and buttocks, and the x-ray tube is centered over the umbilicus. To diminish the distortion produced by the lumbar curvature, the patient can be asked to flex the knees and hips, which may straighten the lumbar curvature to some extent.

The bodies, transverse processes, spinous processes, laminae, and intervertebral disc spaces are clearly seen (Fig. 16-15). The pedicles produce ovoid shadows, and the articular processes and posterior intervertebral joints can be delineated.

Because of the obliquity of the sacroiliac joint, it is visualized as two lines, the lateral one corresponding to the

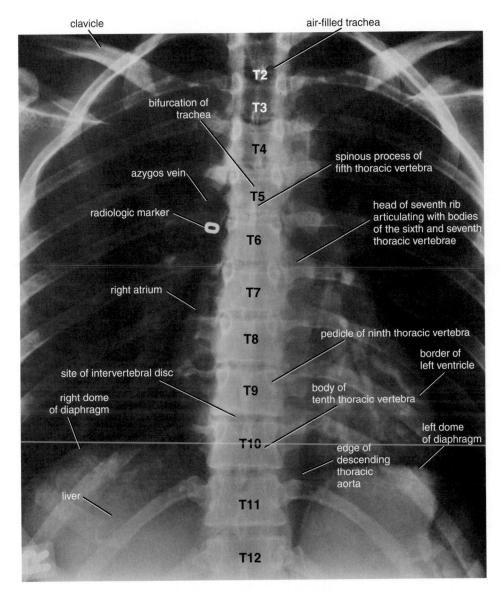

clavicle

air-filled trachea

T2

bifurcation of
trachea

T3

T4

spinous process of
fifth thoracic vertebra

azygos vein

T5

radiologic marker

head of seventh rib
articulating with bodies
of the sixth and seventh
thoracic vertebrae

T6

right atrium

T7

T8

pedicle of ninth thoracic vertebra

border of
left ventricle

site of intervertebral disc

T9

body of
tenth thoracic vertebra

right dome
of diaphragm

left dome
of diaphragm

T10

edge of
descending
thoracic
aorta

liver

T11

**Figure 16-14**   Anteroposterior
radiograph of the thoracic
region of the the vertebral
column.

T12

anterior margin and the medial one to the posterior margin (Fig. 16-15). The lower segments of the sacrum and the coccyx are tilted posteriorly and are usually overlapped by the symphysis pubis. In addition, the presence of gas and fecal material in the rectum and sigmoid colon commonly obscures the sacrum. To demonstrate the sacrum in a more direct anteroposterior view, the x-ray tube may be tilted.

The **lateral view** is taken with the patient lying on the side. If it is desirable to demonstrate the postural curves, the patient assumes the standing position. The film cassette is placed against the side of the lumbar region, and the x-ray tube is directed laterally through the lumbar part of the vertebral column at right angles to the film.

The large vertebral bodies, the intervertebral disc spaces, and the intervertebral foramina are clearly seen

(Fig. 16-16). The pedicles, the articular processes, and the spinous processes are easily visualized. The transverse processes can be identified, but they are superimposed on the sides of the preceding structures. The anterior and posterior surfaces of the vertebral bodies and the posterior wall of the vertebral canal form smooth curved lines that are roughly parallel.

Occasionally, the fifth lumbar vertebra is partly or completely fused with the first sacral vertebra. Not infrequently, the first sacral vertebra is separate from the remainder of the sacrum and has the appearance of a sixth lumbar vertebra.

The **sacrum** on lateral view shows the promontory, the sacral canal, and the fused sacral bodies and spinous processes (Fig. 16-16). Note the localized anterior angulation between the body of the fifth lumbar vertebra and the first sacral vertebra.

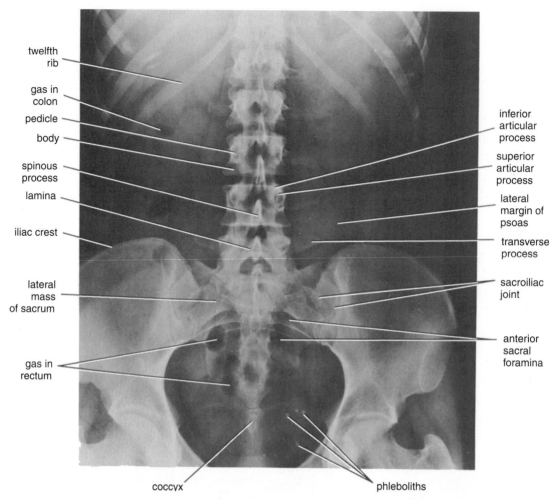

twelfth rib

gas in colon

pedicle

body

spinous process

lamina

iliac crest

lateral mass of sacrum

gas in rectum

inferior articular process

superior articular process

lateral margin of psoas

transverse process

sacroiliac joint

anterior sacral foramina

coccyx

phleboliths

**Figure 16-15**  Anteroposterior radiograph of the lower thoracic, lumbar, and sacral regions of the vertebral column.

# Coccyx

The coccyx is not well shown on routine anteroposterior and lateral radiographs because of its oblique position relative to the film and the presence of gas and feces in the rectum and sigmoid colon. These difficulties may be partially overcome by tilting the x-ray tube and evacuating the contents of the rectum and sigmoid colon.

# Spinal Subarachnoid Space

The subarachnoid space can be studied radiographically by the injection of contrast media into the subarachnoid space by lumbar puncture. Iodized oil has been used with success. This technique is referred to as **myelography** (Figs. 16-17 and 16-18).

If the patient is sitting in the upright position, the oil sinks to the lower limit of the subarachnoid space at the level of the lower border of the second sacral vertebra. By placing the patient on a tilting table, the oil can be made to gravitate gradually to higher levels of the vertebral column.

A normal myelogram will show pointed lateral projections at regular intervals at the intervertebral space levels. This appearance is caused by the opaque medium filling the lateral extensions of the subarachnoid space around each spinal nerve. The presence of a tumor or a prolapsed intervertebral disc may obstruct the movement of the oil from one region to another when the patient is tilted.

# CT and MRI studies

CT and MRI scans are extensively used to detect lesions of the vertebral column, especially those involving the soft tissues. CT scans can concentrate on the intervertebral spaces and reveal the intervertebral disc in transverse slices (Figs. 16-19 and 16-20). The disc has a higher density than the

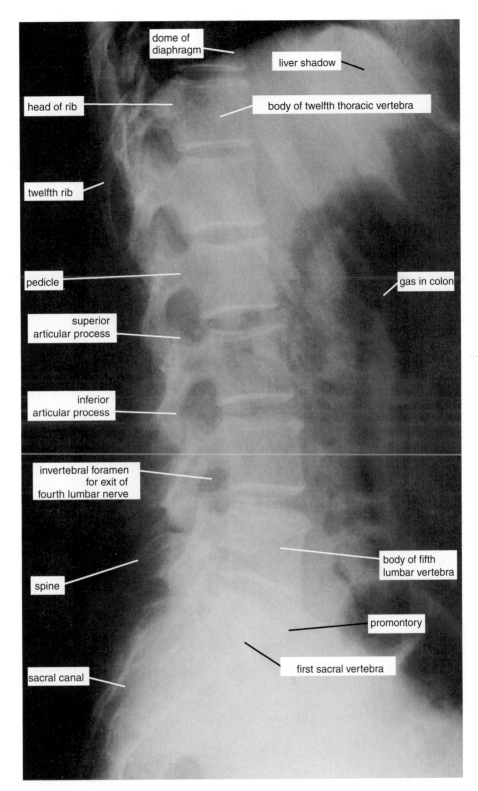

**Figure 16-16**    Lateral radiograph of the lower thoracic, lumbar, and sacral regions of the vertebral column.

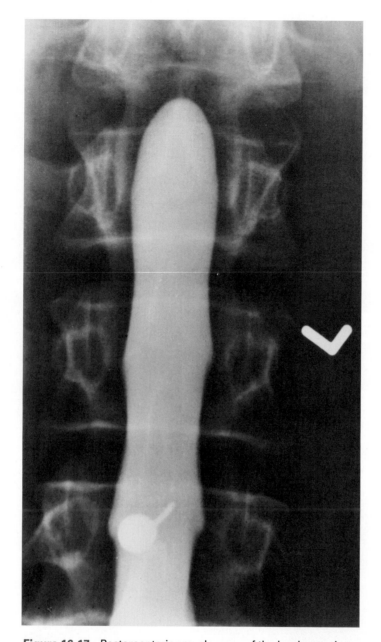

**Figure 16-17**   Posteroanterior myelogram of the lumbar region.

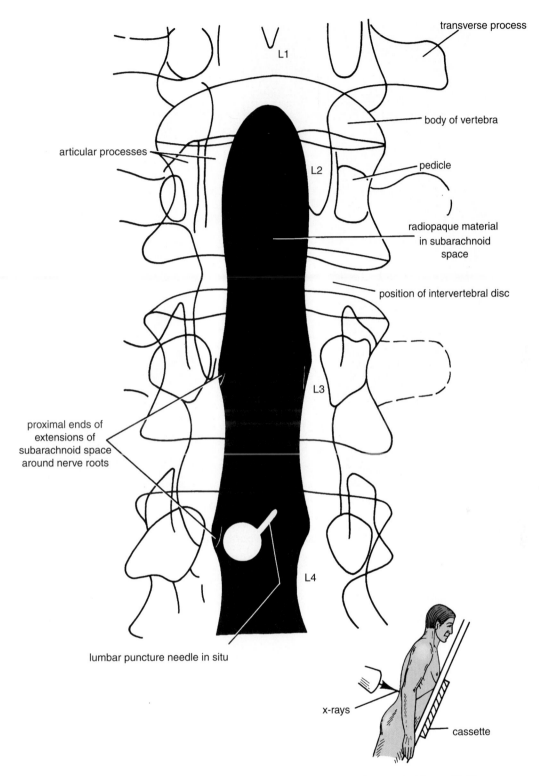

transverse process

body of vertebra

pedicle

radiopaque material
in subarachnoid
space

position of intervertebral disc

articular processes

proximal ends of
extensions of
subarachnoid space
around nerve roots

lumbar puncture needle in situ

x-rays

cassette

**Figure 16-18**   Main features that can be seen in the myelogram in Figure 16-17.

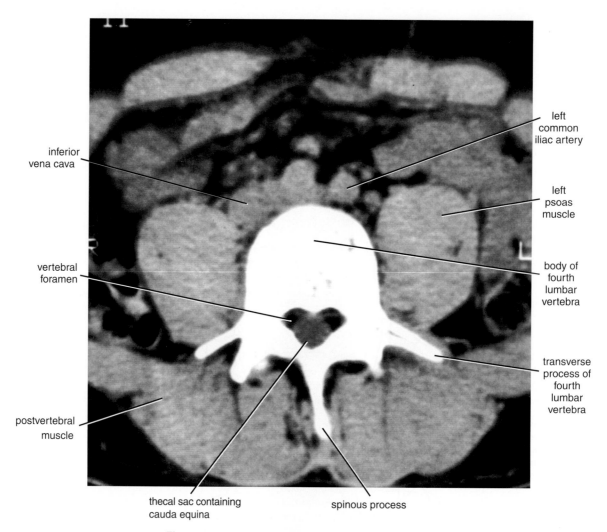

**Figure 16-19**   CT scan of the fourth lumbar vertebra.

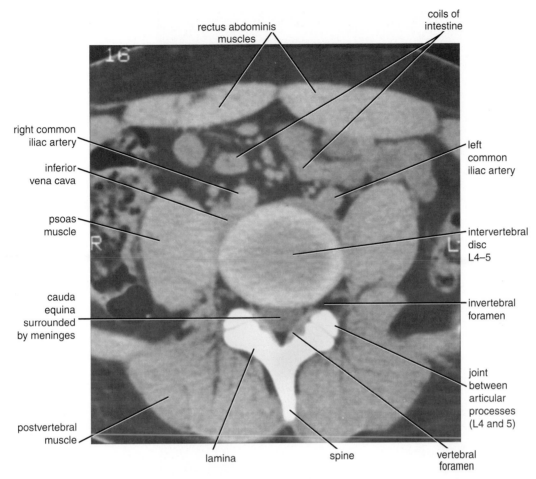

**Figure 16-20** CT scan through the vertebral column at the level of the intervertebral disc between the fourth and fifth lumbar vertebrae. The spine of L4 and the intervertebral foramen on each side are shown. Note the joints between the articular processes.

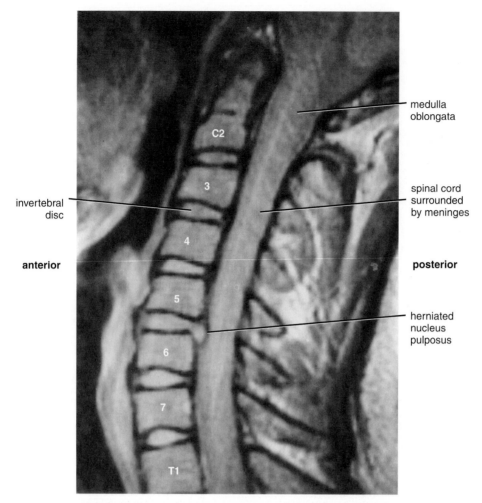

Figure 16-21    Sagittal MRI scan of the cervical part of the vertebral column. A herniated disc between the fifth and sixth vertebrae is shown. Note the position of the spinal cord and its meningeal coverings relative to the herniated disc. (Courtesy of Pait.)

cerebrospinal fluid in the subarachnoid space and the surrounding fat. Fragments of a herniated disc can be identified beyond the boundaries of the anulus fibrosus.

MRI easily defines the intervertebral disc on sagittal section and shows its relationship to the vertebral body and the posterior longitudinal ligament (Fig. 16-21). The herniated fragment of the disc and its relationship to the dural sac can easily be demonstrated. The use of MRI is now largely replacing myelography or CT in this region.

 # SURFACE ANATOMY

The entire posterior aspect of the patient should be examined from head to foot, and the arms should hang loosely at the side.

# Midline Structures

In the midline, the following structures can be palpated from above downward.

## External Occipital Protuberance

The external occipital protuberance lies at the junction of the head and neck (Fig. 16-1). If the index finger is placed on the skin in the midline, it can be drawn downward from the protuberance in the **nuchal groove.**

## Cervical Vertebrae

The most prominent spinous process that can be felt in the neck (Fig. 16-22) is that of the **seventh cervical vertebra (vertebra prominens).** Cervical spines 1 to 6 are covered by

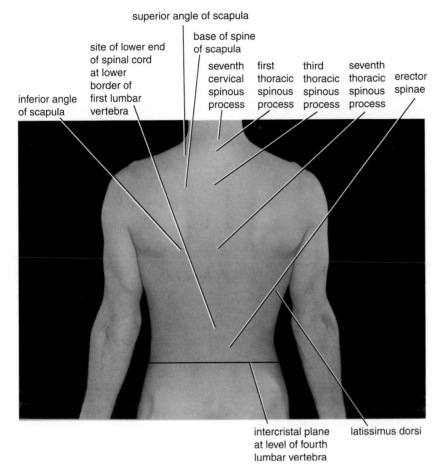

**Figure 16-22** The back of a 27-year-old man.

the **ligamentum nuchae**, a large ligament that runs down the back of the neck connecting the skull to the spinous processes of the cervical vertebrae.

The **transverse processes** are short but easily palpable from the lateral side in a thin neck. The **anterior tubercle of the sixth cervical transverse process (tubercle of Chassaignac)** can be palpated medial to the sternocleidomastoid muscle, and against it, the common carotid artery can be compressed.

## Thoracic and Lumbar Vertebrae

The nuchal groove is continuous below with a furrow that runs down the middle of the back over the tips of the **spines of all the thoracic** and the upper four **lumbar vertebrae** (Fig. 16-22). The most prominent spine is that of the first thoracic vertebra; the others may be easily recognized when the trunk is bent forward.

## Sacrum

The **spines of the sacrum** are fused with each other in the midline to form the **median sacral crest.** The crest can be felt beneath the skin in the uppermost part of the cleft between the buttocks.

The **sacral hiatus** is situated on the posterior aspect of the lower end of the sacrum, and here, the extradural space (epidural space) terminates. The hiatus lies about 2 in (5 cm) above the tip of the coccyx and beneath the skin of the groove between the buttocks.

## Coccyx

The inferior surface and tip of the coccyx can be palpated in the groove between the buttocks about 1 in (2.5 cm) behind the anus (Fig. 16-1). The anterior surface of the coccyx can be palpated with a gloved finger in the anal canal.

# Upper Lateral Part of the Thorax

The upper lateral part of the thorax is covered by the scapula and its associated muscles. The scapula lies posterior to the first to the seventh ribs (Figs. 16-1 and 16-22).

## Scapula

The **medial border** of the scapula forms a prominent ridge, which ends above at the superior angle and below at the inferior angle (Fig. 16-22).

The **superior angle** can be palpated opposite the first thoracic spine, and the **inferior angle** can be palpated opposite the seventh thoracic spine (Figs. 16-1 and 16-22).

The **crest of the spine of the scapula** can be palpated and traced medially to the medial border of the scapula, which it joins at the level of the third thoracic spine (Figs. 16-1 and 16-22).

The **acromion of the scapula** forms the lateral extremity of the spine of the scapula. It is subcutaneous and easily located.

# Lower Lateral Part of the Back

The lower lateral part of the back is formed by the posterior aspect of the upper part of the bony pelvis (false pelvis) and its associated gluteal muscles.

## Iliac Crests

The iliac crests are easily palpable along their entire length (Fig. 16-1). They lie at the level of the fourth lumbar spine and are used as a landmark when performing a lumbar puncture. Each crest ends in front at the **anterior superior iliac spine** and behind at the **posterior superior iliac spine;** the latter lies beneath a skin dimple at the level of the second sacral vertebra and the middle of the sacroiliac joint. The iliac tubercle is a prominence felt on the outer surface of the iliac crest about 2 in (5 cm) posterior to the anterosuperior iliac spine. The iliac tubercle lies at the level of the fifth lumbar spine.

# Spinal Cord and Subarachnoid Space

The **spinal cord** in adults extends down to the level of the lower border of the spine of the first lumbar vertebra (Fig. 16-5). In young children, it may extend to the third lumbar spine.

The **subarachnoid space,** with its **cerebrospinal fluid,** extends down to the lower border of the second sacral vertebra (Fig. 16-5), which lies at the level of the posterosuperior iliac spine.

# Curves of the Vertebral Column

The curves of the vertebral column can be examined by inspecting the lateral contour of the back. Normally, the posterior surface is concave in the cervical region, convex in the thoracic region, and concave in the lumbar region (Fig. 16-2). The anterior surface of the sacrum and coccyx together have an anterior concavity. The lumbar region meets the sacrum at a sharp angle, the **lumbosacral angle.**

Inspection of the posterior surface of the back, with particular reference to the vertical alignment of the vertebral spines, reveals a slight lateral curvature in most normal persons. Right-handed persons, especially those whose work involves extreme and prolonged muscular effort, usually exhibit a lateral thoracic curve to the right; left-handed persons usually exhibit a lateral thoracic curve to the left.

# Review Questions

## Completion Questions

### Select the phrase that BEST completes each statement.

1. The seventh cervical vertebra is characterized by having
   A. the longest spinous process.
   B. a large foramen transversarium.
   C. a heart-shaped body.
   D. a massive body.
   E. an odontoid process.

2. The sixth thoracic vertebra is characterized by
   A. its heart-shaped body.
   B. its bifid spinous process.
   C. its massive body.
   D. having the superior articular processes face medially and the inferior articular processes face laterally.
   E. its thick lamina.

3. The characteristic feature of the fifth lumbar vertebra is its
   A. heart-shaped body.
   B. rounded vertebral foramen.
   C. small pedicles.
   D. massive body.
   E. short and thick transverse process.

4. The cauda equina consists of
   A. a bundle of posterior roots of lumbar, sacral, and coccygeal spinal nerves.
   B. the filum terminale.

C. a bundle of anterior and posterior roots of lumbar, sacral, and coccygeal spinal nerves.

D. a bundle of lumbar, sacral, and coccygeal spinal nerves and the filum terminale.

E. a bundle of anterior and posterior roots of lumbar, sacral, and coccygeal spinal nerves, and the filum terminale.

5. The spinal cord in the adult ends inferiorly at the level of the
   A. L5 vertebra.
   B. L3 vertebra.
   C. S2 to 3 vertebrae.
   D. T12 vertebra.
   E. L1 vertebra.

6. Herniation of the intervertebral disc between the fifth and sixth cervical vertebrae will compress the
   A. fourth cervical nerve root.
   B. sixth cervical nerve root.
   C. fifth cervical nerve root.
   D. seventh and eighth cervical nerve roots.
   E. seventh cervical nerve root.

7. The subarachnoid space ends inferiorly in the adult at the level of
   A. the coccyx.
   B. the lower border of L1.
   C. S2 to 3.
   D. S5.
   E. the promontory of the sacrum.

## Multiple Choice Questions

**Select the BEST answer for each question.**

8. The following statements concerning an intervertebral disc are correct **except** which?
   A. The nucleus pulposus is most likely to herniate in an anterolateral direction.
   B. The discs are the thickest in the lumbar region.
   C. The atlanto-axial joint possesses no disc.
   D. The discs play a major role in the development of the curvatures of the vertebral column.
   E. During aging, the fluid within the nucleus pulposus is replaced by fibrocartilage.

9. The following statements concerning the vertebral column are correct **except** which?
   A. Throughout life, the marrow of the vertebral bodies has a hemopoietic function.
   B. The internal vertebral venous plexus provides a path for the passage of malignant cells from the prostate to the cranial cavity.
   C. The vertebral artery ascends the neck through the foramen transversarium of all the cervical vertebrae.

D. Injection of an anesthetic into the sacral canal can be used to block pain and sensation from the cervix, vagina, and the perineum during childbirth.

E. The atlanto-axial joint permits rotation of the head on the vertebral column.

10. The first cervical vertebra (atlas) has all the following anatomic features **except** which?
    A. Lateral masses
    B. Inferior articular facets
    C. Anterior arch
    D. Spinous process
    E. Superior articular facets

**Read the case history and select the BEST answer to the questions following it.**

An 8-year-old girl was taken to a pediatrician because her mother was concerned about a lateral curvature of the child's spine, which she had noticed since her daughter was 5 months old. The girl was otherwise perfectly healthy and active.

11. The pediatrician performed a thorough physical examination and found the following **except** which?
    A. Both legs were of equal length.
    B. On standing, the heights of the iliac crests were the same on each side.
    C. The left shoulder was lower than the right.
    D. The vertebral column in the midthoracic region showed a sharp curve convex to the right.
    E. There were gentle compensatory curves of the vertebral column above and below the sharp curve in the midthoracic region, with convexities to the right.

The pediatrician performed further clinical examinations and ordered a radiographic examination of the vertebral column.

12. The following statements about this patient are correct **except** which?
    A. The anteroposterior x-ray of the midthoracic region revealed a wedge-shaped vertebra at the level of T5 and fusion of the left fifth and sixth ribs.
    B. Flexion of the vertebral column showed that the sharp curved area was rigid.
    C. The child had a congenital hemivertebra at the level of T5 with compensatory curves above and below that defect.
    D. The condition is caused by a failure in development of one of the three ossification centers that appear in the centrum of the body of each vertebra.
    E. Since the child had no symptoms and the compensatory curves are well balanced, no special treatment is advised.

# Answers and Explanations

1. **A** is correct. The seventh cervical vertebra is characterized by having the longest spinous process (Fig. 16-2).

2. **A** is correct. The sixth thoracic vertebra has a heart-shaped body (Fig. 16-2).

3. **D** is correct. The fifth lumbar vertebra has a massive body; in addition, the pedicles are large, and the transverse processes are long and slender; the vertebral foramen is triangular.

4. **E** is correct. The cauda equina consists of a bundle of anterior and posterior roots of lumbar, sacral, and coccygeal spinal nerves and the filum terminale.

5. **E** is correct. In the adult, the spinal cord ends inferiorly at the level of the first lumbar vertebra.

6. **B** is correct. Herniation of the intervertebral disc between the fifth and sixth cervical vertebrae will compress the sixth cervical nerve root (see CD Fig. 16-1).

7. **C** is correct. The subarachnoid space ends inferiorly in the adult at the level of the lower border of the second sacral vertebra (Fig. 16-5).

8. **A** is incorrect. The nucleus pulposus of an intervertebral disc is most likely to herniate in a posterolateral direction.

9. **C** is incorrect. The vertebral artery ascends through the foramen transversarium of the upper six cervical vertebrae; only the vertebral vein passes through the small foramen transversarium of the seventh cervical vertebra.

10. **D** is incorrect. The atlas does not have a spine but exhibits a posterior tubercle on its posterior arch.

11. **E** is incorrect. The compensatory curves of the vertebral column above and below a sharp curve convex to the right would have their convexities to the left.

12. **D** is incorrect. Normally, the centrum has only two ossification centers and not three as stated. A hemivertebra in the thoracic region often is associated with aplasia or fusion of adjacent ribs.

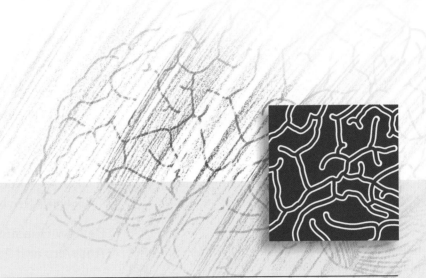

# 17

# The Spinal Nerves and Spinal Nerve Blocks

All clinical material relevant to this chapter can be found on the CD-ROM.

# Chapter Outline

Spinal nerves and their peripheral branches are common sites for disease including knife wounds, damage by bone fragments in fractures, chemical neuropathies, such as diabetes mellitus, and accidental or purposeful damage during surgical procedures. Medical professionals must be in a position to identify the nerve involved when presented with a patient with sensory loss or muscle paralysis. In addition, local anesthesia is extensively used in the emergency departments for the repair of lacerations, the drainage of abscesses, and the reduction of fractures and fracture dislocations.

The purpose of this chapter is to provide an overview of the anatomy of the spinal nerves and their branches that are commonly blocked in an emergency center. Special attention is paid to the important anatomical landmarks used when performing a nerve block.

# BASIC ANATOMY

## Spinal Nerves

A total of 31 pairs of spinal nerves leave the spinal cord and pass through intervertebral foramina in the vertebral column (Figs. 17-1 and 17-2). The spinal nerves are named according to the region of the vertebral column with which they are associated: 8 **cervical**, 12 **thoracic**, 5 **lumbar**, 5 **sacral**, and 1 **coccygeal**. Note that there are 8 cervical nerves and only 7 cervical vertebrae and that there is 1 coccygeal nerve and 4 coccygeal vertebrae.

During development, the spinal cord grows in length more slowly than the vertebral column. In the adult, when growth ceases, the lower end of the spinal cord reaches inferiorly only as far as the lower border of the first lumbar vertebra. To accommodate for this disproportionate growth in length, the length of the roots increases progressively from above downward. In the upper cervical region, the spinal nerve roots are short and run almost horizontally,

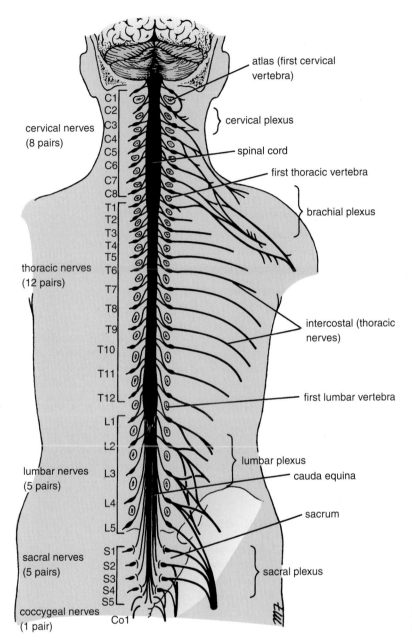

cervical nerves (8 pairs)

C1
C2
C3
C4
C5
C6
C7
C8

thoracic nerves (12 pairs)

T1
T2
T3
T4
T5
T6
T7
T8
T9
T10
T11
T12

lumbar nerves (5 pairs)

L1
L2
L3
L4
L5

sacral nerves (5 pairs)

S1
S2
S3
S4
S5

coccygeal nerves (1 pair)

Co1

atlas (first cervical vertebra)

cervical plexus

spinal cord

first thoracic vertebra

brachial plexus

intercostal (thoracic nerves)

first lumbar vertebra

lumbar plexus

cauda equina

sacrum

sacral plexus

**Figure 17-1**    Brain, spinal cord, spinal nerves, and plexuses of limbs.

but the roots of the lumbar and sacral nerves below the level of the termination of the cord form a vertical bundle of nerves that resembles a horse's tail and is called the **cauda equina** (Fig. 17-1).

Each spinal nerve is connected to the spinal cord by two roots: the **anterior root** and the **posterior root** (Fig. 17-1). The anterior root consists of bundles of nerve fibers carrying nerve impulses away from the central nervous system (Fig. 1-16). Such nerve fibers are called **efferent fibers**. Those efferent fibers that go to skeletal muscle and cause them to contract are called **motor fibers**. Their cells of origin lie in the anterior gray horn of the spinal cord.

The posterior root consists of bundles of nerve fibers that carry impulses to the central nervous system and are called **afferent fibers** (Fig. 1-16). Because these fibers are concerned with conveying information about sensations of touch, pain, temperature, and vibrations, they are called **sensory fibers**. The cell bodies of these nerve fibers are situated in a swelling on the posterior root called the **posterior root ganglion** (Figs. 1-16 and 17-2).

At each intervertebral foramen, the anterior and posterior roots unite to form a spinal nerve (Fig. 17-2). Here, the motor and sensory fibers become mixed together, so that a spinal nerve is made up of a mixture of motor and sensory fibers (Fig. 1-16). On emerging from the foramen, the spinal

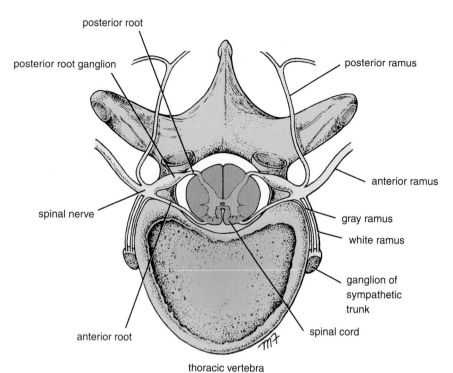

posterior root

posterior root ganglion

posterior ramus

anterior ramus

gray ramus

white ramus

spinal nerve

ganglion of
sympathetic
trunk

anterior root

spinal cord

thoracic vertebra

**Figure 17-2**   The association between
spinal cord, spinal nerves, and
sympathetic trunks.

nerve divides into a large **anterior ramus** and a smaller **posterior ramus**. The posterior ramus passes posteriorly around the vertebral column to supply the muscles and skin of the back (Figs. 1-16 and 17-2). The anterior ramus continues anteriorly to supply the muscles and skin over the anterolateral body wall and all the muscles and skin of the limbs.

In addition to the anterior and posterior rami, spinal nerves give a small **meningeal branch** that supplies the vertebrae and the coverings of the spinal cord (the meninges). Thoracic spinal nerves also have branches, called **rami communicantes,** that are associated with the sympathetic part of the autonomic nervous system (see p. 629).

# Plexuses

At the root of the limbs, the anterior rami join one another to form complicated nerve plexuses (Fig. 17-1). The cervical and brachial plexuses are found at the root of the upper limbs, and the lumbar and sacral plexuses are found at the root of the lower limbs.

# Cervical Plexus

The cervical plexus is formed by the anterior rami of the first four cervical nerves. The rami are joined by connecting branches, which form loops that lie in front of the origins of the levator scapulae and the scalenus medius muscles (Fig. 17-3). The plexus is covered in front by the prevertebral

layer of deep cervical fascia and is related to the internal jugular vein within the carotid sheath. The cervical plexus supplies the skin and the muscles of the head, the neck, and the shoulders.

## Branches

- **Cutaneous Branches**
  - The **lesser occipital nerve** (C2), which supplies the back of the scalp and the auricle.
  - The **greater auricular nerve** (C2 and 3), which supplies the skin over the angle of the mandible.
  - The **transverse cervical nerve** (C2 and 3), which supplies the skin over the front of the neck.
  - The **supraclavicular nerves** (C3 and 4). The medial, intermediate, and lateral branches supply the skin over the shoulder region. These nerves are important clinically because pain may be referred along them from the phrenic nerve (gallbladder disease).
- **Muscular branches to the neck muscles**. Prevertebral muscles, sternocleidomastoid (proprioceptive, C2 and 3), levator scapulae (C3 and 4), and trapezius (proprioceptive, C3 and 4). A branch from C1 joins the hypoglossal nerve. Some of these C1 fibers later leave the hypoglossal as the descending branch, which unites with the **descending cervical nerve** (C2 and 3) to form the **ansa cervicalis** (Fig. 17-4). The first, second, and third cervical nerve fibers within the ansa cer-

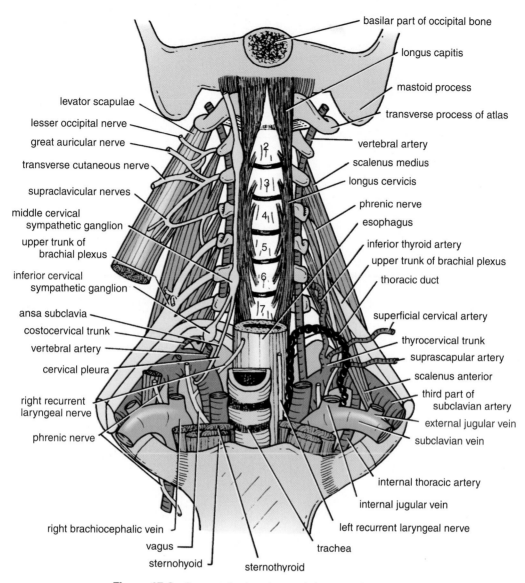

- basilar part of occipital bone
- longus capitis
- mastoid process
- transverse process of atlas
- vertebral artery
- scalenus medius
- longus cervicis
- phrenic nerve
- esophagus
- inferior thyroid artery
- upper trunk of brachial plexus
- thoracic duct
- superficial cervical artery
- thyrocervical trunk
- suprascapular artery
- scalenus anterior
- third part of subclavian artery
- external jugular vein
- subclavian vein
- internal thoracic artery
- internal jugular vein
- left recurrent laryngeal nerve
- trachea
- sternothyroid

levator scapulae
lesser occipital nerve
great auricular nerve
transverse cutaneous nerve
supraclavicular nerves
middle cervical sympathetic ganglion
upper trunk of brachial plexus
inferior cervical sympathetic ganglion
ansa subclavia
costocervical trunk
vertebral artery
cervical pleura
right recurrent laryngeal nerve
phrenic nerve
right brachiocephalic vein
vagus
sternohyoid

**Figure 17-3**  Prevertebral region and the root of the neck.

vicalis supply the omohyoid, sternohyoid, and sternothyroid muscles. Other C1 fibers within the hypoglossal nerve leave it as the nerve to the thyrohyoid and geniohyoid.

■ Muscular branch to the diaphragm. **Phrenic nerve.**

## Phrenic Nerve

The phrenic nerve arises in the neck from the third, fourth, and fifth cervical nerves of the cervical plexus. It runs vertically downward across the front of the scalenus anterior muscle (Fig. 17-3) and enters the thorax by passing in front of the subclavian artery.

The **right phrenic nerve** (Fig. 17-5) descends in the thorax along the right side of the superior vena cava and **in front of** the root of the right lung. It then passes over the pericardium to the diaphragm.

The **left phrenic nerve** (Fig. 17-5) descends along the left side of the left subclavian artery and crosses the left side of the aortic arch and the left vagus nerve. It passes **in front of** the root of the left lung and then descends on the pericardium to the diaphragm.

The phrenic nerve is the **only motor nerve supply** to the diaphragm. It also sends sensory branches to the pericardium, the mediastinal parietal pleura, and the pleura and peritoneum covering the upper and lower surfaces of the central part of the diaphragm.

Table 17-1 summarizes the branches of the cervical plexus and their distribution.

## Brachial Plexus

The brachial plexus is formed in the posterior triangle of the neck by the union of the anterior rami of the fifth, sixth,

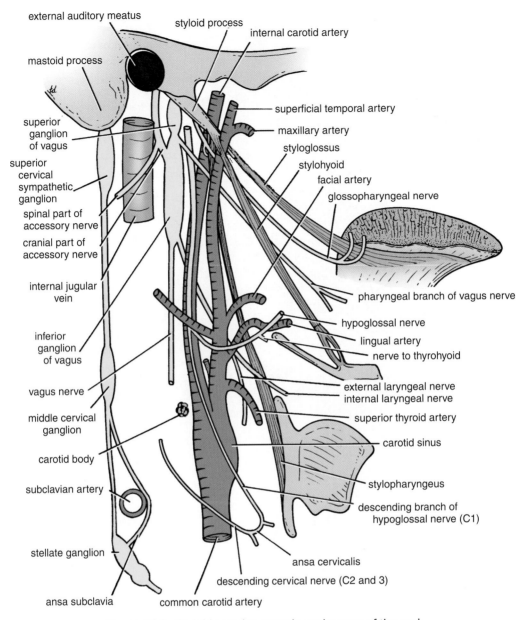

**Figure 17-4** Styloid muscles, vessels, and nerves of the neck.

seventh, and eighth cervical and the first thoracic spinal nerves (Figs. 17-6 and 17-7). This plexus is divided into **roots, trunks, divisions,** and **cords** (Fig. 17-6).

The roots of C5 and 6 unite to form the **upper trunk,** the root of C7 continues as the **middle trunk,** and the roots of C8 and T1 unite to form the **lower trunk.** Each trunk then divides into **anterior** and **posterior divisions.** The anterior divisions of the upper and middle trunks unite to form the **lateral cord,** the anterior division of the lower trunk continues as the **medial cord,** and the posterior divisions of all three trunks join to form the **posterior cord.**

The roots of the brachial plexus enter the base of the neck between the scalenus anterior and the scalenus medius muscles (Fig. 17-3). The trunks and divisions cross the posterior triangle of the neck, and the cords become arranged around the axillary artery in the axilla (Fig. 6-10). Here, the brachial plexus and the axillary artery and vein are enclosed in the **axillary sheath.**

## Branches

The branches of the brachial plexus and their distribution are summarized in Table 17-2.

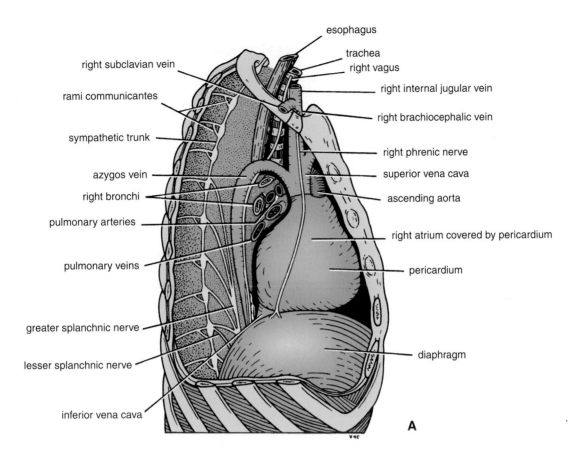

esophagus

trachea
right vagus

right subclavian vein

rami communicantes

right internal jugular vein

right brachiocephalic vein

sympathetic trunk

right phrenic nerve

azygos vein

superior vena cava

right bronchi

ascending aorta

pulmonary arteries

right atrium covered by pericardium

pulmonary veins

pericardium

greater splanchnic nerve

lesser splanchnic nerve

diaphragm

inferior vena cava

**A**

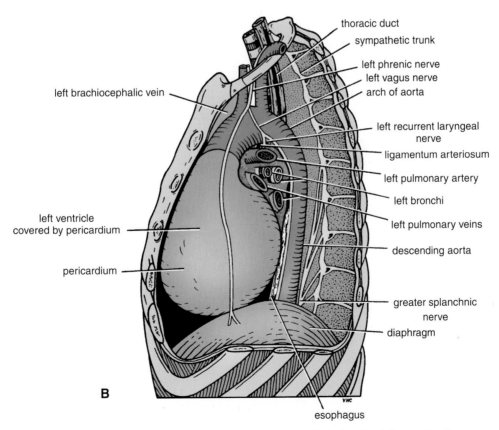

thoracic duct
sympathetic trunk

left phrenic nerve
left vagus nerve
arch of aorta

left brachiocephalic vein

left recurrent laryngeal
nerve

ligamentum arteriosum

left pulmonary artery

left bronchi

left ventricle
covered by pericardium

left pulmonary veins

descending aorta

pericardium

greater splanchnic
nerve

diaphragm

**B**

esophagus

**Figure 17-5** **A.** Right side of the mediastinum. **B.** Left side of the mediastinum.

| Table 17-1 | Summary of the Branches of the Cervical Plexus and Their Distribution |
|---|---|
| **Branches** | **Distribution** |
| Cutaneous | |
|    Lesser occipital | Skin of scalp behind ear |
|    Greater auricular | Skin of parotid salivary gland, auricle, and angle of jaw |
|    Transverse cutaneous | Skin over side and front of neck |
|    Supraclavicular | Skin over upper part of chest and shoulder |
| Muscular | |
|    Segmental | Prevertebral muscles, levator scapulae |
|    Ansa cervicalis (C1, 2, 3) | Omohyoid, sternohyoid, sternothyroid |
|    C1 fibers via hypoglossal nerve | Thyrohyoid, geniohyoid |
|    Phrenic nerve (C3, 4, 5) | Diaphragm (most important muscle of respiration) |
| Sensory | |
|    Phrenic nerve (C3, 4, 5) | Pericardium, mediastinal parietal pleura, and pleura and peritoneum covering central diaphragm |

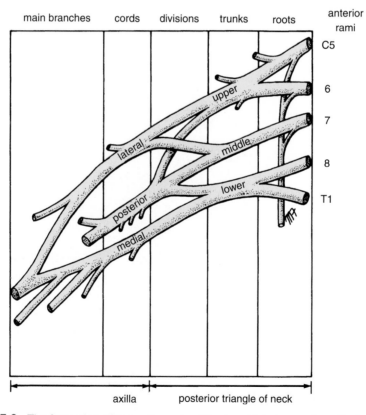

**Figure 17-6** The formation of the main parts of the brachial plexus. Note the locations of the different parts.

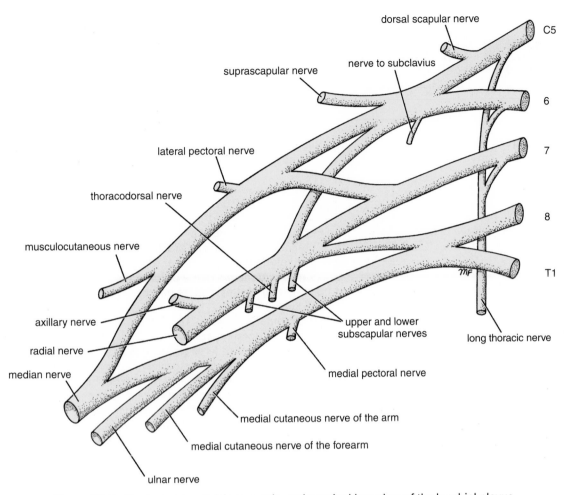

**Figure 17-7** Roots, trunks, divisions, cords, and terminal branches of the brachial plexus.

## Musculocutaneous Nerve

The musculocutaneous nerve (Fig. 17-7) arises from the lateral cord of the brachial plexus (C5, 6, and 7). It pierces the coracobrachialis muscle and then descends between the biceps and the brachialis muscles. In the region of the elbow, it pierces the deep fascia and is distributed to the skin as the **lateral cutaneous nerve of the forearm** (Fig. 17-8). The musculocutaneous nerve supplies the coracobrachialis, both heads of the biceps, and the greater part of the brachialis muscles.

The main branches of the musculocutaneous nerve are summarized in Figure 17-9.

## Median Nerve

The median nerve (Fig. 17-7) arises from the medial and the lateral cords of the brachial plexus (C5, 6, 7, 8, and T1). It descends on the lateral side of the axillary and the brachial arteries (Fig. 17-10). Halfway down the arm, it crosses the brachial artery to reach its medial side. The nerve then descends through the forearm between the two heads of the pronator teres and runs on the posterior surface of the flexor digitorum superficialis (Figs. 17-10 and 17-11). At the wrist, it lies behind the tendon of the palmaris longus (Fig. 17-12). The median nerve enters the palm by passing **behind** the flexor retinaculum (Figs. 17-13 and 17-14) and through the carpal tunnel.

### Branches of the Median Nerve in the Arm.

There are no branches in the arm.

### Branches of the Median Nerve in the Forearm

- **Muscular branches:** Pronator teres, flexor carpi radialis, palmaris longus, and flexor digitorum superficialis muscles.
- **Articular branches:** Elbow joint.
- **Anterior interosseous nerve** (Fig. 17-15)
  - **Muscular branches** to the flexor pollicis longus, pronator quadratus, and lateral half of the flexor digitorum profundus muscles.
  - **Articular branches** to the wrist and carpal joints.
- **Palmar branch:** Skin over the lateral part of the palm (Figs. 17-8 and 17-14).

| Table 17-2 | Summary of the Branches of the Brachial Plexus and Their Distribution |
|---|---|

| Branches | Distribution |
|---|---|
| Roots | |
|    Dorsal scapular nerve (C5) | Rhomboid minor, rhomboid major, levator scapulae muscles |
|    Long thoracic nerve (C5, 6, 7) | Serratus anterior muscle |
| Upper trunk | |
|    Suprascapular nerve (C5, 6) | Supraspinatus and infraspinatus muscles |
|    Nerve to subclavius (C5, 6) | Subclavius |
| Lateral cord | |
|    Lateral pectoral nerve (C5, 6, 7) | Pectoralis major muscle |
|    Musculocutaneous nerve (C5, 6, 7) | Coracobrachialis, biceps brachii, brachialis muscles; supplies skin along lateral border of forearm when it becomes the lateral cutaneous nerve of forearm |
|    Lateral root of median nerve (C5, 6, 7) | See medial root of median nerve |
| Posterior cord | |
|    Upper subscapular nerve (C5, 6) | Subscapularis muscle |
|    Thoracodorsal nerve (C6, 7, 8) | Latissimus dorsi muscle |
|    Lower subscapular nerve (C5, 6) | Subscapularis and teres major muscles |
|    Axillary nerve (C5, 6) | Deltoid and teres minor muscles; upper lateral cutaneous nerve of arm supplies skin over lower half of deltoid muscle |
|    Radial nerve (C5, 6, 7, 8; T1) | Triceps, anconeus, part of brachialis, extensor carpi radialis longus; via deep radial nerve branch supplies extensor muscles of forearm: supinator, extensor carpi radialis brevis, extensor carpi ulnaris, extensor digitorum, extensor digiti minimi, extensor indicis, abductor pollicis longus, extensor pollicis longus, extensor pollicis brevis; skin, lower lateral cutaneous nerve of arm, posterior cutaneous nerve of arm, and posterior cutaneous nerve of forearm; skin on lateral side of dorsum of hand and dorsal surface of lateral three and a half fingers; articular branches to elbow, wrist, and hand |
| Medial cord | |
|    Medial pectoral nerve (C8; T1) | Pectoralis major and minor muscles |
|    Medial cutaneous nerve of arm joined by intercostal brachial nerve from second intercostal nerve (C8; T1, 2) | Skin of medial side of forearm |
| Medial cutaneous nerve of forearm (C8; T1) | Skin of medial side of arm |
| Ulnar nerve (C8; T1) | Flexor carpi ulnaris and medial half of flexor digitorum profundus, flexor digiti minimi, opponens digiti minimi, abductor digiti minimi, adductor pollicis, third and fourth lumbricals, interossei, palmaris brevis, skin of medial half of dorsum of hand and palm, skin of palmar and dorsal surfaces of medial one and a half fingers |
| Medial root of median nerve (with lateral root) forms median nerve (C5, 6, 7, 8; T1) | Pronator teres, flexor carpi radialis, palmaris longus, flexor digitorum superficialis, abductor pollicis brevis, flexor pollicis brevis, opponens pollicis, first two lumbricals (by way of anterior interosseous branch), flexor pollicis longus, flexor digitorum profundus (lateral half), pronator quadratus; palmar cutaneous branch to lateral half of palm and digital branches to palmar surface of lateral three and a half fingers; articular branches to elbow, wrist, and carpal joints |

From Snell RS: Clinical Anatomy. 7th Ed. Philadelphia: Lippincott Williams & Wilkins, 2004, p. 481.

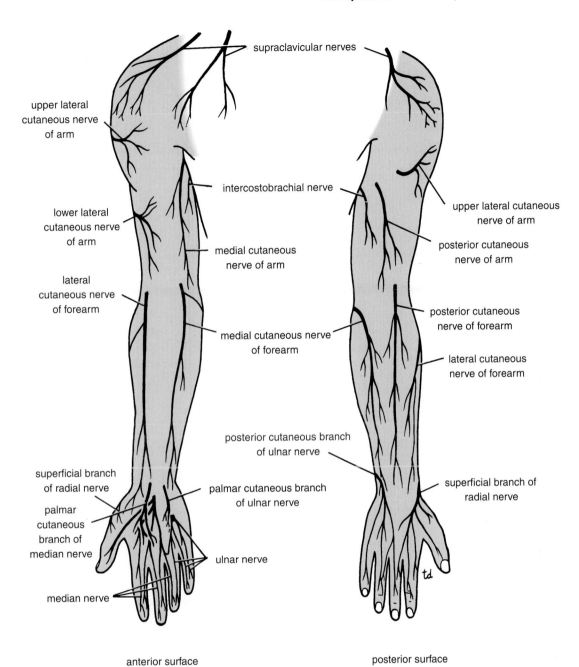

supraclavicular nerves

upper lateral cutaneous nerve of arm

intercostobrachial nerve

upper lateral cutaneous nerve of arm

lower lateral cutaneous nerve of arm

posterior cutaneous nerve of arm

medial cutaneous nerve of arm

lateral cutaneous nerve of forearm

posterior cutaneous nerve of forearm

medial cutaneous nerve of forearm

lateral cutaneous nerve of forearm

posterior cutaneous branch of ulnar nerve

superficial branch of radial nerve

palmar cutaneous branch of ulnar nerve

superficial branch of radial nerve

palmar cutaneous branch of median nerve

ulnar nerve

median nerve

anterior surface

posterior surface

**Figure 17-8** Cutaneous innervation of the upper limb.

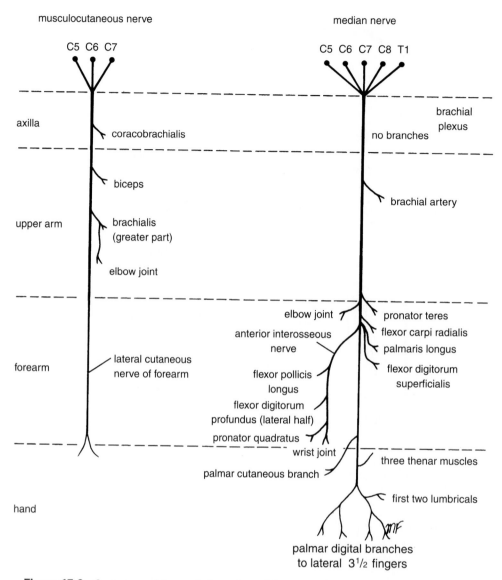

**Figure 17-9** Summary of the main branches of the musculocutaneous and median nerves.

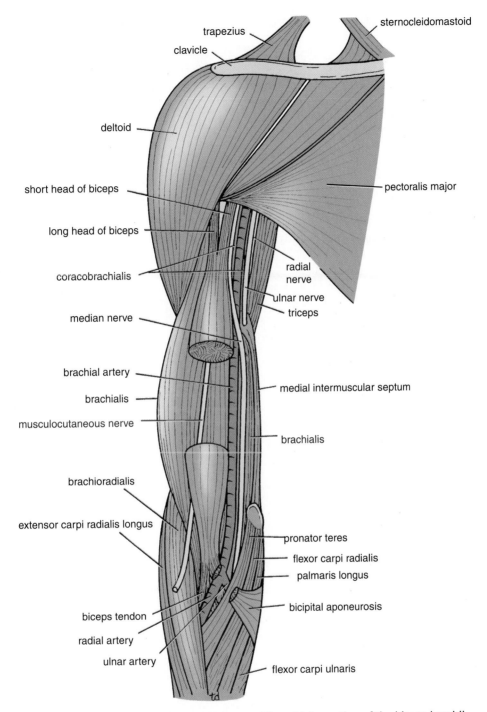

**Figure 17-10**   Anterior view of the upper arm. The middle portion of the biceps brachii has been removed to show the musculocutaneous nerve lying in front of the brachialis.

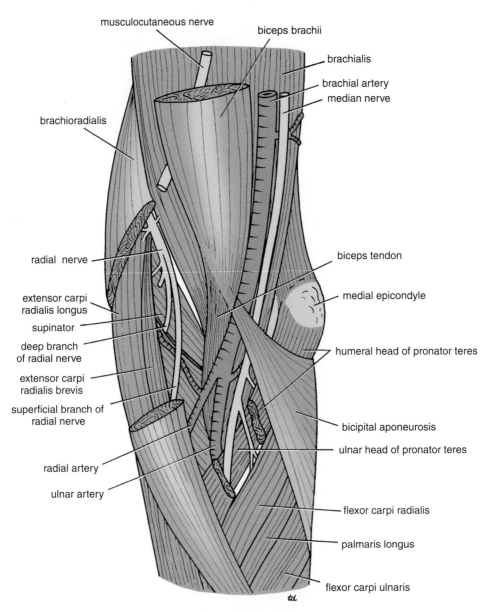

**Figure 17-11** Right cubital fossa.

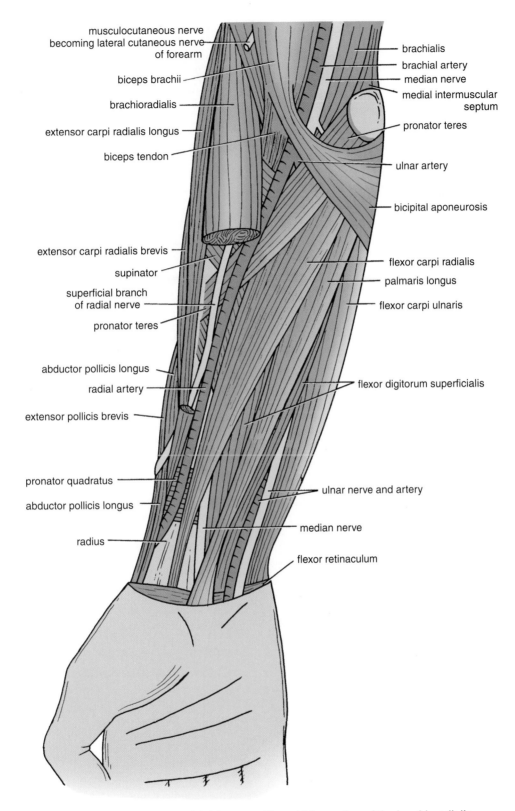

musculocutaneous nerve
becoming lateral cutaneous nerve
of forearm

biceps brachii

brachioradialis

extensor carpi radialis longus

biceps tendon

extensor carpi radialis brevis

supinator

superficial branch
of radial nerve

pronator teres

abductor pollicis longus

radial artery

extensor pollicis brevis

pronator quadratus

abductor pollicis longus

radius

brachialis

brachial artery

median nerve

medial intermuscular
septum

pronator teres

ulnar artery

bicipital aponeurosis

flexor carpi radialis

palmaris longus

flexor carpi ulnaris

flexor digitorum superficialis

ulnar nerve and artery

median nerve

flexor retinaculum

**Figure 17-12** Anterior view of the forearm. The middle portion of the brachioradialis muscle has been removed to display the superficial branch of the radial nerve and the radial artery.

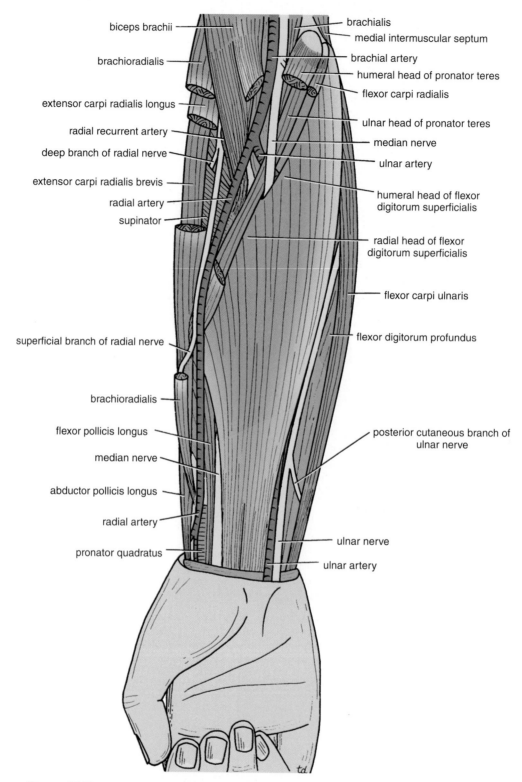

**Figure 17-13** Anterior view of the forearm. Most of the superficial muscles have been removed to display the flexor digitorum superficialis, median nerve, superficial branch of the radial nerve, and radial artery. Note that the ulnar head of the pronator teres separates the median nerve from the ulnar artery.

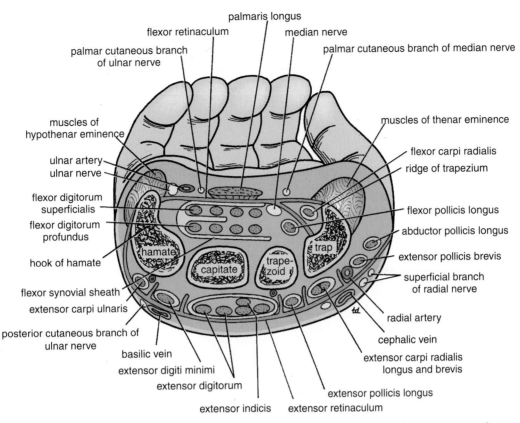

muscles of thenar eminence

palmaris longus

flexor retinaculum

median nerve

palmar cutaneous branch of ulnar nerve

palmar cutaneous branch of median nerve

muscles of hypothenar eminence

ulnar artery

ulnar nerve

flexor digitorum superficialis

flexor digitorum profundus

hook of hamate

flexor synovial sheath

extensor carpi ulnaris

posterior cutaneous branch of ulnar nerve

basilic vein

extensor digiti minimi

extensor digitorum

extensor indicis

extensor retinaculum

extensor pollicis longus

extensor carpi radialis longus and brevis

cephalic vein

radial artery

superficial branch of radial nerve

extensor pollicis brevis

abductor pollicis longus

flexor pollicis longus

ridge of trapezium

flexor carpi radialis

hamate

capitate

trape-zoid

trapezoid

trap

**Figure 17-14** Cross section of the hand showing the relation of the tendons, nerves, and arteries to the flexor and extensor retinacula.

## Branches of the Median Nerve in the Palm (Fig. 17-16)

- **Muscular branches:** Abductor pollicis brevis, flexor pollicis brevis, opponens pollicis, and the first and the second lumbrical muscles.
- **Cutaneous branches:** Palmar aspect of the lateral three and a half fingers and the distal half of the dorsal aspect of each finger as well.

For a summary of the main branches of the median nerve, see Figure 17-9.

## Ulnar Nerve

The ulnar nerve (Fig. 17-7) arises from the medial cord of the brachial plexus (C8, T1). It descends along the medial side of the axillary and the brachial arteries in the anterior compartment of the arm (Fig. 17-10). At the middle of the arm, it pierces the medial intermuscular septum and passes down **behind** the medial epicondyle of the humerus (Fig. 17-17). It then enters the anterior compartment of the forearm and descends behind the flexor carpi ulnaris medial to the ulnar artery. At the wrist, it passes **anterior** to the flexor retinaculum and **lateral** to the **pisiform bone** (Figs. 17-13 and 17-14). It then divides into **superficial** and the **deep terminal branches.**

## Branches of the Ulnar Nerve in the Arm

There are no branches in the arm.

## Branches of the Ulnar Nerve in the Forearm

- **Muscular branches:** Flexor carpi ulnaris and medial half of the flexor digitorum profundus muscles.
- **Articular branches:** Elbow joint.
- **Cutaneous branches.**
  - **Dorsal posterior cutaneous branch** (Figs. 17-8 and 17-18). Supplies the skin over the medial side of the back of the hand and back of the medial one and a half fingers over the proximal phalanges.
  - **Palmar cutaneous branch.** Passes anterior to the flexor retinaculum and supplies the skin over the medial part of the palm (Fig. 17-8).

## Branches of the Ulnar Nerve in the Hand

The **superficial terminal branch** (Fig. 17-16) descends into the palm and gives off the following branches:

- **Muscular branches:** Palmaris brevis muscle.
- **Cutaneous branches:** Supply the skin over the palmar aspect of the medial one and a half fingers (including their nail beds).

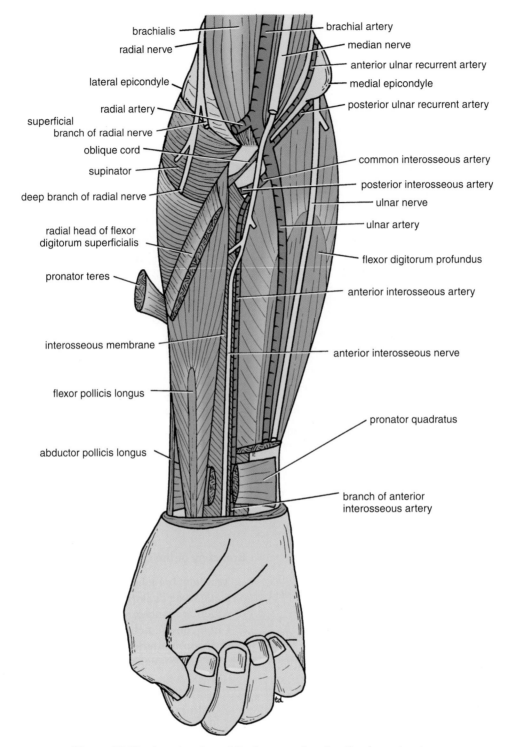

brachialis

radial nerve

lateral epicondyle

radial artery

superficial
branch of radial nerve

oblique cord

supinator

deep branch of radial nerve

radial head of flexor
digitorum superficialis

pronator teres

interosseous membrane

flexor pollicis longus

abductor pollicis longus

brachial artery

median nerve

anterior ulnar recurrent artery

medial epicondyle

posterior ulnar recurrent artery

common interosseous artery

posterior interosseous artery

ulnar nerve

ulnar artery

flexor digitorum profundus

anterior interosseous artery

anterior interosseous nerve

pronator quadratus

branch of anterior
interosseous artery

**Figure 17-15** Anterior view of the forearm showing the deep structures.

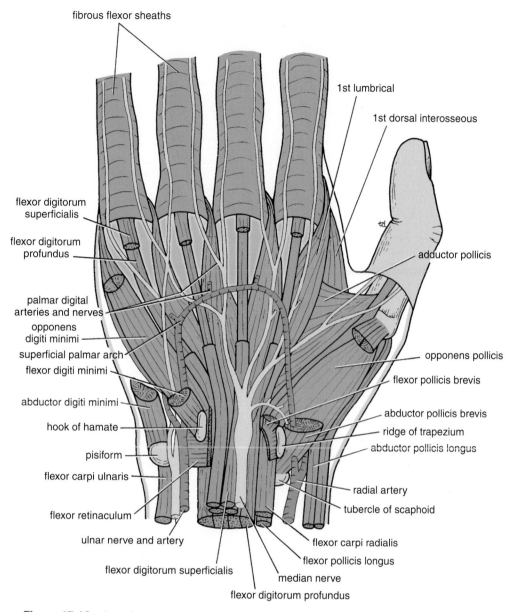

**Figure 17-16**    Anterior view of the palm of the hand. The palmar aponeurosis and the greater part of the flexor retinaculum have been removed to display the superficial palmar arch, the median nerve, and the long flexor tendons. Segments of the tendons of the flexor digitorum superficialis have been removed to show the underlying tendons of the flexor digitorum profundus.

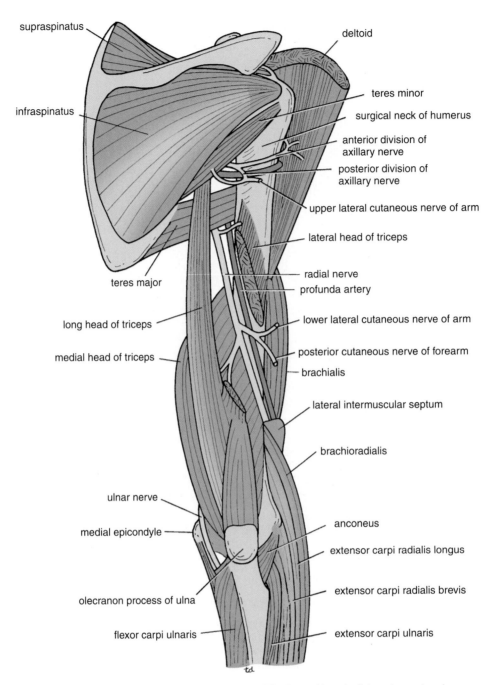

**Figure 17-17**    Posterior view of the upper arm. The lateral head of the triceps has been divided to display the radial nerve and the profunda artery in the spiral groove of the humerus.

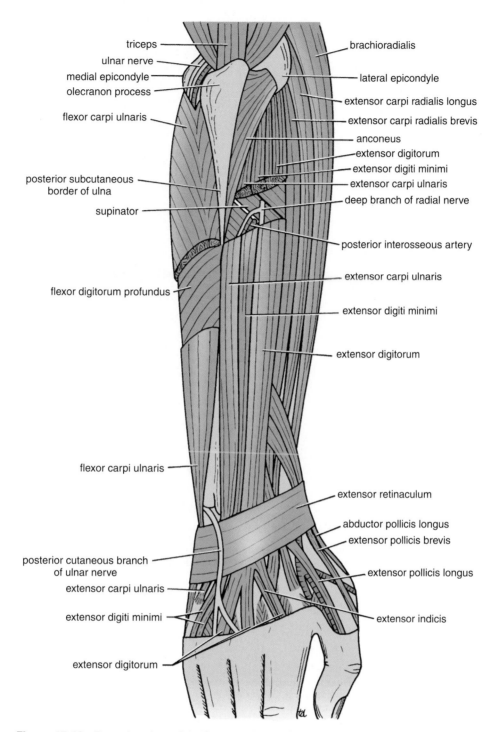

triceps

ulnar nerve

medial epicondyle

olecranon process

flexor carpi ulnaris

posterior subcutaneous
border of ulna

supinator

flexor digitorum profundus

flexor carpi ulnaris

posterior cutaneous branch
of ulnar nerve

extensor carpi ulnaris

extensor digiti minimi

extensor digitorum

brachioradialis

lateral epicondyle

extensor carpi radialis longus

extensor carpi radialis brevis

anconeus

extensor digitorum

extensor digiti minimi

extensor carpi ulnaris

deep branch of radial nerve

posterior interosseous artery

extensor carpi ulnaris

extensor digiti minimi

extensor digitorum

extensor retinaculum

abductor pollicis longus

extensor pollicis brevis

extensor pollicis longus

extensor indicis

**Figure 17-18**   Posterior view of the forearm. Parts of the extensor digitorum, extensor digiti minimi, and extensor carpi ulnaris have been removed to show the deep branch of the radial nerve and the posterior interosseous artery.

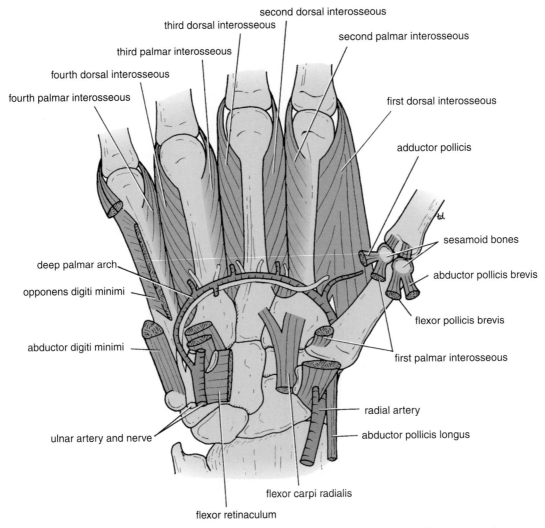

third dorsal interosseous

second dorsal interosseous

third palmar interosseous

second palmar interosseous

fourth dorsal interosseous

fourth palmar interosseous

first dorsal interosseous

adductor pollicis

sesamoid bones

abductor pollicis brevis

deep palmar arch

opponens digiti minimi

flexor pollicis brevis

first palmar interosseous

abductor digiti minimi

radial artery

abductor pollicis longus

ulnar artery and nerve

flexor carpi radialis

flexor retinaculum

**Figure 17-19**   Anterior view of the palm of the hand showing the deep palmar arch and the deep terminal branch of the ulnar nerve. The interossei are also shown.

The **deep terminal branch** (Fig. 17-19) runs backward between the abductor digiti minimi and the flexor digiti minimi muscles, pierces the opponens digiti minimi muscle, and gives off the following branches.

■ **Muscular branches:** Abductor digiti minimi, flexor digiti minimi, opponens digiti minimi, all palmar and all dorsal interossei, third and fourth lumbricals, and adductor pollicis muscles.
■ **Articular branches:** Carpal joints.

The main branches of the ulnar nerve are summarized in Figure 17-20.

## Radial Nerve

The radial nerve (Fig. 17-7) arises from the posterior cord of the brachial plexus (C5, 6, 7, 8, and T1). It descends behind the axillary and the brachial arteries, and it enters the posterior compartment of the arm. The radial nerve winds around the back of the humerus in the spiral groove with the profunda artery (Fig. 17-17). Piercing the lateral intermuscular septum just above the elbow, it descends in front of the lateral epicondyle and divides into the superficial and the deep terminal branches (Fig. 17-15).

### Branches of the Radial Nerve in the Axilla

■ **Muscular branches:** Long and medial heads of the triceps muscle.
■ **Cutaneous branch:** Posterior cutaneous nerve of the arm (Fig. 17-8).

### Branches of the Radial Nerve in the Spiral Groove behind the Humerus

■ **Muscular branches:** Lateral and medial heads of the triceps muscle and the anconeus muscle (Fig. 17-17).
■ **Cutaneous branches:** Lower lateral cutaneous nerve of the arm, posterior cutaneous nerve of the forearm (Figs. 17-8 and 17-17).

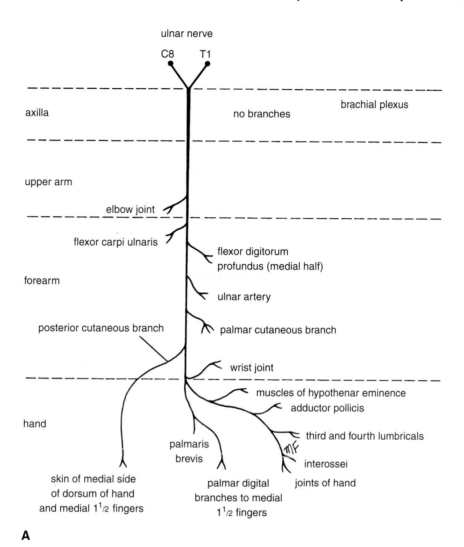

**A**

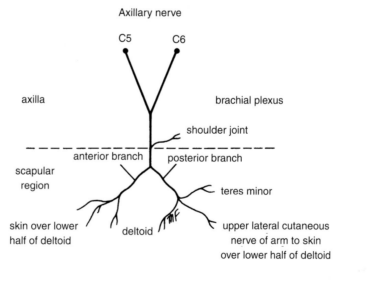

**B**

**Figure 17-20**  **A.** Summary of the main branches of the ulnar nerve. **B.** Summary of the main branches of the axillary nerve.

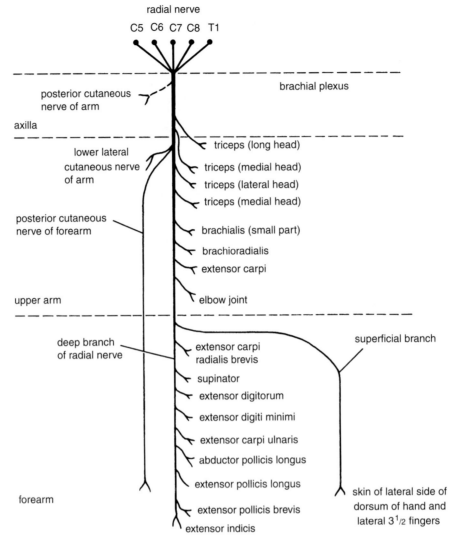

**Figure 17-21** Summary of the main branches of the radial nerve.

### Branches of the Radial Nerve in the Anterior Compartment of the Arm Close to the Lateral Epicondyle

▓ **Muscular branches:** Brachialis, brachioradialis, and extensor carpi radialis longus muscles (Fig. 17-11).

▓ **Articular branches:** Elbow joint.

#### Superficial Branch of Radial Nerve

The superficial branch of the radial nerve descends under cover of the brachioradialis muscle on the lateral side of the radial artery (Figs. 17-12 and 17-13). It emerges from beneath the brachioradialis tendon and then descends on the back of the hand (Fig. 17-8).

#### Deep Branch of the Radial Nerve

The deep branch of the radial nerve winds around the lateral side of the neck of the radius within the supinator muscle (Fig. 17-15). It enters the posterior compartment of the

forearm, and it descends between the muscles. It gives off the following branches:

▓ **Muscular branches:** Extensor carpi radialis brevis, supinator, extensor carpi ulnaris, abductor pollicis longus, extensor pollicis brevis, extensor pollicis longus, and extensor indicis muscles.

▓ **Articular branches.** Wrist and carpal joints.

The main branches of the radial nerve are summarized in Figure 17-21.

### Axillary Nerve

The axillary nerve (Fig. 17-7) arises from the posterior cord of the brachial plexus (C5 and 6). It passes backward through the quadrangular space below the shoulder joint with the posterior circumflex humeral vessels (Fig. 17-17).

### Branches of the Axillary Nerve

■ **Articular branch,** which supplies the shoulder joint.
■ **Anterior terminal branch,** which winds around the surgical neck of the humerus and supplies the deltoid muscle and the skin that covers its lower half. (Supraclavicular nerves supply the skin over the upper half of the deltoid muscle.)
■ **Posterior terminal branch,** which supplies the teres minor and the deltoid muscles and then becomes the **upper lateral cutaneous nerve of the arm** (Fig. 17-8), which also supplies the skin over the lower part of the deltoid muscle.

The main branches of the axillary nerve are summarized in Figure 17-20.

## Long Thoracic Nerve

The long thoracic nerve (C5, 6, and 7) is a branch of the brachial plexus (Fig. 17-7) that is clinically important. (It is vulnerable in operations involving the axilla, especially a radical mastectomy). It arises from the roots of the brachial plexus in the neck and enters the axilla by passing down over the lateral border of the first rib behind the axillary vessels and brachial plexus (Fig. 7-3). It descends over the lateral surface of the serratus anterior muscle, which it supplies

# Intercostal Nerves

The intercostal nerves are the anterior rami of the first eleven thoracic spinal nerves (Fig. 17-22). The anterior ramus of the twelfth thoracic nerve lies in the abdomen and runs forward in the abdominal wall as the subcostal nerve.

Each intercostal nerve enters an intercostal space between the parietal pleura and the posterior intercostal membrane (Fig. 17-23). It then runs forward inferiorly to the intercostal vessels in the subcostal groove of the corresponding rib, between the innermost intercostal and internal intercostal muscle.

The first six nerves are distributed within their intercostal spaces. The seventh to ninth intercostal nerves leave the anterior ends of their intercostal spaces by passing deep to the costal cartilages, to enter the anterior abdominal wall. The tenth and eleventh nerves, because the corresponding ribs are floating, pass directly into the abdominal wall.

## Branches of the Intercostal Nerves

■ **Rami communicantes** connect the intercostal nerve to a ganglion of the sympathetic trunk (Fig. 17-2). The gray

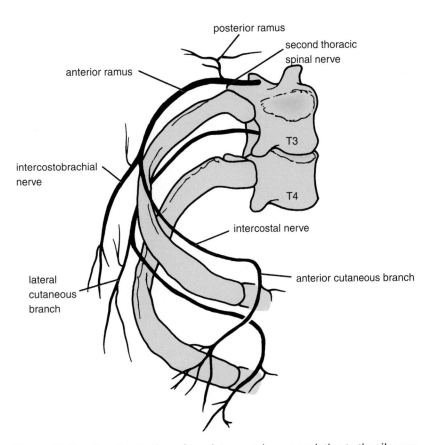

**Figure 17-22** The distribution of two intercostal nerves relative to the rib cage.

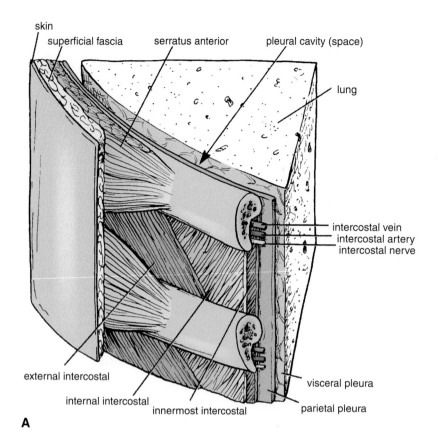

**A**

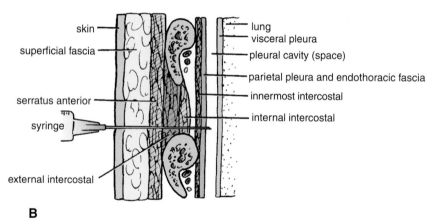

**B**

**Figure 17-23** **A.** Section through an intercostal space. **B.** Structures penetrated by a needle when it passes from skin surface to pleural cavity. Depending on the site of penetration, the pectoral muscles will be pierced in addition to the serratus anterior muscle.

ramus joins the nerve medial at the point at which the white ramus leaves it.

- **Collateral branch** runs forward below the main nerve.
- **Lateral cutaneous branch** reaches the skin on the side of the chest. It divides into an anterior and a posterior branch (Fig. 17-22).
- **Anterior cutaneous branch,** which is the terminal portion of the main trunk, reaches the skin near the midline. It divides into a medial and a lateral branch (Fig. 17-22).
- **Muscular branches** run to the intercostal muscles.
- **Pleural sensory branches** run to the parietal pleura.

- **Peritoneal sensory branches** (seventh to eleventh intercostal nerves only) run to the parietal peritoneum.

It should be noted that the seventh to the eleventh intercostal nerves supply the skin and parietal peritoneum covering the outer and inner surfaces of the anterior abdominal wall, respectively. The seventh to eleventh intercostal nerves also supply the anterior abdominal muscles (external and internal oblique, transversus abdominis, and rectus abdominis muscles).

The **first** and **second intercostal nerves,** however, are exceptions.

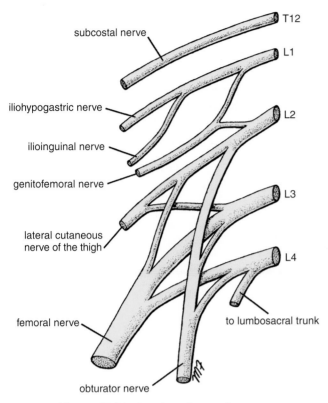

**Figure 17-24**    Lumbar plexus of nerves.

The **first intercostal nerve** gives rise to a large branch (equivalent to the lateral cutaneous branch of a typical intercostal nerve) that joins the anterior ramus of the eighth cervical nerve to form the lower trunk of the brachial plexus (Fig. 17-7). The remainder of the first intercostal nerve is small, and there is no cutaneous branch.

The **second intercostal nerve** is joined to the medial cutaneous nerve of the arm by the **intercostobrachial nerve** (Fig. 17-8). The second intercostal nerve therefore supplies the skin of the armpit and the upper medial side of the arm.

# Lumbar Plexus

The lumbar plexus, which is one of the main nervous pathways supplying the lower limb, is formed in the psoas muscle from the anterior rami of the upper four lumbar nerves (Fig. 17-24). The anterior rami receive gray rami communicantes from the sympathetic trunk, and the upper two give off white rami communicantes to the sympathetic trunk. The branches of the plexus emerge from the lateral and medial borders of the muscle and from its anterior surface (Fig. 17-25). The courses of the more important nerves and their branches are as follows:

The iliohypogastric nerve, ilioinguinal nerve, lateral cutaneous nerve of the thigh, and femoral nerve emerge from

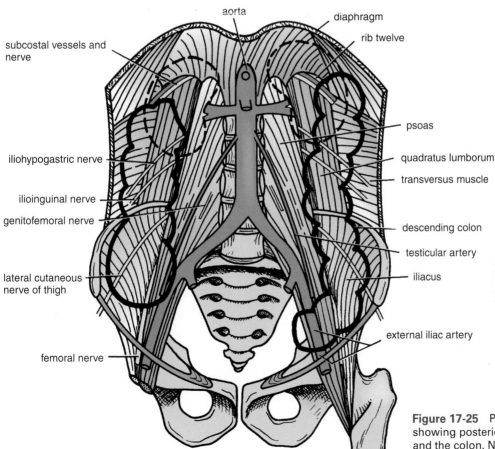

**Figure 17-25**    Posterior abdominal wall showing posterior relations of the kidneys and the colon. Note the branches of the lumbar plexus.

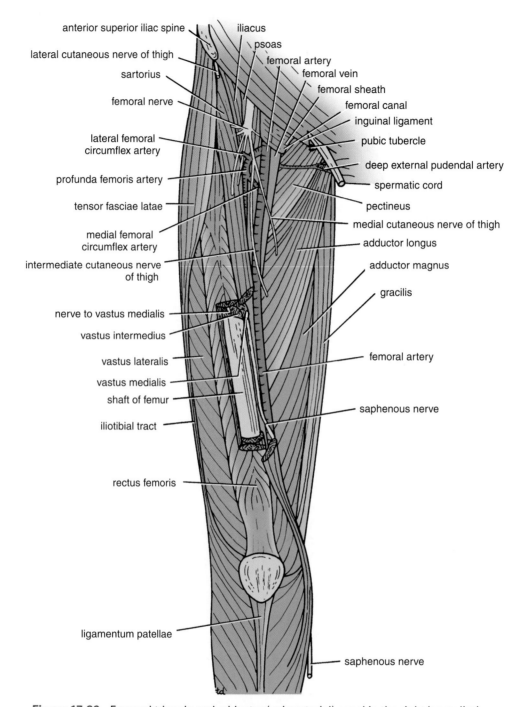

**Figure 17-26**   Femoral triangle and adductor (subsartorial) canal in the right lower limb.

the lateral border of the psoas, in that order, from above downward (Fig. 17-25). The obturator nerve and the lumbosacral trunk emerge from the medial border. The genitofemoral nerve emerges from the anterior surface (Fig. 17-25).

## Iliohypogastric Nerve

The iliohypogastric nerve (L1) emerges from the lateral border of the psoas (Fig. 17-25). It runs forward to supply the transversus abdominis, the internal and external oblique muscles, and the skin above the inguinal ligament.

## Ilioinguinal Nerve

The ilioinguinal nerve (L1) emerges from the lateral border of the psoas (Fig. 17-25). It runs forward through the inguinal canal and exits through the superficial inguinal ring. It supplies the transversus abdominis, internal oblique, and external oblique muscles. It also supplies the skin just above the symphysis pubis and the scrotum or labium majus.

## Lateral Cutaneous Nerve of the Thigh

The lateral cutaneous nerve of the thigh (L2 and 3) emerges from the lateral border of the psoas, crosses the iliacus muscle, and enters the thigh behind the lateral end of the inguinal ligament (Fig. 17-25). It divides into anterior and posterior branches and supplies the skin of the lateral aspect of the thigh and knee and the lower part of the buttock.

## Femoral Nerve

This is the largest branch of the lumbar plexus and arises from L2, 3, and 4 lumbar nerves (Fig. 17-24). It emerges from the lateral border of the psoas muscle in the abdomen and descends between the psoas and the iliacus muscles (Fig. 17-25). It enters the thigh behind the inguinal ligament and lies lateral to the femoral vessels and the femoral sheath. About 1½ in (4 cm) below the inguinal ligament, it terminates by dividing into anterior and posterior divisions (Fig. 17-26).

### Branches of the Femoral Nerve in the Abdomen

**Muscular branches** to the iliacus.

### Branches of the Femoral Nerve in the Thigh

- **Cutaneous branches:** The **medial cutaneous nerve of the thigh** supplies the skin on the medial side of the thigh (Fig. 17-27). The **intermediate cutaneous nerve of the thigh** supplies the skin on the anterior surface of the thigh (Fig. 17-27). The **saphenous nerve** descends through the femoral triangle into the adductor canal, crossing the femoral artery on its anterior surface (Fig.

17-26). The nerve emerges on the medial side of the knee joint between the tendons of the sartorius and gracilis muscles. It descends along the medial side of the leg in company with the great saphenous vein. It passes anterior to the medial malleolus and along the medial border of the foot as far as the ball of the big toe (Figs. 17-27 and 17-28).
- **Muscular branches:** Sartorius, pectineus, and quadriceps femoris muscles.
- **Articular branches** to the hip and knee joints.

The main branches of the femoral nerve are summarized in Figure 17-29.

**Patellar plexus:** This small plexus lies in front of the knee and is formed from the terminal branches of the lateral, intermediate, and medial cutaneous nerves of the thigh and the infrapatellar branch of the saphenous nerve (Fig. 17-27).

## Obturator Nerve

This nerve arises from the lumbar plexus (L2, 3, and 4) (Fig. 17-24). It emerges on the medial border of the psoas muscle within the abdomen. The nerve descends and crosses the pelvic brim anterior to the sacroiliac joint and posterior to the common iliac vessels. It then runs downward and forward on the lateral wall of the pelvis in the angle between the internal and external iliac vessels (Fig. 17-30). Here, it is accompanied by the obturator vessels. On reaching the obturator canal (upper part of the obturator foramen of the hip bone), it divides into anterior and posterior divisions.

## Branches of the Obturator Nerve

- **Parietal peritoneum:** Sensory fibers to parietal peritoneum on the lateral wall of the pelvis.
- **Anterior division:** Descends into the thigh anterior to the obturator externus and adductor brevis muscles (Figs. 17-31 and 17-32).
  - **Muscular branches** to the gracilis, adductor brevis, adductor longus, and sometimes the pectineus muscles.
  - **Sensory branch** to the skin on the medial side of the thigh (Fig. 17-27).
  - **Articular branch** to the hip joint.
- **Posterior division:** Descends through the obturator externus muscle and passes behind the adductor brevis and in front of the adductor magnus muscle (Figs. 17-31 and 17-32).
  - **Muscular branches** to the obturator externus, adductor magnus (adductor part), and sometimes the adductor brevis muscles.
- **Articular branch** to the knee joint.

The main branches of the obturator nerve are summarized in Figure 17-33.

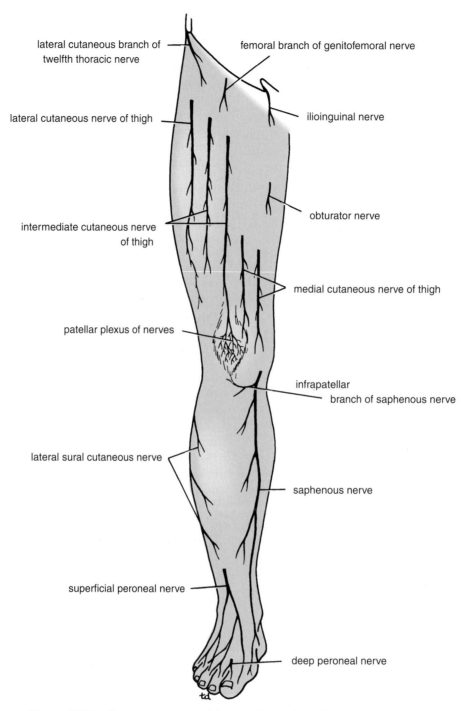

**Figure 17-27** Cutaneous nerves of the anterior surface of the right lower limb.

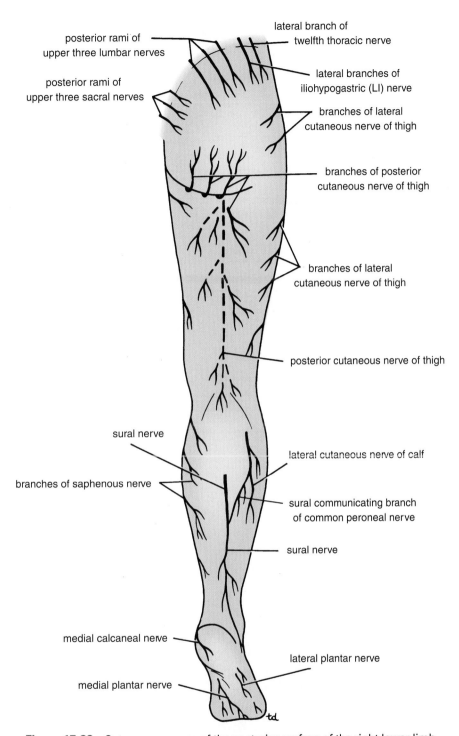

**Figure 17-28** Cutaneous nerves of the posterior surface of the right lower limb.

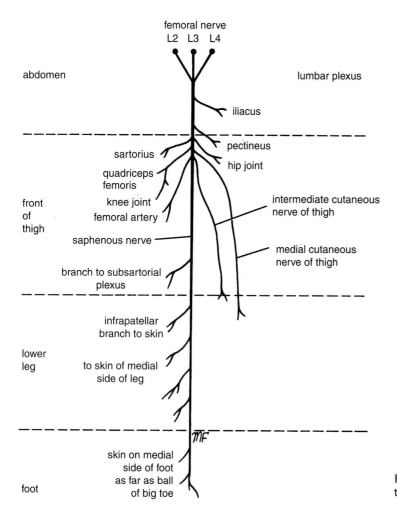

**Figure 17-29** Summary of the main branches of the femoral nerve.

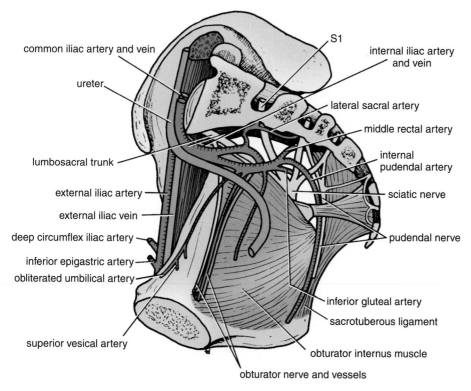

**Figure 17-30** Lateral wall of the pelvis.

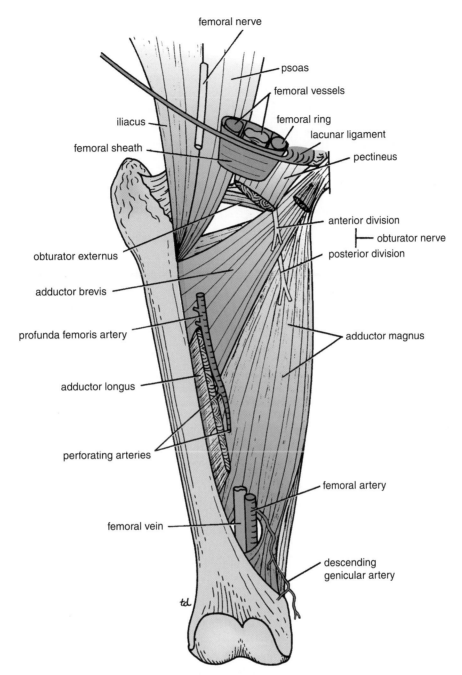

**Figure 17-31** Relationship between the obturator nerve and the adductor muscles in the right lower limb.

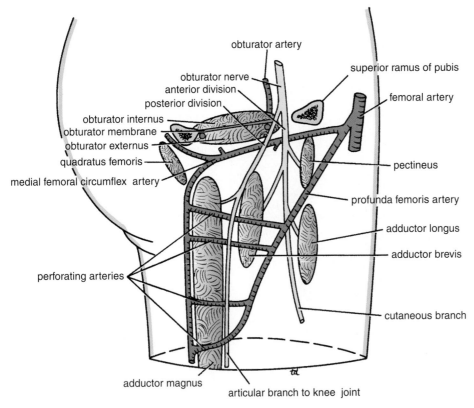

**Figure 17-32** Vertical section of the medial compartment of the thigh. Note the courses taken by the obturator nerve and its divisions and the profunda femoris artery and its branches. Note also the anastomosis between the perforating arteries and the medial femoral circumflex artery.

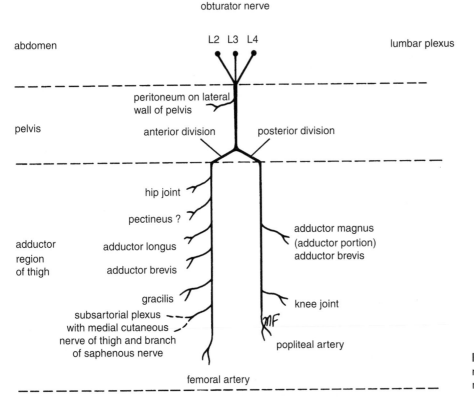

**Figure 17-33** Summary of the main branches of the obturator nerve.

## Genitofemoral Nerve

The genitofemoral nerve (L1 and 2) is a small nerve that emerges from the anterior surface of the psoas (Fig. 17-24). It runs downward in front of the muscle and divides into a **genital** branch, which enters the spermatic cord and supplies the cremaster muscle, and a **femoral branch**, which supplies a small area of skin of the thigh (Fig. 17-27).

### Cremasteric Reflex

The genitofemoral nerve is the nervous pathway involved in the cremasteric reflex contraction of the cremaster muscle and the drawing upward of the testis within the scrotum.

## Lumbosacral Trunk

The lumbosacral trunk is formed from the union of a branch of the anterior ramus of the fourth lumbar nerve and that of the fifth lumbar nerve (Fig. 7-24). The trunk passes down along the medial border of the psoas muscle in front of the sacroiliac joint and enters the pelvis, where it joins the sacral nerves to form the sacral plexus (Fig. 17-34).

The branches of the lumbar plexus and their distribution are summarized in Table 17-3.

# Sacral Plexus

The sacral plexus lies on the posterior pelvic wall in front of the piriformis muscle (Fig. 17-34). It is formed from the anterior rami of the fourth and fifth lumbar nerves and the anterior rami of the first, second, third, and fourth sacral nerves (Fig. 17-35). The contribution from the fourth lumbar nerve joins the fifth lumbar nerve to form the **lumbosacral trunk,** which passes down into the pelvis and joins the sacral nerves as they emerge from the anterior sacral foramina.

# Branches of the Sacral Plexus to the Lower Limb

Branches of the sacral plexus to the lower limb that leave the pelvis through the greater sciatic foramen are as follows (Fig. 17-29):

- **Sciatic nerve** (L4 and 5; S1, 2, and 3). This is the largest branch of the plexus and the largest nerve in the body (Fig. 17-35).
- **Superior gluteal nerve,** which supplies the gluteus medius and minimus and the tensor fasciae latae muscles (Fig. 17-36).
- **Inferior gluteal nerve,** which supplies the gluteus maximus muscle (Fig. 17-36).

| Table 17-3 | Summary of the Branches of the Lumbar Plexus and Their Distribution |
|---|---|
| **Branches** | **Distribution** |
| Iliohypogastric nerve | External oblique, internal oblique, transversus abdominis muscles of anterior abdominal wall; skin over lower anterior abdominal wall and buttock |
| Ilioinguinal nerve | External oblique, internal oblique, transversus abdominis muscles of anterior abdominal wall; skin of upper medial aspect of thigh, root of penis and scrotum in the male, mons pubis and labia majora in the female |
| Lateral cutaneous nerve of the thigh | Skin of anterior and lateral surfaces of the thigh |
| Genitofemoral nerve (L1, 2) | Cremaster muscle in scrotum in male; skin over anterior surface of thigh; nervous pathway for cremasteric reflex |
| Femoral nerve (L2, 3, 4) | Iliacus, pectineus, sartorius, quadriceps femoris muscles, and intermediate cutaneous branches to the skin of the anterior surface of the thigh and by saphenous branch to the skin of the medial side of the leg and foot; articular, branches to hip and knee joints |
| Obturator nerve (L2, 3, 4) | Gracilis, adductor brevis, adductor longus, obturator externus, pectineus, adductor magnus (adductor portion), and skin on medial surface of thigh; articular branches to hip and knee joints |
| Segmental branches | Quadratus lumborum and psoas muscles |

From Snell RS: Clinical Anatomy. 7th Ed. Philadelphia: Lippincott Williams & Wilkins, 2004, p. 299.

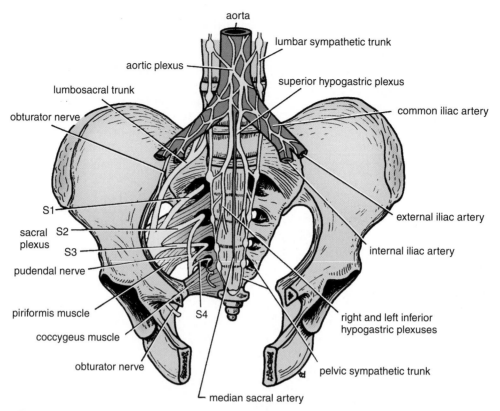

**Figure 17-34** Posterior pelvic wall showing the sacral plexus, superior hypogastric plexus, and right and left inferior hypogastric plexuses. Pelvic parts of the sympathetic trunks are also shown.

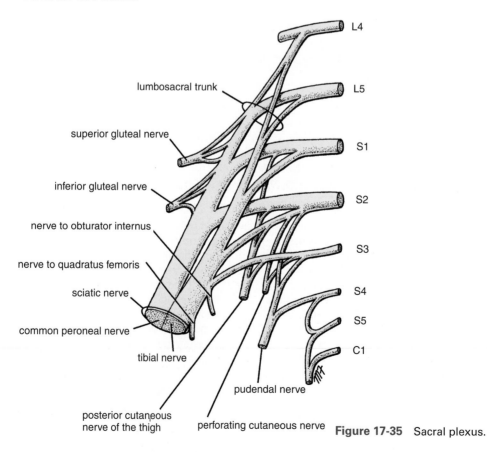

**Figure 17-35** Sacral plexus.

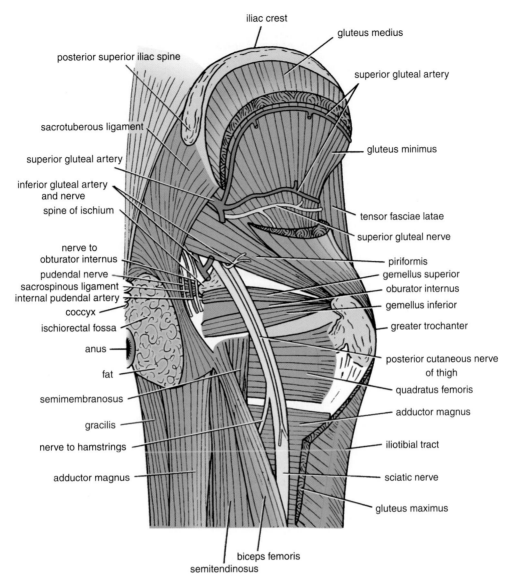

**Figure 17-36**   Structures in the right gluteal region. The greater part of the gluteus maximus and part of the gluteus medius have been removed.

■ **Nerve to the quadratus femoris muscle,** which also supplies the inferior gemellus muscle.

■ **Nerve to the obturator internus muscle,** which also supplies the superior gemellus muscle (Fig. 17-36).

■ **Posterior cutaneous nerve of the thigh,** which supplies the skin of the buttock and the back of the thigh (Figs. 17-28 and 17-36).

# Branches of the Sacral Plexus to the Pelvic Muscles, Pelvic Viscera, and Perineum

■ **Pudendal nerve** (S2, 3, and 4) leaves the pelvis through the greater sciatic foramen and enters the perineum through the lesser sciatic foramen (Figs. 17-29 and 17-36).

■ **Nerves to the piriformis muscle.**

■ **Pelvic splanchnic nerves,** which constitute the sacral part of the parasympathetic system and arise from the second, third, and fourth sacral nerves. They are distributed to the pelvic viscera.

The branches of the sacral plexus and their distribution are summarized in Table 17-4.

## Sciatic Nerve

The sciatic nerve (Fig. 17-35) is the largest nerve in the body and arises from L4 and 5 and S1, 2, and 3. It passes out of the pelvis and into the gluteal region through the greater sciatic foramen (Figs. 17-30 and 17-36). The nerve appears below the piriformis muscle and is covered by the gluteus maximus muscle. It descends through the gluteal region, and it enters the posterior compartment of the thigh. In the

| Table 17-4 | Summary of the Branches of the Sacral Plexus and Their Distribution |
|---|---|

| Branches | Distribution |
|---|---|
| Superior gluteal nerve | Gluteus medius, gluteus minimus, and tensor fasciae latae muscles |
| Inferior gluteal nerve | Gluteus maximus muscle |
| Nerve to piriformis | Piriformis muscle |
| Nerve to obturator internus | Obturator internus and superior gemellus muscles |
| Nerve to quadratus femoris | Quadratus femoris and inferior gemellus muscles |
| Perforating cutaneous nerve | Skin over medial aspect of buttock |
| Posterior cutaneous nerve of thigh | Skin over posterior surface of thigh and popliteal fossa, also over lower part of buttock, scrotum, or labium majus |
| Sciatic nerve (L4, 5; S1, 2, 3) Tibial portion | Hamstring muscles (semitendinosus, biceps femoris [long head], adductor magnus [hamstring part]), gastrocnemius, soleus, plantaris, popliteus, tibialis posterior, flexor digitorum longus, flexor hallucis longus, and via medial and lateral plantar branches to muscles of sole of foot; sural branch supplies skin on lateral side of leg and foot |
| Common peroneal portion | Biceps femoris muscle (short head) and via deep peroneal branch: tibialis anterior, extensor hallucis longus, extensor digitorum longus, peroneus tertius, and extensor digitorum brevis muscles; skin over cleft between first and second toes; the superficial peroneal branch supplies the peroneus longus and brevis muscles and skin over lower third of anterior surface of leg and dorsum of foot |
| Pudendal nerve | Muscles of perineum including the external anal sphincter, mucous membrane of lower half of anal canal, perianal skin, skin of penis, scrotum, clitoris, and labia majora and minora |

From Snell RS: Clinical Anatomy. 7th Ed. Philadelphia: Lippincott Williams & Wilkins, 2004, p. 349.

lower third of the thigh (and occasionally at a higher level), it ends by dividing into the tibial and the common peroneal nerves (Fig. 17-37).

## Branches of the Sciatic Nerve

- **Muscular branches:** Biceps femoris (long head), semitendinosus, semimembranosus, and hamstring part of the adductor magnus muscles.
- **Articular branches:** Hip joint.
- **Terminal branches:** Tibial and common peroneal nerves.

## Tibial Nerve

The tibial nerve descends through the popliteal fossa and the posterior compartment of the leg (Fig. 17-38). It lies deep to the gastrocnemius and soleus muscles, and it reaches the interval between the medial malleolus and the heel (Fig. 17-39). It is covered by the flexor retinaculum and divides into the medial and the lateral plantar nerves.

## Branches of the Tibial Nerve

- **Cutaneous branches: Sural nerve** (joined by communicating branch of the common peroneal nerve), which supplies the skin of the calf, the back of the leg, the lateral border of the foot, and the lateral side of the little toe (Fig. 17-28); **medial calcaneal nerve,** which supplies the skin over the medial surface of the heel.
- **Muscular branches:** Gastrocnemius, plantaris, soleus, popliteus, flexor digitorum longus, flexor hallucis longus, and tibialis posterior muscles.
- **Articular branches:** Knee and ankle joints.
- **Terminal branches.**
  - **Medial plantar nerve:** Runs forward deep to the abductor hallucis muscle with the medial plantar artery (Figs. 17-40 and 17-41).
    - **Cutaneous branch,** which supplies the medial part of the sole and the medial three and a half toes and nail beds.
    - **Muscular branch,** which supplies the abductor hallucis, flexor digitorum brevis, flexor hallucis brevis, and first lumbrical muscles.

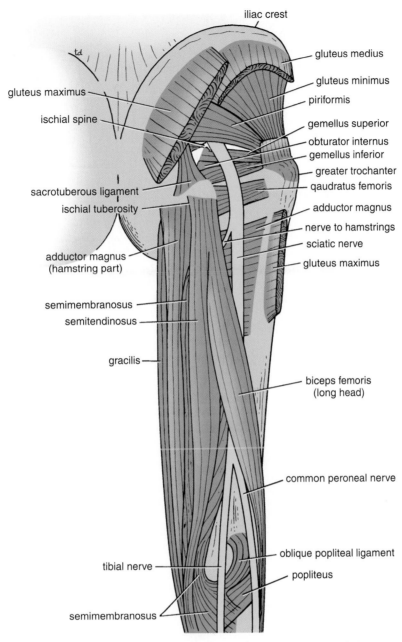

**Figure 17-37**   Structures in the posterior aspect of the right thigh.

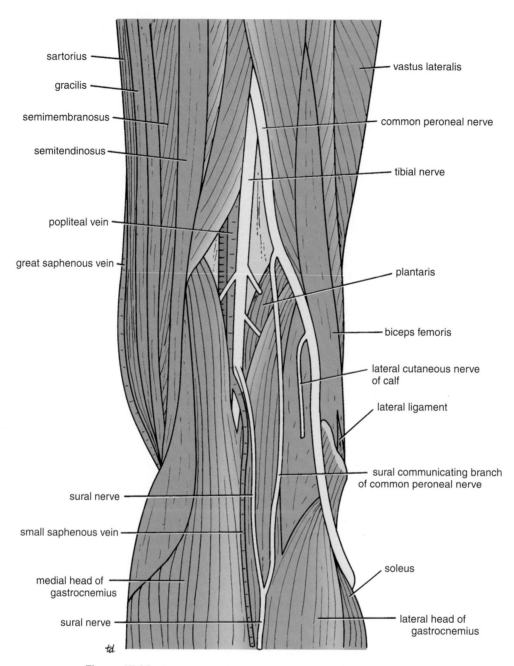

sartorius

gracilis

semimembranosus

semitendinosus

popliteal vein

great saphenous vein

sural nerve

small saphenous vein

medial head of
gastrocnemius

sural nerve

vastus lateralis

common peroneal nerve

tibial nerve

plantaris

biceps femoris

lateral cutaneous nerve
of calf

lateral ligament

sural communicating branch
of common peroneal nerve

soleus

lateral head of
gastrocnemius

**Figure 17-38** Boundaries and contents of the right popliteal fossa.

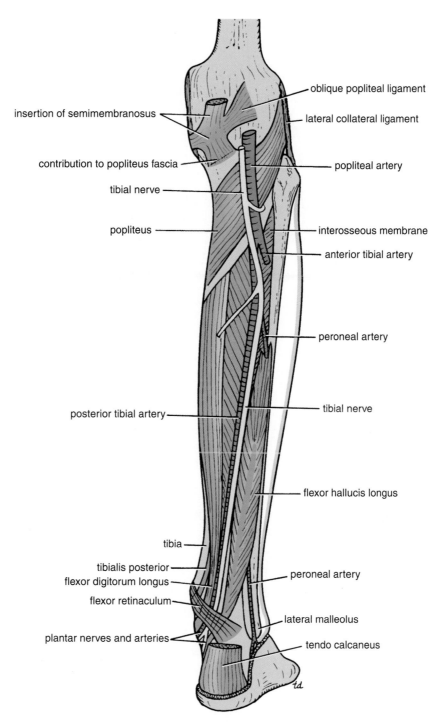

insertion of semimembranosus

oblique popliteal ligament

lateral collateral ligament

contribution to popliteus fascia

popliteal artery

tibial nerve

popliteus

interosseous membrane

anterior tibial artery

peroneal artery

posterior tibial artery

tibial nerve

flexor hallucis longus

tibia

tibialis posterior
flexor digitorum longus

peroneal artery

flexor retinaculum

plantar nerves and arteries

lateral malleolus

tendo calcaneus

**Figure 17-39** Deep structures in the posterior aspect of the right leg.

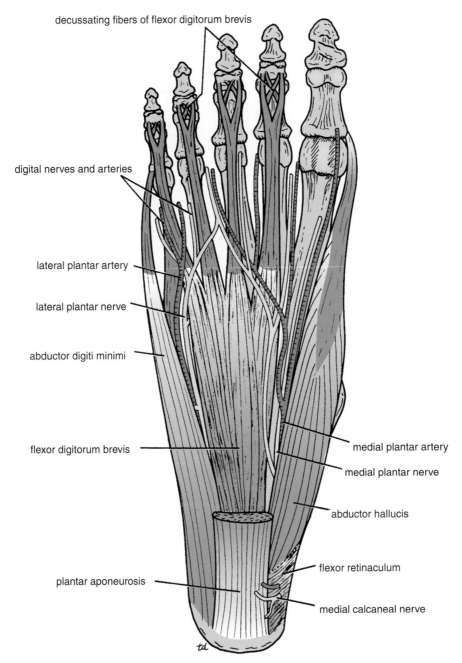

decussating fibers of flexor digitorum brevis

digital nerves and arteries

lateral plantar artery

lateral plantar nerve

abductor digiti minimi

flexor digitorum brevis

medial plantar artery

medial plantar nerve

abductor hallucis

flexor retinaculum

plantar aponeurosis

medial calcaneal nerve

**Figure 17-40** First layer of the plantar muscles of the right foot. Medial and lateral plantar arteries and nerves are also shown.

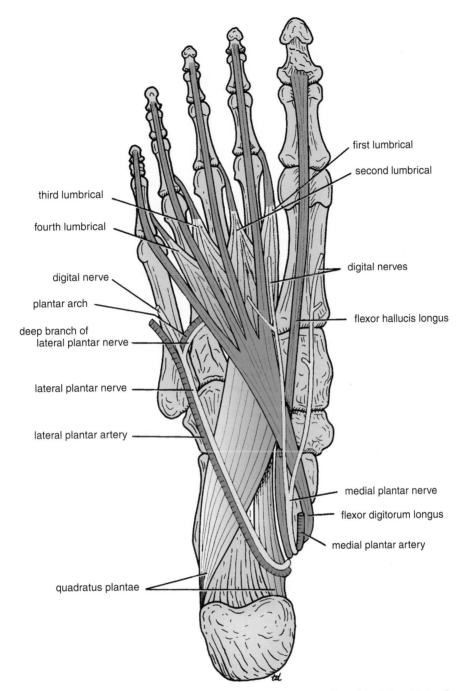

first lumbrical

second lumbrical

third lumbrical

fourth lumbrical

digital nerve

plantar arch

deep branch of
lateral plantar nerve

lateral plantar nerve

lateral plantar artery

digital nerves

flexor hallucis longus

medial plantar nerve

flexor digitorum longus

medial plantar artery

quadratus plantae

**Figure 17-41**  Second layer of the plantar muscles of the right foot. Medial and lateral plantar arteries and nerves are also shown.

■ **Lateral plantar nerve:** Runs forward deep to the abductor hallucis and flexor digitorum brevis muscles in company with the lateral plantar artery (Figs. 17-40 and 17-41).

■ **Cutaneous branch,** which supplies the lateral part of the sole and the lateral one and a half toes and nail beds.

■ **Muscular branch,** which supplies the flexor digitorum accessorius, abductor digiti minimi, flexor digiti minimi brevis, adductor hallucis, interosseous muscles, second lumbrical, third lumbrical, and fourth lumbrical muscles.

The main branches of the tibial nerve are summarized in Figure 17-42.

## Common Peroneal Nerve

The common peroneal nerve descends through the popliteal fossa (Fig. 17-38). It then passes laterally around the neck of the fibula, pierces the peroneus longus muscle,

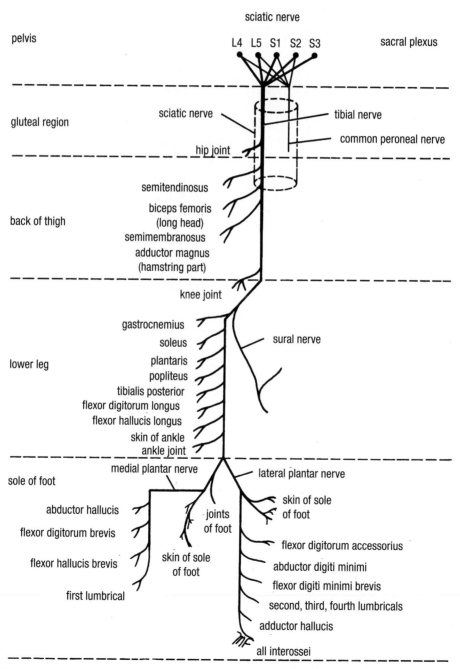

**Figure 17-42** Summary of the origin of the sciatic nerve and the main branches of the tibial nerve.

and divides into the superficial and the deep peroneal nerves (Figs. 17-43 and 17-44).

## Branches of the Common Peroneal Nerve

■ **Cutaneous branches: Sural communicating branch** (Fig. 17-28), which joins the sural nerve (see Branches of the Tibial Nerve); **lateral cutaneous nerve of the calf,** which supplies the skin on the lateral side of the back of the leg (Fig. 17-28).

■ **Muscular branch:** Short head of the biceps femoris muscle.

■ **Articular branch:** Knee joint.

■ **Terminal branches.**

■ **Superficial peroneal nerve:** Descends between the peroneus longus and brevis muscles in the lateral facial compartment and becomes subcutaneous (Figs. 17-43 and 17-44).

■ **Cutaneous branch** to the skin on the front of the lower leg and dorsum of the foot (except for

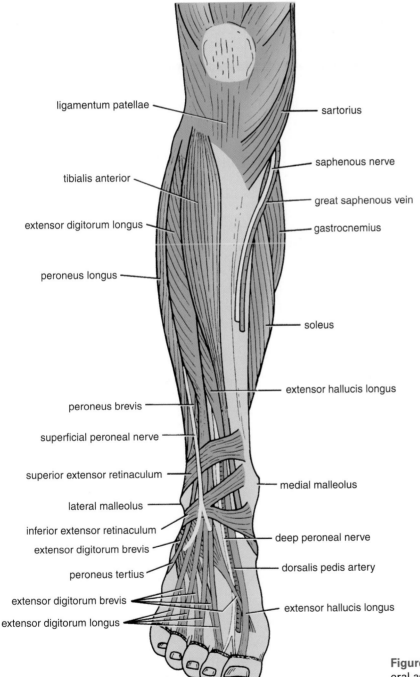

**Figure 17-43**  Structures in the anterior and lateral aspects of the right leg and the dorsum of the foot.

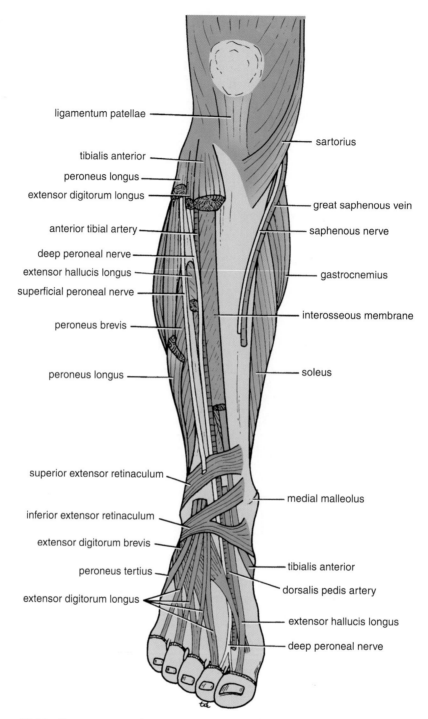

**Figure 17-44** Deep structures in the anterior and lateral aspects of the right leg and the dorsum of the foot.

the cleft between the big and the second toes, which is innervated by the deep peroneal nerve) (Fig. 17-27).

■ **Muscular branch,** which supplies the peroneus longus and brevis muscles.

■ **Deep peroneal nerve:** Descends in the anterior fascial compartment deep to the extensor digitorum

longus muscle and on the interosseous membrane (Fig. 17-44). It is accompanied by the anterior tibial vessels, and on the dorsum of the foot (Fig. 17-45), it divides into the medial and the lateral terminal branches.

■ **Cutaneous branch,** supplies the adjacent side of the big and the second toes (Fig. 17-27).

■ **Muscular branch,** which supplies the tibialis anterior, extensor digitorum longus, peroneus tertius, extensor hallucis longus, and extensor digitorum brevis muscles.

■ **Articular branch,** which supplies the ankle and tarsal joints.

The branches of the common peroneal nerve are summarized in Figure 17-46. Dermatomal charts for the ante-

rior and the posterior surfaces of the body are shown in CD Figures 1-2 and 1-3.

## Pudendal Nerve

The pudendal nerve arises from S2, 3, and 4 spinal nerves (Fig. 17-35). It leaves the pelvis through the greater sciatic foramen and, after a brief course in the gluteal region (Fig. 17-36), enters the perineum through the lesser sciatic fora-

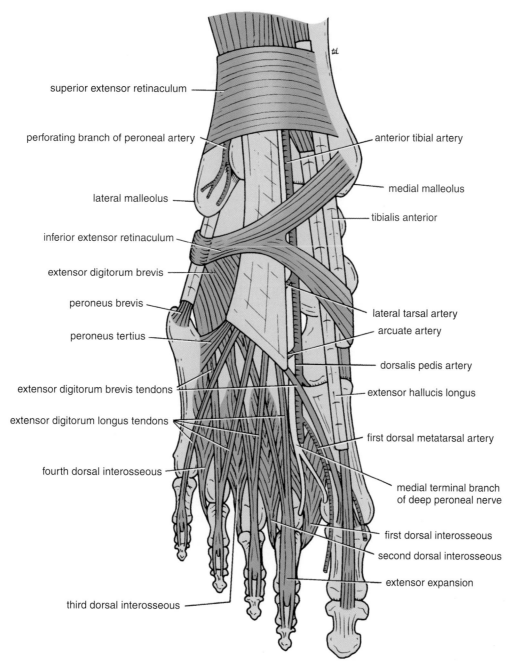

**Figure 17-45** Structures in the dorsal aspect of the right foot.

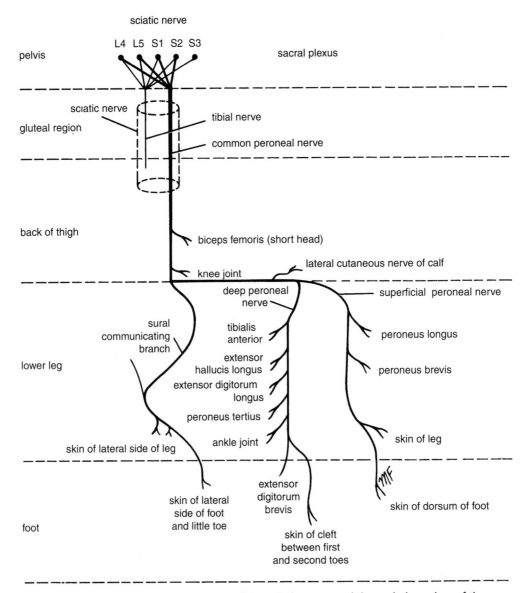

**Figure 17-46** Summary of the origin of the sciatic nerve and the main branches of the common peroneal nerve.

men. The nerve passes forward in the **pudendal canal** with the internal pudendal vessels on the lateral wall of the is-chiorectal fossa. It ends by dividing into the perineal nerve and the dorsal nerve of the penis (clitoris).

## Branches of the Pudendal Nerve

■ **Inferior rectal nerve:** Crosses the ischiorectal fossa to the anal canal and perianal skin.
   ■ **Sensory** to the mucous membrane of the lower half of the anal canal and perianal skin.
   ■ **Muscular** to the external anal sphincter.
■ **Perineal nerve**
   ■ **Cutaneous: Posterior scrotal (labial) nerves** to the posterior surface of the scrotum or labia majora.

■ **Muscular** to the superficial and deep transverse perineal muscles, bulbospongiosus, ischiocavernosus muscles, the sphincter urethrae, and the levator ani.
■ **Dorsal nerve of the penis (clitoris)** supplies the skin and deeper structures of the penis (clitoris).

The branches of the sacral plexus and their distribution are shown in Table 17-4.

# Coccygeal Nerve

The anterior ramus of the coccygeal nerve joins with branches of the anterior rami of the fourth and fifth sacral nerves to form the **coccygeal plexus.** Branches from this plexus supply the skin over the coccyx.

# Review Questions

## Fill in the Blank

**Fill in the blank with the BEST answer.**

1. The phrenic nerve arises from _____.
   A. C1, 2, and 3 spinal nerves.
   B. C8, T1, T2, and T3 spinal nerves.
   C. C3, 4, and 5 posterior rami of the spinal nerves.
   D. C5, 6, and 7 spinal nerves.
   E. C3, 4, and 5 anterior rami of the spinal nerves.

2. The motor nerve fibers of the phrenic nerve supply the _____.
   A. scalenus anterior muscle
   B. diaphragm
   C. prevertebral muscles
   D. intercostal muscles
   E. rectus abdominis muscle

3. The supraclavicular nerves arise from the _____.
   A. brachial plexus
   B. C5 and 6 spinal nerves
   C. C3 and 4 spinal nerves
   D. C7 and 8 spinal nerves
   E. C8 and T1 spinal nerves

4. The supraclavicular nerves supply the _____.
   A. supraspinatus muscle
   B. deltoid muscle
   C. subclavius muscle
   D. skin over the shoulder
   E. skin over the lower half of the deltoid muscle

5. The cervical plexus emerges into the neck between the _____.
   A. scalenus anterior and the scalenus medius muscles
   B. scalenus medius and the scalenus posterior muscles
   C. posterior belly of the digastric and the sternocleidomastoid muscles
   D. levator scapulae and the scalenus anterior muscles
   E. trapezius and the levator scapulae muscles

6. An inability to oppose the thumb to the little finger can result from damage to the _____ nerve.
   A. anterior interosseous
   B. posterior interosseous
   C. radial
   D. ulnar
   E. median

7. The sensory innervation of the nail bed of the index finger is the _____.
   A. median nerve
   B. radial nerve
   C. dorsal cutaneous branch of the ulnar nerve
   D. superficial branch of the ulnar nerve
   E. palmar cutaneous branch of the ulnar nerve

8. The sensory innervation of the medial side of the palm is the _____.
   A. radial nerve
   B. palmar cutaneous branch of the ulnar nerve
   C. dorsal cutaneous branch of the ulnar nerve
   D. median nerve
   E. superficial branch of the ulnar nerve

9. The sensory innervation of the dorsal surface of the root of the thumb is the _____.
   A. median nerve
   B. radial nerve
   C. superficial branch of the ulnar nerve
   D. dorsal cutaneous branch of the ulnar nerve
   E. posterior interosseous nerve

10. The sensory innervation of the medial side of the palmar aspect of the ring finger is the _____.
    A. radial nerve
    B. posterior interosseous nerve
    C. dorsal cutaneous branch of the ulnar nerve
    D. median nerve
    E. superficial branch of the ulnar nerve

11. The musculocutaneous nerve originates from the _____ of the brachial plexus.
    A. posterior cord
    B. lateral cord
    C. both medial and lateral cords
    D. upper trunk
    E. medial cord

12. The suprascapular nerve originates from the _____ of the brachial plexus.
    A. medial cord
    B. lower trunk
    C. posterior cord
    D. lateral cord
    E. upper trunk

13. The median nerve originates from the _____ of the brachial plexus.
    A. both medial and lateral cords
    B. medial cord
    C. posterior cord
    D. upper and lower trunk
    E. lateral cord

14. The thoracodorsal nerve originates from the _____ of the brachial plexus.
    A. lateral cord
    B. posterior cord
    C. medial cord
    D. medial and posterior cords
    E. lower trunk

15. The axillary nerve originates from the _____ of the brachial plexus.
    A. posterior cord
    B. middle trunk
    C. lateral cord
    D. lower trunk
    E. medial cord

## Multiple Choice Questions

**Select the BEST answer for each question.**

16. All the following statements concerning the brachial plexus are correct **except** which?
    A. The roots C8 and T1 join to form the lower trunk.
    B. The cords are named according to their position relative to the first part of the axillary artery.
    C. The nerve that innervates the levator scapulae is a branch of the upper trunk.
    D. The roots, trunks, and divisions are not located in the axilla.
    E. No nerves originate as branches from the individual divisions of the brachial plexus.

17. The radial nerve gives off the following branches in the posterior compartment of the arm **except** which?
    A. Lateral head of the triceps
    B. Lower lateral cutaneous nerve of the arm
    C. Medial head of the triceps
    D. Brachioradialis
    E. Anconeus

18. The following statements concerning structures in the intercostal space are correct **except** which?
    A. The anterior intercostal arteries of the upper six intercostal spaces are branches of the internal thoracic artery.
    B. The intercostal nerves travel forward in an intercostal space between the internal intercostal and innermost intercostal muscles.
    C. The intercostal blood vessels and nerves are positioned in the order of vein, nerve, and artery from superior to inferior in a subcostal groove.
    D. The lower five intercostal nerves supply sensory innervation to the skin of the lateral thoracic and anterior abdominal walls.
    E. The posterior intercostal veins drain backward into the azygos and hemiazygos veins.

A 30-year-old man was seen in the emergency department with a stab wound in the right inguinal region.

19. Which of the following nerves supplies the skin of the inguinal region?
    A. The eleventh thoracic nerve
    B. The tenth thoracic nerve
    C. The twelfth thoracic nerve
    D. The first lumbar nerve
    E. The femoral nerve

20. Which of the following nerves innervates at least one muscle that acts on both the hip and knee joints?
    A. Ilioinguinal nerve
    B. Femoral nerve
    C. Saphenous nerve
    D. Common peroneal nerve
    E. Superficial peroneal nerve

21. The following statements concerning the lumbar plexus are correct **except** which?
    A. The plexus lies within the psoas muscle.
    B. The plexus is formed from the posterior rami of the upper four lumbar nerves.
    C. The femoral nerve emerges from the lateral border of the psoas muscle.
    D. The obturator nerve emerges from the medial border of the psoas muscle.
    E. The iliohypogastric nerve emerges from the lateral border of the psoas muscle.

22. The following statements concerning the nerves of the pelvic cavity are correct **except** which?
    A. The inferior hypogastric plexus contains both sympathetic and parasympathetic nerves.
    B. The sacral plexus lies behind the rectum.
    C. The pelvic part of the sympathetic trunk possesses both white and gray rami communicantes.
    D. The superior hypogastric plexus is formed from the aortic sympathetic plexus and branches of the lumbar sympathetic ganglia.
    E. The anterior rami of the upper four sacral nerves emerge into the pelvis through the anterior sacral foramina.

23. The statements concerning the segmental origin of the following nerves are correct **except** which?
    A. The sciatic nerve is derived from the segments L4 and 5; S1, 2, and 3.
    B. The pudendal nerve is derived from the segments L3, 4, and 5.
    C. The pelvic splanchnic nerve is derived from the segments S2, 3, and 4.
    D. The obturator nerve is derived from the segments L2, 3, and 4.
    E. The lumbosacral trunk is derived from the segments L4 and 5.

24. The following structures receive innervation from branches of the pudendal nerve **except** which?
    A. Labia minora
    B. Urethral sphincter
    C. The posterior fornix of the vagina
    D. Ischiocavernosus muscles
    E. Skin of the penis or clitoris

25. The statements concerning the motor nerve supply of the muscles of the pelvic walls are correct **except** which?
    A. The sacral nerves or plexus supply the obturator internus muscle.
    B. The obturator nerve supplies the piriformis muscle.
    C. The sacral nerves, or plexus, supply the iliococcygeus muscle.
    D. The sacral nerves, or plexus, supply the coccygeus muscle.
    E. The perineal branch of the fourth sacral nerve and the perineal branch of the pudendal nerve supply the levator ani muscle.

# Answers and Explanations

1. **E** is correct. The phrenic nerve arises from the cervical plexus (C3, 4, and 5).

2. **B** is correct. The motor fibers of the phrenic nerve supply all the muscle of the diaphragm; the sensory fibers supply the pleura and peritoneum on the upper and lower surfaces of the central part of the diaphragm.

3. **C** is correct. The supraclavicular nerves arise from the cervical plexus (C3 and 4).

4. **D** is correct. The supraclavicular nerves are sensory nerves and supply the skin over the shoulder down as far as the middle of the deltoid muscle (Fig. 17-8).

5. **A** is correct. The cervical plexus emerges into the neck between the scalenus anterior and scalenus medius muscles (Fig. 17-3).

6. **E** is correct. The opponens pollicis muscle, which is responsible for pulling the thumb medially and forward across the palm so that the palmar surface of the tip of the thumb may come into contact with the palmar surface of the tips of the other fingers, is supplied by the median nerve.

7. **A** is correct. The sensory innervation of the nail bed of the index finger is the median nerve.

8. **B** is correct. The sensory innervation of the medial side of the palm is the palmar cutaneous branch of the ulnar nerve (Fig. 17-8).

9. **B** is correct. The radial nerve gives sensory innervation to the dorsal surface of the root of the thumb (Fig. 17-8).

10. **E** is correct. The superficial branch of the ulnar nerve gives the sensory innervation to the medial side of the palmar aspect of the ring finger (Fig. 17-8).

11. **B** is correct. The musculocutaneous nerve originates from the lateral cord of the brachial plexus (Fig. 17-7).

12. **E** is correct. The suprascapular nerve originates from the upper trunk of the brachial plexus (Fig. 17-7).

13. **A** is correct. The median nerve originates from the medial and lateral cords of the brachial plexus (Fig. 17-7).

14. **B** is correct. The thoracodorsal nerve originates from the posterior cord of the brachial plexus (Fig. 17-7).

15. **A** is correct. The axillary nerve originates from the posterior cord of the brachial plexus (Fig. 17-7).

16. **B** is incorrect. The cords of the brachial plexus are named according to their relative position to the second part of the axillary artery as it lies behind the pectoralis minor muscle.

17. **D** is incorrect. The branch from the radial nerve to the brachioradialis muscle leaves the nerve after it has left the posterior compartment of the arm by piercing the lateral intermuscular septum.

18. **C** is incorrect. The order from superior to inferior is intercostal vein, artery, and nerve (Fig. 17-23).

19. **D** is correct. The first lumbar nerve, represented by the iliohypogastric and ilioinguinal neves, supplies the skin just above the inguinal ligament and the symphysis pubis.

20. **B** is correct. The femoral nerve innervates the rectus femoris muscle, which is a flexor of the hip joint and an extensor of the knee joint.

21. **B** is incorrect. The lumbar plexus is formed from the anterior rami of the upper four lumbar spinal nerves (Fig. 17-24).

22. **C** is incorrect. The pelvic part of the sympathetic trunk gives rise to only gray postganglionic nerve fibers, which are distributed to the pelvic visceral and blood vessels.

23. **B** is incorrect. The pudendal nerve is a branch of the sacral plexus and is derived from S2, 3, and 4.

24. **C** is incorrect. The posterior fornix of the vagina is innervated by the inferior hypogastric plexuses.

25. **B** is incorrect. The piriformis muscle receives its motor nerve supply from the sacral plexus.

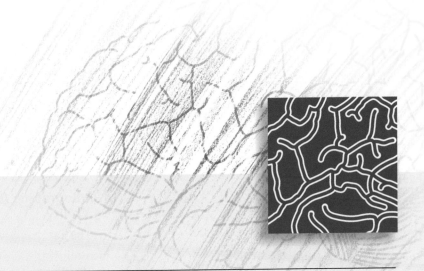

# 18

# The Eye and the Ear

All clinical material relevant to this chapter can be found on the CD-ROM.

# Chapter Outline

E ye problems range from minor to major. Conjunctivitis, styes, corneal abrasions, corneal foreign bodies, acute iritis, and eyelid lacerations. Sudden vision loss, chemical trauma, hyphema (hemorrhage in the anterior chamber of the eye), globe perforations, blow out fractures of the maxillary sinus, and acute glaucoma may be confronted by the medical professional in emergency situations. Preexisting eye problems and the ocular manifestations of systemic disease also have to be recognized.

Ear problems also show wide variations. Trauma to the external ear, deafness, excessive wax in the external auditory meatus, otitis media and its complications, and diseases involving the labyrinth are just a few of the medical conditions requiring treatment.

In this chapter, the normal anatomy of the eye and ear are reviewed, including the arrangement and action of both the intraocular and extraocular muscles and the clinical anatomy of the external and middle ear.

# BASIC ANATOMY

## Eyelids

The eyelids protect the eye from injury and excessive light by their closure (Fig. 18-1). The upper eyelid is larger and more mobile than the lower eyelid, and they meet each other at the **medial** and **lateral angles**. The **palpebral** fissure is the elliptical opening between the eyelids and is the entrance into the conjunctival sac. When the eye is closed, the upper eyelid completely covers the cornea of the eye. When the eye is open and looking straight ahead, the upper lid just covers the upper margin of the cornea. The lower lid lies just below the cornea when the eye is open and rises only slightly when the eye is closed.

The superficial surface of the eyelids is covered by skin, and the deep surface is covered by a mucous membrane, called the **conjunctiva**.

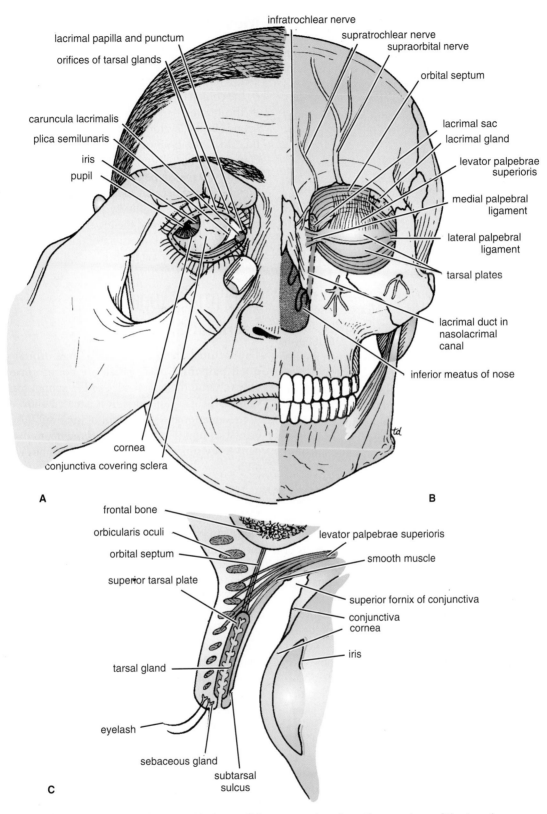

**Figure 18-1    A.** Right eye, with the eyelids separated to show the openings of the tarsal glands, plica semilunaris, caruncula lacrimalis, and puncta lacrimalia. **B.** Left eye, showing the superior and inferior tarsal plates and the lacrimal gland, sac, and duct. Note that a small window has been cut in the orbital septum to show the underlying lacrimal gland and fat (*yellow*). **C.** Sagittal section through the upper eyelid and the superior fornix of the conjunctiva. Note the presence of smooth muscle in the levator palpebrae superioris.

The **eyelashes** are short, curved hairs on the free edges of the eyelids (Fig. 18-1). They are arranged in double or triple rows at the mucocutaneous junction. The sebaceous glands (glands of Zeis) open directly into the eyelash follicles. The **ciliary glands** (glands of Moll) are modified sweat glands that open separately between adjacent lashes. The **tarsal glands** are long, modified sebaceous glands that pour their oily secretion onto the margin of the lid; their openings lie behind the eyelashes (Fig. 18-1).

---

### PHYSIOLOGIC NOTE

### Function of the Tarsal Glands

The oily secretion of the tarsal glands prevents the overflow of tears and helps make the closed eyelids airtight.

---

The more rounded medial angle of the palpebral fissure is separated from the eyeball by a small space, the **lacus lacrimalis,** in the center of which is a small, reddish-yellow elevation, the **caruncula lacrimalis** (Fig. 18-1). A reddish semilunar fold, called the **plica semilunaris,** lies on the lateral side of the caruncle.

Near the medial angle of the eye, there is a small elevation on the eyelid called the **papilla lacrimalis.** On the summit of the papilla is a small hole, the **punctum lacrimale,** which leads into the **canaliculus lacrimalis** (Fig. 18-1). The papilla lacrimalis projects into the lacus, and the punctum and canaliculus carry tears down into the nose.

The **conjunctiva** is a thin mucous membrane that lines the eyelids and is reflected at the **superior** and **inferior fornices** onto the anterior surface of the eyeball (Fig. 18-1). Its epithelium is continuous with that of the cornea. The upper lateral part of the superior fornix is pierced by the ducts of the lacrimal gland (see next column). The conjunctiva thus forms a potential space, the **conjunctival sac,** which is open at the palpebral fissure.

Beneath the eyelid is a groove, the **subtarsal sulcus,** which runs close to and parallel with the margin of the lid (Fig. 18-1).

The framework of the eyelids is formed by a fibrous sheet, the **orbital septum** (Fig. 18-1). This is attached to the periosteum at the orbital margins. The orbital septum is thickened to form the superior and inferior **tarsal plates.** The lateral ends of the plates are attached by a band, the **lateral palpebral ligament,** to a bony tubercle just within the orbital margin. The medial ends of the plates are attached by a similar band, the **medial palpebral ligament,** to the crest of the lacrimal bone (Fig. 18-1). The tarsal glands are embedded in the posterior surface of the tarsal plates.

The superficial surface of the tarsal plates and the orbital septum are covered by the palpebral fibers of the **orbicularis oculi muscle** (Table 18-1). The aponeurosis of insertion of the **levator palpebrae superioris muscle** pierces the orbital septum to reach the anterior surface of the superior tarsal plate and the skin (Fig. 18-1).

## Movements of the Eyelids

The position of the eyelids at rest depends on the tone of the **orbicularis oculi** and the **levator palpebrae superioris muscles** and the position of the eyeball. The eyelids are closed by the contraction of the orbicularis oculi and the relaxation of the levator palpebrae superioris muscles. The eye is opened by the levator palpebrae superioris raising the upper lid. On looking upward, the levator palpebrae superioris contracts, and the upper lid moves with the eyeball. On looking downward, both lids move, the upper lid continues to cover the upper part of the cornea, and the lower lid is pulled downward slightly by the conjunctiva, which is attached to the sclera and the lower lid.

The origins and insertions of the muscles of the eyelids are summarized in Table 18-1.

# Lacrimal Apparatus

## Lacrimal Gland

The lacrimal gland consists of a large **orbital part** and a small **palpebral part,** which are continuous with each other around the lateral edge of the aponeurosis of the levator palpebrae superioris. It is situated above the eyeball in the anterior and upper part of the orbit posterior to the orbital septum (Fig. 18-1). The gland opens into the lateral part of the superior fornix of the conjunctiva gland by 12 ducts.

## Nerve Supply

**Parasympathetic secretomotor nerve supply** is derived from the **lacrimal nucleus** of the facial nerve. The preganglionic fibers reach the pterygopalatine ganglion (sphenopalatine ganglion) via the nervus intermedius and its great petrosal branch and via the nerve of the pterygoid canal. The postganglionic fibers leave the ganglion and join the maxillary nerve. They then pass into its zygomatic branch and the zygomaticotemporal nerve. They reach the lacrimal gland within the lacrimal nerve.

**Sympathetic postganglionic nerve supply** is derived from the internal carotid plexus and travels via the deep petrosal nerve, the nerve of the pterygoid canal, the maxillary nerve, the zygomatic nerve, the zygomaticotemporal nerve, and finally, the lacrimal nerve.

## Lacrimal Ducts

The tears circulate across the cornea and accumulate in the **lacus lacrimalis.** From here, the tears enter the **canaliculi lacrimales** through the **puncta lacrimalia.** The canaliculi lacrimales pass medially and open into the **lacrimal sac** (Fig. 18-1), which lies in the lacrimal groove behind the medial palpebral ligament and is the upper blind end of the nasolacrimal duct.

| Table 18-1 | Muscles of the Eyeball and Eyelids | | | |
|---|---|---|---|---|
| **Muscle** | **Origin** | **Insertion** | **Nerve Supply** | **Action** |
| **Extrinsic Muscles of Eyeball (Striated Skeletal Muscle)** | | | | |
| Superior rectus | Tendinous ring on posterior wall of orbital cavity | Superior surface of eyeball just posterior to corneoscleral junction | Oculomotor nerve (third cranial nerve) | Raises cornea upward and medially |
| Inferior rectus | Tendinous ring on posterior wall of orbital cavity | Inferior surface of eyeball just posterior to corneoscleral junction | Oculomotor nerve (third cranial nerve) | Depresses cornea downward and medially |
| Medial rectus | Tendinous ring on posterior wall of orbital cavity | Medial surface of eyeball just posterior to corneoscleral junction | Oculomotor nerve (third cranial nerve) | Rotates eyeball so that cornea looks medially |
| Lateral rectus | Tendinous ring on posterior wall of orbital cavity | Lateral surface of eyeball just posterior to corneoscleral junction | Abducent nerve (sixth cranial nerve) | Rotates eyeball so that cornea looks laterally |
| Superior oblique | Posterior wall of orbital cavity | Passes through pulley and is attached to superior surface of eyeball beneath superior rectus | Trochlear nerve (fourth cranial nerve) | Rotates eyeball so that cornea looks downward and laterally |
| Inferior oblique | Floor of orbital cavity | Lateral surface of eyeball deep to lateral rectus | Oculomotor nerve (third cranial nerve) | Rotates eyeball so that cornea looks upward and laterally |
| **Intrinsic Muscles of Eyeball (Smooth Muscle)** | | | | |
| Sphincter pupillae of iris | | | Parasympathetic via oculomotor nerve | Constricts pupil |
| Dilator pupillae of iris | | | Sympathetic | Dilates pupil |
| Ciliary muscle | | | Parasympathetic via oculomotor nerve | Controls shape of lens; in accommodation, makes lens more globular |
| **Muscles of Eyelids** | | | | |
| Orbicularis oculi (Table 13-2) | | | | |
| Levator palpebrae superioris | Back of orbital cavity | Anterior surface and upper margin of superior tarsal plate | Striated muscle oculomotor nerve, smooth muscle sympathetic | Raises upper lid |

From Snell RS: Clinical Anatomy. 7th Ed. Philadelphia: Lippincott Williams & Wilkins, 2004, p. 828.

The **nasolacrimal duct** is about 0.5 in (1.3 cm) long and emerges from the lower end of the lacrimal sac (Fig. 18-1). The duct descends downward, backward, and laterally in a bony canal and opens into the inferior meatus of the nose. The opening is guarded by a fold of mucous membrane, the **lacrimal fold.** This prevents air from being forced up the duct into the lacrimal sac on blowing the nose.

# The Orbit

## Orbital Margin

The orbital margin is formed by the frontal, maxilla, and zygomatic bones.

## Orbital Cavity

The orbital cavity is pyramidal, with its base in front and its apex behind (Fig. 18-2). The orbital walls are shown in Figure 18-2.

**Roof:** Formed by the orbital plate of the frontal bone, which separates the orbital cavity from the anterior cranial fossa and the frontal lobe of the cerebral hemisphere.

**Floor:** Formed by the orbital plate of the maxilla, which separates the orbital cavity from the maxillary sinus.

**Lateral wall:** Formed by the zygomatic bone and the greater wing of the sphenoid (Fig. 18-2).

**Medial wall:** Formed from before backward by the frontal process of the maxilla, the lacrimal bone, the orbital plate of the ethmoid (which separates the orbital cavity from the ethmoid sinuses), and the body of the sphenoid.

### Openings into the Orbital Cavity

The openings into the orbital cavity are shown in Figure 18-2.

**Orbital opening:** Lies anteriorly (Fig. 18-2). About one sixth of the eye is exposed; the remainder is protected by the walls of the orbit.

**Supraorbital notch (foramen):** The supraorbital notch is situated on the superior orbital margin (Fig. 18-2). It transmits the supraorbital nerve and blood vessels.

**Infraorbital groove and canal:** Situated on the floor of the orbit in the orbital plate of the maxilla (see Fig. 18-4); they transmit the infraorbital nerve (a continuation of the maxillary nerve) and blood vessels.

**Nasolacrimal canal:** Located anteriorly on the medial wall; it communicates with the inferior meatus of the nose (Fig. 18-1). It transmits the nasolacrimal duct.

**Inferior orbital fissure:** Located posteriorly between the maxilla and the greater wing of the sphenoid (Fig. 18-2); it communicates with the pterygopalatine fossa. It transmits the maxillary nerve and its zygomatic branch, the inferior ophthalmic vein, and sympathetic nerves.

**Superior orbital fissure:** Located posteriorly between the greater and lesser wings of the sphenoid (Fig. 18-2); it communicates with the middle cranial fossa. It transmits the lacrimal nerve, the frontal nerve, the trochlear nerve, the oculomotor nerve (upper and lower divisions), the abducent nerve, the nasociliary nerve, and the superior ophthalmic vein.

**Optic canal:** Located posteriorly in the lesser wing of the sphenoid (Fig. 18-2). It communicates with the middle cranial fossa and transmits the optic nerve and the ophthalmic artery.

## Orbital Fascia

The orbital fascia is the periosteum of the bones that form the walls of the orbit. It is loosely attached to the bones and is continuous through the foramina and fissures with the periosteum covering the outer surfaces of the bones. The muscle of Müller, or orbitalis muscle, is a thin layer of smooth muscle that bridges the inferior orbital fissure. It is supplied by sympathetic nerves, and its function is unknown.

## Nerves of the Orbit

### Optic Nerve

The optic nerve enters the orbit from the middle cranial fossa by passing through the optic canal (Fig. 18-3). It is accompanied by the ophthalmic artery, which lies on its lower lateral side. The nerve is surrounded by sheaths of pia mater, arachnoid mater, and dura mater (see Fig. 18-9). It runs forward and laterally within the cone of the recti muscles and pierces the sclera at a point medial to the posterior pole of the eyeball. Here, the meninges fuse with the sclera so that the subarachnoid space with its contained cerebrospinal fluid extends forward from the middle cranial fossa, around the optic nerve, and through the optic canal, as far as the eyeball. A rise in pressure of the cerebrospinal fluid within the cranial cavity, therefore, is transmitted to the back of the eyeball.

### Lacrimal Nerve

The lacrimal nerve arises from the ophthalmic division of the trigeminal nerve. It enters the orbit through the upper part of the superior orbital fissure (Fig. 18-2) and passes forward along the upper border of the lateral rectus muscle (Fig. 18-3). It is joined by a branch of the zygomaticotemporal nerve, which later leaves it to enter the lacrimal gland (parasympathetic secretomotor fibers). The lacrimal nerve ends by supplying the skin of the lateral part of the upper lid.

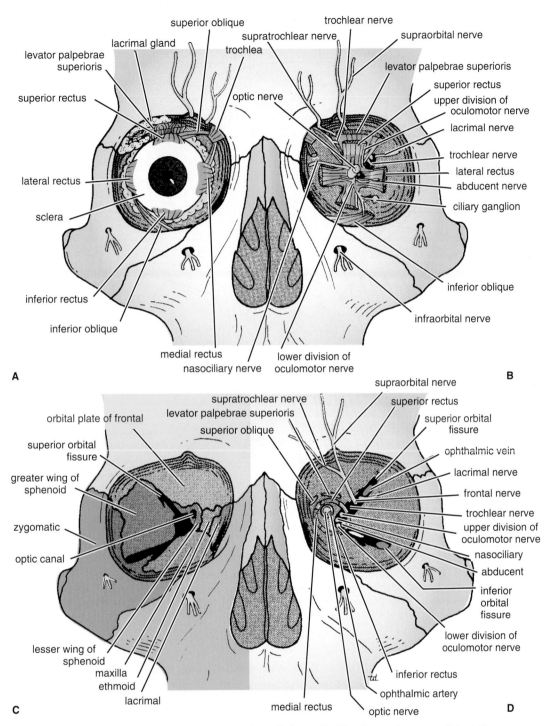

**Figure 18-2  A.** Right eyeball exposed from in front. **B.** Muscles and nerves of the left orbit as seen from in front. **C.** Bones forming the walls of the right orbit. **D.** The optic canal and the superior and inferior orbital fissures on the left side.

## Frontal Nerve

The frontal nerve arises from the ophthalmic division of the trigeminal nerve. It enters the orbit through the upper part of the superior orbital fissure (Fig. 18-2) and passes forward on the upper surface of the levator palpebrae superioris beneath the roof of the orbit (Fig. 18-3). It divides into the **supratrochlear** and **supraorbital nerves**, which wind around the upper margin of the orbital cavity to supply the skin of the forehead; the supraorbital nerve also supplies the mucous membrane of the frontal air sinus.

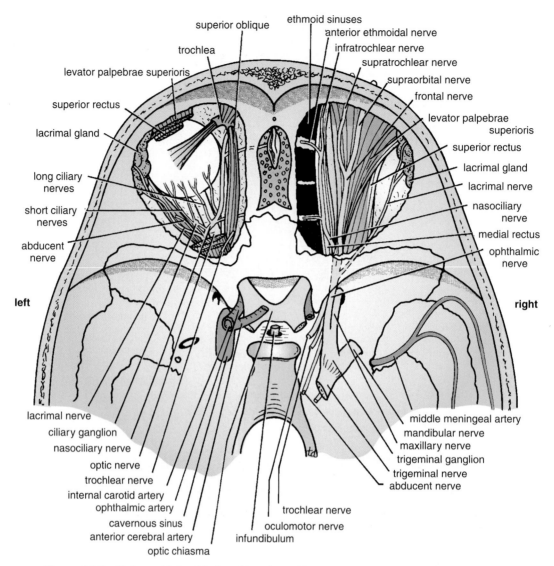

**Figure 18-3**   Right and left orbital cavities viewed from above. The roof of the orbit, formed by the orbital plate of the frontal bone, has been removed from both sides. On the left side, the levator palpebrae superioris and the superior rectus muscles have also been removed to expose the underlying structures.

## Trochlear Nerve

The trochlear nerve enters the orbit through the upper part of the superior orbital fissure (Fig. 18-2). It runs forward and supplies the superior oblique muscle (Fig. 18-3).

## Oculomotor Nerve

The **superior ramus** of the oculomotor nerve enters the orbit through the lower part of the superior orbital fissure (Fig. 18-2). It supplies the superior rectus muscle, then pierces it, and supplies the levator palpebrae superioris muscle (Fig. 18-2).

The **inferior ramus** of the oculomotor nerve enters the orbit in a similar manner and supplies the inferior rectus,

the medial rectus, and the inferior oblique muscles. The nerve to the inferior oblique muscle gives off a branch (Fig. 18-4) that passes to the ciliary ganglion and carries parasympathetic fibers to the sphincter pupillae and the ciliary muscle.

## Nasociliary Nerve

The nasociliary nerve arises from the ophthalmic division of the trigeminal nerve. It enters the orbit through the lower part of the superior orbital fissure (Fig. 18-2). It crosses above the optic nerve, runs forward along the upper margin of the medial rectus muscle, and ends by dividing into the **anterior ethmoidal** and **infratrochlear nerves** (Fig. 18-3).

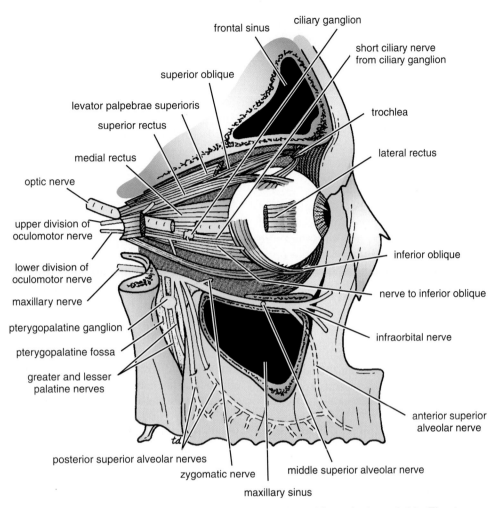

frontal sinus | ciliary ganglion | short ciliary nerve from ciliary ganglion
superior oblique
levator palpebrae superioris
superior rectus
medial rectus
optic nerve
upper division of oculomotor nerve
lower division of oculomotor nerve
maxillary nerve
pterygopalatine ganglion
pterygopalatine fossa
greater and lesser palatine nerves
posterior superior alveolar nerves
zygomatic nerve
maxillary sinus
middle superior alveolar nerve
anterior superior alveolar nerve
infraorbital nerve
nerve to inferior oblique
inferior oblique
lateral rectus
trochlea

**Figure 18-4** Muscles and nerves of the right orbit viewed from the lateral side. The maxillary nerve and the pterygopalatine ganglion are also shown.

**Branches of the Nasociliary Nerve**

- The **communicating branch to the ciliary ganglion** is a sensory nerve. The sensory fibers from the eyeball pass to the ciliary ganglion via the short ciliary nerves, pass through the ganglion without interruption, and then join the nasociliary nerve by means of the communicating branch.
- The long **ciliary nerves,** two or three in number, arise from the nasociliary nerve as it crosses the optic nerve (Fig. 18-3). They contain sympathetic fibers for the dilator pupillae muscle. The nerves pass forward with the short ciliary nerves and pierce the sclera of the eyeball. They continue forward between the sclera and the choroid to reach the iris.
- The **posterior ethmoidal nerve** supplies the ethmoidal and sphenoidal air sinuses (Fig. 18-3).
- The **infratrochlear nerve** passes forward below the pulley of the superior oblique muscle and supplies the skin of the medial part of the upper eyelid and the adjacent part of the nose (Fig. 18-1).

- The **anterior ethmoidal nerve** passes through the anterior ethmoidal foramen and enters the anterior cranial fossa on the upper surface of the cribriform plate of the ethmoid (Fig. 18-3). It enters the nasal cavity through a slitlike opening alongside the crista galli. After supplying an area of mucous membrane, it appears on the face as the **external nasal branch** at the lower border of the nasal bone and supplies the skin of the nose down as far as the tip.

## Abducent Nerve

The abducent nerve enters the orbit through the lower part of the superior orbital fissure (Fig. 18-2). It supplies the lateral rectus muscle.

## Ciliary Ganglion

The ciliary ganglion is a parasympathetic ganglion about the size of a pinhead (Fig. 18-4) and situated in the posterior part of the orbit. It receives its preganglionic parasympathetic fibers from the oculomotor nerve via the nerve to the inferior oblique. The postganglionic fibers leave the ganglion in the **short ciliary nerves,** which enter the back of the eyeball and supply the sphincter pupillae and the ciliary muscle.

A number of sympathetic fibers pass from the internal carotid plexus into the orbit and run through the ganglion without interruption.

## Blood Vessels and Lymph Vessels of the Orbit

### Ophthalmic Artery

The ophthalmic artery is a branch of the internal carotid artery after the artery emerges from the cavernous sinus. It enters the orbit through the optic canal with the optic nerve (Fig. 18-3). It runs forward and crosses the optic nerve to reach the medial wall of the orbit. It gives off numerous branches, which accompany the nerves in the orbital cavity.

#### Branches of the Ophthalmic Artery

- The **central artery of the retina** is a small branch that pierces the meningeal sheaths of the optic nerve to gain entrance to the nerve (see Fig. 18-9). It runs in the substance of the optic nerve and enters the eyeball at the center of the **optic disc.** Here, it divides into branches, which may be studied in a patient through an ophthalmoscope. The branches are end arteries.
- The **muscular branches.**
- The **ciliary arteries** can be divided into anterior and posterior groups. The former group enters the eyeball near the corneoscleral junction; the latter group enters near the optic nerve.
- The **lacrimal artery** to the lacrimal gland.

### Ophthalmic Veins

The **superior ophthalmic vein** communicates in front with the facial vein. The **inferior ophthalmic vein** communicates through the inferior orbital fissure with the pterygoid venous plexus. Both veins pass backward through the superior orbital fissure and drain into the cavernous sinus.

### Lymph Vessels

No lymph vessels or nodes are present in the orbital cavity.

# Eye

## Movements of the Eyeball

### Terms Used in Describing Eye Movements

The center of the cornea or the center of the pupil is used as the anatomic "anterior pole" of the eye. All movements of the eye are then related to the direction of the movement of the anterior pole as it rotates on any one of the three axes (horizontal, vertical, and sagittal). The terminology then becomes as follows: **elevation** is the rotation of the eye upward, **depression** is the rotation of the eye downward, **abduction** is the rotation of the eye laterally, and **adduction** is the rotation of the eye medially.

Rotatory movements of the eyeball use the upper rim of the cornea (or pupil) as the marker. The eye either rotates medially or laterally.

### Extrinsic Muscles Producing Movement of the Eye

There are six voluntary muscles that run from the posterior wall of the orbital cavity to the eyeball (Fig. 18-2). These are the **superior rectus,** the **inferior rectus,** the **medial rectus,** the **lateral rectus,** and the **superior** and **inferior oblique muscles.**

Because the superior and the inferior recti are inserted on the medial side of the vertical axis of the eyeball, they not only raise and depress the cornea, respectively, but they also **rotate it medially** (Fig. 18-5). For the superior rectus muscle to raise the cornea directly upward, the inferior oblique muscle must assist; for the inferior rectus to depress the cornea directly downward, the superior oblique muscle must assist (Figs. 18-6 and 18-7). Note that the tendon of the superior oblique muscle passes through a fibrocartilaginous pulley (trochlea) attached to the frontal bone. The tendon now turns backward and laterally and is inserted into the sclera beneath the superior rectus muscle.

The origins, insertions, nerve supply, and actions of the muscles of the eyeball are summarized in Table 18-1. Study carefully Figure 18-8.

### Intrinsic Muscles

The involuntary intrinsic muscles are the **ciliary muscle** and the **constrictor** and the **dilator pupillae of the iris;** these muscles take no part in the movement of the eyeball and are discussed later.

## Fascial Sheath of the Eyeball

The fascial sheath surrounds the eyeball from the optic nerve to the corneoscleral junction (Fig. 18-9). It separates the eyeball from the orbital fat and provides it with a socket for free movement. It is perforated by the tendons of the orbital muscles and is reflected onto each of them as a tubular sheath. The sheaths for the tendons of the medial and lateral recti are attached to the medial and lateral walls of the orbit by triangular ligaments called the **medial and lateral check ligaments.** The lower part of the fascial sheath, which passes beneath the eyeball and connects the check ligaments, is thickened and serves to suspend the eyeball; it is called the **suspensory ligament of the eye** (Fig. 18-9).

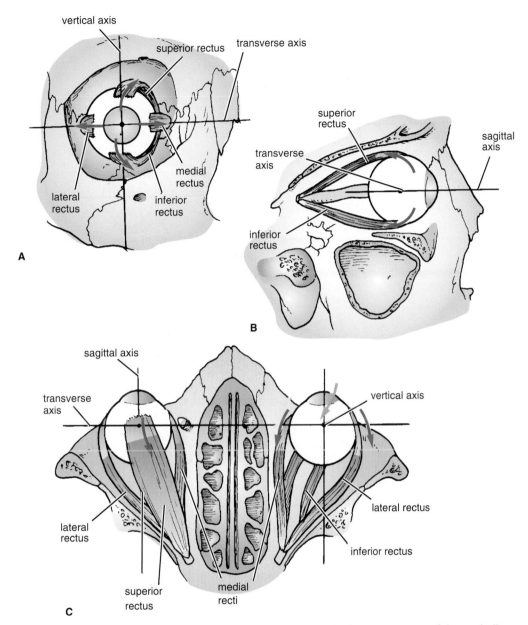

**Figure 18-5** The actions of the four recti muscles in producing movements of the eyeball.

By this means, the eye is suspended from the medial and lateral walls of the orbit, as if in a hammock.

## Structure of the Eye

The eyeball (Fig. 18-9) is embedded in orbital fat but is separated from it by the fascial sheath of the eyeball. The eyeball consists of three coats, which, from without inward, are the fibrous coat, the vascular pigmented coat, and the nervous coat.

## Coats of the Eyeball

### Fibrous Coat

The fibrous coat is made up of a posterior opaque part, the sclera, and an anterior transparent part, the cornea (Fig. 18-9).

### Sclera

The opaque **sclera** is composed of dense fibrous tissue and is white. Posteriorly, it is pierced by the optic nerve and is fused with the dural sheath of that nerve (Fig. 18-9). The **lamina cribrosa** is the area of the sclera that is pierced by the nerve fibers of the optic nerve.

The sclera is also pierced by the ciliary arteries and nerves and their associated veins, the venae vorticosae. The sclera is directly continuous in front with the cornea at the corneoscleral junction, or limbus.

### Cornea

The transparent **cornea** is largely responsible for the refraction of the light entering the eye (Fig. 18-9). It is in contact posteriorly with the aqueous humor.

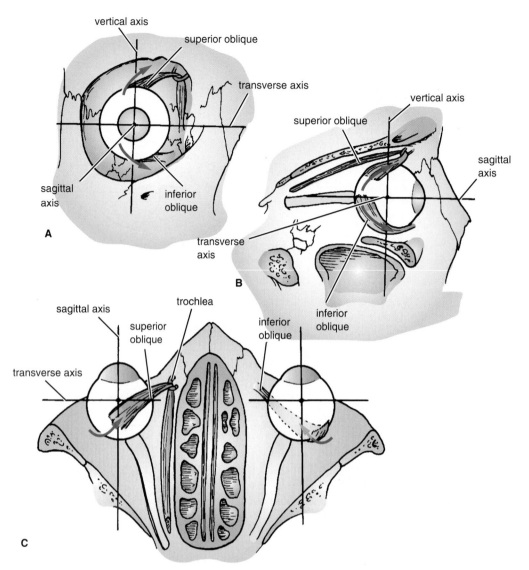

**Figure 18-6**   The actions of the superior and inferior oblique muscles in producing movements of the eyeball.

**Blood supply:** The cornea is avascular and devoid of lymphatic drainage. It is nourished by diffusion from the aqueous humor and from the capillaries at its edge.

**Nerve supply:** Long ciliary nerves from the ophthalmic division of the trigeminal nerve.

---

### PHYSIOLOGIC NOTE

### Function of the Cornea

The cornea is the most important refractive medium of the eye. This refractive power occurs on the anterior surface of the cornea, where the refractive index of the cornea (1.38) differs greatly from that of the air. The importance of the tear film in maintaining the normal environment for the corneal epithelial cells should be stressed.

## Vascular Pigmented Coat

The vascular pigmented coat consists, from behind forward, of the choroid, the ciliary body, and the iris.

### The Choroid

The choroid is composed of an outer pigmented layer and an inner, highly vascular layer.

### The Ciliary Body

The **ciliary body** is continuous posteriorly with the choroid, and anteriorly, it lies behind the peripheral margin of the iris (Fig. 18-9). It is composed of the ciliary ring, the ciliary processes, and the ciliary muscle.

The **ciliary ring** is the posterior part of the body, and its surface has shallow grooves, the ciliary striae.

The **ciliary processes** are radially arranged folds, or ridges, to the posterior surfaces of which are connected the suspensory ligaments of the lens.

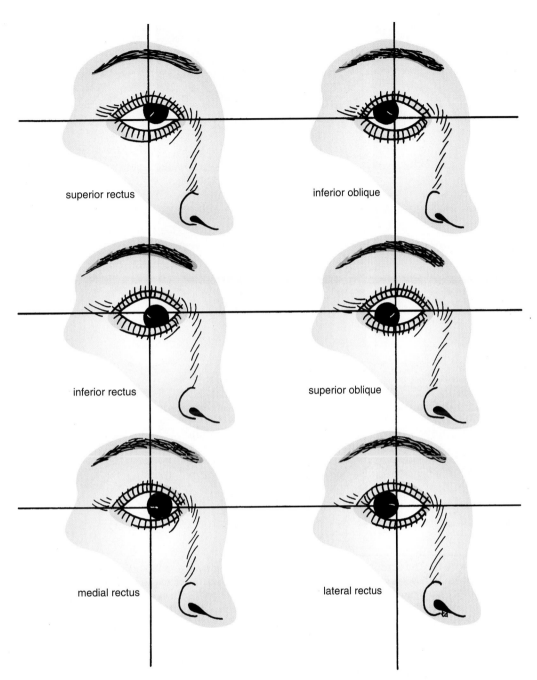

superior rectus

inferior oblique

inferior rectus

superior oblique

medial rectus

lateral rectus

**Figure 18-7** Actions of the four recti and two oblique muscles of the right orbit, *assuming that each muscle is acting alone.* The position of the pupil in relation to the vertical and horizontal planes should be noted in each case. The actions of the superior and inferior recti and the oblique muscles in the living intact eye are tested clinically.

The **ciliary muscle** (Fig. 18-9) is composed of meridianal and circular fibers of smooth muscle. The meridianal fibers run backward from the region of the corneoscleral junction to the ciliary processes. The circular fibers are fewer in number and lie internal to the meridianal fibers.

■ Nerve supply: The ciliary muscle is supplied by the parasympathetic fibers from the oculomotor nerve. After synapsing in the ciliary ganglion, the postganglionic fibers pass forward to the eyeball in the short ciliary nerves.

■ **Action:** Contraction of the ciliary muscle, especially the meridianal fibers, pulls the ciliary body forward. This relieves the tension in the suspensory ligament, and the elastic lens becomes more convex. This increases the refractive power of the lens.

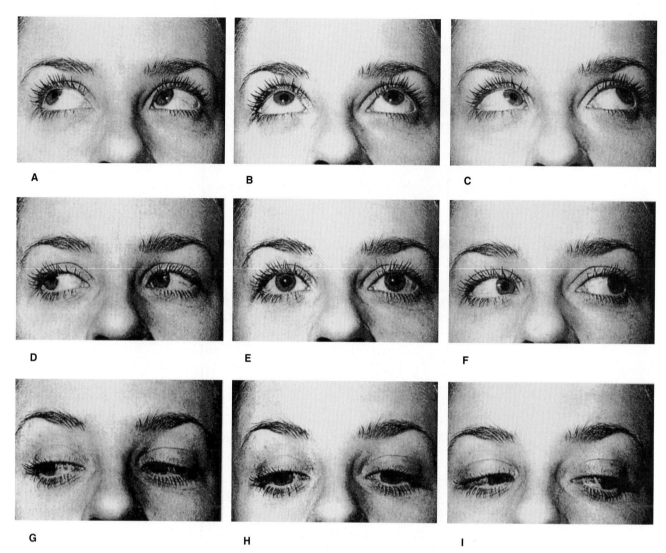

**Figure 18-8** The cardinal positions of the right and left eyes and the actions of the recti and oblique muscles *principally* responsible for the movements of the eyes. **A.** Right eye, superior rectus muscle; left eye, inferior oblique muscle. **B.** Both eyes, superior recti and inferior oblique muscles. **C.** Right eye, inferior oblique muscle; left eye, superior rectus muscle. **D.** Right eye, lateral rectus muscle; left eye, medial rectus muscle. **E.** Primary position, with the eyes fixed on a distant fixation point. **F.** Right eye, medial rectus muscle; left eye, lateral rectus muscle. **G.** Right eye, inferior rectus muscle; left eye, superior oblique muscle. **H.** Both eyes, inferior recti and superior oblique muscles. **I.** Right eye, superior oblique muscle; left eye, inferior rectus muscle.

### The Iris and Pupil

The iris is a thin, contractile, pigmented diaphragm with a central aperture, the pupil (Fig. 18-9). It is suspended in the aqueous humor between the cornea and the lens. The periphery of the iris is attached to the anterior surface of the ciliary body. It divides the space between the lens and the cornea into an **anterior** and a **posterior chamber.**

The muscle fibers of the iris are involuntary and consist of circular and radiating fibers. The circular fibers form the **sphincter pupillae** and are arranged around the margin of

the pupil. The radial fibers form the **dilator pupillae** and consist of a thin sheet of radial fibers that lie close to the posterior surface.

▪ **Nerve supply:** The sphincter pupillae is supplied by parasympathetic fibers from the oculomotor nerve. After synapsing in the ciliary ganglion, the postganglionic fibers pass forward to the eyeball in the short ciliary nerves. The dilator pupillae is supplied by sympathetic fibers, which pass forward to the eyeball in the long ciliary nerves.

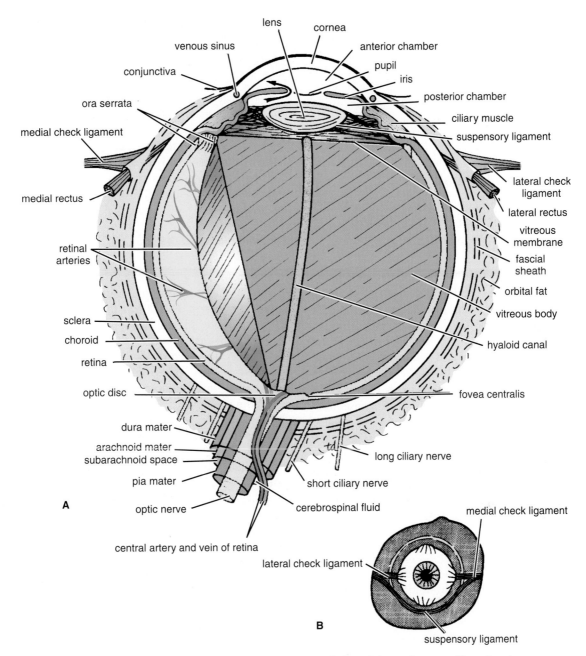

**Figure 18-9   A.** Horizontal section through the eyeball and the optic nerve. Note that the central artery and vein of the retina cross the subarachnoid space to reach the optic nerve. **B.** Check ligaments and suspensory ligament of the eyeball.

■ **Action:** The sphincter pupillae constricts the pupil in the presence of bright light and during accommodation. The dilator pupillae dilates the pupil in the presence of light of low intensity or in the presence of excessive sympathetic activity such as occurs in fright.

### Nervous Coat: The Retina
The retina consists of an **outer pigmented layer** and an **inner nervous layer**. Its outer surface is in contact with the choroid, and its inner surface is in contact with the vitreous body (Fig.

18-9). The posterior three fourths of the retina is the receptor organ. Its anterior edge forms a wavy ring, the **ora serrata**, and the nervous tissues end here. The anterior part of the retina is nonreceptive and consists merely of pigment cells, with a deeper layer of columnar epithelium. This anterior part of the retina covers the ciliary processes and the back of the iris.

At the center of the posterior part of the retina is an oval, yellowish area, the **macula lutea**, which is the area of the retina for the most distinct vision. It has a central depression, the **fovea centralis** (Fig. 18-9).

The optic nerve leaves the retina about 3 mm to the medial side of the macula lutea by the optic disc. The **optic disc** is slightly depressed at its center, where it is pierced by the **central artery of the retina** (Fig.18-9). At the optic disc is a complete absence of **rods** and **cones**, so that the optic disk is insensitive to light and is referred to as the "**blind spot.**" On ophthalmoscopic examination, the optic disc is seen to be pale pink in color and much paler than the surrounding retina.

## Contents of the Eyeball

The contents of the eyeball consist of the refractive media, the aqueous humor, the vitreous body, and the lens.

## Aqueous Humor

The aqueous humor is a clear fluid that fills the anterior and posterior chambers of the eyeball (Fig. 18-9). It is a secretion from the ciliary processes, from which it enters the posterior chamber. It then flows into the anterior chamber through the pupil and is drained away through the spaces at the iridocorneal angle into the **canal of Schlemm.** Obstruction to the draining of the aqueous humor results in a rise in intraocular pressure called **glaucoma.** This can produce degenerative changes in the retina, with consequent blindness.

The function of the aqueous humor is to support the wall of the eyeball by exerting internal pressure and thus maintaining its optical shape. It also nourishes the cornea and the lens and removes the products of metabolism; these functions are important because the cornea and the lens do not possess a blood supply.

## Vitreous Body

The vitreous body fills the eyeball behind the lens (Fig. 18-9) and is a transparent gel. The **hyaloid canal** is a narrow channel that runs through the vitreous body from the optic disc to the posterior surface of the lens; in the fetus, it is filled by the hyaloid artery, which disappears before birth.

The function of the vitreous body is to contribute slightly to the magnifying power of the eye. It supports the posterior surface of the lens and assists in holding the neural part of the retina against the pigmented part of the retina.

## Lens

The lens (Fig. 18-9) is a transparent, biconvex structure enclosed in a transparent capsule. It is situated behind the iris and in front of the vitreous body and is encircled by the ciliary processes.

The lens consists of an elastic **capsule,** which envelops the structure; a **cuboidal epithelium,** which is confined to the anterior surface of the lens; and **lens fibers,** which are formed from the cuboidal epithelium at the equator of the lens. The lens fibers make up the bulk of the lens.

**Changes in the Shape of the Lens and Accommodation**

The elastic lens capsule is under tension, causing the lens constantly to endeavor to assume a globular rather than a disc shape. The equatorial region, or circumference, of the lens is attached to the ciliary processes of the ciliary body by the suspensory ligament. The pull of the radiating fibers of the suspensory ligament tends to keep the elastic lens flattened so that the eye can be focused on distant objects.

**Accommodation of the Eye**

To accommodate the eye for close objects, the ciliary muscle contracts and pulls the ciliary body forward and inward so that the radiating fibers of the suspensory ligament are relaxed. This allows the elastic lens to assume a more globular shape.

With advancing age, the lens becomes denser and less elastic, and, as a result, the ability to accommodate is lessened (presbyopia). This disability can be overcome by the use of an additional lens in the form of glasses to assist the eye in focusing on nearby objects.

**Constriction of the Pupil During Accommodation of the Eye**

To ensure that the light rays pass through the central part of the lens so that spherical aberration is diminished during accommodation for near objects, the sphincter pupillae muscle contracts so the pupil becomes smaller.

**Convergence of the Eyes During Accommodation of the Lens**

In humans, the retinae of both eyes focus on only one set of objects (single binocular vision). When an object moves from a distance toward an individual, the eyes converge so that a single object, not two, is seen. Convergence of the eyes results from the coordinated contraction of the medial rectus muscles.

# RADIOGRAPHIC APPEARANCES OF THE ORBITAL CAVITY

Radiographic appearances of the orbital cavity are shown in Figures 11-11 through 11-14, 18-10, and 18-11.

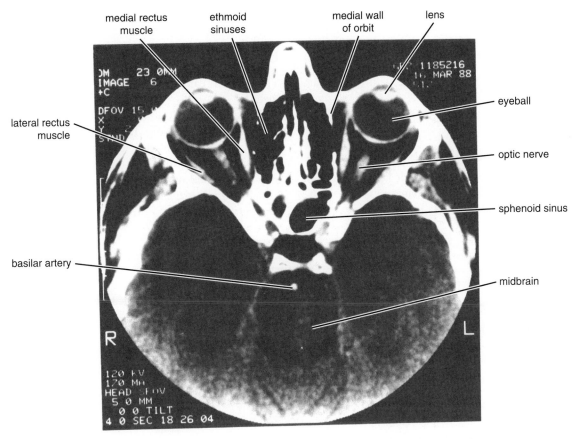

**Figure 18-10** CT scan of the skull showing the walls of the orbit and the eyeball.

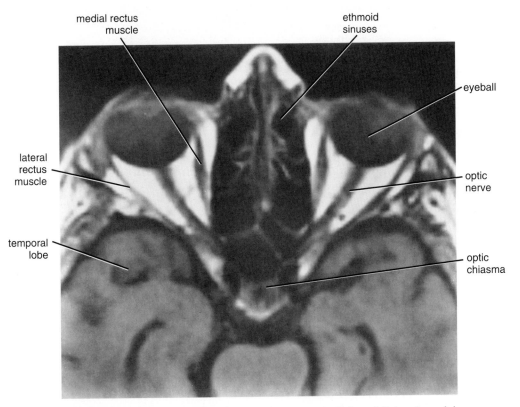

**Figure 18-11** Axial (horizontal) MRI showing the contents of the orbital and cranial cavities. Note that the eyeballs, the optic nerves, the optic chiasma, and the extraocular muscles can be identified.

## Surface Anatomy of the Eye

The main surface landmarks associated with the eye are shown in Figure 18-12.

# The Ear

The ear consists of the external ear, the middle ear, or tympanic cavity, and the internal ear, or labyrinth, which contains the organs of hearing and balance.

## External Ear

The external ear has an auricle and an external auditory meatus.

The **auricle** has a characteristic shape (Fig. 18-13A) and collects air vibrations. It consists of a thin plate of elastic cartilage covered by skin. It possesses both extrinsic and intrinsic muscles, which are supplied by the facial nerve.

The **external auditory meatus** is a curved tube that leads from the auricle to the tympanic membrane (Figs. 18-13 and

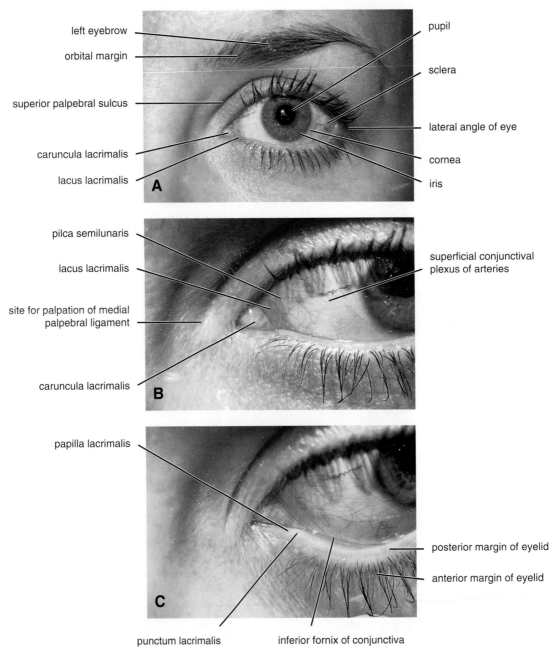

**Figure 18-12**   Left eye of a 29-year-old woman. **A.** The main structures seen in examining the eye. **B.** An enlarged view of the medial angle between the eyelids. **C.** The lower eyelid, pulled downward and slightly everted to reveal the punctum lacrimale.

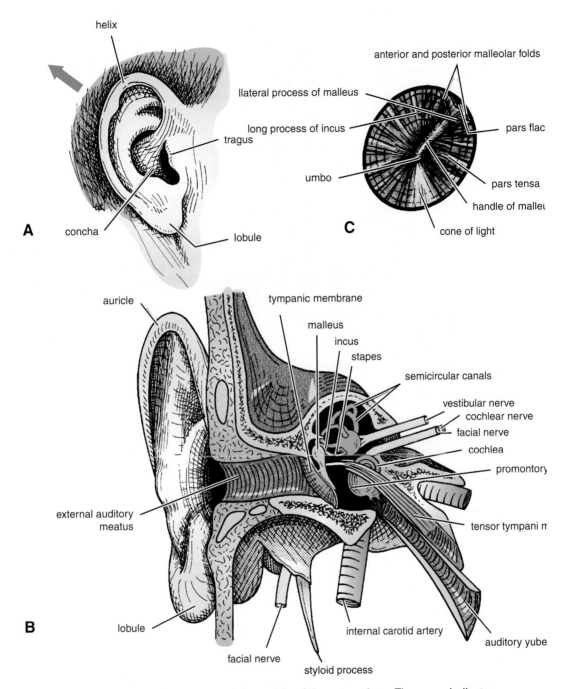

helix

anterior and posterior malleolar folds

llateral process of malleus

long process of incus

tragus

pars flac

umbo

pars tensa

handle of mallet

cone of light

**A**    concha    lobule    **C**

auricle    tympanic membrane

malleus

incus

stapes

semicircular canals

vestibular nerve

cochlear nerve

facial nerve

cochlea

promontory

external auditory meatus

tensor tympani m

**B**    lobule    internal carotid artery

auditory yube

facial nerve    styloid process

**Figure 18-13    A.** Different parts of the auricle of the external ear. The *arrow* indicates the direction that the auricle should be pulled to straighten the external auditory meatus before insertion of the otoscope in the adult. **B.** External and middle portions of the right ear viewed from in front. **C.** The right tympanic membrane as seen through the otoscope.

18-14). It conducts sound waves from the auricle to the tympanic membrane.

The framework of the outer third of the meatus is elastic cartilage, and the inner two thirds is bone, formed by the tympanic plate. The meatus is lined by skin, and its outer third is provided with **hairs** and **sebaceous** and **ceruminous glands.** The latter are modified sweat glands that secrete a yellowish brown wax. The hairs and the wax provide a sticky barrier that prevents the entrance of foreign bodies.

The **sensory nerve** supply of the lining skin is derived from the auriculotemporal nerve and the auricular branch of the vagus nerve.

The **lymph drainage** is to the superficial parotid, mastoid, and superficial cervical lymph nodes.

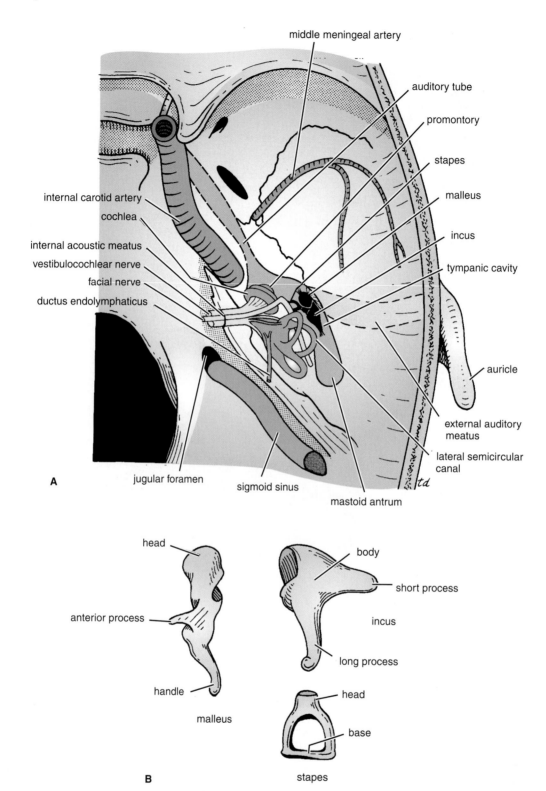

**Figure 18-14   A.** Parts of the right ear in relation to the temporal bone viewed from above. **B.** The auditory ossicles.

# Middle Ear (Tympanic Cavity)

The middle ear is an air-containing cavity in the petrous part of the temporal bone (Fig. 18-13). It is a narrow, slitlike cavity lined with mucous membrane. It contains the auditory ossicles, whose function is to transmit the vibrations of the tympanic membrane (eardrum) to the perilymph of the internal ear. The middle ear communicates in front through the auditory tube with the nasopharynx and behind with the mastoid antrum.

The middle ear has a roof, floor, anterior wall, posterior wall, lateral wall, and medial wall.

- **Roof** is formed by a thin plate of bone, the **tegmen tympani,** which is part of the petrous temporal bone (Figs. 18-15 and 18-16). It separates the tympanic cavity from the meninges and the temporal lobe of the brain in the middle cranial fossa.
- **Floor** is formed by a thin plate of bone, which may be partly replaced by fibrous tissue. It separates the tympanic cavity from the superior bulb of the internal jugular vein (Fig. 18-16).
- **Anterior wall** is formed below by a thin plate of bone that separates the tympanic cavity from the internal carotid artery (Fig. 18-16). At the upper part of the anterior wall are the openings into two canals. The lower and larger of these leads into the auditory tube, and the upper and smaller is the entrance into the canal for the tensor tympani muscle (Fig. 18-15). The thin, bony septum, which separates the canals, is prolonged backward on the medial wall, where it forms a shelflike projection.
- **Posterior wall** has in its upper part a large, irregular opening, the **aditus to the mastoid antrum** (Figs. 18-15 and 18-16). Below this is a small, hollow, conical projection, the **pyramid,** from whose apex emerges the tendon of the **stapedius muscle.**
- **Lateral wall** is largely formed by the tympanic membrane (Figs. 18-13 and 18-15).
- **Medial wall** is formed by the lateral wall of the inner ear. The greater part of the wall shows a rounded projection, called the **promontory,** which results from the underlying first turn of the cochlea (Figs. 18-13 and 18-15). Above and behind the promontory lies the **fenestra vestibuli,** which is oval shaped and closed by the base of the stapes. On the medial side of the window is the perilymph of the scala vestibuli of the internal ear. Below the posterior end of the promontory lies the **fenestra cochleae,** which is round and closed by the **secondary tympanic membrane.** On the medial side of this window is the perilymph of the blind end of the scala tympani (see p 683).

The bony shelf derived from the anterior wall extends backward on the medial wall above the promontory and above the fenestra vestibuli. It supports the tensor tympani muscle. Its posterior end is curved upward and forms a pulley, the **processus cochleariformis,** around which the tendon of the tensor tympani bends laterally to reach its insertion on the handle of the malleus (Fig. 18-16).

A rounded ridge runs horizontally backward above the promontory and the fenestra vestibuli and is known as the **prominence of the facial nerve canal** (contains the facial nerve). On reaching the posterior wall, it curves downward behind the pyramid.

## Tympanic Membrane

The tympanic membrane (Fig. 18-13) is a thin, fibrous membrane that is pearly gray. It is obliquely placed, facing downward, forward, and laterally. The membrane is concave laterally, and at the depth of the concavity is a small depression, the **umbo,** produced by the tip of the handle of the malleus. When the membrane is illuminated through an otoscope, the concavity produces a "cone of light," which radiates anteriorly and inferiorly from the umbo.

The tympanic membrane is circular and measures about 1 cm in diameter. The circumference is thickened and is slotted into a groove in the bone. The groove, or **tympanic sulcus,** is deficient superiorly, which forms a notch. From the sides of the notch, two bands, termed the **anterior** and **posterior malleolar folds,** pass to the lateral process of the malleus. The small triangular area on the tympanic membrane that is bounded by the folds is slack and is called the **pars flaccida** (Fig. 18-13). The remainder of the membrane is tense and is called the **pars tensa.** The handle of the malleus is bound down to the inner surface of the tympanic membrane by the mucous membrane.

The tympanic membrane is extremely sensitive to pain and is innervated on its outer surface by the auriculotemporal nerve and the auricular branch of the vagus.

## Auditory Ossicles

The auditory ossicles are the malleus, incus, and stapes (Figs. 18-14 and 18-15).

- **Malleus** is the largest ossicle and possesses a head, a neck, a long process or handle, an anterior process, and a lateral process. The **head** is rounded and articulates posteriorly with the incus. The **neck** is the constricted part below the head. The **handle** passes downward and backward and is firmly attached to the medial surface of the tympanic membrane. It can be seen through the tympanic membrane on otoscopic examination. The **anterior process** is a spicule of bone that is connected to the anterior wall of the tympanic cavity by a ligament. The lateral process projects laterally and is attached to the anterior and posterior malleolar folds of the tympanic membrane.
- **Incus** possesses a large body and two processes (Fig. 18-14). The **body** is rounded and articulates anteriorly with

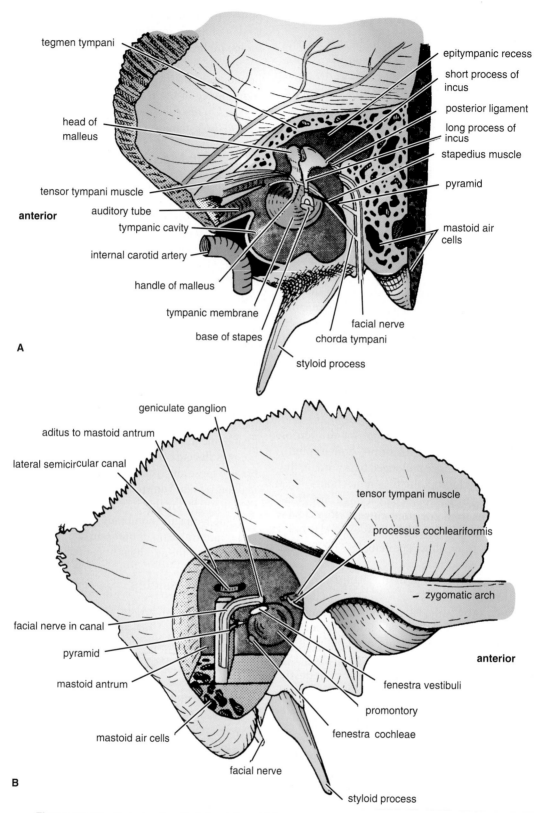

**Figure 18-15** **A.** Lateral wall of the right middle ear viewed from the medial side. Note the position of the ossicles and the mastoid antrum. **B.** Medial wall of the right middle ear viewed from the lateral side. Note the position of the facial nerve in its bony canal.

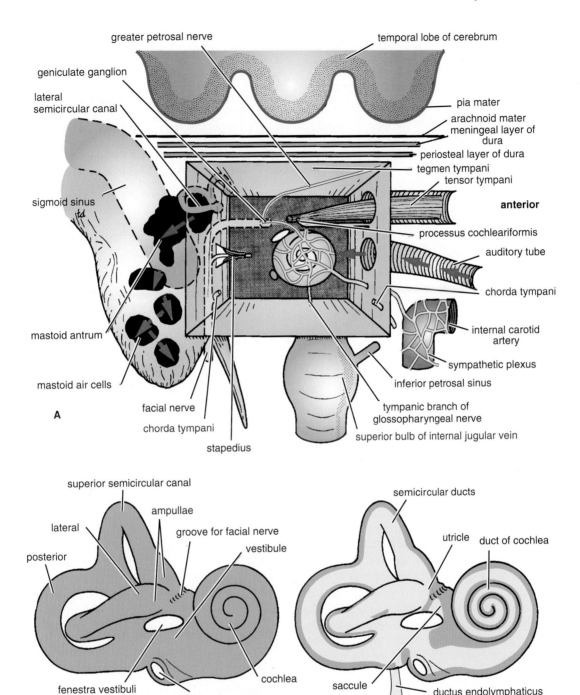

greater petrosal nerve

temporal lobe of cerebrum

geniculate ganglion

lateral
semicircular canal

pia mater
arachnoid mater
meningeal layer of
dura
periosteal layer of dura
tegmen tympani
tensor tympani

sigmoid sinus

**anterior**

processus cochleariformis

auditory tube

chorda tympani

mastoid antrum

internal carotid
artery

mastoid air cells

sympathetic plexus

inferior petrosal sinus

tympanic branch of
glossopharyngeal nerve

facial nerve

chorda tympani

superior bulb of internal jugular vein

stapedius

A

superior semicircular canal

semicircular ducts

ampullae

lateral

groove for facial nerve

utricle  duct of cochlea

posterior

vestibule

fenestra vestibuli

cochlea

saccule

ductus endolymphaticus

fenestra cochleae

saccus endolymphaticus

B

C

**Figure 18-16** **A.** The middle ear and its relations. Bony (**B**) and membranous (**C**) labyrinths.

the head of the malleus. The **long process** descends behind and parallel to the handle of the malleus. Its lower end bends medially and articulates with the head of the stapes. Its shadow on the tympanic membrane can sometimes be recognized on otoscopic examination. The **short process** projects backward and is attached to the posterior wall of the tympanic cavity by a ligament.

■ **Stapes** has a head, a neck, two limbs, and a base (Fig. 18-14). The **head** is small and articulates with the long process of the incus. The **neck** is narrow and receives the insertion of the stapedius muscle. The **two limbs** diverge from the neck and are attached to the oval **base.** The edge of the base is attached to the margin of the fenestra vestibuli by a ring of fibrous tissue, the **anular ligament.**

| Table 18-2 | Muscles of the Middle Ear | | | |
|---|---|---|---|---|
| **Muscle** | **Origin** | **Insertion** | **Nerve Supply** | **Action** |
| Tensor tympani | Wall of auditory tube and wall of its own canal | Handle of malleus | Mandibular division of trigeminal nerve | Dampens down vibrations of tympanic membrane |
| Stapedius | Pyramid (bony projection on posterior wall of middle ear) | Neck of stapes | Facial nerve | Dampens down vibrations of stapes |

From Snell RS: Clinical Anatomy. 7th Ed. Philadelphia: Lippincott Williams & Wilkins, 2004, p. 839.

## Muscles of the Ossicles

The muscles of the ossicles are the **tensor tympani** and the **stapedius muscles.**

The muscles of the ossicles, their nerve supply, and their actions are summarized in Table 18-2.

### PHYSIOLOGIC NOTE

### Movements of the Auditory Ossicles

The malleus and incus rotate on an anteroposterior axis that runs through the ligament connecting the anterior process of the malleus to the anterior wall of the tympanic cavity, the anterior process of the malleus and the short process of the incus, and the ligament connecting the short process of the incus to the posterior wall of the tympanic cavity.

When the tympanic membrane moves medially (Fig. 18-17), the handle of the malleus also moves medially. The head of the malleus and the body of the incus move laterally. The long process of the incus moves medially with the stapes. The base of the stapes is pushed medially in the fenestra vestibuli, and the motion is communicated to the perilymph in the scala vestibuli. Liquid being incompressible, the perilymph causes an outward bulging of the secondary tympanic membrane in the fenestra cochleae at the lower end of the scala tympani (Fig. 18-17). The above movements are reversed if the tympanic membrane moves laterally. Excessive lateral movements of the head of the malleus cause a temporary separation of the articular surfaces between the malleus and incus so that the base of the stapes is not pulled laterally out of the fenestra vestibuli.

During passage of the vibrations from the tympanic membrane to the perilymph via the small ossicles, the leverage increases at a rate of 1.3 to 1. Moreover, the area of the tympanic membrane is about 17 times greater than the area of the base of the stapes, causing the effective pressure on the perilymph to increase by a total of 22 to 1.

## Auditory Tube

The auditory tube connects the anterior wall of the tympanic cavity to the nasal pharynx (Fig. 18-13). Its posterior third is bony, and its anterior two thirds is cartilaginous. As the tube descends, it passes over the upper border of the superior constrictor muscle. It serves to equalize air pressures in the tympanic cavity and the nasal pharynx.

## Mastoid Antrum

The mastoid antrum lies behind the middle ear in the petrous part of the temporal bone (Fig. 18-14). It communicates with the middle ear by the aditus (Fig. 18-15).

### Relations of the Mastoid Antrum

The relations of the mastoid antrum are important in understanding the spread of infection.

- **Anterior wall** is related to the middle ear and contains the aditus to the mastoid antrum (Fig. 18-16).
- **Posterior wall** separates the antrum from the sigmoid venous sinus and the cerebellum (Fig. 18-16).
- **Lateral wall** is 1.5 cm thick and forms the floor of the suprameatal triangle (see Fig. 11-5).
- **Medial wall** is related to the posterior semicircular canal (Fig. 18-16).
- **Superior wall** is the thin plate of bone, the tegmen tympani, that is related to the meninges of the middle cranial fossa and the temporal lobe of the brain (Fig. 18-16).
- **Inferior wall** is perforated with holes, through which the antrum communicates with the mastoid air cells (Fig. 18-16).

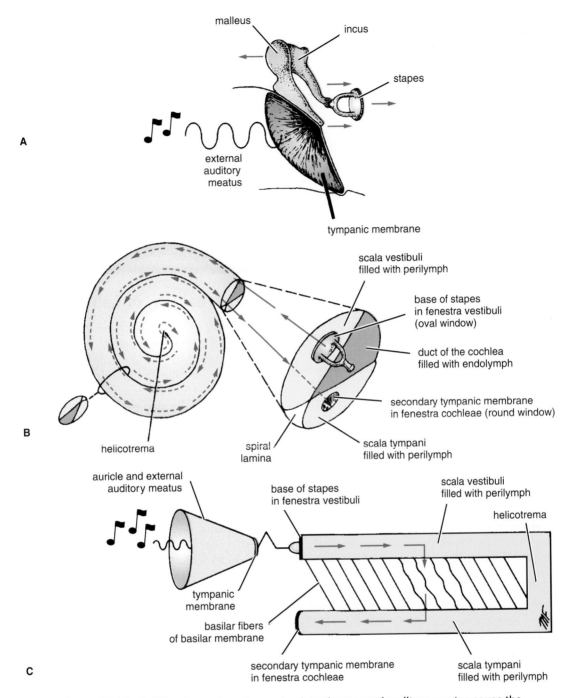

**Figure 18-17  A.** Vibrations of music passing into the external auditory meatus cause the tympanic membrane to move medially; the head of the malleus and incus move laterally, and the long process of the incus, with the stapes, moves laterally. **B.** The medial movement of the base of the stapes in the fenestra vestibuli causes motion (*arrows*) in the perilymph in the scala vestibuli. At the apex of the cochlea (the helicotrema), the compression wave in the perilymph passes down the scala tympani, causing a lateral bulging of the secondary tympanic membrane in the fenestra cochleae. **C.** Movement of the perilymph (*arrows*) after movement of the base of the stapes. Note the position of the basilar fibers of the basilar membrane.

## Mastoid Air Cells

The mastoid process begins to develop during the second year of life. The mastoid air cells are communicating cavities within the process that are continuous above with the antrum and the middle ear (Fig. 18-16). They are lined with mucous membrane.

# Facial Nerve

The entire course of the facial nerve is described on page 567. On reaching the bottom of the internal acoustic meatus, the facial nerve enters the facial canal (Fig. 18-14). The nerve runs laterally above the vestibule of the internal ear until it reaches the medial wall of the middle ear. Here, the nerve expands to form the sensory **geniculate ganglion** (Figs. 18-15 and 18-16). The nerve then bends sharply backward above the promontory.

On arriving at the posterior wall of the middle ear, it curves downward on the medial side of the aditus of the mastoid antrum (Fig. 18-16). It descends in the posterior wall of the middle ear, behind the pyramid, and finally emerges through the stylomastoid foramen into the neck.

## Important Branches of the Intrapetrous Part of the Facial Nerve

- The **greater petrosal nerve** arises from the facial nerve at the geniculate ganglion (Fig. 18-16). It contains preganglionic parasympathetic fibers that pass to the pterygopalatine ganglion and are there relayed through the zygomatic and lacrimal nerves to the lacrimal gland; other postganglionic fibers pass through the nasal and palatine nerves to the glands of the mucous membrane of the nose and palate. The great petrosal nerve also contains many taste fibers from the mucous membrane of the palate. The nerve emerges on the superior surface of the petrous part of the temporal bone and is eventually joined by the deep petrosal nerve from the sympathetic plexus on the internal carotid artery and forms the **nerve of the pterygoid canal.** This passes forward and enters the pterygopalatine fossa, where it ends in the pterygopalatine ganglion.
- The **nerve to the stapedius** arises from the facial nerve as it descends in the facial canal behind the pyramid (Fig. 18-16). It supplies the muscle within the pyramid.
- The **chorda tympani** arises from the facial nerve just above the stylomastoid foramen (Fig. 18-15). It enters the middle ear close to the posterior border of the tympanic membrane. It then runs forward over the upper part of the tympanic membrane and crosses the root of the handle of the malleus (Fig. 18-15). It lies in the interval between the mucous membrane and the fibrous layers of the tympanic membrane. The nerve leaves the middle ear through the petrotympanic fissure and enters the infratemporal fossa, where it joins the lingual nerve. The chorda tympani contains:

- **Taste fibers** from the mucous membrane covering the anterior two thirds of the tongue (not the vallate papillae) and the floor of the mouth. The taste fibers are the peripheral processes of the cells in the geniculate ganglion.
- **Preganglionic parasympathetic secretomotor fibers** that reach the submandibular ganglion and are there relayed to the submandibular and sublingual salivary glands.

## Tympanic Nerve

The tympanic nerve arises from the glossopharyngeal nerve, just below the jugular foramen (see p 569). It passes through the floor of the middle ear and onto the promontory (Fig. 18-16). Here, it forms the **tympanic plexus.** The tympanic plexus supplies the lining of the middle ear and gives off the **lesser petrosal nerve,** which sends secretomotor fibers to the parotid gland via the otic ganglion.

## The Internal Ear, or Labyrinth

The labyrinth is situated in the petrous part of the temporal bone, medial to the middle ear (Fig. 18-14). It consists of the bony labyrinth, comprising a series of cavities within the bone, and the membranous labyrinth, comprising a series of membranous sacs and ducts contained within the bony labyrinth. For a detailed description of the microscopic structure of the labyrinth, a textbook of histology should be consulted.

## Bony Labyrinth

The bony labyrinth consists of three parts: the vestibule, the semicircular canals, and the cochlea (Fig. 18-16). These are cavities situated in the substance of dense bone. They are lined by endosteum and contain a clear fluid, the **perilymph,** in which is suspended the membranous labyrinth (Fig. 18-17).

The **vestibule,** the central part of the bony labyrinth, lies posterior to the cochlea and anterior to the semicircular canals. In its lateral wall are the **fenestra vestibuli,** which is closed by the base of the stapes and its anular ligament, and the **fenestra cochleae,** which is closed by the **secondary tympanic membrane** (Fig. 18-17). Lodged within the vestibule are the **saccule** and **utricle** of the membranous labyrinth (Fig. 18-16).

The three **semicircular canals—superior, posterior,** and **lateral**—open into the posterior part of the vestibule. Each canal has a swelling at one end called the **ampulla.** The canals open into the vestibule by five orifices, one of which is common to two of the canals. Lodged within the canals are the **semicircular ducts** (Fig. 18-16).

The superior semicircular canal is vertical and placed at right angles to the long axis of the petrous bone. The posterior canal is also vertical but is placed parallel with the long axis of the petrous bone. The lateral canal is set in a horizontal position, and it lies in the medial wall of the aditus to the mastoid antrum, above the facial nerve canal.

The **cochlea** resembles a snail shell. It opens into the anterior part of the vestibule (Fig. 18-16). Basically, it consists of a central pillar, the **modiolus,** around which a hollow bony tube makes two and one half spiral turns. Each successive turn is of decreasing radius so that the whole structure is conical. The apex faces anterolaterally and the base faces posteromedially. The first basal turn of the cochlea is responsible for the promontory seen on the medial wall of the middle ear.

The modiolus has a broad base, which is situated at the bottom of the internal acoustic meatus. It is perforated by branches of the cochlear nerve. A spiral ledge, the **spiral lamina,** winds around the modiolus and projects into the interior of the canal and partially divides it. The **basilar membrane** stretches from the free edge of the spiral lamina to the outer bony wall, thus dividing the cochlear canal into the **scala vestibuli** above and the **scala tympani** below. The perilymph within the scala vestibuli is separated from the middle ear by the base of the stapes and the anular ligament at the fenestra vestibuli. The perilymph in the scala tympani is separated from the middle ear by the secondary tympanic membrane at the fenestra cochleae.

## Membranous Labyrinth

The membranous labyrinth is lodged within the bony labyrinth (Fig. 18-16). It is filled with endolymph and surrounded by perilymph. It consists of the utricle and saccule, which are lodged in the bony vestibule; the three semicircular ducts, which lie within the bony semicircular canals; and the duct of the cochlea, which lies within the bony cochlea. All these structures freely communicate with one another.

The **utricle** is the larger of the two vestibular sacs. It is indirectly connected to the saccule and the **ductus endolymphaticus** by the **ductus utriculosaccularis.**

The **saccule** is globular and is connected to the utricle, as described previously. The ductus endolymphaticus, after being joined by the ductus utriculosaccularis, passes on to end in a small blind pouch, the **saccus endolymphaticus** (Fig. 18-16). This lies beneath the dura on the posterior surface of the petrous part of the temporal bone.

Located on the walls of the utricle and saccule are specialized sensory receptors, which are sensitive to the orientation of the head to gravity or other acceleration forces.

The **semicircular ducts,** although much smaller in diameter than the semicircular canals, have the same configuration. They are arranged at right angles to each other so that all three planes are represented. Whenever the head begins or ceases to move, or whenever a movement of the head accelerates or decelerates, the endolymph in the semicircular ducts changes its speed of movement relative to that of the walls of the semicircular ducts. This change is detected in the sensory receptors in the ampullae of the semicircular ducts.

The **duct of the cochlea** is triangular in cross section and is connected to the saccule by the **ductus reuniens.** The highly specialized epithelium that lies on the **basilar membrane** forms the spiral organ of Corti and contains the sensory receptors for hearing. For a detailed description of the spiral organ, a textbook of histology should be consulted.

# Vestibulocochlear Nerve

On reaching the bottom of the internal acoustic meatus (see p 567), the nerve divides into vestibular and cochlear portions (Fig. 18-13).

The **vestibular nerve** is expanded to form the **vestibular ganglion.** The branches of the nerve then gain entrance to the membranous labyrinth, where they supply the utricle, the saccule, and the ampullae of the semicircular ducts.

The **cochlear nerve** divides into branches, which enter the base of the modiolus. The sensory ganglion of this nerve takes the form of an elongated **spiral ganglion** that is lodged in a canal winding around the modiolus in the base of the spiral lamina. The peripheral branches of this nerve pass from the ganglion to the spiral organ of Corti.

# Review Questions

## Completion Questions

**Select the phrase that BEST completes each statement.**

1. The levator palpebrae superioris muscle is innervated by the
   A. facial nerve.
   B. trochlear nerve.
   C. trigeminal nerve.
   D. oculomotor nerve.
   E. abducent nerve.

2. The inferior oblique muscle of the eye is innervated by the
   A. abducent nerve.
   B. trigeminal nerve.
   C. oculomotor nerve.
   D. facial nerve.
   E. trochlear nerve.

3. The lateral rectus muscle of the eye is innervated by the
   A. optic nerve.
   B. trochlear nerve.
   C. oculomotor nerve.
   D. facial nerve.
   E. abducent nerve.

4. The superior oblique muscle of the eye is innervated by the
   A. trigeminal nerve.
   B. trochlear nerve.
   C. abducent nerve.
   D. chorda tympani nerve.
   E. oculomotor nerve.

5. The orbicularis oculi muscle is innervated by the
   A. facial nerve.
   B. lacrimal nerve.
   C. maxillary nerve.
   D. nasociliary nerve.
   E. frontal nerve.

6. The optic canal is an opening in the
   A. lesser wing of the sphenoid bone.
   B. occipital bone.
   C. petrous part of the temporal bone.
   D. frontal bone.
   E. squamous part of the temporal bone.

7. The internal acoustic meatus in the skull
   A. is located in the body of the sphenoid bone.
   B. is located in the mastoid bone.
   C. allows passage of the glossopharyngeal nerve.
   D. allows passage of the vestibulocochlear nerve only.
   E. allows passage of the facial nerve and the vestibulocochlear nerve.

## Multiple Choice Questions

**Select the BEST answer for each question.**

8. The muscles and nerves that are responsible for adducting the eyeball (rotating the cornea medially) include the following **except** which?
   A. The superior rectus
   B. The medial rectus muscle

   C. The oculomotor nerve
   D. The inferior oblique muscle
   E. The inferior rectus muscle

9. Infection of the middle ear can spread along all the following pathways **except** which?
   A. Through the tegmen tympani to the middle cranial fossa
   B. Through the medial wall into the labyrinth
   C. Through the canal for the tensor tympani muscle into the internal carotid artery
   D. Through the floor into the internal jugular vein
   E. Through the aditus to the mastoid antrum into the mastoid air cells

10. The following general statements concerning the tympanic membrane are correct **except** which?
    A. It is pearly gray in color.
    B. It is concave laterally.
    C. It is crossed by the chorda tympani over the medial surface of the inferior part of the membrane.
    D. It is best visualized in the adult by pulling the auricle upward and backward.
    E. The inner surface is covered with mucous membrane.

11. The following statements concerning the chorda tympani are correct **except** which?
    A. It contains parasympathetic postganglionic fibers.
    B. It contains special sensory (taste) fibers.
    C. It joins the lingual nerve in the infratemporal fossa.
    D. It is a branch of the facial nerve in the temporal bone.
    E. It carries secretomotor fibers to the submandibular and sublingual salivary glands.

12. Assuming that the patient's eyesight is normal, in which cranial nerve is there likely to be a lesion when the direct and consensual light reflexes are absent?
    A. Trochlear nerve
    B. Optic nerve
    C. Abducent nerve
    D. Oculomotor nerve
    E. Trigeminal nerve

# Answers and Explanations

1. **D** is correct. The levator palpebrae superioris is innervated by the oculomotor nerve. However, the smooth muscle fibers of the levator palpebrae superioris are innervated by the sympathetic nerves. (In Horner's syndrome, where there is a lesion of the sympathetic nerves, mild ptosis occurs). The striated muscle, which forms the greater part of the muscle, receives its innervation from the oculomotor nerve. Division of the oculomotor nerve causes severe ptosis.

2. **C** is correct. The inferior oblique muscle of the eye is innervated by the oculomotor nerve.

3. **E** is correct. The lateral rectus muscle of the eye is innervated by the abducent nerve (Fig. 18-2).

4. **B** is correct. The superior oblique muscle of the eye is innervated by the trochlear nerve (Fig. 18-3).

5. **A** is correct. The orbicularis oculi muscle is a muscle of facial expression and is innervated by the facial nerve.

6. **A** is correct. The optic canal is an opening in the lesser wing of the sphenoid bone (Fig. 18-2). Through the canal pass the optic nerve and its surrounding tubular sheath of meninges and cerebrospinal fluid and the ophthalmic artery.

7. **E** is correct. The internal acoustic meatus in the skull allows passage of the facial and vestibulocochlear nerves from the posterior cranial fossa to the ear and beyond, in the case of the facial nerve (Fig. 18-14).

8. **D** is incorrect. The oblique muscles turn the eyeball laterally. (In addition, the superior oblique muscle turns the eye downward, and the inferior oblique muscle turns the eye upward.) The superior rectus muscle turns the eye medially as well as upward, and the inferior rectus turns the eye medially as well as downward (because these muscles take origin from the back of the orbit medial to the vertical axis of the eyeball). The oculomotor nerve supplies the medial, superior, and inferior recti and the inferior oblique muscles.

9. **C** is incorrect. The canal for the tensor tympani muscle is closed at its deep end and is filled by the origin of the tensor tympani muscle.

10. **C** is incorrect. The chorda tympani crosses the medial surface of the superior part of the tympanic membrane (Fig. 18-15).

11. **A** is incorrect. The chorda tympani contains parasympathetic preganglionic fibers.

12. **D** is correct. The oculomotor nerve contains the parasympathetic fibers that supply the constrictor pupillae muscle, which is necessary for the constriction of the pupil in the direct and consensual light reflexes.

# The Digestive System

# 19

# The Abdominal Wall, the Peritoneal Cavity, the Retroperitoneal Space, and the Alimentary Tract

All clinical material relevant to this chapter can be found on the CD-ROM.

# Chapter Outline

cute abdominal pain, blunt and penetrating trauma to the abdominal wall, and gastrointestinal bleeding are common problems facing the medical professional. The problems are complicated by the fact that the abdomen contains multiple organ systems, and in many patients, more than one system is involved.

Knowledge of spatial relationships of different abdominal organs is essential to making an accurate and complete diagnosis. Children with abdominal pain present a special diagnostic challenge; many diseases of childhood produce the symptoms of abdominal pain.

The object of this chapter is to provide an overview of the basic anatomy of the abdomen with special reference to the abdominal wall, alimentary tract, and peritoneal cavity. The liver and bile ducts, pancreas, and spleen, which are so closely associated with the gastrointestinal tract, will be discussed in the next chapter.

# BASIC ANATOMY

## Abdominal Wall

The anterior abdominal wall is made up of skin, superficial fascia, deep fascia, muscles, extraperitoneal fat and parietal peritoneum.

### Skin

The skin is loosely attached to the underlying structures except at the umbilicus, where it is tethered to the scar tissue of the umbilicus. The natural lines of cleavage in the skin run downward and forward almost horizontally around the trunk.

The umbilicus is a consolidated scar representing the site of attachment of the umbilical cord in the fetus; it is situated in the linea alba (see p. 696).

### Nerve Supply

The cutaneous nerve supply to the anterior abdominal wall is derived from the anterior rami of the lower six thoracic and the first lumbar nerves (Fig. 19-1). The thoracic nerves are the lower five intercostal and the subcostal nerves; the first lumbar nerve is represented by the iliohypogastric and the ilioinguinal nerves.

The dermatome of T7 is located in the epigastrium over the xiphoid process. The dermatome of T10 includes the umbilicus, and that of L1 lies just above the inguinal ligament and the symphysis pubis. The dermatomes and distribution of cutaneous nerves are shown in Figure 19-1.

### Blood Supply

**Arteries:** The skin near the midline is supplied by branches of the superior and the inferior epigastric arteries. The skin of the flanks is supplied by branches of the intercostal, lumbar, and deep circumflex iliac arteries (Fig. 19-2). In addition, the skin in the inguinal region is supplied by the superficial epigastric, superficial circumflex iliac, and the superficial external pudendal arteries, which are branches of the femoral artery.

**Veins:** The venous drainage passes above mainly into the axillary vein via the lateral thoracic vein and below into the femoral vein via the superficial epigastric and the great saphenous veins (Fig. 19-2).

### Lymph Drainage

The cutaneous lymph vessels above the level of the umbilicus drain upward into the anterior axillary lymph nodes. The vessels below this level drain downward into the superficial inguinal nodes (Fig. 19-3).

### Superficial Fascia

The superficial fascia is divided into the superficial fatty layer (fascia of Camper) and the deep membranous layer (Scarpa's fascia) (Fig. 19-4).

The fatty layer is continuous with the superficial fascia over the rest of the body. The membranous layer fades out laterally and above. Inferiorly, the membranous layer passes over the inguinal ligament to fuse with the deep fascia of the thigh (fascia lata) approximately one fingerbreadth below the inguinal ligament. In the midline, the membranous layer is not attached to the pubis but, instead, forms a tubular sheath for the penis (clitoris). In the perineum, it is attached on each

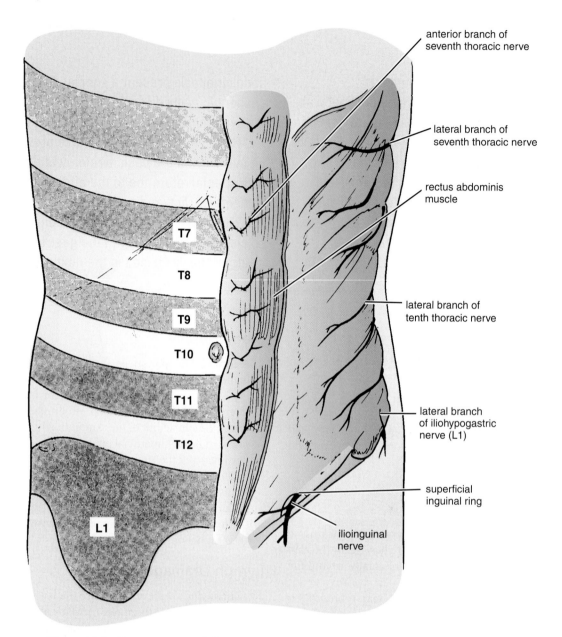

anterior branch of
seventh thoracic nerve

lateral branch of
seventh thoracic nerve

rectus abdominis
muscle

T7

T8

T9

T10

T11

T12

L1

lateral branch of
tenth thoracic nerve

lateral branch
of iliohypogastric
nerve (L1)

superficial
inguinal ring

ilioinguinal
nerve

**Figure 19-1**   Dermatomes and distribution of cutaneous nerves on the anterior
abdominal wall.

side to the margins of the pubic arch and is known as Colles'
fascia. Posteriorly, it fuses with the perineal body and the
posterior margin of the perineal membrane.

## Deep Fascia

In the anterior abdominal wall, the deep fascia is a thin layer
of areolar tissue covering the muscles.

## Muscles of the Anterior Abdominal Wall

The muscles of the anterior abdominal wall consist mainly
of three broad, thin sheets that are aponeurotic in front.

From exterior to interior, these sheets are the **external
oblique,** the **internal oblique,** and the **transversus** (Fig. 19-
5). In addition, on either side of the midline anteriorly, there
is a wide, strap-like, vertical muscle called the **rectus abdominis** (Fig. 19-6). As the aponeuroses of the three sheets
pass forward, they enclose the rectus abdominis to form the
**rectus sheath.**

In the lower part of the rectus sheath, there may be a
small muscle called the **pyramidalis.**

The **cremaster muscle** forms part of the covering of the
spermatic cord and is derived from the lower fibers of the internal oblique; it passes inferiorly and enters the scrotum
(see Fig. 19-11).

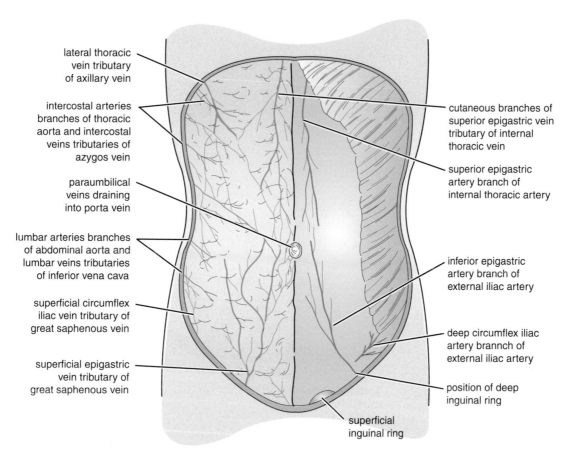

lateral thoracic
vein tributary
of axillary vein

intercostal arteries
branches of thoracic
aorta and intercostal
veins tributaries of
azygos vein

paraumbilical
veins draining
into porta vein

lumbar arteries branches
of abdominal aorta and
lumbar veins tributaries
of inferior vena cava

superficial circumflex
iliac vein tributary of
great saphenous vein

superficial epigastric
vein tributary of
great saphenous vein

cutaneous branches of
superior epigastric vein
tributary of internal
thoracic vein

superior epigastric
artery branch of
internal thoracic artery

inferior epigastric
artery branch of
external iliac artery

deep circumflex iliac
artery brannch of
external iliac artery

position of deep
inguinal ring

superficial
inguinal ring

**Figure 19-2**   On the left, arterial and venous drainage of the anterior abdominal wall. On the right, arterial supply to the anterior abdominal wall.

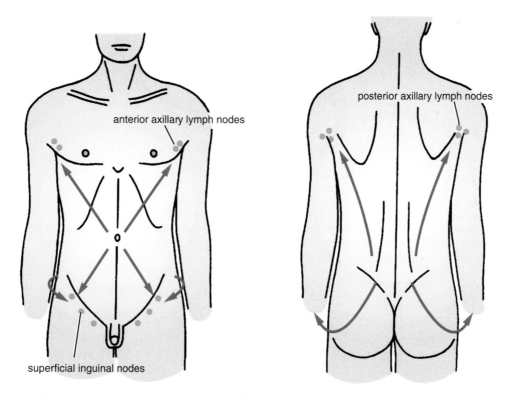

anterior axillary lymph nodes

posterior axillary lymph nodes

superficial inguinal nodes

**Figure 19-3**   Lymph drainage of the skin of the anterior and posterior abdominal walls.

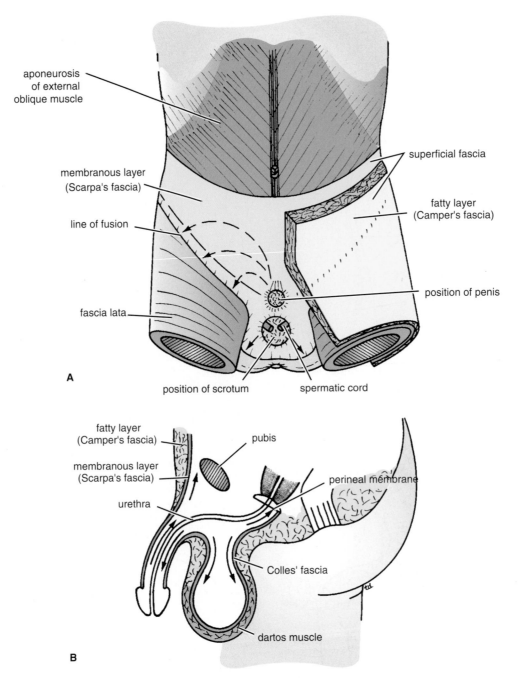

aponeurosis
of external
oblique muscle

membranous layer
(Scarpa's fascia)

line of fusion

fascia lata

superficial fascia

fatty layer
(Camper's fascia)

position of penis

position of scrotum        spermatic cord

A

fatty layer
(Camper's fascia)

membranous layer
(Scarpa's fascia)

urethra

pubis

perineal membrane

Colles' fascia

dartos muscle

B

**Figuer 19-4**  **A.** Arrangement of the fatty layer and the membranous layer of the superficial fascia in the lower part of the anterior abdominal wall. Note the line of fusion between the membranous layer and the deep fascia of the thigh (fascia lata). **B.** Note the attachment of the membranous layer to the posterior margin of the perineal membrane. *Arrows* indicate paths taken by urine in cases of ruptured urethra.

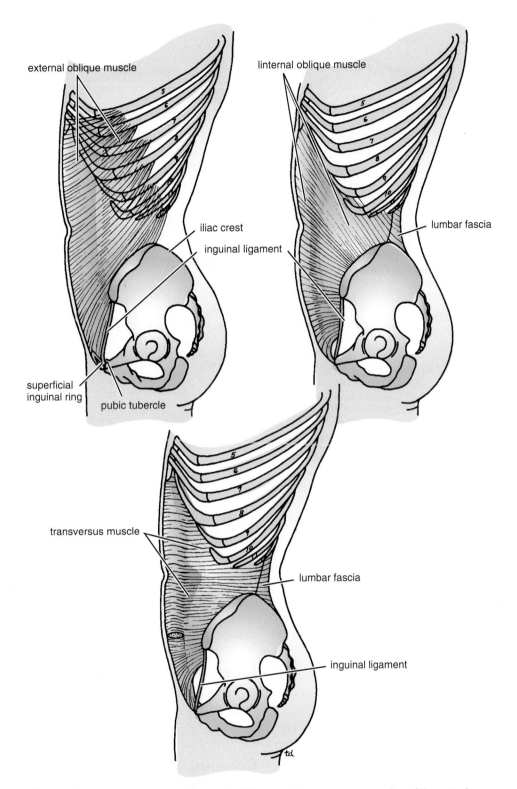

**Figure 19-5**   External oblique, internal oblique, and transversus muscles of the anterior abdominal wall.

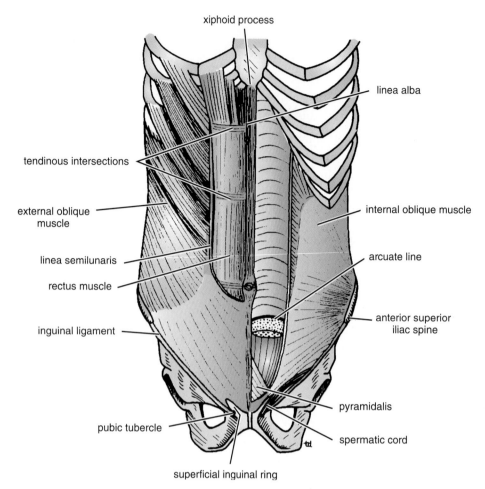

xiphoid process

linea alba

tendinous intersections

external oblique
muscle

internal oblique muscle

linea semilunaris

arcuate line

rectus muscle

anterior superior
iliac spine

inguinal ligament

pyramidalis

spermatic cord

pubic tubercle

superficial inguinal ring

**Figure 19-6** Anterior view of the rectus abdominis muscle and the rectus sheath. *Left:* The anterior wall of the sheath has been partly removed, revealing the rectus muscle with its tendinous intersections. *Right:* The posterior wall of the rectus sheath is shown. The edge of the arcuate line is shown at the level of the anterior superior iliac spine.

The muscles of the anterior abdominal wall are summarized in Table 19-1.

## Rectus Sheath

The rectus sheath (Figs. 19-7 and 19-8) is a long fibrous sheath that encloses the rectus abdominis muscle and pyramidalis muscle (if present) and contains the anterior rami of the lower six thoracic nerves and the superior and inferior epigastric vessels and lymph vessels. It is formed by the aponeuroses of the three lateral abdominal muscles. The internal oblique aponeurosis splits at the lateral edge of the rectus abdominis to form two laminae; one passes anteriorly and one passes posteriorly to the rectus. The aponeurosis of the external oblique fuses with the anterior lamina, and the transversus aponeurosis fuses with the posterior lamina. At the level of the anterior superior iliac spines, all three aponeuroses pass anteriorly to the rectus muscle, leaving the sheath deficient posteriorly below this level. The lower, crescent-shaped edge of the posterior wall of the sheath is called the arcuate line. All three aponeuroses fuse with each other and with their fellows of the opposite side in the midline between the right and the left recti muscles to form a fibrous band called the **linea alba**, which extends from the xiphoid process above to the pubic symphysis below.

The posterior wall of the sheath is not attached to the rectus muscle. The transverse tendinous intersections, which divide the rectus abdominis muscle into segments, are usually three in number: One at the level of the xiphoid process, one at the level of the umbilicus, and one between these two (Fig. 19-7). These intersections are strongly attached to the anterior wall of the rectus sheath.

## Linea Semilunaris

The linea semilunaris is the lateral edge of the rectus abdominis muscle (Fig. 19-6). It crosses the costal margin at the tip of the ninth costal cartilage.

## Conjoint Tendon

The internal oblique muscle (Fig. 19-5) has a lower, free border that arches over the spermatic cord (or the round ligament of the uterus) and then descends behind and attaches to the pubic crest and the pectineal line (Fig. 19-10). Near their insertion, the lowest tendinous fibers are joined by similar fibers from the transversus abdominis to form the conjoint tendon, which strengthens the medial half of the posterior wall of the inguinal canal.

| Table 19-1 | Muscles of the Anterior Abdominal Wall | | | |
|---|---|---|---|---|
| **Name of Muscle** | **Origin** | **Insertion** | **Nerve Supply** | **Action** |
| External oblique | Lower eight ribs | Xiphoid process, linea alba, pubic crest, pubic tubercle, iliac crest | Lower six thoracic nerves and iliohypogastric and ilioinguinal nerves (L1) | Supports abdominal contents; compresses abdominal contents; assists in flexing and rotation of trunk; assists in forced expiration, micturition, defecation, parturition, and vomiting |
| Internal oblique | Lumbar fascia, iliac crest, lateral two thirds of inguinal ligament | Lower three ribs and costal cartilages, xiphoid process, linea alba, symphysis pubis | Lower six thoracic nerves and iliohypogastric and ilioinguinal nerves (L1) | As above |
| Transversus | Lower six costal cartilages, lumbar fascia, iliac crest, lateral third of inguinal ligament | Xiphoid process, linea alba, symphysis pubis | Lower six thoracic nerves and iliohypogastric and ilioinguinal nerves (L1) | Compresses abdominal contents |
| Rectus abdominis | Symphysis pubis and pubic crest | Fifth, sixth, and seventh costal cartilages and xiphoid process | Lower six thoracic nerves | Compresses abdominal contents and flexes vertebral column; accessory muscle of expiration |
| Pyramidalis (if present) | Anterior surface of pubis | Linea alba | Twelfth thoracic nerve | Tenses the linea alba |

From Snell RS: Clinical Anatomy. 7th Ed. Philadelphia: Lippincott Williams & Wilkins, 2004, p 167.

## Inguinal Ligament

The inguinal ligament (Fig. 19-10) connects the anterior superior iliac spine with the pubic tubercle. This ligament is formed by the lower border of the aponeurosis of the external oblique, which is folded back on itself (Fig. 19-6). From the medial end of the ligament, the lacunar ligament (Fig. 19-10) extends backward and upward to the pectineal line of the superior ramus of the pubis, where it becomes continuous with the pectineal ligament (a thickening of the periosteum, Fig. 19-10). The lower border of the inguinal ligament is attached to the deep fascia of the thigh (the fascia lata).

## Fascia Transversalis

The fascia transversalis is a thin layer of fascia that lines the transversus muscle (Fig. 19-7) and is continuous with a similar layer lining the diaphragm and the iliacus muscle. The femoral sheath of the femoral vessels is formed from the fascia transversalis and the fascia iliaca.

## Extraperitoneal Fat

The extraperitoneal fat is a thin layer of connective tissue that contains a variable amount of fat and lies between the fascia transversalis and the parietal peritoneum (Fig. 19-8).

## Parietal Peritoneum

The walls of the abdomen are lined with parietal peritoneum (Fig. 19-7). This is a thin serous membrane and is continuous below with the parietal peritoneum lining the pelvis.

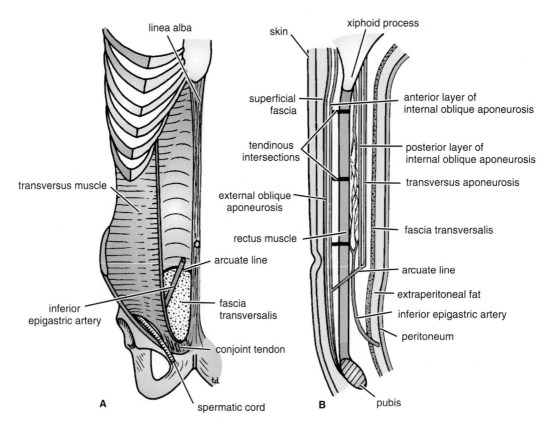

**Figure 19-7** Rectus sheath in anterior view (**A**) and in sagittal section (**B**). Note the arrangement of the aponeuroses forming the rectus sheath.

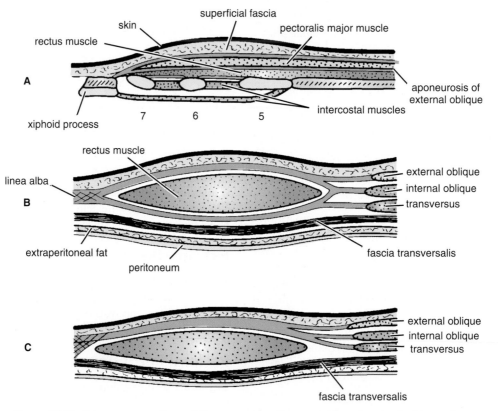

**Figure 19-8** Transverse sections of the rectus sheath seen at three levels. **A.** Above the costal margin. **B.** Between the costal margin and the level of the anterior superior iliac spine. **C.** Below the level of the anterior superior iliac spine and above the pubis.

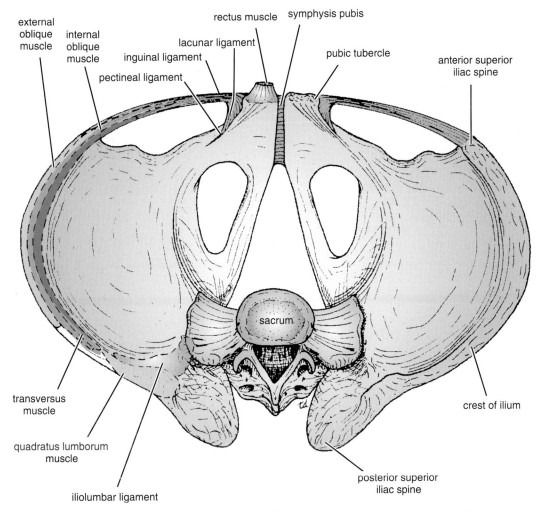

**Figure 19-9**  Bony pelvis viewed from above. Note the attachments of the inguinal, lacunar, and pectineal ligaments.

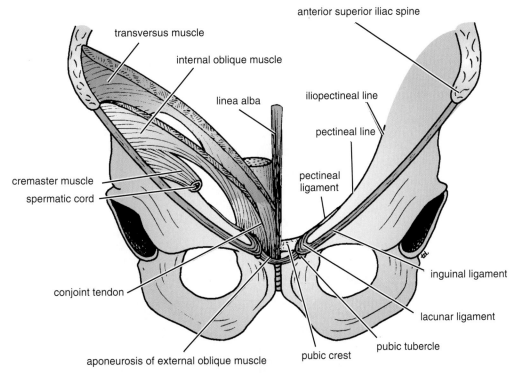

**Figure 19-10**  Anterior view of the pelvis showing the attachment of the conjoint tendon to the pubic crest and the adjoining part of the pectineal line.

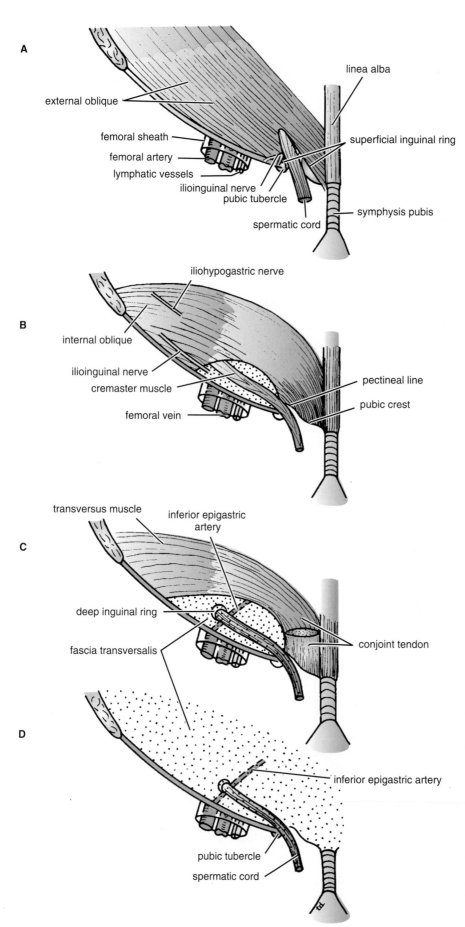

**Figure 19-11** Inguinal canal showing the arrangement of the external oblique muscle (**A**), the internal oblique muscle (**B**), the transversus muscle (**C**), and the fascia transversalis (**D**). Note that the anterior wall of the canal is formed by the external oblique and the internal oblique muscles and the posterior wall is formed by the fascia transversalis and the conjoint tendon. The deep inguinal ring lies lateral to the inferior epigastric artery.

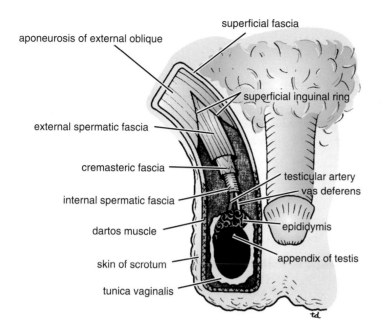

**Figure 19-12**   Scrotum dissected from in front. Note the spermatic cord and its coverings.

## Inguinal Canal

The inguinal canal (Fig. 19-11) is an oblique passage through the lower part of the anterior abdominal wall. In males, it allows structures to pass to and from the testis to the abdomen. In females, it allows the round ligament of the uterus to pass from the uterus to the labium majus.

The canal is approximately 1.5 in. (4 cm) in length in adults and extends from the deep inguinal ring downward and medially to the superficial inguinal ring. It lies parallel to and immediately above the inguinal ligament.

The **deep inguinal ring*** is an oval opening in the fascia transversalis and lies approximately 0.5 in. (1.3 cm) above the inguinal ligament (Fig. 19-11). The margins of this ring give attachment to the **internal spermatic fascia.**

The **superficial inguinal ring** is a triangular-shaped defect in the aponeurosis of the external oblique muscle and **lies immediately above and medial to the pubic tubercle** (Fig. 19-11). The margins of this ring give attachment to the **external spermatic fascia** (Fig. 19-12).

### Walls of the Inguinal Canal

**Anterior wall:** External oblique aponeurosis, reinforced laterally by the origin of the internal oblique from the in-

guinal ligament (Figs. 19-6 and 19-11).
**Posterior wall:** Conjoint tendon medially; fascia transversalis laterally (Figs. 19-11 and 19-13).
**Roof or superior wall:** Arching fibers of the internal oblique and transversus muscles (Fig. 19-10).
**Floor or inferior wall:** Upturned lower edge of the inguinal ligament and the lacunar ligaments (Fig. 19-10).

---

### PHYSIOLOGIC NOTE

### Function of the Inguinal Canal

In males, the inguinal canal allows structures to pass to and from the testis to the abdomen. (Normal spermatogenesis occurs only if the testis leaves the abdominal cavity and enters a cooler environment in the scrotum.) In females, the smaller canal allows the round ligament of the uterus to pass from the uterus to the labium majus. In both sexes, the canal also transmits the ilioinguinal nerve.

---

### PHYSIOLOGIC NOTE

### Mechanics of the Inguinal Canal

The inguinal canal is a site of potential weakness in both sexes. On coughing and straining (as in micturition, defecation, and parturition), the arching lowest fibers of the internal oblique and transversus abdominis muscles contract and flatten the arch (Fig. 19-14). In turn, this lowers the roof of the canal toward the floor and virtually closes the canal.

---

* A common frustration for students is the inability to observe these rings as openings. One must remember that the internal spermatic fascia is attached to the margins of the deep inguinal ring and the external spermatic fascia is attached to the margins of the superficial inguinal ring so that the edges of the rings cannot be observed externally. Compare this arrangement with the openings for the fingers seen inside a glove with the absence of openings for the fingers when the glove is viewed from the outside.

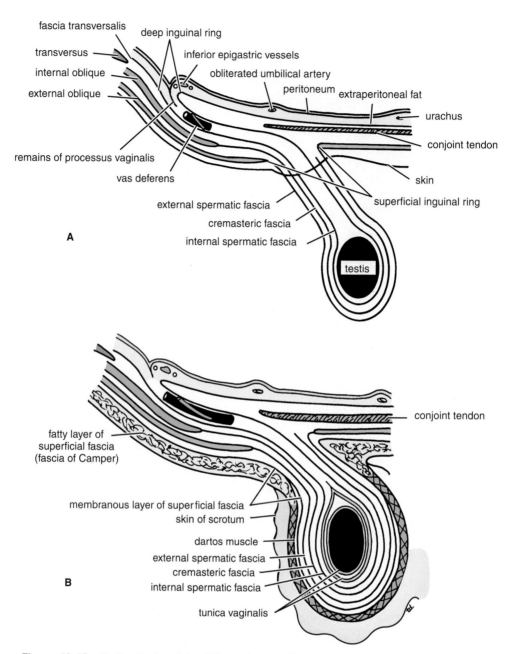

**Figure 19-13   A.** Continuity of the different layers of the anterior abdominal wall with coverings of the spermatic cord. **B.** The skin and superficial fascia of the abdominal wall and scrotum have been included, and the tunica vaginalis is shown.

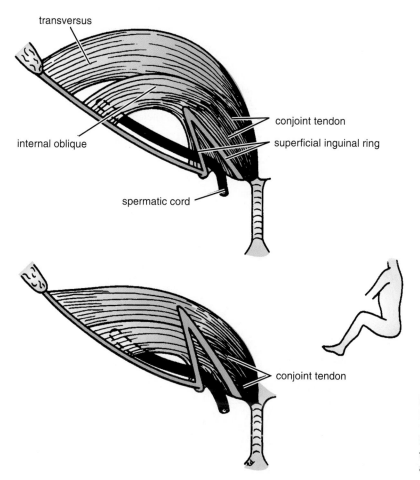

transversus

conjoint tendon

internal oblique

superficial inguinal ring

spermatic cord

conjoint tendon

**Figure 19-14**  Action of the muscles on the inguinal canal. Note that the canal is "obliterated" when the muscles contract. Note also that the anterior surface of the thigh protects the inguinal region when one assumes the squatting position.

# Spermatic Cord

The spermatic cord is a collection of structures that pass through the inguinal canal to and from the testis (Fig. 19-15). These structures include the following:

- The vas deferens.
- The testicular artery.
- Testicular veins (pampiniform plexus).
- Testicular lymph vessels.
- Autonomic nerves.
- Remains of the processus vaginalis.
- The cremasteric artery.
- The genital branch of the genitofemoral nerve, which supplies the cremaster muscle.

## Coverings of the Spermatic Cord

There are three concentric layers of fascia derived from the layers of the anterior abdominal wall (Figs. 19-12 and 19-13):

**External spermatic fascia** derived from the external oblique muscle and attached to the margins of the superficial inguinal ring.

**Cremasteric fascia** derived from the internal oblique muscle.

**Internal spermatic fascia** derived from the fascia transversalis that lines the abdominal muscles; it is attached to the margins of the deep inguinal ring.

To understand the coverings of the spermatic cord, one must first consider the development of the inguinal canal.

---

**EMBRYOLOGIC NOTE**

### Development of the Inguinal Canal

Before the descent of the testis and the ovary from their site of origin high on the posterior abdominal wall (L1), a peritoneal diverticulum called the **processus vaginalis** is formed (Fig. 19-16). The processus vaginalis passes through the layers of the lower part of the anterior abdominal wall and, as it does so, it acquires a tubular covering from each layer. It traverses the fascia transversalis at the deep inguinal ring and acquires a tubular covering, the **internal spermatic fascia** (Fig. 19-13). As it passes through the lower part of the internal oblique muscle, it takes with it some of its l▮

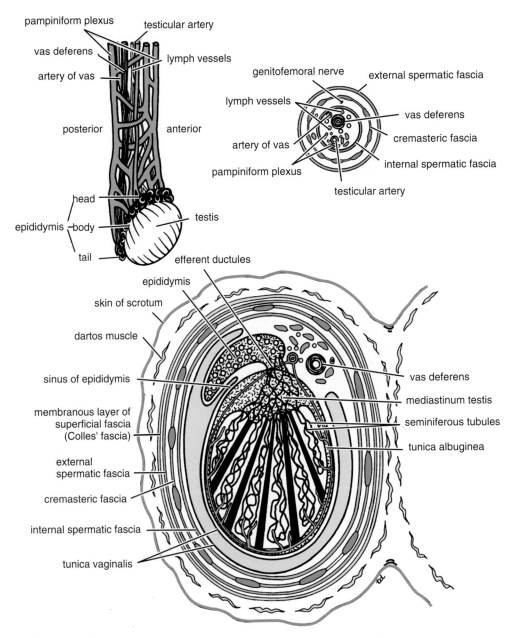

**Figure 19-15** Testis and epididymis, spermatic cord, and scrotum. Also shown is the testis and epididymis cut across in horizontal section.

fibers, which form the **cremaster muscle.** The muscle fibers are embedded in fascia, and thus the second tubular sheath is known as the **cremasteric fascia** (Fig. 19-13). The processus vaginalis passes under the arching fibers of the transversus abdominis muscle and, therefore, does not acquire a covering from this abdominal layer. On reaching the aponeurosis of the external oblique, it evaginates this to form the superficial inguinal ring and acquires a third tubular fascial coat, the **external spermatic fascia** (Figs. 19-12 and 19-13). It is in this manner that the inguinal canal is formed in both sexes. (In the female, the term **spermatic** fas-

cia should be replaced by **the covering of the round ligament of the uterus.**)

Meanwhile, a band of mesenchyme, extending from the lower pole of the developing gonad through the inguinal canal to the labioscrotal swelling, has condensed to form the **gubernaculum** (Fig. 19-16).

In the male, the testis descends through the pelvis and inguinal canal during the seventh and eighth months of fetal life. The normal stimulus for the descent of the testis is testosterone, which is secreted by the fetal testes. The testis follows the gubernaculum and descends behind the peritoneum on the posterior abdominal wall. The testis then passes behind the processus vaginalis and pulls down its duct, blood vessels, nerves, and lymph vessels. The testis takes up its final position in the developing scrotum by the end of the eighth month.

Because the testis and its accompanying vessels, ducts, and so on follow the course previously taken by the processus vaginalis, they acquire the same three coverings as they pass down the inguinal canal. Thus, the spermatic cord is covered by three concentric layers of fascia: the external spermatic fascia, the cremasteric fascia, and the internal spermatic fascia.

In the female, the ovary descends into the pelvis following the gubernaculum (Fig. 19-16). The gubernaculum becomes attached to the side of the developing uterus, and the gonad descends no farther. That part of the gubernaculum extending from the uterus into the de-

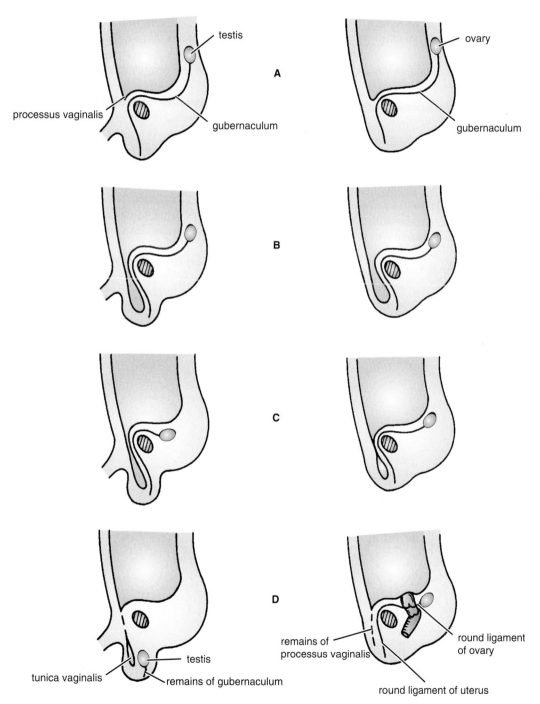

**Figure 19-16**  Origin, development, and fate of the processus vaginalis in the two sexes. Note the descent of the testis into the scrotum and the descent of the ovary into the pelvis.

veloping labium majus persists as the **round ligament of the uterus.** Thus, in the female, the only structures that pass through the inguinal canal from the abdominal cavity are the round ligament of the uterus and a few lymph vessels. The lymph vessels convey a small amount of lymph from the body of the uterus to the superficial inguinal nodes.

# Scrotum

The scrotum is an outpouching of the lower part of the anterior abdominal wall. It contains the testes, the epididymides, and the lower ends of the spermatic cords (Figs. 19-13 and 19-15).

The wall of the scrotum has the following layers:

- **Skin:** The skin of the scrotum is thin, wrinkled, and pigmented and forms a single pouch. A slightly raised ridge in the midline indicates the line of fusion of the two lateral labioscrotal swellings. (In the female, the swellings remain separate and form the labia majora).
- **Superficial fascia:** This is continuous with the fatty and membranous layers of the anterior abdominal wall; the fat is, however, replaced by smooth muscle called the **dartos muscle.** This is innervated by sympathetic nerve fibers and is responsible for the wrinkling of the overlying skin. The membranous layer of the superficial fascia (often referred to as Colles' fascia) is continuous in front with the membranous layer of the anterior abdominal wall (Scarpa's fascia), and behind, it is attached to the perineal body and the posterior edge of the perineal membrane (Fig. 19-4). At the sides, it is attached to the ischiopubic rami. Both layers of superficial fascia contribute to a median partition that crosses the scrotum and separates the testes from each other.
- **Spermatic fasciae:** These three layers lie beneath the superficial fascia and are derived from the three layers of the anterior abdominal wall on each side, as previously explained. The **cremaster muscle** in the cremasteric fascia can be made to contract by stroking the skin on the medial aspect of the thigh. This is called the **cremasteric reflex.** The afferent fibers of this reflex arc travel in the femoral branch of the genitofemoral nerve (L1 and 2), and the efferent motor nerve fibers travel in the genital branch of the genitofemoral nerve. The function of the cremaster muscle is to raise the testis and the scrotum upward for warmth and for protection against injury. For testicular temperature and fertility, see page 701.
- **Tunica vaginalis** (Figs. 19-12, 19-13, and 19-15): This lies within the spermatic fasciae and covers the anterior, medial, and lateral surfaces of each testis. It is the lower expanded part of the processus vaginalis; normally, just before birth, it becomes shut off from the upper part of the processus and the peritoneal cavity. The tunica vaginalis is thus a closed sac, invaginated from behind by the testis.

## Lymph Drainage of the Scrotum

Lymph from the skin and fascia, including the tunica vaginalis, drains into the superficial inguinal lymph nodes (Fig. 19-17).

The testes and epididymides are considered in Chapter 22.

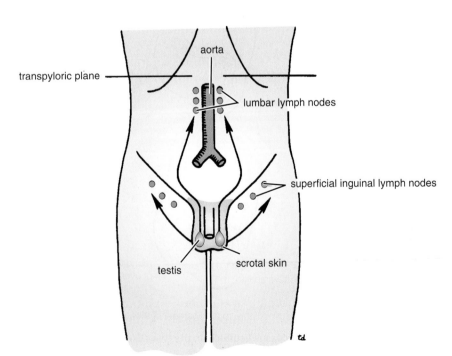

**Figure 19-17** Lymph drainage of the testis and the skin of the scrotum.

# Labia Majora

The labia majora are prominent, hair-bearing folds of skin formed by the enlargement of the genital swellings in the fetus. (In the male, the genital swellings fuse in the midline to form the scrotum.) Within the labia are a large amount of adipose tissue and the terminal strands of the round ligaments of the uterus.

# Nerves of the Anterior Abdominal Wall

The nerves of the anterior abdominal wall are the anterior rami of the lower six thoracic and the first lumbar nerves (Fig. 19-1). These nerves run downward and forward between the internal oblique and the transversus muscles (Fig. 19-18). They supply the skin, the muscles, and the parietal peritoneum of the anterior abdominal wall. The lower six thoracic nerves pierce the posterior wall of the rec-

tus sheath. The first lumbar nerve is represented by the iliohypogastric and the ilioinguinal nerves, which do not enter the rectus sheath. Instead, the iliohypogastric nerve pierces the external oblique aponeurosis above the superficial inguinal ring, and the ilioinguinal nerve passes through the inguinal canal to emerge through the ring.

# Blood Supply of the Anterior Abdominal Wall

**Arteries:** These include the **superior and inferior epigastric arteries**, the **deep circumflex iliac artery**, the **lower two posterior intercostal arteries**, and the **four lumbar arteries** (Fig. 19-2). The **superficial epigastric artery**, the **superficial circumflex iliac artery**, and the **superficial external pudendal artery** also supply the lower part of the anterior abdominal wall.

**Veins:** The veins have the same names as the arteries and follow them to drain into the internal thoracic and the ex-

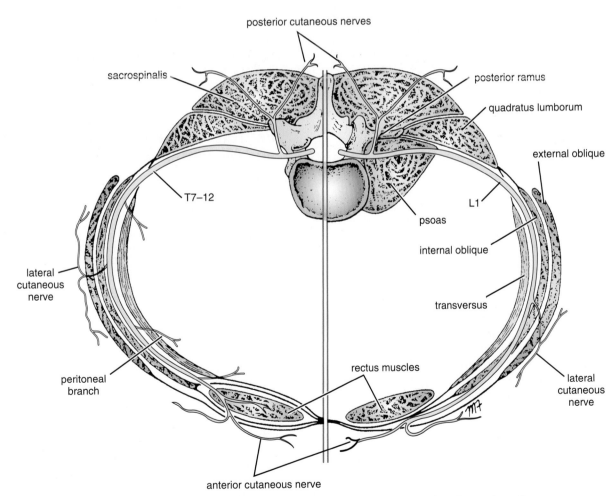

**Figure 19-18** Cross section of the abdomen showing the courses of the lower thoracic and first lumbar nerves.

ternal iliac veins, the azygos veins, and the inferior vena cava (Fig. 19-2).

The superficial epigastric, the superficial circumflex iliac, and the superficial external pudendal veins drain into the great saphenous vein and, from there, into the femoral vein. The thoracoepigastric vein is the name given to the anastomoses between the lateral thoracic vein, a tributary of the axillary vein, and the superficial epigastric vein, a tributary of the great saphenous vein. This vein provides an alternative path for the venous blood should the superior or inferior vena cava become obstructed.

---

**E M B R Y O L O G I C   N O T E**

### Development of the Abdominal Wall

Following segmentation of the mesoderm, the lateral mesoderm splits into a somatic and a splanchnic layer associated with ectoderm and entoderm, respectively (Fig. 19-19). The muscles of the anterior abdominal wall are derived from the somatopleuric mesoderm and retain their segmental innervation from the anterior rami of the spinal nerves. Unlike the thorax, the segmental arrangement becomes lost due to the absence of ribs, and the mesenchyme fuses to form large sheets of muscle. The rectus abdominis retains indications of its segmental origin, as seen by the presence of the tendinous intersections. The somatopleuric mesoderm becomes split tangentially into three layers, which form the external oblique, internal oblique, and transversus abdominis muscles. The anterior body wall finally closes in the midline at 3 months, when the right and left sides meet in the midline and fuse. The line of fusion of the mesenchyme forms the linea alba; and on either side of this, the rectus muscles come to lie within their rectus sheaths.

### Development of the Umbilical Cord and the Umbilicus

As the tail fold of the embryo develops, the embryonic attachment of the body stalk to the caudal end of the embryonic disc comes to lie on the anterior surface of the embryo, close to the remains of the yolk sac (Fig. 19-20). The amnion and chorion now fuse, so that the amnion encloses the body stalk and the yolk sac with their blood vessels to form the tubular umbilical cord. The mesenchymal core of the cord forms the loose connective tissue called **Wharton's jelly**. Embedded in this are the remains of the yolk sac, the vitelline duct, the remains of the allantois, and the umbilical blood vessels.

The umbilical vessels consist of two arteries that carry deoxygenated blood from the fetus to the chorion (later the placenta). The two umbilical veins convey oxygenated blood from the placenta to the fetus. The right vein soon disappears (Fig. 19-20).

The umbilical cord is a twisted tortuous structure that measures about 0.75 in. (2 cm) in diameter. It increases in length until, at the end of pregnancy, it is about 20 in. (50 cm) long—that is, about the same length as the child.

# Peritoneum and Peritoneal Cavity

The peritoneum is the serous membrane that lines the abdominal and the pelvic cavities and that clothes the viscera (Figs. 19-21 and 19-22). The peritoneum can be regarded as a balloon against which organs are pressed from the outside. The **parietal layer** lines the walls of the abdominal and the pelvic cavities, and the **visceral layer** covers the organs. The potential space between the parietal and the visceral layer is called the **peritoneal cavity.** In males, this is a closed cavity, but in females, there is communication with the exterior through the uterine tubes, the uterus, and the vagina.

The peritoneal cavity is divided into two parts: the greater sac and the lesser sac (Figs. 19-21 and 19-22). The **greater sac** is the main compartment and extends from the diaphragm down into the pelvis. The **lesser sac** is smaller and lies behind the stomach. The greater and the lesser sacs are in free communication with one another through the **epiploic foramen**. The peritoneum secretes a small amount of serous fluid that lubricates the peritoneal surfaces and facilitates free movement between the viscera.

## Peritoneal Ligaments, Omenta, and Mesenteries

The peritoneal ligaments, omenta, and mesenteries permit blood, lymph vessels, and nerves to reach the viscera.

### Peritoneal Ligaments

Peritoneal ligaments are two-layered folds of peritoneum that connect solid viscera with the abdominal walls. The liver, for example, is connected to the diaphragm by the **falciform ligament**, the **coronary ligament**, and the **right** and the **left triangular ligaments** (see Fig. 20-6).

### Omenta

Omenta are two-layered folds of peritoneum that connect the stomach with another viscus. The **greater omentum** connects the greater curvature of the stomach with the transverse colon (Fig. 19-21). It hangs down like an apron in front of the coils of the small intestine and is folded back on itself. The **lesser omentum** suspends the lesser curvature of the stomach to the fissure for the ligamentum venosum and the porta hepatis of the liver (Fig. 19-21). The **gastrosplenic omentum** (ligament) connects the stomach to the hilus of the spleen(Fig. 19-22).

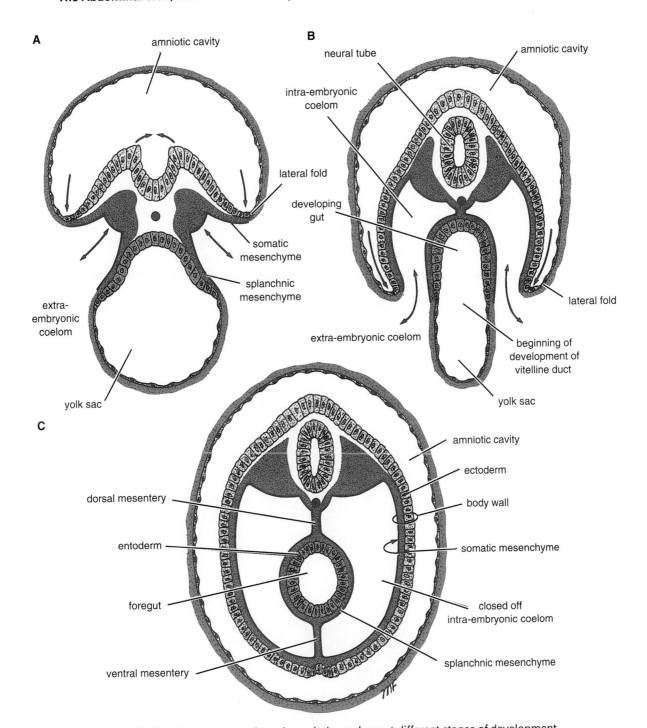

**Figure 19-19**   Transverse sections through the embryo at different stages of development showing the formation of the abdominal wall and peritoneal cavity. **A.** The intraembryonic coelom in free communication with the extraembryonic coelom (*double-headed arrows*). **B.** The development of the lateral folds of the embryo and the beginning of the closing off of the intraembryonic coelom. **C.** The lateral folds of the embryo finally fused in the midline and closing off the intraembryonic coelom or future peritoneal cavity. Most of the ventral mesentery will break down and disappear.

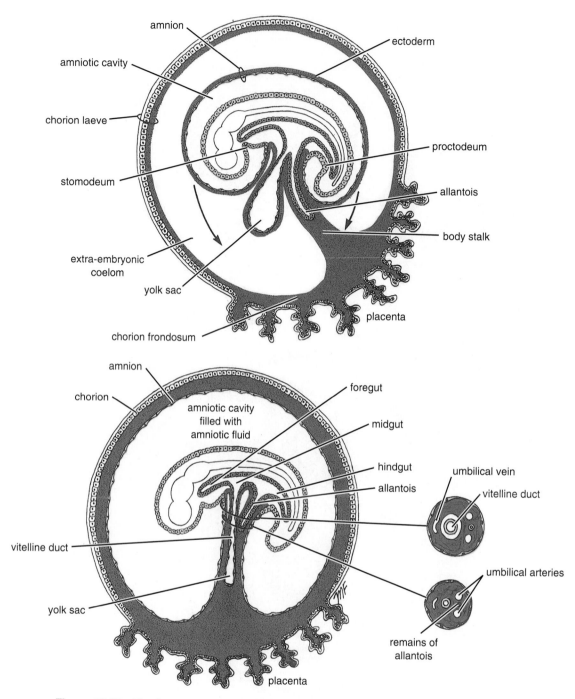

**Figure 19-20** The formation of the umbilical cord. Note the expansion of the amniotic cavity (*arrows*) so that the cord becomes covered with amnion. Note also that the umbilical vessels have been reduced to one vein and two arteries.

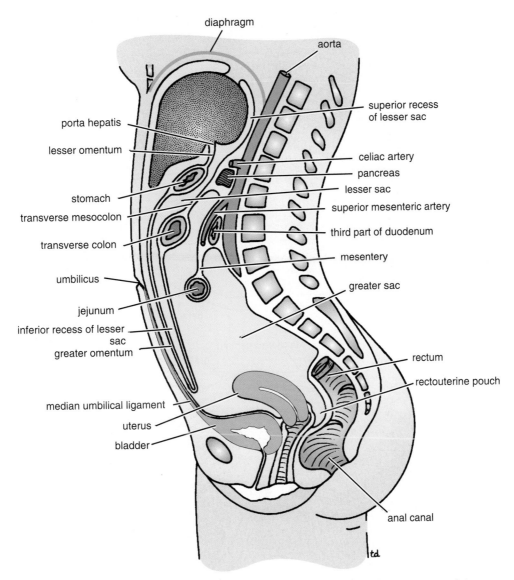

**Figure 19-21** Sagittal section of the female abdomen showing the arrangement of the peritoneum.

## Mesenteries

Mesenteries are two-layered folds of peritoneum connecting parts of the intestines with the posterior abdominal wall (e.g., the **mesentery of the small intestine,** the **transverse mesocolon,** and the **sigmoid mesocolon**) (Fig. 19-21).

The extent of the peritoneum and the peritoneal cavity should be studied in the transverse and sagittal sections of the abdomen seen in Figures 19-21 and 19-22.

## Lesser Sac

The lesser sac lies behind the stomach and the lesser omentum (Figs. 19-21 and 19-22). It extends upward as far as the diaphragm and downward between the layers of the greater omentum. The left margin of the sac is formed by the spleen, the gastrosplenic omentum, and the splenicorenal ligament. The right margin opens into the greater sac (the main part of the peritoneal cavity) through the **epiploic foramen** (Figs. 19-23 and 19-24).

## Boundaries of the Epiploic Foramen

- **Anteriorly:** Free border of the lesser omentum, the bile duct, the hepatic artery, and the portal vein.
- **Posteriorly:** Inferior vena cava.
- **Superiorly:** Caudate process of the caudate lobe of the liver.
- **Inferiorly:** First part of the duodenum.

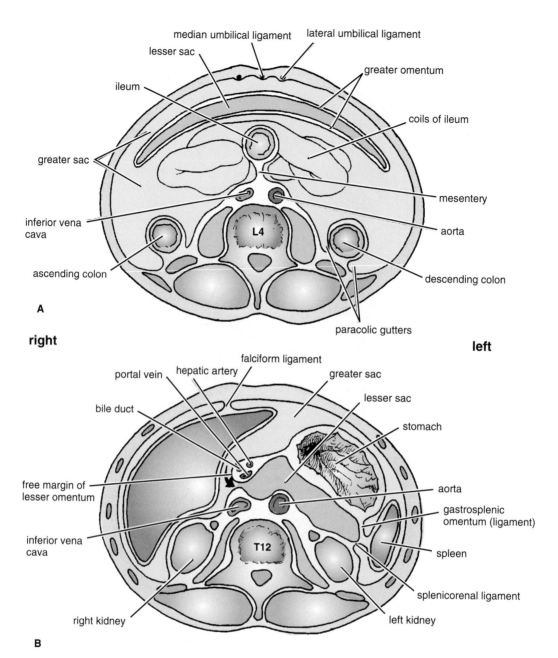

**right**                                                                    **left**

**Figure 19-22**   Transverse sections of the abdomen showing the arrangement of the peritoneum. The *arrow* in B indicates the position of the opening of the lesser sac. These sections are viewed from below.

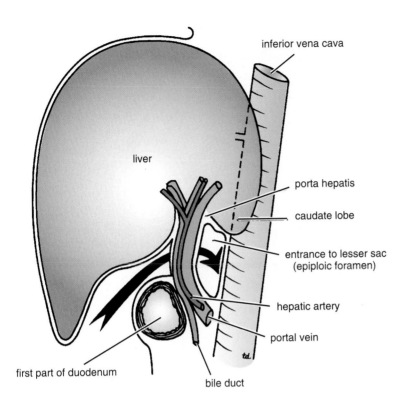

**Figure 19-23** Sagittal section through the entrance into the lesser sac showing the important structures that form boundaries to the opening. (Note the *arrow* passing from the greater sac through the epiploic foramen into the lesser sac.)

## Peritoneal Recesses, Spaces, and Gutters

### Duodenal Recesses

Close to the duodenojejunal junction, there may be four small pouches of peritoneum called the **superior duodenal recess,** the **inferior duodenal recess,** the **paraduodenal recess,** and the **retroduodenal recess** (Fig. 19-25).

### Cecal Recesses

Folds of peritoneum close to the cecum produce three peritoneal recesses called the **superior ileocecal recess,** the **inferior ileocecal recess,** and the **retrocecal recess** (Fig. 19-26).

### Intersigmoid Recess

The intersigmoid recess is situated at the apex of the inverted, V-shaped root of the sigmoid mesocolon (Fig. 19-26). Its mouth opens downward and lies in front of the left ureter.

### Subphrenic Spaces

The **right and left anterior subphrenic spaces** lie between the diaphragm and the liver on each side of the falciform ligament (Fig. 19-27). The **right posterior subphrenic space** lies between the right lobe of the liver, the right kidney, and the right colic flexure. The **right extraperitoneal space** lies

between the layers of the coronary ligament and is, therefore, situated between the liver and the diaphragm.

## Paracolic Gutters

The paracolic gutters lie on the lateral and medial sides of the ascending and descending colons, respectively (Figs. 19-22 and 19-27).

The subphrenic spaces and the paracolic gutters are clinically important because they may be sites for the collection and movement of infected peritoneal fluid.

## Nerve Supply of the Peritoneum

The **parietal peritoneum** is supplied for the sensations of pain, temperature, touch, and pressure by the lower six thoracic and first lumbar nerves. The parietal peritoneum in the pelvis is mainly supplied by the obturator nerve.

The visceral peritoneum is supplied for the sensation of stretch only by autonomic nerves that supply the viscera or that are traveling in the mesenteries.

**PHYSIOLOGIC NOTE**

### Functions of the Peritoneum

The peritoneal fluid, which is pale yellow and somewhat viscid, contains leukocytes. It is secreted by the peritoneum and ensures that the mobile viscera glide easily

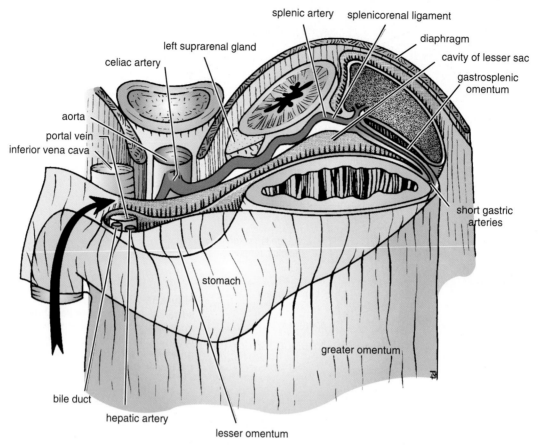

**Figure 19-24** Transverse section of the lesser sac showing the arrangement of the peritoneum in the formation of the lesser omentum, the gastrosplenic omentum, and the splenicorenal ligament. *Arrow* indicates the position of the opening of the lesser sac.

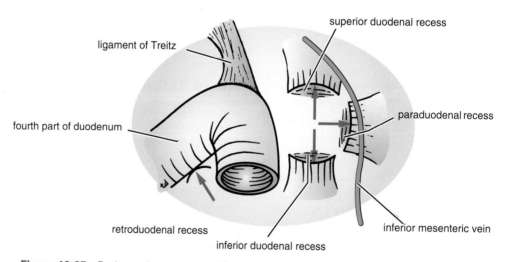

**Figure 19-25** Peritoneal recesses, which may be present in the region of the duodenojejunal junction. Note the presence of the inferior mesenteric vein in the peritoneal fold, forming the paraduodenal recess.

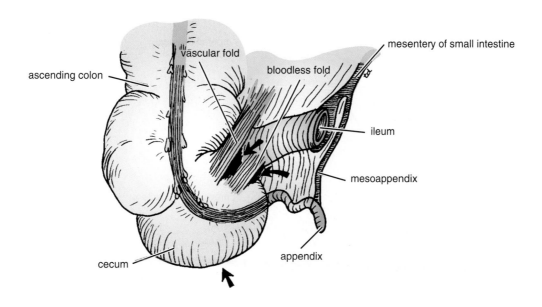

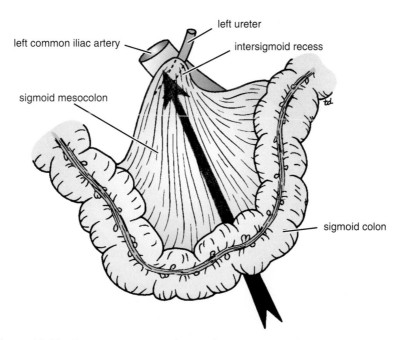

**Figure 19-26**   Peritoneal recesses (*arrows*) in the region of the cecum and the recess related to the sigmoid mesocolon.

on one another. As a result of the movements of the diaphragm and the abdominal muscles, together with the peristaltic movements of the intestinal tract, the peritoneal fluid is not static. Experimental evidence has shown that particulate matter introduced into the lower part of the peritoneal cavity reaches the subphrenic peritoneal spaces rapidly, whatever the position of the body. It seems that intraperitoneal movement of fluid toward the diaphragm is continuous (Fig. 19-27), and there, it is

quickly absorbed into the subperitoneal lymphatic capillaries. This can be explained on the basis that the area of peritoneum is extensive in the region of the diaphragm and the respiratory movements of the diaphragm aid lymph flow in the lymph vessels.

The peritoneal coverings of the intestine tend to stick together in the presence of infection. The greater omentum, which is kept constantly on the move by the peristalsis of the neighboring intestinal tract, may adhere to

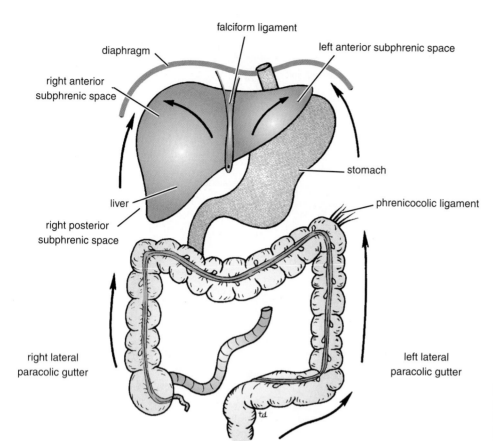

right anterior
subphrenic space

diaphragm

falciform ligament

left anterior subphrenic space

liver

right posterior
subphrenic space

stomach

phrenicocolic ligament

right lateral
paracolic gutter

left lateral
paracolic gutter

**Figure 19-27**   Normal direction
of flow of the peritoneal fluid
from different parts of the
peritoneal cavity to the
subphrenic spaces.

other peritoneal surfaces around a focus of infection. In this manner, many of the intraperitoneal infections are sealed off and remain localized.

The peritoneal folds play an important part in suspending the various organs within the peritoneal cavity and serve as a means of conveying the blood vessels, lymphatics, and nerves to these organs.

Large amounts of fat are stored in the peritoneal ligaments and mesenteries, and especially large amounts can be found in the greater omentum.

## E M B R Y O L O G I C   N O T E

### Development of the Peritoneum and the Peritoneal Cavity

Once the lateral mesoderm has split into somatic and splanchnic layers, a cavity is formed between the two, called the intraembryonic coelom. The peritoneal cavity is derived from that part of the embryonic coelom situated caudal to the septum transversum. In the earliest stages, the peritoneal cavity is in free communication with the extraembryonic coelom on each side (Fig. 19-19B). Later, with the development of the head, tail, and lateral folds of

the embryo, this wide area of communication becomes restricted to the small area within the umbilical cord.

Early in development, the peritoneal cavity is divided into right and left halves by a central partition formed by the dorsal mesentery, the gut, and small ventral mesentery (Fig. 19-28). However, the ventral mesentery extends only for a short distance along the gut (see below), so that below this level, the right and left halves of the peritoneal cavity are in free communication (Fig. 19-28). As a result of the enormous growth of the liver and the enlargement of the developing kidneys, the capacity of the abdominal cavity becomes greatly reduced at about the 6th week of development. It is at this time that the small remaining communication between the peritoneal cavity and extraembryonic coelom becomes important. An intestinal loop is forced out of the abdominal cavity through the umbilicus into the umbilical cord. This physiologic herniation of the midgut takes place during the 6th week of development.

### Formation of the Peritoneal Ligaments and Mesenteries

The peritoneal ligaments are developed from the ventral and dorsal mesenteries. The ventral mesentery is formed from the mesoderm of the septum transversum (derived from the cervical somites, which migrate downward).

The ventral mesentery forms the falciform ligament, the lesser omentum, and the coronary and triangular ligaments of the liver (Fig. 19-28).

The dorsal mesentery is formed from the fusion of the splanchnopleuric mesoderm on the two sides of the embryo. It extends from the posterior abdominal wall to the posterior border of the abdominal part of the gut (Figs. 19-19 and 19-28). The dorsal mesentery forms the gastrophrenic ligament, the gastrosplenic omentum, the splenicorenal ligament, the greater omentum, and the mesenteries of the small and large intestines.

### Formation of the Lesser and Greater Peritoneal Sacs

The extensive growth of the right lobe of the liver pulls the ventral mesentery to the right and causes rotation of the stomach and duodenum (Fig. 19-29). By this means, the upper right part of the peritoneal cavity becomes incorporated into the lesser sac. The right free border of the ventral mesentery becomes the right border of the lesser omentum and the anterior boundary of the entrance into the lesser sac.

The remaining part of the peritoneal cavity, which is not included in the lesser sac, is called the greater sac, and the two sacs are in communication through the **epiploic foramen.**

### Formation of the Greater Omentum

The spleen is developed in the upper part of the dorsal mesentery, and the greater omentum is formed as a result of the rapid and extensive growth of the dorsal mesentery caudal to the spleen. To begin with, the greater omentum extends from the greater curvature of the stomach to the posterior abdominal wall superior to the transverse mesocolon. With continued growth, it reaches inferiorly as an apron-like double layer of peritoneum anterior to the transverse colon.

Later, the posterior layer of the omentum fuses with the transverse mesocolon; as a result, the greater omentum becomes attached to the anterior surface of the transverse colon (Fig. 19-29). As development proceeds, the omentum becomes laden with fat. The inferior recess of the lesser sac extends inferiorly between the anterior and the posterior layers of the fold of the greater omentum.

## Retroperitoneal Space

The retroperitoneal space lies on the posterior abdominal wall behind the parietal peritoneum. It extends from the twelfth thoracic vertebra and the twelfth rib to the sacrum and the iliac crests below (Fig. 19-30).

The floor or posterior wall of the space is formed from medial to lateral by the psoas and quadratus lumborum muscles and the origin of the transversus abdominis muscle. Each of these muscles is covered on the anterior surface by a definite layer of fascia. In front of the fascial layers is a variable amount of fatty connective tissue that forms a bed for the suprarenal glands, the

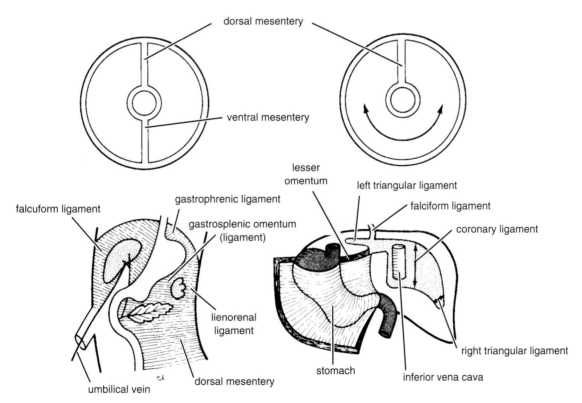

**Figure 19-28** Ventral and dorsal mesenteries and the organs that develop within them.

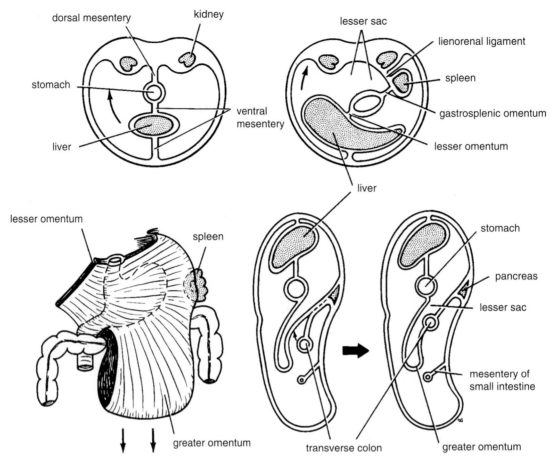

**Figure 19-29** The rotation of the stomach and the formation of the greater omentum and lesser sac.

kidneys, the ascending and descending parts of the colon, and the duodenum. The retroperitoneal space also contains the ureters and the renal and gonadal blood vessels.

# The Alimentary Tract

The mouth cavity, pharynx, and related structures are described in Chapter 2.

## Esophagus

The esophagus is a muscular tube about 10 in. (25 cm) long, extending from the pharynx to the stomach (Fig. 19-31). It begins at the level of the cricoid cartilage in the neck and descends in the midline behind the trachea. In the thorax, it passes downward through the mediastinum and enters the abdominal cavity by piercing the diaphragm at the level of the tenth thoracic vertebra. The esophagus has a short course of about ½ in. (1.25 cm) before it enters the right side of the stomach.

## Esophagus in the Neck

### Relations (Fig. 19-32)
**Anteriorly:** Trachea, recurrent laryngeal nerves.
**Posteriorly:** The prevertebral muscles and the vertebral column.
**Laterally:** The thyroid gland, carotid sheath (common carotid artery, internal jugular vein, and vagus nerve), and on the left side, the thoracic duct.

### Blood Supply
**Arteries:** Inferior thyroid arteries.
**Veins:** Inferior thyroid veins.

### Lymph Drainage
Deep cervical lymph nodes.

### Nerve Supply
Recurrent laryngeal nerves and branches from the sympathetic trunks.

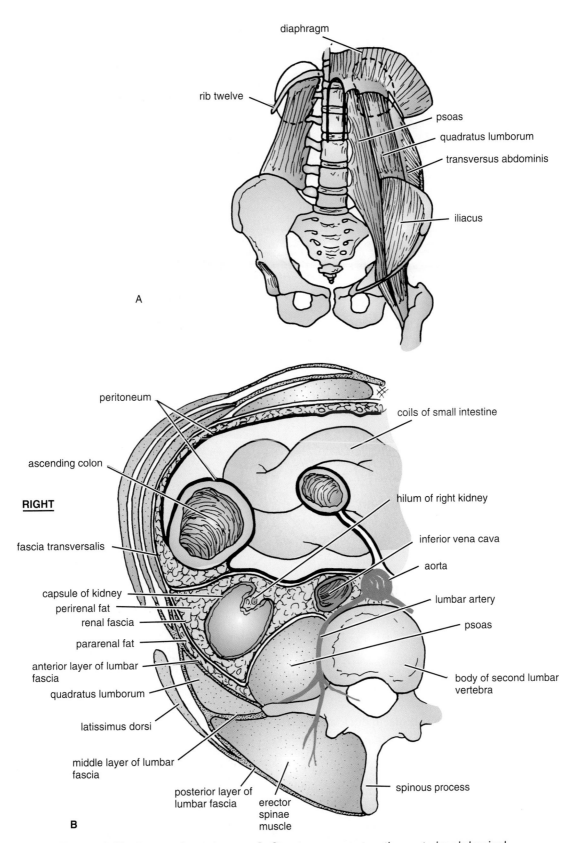

**Figure 19-30**  Retroperitoneal space. **A.** Structures present on the posterior abdominal wall behind the peritoneum. **B.** Transverse section of the posterior abdominal wall showing structures in the retroperitoneal space as seen from below.

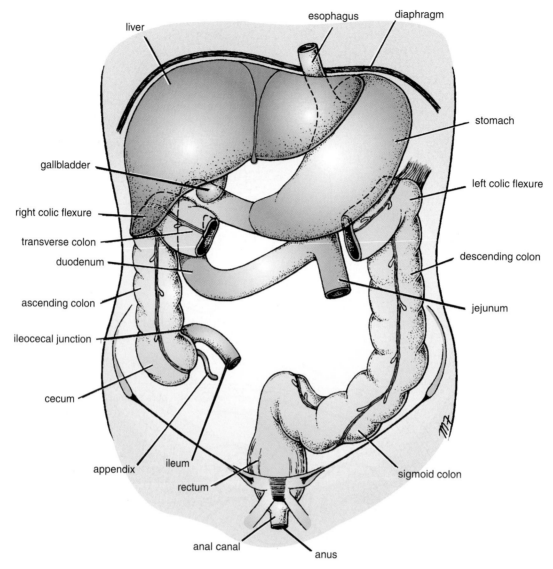

**Figure 19-31**    General arrangement of abdominal viscera.

## Esophagus in the Thorax

### Relations (Fig. 19-33)

**Anteriorly:** Trachea, left recurrent laryngeal nerve, left bronchus, left atrium of the heart.

**Posteriorly:** Vertebral column, thoracic duct, azygos veins, right posterior intercostal arteries, descending thoracic aorta.

**Laterally, right side:** Mediastinal pleura, azygos vein.

**Left side:** Aortic arch, left subclavian artery, thoracic duct, mediastinal pleura.

### Blood Supply

**Arteries:** Upper part from the descending thoracic aorta, lower third from the left gastric artery.

**Veins:** These drain into the azygos veins, and from the lower third, they drain into the **left gastric vein, a tributary of the portal vein.**

### Lymph Drainage

Upper part into the superior and posterior mediastinal nodes and, from the lower third, into nodes along the left gastric blood vessels and the celiac nodes in the abdomen.

### Nerve Supply

Vagal trunks (left vagus lies anterior and right vagus lies posterior), esophageal plexus, sympathetic trunks, greater splanchnic nerves.

## Esophagus in the Abdomen

The esophagus enters the abdomen through an opening in the right crus of the diaphragm (Fig. 19-31). After a course of about 0.5 in. (1.25 cm), it enters the stomach on its right side.

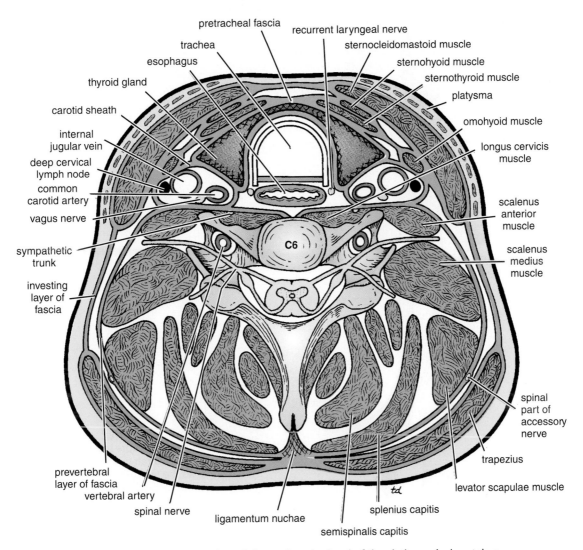

**Figure 19-32** Cross section of the neck at the level of the sixth cervical vertebra.

## Relations

**Anteriorly:** The esophagus lies posterior to the left lobe of the liver and in front of the left crus of the diaphragm. The left and right vagi lie on its anterior and posterior surfaces, respectively.

## Blood Supply

**Arteries:** Branches from the left gastric artery (see Fig. 19-39).

**Veins:** The left gastric vein, a tributary of the portal vein (see portal–systemic anastomosis).

## Lymph Drainage

The lymph vessels follow the arteries into the left gastric nodes.

## Nerve Supply

Anterior and posterior gastric nerves (vagi) and sympathetic branches of the thoracic part of the sympathetic trunk.

---

**PHYSIOLOGIC NOTE**

### Function of the Esophagus

The esophagus conducts food from the pharynx into the stomach. Wave-like contractions of the muscular coat, called peristalsis, propel the food onward.

---

**PHYSIOLOGIC NOTE**

### Gastroesophageal Sphincter

No anatomic sphincter exists at the lower end of the esophagus. However, the circular layer of smooth muscle in this region serves as a physiologic sphincter. As the food descends through the esophagus, relaxation of the muscle at the lower end occurs ahead of the peristaltic wave so that the food enters the stomach. The tonic contraction of this sphincter prevents the stomach contents from regurgitating into the esophagus.

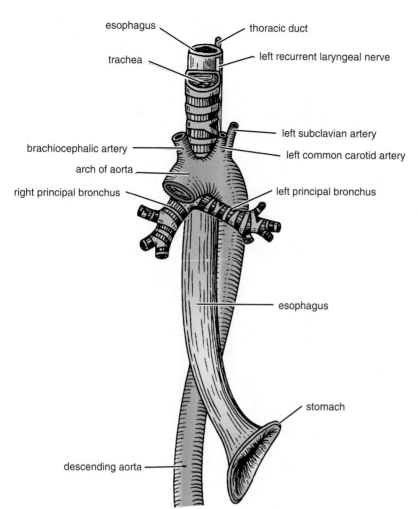

esophagus

thoracic duct

trachea

left recurrent laryngeal nerve

left subclavian artery

brachiocephalic artery

left common carotid artery

arch of aorta

right principal bronchus

left principal bronchus

esophagus

stomach

descending aorta

**Figure 19-33** Thoracic part of the esophagus. Note the position of the trachea and aorta relative to the esophagus. Note also the left principal bronchus crossing the anterior surface of the esophagus below the aortic arch.

The closure of the sphincter is under vagal control, and this can be augmented by the hormone gastrin and reduced in response to secretin, cholecystokinin, and glucagon.

# RADIOGRAPHIC APPEARANCES OF THE ESOPHAGUS

The radiographic appearances of the esophagus are shown in Figures 19-34 and 19-35.

## Stomach

The stomach is a dilated portion of the alimentary canal situated in the upper part of the abdomen (Fig. 19-31). It is roughly J shaped (Fig. 19-36), and it has two openings (the **cardiac** and **pyloric orifices**), two curvatures (the **greater** and the **lesser curvatures**), and two surfaces (an **anterior** and a **posterior surface**).

### PHYSIOLOGIC NOTE

### Function of the Stomach

The stomach is concerned with the storage and digestion of food.

The stomach may be divided into the following parts:

- **Fundus:** This is dome shaped and projects upward and to the left of the cardiac orifice. It is usually full of gas.
- **Body:** This extends from the cardiac orifice to the **incisura angularis** (a constant notch in the lower part of the lesser curvature).
- **Pyloric antrum:** This extends from the incisura angularis to the pylorus (Fig. 19-36).
- **Pylorus:** This is the most tubular part of the stomach. The thick, muscular wall is called the **pyloric sphincter**, and the cavity of the pylorus is called the **pyloric canal** (Fig. 19-36).

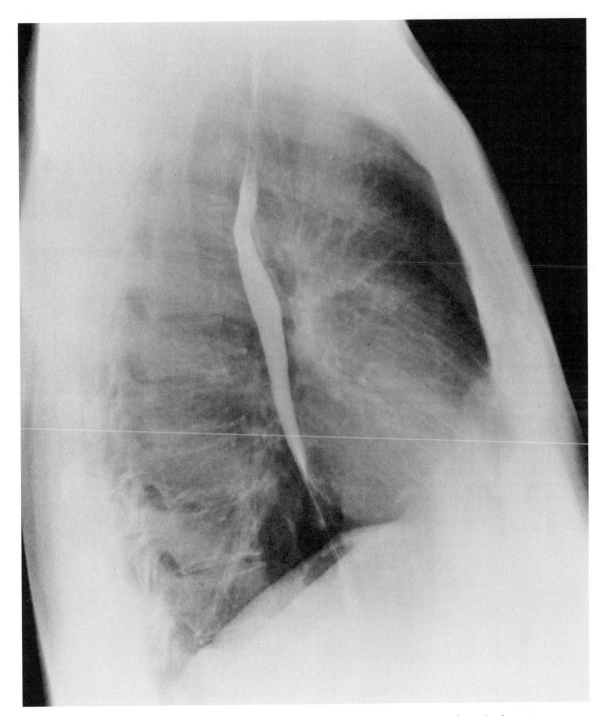

**Figure 19-34**   Left lateral radiograph of the chest of a normal adult man after a barium swallow.

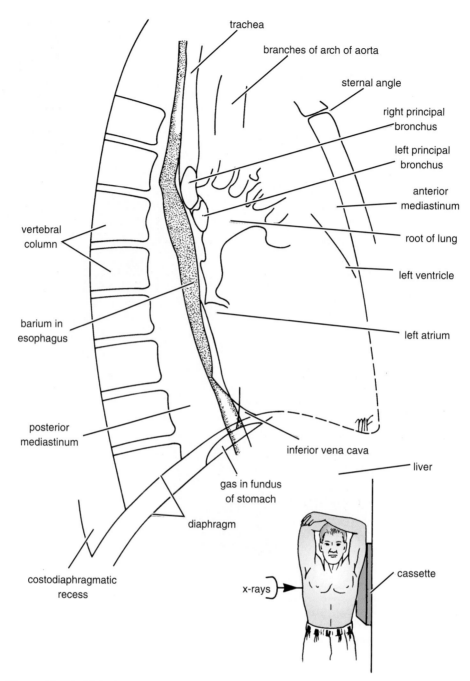

**Figure 19-35**   Main features observable in a left lateral radiograph of the chest shown in Figure 19-34. Note the position of the patient in relation to the x-ray source and cassette holder.

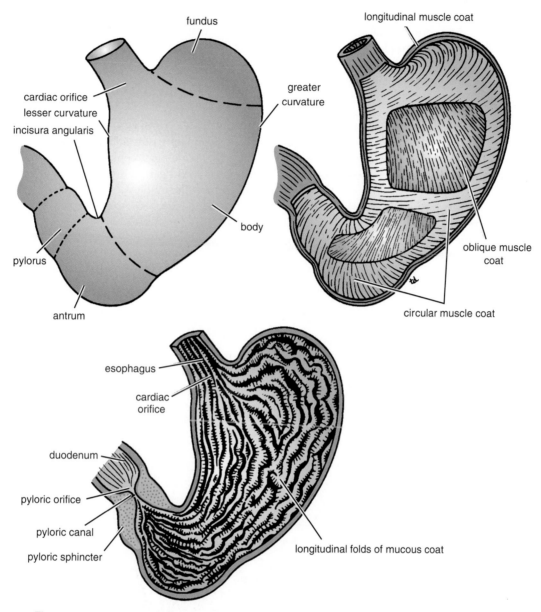

**Figure 19-36**  Stomach showing the parts, muscular coats, and mucosal lining. Note the increased thickness of the circular muscle forming the pyloric sphincter.

## PHYSIOLOGIC NOTE

### The Pyloric Sphincter

The pyloric sphincter controls the outflow of gastric contents into the duodenum. The sphincter receives motor fibers from the sympathetic system and inhibitory fibers from the vagi. In addition, the pylorus is controlled by local nervous and hormonal influences from the stomach and duodenal walls. For example, the stretching of the stomach due to filling will stimulate the myenteric nerve plexus in its wall and reflexly cause relaxation of the sphincter. The release of the hormone gastrin from the antral mucosa stimulates peristalsis in the stomach wall and thus promotes emptying of the stomach. In the duodenum, the presence of gastric contents stimulates local enteric reflexes, which inhibit the relaxation of the sphincter. The presence of fat in the duodenum causes the release of hormones, such as cholecystokinin, from the mucosa in the duodenum and jejunum, which inhibit gastric motility and thus slow the emptying of gastric contents into the duodenum.

The **lesser curvature** forms the right border of the stomach and is connected to the liver by the lesser omentum (Fig. 19-21). The **greater curvature** is much longer than the lesser curvature, and it extends from the left of the cardiac orifice over the dome of the fundus and along the left border of the stomach. The gastrosplenic omentum (ligament) extends from the upper part of the greater curvature to the spleen (Fig. 19-24). The greater omentum extends from the lower part of the greater curvature to the transverse colon (Fig. 19-21).

The esophagus enters the stomach at the **cardiac orifice** (Fig. 19-36). No anatomic sphincter can be demonstrated here, but a physiologic mechanism prevents the regurgitation of stomach contents into the esophagus (see p 721).

The **pyloric orifice** is formed by the pyloric canal (Fig. 19-36). The circular muscle coat of the stomach is much thicker here and forms the anatomic and physiologic pyloric sphincter.

## Relations of the Stomach

These vary with the degree of filling.

**Anteriorly:** Left costal margin, anterior abdominal wall, diaphragm, left pleura, base of the left lung, pericardium, quadrate and left lobes of liver (Fig. 19-37).

**Posteriorly:** Lesser sac, pancreas (body and tail), splenic artery, diaphragm, left suprarenal gland, and upper part of the left kidney, spleen, and transverse mesocolon (Fig. 19-38).

## Blood Supply

### Arteries
The right and left gastric arteries supply the lesser curvature. The right and left gastroepiploic arteries supply the greater curvature. Short gastric arteries derived from the splenic artery supply the fundus (Fig. 19-39).

### Veins
The veins drain into the portal circulation. The right and left gastric veins drain into the portal vein. The short gastric and the left gastroepiploic veins drain into the splenic vein, and the right gastroepiploic vein drains into the superior mesenteric vein.

## Lymph Drainage

The lymph vessels follow the arteries into the left and right gastric nodes, the left and right gastroepiploic nodes, and the

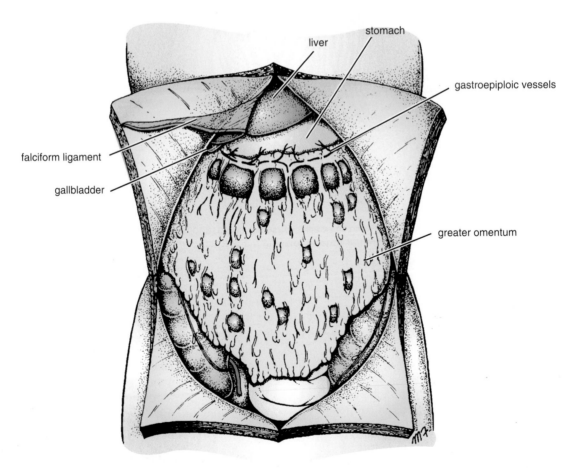

**Figure 19-37**   Abdominal organs in situ. Note that the greater omentum hangs down in front of the small and large intestines.

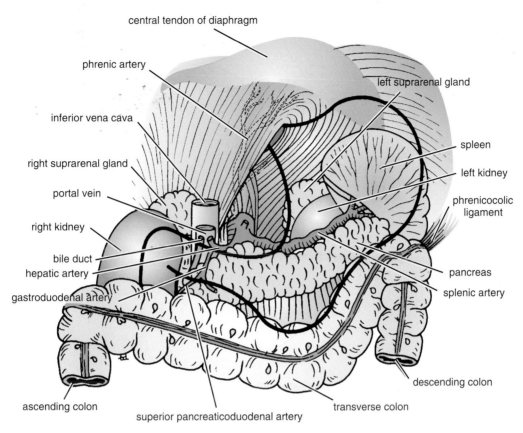

**Figure 19-38**  Structures situated on the posterior abdominal wall behind the stomach.

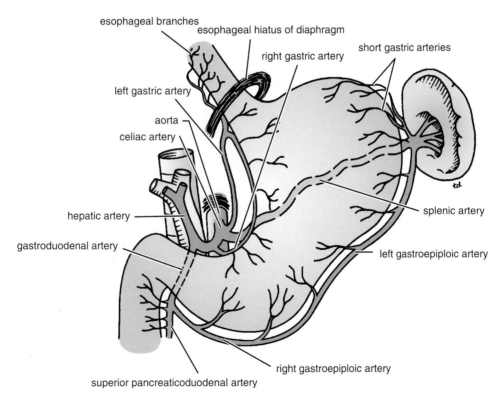

**Figure 19-39**  Arteries that supply the stomach. Note that all the arteries are derived from branches of the celiac artery.

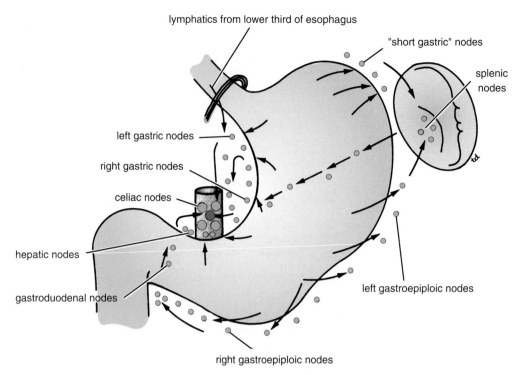

lymphatics from lower third of esophagus

"short gastric" nodes

splenic nodes

left gastric nodes

right gastric nodes

celiac nodes

hepatic nodes

gastroduodenal nodes

left gastroepiploic nodes

right gastroepiploic nodes

**Figure 19-40** Lymph drainage of the stomach. Note that all the lymph eventually passes through the celiac lymph nodes.

short gastric nodes. All lymph from the stomach eventually passes to the celiac nodes (Fig. 19-40).

## Nerve Supply

The sympathetic nerve supply is from the celiac plexus, and parasympathetic nerve supply is from the vagus nerves (Fig. 19-41).

# Cross-Sectional Anatomy of the Abdomen

To assist in interpretation of CT scans of the abdomen, study the labeled cross sections of the abdomen shown in Figures 19-42 and 19-43. The sections have been photographed on their **inferior surfaces.**

# RADIOGRAPHIC APPEARANCES OF THE STOMACH

The radiographic appearances of the stomach are shown in Figures 19-45 and 19-46. See Figure 19-44 for CT scan.

## Small Intestine

The small intestine extends from the pylorus of the stomach to the ileocecal junction (Fig. 19-31). It is divided into three parts: the duodenum, the jejunum, and the ileum.

---

**P H Y S I O L O G I C  N O T E**

**Function of the Small Intestine**

The greater part of digestion and food absorption occurs in the small intestine.

---

## Duodenum

The duodenum is a C-shaped tube approximately 10 in. (25 cm) in length that curves around the head of the pancreas (Fig. 19-47). The duodenum begins at the pyloric sphincter of the stomach, and it ends by becoming continuous with the jejunum. The first inch of the duodenum has the lesser omentum attached to its upper border and the greater omentum attached to its lower border. The remainder of the duodenum is retroperitoneal.

The duodenum is divided into four parts:

■ The **first part** runs upward and backwards on the transpyloric plane at the level of the first lumbar vertebra (Fig. 19-48).

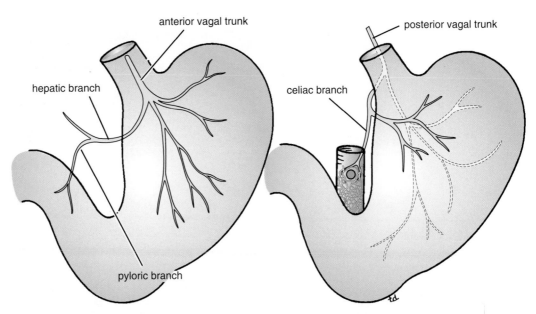

**Figure 19-41**    Distribution of the anterior and posterior vagal trunks within the abdomen. Note that the celiac branch of the posterior vagal trunk is distributed with the sympathetic nerves as far down the intestinal tract as the left colic flexure.

- The **second part** runs vertically downward (Fig. 19-48). The bile and main pancreatic ducts pierce the medial wall approximately halfway down, and they unite to form an ampulla that opens on the summit of a **major duodenal papilla** (Fig. 19-49). The accessory pancreatic duct (if present) opens into the duodenum on a **minor duodenal papilla**, approximately 0.75 in. (1.9 cm) above the major duodenal papilla.
- The **third part** passes horizontally in front of the vertebral column. The root of the mesentery of the small intestine and the superior mesenteric vessels cross this part anteriorly (Fig. 19-48).
- The **fourth part** runs upward and to the left to the **duodenojejunal flexure**. The flexure is held in position by the **ligament of Treitz**, which is attached to the right crus of the diaphragm.

### Relations

**First part:** Anteriorly: Quadrate lobe of the liver, gallbladder (Fig. 19-50). Posteriorly: lesser sac (first inch only), gastroduodenal artery, bile duct, portal vein, inferior vena cava (Fig. 19-48).

**Second part:** Anteriorly: fundus of the gallbladder (Fig. 19-50), right lobe of the liver, transverse colon, coils of the small intestine. Posteriorly: hilus of the right kidney (Fig. 19-48). Medially: head of the pancreas, bile duct, and pancreatic ducts (Figs. 19-47 and 19-48).

**Third part:** Anteriorly: root of the mesentery of the small intestine, superior mesenteric vessels, coils of the jejunum (Fig. 19-48). Posteriorly: Right ureter, inferior vena cava, aorta (Fig. 19-48). Superiorly: head of the pancreas (Fig. 19-47).

**Fourth part:** Anteriorly: beginning of the root of the mesentery and coils of the jejunum. Posteriorly: left margin of the aorta (Fig. 19-48).

### Blood Supply

#### Arteries

The upper half of the duodenum is supplied by the superior pancreaticoduodenal artery, which is a branch of the gastroduodenal artery. The lower half is supplied by the inferior pancreaticoduodenal artery, which is a branch of the superior mesenteric artery (Fig. 19-47).

#### Veins

The superior pancreaticoduodenal vein joins the portal vein. The inferior pancreaticoduodenal vein joins the superior mesenteric vein.

### Lymph Drainage

The lymph vessels drain upward via the pancreaticoduodenal nodes to the gastroduodenal nodes and the celiac nodes. They drain downward via the pancreaticoduodenal nodes to the superior mesenteric nodes.

### Nerve Supply

The duodenum is supplied by the sympathetic and vagus nerves via the celiac and the superior mesenteric plexuses.

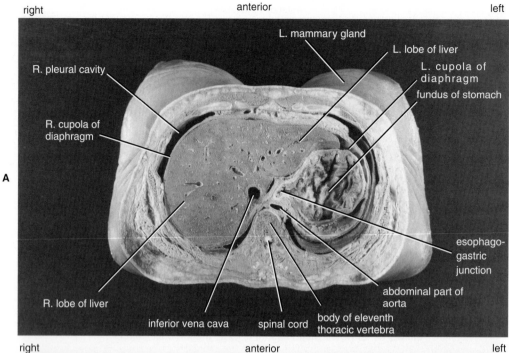

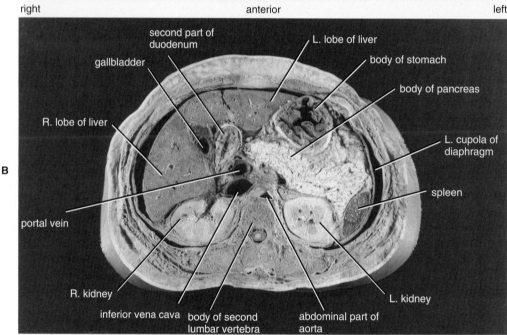

**Figure 19-42  A.** Cross section of the abdomen at the level of the body of the eleventh thoracic vertebra, viewed from below. Note that the large size of the pleural cavity is an artifact caused by the embalming process. **B.** Cross section of the abdomen at the level of the body of the second lumbar vertebra, viewed from below.

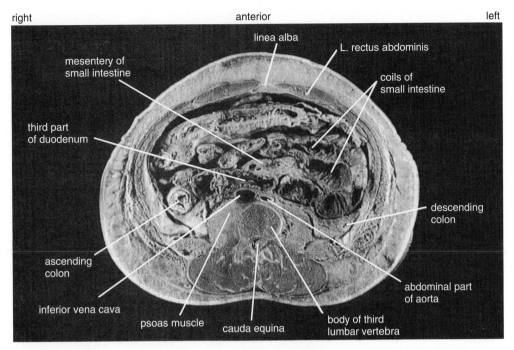

**Figure 19-43** Cross section of the abdomen at the level of the body of the third lumbar vertebra, viewed from below.

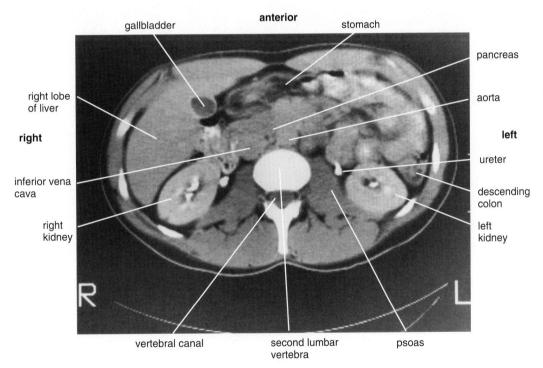

**Figure 19-44** CT scan of the abdomen at the level of the second lumbar vertebra after intravenous pyelography. The radiopaque material can be seen in the renal pelvis and the ureters. The section is viewed from below.

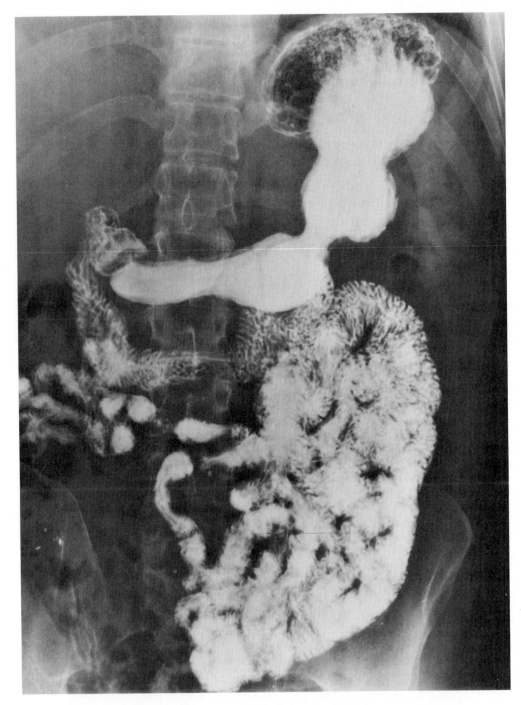

**Figure 19-45**  Anteroposterior radiograph of the stomach and the small intestine after ingestion of barium meal.

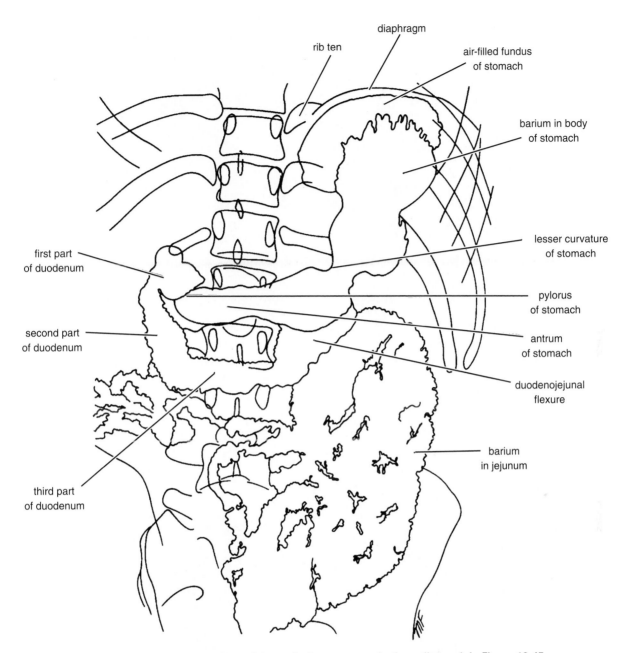

**Figure 19-46** Representation of the main features seen in the radiograph in Figure 19-45.

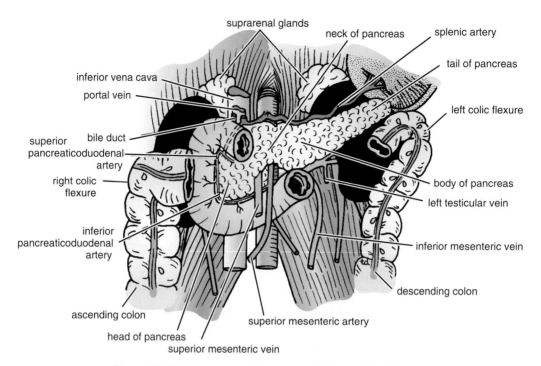

**Figure 19-47** Pancreas and anterior relations of the kidneys.

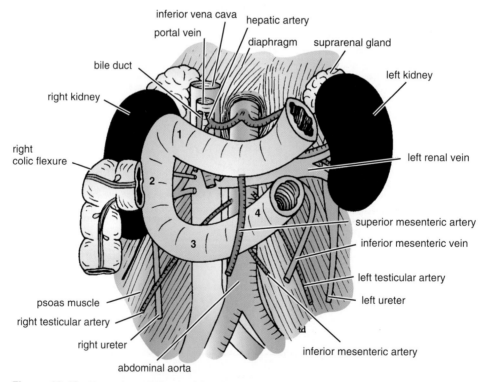

**Figure 19-48** Posterior relations of the duodenum and the pancreas. The *numbers* represent the four parts of the duodenum.

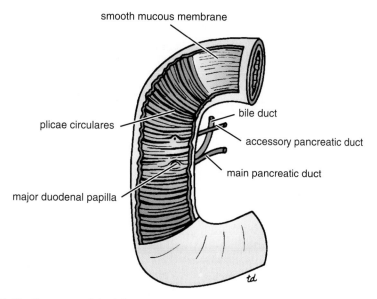

**Figure 19-49** Entrance of the bile duct and the main and accessory pancreatic ducts into the second part of the duodenum. Note the smooth lining of the first part of the duodenum, the plicae circulares of the second part, and the major duodenal papilla.

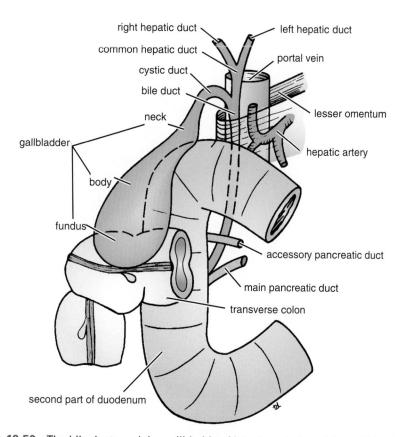

**Figure 19-50** The bile ducts and the gallbladder. Note the relation of the gallbladder to the transverse colon and the duodenum.

# RADIOGRAPHIC APPEARANCES OF THE DUODENUM

The radiographic appearances of the duodenum are shown in Figures 19-45, 19-46, and 19-51.

## Jejunum and Ileum

The jejunum measures approximately 8 ft (2.5 m) long, and the ileum measures approximately 12 ft (3.6 m) long. The jejunum begins at the duodenojejunal flexure (Figs. 19-31 and 19-52) in the upper part of the abdominal cavity and to the left of the midline. It is wider in diameter, thicker walled, and redder in color (more vascular) than the ileum.

The coils of the ileum occupy the lower right part of the abdominal cavity (Fig. 19-53) and tend to hang down into the pelvis. The ileum ends at the ileocecal junction.

The coils of the jejunum and the ileum are suspended from the posterior abdominal wall by a fan-shaped fold of peritoneum called the **mesentery of the small intestine** (Fig. 19-52).

### Relations
**Anteriorly:** Anterior abdominal wall and greater omentum, which usually covers over the coils (Fig. 19-53).
**Posteriorly:** Posterior abdominal wall and retroperitoneal structures.

### Blood Supply

#### Arteries
Branches of the superior mesenteric artery (Fig. 19-54).

### Veins
The veins drain into the superior mesenteric vein.

### Lymph Drainage
The lymph passes to the superior mesenteric nodes via intermediate nodes.

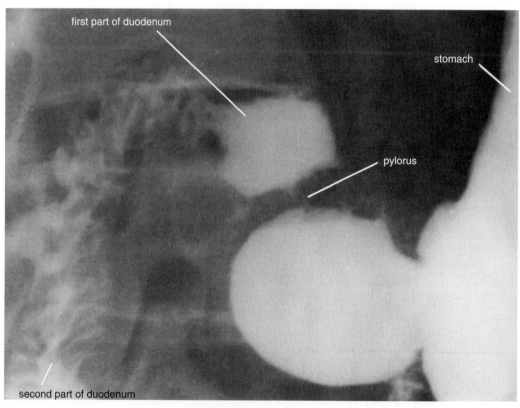

**Figure 19-51** Anteroposterior radiograph of the duodenum after ingestion of barium meal.

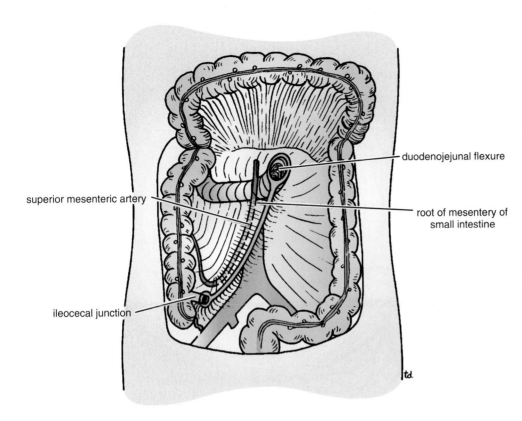

duodenojejunal flexure

superior mesenteric artery

root of mesentery of small intestine

ileocecal junction

**Figure 19-52** Attachment of the root of the mesentery of the small intestine to the posterior abdominal wall. Note that it extends from the duodenojejunal flexure on the left of the aorta downward and to the right to the ileocecal junction. The superior mesenteric artery lies in the root of the mesentery.

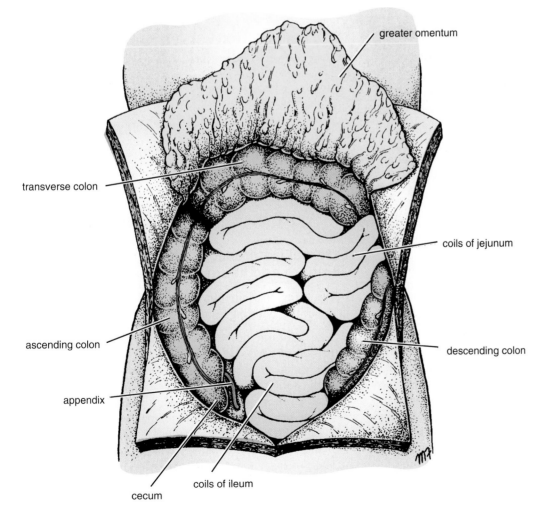

greater omentum

transverse colon

coils of jejunum

ascending colon

descending colon

appendix

cecum

coils of ileum

**Figure 19-53** Abdominal contents after the greater omentum has been reflected upward. Coils of small intestine occupy the central part of the abdominal cavity, whereas ascending, transverse, and descending parts of the colon are located at the periphery.

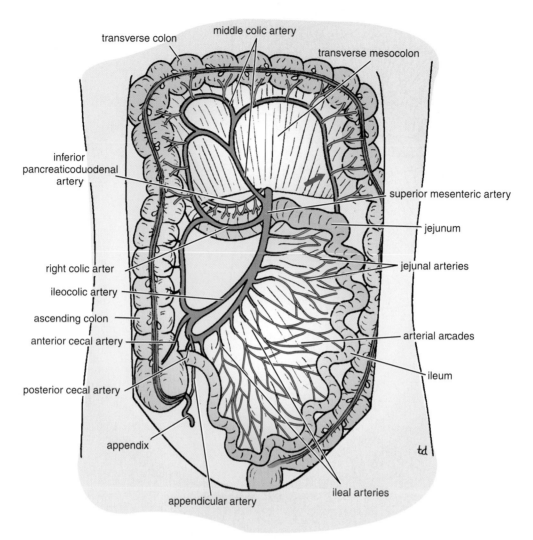

**Figure 19-54** Superior mesenteric artery and its branches. Note that this artery supplies blood to the gut from halfway down the second part of the duodenum to the distal third of the transverse colon (*arrow*).

### Nerve Supply

Sympathetic and vagus nerve fibers arise from the superior mesenteric plexus.

## Mesentery of the Small Intestine

The coils of jejunum and ileum are freely mobile and are attached to the posterior abdominal wall by a fan-shaped fold of peritoneum known as the mesentery of the small intestine (Fig. 19-52). The long free edge of the mesentery is attached to the mobile intestine. The short fixed root of the mesentery is attached to the peritoneum on the posterior abdominal wall along a line that extends downward and to the right from the left side of the second lumbar vertebra to the region of the right sacroiliac joint. The root of the mesentery permits the entrance and exit of the branches of the superior mesenteric artery and vein, lymph vessels, and nerves into the mesentery.

## External Differences between the Jejunum and Ileum

In the living, the jejunum can be distinguished from the ileum by the following features:

- The jejunum lies coiled in the upper part of the peritoneal cavity below the left side of the transverse mesocolon; the ileum is in the lower part of the cavity and in the pelvis (Fig. 19-53).
- The jejunum is wider bored, thicker walled, and redder than the ileum. The jejunal wall feels thicker because the permanent infoldings of the mucous membrane, the plicae circulares, are larger, more numerous, and closely set in the jejunum; whereas in the upper part of the ileum, they are smaller and more widely separated, and in the lower part, they are absent (Fig. 19-55).

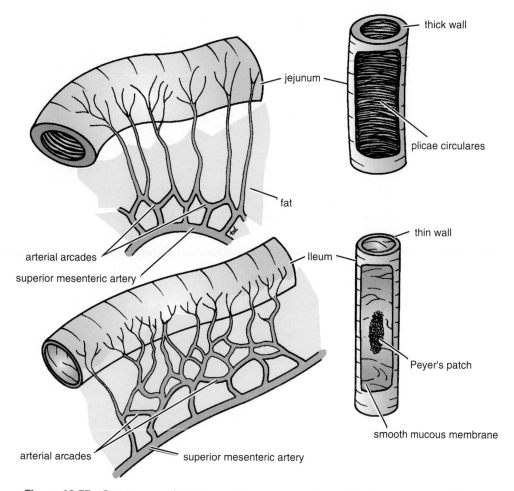

**Figure 19-55** Some external and internal differences between the jejunum and the ileum.

- The jejunal mesentery is attached to the posterior abdominal wall above and to the left of the aorta, whereas the ileal mesentery is attached below and to the right of the aorta.
- The jejunal mesenteric vessels form only one or two arcades, with long and infrequent branches passing to the intestinal wall. The ileum receives numerous short terminal vessels that arise from a series of three or four or even more arcades (Fig. 19-55).
- At the jejunal end of the mesentery, the fat is deposited near the root and is scanty near the intestinal wall. At the ileal end of the mesentery, the fat is deposited throughout so that it extends from the root to the intestinal wall (Fig. 19-55).
- Aggregations of lymphoid tissue (Peyer's patches) are present in the mucous membrane of the lower ileum along the antimesenteric border (Fig. 19-55). In the living, these may be visible through the wall of the ileum from the outside.

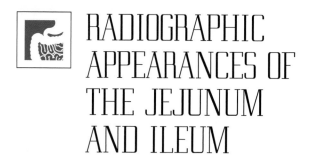

# RADIOGRAPHIC APPEARANCES OF THE JEJUNUM AND ILEUM

The radiographic appearances of the jejunum and ileum are shown in Figure 19-56.

## Large Intestine

The large intestine extends from the ileum to the anus (Fig. 19-31). It is divided into the cecum, the appendix, the ascending colon, the transverse colon, the descending colon, the sigmoid colon, the rectum, and the anal canal.

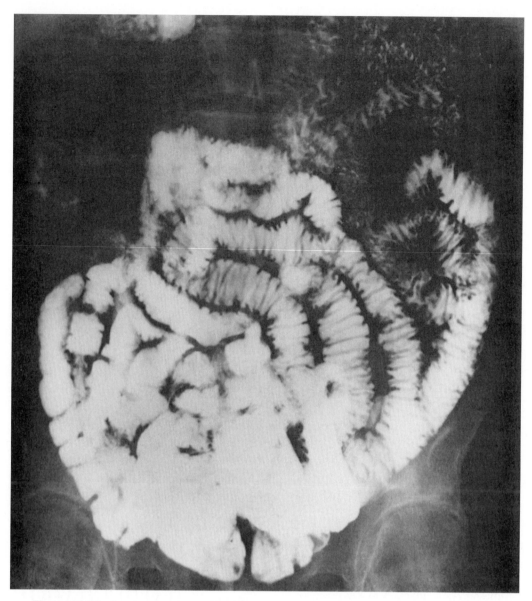

**Figure 19-56**   Anteroposterior radiograph of the small intestine after ingestion of barium meal.

### Function of the Large Intestine

The main functions of the large intestine include the absorption of water, production of certain vitamins, storage of undigested food materials, and formation and excretion of feces from the body.

## Cecum

The cecum is a blind-ended pouch within the right iliac fossa and is completely covered with peritoneum (Fig. 19-57). At the junction of the cecum and the ascending colon, it is joined on the left side by the terminal part of the ileum. The appendix is attached to its posteromedial surface (Fig. 19-58).

### Relations

**Anteriorly:** Anterior abdominal wall in the right iliac region, coils of small intestine.
**Posteriorly:** Iliopsoas muscle (Fig. 19-59).

### Blood Supply

#### Arteries

Anterior and posterior cecal arteries from the ileocolic artery (Fig. 19-58), which is a branch of the superior mesenteric artery.

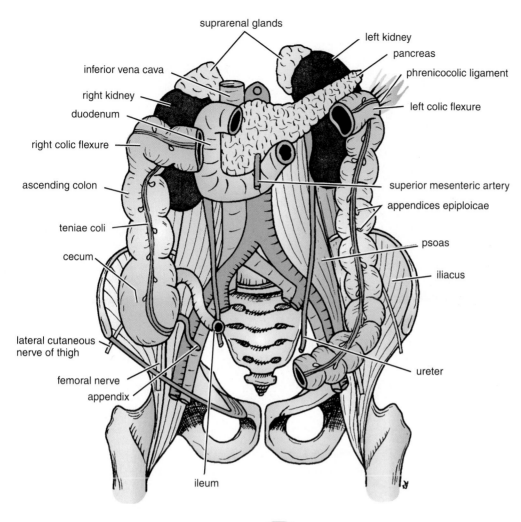

suprarenal glands

left kidney

pancreas

phrenicocolic ligament

inferior vena cava

right kidney

duodenum

left colic flexure

right colic flexure

ascending colon

superior mesenteric artery

appendices epiploicae

teniae coli

psoas

cecum

iliacus

lateral cutaneous
nerve of thigh

ureter

femoral nerve

appendix

ileum

**Figure 19-57** Abdominal cavity showing the terminal part of the ileum, the cecum, the appendix, the ascending colon, the right colic flexure, the left colic flexure, and the descending colon. Note the teniae coli and the appendices epiploicae.

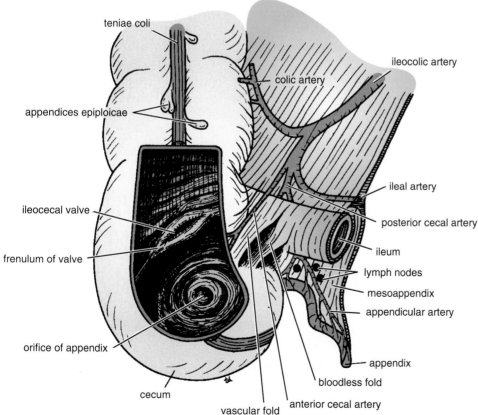

teniae coli

ileocolic artery

colic artery

appendices epiploicae

ileal artery

ileocecal valve

posterior cecal artery

ileum

frenulum of valve

lymph nodes

mesoappendix

appendicular artery

orifice of appendix

appendix

cecum

bloodless fold

vascular fold

anterior cecal artery

**Figure 19-58** Cecum and appendix. Note that the appendicular artery is a branch of the posterior cecal artery. The edge of the mesoappendix has been cut to show the peritoneal layers.

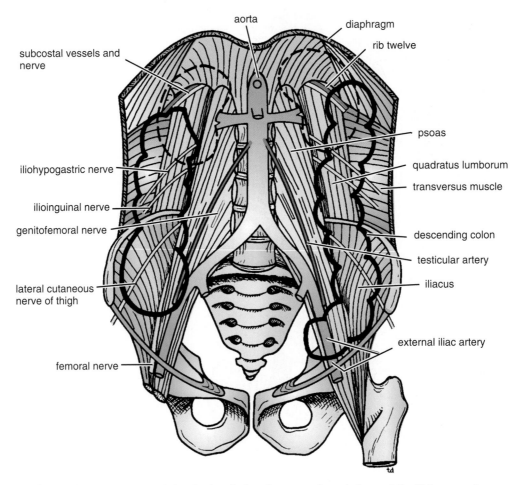

**Figure 19-59**  Posterior abdominal wall showing posterior relations of the kidneys and the colon.

*Veins*
The veins drain into the superior mesenteric vein.

**Lymph Drainage**
The lymph drains into the mesenteric and superior mesenteric nodes.

**Nerve Supply**
Sympathetic and vagus nerves, via the superior mesenteric plexus, supply the cecum.

## Ileocecal Valve

A rudimentary structure, the ileocecal valve consists of two horizontal folds of mucous membrane that project around the orifice of the ileum (Fig. 19-58).

---

**PHYSIOLOGIC NOTE**

### Function of the Ileocecal Valve

The ileocecal valve plays little or no part in preventing reflux of cecal contents into the ileum. The circular muscle at the lower end of the ileum (the ileocecal sphincter) serves as a sphincter and controls the flow of contents from the ileum into the colon. The smooth muscle tone is reflexly increased when the cecum is distended; the hormone gastrin, which is produced by the stomach, causes relaxation of the muscle tone.

# RADIOGRAPHIC APPEARANCES OF THE CECUM

The radiographic appearances of the cecum are shown in Figure 19-60.

## Appendix

The appendix (Figs. 19-57 and 19-58) is a narrow, muscular tube with a large amount of lymphoid tissue in its wall.

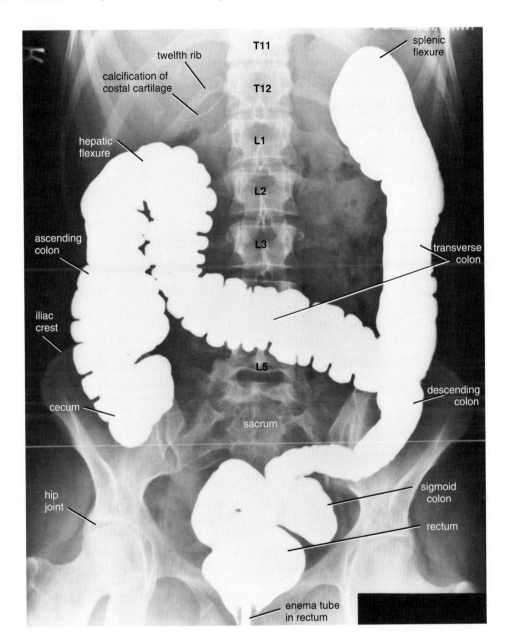

**Figure 19-60**   Anteroposterior radiograph of the large intestine after a barium enema.

It is attached to the posteromedial surface of the cecum approximately 1 in. (2.5 cm) below the ileocecal junction. It has a complete peritoneal covering, which is attached to the mesentery of the small intestine by a short mesentery of its own called the **mesoappendix**. The mesoappendix contains the appendicular vessels and nerves (Fig. 19-58).

The appendix lies in the right iliac fossa, and in relation to the anterior abdominal wall, its base is situated one third of the way up the line joining the right anterior superior iliac spine to the umbilicus (McBurney's point). Inside the abdomen, the base of the appendix is easily recognized

tracing the teniae coli of the cecum and then following them to the appendix, where they converge to form a continuous muscle coat.

### Blood Supply

#### Arteries
Appendicular artery is a branch of the posterior cecal artery (Fig. 19-58).

#### Veins
The veins drain into the posterior cecal vein.

### Lymph Drainage

The lymph drains into nodes in the mesoappendix and eventually into the superior mesenteric lymph nodes.

### Nerve Supply

The appendix is supplied by the sympathetic and vagus nerves from the superior mesenteric plexus. Afferent nerve fibers concerned with the conduction of visceral pain from the appendix accompany the sympathetic nerves and enter the spinal cord at the level of the tenth thoracic segment.

## Ascending Colon

The ascending colon is approximately 5 in. (13 cm) in length and extends upward from the cecum to the inferior surface of the right lobe of the liver (Figs. 19-31 and 19-57). Here, it turns to the left (forming the **right colic flexure**) and becomes continuous with the transverse colon. The peritoneum covers the front and the sides of the ascending colon, binding it to the posterior abdominal wall. The ascending colon lies posteriorly on the iliacus, quadratus lumborum, and the lower pole of the right kidney (Fig. 19-59).

### Blood Supply

#### Arteries

The ascending colon is supplied by the ileocolic and the right colic branches of the superior mesenteric artery (Fig. 19-54).

#### Veins

The veins drain into the superior mesenteric vein.

### Lymph Drainage

The lymph drains into the colic and superior mesenteric lymph nodes.

### Nerve Supply

Sympathetic and vagus nerves from the superior mesenteric plexus supply the ascending colon.

## Transverse Colon

The transverse colon is approximately 15 in. (38 cm) in length and passes across the abdomen, occupying the umbilical and the hypogastric regions (Fig. 19-31). It begins at the right colic flexure below the right lobe of the liver and hangs downward, suspended by the transverse mesocolon from the pancreas. It then ascends to the left colic flexure below the spleen. The **left colic flexure** is higher than the right colic flexure and is held in position by the **phrenicocolic ligament**. The **transverse mesocolon** (or mesentery of the transverse colon) is attached to the superior border of the transverse colon and suspends it from the pancreas (Fig. 19-21); the posterior layers of the greater omentum are attached to the inferior border.

### Blood Supply

#### Arteries

The proximal two thirds of the transverse colon is supplied by the middle colic artery (Fig. 19-54), which is a branch of the superior mesenteric artery. The distal one third is supplied by the left colic artery, which is a branch of the inferior mesenteric artery (Fig. 19-61).

#### Veins

The veins drain into the superior and the inferior mesenteric veins.

### Lymph Drainage

The proximal two thirds drain into the colic nodes and into the superior mesenteric nodes. The distal one third drains into the colic nodes and then the inferior mesenteric nodes.

### Nerve Supply

The proximal two thirds are innervated by the sympathetic and the vagal nerves through the superior mesenteric plexus. The distal one third is innervated by the sympathetic and the parasympathetic pelvic splanchnic nerves through the inferior mesenteric plexus.

## Descending Colon

The descending colon is approximately 10 in. (25 cm) in length and extends downward from the left colic flexure to the pelvic brim, where it becomes continuous with the sigmoid colon (Fig. 19-31). The peritoneum covers the front and the sides and also binds it to the posterior abdominal wall. The descending colon lies posteriorly on the left kidney, the quadratus lumborum, and the iliacus muscles (Fig. 19-59).

### Blood Supply

#### Arteries

The left colic branch and sigmoid branches of the inferior mesenteric artery (Fig. 19-61) supply the descending colon.

#### Veins

The veins drain into the inferior mesenteric vein.

### Lymph Drainage

The lymph passes to the colic and inferior mesenteric nodes.

### Nerve Supply

Sympathetic and parasympathetic pelvic splanchnic nerves through the inferior mesenteric plexus supply the descending colon.

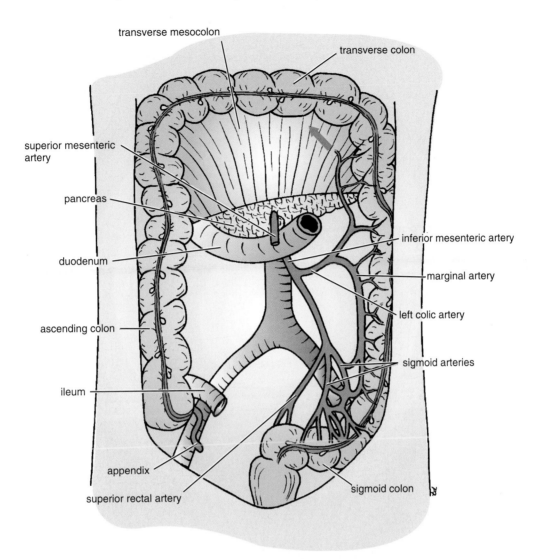

**Figure 19-61**  Inferior mesenteric artery and its branches. Note that this artery supplies the large bowel from the distal third of the transverse colon to halfway down the anal canal. It anastomoses with the middle colic branch of the superior mesenteric artery (*arrow*).

## Sigmoid Colon

The sigmoid colon is 10 to 15 in. (25 to 38 cm) in length and begins as a continuation of the descending colon in front of the pelvic brim (Fig. 19-31). Below, it becomes continuous with the rectum in front of the third sacral vertebra. It hangs down into the pelvic cavity in the form of a loop and is attached to the posterior pelvic wall by the fan-shaped **sigmoid mesocolon** (Fig. 19-62).

### Blood Supply

#### Arteries
Sigmoid branches of the inferior mesenteric artery (Fig. 19-61) supply the sigmoid colon.

#### Veins
The veins drain into the inferior mesenteric vein.

### Lymph Drainage
The lymph drains into the colic and inferior mesenteric nodes.

### Nerve Supply
Sympathetic and parasympathetic nerves through the inferior hypogastric plexuses supply the sigmoid colon.

## Rectum

The rectum is about 5 in. (13 cm) long and begins in front of the third sacral vertebra as a continuation of the sigmoid

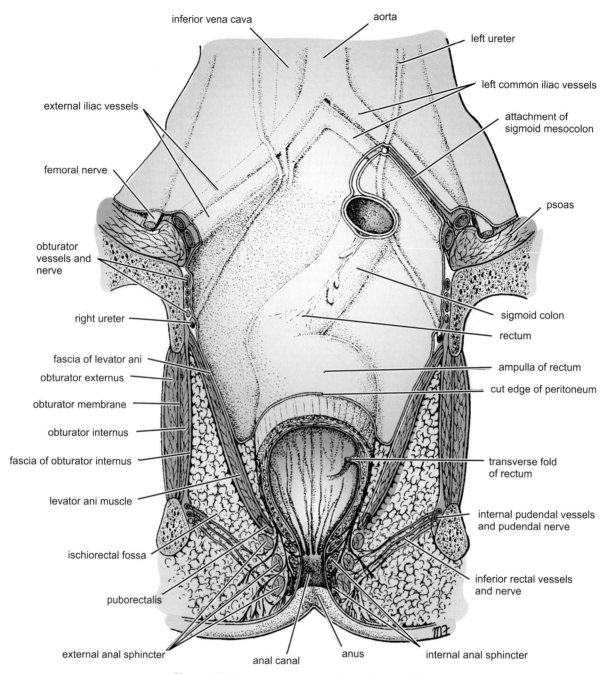

**Figure 19-62** Coronal section through the pelvis.

colon (Figs. 19-62, 19-63, and 19-64). It passes downward, following the curve of the sacrum and the coccyx, and ends in front of the tip of the coccyx by piercing the pelvic floor and becomes continuous with the anal canal. The lower part of the rectum lies immediately above the pelvic floor and is dilated to form the **rectal ampulla** (Fig. 19-63). The peritoneum covers only the upper two thirds of the rectum. The teniae coli of the sigmoid colon come together, so that the longitudinal muscle fibers form a broad band on the anterior and posterior surfaces of the rectum.

The mucous membrane of the rectum, together with the circular muscle layer, form three semicircular folds; two are placed on the left rectal wall, and one is placed on the right wall. They are called the **transverse folds of the rectum** (Fig. 19-63).

### Relations

**Anteriorly in the male:** Rectovesical pouch, sigmoid colon, coils of ileum, bladder, vas deferens, seminal vesicles, prostate (Fig. 19-64).

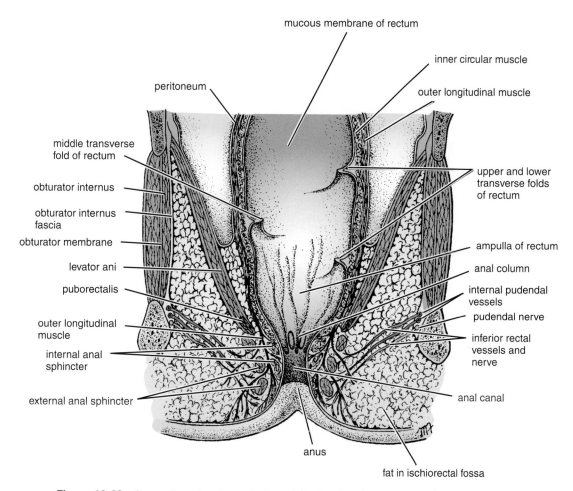

**Figure 19-63** Coronal section through the pelvis showing the rectum and the pelvic floor.

**Anteriorly in the female:** Rectouterine pouch (pouch of Douglas), vagina (Fig. 19-65).

**Posteriorly:** Sacrum, coccyx, piriformis, and coccygeus muscles, levatores ani muscles, sacral plexus, sympathetic trunks.

### Blood Supply

#### Arteries
The superior rectal artery, a branch of the inferior mesenteric artery, is the chief artery and supplies the mucous membrane (Fig. 19-61); the middle rectal artery, a branch of the internal iliac artery, supplies the muscle wall; the inferior rectal artery, a branch of the internal pudendal artery, supplies the muscle wall (Fig. 19-63).

#### Veins
The superior rectal vein drains into the inferior mesenteric vein and is a tributary of the portal circulation. The middle and inferior rectal veins drain into the internal iliac and internal pudendal veins, respectively. The anastomosis between the rectal veins is an important **portal–systemic anastomosis.**

### Lymph Drainage
The lymph passes to pararectal nodes and then upward to the inferior mesenteric nodes. Some lymph vessels pass to the internal iliac nodes.

### Nerve Supply
Sympathetic and parasympathetic pelvic splanchnic nerves through the hypogastric plexuses.

## Anal Canal

The anal canal is about 1 ½ in. (4 cm) long and passes downward and backward from the rectal ampulla to open on the surface at the anus (Figs. 19-64 and 19-65). Except during defecation, its lateral walls are kept in apposition by the levator ani muscles and the anal sphincters.

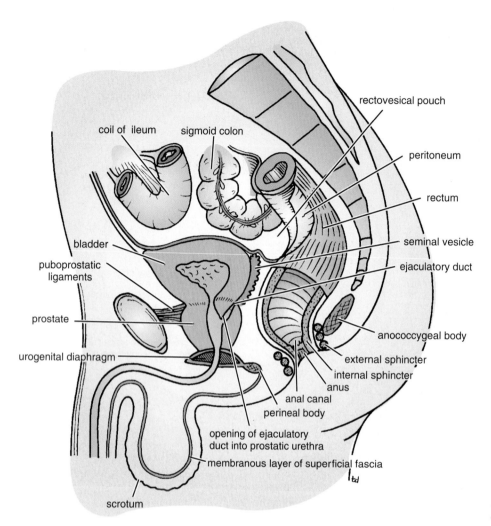

**Figure 19-64** Sagittal section of the male pelvis.

### Relations

**Posteriorly:** Anococcygeal body, coccyx (Fig. 19-64).

**Anteriorly in the male:** The perineal body, the urogenital diaphragm, the membranous part of the urethra, the bulb of the penis.

**Anteriorly in the female:** The perineal body, the urogenital diaphragm, the lower part of the vagina (Fig. 19-65).

**Laterally:** Fat-filled ischiorectal fossa (Fig. 19-66).

The **mucous membrane** of the upper half of the anal canal shows vertical folds called **anal columns** (Figs. 19-66 and 19-67). These are connected together at their lower ends by small semilunar folds called **anal valves.** The mucous membrane of the lower half of the anal canal is smooth and merges with the skin at the anus. The **pectinate line** indicates the level where the upper half of the anal canal joins the lower half (Fig. 19-67).

The **muscular coat**, as in other parts of the intestinal tract, is divided into an outer longitudinal and an inner circular layer of smooth muscle (Figs. 19-66 and 19-67). The circular coat is thickened at the upper end of the canal to form the **involuntary internal sphincter.** Surrounding the internal sphincter of smooth muscle is a collar of striped muscle called the **voluntary external sphincter.** The external sphincter is divided into three parts: subcutaneous, superficial, and deep. The attachments of these parts are given in Table 19-2.

The **puborectalis** fibers of the levatores ani muscles form a sling, which is attached anteriorly to the pubic bones. The muscular sling passes backward around the junction of the rectum and the anal canal pulling them forward so that the rectum joins the anal canal at an acute angle (Fig. 19-68).

At the junction of the rectum and the anal canal, the internal sphincter, the deep part of the external sphincter, and the puborectalis form a distinct ring called the anorectal ring, which can be palpated on rectal examination.

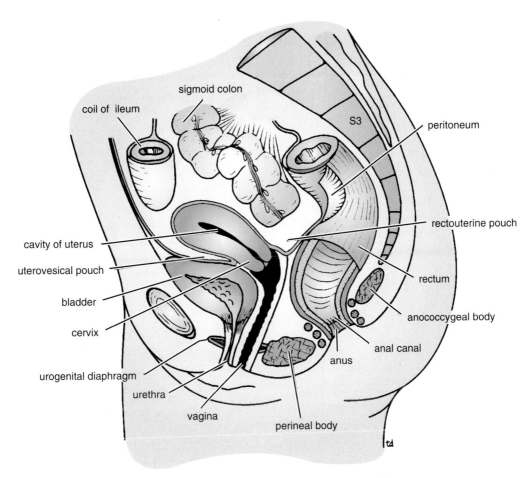

**Figure 19-65**    Sagittal section of the female pelvis.

## Blood Supply

### Arteries

The superior rectal artery supplies the upper half of the rectum, and the inferior rectal artery supplies the lower half of the rectum (Fig. 19-68).

### Veins

The upper half is drained by the superior rectal vein into the inferior mesenteric vein; the lower half is drained by the inferior rectal vein into the systemic circulation (Fig. 19-68). The anastomosis between the rectal veins forms an important **portal–systemic anastomosis.**

## Lymphatic Drainage

Lymph from the upper half of the anal canal ascends to the pararectal nodes and joins the inferior mesenteric nodes. Lymph from the lower half of the canal drains into the medial group of superficial inguinal nodes (Fig. 19-68).

## Nerve Supply

The mucous membrane of the upper half of the anal canal is sensitive to stretch and is innervated by fibers that ascend through the hypogastric plexuses. The lower half is sensitive to pain, temperature, touch, and pressure and is innervated by the inferior rectal nerves.

The internal anal sphincter is supplied by sympathetic fibers from the inferior hypogastric plexus. The voluntary external sphincter is supplied by the inferior rectal nerves.

### PHYSIOLOGIC NOTE

## Defecation

The time, place, and frequency of defecation are a matter of habit. Some adults defecate once a day, some defecate several times a day, and some perfectly normal people defecate only once in several days.

The desire to defecate is initiated by stimulation of the stretch receptors in the wall of the rectum by the presence of feces in the lumen. The act of defecation involves a coordinated reflex that results in the emptying of the descending colon, sigmoid colon, rectum, and anal canal. It is assisted by a rise in intra-abdominal pressure brought about by contraction of the muscles of the anterior abdominal wall. The tonic contraction of the internal and

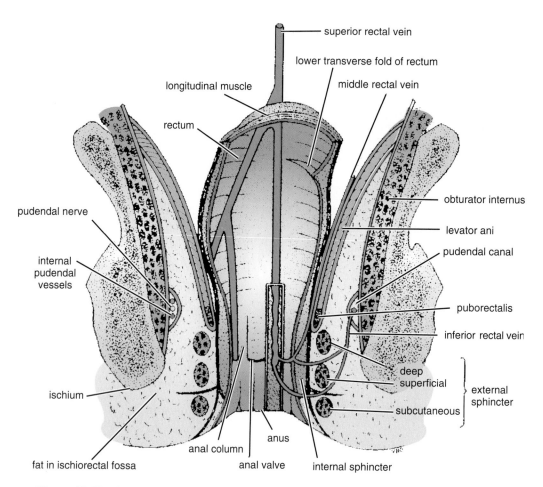

**Figure 19-66**   Coronal section of the pelvis and the perineum showing venous drainage of the anal canal.

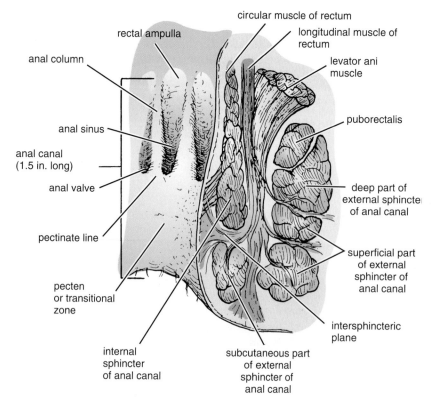

**Figure 19-67**   Coronal section of the anal canal showing the detailed anatomy of the mucous membrane and the arrangement of the internal and external anal sphincters. Note that the terms *pectinate line* (the line at the level of the anal valves) and *pecten* (the transitional zone between the skin and the mucous membrane) are sometimes used by clinicians.

## Table 19-2    Sphincters of the Anal Canal

| Muscle | Origin | Insertion | Nerve Supply | Action |
|---|---|---|---|---|
| External anal sphincter Subcutaneous part | Encircles anal canal, no bony attachments | | Inferior rectal nerve and perineal branch of fourth sacral nerve | Together with puborectalis muscle, forms voluntary sphincter of anal canal |
| Superficial part Deep part | Perineal body Encircles anal canal, no bony attachments | Coccyx | | |
| Puborectalis (part of levator ani) | Pubic bones | Sling around junction of rectum and anal canal | Perineal branch of fourth sacral nerve and from perineal branch of pudendal nerve | Together with external anal sphincter, forms voluntary sphincter for anal canal |

external anal sphincters, including the puborectalis muscles, is now voluntarily inhibited, and the feces are evacuated through the anal canal. Depending on the laxity of the submucous coat, the mucous membrane of the lower part of the anal canal is extruded through the anus ahead of the fecal mass. At the end of the act, the mucosa is returned to the anal canal by the tone of the longitudinal fibers of the anal walls and the contraction and upward pull of the puborectalis muscle. The empty lumen of the anal canal is now closed by the tonic contraction of the anal sphincters.

## Differences between the Small and Large Intestines

### External Differences (Fig. 19-69)

▨ The small intestine (with the exception of the duodenum) is mobile, whereas the ascending and descending parts of the colon are fixed.

▨ The caliber of the full small intestine is normally smaller than that of the filled large intestine.

▨ The small intestine (with the exception of the duodenum) has a mesentery that passes downward across the midline into the right iliac fossa.

▨ The longitudinal muscle of the small intestine forms a continuous layer around the gut. In the large intestine (with the exception of the appendix), the longitudinal muscle is collected into three bands, the teniae coli.

▨ The small intestine has no fatty tags attached to its wall. The large intestine has fatty tags, called the **appendices epiploicae.**

▨ The wall of the small intestine is smooth, whereas that of the large intestine is sacculated.

### Internal Differences (Fig. 19-69)

▨ The mucous membrane of the small intestine has permanent folds, called **plicae circulares,** which are absent in the large intestine.

▨ The mucous membrane of the small intestine has villi, which are absent in the large intestine.

▨ Aggregations of lymphoid tissue, called Peyer's patches, are found in the mucous membrane of the small intestine; these are absent in the large intestine.

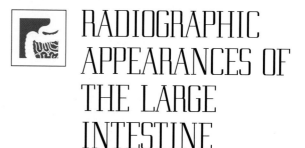

# RADIOGRAPHIC APPEARANCES OF THE LARGE INTESTINE

The radiographic appearances of the large intestine are shown in Figures 19-60 and 19-70.

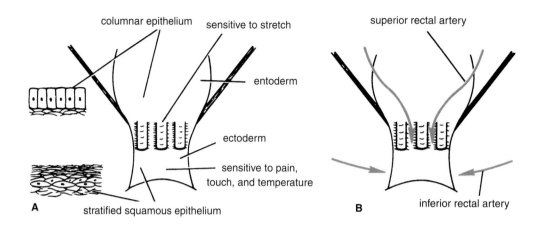

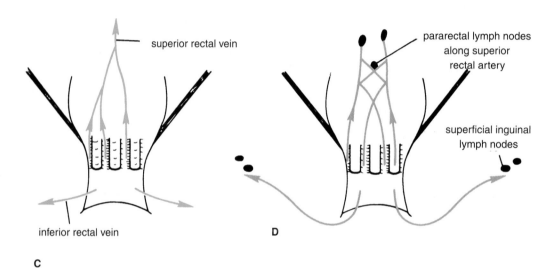

**Figure 19-68**   Upper and lower halves of the anal canal showing their embryologic origin and lining epithelium (**A**), their arterial supply (**B**), their venous drainage (**C**), and their lymph drainage (**D**). **E.** Arrangement of the muscle fibers of the puborectalis muscle and different parts of the external anal sphincter.

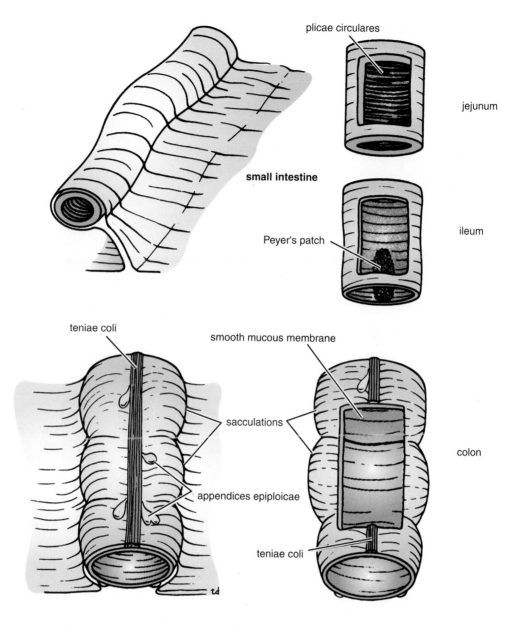

**Figure 19-69** Some external and internal differences between the small and the large intestine.

**E M B R Y O L O G I C    N O T E**

## Development of the Alimentary Tract

The digestive tube is formed from the yolk sac. The entoderm forms the epithelial lining, and the splanchnic mesenchyme forms the surrounding muscle and serous coats. The developing gut is divided into the foregut, midgut, and hindgut (Fig. 19-71).

### Development of the Esophagus

The esophagus develops from the narrow part of the foregut that succeeds the pharynx (Fig. 19-71). At first, it

is a short tube, but when the heart and diaphragm descend, it elongates rapidly.

### Development of the Stomach

The stomach develops as a dilatation of the foregut (Fig. 19-72). To begin with, it has a ventral and dorsal mesentery. Very active growth takes place along the dorsal border, which becomes convex and forms the greater curvature. The anterior border becomes concave and forms the lesser curvature. The fundus appears as a dilatation at the upper end of the stomach. At this stage, the stomach has a right and left surface to which the right and left vagus nerves are attached, respectively (Fig. 19-72).

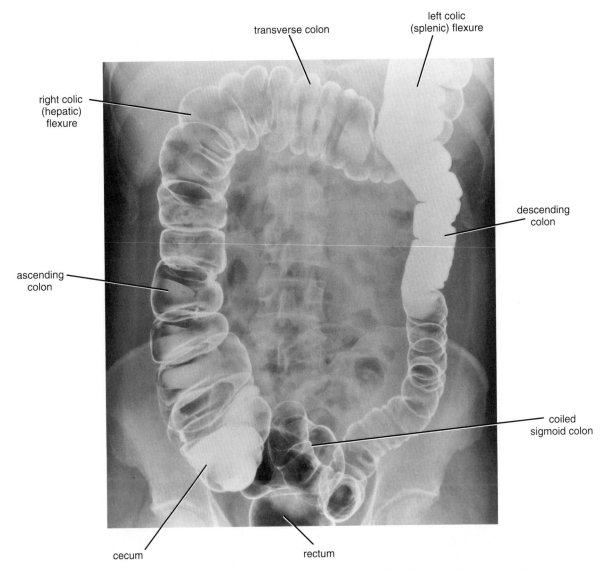

right colic
(hepatic)
flexure

transverse colon

left colic
(splenic) flexure

descending
colon

ascending
colon

coiled
sigmoid colon

cecum

rectum

**Figure 19-70**   Anteroposterior radiograph of the large intestine after a barium enema. Air
has been introduced into the intestine through the enema tube after evacuation of most of
the barium. This procedure is referred to as a contrast enema.

With the great growth of the right lobe of the liver, the stomach is gradually rotated to the right so that the left surface becomes anterior and the right surface becomes posterior. The ventral and dorsal mesenteries now change position as the result of rotation of the stomach, and they form the omenta and various peritoneal ligaments.

The pouch of peritoneum behind the stomach is known as the lesser sac.

### Development of the Duodenum

The duodenum is formed from the most caudal portion of the foregut and the most cephalic end of the midgut. This region rapidly grows to form a loop. At this time, the duodenum has a mesentery that extends to the posterior abdominal wall and is part of the dorsal mesentery. A small part of the ventral mesentery is also attached to the ventral border of the first part of the duodenum and the upper half of the second part of the duodenum. When the stomach rotates, the duodenal loop is forced to rotate to the right, where the second, third, and fourth parts adhere to the posterior abdominal wall. Now the peritoneum behind the duodenum disappears. However, some smooth muscle and fibrous tissue that belong to the dorsal mesentery remain as the suspensory ligament of the duodenum (ligament of Treitz), and this fixes the terminal part of the duodenum and prevents it from moving inferiorly (Fig. 19-73). The liver and pancreas arise as entodermal buds from the developing duodenum.

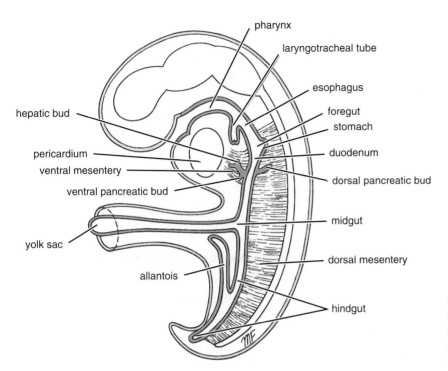

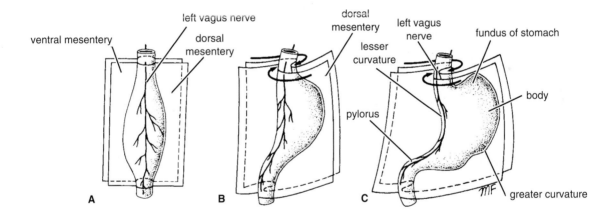

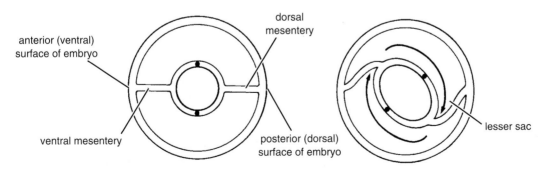

**Figure 19-71** The foregut, midgut, and hindgut. The positions of the ventral and dorsal mesenteries, the hepatic bud, and the ventral and dorsal pancreatic buds are also shown.

**Figure 19-72** Development of the stomach in relation to the ventral and dorsal mesenteries. Note how the stomach rotates so that the left vagus nerve comes to lie on the anterior surface of the stomach. Note also the position of the lesser sac.

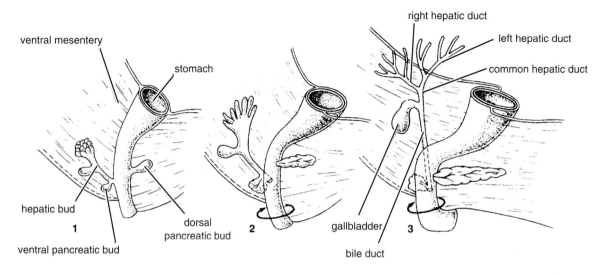

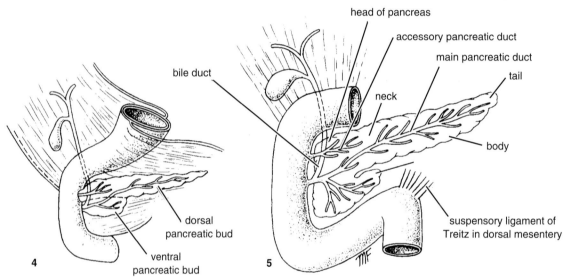

**Figure 19-73** The development of the pancreas and the extrahepatic biliary apparatus.

### Development of the Jejunum, Ileum, Cecum, Appendix, Ascending Colon, and Proximal Two Thirds of the Transverse Colon

Distal to the duodenum, the small intestine and the large intestine, as far as the distal third of the transverse colon, develop from the midgut. The midgut increases rapidly in length and forms a loop to the apex, on which is attached the **vitelline duct;** this duct passes through the widely open umbilicus (Fig. 19-71). At the same time, the dorsal mesentery elongates, and passing through it from the aorta to the yolk sac are the **vitelline arteries.** These arteries now fuse to form the **superior mesenteric artery,** which supplies the midgut and its derivatives. The rapidly growing liver and kidneys now encroach on the abdominal cavity, causing the midgut loop to herniate into the umbilical cord.

A diverticulum appears at the caudal end of the bowel loop, and this forms the cecum. At first, the diver-

ticulum is conical; later, the upper part expands and forms the **cecum,** while the lower part remains rudimentary and forms the **appendix** (Fig. 19-74). After birth, the wall of the cecum grows unequally, and the appendix comes to lie on its medial side.

While the loop of gut is in the umbilical cord, its cephalic limb becomes greatly elongated and coiled and forms the future **jejunum** and greater part of the **ileum.** The caudal limb of the loop also increases in length, but it remains uncoiled and forms the future distal part of the ileum, the cecum, the appendix, the **ascending colon,** and the **proximal two thirds of the transverse colon.**

### Rotation of the Midgut Loop in the Umbilical Cord and Its Return to the Abdominal Cavity

While in the umbilical cord, the midgut rotates around an axis formed by the superior mesenteric artery and the vitelline duct. As one views the embryo from the anterior

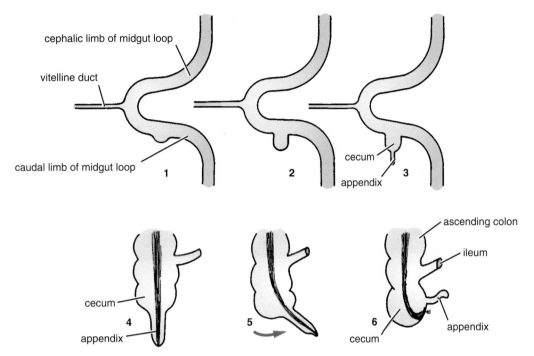

**Figure 19-74** Stages in the development of the cecum and appendix. The final stages of development (stages 4, 5, and 6) take place after birth.

aspect, a counterclockwise rotation of approximately 90° occurs (Fig. 19-75). Later, as the gut returns to the abdominal cavity, the midgut rotates counterclockwise an additional 180°. Thus, a total rotation of 270° counterclockwise has occurred (Fig. 19-76).

The rotation of the gut results in part of the large intestine (transverse colon) coming in front of the superior mesenteric artery and the second part of the duodenum; the third part of the duodenum comes to lie behind the artery. The cecum and appendix come into close contact

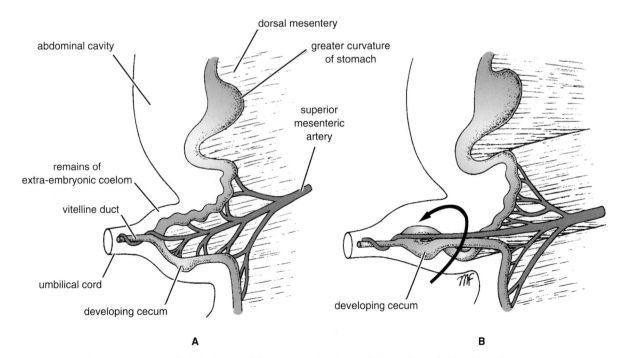

**Figure 19-75** Left side views of the counterclockwise 90° rotation of the midgut loop while it is in the extraembryonic coelom in the umbilical cord.

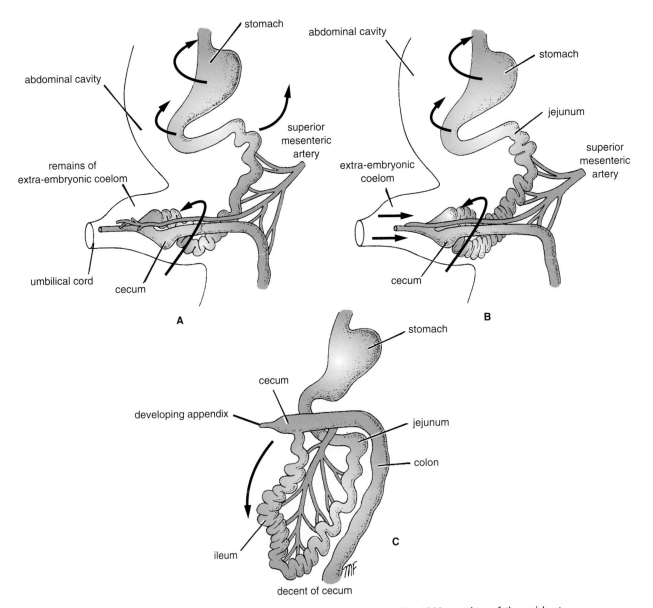

**Figure 19-76** Left side views (**A, B**) of the counterclockwise 180° rotation of the midgut loop as it is withdrawn into the abdominal cavity. **C.** The descent of the cecum takes place later.

with the right lobe of the liver. Later, the cecum and appendix descend into the right iliac fossa so that the ascending colon and right colic flexure are formed. Thus, the rotation of the gut has resulted in the large gut coming to lie laterally and encircle the centrally placed small gut.

The primitive mesenteries of the duodenum and ascending and descending colons now fuse with the parietal peritoneum on the posterior abdominal wall. This explains how these parts of the developing gut become retroperitoneal. The primitive mesenteries of the jejunum and ileum, the transverse colon, and the sigmoid colon persist as the mesentery of the small intestine, the transverse mesocolon, and the sigmoid mesocolon, respectively.

The rotation of the stomach and duodenum to the right is largely brought about by the great growth of the right lobe of the liver. The left surface of the stomach becomes anterior, and the right surface becomes posterior. A pouch of peritoneum becomes located behind the stomach and is called the **lesser sac**.

### Fate of the Vitelline Duct

The midgut is at first connected with the yolk sac by the vitelline duct. By the time the gut returns to the abdominal cavity, the duct becomes obliterated and severs its connection with the gut.

### Development of the Left Colic Flexure, Descending Colon, Sigmoid Colon, Rectum, and Upper Half of the Anal Canal

The left colic flexure, descending colon, sigmoid colon, rectum, and the upper half of the anal canal are developed from the hindgut. Distally, this terminates as a blind sac of entoderm, which is in contact with a shallow ectodermal depression called the **proctodeum.** The apposed layers of ectoderm and entoderm form the **cloacal membrane,** which separates the cavity of the hindgut from the surface (Fig. 19-77). The hindgut sends off a diverticulum, the **allantois,** which passes into the umbilical cord. Distal to the allantois, the hindgut dilates to form the **entodermal cloaca** (Fig. 19-77). In the interval between the allantois and the hindgut, a wedge of mesenchyme invaginates the entoderm. With continued proliferation of the mesenchyme, a septum is formed that grows inferiorly and divides the cloaca into anterior and posterior parts. The septum is called the **urorectal septum;** the anterior part of the cloaca becomes the **primitive bladder** and the **urogenital sinus,** and the posterior part of the cloaca forms the **anorectal canal.** On reaching the cloacal membrane, the urorectal septum fuses with it and forms the future **perineal body** (Fig. 19-77). The fates of the primitive bladder and the urogenital sinus in both sexes are considered in detail on page 823.

The anorectal canal forms the **rectum** and the superior half of the **anal canal.** The lining of the inferior half of the anal canal is formed from the ectoderm of the proctodeum (Fig. 19-68). The posterior part of the cloacal membrane breaks down so that the gut opens onto the surface of the embryo.

### Hindgut Artery

The hindgut, which extends from the left colic flexure to halfway down the anal canal, is supplied by the **inferior mesenteric artery** (Fig. 19-78). Here, a number of ventral branches of the aorta fuse to form a single artery.

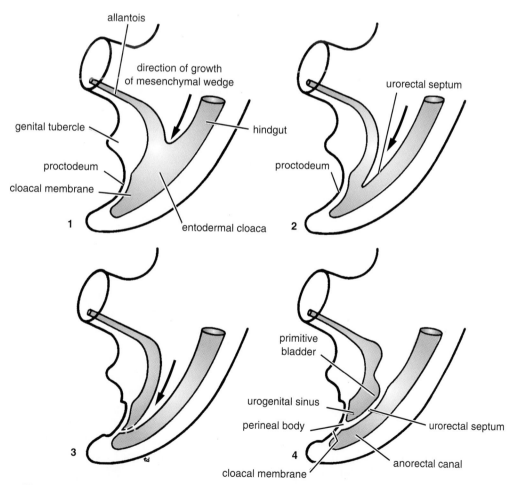

**Figure 19-77** Progressive stages (1-4) in the formation of the urorectal septum, which divides the cloaca into an anterior part (the primitive bladder and the urogenital sinus) and a posterior part (the anorectal canal).

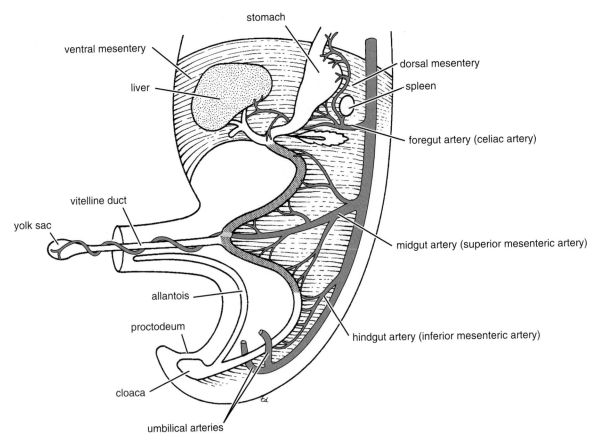

**Figure 19-78** Formation of the midgut loop (*shaded*). Note how the superior mesenteric artery and vitelline duct form an axis for the future rotation of the midgut loop.

### Meconium

At full term, the large intestine is filled with a mixture of intestinal gland secretions, bile, and amniotic fluid. This substance is dark green in color and is called meconium. It starts to accumulate at 4 months and reaches the rectum at the 5th month.

### Development of the Anal Canal

The distal end of the hindgut terminates as a blind sac of entoderm called the **cloaca** (Fig. 19-77). The cloaca lies in contact with a shallow ectodermal depression called the **proctodeum.** The apposed layers of ectoderm and entoderm form the **cloacal membrane,** which separates the cavity of the hindgut from the surface (Fig. 19-77). The cloaca becomes divided into anterior and posterior parts by the **urorectal septum;** the posterior part of the cloaca is called the **anorectal canal.** The anorectal canal forms the rectum and the upper half of the anal canal. The lining of the superior half of the anal canal is formed from entoderm, and the lining of the inferior half of the anal canal is formed

from the ectoderm of the proctodeum (Fig. 19-77). The sphincters of the anal canal are formed from the surrounding mesenchyme. The posterior part of the cloacal membrane breaks down so that the gut opens onto the surface of the embryo.

### Arterial Supply of the Developing Alimentary Tract

The arterial supply to the gut and its relationship to the development of the different parts of the gut are illustrated diagrammatically in Figure 19-78. The celiac artery is the artery of the foregut and supplies the gastrointestinal tract from the lower one third of the esophagus down as far as the middle of the second part of the duodenum. The superior mesenteric artery is the artery of the midgut and supplies the gastrointestinal tract from the middle of the second part of the duodenum as far as the distal one third of the transverse colon. The inferior mesenteric artery is the artery of the hindgut and supplies the large intestine from the distal one third of the transverse colon to halfway down the anal canal.

# SURFACE ANATOMY OF THE ABDOMINAL WALL AND THE GASTROINTESTINAL TRACT

## Surface Landmarks of the Abdominal Wall

### Xiphoid Process

The xiphoid process is the thin cartilaginous lower part of the sternum. It is easily palpated in the depression where the costal margins meet in the upper part of the anterior abdominal wall (Figs. 19-79 and 19-80). The xiphisternal junction is identified by feeling the lower edge of the body of the sternum, and it lies opposite the body of the ninth thoracic vertebra.

### Costal Margin

The costal margin is the curved lower margin of the thoracic wall and is formed in front by the cartilages of the seventh, eighth, ninth, and tenth ribs (Figs. 19-79 and 19-80) and behind by the cartilages of the eleventh and twelfth ribs. The costal margin reaches its lowest level at the tenth costal cartilage, which lies opposite the body of the third lumbar vertebra. The twelfth rib may be short and difficult to palpate.

### Iliac Crest

The iliac crest can be felt along its entire length and ends in front at the anterior superior iliac spine (Figs. 19-79 and 19-80) and behind at the posterior superior iliac spine (Fig. 19-82). Its highest point lies opposite the body of the fourth lumbar vertebra.

About 2 in. (5 cm) posterior to the anterior superior iliac spine, the outer margin of the iliac crest projects to form the tubercle of the crest. The tubercle lies at the level of the body of the fifth lumbar vertebra.

### Pubic Tubercle

The pubic tubercle is an important surface landmark. It may be identified as a small protuberance along the superior surface of the pubis (Figs. 19-6 and 19-80).

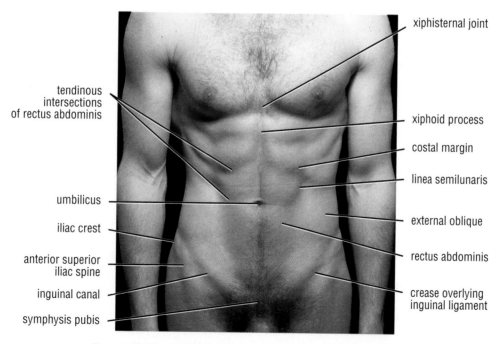

**Figure 19-79**   Anterior abdominal wall of a 27-year-old man.

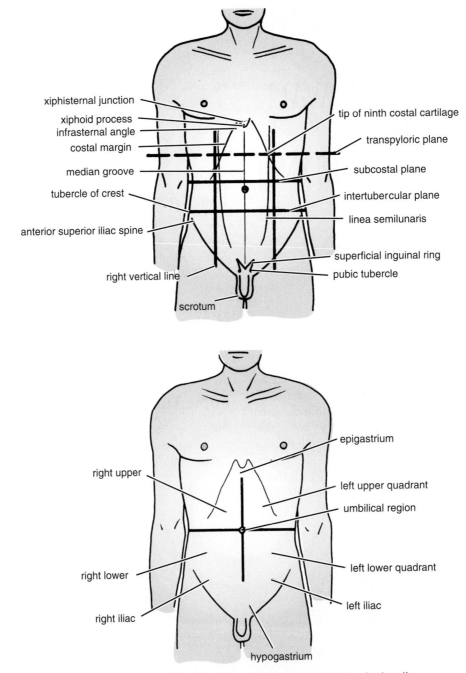

**Figure 19-80** Surface landmarks and regions of the anterior abdominal wall.

## Symphysis Pubis

The symphysis pubis is the cartilaginous joint that lies in the midline between the bodies of the pubic bones (Fig. 19-79). It is felt as a solid structure beneath the skin in the midline at the lower extremity of the anterior abdominal wall. The pubic crest is the name given to the ridge on the superior surface of the pubic bones medial to the pubic tubercle (Fig. 19-10).

## Inguinal Ligament

The inguinal ligament lies beneath a skin crease in the groin. It is the rolled-under inferior margin of the aponeurosis of the external oblique muscle (Fig. 19-79). It is attached laterally to the anterior superior iliac spine and curves downward and medially, to be attached to the pubic tubercle.

## Superficial Inguinal Ring

The superficial inguinal ring is a triangular aperture in the aponeurosis of the external oblique muscle and is situated above and medial to the pubic tubercle (Figs. 19-11, 19-12, and 19-80). In the adult male, the margins of the ring can be felt by invaginating the skin of the upper part of the scrotum with the tip of the little finger. The soft tubular spermatic cord can be felt emerging from the ring and descending over or medial to the pubic tubercle into the scrotum (Figs. 19-11 and 19-12). Palpate the spermatic cord in the upper part of the scrotum between the finger and thumb and note the presence of a firm cordlike structure in its posterior part called the vas deferens (Figs. 19-12 and 19-15).

In the female, the superficial inguinal ring is smaller and difficult to palpate; it transmits the round ligament of the uterus.

## Scrotum

The scrotum is a pouch of skin and fascia containing the testes, the epididymides, and the lower ends of the spermatic cords. The skin of the scrotum is wrinkled and is covered with sparse hairs. The bilateral origin of the scrotum is indicated by the presence of a dark line in the midline, called the scrotal raphe, along the line of fusion. The testis on each side is a firm ovoid body surrounded on its lateral, anterior, and medial surfaces by the two layers of the tunica vaginalis (Fig. 19-15). The testis should therefore lie free and not tethered to the skin or subcutaneous tissue. Posterior to the testis is an elongated structure, the epididymis (Fig. 19-15). It has an enlarged upper end called the head, a body, and a narrow lower end, the tail. The vas deferens emerges from the tail and ascends medial to the epididymis to enter the spermatic cord at the upper end of the scrotum.

## Linea Alba

The linea alba is a vertically running fibrous band that extends from the symphysis pubis to the xiphoid process and lies in the midline (Fig. 19-6). It is formed by the fusion of the aponeuroses of the muscles of the anterior abdominal wall and is represented on the surface by a slight median groove (Figs. 19-79 and 19-80).

## Umbilicus

The umbilicus lies in the linea alba and is inconstant in position. It is a puckered scar and is the site of attachment of the umbilical cord in the fetus.

## Rectus Abdominis

The rectus abdominis muscles lie on either side of the linea alba (Fig. 19-79) and run vertically in the abdominal wall; they can be made prominent by asking the patient to raise the shoulders while in the supine position without using the arms.

## Tendinous Intersections of the Rectus Abdominis

The tendinous intersections are three in number and run across the rectus abdominis muscle. In muscular individuals, they can be palpated as transverse depressions at the level of the tip of the xiphoid process, at the umbilicus, and halfway between the two (Fig. 19-79).

## Linea Semilunaris

The linea semilunaris is the lateral edge of the rectus abdominis muscle and crosses the costal margin at the tip of the ninth costal cartilage (Figs. 19-79 and 19-80). To accentuate the semilunar lines, the patient is asked to lie on the back and raise the shoulders off the couch without using the arms. To accomplish this, the patient contracts the rectus abdominis muscles so that their lateral edges stand out.

## Abdominal Lines and Planes

Vertical lines and horizontal planes (Fig. 19-80) are commonly used to facilitate the description of the location of diseased structures or the performing of abdominal procedures.

### Vertical Lines

Each vertical line (right and left) passes through the midpoint between the anterior superior iliac spine and the symphysis pubis.

### Transpyloric Plane

The horizontal transpyloric plane passes through the tips of the ninth costal cartilages on the two sides—that is, the point where the lateral margin of the rectus abdominis (linea semilunaris) crosses the costal margin (Fig. 19-80). It lies at the level of the body of the first lumbar vertebra. This plane passes through the pylorus of the stomach, the duodenojejunal junction, the neck of the pancreas, and the hila of the kidneys.

### Subcostal Plane

The horizontal subcostal plane joins the lowest point of the costal margin on each side—that is, the tenth costal cartilage (Fig. 19-80). This plane lies at the level of the third lumbar vertebra.

### Intercristal Plane

The intercristal plane passes across the highest points on the iliac crests and lies on the level of the body of the fourth lumbar vertebra. This is commonly used as a surface landmark when performing a lumbar spinal tap.

## Intertubercular Plane

The horizontal intertubercular plane joins the tubercles on the iliac crests (Fig. 19-80) and lies at the level of the fifth lumbar vertebra.

## Abdominal Quadrants

It is common practice to divide the abdomen into quadrants by using a vertical and a horizontal line that intersect at the umbilicus (Fig. 19-80). The quadrants are the upper right, upper left, lower right, and lower left. The terms **epigastrium** and **periumbilical** are loosely used to indicate the area below the xiphoid process and above the umbilicus and the area around the umbilicus, respectively.

# Surface Landmarks of the Abdominal Viscera

It must be emphasized that the positions of most of the abdominal viscera show individual variations as well as variations in the same person at different times. Posture and respiration have a profound influence on the position of viscera.

The following organs are more or less fixed, and their surface markings are of clinical value.

## Liver

The liver lies under cover of the lower ribs, and most of its bulk lies on the right side (Fig. 19-81). In infants, until about the end of the 3rd year, the lower margin of the liver extends one or two fingerbreadths below the costal margin (Fig. 19-81). In the adult who is obese or has a well-developed right rectus abdominis muscle, the liver is not palpable. In a thin adult, the lower edge of the liver may be felt a fingerbreadth below the costal margin. It is most easily felt when the patient inspires deeply and the diaphragm contracts and pushes down the liver.

## Gallbladder

The fundus of the gallbladder lies opposite the tip of the right ninth costal cartilage—that is, where the lateral edge of the right rectus abdominis muscle crosses the costal margin (Fig. 19-81).

## Spleen

The spleen is situated in the left upper quadrant and lies under cover of the ninth, tenth, and eleventh ribs (Fig. 19-81). Its long axis corresponds to that of the tenth rib, and in the adult, it does not normally project forward in front of the midaxillary line. In infants, the lower pole of the spleen may just be felt (Fig. 19-81).

## Pancreas

The pancreas lies across the transpyloric plane. The head lies below and to the right, the neck lies on the plane, and the body and tail lie above and to the left.

## Kidneys

The right kidney lies at a slightly lower level than the left kidney (because of the bulk of the right lobe of the liver), and the lower pole can be palpated in the right lumbar region at the end of deep inspiration in a person with poorly developed abdominal muscles. Each kidney moves about 1 in. (2.5 cm) in a vertical direction during full respiratory movement of the diaphragm. The normal left kidney, which is higher than the right kidney, is not palpable.

On the anterior abdominal wall, the hilum of each kidney lies on the transpyloric plane, about three fingerbreadths from the midline (Fig. 19-82). On the back, the kidneys extend from the twelfth thoracic spine to the third lumbar spine, and the hili are opposite the first lumbar vertebra (Fig. 19-82).

## Stomach

The **cardioesophageal junction** lies about three fingerbreadths below and to the left of the xiphisternal junction (the esophagus pierces the diaphragm at the level of the tenth thoracic vertebra).

The **pylorus** lies on the transpyloric plane just to the right of the midline. The **lesser curvature** lies on a curved line joining the cardioesophageal junction and the pylorus. The **greater curvature** has an extremely variable position in the umbilical region or below.

## Duodenum (First Part)

The duodenum lies on the transpyloric plane about four fingerbreadths to the right of the midline.

## Cecum

The cecum is situated in the right lower quadrant. It is often distended with gas and gives a resonant sound when percussed. It can be palpated through the anterior abdominal wall.

## Appendix

The appendix lies in the right lower quadrant. The base of the appendix is situated one third of the way up the line, joining the anterior superior iliac spine to the umbilicus (McBurney's point). The position of the free end of the appendix is variable.

## Ascending Colon

The ascending colon extends upward from the cecum on the lateral side of the right vertical line and disappears under

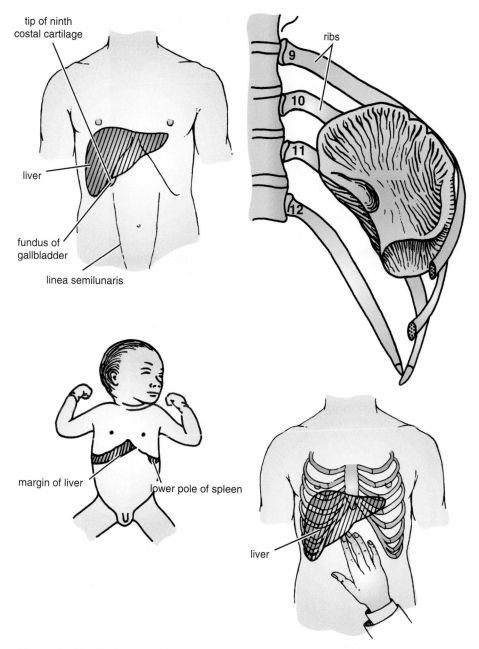

**Figure 19-81** Surface markings of the fundus of the gallbladder, spleen, and liver. In a young child, the lower margin of the normal liver and the lower pole of the normal spleen can be palpated. In a thin adult, the lower margin of the normal liver may just be felt at the end of deep inspiration.

the right costal margin. It can be palpated through the anterior abdominal wall.

## Transverse Colon

The transverse colon extends across the abdomen, occupying the umbilical region. It arches downward with its concavity directed upward. Because it has a mesentery, its position is variable.

## Descending Colon

The descending colon extends downward from the left costal margin on the lateral side of the left vertical line. In the left lower quadrant, it curves medially and downward to become continuous with the sigmoid colon. The descending colon has a smaller diameter than the ascending colon and can be palpated through the anterior abdominal wall.

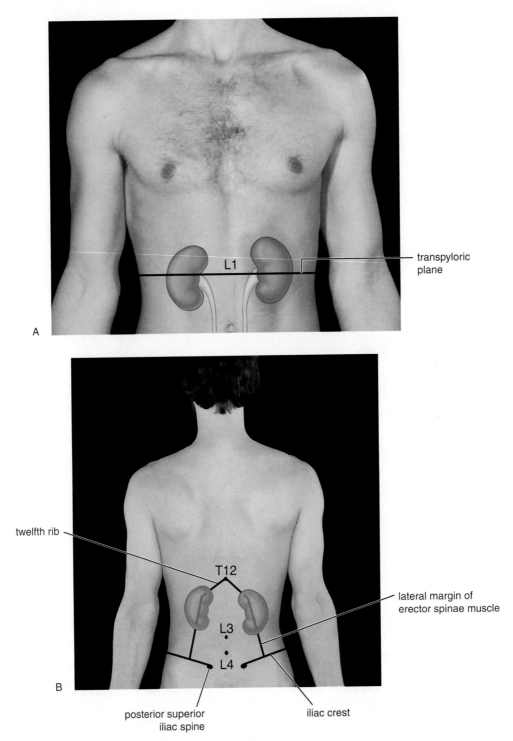

**Figure 19-82   A.** Surface anatomy of the kidneys and ureters on the anterior abdominal wall. Note the relationship of the hilum of each kidney to the transpyloric plane. **B.** Surface anatomy of the kidneys on the posterior abdominal wall.

# Review Questions

## Multiple Choice Questions

**Select the BEST answer for each question.**

1. The following structures form the walls of the inguinal canal **except** which?
   A. The conjoint tendon
   B. The aponeurosis of the external oblique muscle
   C. The internal oblique muscle
   D. The lacunar ligament
   E. The fascia transversalis

2. The following statements concerning the walls of the inguinal canal are correct **except** which?
   A. The inguinal ligament is made tense by flexing the hip joint.
   B. The contracting internal oblique muscle reinforces the anterior wall of the canal in front of the weak deep inguinal ring.
   C. The strong conjoint tendon reinforces the posterior wall of the canal behind the weak superficial inguinal ring.
   D. Contraction of the arching fibers of the internal oblique and transversus abdominis muscles lowers the roof of the canal so that the canal is practically obliterated.
   E. After birth, as the result of growth, the deep inguinal ring moves away from the superficial ring so that the canal becomes oblique and the two rings no longer lie opposite one another.

3. In the female, the inguinal canal contains the following structures **except** which?
   A. Ilioinguinal nerve
   B. Remnant of the processus vaginalis
   C. Round ligament of the uterus
   D. Inferior epigastric artery
   E. Lymph vessels from the fundus of the uterus

4. The following statements concerning the spermatic cord are correct **except** which?
   A. It extends from the deep inguinal ring to the scrotum.
   B. It contains the testicular artery.
   C. It is covered by five layers of spermatic fascia.
   D. It contains the pampiniform plexus.
   E. It contains lymph vessels that drain the testis.

5. The following structures are present in the inguinal canal in the male **except** which?
   A. Internal spermatic fascia
   B. Genital branch of the genitofemoral nerve
   C. Testicular vessels
   D. Deep circumflex iliac artery
   E. Ilioinguinal nerve

6. The following statements concerning the conjoint tendon are correct **except** which?
   A. It is attached to the pubic crest and the pectineal line.
   B. It is formed by the fusion of the aponeuroses of the transversus abdominis and internal oblique muscles.
   C. It is attached medially to the linea alba.
   D. It is continuous with the inguinal ligament.
   E. It may bulge forward in a direct inguinal hernia.

7. To pass a needle into the cavity of the tunica vaginalis in the scrotum, the following structures have to be pierced **except** which?
   A. Skin
   B. Dartos muscle and Colles' fascia
   C. Tunica albuginea
   D. Internal spermatic fascia
   E. Cremasteric fascia

## Matching Questions

**Match each structure listed below with the region on the anterior abdominal wall in which it is located. Each lettered answer may be used more than once.**

8. Appendix

9. Gallbladder

10. Cecum

11. Left colic flexure
    A. Right upper quadrant
    B. Left lower quadrant
    C. Right lower quadrant
    D. None of the above

**Match each structure listed below with the structure with which it is most closely associated.**

12. External spermatic fascia

13. Round ligament of the uterus

14. Cremasteric fascia

15. Internal spermatic fascia

16. Deep inguinal ring
    A. Internal oblique
    B. Fascia transversalis
    C. Gubernaculum
    D. External oblique
    E. None of the above

**Match each structure listed below with the group of lymph nodes that drain it.**

17. Testis

18. Skin of anterior abdominal wall below level of the umbilicus

19. Epididymis

20. Skin of the scrotum
    A. Anterior axillary lymph nodes
    B. Para-aortic or lumbar lymph nodes
    C. Superficial inguinal lymph nodes
    D. External iliac nodes
    E. None of the above

## Multiple Choice Questions

**Read the case histories and select the BEST answer to the question following them.**

A 30-year-old man was seen in the emergency department with a stab wound in the right inguinal region.

21. Which of the following nerves supplies the skin of the inguinal region?
    A. The eleventh thoracic nerve
    B. The tenth thoracic nerve
    C. The twelfth thoracic nerve
    D. The first lumbar nerve
    E. The femoral nerve

Immediately after delivery, it was noted that a 7.5-lb male neonate had a large swelling on the anterior abdominal wall. The swelling consisted of a large sac, the walls of which were translucent and soft. The umbilical cord was attached to the apex of the sac, and the umbilical arteries and vein ran within its walls.

22. The following statements concerning this case are probably correct **except** which?
    A. On closer examination, it was possible to see within the sac coils of small intestine and the lower margin of the liver.

B. As the baby cried and started to swallow air, the sac became larger.

C. Failure of the formation of adequate head and tail folds of the embryonic disc causes a defect in the anterior abdominal wall in the umbilical region.

D. The defect in the anterior abdominal wall is filled with thin amnion, which forms the wall of the sac.

E. The condition is known as exomphalos or omphalocele.

**Select the BEST answer for each question.**

23. The following statements concerning the ileum are correct **except** which?
    A. The circular smooth muscle of the lower end of the ileum serves as a sphincter at the junction of the ileum and the cecum.
    B. The branches of the superior mesenteric artery serving the ileum form more arcades than those serving the jejunum.
    C. Peyer's patches are present in the mucous membrane of the lower ileum along the antimesenteric border.
    D. The plicae circulares are more prominent at the distal end of the ileum than in the jejunum.
    E. The parasympathetic innervation of the ileum is from the vagus nerves.

24. The following structures are present within the lesser omentum **except** which?
    A. The portal vein
    B. The bile duct
    C. The inferior vena cava
    D. The hepatic artery
    E. The lymph nodes

## Matching Questions

**Match the numbered structures shown on the posteroanterior radiograph of the stomach and small** intestine—after ingestion of a barium meal—with the appropriate lettered structures (38-year-old male).

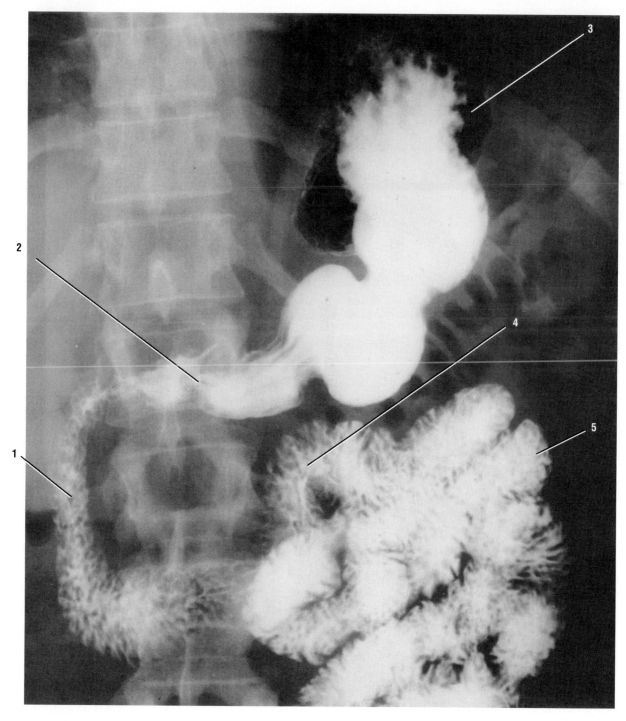

25. Structure 1

26. Structure 2

27. Structure 3

28. Structure 4

29. Structure 5

A. First part of duodenum
B. Second part of duodenum
C. Third part of duodenum
D. Air-filled fundus of stomach
E. Jejunum
F. Pylorus of stomach
G. None of the above

Match the numbered structures shown on the posteroanterior radiograph of the large intestine—after evacuation of a barium enema—with the appropriate lettered lymphatic drainage (20-year-old female).

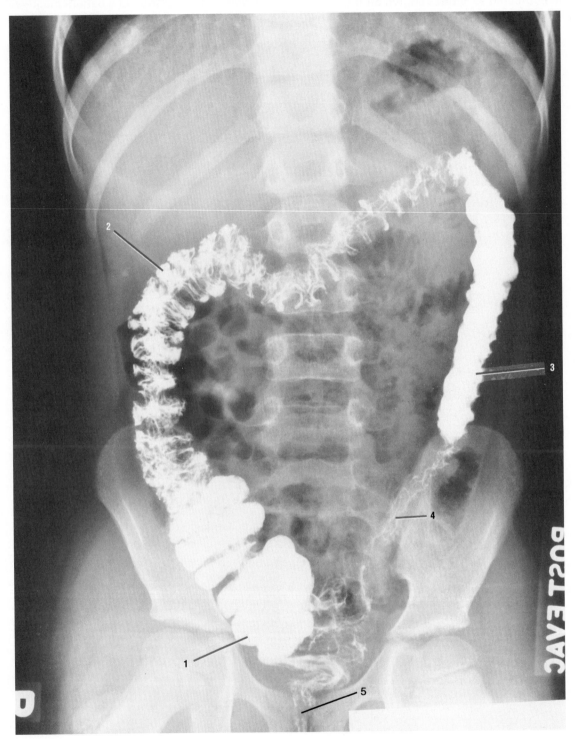

30. Structure 1

31. Structure 2

32. Structure 3

33. Structure 4

34. Structure 5

A. Appendix
B. Splenic flexure
C. Transverse colon
D. Cecum
E. Rectum
F. Sigmoid colon
G. Descending colon
H. None of the above

## Multiple Choice Questions

### Select the BEST answer for each question.

35. The following statements concerning the anal canal are correct **except** which?
    A. It is about 1.5 in. (3.8 cm) long.
    B. It pierces the urogenital diaphragm.
    C. It is related laterally to the external anal sphincter.
    D. It is the site of an important portal–systemic anastomosis.
    E. The mucous membrane of the lower half receives its arterial supply from the inferior rectal artery.

36. The following statements concerning the subcutaneous part of the external anal sphincter are correct **except** which?
    A. It encircles the anal canal.
    B. It is not attached to the anococcygeal body.

    C. It is composed of striated muscle fibers.
    D. It is not responsible for causing the anal canal and rectum to join at an acute angle.
    E. It is innervated by the middle rectal nerve.

37. The following statements concerning defecation are correct **except** which?
    A. The act is often preceded by the entrance of feces into the rectum, which gives rise to the desire to defecate.
    B. The muscles of the anterior abdominal wall contract.
    C. The external anal sphincters and the puborectalis relax.
    D. The internal sphincter contracts and causes the evacuation of the feces.
    E. The mucous membrane of the lower part of the anal canal is extruded through the anus ahead of the fecal mass.

# Answers and Explanations

1. **D** is incorrect. The lacunar ligament does not form part of the walls of the inguinal canal (Fig. 19-9).

2. **A** is incorrect. The inguinal ligament is made tense by extending the hip joint, since the fascia lata of the thigh is attached to the lower margin of the ligament and pulls it downwards in this movement.

3. **D** is incorrect. The inferior epigastric artery lies outside the inguinal canal behind the fascia transversalis (Figs. 19-11 and 19-13).

4. **C** is incorrect. The spermatic cord is covered by three layers of spermatic fascia, which are derived from the three layers of the anterior abdominal wall. The external spermatic fascia is from the aponeurosis of the external oblique muscle, the cremasteric fascia is derived from the internal oblique muscle, and the internal spermatic fascia is formed from the fascia transversalis (Fig. 19-11).

5. **D** is incorrect. The deep circumflex iliac artery is a branch of the external iliac artery and runs upward and laterally toward the anterior superior iliac spine away from the inguinal canal (Fig. 19-2).

6. **D** is incorrect. The conjoint tendon is not continuous with the inguinal ligament (Fig. 19-10).

7. **C** is incorrect. The tunica albuginea is the outer tough fibrous capsule of the testis (Fig. 19-15).

8. **C** is correct. The appendix is located in the right lower quadrant of the abdomen (Fig. 19-31).

9. **A** is correct. The gallbladder is located in the upper right quadrant of the abdomen (Fig. 19-31).

10. **C** is correct. The cecum is located in the lower right quadrant of the abdomen (Fig. 19-31).

11. **D** is correct. The left colic flexure is located in the upper left quadrant of the abdomen (Fig. 19-31).

12. **D** is correct. The external spermatic fascia is attached to the external oblique aponeurosis at the superficial inguinal ring (Fig. 19-12).

13. **C** is correct. The round ligament of the uterus is derived embryologically from the gubernaculum.

14. **A** is correct. The cremasteric fascia is formed from the lower margin of the internal oblique muscle (Fig. 19-11).

15. **B** is correct. The internal spermatic fascia is formed from the fascia transversalis at the deep inguinal ring (Fig. 19-13).

16. **B** is correct. The deep inguinal ring is an opening in the fascia transversalis (Fig. 19-17).

17. **B** is correct. The testis is drained into the para-aortic (lumbar) lymph nodes.

18. **C** is correct. The skin of the anterior abdominal wall below the level of the umbilicus drains into the superficial inguinal lymph nodes.

19. **B** is correct. The epididymis drains into the para-aortic (lumbar) lymph nodes.

20. **C** is correct. The skin of the scrotum drains into the superficial inguinal lymph nodes.

21. **D** is correct. The first lumbar nerve, represented by the iliohypogastric and ilioinguinal nerves, supplies the skin just above the inguinal ligament and the symphysis pubis (Fig. 19-1).

22. **C** is incorrect. The defect is caused by a failure of the formation of adequate lateral folds in the umbilical region, which is filled in by amnion only. During the first 24 hours after birth, the wall of the sac becomes dry and opaque and may rupture, causing evisceration. Bacteria at once gain entrance to the peritoneal cavity, producing peritonitis. The sac of amnion should be surgically excised as soon as possible after birth, and the contained viscera should be returned to the abdominal cavity. The defect in the anterior abdominal wall should then be closed.

23. **D** is incorrect. The plicae circulares are absent from the distal end of the ileum.

24. **C** is incorrect. The inferior vena cava lies on the posterior abdominal wall behind the parietal peritoneum. It is separated from the lesser omentum by the epiploic foramen (Fig. 19-23).

25. **B** is correct. Structure 1 is the second part of the duodenum.

26. **F** is correct. Structure 2 is the pylorus of the stomach.

27. **D** is correct. Structure 3 is the air-filled fundus of stomach.

28. **G** is correct. Structure 4 is the duodenojejunal junction.

29. **E** is correct. Structure 5 is the jejunum.

30. **D** is correct. Structure 1 is the cecum.

31. **H** is correct. Structure 2 is the right colic flexure.

32. **G** is correct. Structure 3 is the descending colon.

33. **F** is correct. Structure 4 is the sigmoid colon.

34. **E** is correct. Structure 5 is the rectum.

35. **B** is incorrect. The anal canal lies posterior to the urogenital diaphragm and, therefore, does not pierce it.

36. **E** is incorrect. The subcutaneous part of the external anal sphincter is innervated by the inferior rectal nerve, which is a branch of the pudendal nerve.

37. **D** is incorrect. The internal anal sphincter is relaxed during defecation.

# 20 The Viscera Associated with the Alimentary Tract: The Liver, the Pancreas, and the Spleen

All clinical material relevant to this chapter can be found on the CD-ROM.

# Chapter Outline

L iver disorders can greatly disturb the body's homeostasis. Unfortunately, the liver is susceptible to numerous metabolic, circulatory, toxic, microbial, and neoplastic diseases. Even though the liver is protected by the lower ribs, it is subject to trauma from crush injuries, gunshot, and knife wounds. For these reasons, the gross anatomy of the liver must be known and its relationship to the diaphragm and surrounding structures understood. In this chapter, the clinical anatomy of the liver and the biliary apparatus is reviewed.

The pancreas, which is both an exocrine and endocrine gland, is regarded clinically as virtually a hidden organ on the posterior abdominal wall. This fact places the medical professional at a considerable disadvantage. Signs and symptoms only become apparent when the disease has spread to involve neighboring structures such as the bile duct, the vertebral column, or stomach and intestines. Some of the most common diseases include cystic fibrosis, pancreatitis, and carcinoma. The objective in this chapter is to review the gross structure of the pancreas in relation to its neighboring organs so that pancreatic disease may be diagnosed and the appropriate treatment instituted.

The spleen is clearly part of the lymphatic and circulatory systems. Its function is to remove foreign particles, including worn out red cells, and create antibodies to blood borne antigens. Nevertheless, it is closely related to the digestive system in the upper left part of the abdomen and is often involved in the disease of that system. The purpose of its inclusion here is to emphasize this relationship and assist the student in the understanding of the peritoneal ligaments of the spleen and the closeness of the spleen and its blood supply to the stomach, pancreas, kidney, and colon.

 # BASIC ANATOMY

## Liver

The liver is the largest organ in the body. It is soft and pliable and occupies the upper part of the abdominal cavity just beneath the diaphragm (Fig. 20-1). The greater part of the liver is situated under cover of the right costal margin, and the right hemidiaphragm separates it from the pleura, lungs, pericardium, and heart. The liver extends to the left to reach the left hemidiaphragm. The convex upper surface of the liver is molded to the undersurface of the domes of the diaphragm. The posteroinferior, or visceral, surface is molded to adjacent viscera and is, therefore, irregular in shape; it lies in contact with the abdominal part of the esophagus, the stomach, the duodenum, the right colic flexure, the right kidney and suprarenal gland, and the gallbladder.

The liver may be divided into a large **right lobe** and a small **left lobe** by the attachment of the peritoneum of the falciform ligament (Fig. 20-2). The right lobe is further divided into a **quadrate lobe** and a **caudate lobe** by the presence of the

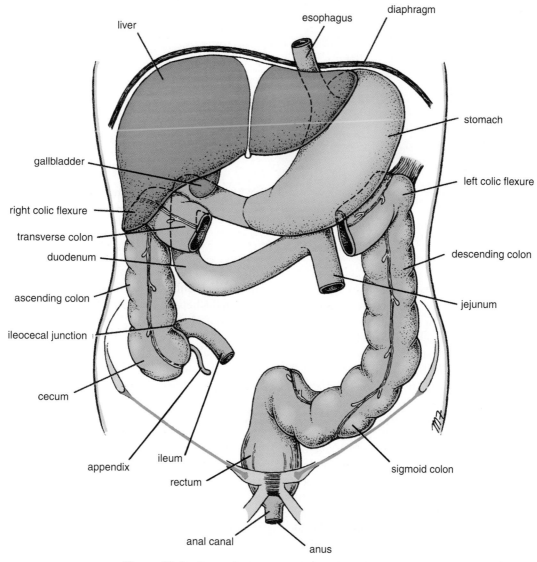

**Figure 20-1**   General arrangement of abdominal viscera.

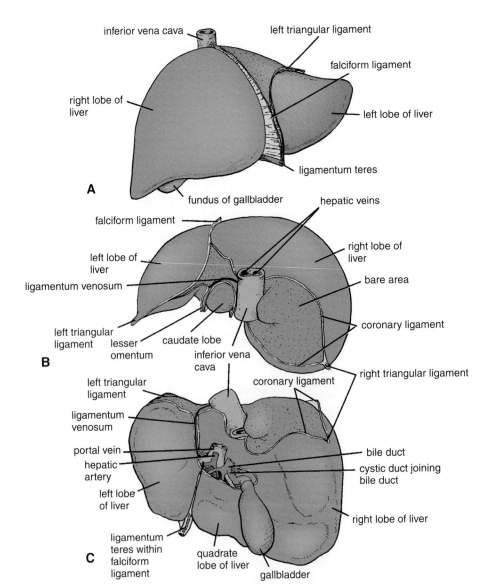

**Figure 20-2** Liver as seen from in front (**A**), from above (**B**), and from behind (**C**). Note the position of the peritoneal reflections, the bare areas, and the peritoneal ligaments.

gallbladder, the fissure for the ligamentum teres, the inferior vena cava, and the fissure for the ligamentum venosum. Experiments have shown that, in fact, the quadrate and caudate lobes are a functional part of the left lobe of the liver. Thus, the right and left branches of the hepatic artery and portal vein, and the right and left hepatic ducts, are distributed to the right lobe and the left lobe (plus quadrate plus caudate lobes), respectively. Apparently, the two sides overlap very little.

The **porta hepatis**, or hilum of the liver, is found on the posteroinferior surface and lies between the caudate and quadrate lobes (Fig. 20-2). The upper part of the free edge of the lesser omentum is attached to its margins. In it lie the right and left hepatic ducts, the right and left branches of the hepatic artery, the portal vein, and sympathetic and parasympathetic nerve fibers (Fig. 20-3). A few hepatic lymph nodes lie here; they drain the liver and gallbladder and send their efferent vessels to the celiac lymph nodes.

The liver is completely surrounded by a fibrous capsule but only partially covered by peritoneum. The liver is made up of **liver lobules**. The **central vein** of each lobule is a tributary of the hepatic veins. In the spaces between the lobules are the **portal canals**, which contain branches of the hepatic artery, portal vein, and a tributary of a bile duct (portal triad). The arterial and venous blood passes between the liver cells by means of **sinusoids** and drains into the central vein.

---

### P H Y S I O L O G I C   N O T E

### Summary of Liver Functions

The liver is the largest gland in the body and has a wide variety of functions. Three of its basic functions are production and secretion of bile, which is passed into the intestinal tract; involvement in many metabolic activities related to carbohydrate, fat, and protein metabolism; and filtration of the blood, removing bacteria and other foreign particles that have gained entrance to the blood from the lumen of the intestine.

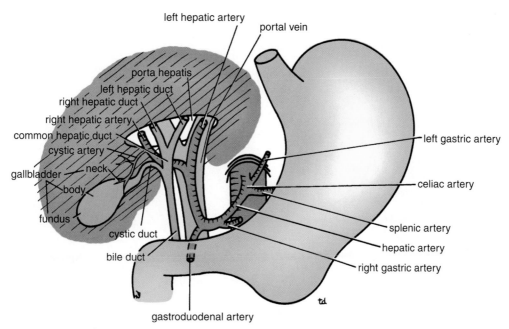

**Figure 20-3**  Structures entering and leaving the porta hepatis.

The liver synthesizes heparin, an anticoagulant substance, and has an important detoxicating function. It produces bile pigments from the hemoglobin of worn-out red blood corpuscles and secretes bile salts; these together are conveyed to the duodenum by the biliary ducts.

It is interesting to note that the liver has a vast reserve and regenerative capacity. It has been estimated that a patient with a normal liver can survive resection of about 85% of its total volume.

## Important Relations of the Liver

▪ **Anteriorly:** Diaphragm, right and left costal margins, right and left pleura and lower margins of both lungs, xiphoid process, and anterior abdominal wall in the subcostal angle.

▪ **Posteriorly:** Diaphragm, right kidney, hepatic flexure of the colon, duodenum, gallbladder, inferior vena cava, and esophagus and fundus of the stomach.

## Peritoneal Ligaments of the Liver

The **falciform ligament,** which is a two-layered fold of the peritoneum, ascends from the umbilicus to the liver (Figs. 20-2 and 20-4). It has a sickle-shaped free margin that contains the ligamentum teres, the remains of the umbilical vein. The falciform ligament passes on to the anterior and then the superior surfaces of the liver and then splits into two layers. The right layer forms the upper layer of the **coronary ligament;** the left layer forms the upper layer of the **left triangular ligament** (Fig. 20-2). The right extremity of the coronary ligament is known as the **right triangular ligament** of the liver. It should be noted that the peritoneal layers forming the coronary ligament are widely separated,

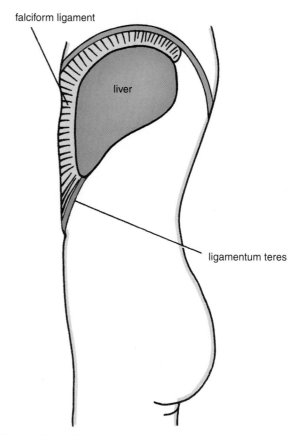

**Figure 20-4**  Shows the falciform ligament connecting the liver to the anterior abdominal wall and the diaphragm.

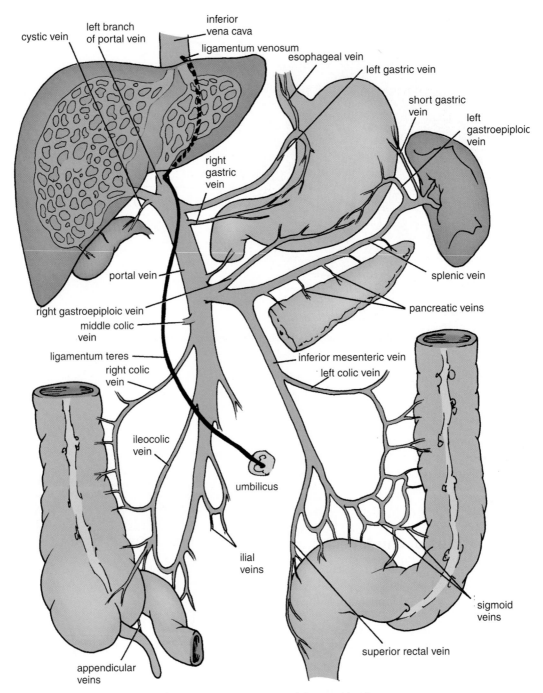

**Figure 20-5** Tributaries of the portal vein.

leaving an area of liver devoid of peritoneum. Such an area is referred to as a **bare area of the liver** (Fig. 20-2).

The **ligamentum teres** passes into a fissure on the visceral surface of the liver and joins the left branch of the portal vein in the porta hepatis (Figs. 20-2 and 20-5). The **ligamentum venosum,** a fibrous band that is the remains of the **ductus venosus,** is attached to the left branch of the portal vein and ascends in a fissure on the visceral surface of the liver to be attached above to the inferior vena cava

(Figs. 20-2 and 20-5). In the fetus, oxygenated blood is brought to the liver in the umbilical vein (ligamentum teres). The greater proportion of the blood bypasses the liver in the ductus venosus (ligamentum venosum) and joins the inferior vena cava. At birth, the umbilical vein and ductus venosus close and become fibrous cords.

The **lesser omentum** arises from the edges of the porta hepatis and the fissure for the ligamentum venosum and passes down to the lesser curvature of the stomach (Fig. 20-6).

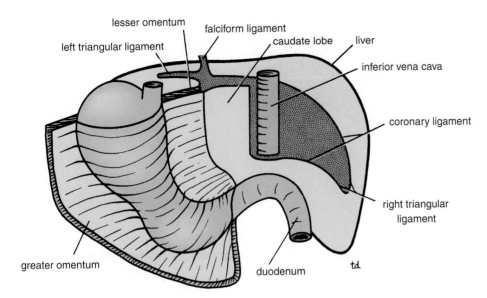

**Figure 20-6** Attachment of the lesser omentum to the stomach and the posterior surface of the liver.

## Blood Supply

### Arteries

The hepatic artery, a branch of the celiac artery, divides into right and left terminal branches that enter the porta hepatis (Fig. 20-3).

### Veins

The portal vein divides into right and left terminal branches that enter the porta hepatis behind the arteries (Fig. 20-3). The **hepatic veins** (three or more) emerge from the posterior surface of the liver and drain into the inferior vena cava.

### Blood Circulation through the Liver

The blood vessels (Fig. 20-3) conveying blood to the liver are the hepatic artery (30%) and portal vein (70%). The hepatic artery brings oxygenated blood to the liver, and the portal vein brings venous blood rich in the products of digestion, which have been absorbed from the gastrointestinal tract. The arterial and venous blood is conducted to the central vein of each liver lobule by the liver sinusoids. The central veins drain into the right and left hepatic veins, and these leave the posterior surface of the liver and open directly into the inferior vena cava.

## Lymph Drainage

The liver produces a large amount of lymph—about one third to one half of all body lymph. The lymph vessels leave the liver and enter several lymph nodes in the porta hepatis. The efferent vessels pass to the celiac nodes. A few vessels pass from the bare area of the liver through the diaphragm to the posterior mediastinal lymph nodes.

## Nerve Supply

Sympathetic and parasympathetic nerves form the celiac plexus. The anterior vagal trunk gives rise to a large hepatic branch, which passes directly to the liver.

# Portal Vein

The portal vein enters the liver and breaks up into sinusoids, from which blood passes into the hepatic veins that join the inferior vena cava. The portal vein is about 2 in. (5 cm) long and is formed behind the neck of the pancreas by the union of the superior mesenteric and splenic veins (Fig. 20-7). It ascends to the right, behind the first part of the duodenum, and enters the lesser omentum (Fig. 20-8; see also Fig. 20-10). It then runs upward in front of the opening into the lesser sac to the porta hepatis, where it divides into right and left terminal branches.

The portal vein (Fig. 20-5) drains blood from the abdominal part of the gastrointestinal tract from the lower third of the esophagus to halfway down the anal canal; it also drains blood from the spleen, pancreas, and gallbladder.

The portal circulation begins as a capillary plexus in the organs it drains and ends by emptying its blood into sinusoids within the liver.

For further details concerning the basic anatomy of the portal vein, see page 233.

# Bile Ducts of the Liver

Bile is secreted by the liver cells at a constant rate of about 40 mL per hour. When digestion is not taking place, the bile is stored and concentrated in the gallbladder; later, it is delivered to the duodenum. The bile ducts of the liver consist of the **right** and **left hepatic ducts,** the **common hepatic duct,** the **bile duct,** the **gallbladder,** and the **cystic duct.**

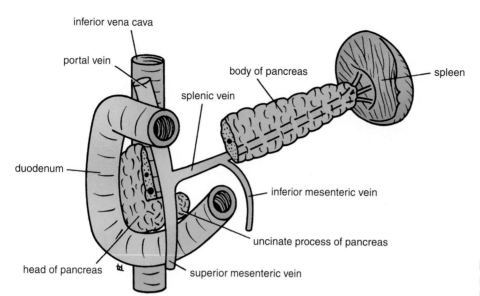

inferior vena cava

portal vein

body of pancreas

spleen

splenic vein

duodenum

inferior mesenteric vein

uncinate process of pancreas

head of pancreas

superior mesenteric vein

**Figure 20-7**   Formation of the portal vein behind the neck of the pancreas.

The smallest interlobular tributaries of the bile ducts are situated in the portal canals of the liver; they receive the bile canaliculi. The interlobular ducts join one another to form progressively larger ducts and, eventually, at the porta hepatis, form the right and left hepatic ducts. The right hepatic duct drains the right lobe of the liver and the left duct drains the left lobe, caudate lobe, and quadrate lobe.

## Hepatic Ducts

The right and left hepatic ducts emerge from the right and left lobes of the liver in the porta hepatis (Fig. 20-3). After a short course, the hepatic ducts unite to form the common hepatic duct (Fig. 20-9).

The **common hepatic duct** is about 1.5 in. (4 cm) long and descends within the free margin of the lesser omentum. It is joined on the right side by the cystic duct from the gall-bladder to form the bile duct (Fig. 20-9).

## Bile Duct

The bile duct (common bile duct) is about 3 in. (8 cm) long. In the first part of its course, it lies in the right free margin of the lesser omentum in front of the opening into the lesser sac. Here, it lies in front of the right margin of the portal vein and on the right of the hepatic artery (Fig. 20-10). In the second part of its course, it is situated behind the first part of the duodenum (Fig. 20-8) to the right of the gastroduodenal artery.

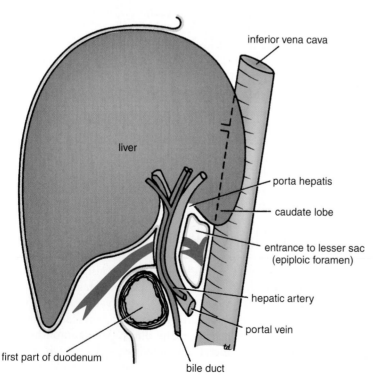

inferior vena cava

liver

porta hepatis

caudate lobe

entrance to lesser sac (epiploic foramen)

hepatic artery

portal vein

first part of duodenum

bile duct

**Figure 20-8**   Sagittal section through the entrance into the lesser sac showing the important structures that form boundaries to the opening. (Note the *arrow* passing from the greater sac through the epiploic foramen into the lesser sac.)

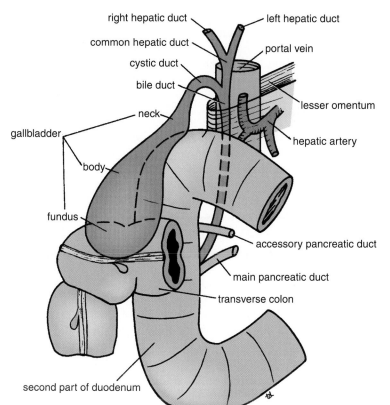

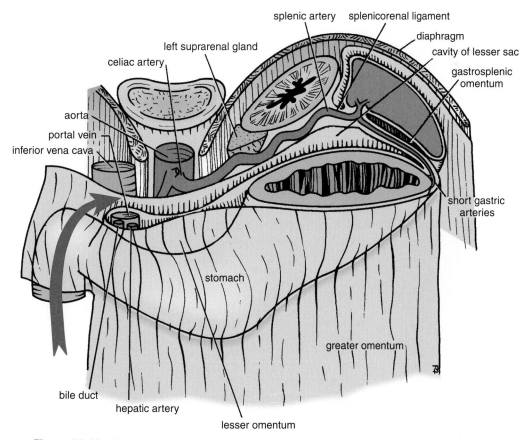

**Figure 20-9**   The bile ducts and the gallbladder. Note the relation of the gallbladder to the transverse colon and the duodenum.

**Figure 20-10**   Transverse section of the lesser sac showing the arrangement of the peritoneum in the formation of the lesser omentum, the gastrosplenic omentum, and the splenicorenal ligament. *Arrow* indicates the position of the opening of the lesser sac.

In the third part of its course, it lies in a groove on the posterior surface of the head of the pancreas (Fig. 20-9). Here, the bile duct comes into contact with the main pancreatic duct.

The bile duct ends below by piercing the medial wall of the second part of the duodenum about halfway down its length (Fig. 20-11). It is usually joined by the main pancreatic duct, and together, they open into a small ampulla in the duodenal wall, called the **hepatopancreatic ampulla (ampulla of Vater)**. The ampulla opens into the lumen of the duodenum by means of a small papilla, the **major duodenal papilla** (Fig. 20-11). The terminal parts of both ducts and the ampulla are surrounded by a circular muscle, known as the **sphincter of the hepatopancreatic ampulla (sphincter of Oddi)** (Fig. 20-11). Occasionally, the bile and pancreatic ducts open separately into the duodenum. The common variations of this arrangement are shown in Figure 20-12.

# Gallbladder

## Location and Description

The gallbladder is a pear-shaped sac lying on the undersurface of the liver (Figs. 20-2 and 20-9). It has a capacity of 30 to 50 mL and stores bile, which it concentrates by absorbing water. For descriptive purposes, the gallbladder is divided into the fundus, body, and neck. The **fundus** is rounded and usually projects below the inferior margin of the liver, where it comes in contact with the anterior abdominal wall at the level of the tip of the ninth right costal cartilage. The **body** lies in contact with the visceral surface of the liver and is directed upward, backward, and to the left. The **neck** becomes continuous with the cystic duct, which turns into the lesser omentum to join the right side of the common hepatic duct, to form the bile duct (Fig. 20-9).

The peritoneum completely surrounds the fundus of the gallbladder and binds the body and neck to the visceral surface of the liver.

## Relations

- **Anteriorly:** The anterior abdominal wall and the inferior surface of the liver (Fig. 20-13).
- **Posteriorly:** The transverse colon and the first and second parts of the duodenum (Fig. 20-9).

---

**PHYSIOLOGIC NOTE**

### Function of the Gallbladder

When digestion is not taking place, the sphincter of Oddi remains closed, and bile accumulates in the gallbladder. The gallbladder concentrates bile; stores bile; selectively absorbs bile salts, keeping the bile acid; excretes cholesterol; and secretes mucus. To aid in these functions, the mucous membrane is thrown into permanent folds that unite with each other, giving the surface a honeycombed appearance. The columnar cells lining the surface also have numerous microvilli on their free surface.

Bile is delivered to the duodenum as the result of contraction and partial emptying of the gallbladder. This mechanism is initiated by the entrance of fatty foods into the duodenum. The fat causes release of the hormone **cholecystokinin** from the mucous membrane of the duodenum; the hormone then enters the blood, causing the gallbladder to contract. At the same time, the smooth

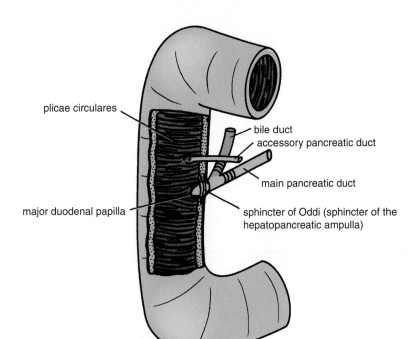

plicae circulares

bile duct
accessory pancreatic duct

main pancreatic duct

major duodenal papilla

sphincter of Oddi (sphincter of the hepatopancreatic ampulla)

**Figure 20-11** Terminal parts of the bile and pancreatic ducts as they enter the second part of the duodenum. Note the sphincter of Oddi and the smooth muscle around the ends of the bile duct and the main pancreatic duct.

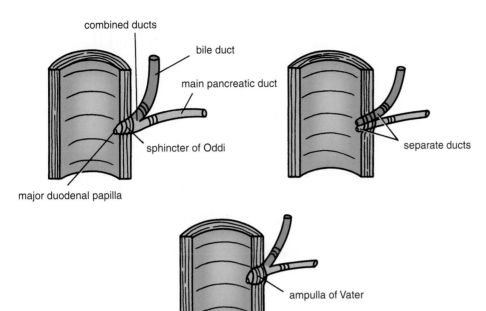

**Figure 20-12**  Three common variations of terminations of the bile and main pancreatic ducts as they enter the second part of the duodenum.

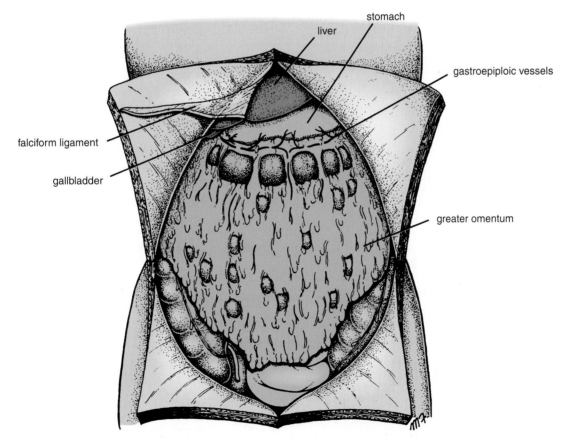

**Figure 20-13**  Abdominal organs in situ. Note that the greater omentum hangs down in front of the small and large intestines.

muscle around the distal end of the bile duct and the ampulla is relaxed, thus allowing the passage of concentrated bile into the duodenum. The bile salts in the bile are important in emulsifying the fat in the intestine and in assisting with its digestion and absorption.

## Blood Supply

The **cystic artery,** a branch of the right hepatic artery (Fig. 20-3), supplies the gallbladder. The **cystic vein** drains directly into the portal vein. Several very small arteries and veins also run between the liver and gallbladder.

## Lymph Drainage

The lymph drains into a **cystic lymph node** situated near the neck of the gallbladder. From here, the lymph vessels pass to the hepatic nodes along the course of the hepatic artery and then to the celiac nodes.

## Nerve Supply

Sympathetic and parasympathetic vagal fibers form the celiac plexus. The gallbladder contracts in response to the hormone cholecystokinin, which is produced by the mucous membrane of the duodenum on the arrival of fatty food from the stomach.

## Cystic Duct

The **cystic duct** is about 1.5 in. (3.8 cm) long and connects the neck of the gallbladder to the common hepatic duct to form the bile duct (Fig. 20-9). It usually is somewhat S shaped and descends for a variable distance in the right free margin of the lesser omentum.

The mucous membrane of the cystic duct is raised to form a spiral fold that is continuous with a similar fold in the neck of the gallbladder. The fold is commonly known as the "spiral valve." The function of the spiral valve is to keep the lumen constantly open.

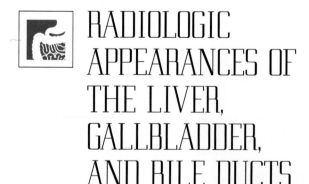

# RADIOLOGIC APPEARANCES OF THE LIVER, GALLBLADDER, AND BILE DUCTS

The radiologic appearances of the liver, gallbladder, and bile ducts are shown in Figures 20-14, 20-15, 20-16, and 20-17.

# SONOGRAM OF THE GALLBLADDER

A sonogram of the gallbladder is shown in Figure 20-18.

---

**EMBRYOLOGIC    NOTE**

### Development of the Liver and Bile Ducts

**Liver**

The liver arises from the distal end of the foregut as a solid bud of entodermal cells (Figs. 20-19 and 20-20). The site of origin lies at the apex of the loop of the developing duodenum and corresponds to a point halfway along the second part of the fully formed duodenum. The **hepatic bud** grows anteriorly into the mass of splanchnic mesoderm called the septum transversum. The end of the bud now divides into right and left branches, from which columns of entodermal cells grow into the vascular mesoderm. The paired vitelline veins and umbilical veins that course through the septum transversum become broken up by the invading columns of liver cells and form the **liver sinusoids.** The columns of entodermal cells form the liver cords. The connective tissue of the liver is formed from the mesenchyme of the septum transversum.

The main hepatic bud and its right and left terminal branches now become canalized to form the **common hepatic duct** and the **right** and **left hepatic ducts.** The liver grows rapidly in size and comes to occupy the greater part of the abdominal cavity; the right lobe becomes much larger than the left lobe.

**Gallbladder and Cystic Duct**

The gallbladder develops from the hepatic bud as a solid outgrowth of cells (Fig. 20-19). The end of the outgrowth expands to form the gallbladder, while the narrow stem remains as the cystic duct. Later, the gallbladder and cystic duct become canalized. The cystic duct now opens into the **common hepatic duct** to form the **bile duct.**

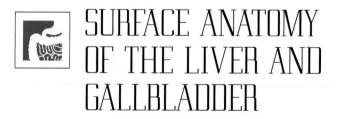

# SURFACE ANATOMY OF THE LIVER AND GALLBLADDER

## Liver

The liver lies under cover of the lower ribs, and most of its bulk lies on the right side (Fig. 20-21). In infants, until

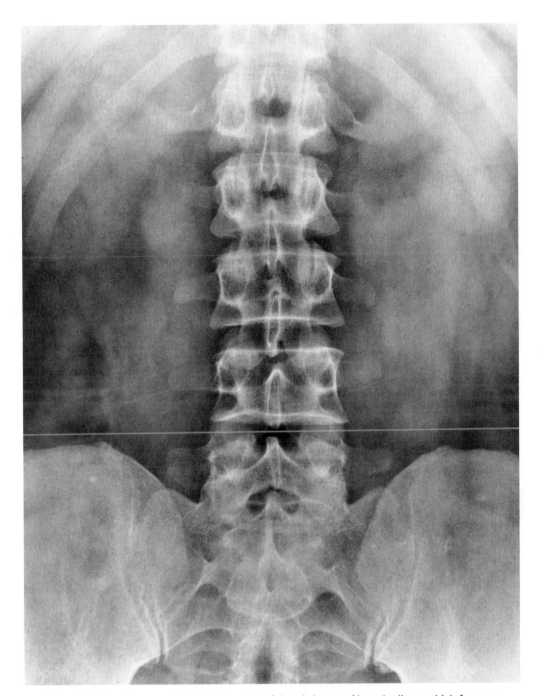

**Figure 20-14** Anteroposterior radiograph of the abdomen. Note the liver, which forms a homogenous opacity in the upper part of the abdomen.

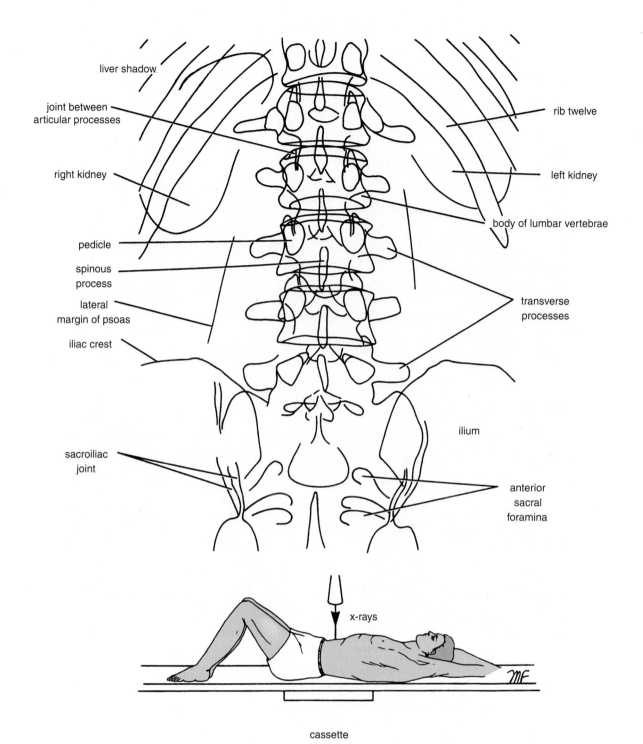

**Figure 20-15**   Representation of the main features seen in the anteroposterior radiograph in Figure 20-14.

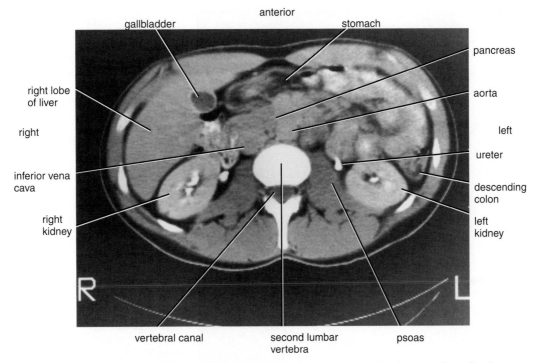

anterior

gallblader

stomach

pancreas

right lobe
of liver

aorta

right

left

ureter

inferior vena
cava

descending
colon

right
kidney

left
kidney

R

L

vertebral canal

second lumbar
vertebra

psoas

**Figure 20-16** CT scan of the abdomen at the level of the second lumbar vertebra after in-travenous pyelography. The radiopaque material can be seen in the renal pelvis and the ureters. The section is viewed from below.

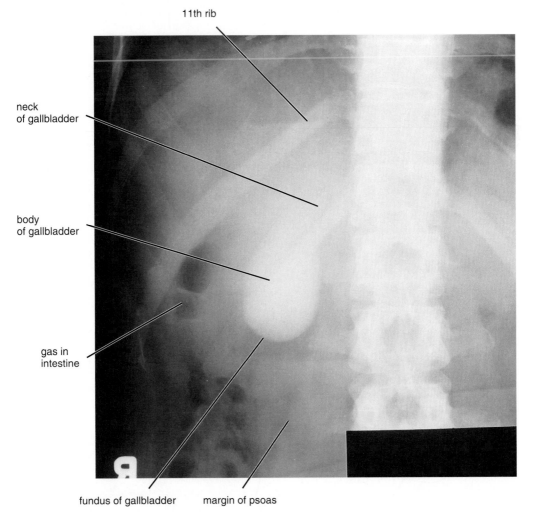

11th rib

neck
of gallbladder

body
of gallbladder

gas in
intestine

fundus of gallbladder    margin of psoas

**Figure 20-17** A cholecystogram. An anteroposterior radiograph of the gallbladder after administration of an iodine-containing compound.

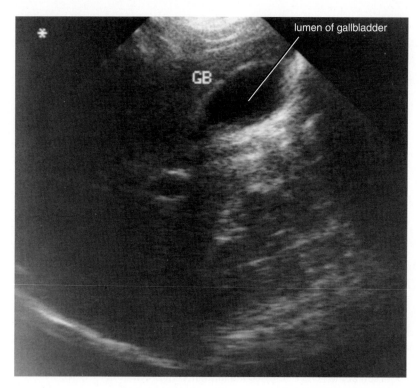

lumen of gallbladder

GB

Figure 20-18 Longitudinal sonogram of the upper part of the abdomen showing the lumen of the gallbladder. (Courtesy of Dr. M.C. Hill.)

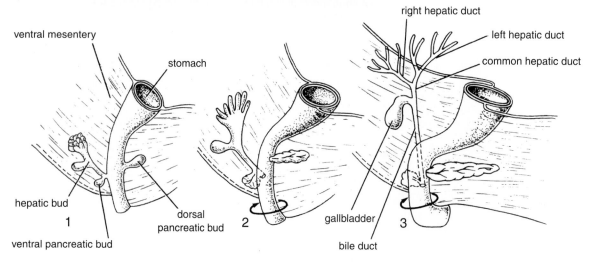

ventral mesentery

stomach

right hepatic duct

left hepatic duct

common hepatic duct

hepatic bud

ventral pancreatic bud

dorsal pancreatic bud

1

2

gallbladder

bile duct

3

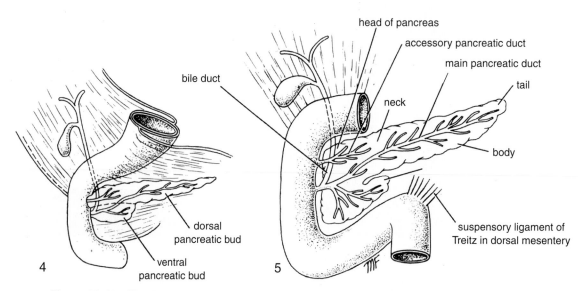

bile duct

head of pancreas

accessory pancreatic duct

main pancreatic duct

tail

neck

body

dorsal pancreatic bud

ventral pancreatic bud

4

5

suspensory ligament of Treitz in dorsal mesentery

Figure 20-19 The development of the pancreas and the extrahepatic biliary apparatus.

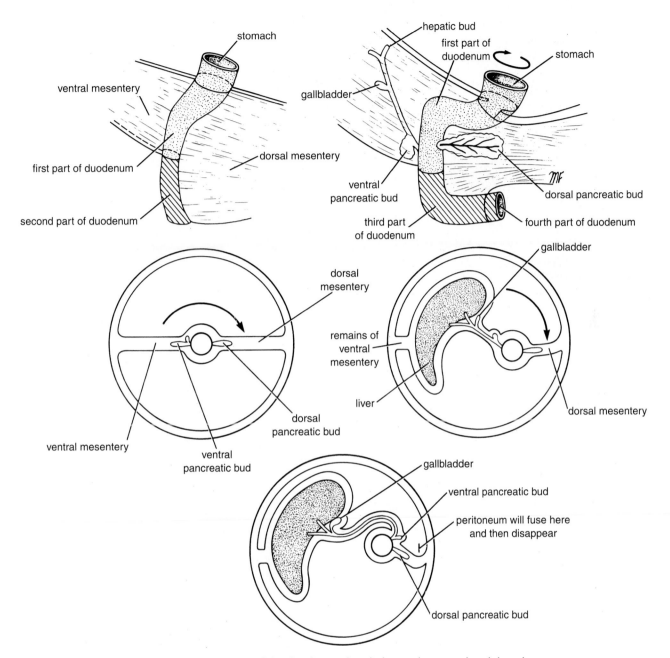

**Figure 20-20**   Development of the duodenum in relation to the ventral and dorsal mesenteries. *Stippled area,* foregut; *crosshatched area,* midgut.

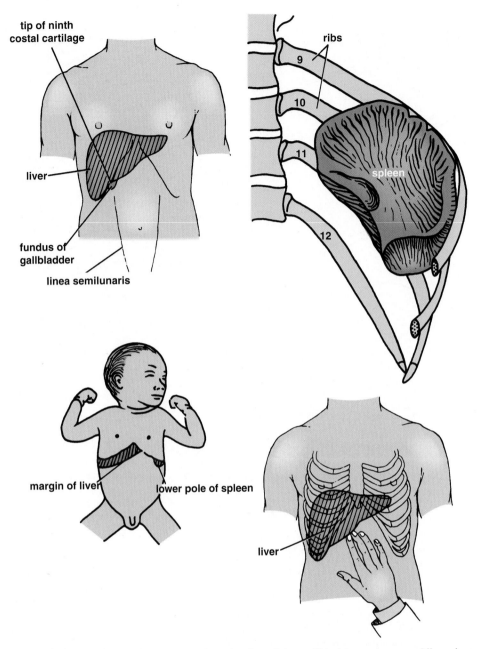

**Figure 20-21** Surface markings of the fundus of the gallbladder, spleen, and liver. In a young child, the lower margin of the normal liver and the lower pole of the normal spleen can be palpated. In a thin adult, the lower margin of the normal liver may just be felt at the end of deep inspiration.

about the end of the 3rd year, the lower margin of the liver extends one or two fingerbreadths below the costal margin (Fig. 20-21). In the adult who is obese or has a well-developed right rectus abdominis muscle, the liver is not palpable. In a thin adult, the lower edge of the liver may be felt a fingerbreadth below the costal margin. It is most easily felt when the patient inspires deeply and the diaphragm contracts and pushes down the liver. The normal edge of the liver is smooth, yielding, and devoid of nodularity.

# Gallbladder

The fundus of the gallbladder lies opposite the tip of the right ninth costal cartilage—that is, where the lateral edge of the right rectus abdominis muscle crosses the costal margin (Fig. 20-21).

# Pancreas

## Location and Description

The pancreas is an elongated structure that lies in the epigastrium and the left upper quadrant. It is soft and lob-ulated and situated on the posterior abdominal wall behind the peritoneum. It crosses the transpyloric plane. The pancreas is divided into a head, neck, body, and tail (Fig. 20-22).

The **head** of the pancreas is disc shaped and lies within the concavity of the duodenum (Fig. 20-22). A part of the head extends to the left behind the superior mesenteric vessels and is called the **uncinate process.**

The **neck** is the constricted portion of the pancreas and connects the head to the body. It lies in front of the beginning of the portal vein and the origin of the superior mesenteric artery from the aorta.

The **body** runs upward and to the left across the midline (Fig. 20-22). It is somewhat triangular in cross section.

The **tail** passes forward in the splenicorenal ligament and comes in contact with the hilum of the spleen (Fig. 20-22).

---

### PHYSIOLOGIC NOTE

#### Functions of the Pancreas

The pancreas is both an exocrine and an endocrine gland. The exocrine portion of the gland produces a

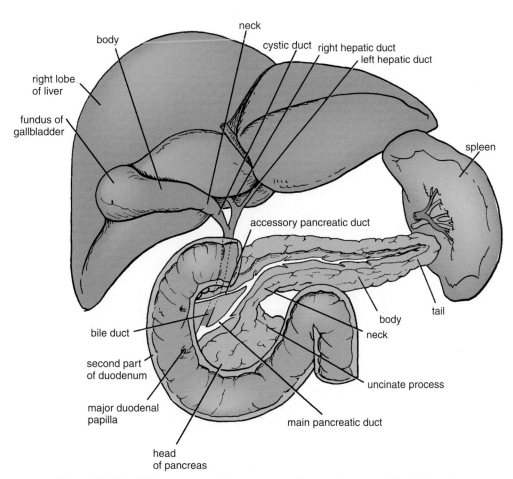

**Figure 20-22**  Different parts of the pancreas dissected to reveal the duct system.

secretion that contains enzymes capable of hydrolyzing proteins, fats, and carbohydrates. The endocrine portion of the gland, the pancreatic islets (islets of Langerhans), produces the hormones insulin and glucagon, which play a key role in carbohydrate metabolism.

## Relations

■ **Anteriorly:** From right to left, the transverse colon and the attachment of the transverse mesocolon, the lesser sac, and the stomach (Fig. 20-23; see also Fig. 19-21).

■ **Posteriorly:** From right to left, the bile duct, the portal and splenic veins, the inferior vena cava, the aorta, the origin of the superior mesenteric artery, the left psoas muscle, the left suprarenal gland, the left kidney, and the hilum of the spleen (Figs. 20-23 and 20-24).

## Pancreatic Ducts

The **main duct of the pancreas** begins in the tail and runs the length of the gland, receiving numerous tributaries on the way (Fig. 20-22). It opens into the second part of the duodenum at about its middle with the bile duct on the **major duodenal papilla** (Fig. 20-11). Sometimes the main duct drains separately into the duodenum (Fig. 20-12).

The **accessory duct** of the pancreas, when present, drains the upper part of the head and then opens into the duodenum a short distance above the main duct on the **minor duodenal papilla** (Figs. 20-11 and 20-22). The accessory duct frequently communicates with the main duct.

## Blood Supply

### Arteries

The splenic and the superior and inferior pancreaticoduodenal arteries (see Fig. 19-47) supply the pancreas.

### Veins

The corresponding veins drain into the portal system.

## Lymph Drainage

Lymph nodes situated along the arteries that supply the gland. The efferent vessels of these nodes ultimately drain into the celiac and superior mesenteric lymph nodes.

## Nerve Supply

Sympathetic and parasympathetic (vagal) nerve fibers supply the area.

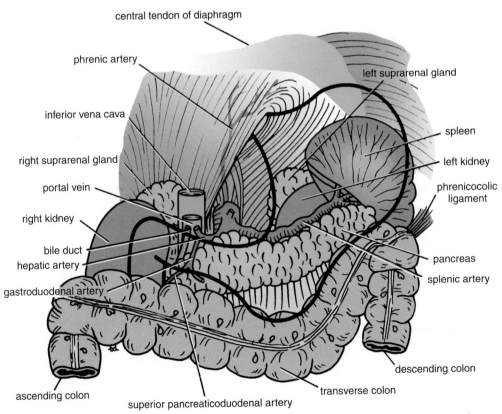

**Figure 20-23**   Structures situated on the posterior abdominal wall behind the stomach.

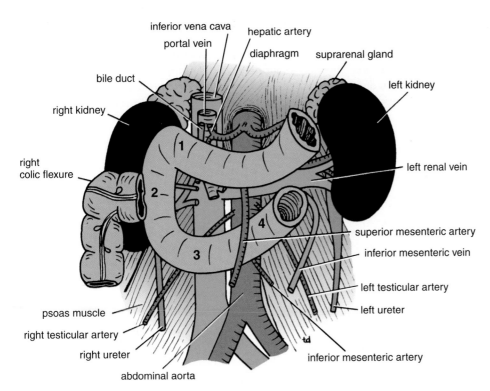

**Figure 20-24** Posterior relations of the duodenum and the pancreas. The *numbers* represent the four parts of the duodenum.

---

**E M B R Y O L O G I C   N O T E**

### Development of the Pancreas

The pancreas develops from a **dorsal** and **ventral bud** of entodermal cells that arise from the foregut. The dorsal bud originates a short distance above the ventral bud and grows into the dorsal mesentery. The ventral bud arises in common with the hepatic bud, close to the junc-

tion of the foregut with the midgut (Fig. 20-19). A canalized duct system now develops in each bud. The rotation of the stomach and duodenum, together with the rapid growth of the left side of the duodenum, results in the ventral bud's coming into contact with the dorsal bud, and fusion occurs (Fig. 20-25).

Fusion also occurs between the ducts, so that the **main pancreatic duct** is derived from the entire ventral

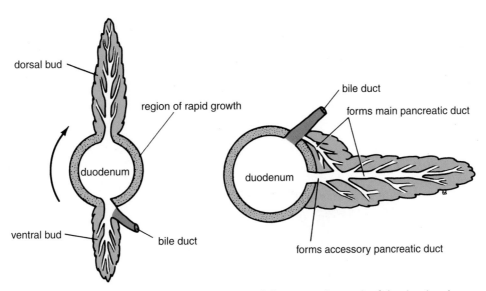

**Figure 20-25** The rotation of the duodenum and the unequal growth of the duodenal wall lead to the fusing of the ventral and dorsal pancreatic buds.

pancreatic duct and the distal part of the dorsal pancreatic duct. The main pancreatic duct joins the bile duct and enters the second part of the duodenum. The proximal part of the dorsal pancreatic duct may persist as an **accessory duct,** which may or may not open into the duodenum about 0.75 in. (2 cm) above the opening of the main duct.

Continued growth of the entodermal cells of the now-fused ventral and dorsal pancreatic buds extend into the surrounding mesenchyme as columns of cells. These columns give off side branches, which later become canalized to form collecting ducts. Secretory acini appear at the ends of the ducts.

The **pancreatic islets** arise as small buds from the developing ducts. Later, these cells sever their connection with the duct system and form isolated groups of cells that start to secrete **insulin** and **glucagon** at about the 5th month.

The inferior part of the head and the uncinate process of the pancreas are formed from the ventral pancreatic bud; the superior part of the head, the neck, the body, and the tail of the pancreas are formed from the dorsal pancreatic bud (Fig. 20-25).

### Entrance of the Bile Duct and Pancreatic Duct into the Duodenum

As seen from development, the bile duct and the main pancreatic duct are joined to one another. They pass obliquely through the wall of the second part of the duodenum to open on the summit of the **major duodenal papilla,** which is surrounded by the **sphincter of Oddi** (Fig. 20-26). In some individuals, they pass separately through the duodenal wall, although in close contact, and open separately on the summit of the duodenal papilla. In other individuals, the two ducts join and form a common dilatation, the **hepatopancreatic ampulla (ampulla of Vater).** This opens on the summit of the duodenal papilla.

# RADIOLOGIC APPEARANCES OF THE PANCREAS

The radiologic appearances of the pancreas are seen in Figure 20-16.

# SURFACE ANATOMY OF THE PANCREAS

Despite the fact that the pancreas lies on the posterior abdominal wall, it can be represented on the anterior abdominal wall. The pancreas lies across the transpyloric plane. The head lies below and to the right, the neck lies on the plane, and the body and tail lie above and to the left.

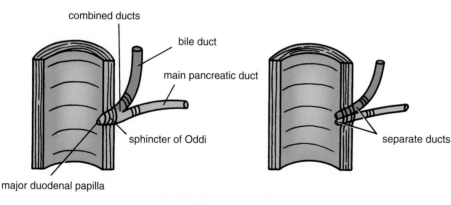

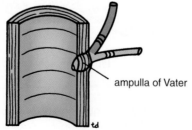

**Figure 20-26**  Three common variations of terminations of the bile and main pancreatic ducts as they enter the second part of the duodenum.

# Spleen

## Location and Description

The spleen is reddish and is the largest single mass of lymphoid tissue in the body. It is oval shaped and has a notched anterior border. It lies just beneath the left half of the diaphragm close to the ninth, tenth, and eleventh ribs. The long axis lies along the shaft of the tenth rib, and its lower pole extends forward only as far as the midaxillary line and cannot be palpated on clinical examination (Fig. 20-27).

The spleen is surrounded by peritoneum (Figs. 20-10 and 20-27), which passes from it at the hilum as the gastrosplenic omentum (ligament) to the greater curvature of the stomach (carrying the short gastric and left gastroepiploic vessels). The peritoneum also passes to the left kidney as the splenicorenal ligament (carrying the splenic vessels and the tail of the pancreas).

## Relations

- **Anteriorly:** The stomach, tail of the pancreas, and left colic flexure. The left kidney lies along its medial border (Figs. 20-10 and 20-23).
- **Posteriorly:** The diaphragm; left pleura (left costodiaphragmatic recess); left lung; and ninth, tenth, and eleventh ribs (Figs. 20-10 and 20-27).

## Blood Supply

### Arteries

The large splenic artery is the largest branch of the celiac artery. It has a tortuous course as it runs along the upper border of the pancreas. The splenic artery then divides into about six branches, which enter the spleen at the hilum.

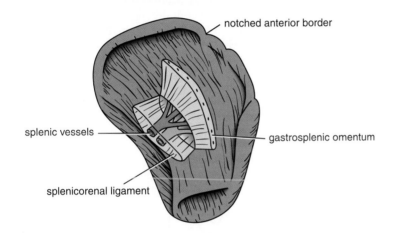

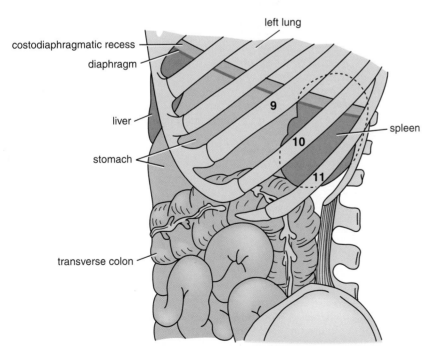

**Figure 20-27** Spleen, with its notched anterior border, and its relation to adjacent structures.

## Veins

The splenic vein leaves the hilum and runs behind the tail and the body of the pancreas. Behind the neck of the pancreas, the splenic vein joins the superior mesenteric vein to form the portal vein.

## Lymph Drainage

The lymph vessels emerge from the hilum and pass through a few lymph nodes along the course of the splenic artery and then drain into the celiac nodes.

## Nerve Supply

The nerves accompany the splenic artery and are derived from the celiac plexus.

---

### E M B R Y O L O G I C   N O T E

#### Development of the Spleen

The spleen develops as a thickening of the mesenchyme in the dorsal mesentery (Fig. 20-28). In the earliest stages, the spleen consists of a number of mesenchymal masses that later fuse. The notches along its anterior border are permanent and indicate that the mesenchymal masses never completely fuse.

The part of the dorsal mesentery that extends between the hilum of the spleen and the greater curvature of the stomach is called the **gastrosplenic omentum;** the part that extends between the spleen and the left kidney on the posterior abdominal wall is called the **splenicorenal ligament.** The spleen is supplied by a branch of the foregut artery (celiac artery), the **splenic artery.**

# SURFACE ANATOMY OF THE SPLEEN

The spleen is situated in the left upper quadrant and lies under cover of the ninth, tenth, and eleventh ribs (Fig. 20-21). Its long axis corresponds to that of the tenth rib, and in the adult, it does not normally project forward in front of the midaxillary line. In infants, the lower pole of the spleen may just be felt (Fig. 20-21).

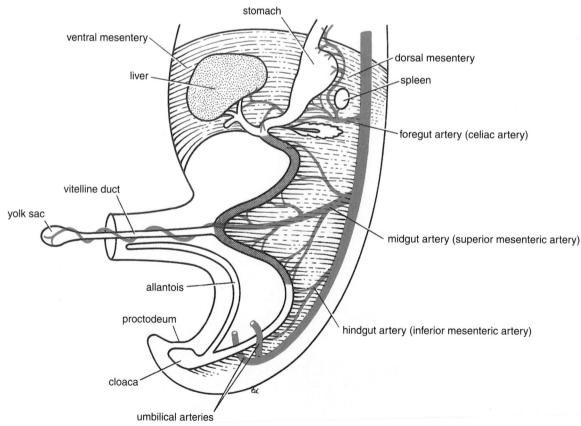

**Figure 20-28**   Formation of the midgut loop (*shaded*). Note how the superior mesenteric artery and vitelline duct form an axis for the future rotation of the midgut loop. Note also the formation of the spleen in the dorsal mesentery.

# Review Questions

## Multiple Choice Questions

**Select the BEST answer for each question.**

1. The following statements concerning the liver are correct **except** which?
   A. The quadrate lobe drains into the right hepatic duct.
   B. The lesser omentum suspends the stomach from the visceral surface of the liver.
   C. The left triangular ligament of the liver lies anterior to the abdominal part of the esophagus.
   D. The attachment of the hepatic veins to the inferior vena cava is one of the most important supports of the liver.
   E. The ligamentum venosum is attached to the left branch of the portal vein in the porta hepatis.

2. The following statements concerning the pancreas are correct **except** which?
   A. The pancreas is located on the posterior abdominal wall behind the peritoneum.
   B. The main pancreatic duct opens into the third part of the duodenum.
   C. The uncinate process of the pancreas projects from the head of the pancreas.
   D. The bile duct (common bile duct) lies posterior to the head of the pancreas.
   E. The transverse mesocolon is attached to the anterior border of the pancreas.

3. The following veins form important portal–systemic anastomoses **except** which?
   A. Esophageal branches of the left gastric vein and tributaries of the azygos veins
   B. Superior rectal vein and inferior vena cava
   C. Paraumbilical veins and superficial veins of the anterior abdominal wall
   D. Veins of the ascending and descending parts of the colon with the lumbar veins
   E. Veins from the bare areas of the liver with the phrenic veins

4. The following statements concerning the liver are correct **except** which?
   A. Its lymph drainage is to the celiac nodes.
   B. The quadrate and the caudate lobes are functionally part of the left lobe.
   C. Its parasympathetic innervation is from the vagus nerve.
   D. It receives highly oxygenated blood from the portal vein.
   E. The triangular ligaments connect the liver to the diaphragm.

5. The following statements concerning the gallbladder are correct **except** which?
   A. The arterial supply is from the cystic artery, which is a branch of the right hepatic artery.
   B. The fundus of the gallbladder is located just beneath the tip of the right ninth costal cartilage.
   C. The peritoneum completely surrounds the fundus, the body, and the neck.
   D. The nerves of the gallbladder are derived from the celiac plexus.
   E. Pain sensation from gallbladder disease may be referred along the phrenic nerve and the supraclavicular nerves to the skin over the shoulder.

6. The following structures are present in the porta hepatis **except** which?
   A. Lymph nodes
   B. The right and left branches of the portal vein
   C. The right and left hepatic ducts
   D. The right and left hepatic veins
   E. The right and left branches of the hepatic artery

7. The following statements concerning the spleen are correct **except** which?
   A. The spleen is situated in the left upper quadrant of the abdomen.
   B. The lower pole of the spleen can be easily palpated in a normal thin adult.
   C. The splenic artery runs along the upper border of the pancreas to reach the spleen.
   D. The spleen is attached to the stomach by the gastrosplenic omentum.
   E. The spleen is completely surrounded by peritoneum.

8. Which of the following arteries forms the main blood supply to the pancreas?
   A. The splenic artery
   B. The aorta
   C. The right renal artery
   D. The left suprarenal artery
   E. The superior mesenteric artery

9. Which of the following structures lies within the splenicorenal ligament?
   A. The left gastroepiploic artery
   B. The tail of the pancreas
   C. The left vagus nerve
   D. The lymphatic vessels from the greater curvature of the stomach
   E. The portal vein

# Answers and Explanations

1. **A** is incorrect. The quadrate lobe and the caudate lobe of the liver are, in fact, parts of the left lobe. Thus, the right and left branches of the hepatic artery and portal vein and the right and left hepatic ducts are distributed to the right lobe and the left lobe plus the quadrate and caudate lobes.

2. **B** is incorrect. The main pancreatic duct opens into the second part of the duodenum, at about its middle, with the bile duct on the major duodenal papilla. Sometimes, the main duct drains separately into the duodenum.

3. **B** is incorrect. The superior rectal veins (tributaries of the portal vein) anastomose with the middle and inferior rectal veins (systemic tributaries).

4. **D** is incorrect. The liver receives highly oxygenated blood via the hepatic artery.

5. **C** is incorrect. The fundus is the only part of the gallbladder completely surrounded by peritoneum.

6. **D** is incorrect. The hepatic veins (three or more in number) leave the posterior surface of the liver and drain directly into the inferior vena cava.

7. **B** is incorrect. The spleen is situated in the left upper quadrant and lies beneath the ninth, tenth, and eleventh ribs (Fig. 20-27). Its long axis corresponds to the tenth rib. In the adult, the lower pole of the spleen does not normally project forward in front of the midaxillary line and therefore cannot be palpated. In infants, the lower pole of the normal spleen may just be felt.

8. **A** is correct. Multiple small branches of the splenic artery supply the neck, body, and tail of the pancreas. The head of the pancreas is supplied by the superior and inferior pancreaticoduodenal arteries.

9. **B** is correct. The tail of the pancreas passes upward and to the left to the hilum of the spleen in the splenicorenal ligament. It is here that the pancreas may be damaged during a splenectomy operation.

# The Urinary System

# 21

# The Kidneys, Ureters, Bladder, and Urethra

All clinical material relevant to this chapter can be found on the CD-ROM.

# Chapter Outline

C linical problems involving the urinary system are common and may involve the kidney, ureter, urinary bladder, and urethra. The patient may present with diverse symptoms, ranging from excruciating pain, to painless hematuria, to failure to void urine. Traumatic injury to the renal system occurs in about 10% of patients with abdominal injury.

The purpose of this chapter is to review the significant anatomy of the urinary system relative to medical problems. Emphasis will be placed on age and sexual differences; for example, in young children, the urinary bladder is an abdominal organ rather than a pelvic organ and is consequently more prone to abdominal injuries than in adults; in females, cystitis is much more common than in males because the urethra is much shorter, and ascending infection is more likely.

# BASIC ANATOMY

The urinary system consists of two kidneys situated on the posterior abdominal wall; two ureters, which run down on the posterior abdominal wall and enter the pelvis; one urinary bladder located within the pelvis; and one urethra, which passes through the perineum (Fig. 21-1).

The urethra in the male not only conducts urine to the surface but is an excretory duct for the reproductive system, conveying the semen to the exterior.

# Kidneys

## Location and Description

The two kidneys are reddish brown and lie behind the peritoneum high up on the posterior abdominal wall on either side of the vertebral column; they are largely under cover of the costal margin (Fig. 21-1). The right kidney lies slightly lower than the left kidney because of the large size of the right lobe of the liver. With contraction of the diaphragm during respiration, both kidneys move downward in a vertical direction by as much as 1 in. (2.5 cm). On the medial concave border of each kidney is a vertical slit that is bounded by thick lips of renal substance and is called the **hilum** (Fig. 21-2). The hilum extends into a large cavity called the **renal sinus**. The hilum transmits, from the front backward, the renal vein, two branches of the renal artery, the ureter, and the third branch of the renal artery (V.A.U.A.). Lymph vessels and sympathetic fibers also pass through the hilum.

---

### P H Y S I O L O G I C   N O T E

#### Functions of the Kidneys

The two kidneys function to excrete most of the waste products of metabolism. They play a major role in controlling the water and electrolyte balance within the body and in maintaining the acid–base balance of the blood. The waste products leave the kidneys as **urine,** which passes down the **ureters** to the **urinary bladder,** located within the pelvis. The urine leaves the body in the **urethra.**

---

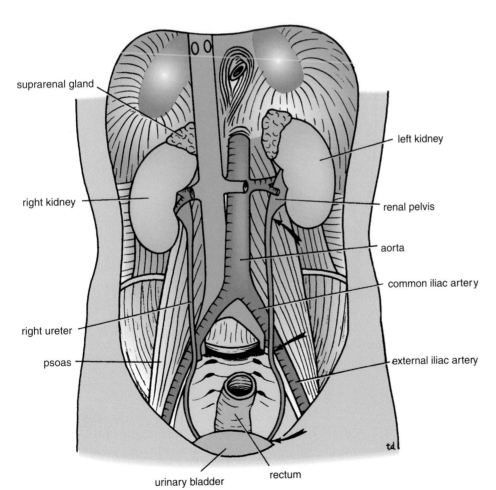

**Figure 21-1** Posterior abdominal wall showing the kidneys and the ureters in situ. *Arrows* indicate three sites where the ureter is narrowed.

suprarenal gland

left kidney

right kidney

renal pelvis

aorta

common iliac artery

right ureter

psoas

external iliac artery

urinary bladder    rectum

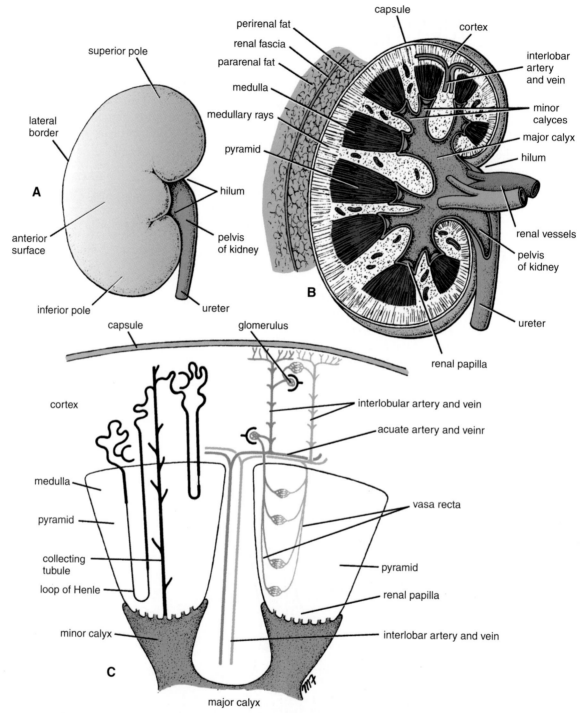

**Figure 21-2** **A.** Right kidney, anterior surface. **B.** Right kidney, coronal section showing the cortex, medulla, pyramids, renal papillae, and calyces. **C.** Section of the kidney showing the position of the nephrons and the arrangement of the blood vessels within the kidney.

## Coverings of the Kidneys

The kidneys have the following coverings (Fig. 21-2):

■ **Fibrous capsule:** This surrounds the kidney and is closely applied to its outer surface.

■ **Perirenal fat:** This is fat that covers the fibrous capsule.

■ **Renal fascia:** This is a condensation of connective tissue that lies outside the perirenal fat and encloses the kidneys and suprarenal glands; it is continuous laterally with the fascia transversalis.

■ **Pararenal fat:** This lies external to the renal fascia and is often in large quantity. It forms part of the retroperitoneal fat.

The perirenal fat, renal fascia, and pararenal fat support the kidneys and hold them in position on the posterior abdominal wall.

## Renal Structure

Each kidney has a dark brown outer **cortex** and a light brown inner **medulla.** The medulla is composed of about a dozen **renal pyramids,** each having its base oriented toward the cortex and its apex, the **renal papilla,** projecting medially (Fig. 21-2). The cortex extends into the medulla between adjacent pyramids as the **renal columns.** Extending from the bases of the renal pyramids into the cortex are striations known as **medullary rays.**

The renal sinus, which is the space within the hilum, contains the upper expanded end of the ureter, the **renal pelvis.** This divides into two or three **major calyces,** each of which divides into two or three **minor calyces** (Fig. 21-2). Each minor calyx is indented by the apex of the renal pyramid, the **renal papilla.**

## Important Relations: Right Kidney

■ **Anteriorly:** The suprarenal gland, the liver, the second part of the duodenum, and the right colic flexure (Figs. 21-3 and 21-4).
■ **Posteriorly:** The diaphragm; the costodiaphragmatic recess of the pleura; the twelfth rib; and the psoas, quadratus lumborum, and transversus abdominis muscles. The subcostal (T12), iliohypogastric, and ilioinguinal nerves (L1) run downward and laterally (Fig. 21-5).

## Important Relations: Left Kidney

■ **Anteriorly:** The suprarenal gland, the spleen, the stomach, the pancreas, the left colic flexure, and coils of jejunum (Figs. 21-3 and 21-4).
■ **Posteriorly:** The diaphragm; the costodiaphragmatic recess of the pleura; the eleventh (the left kidney is higher) and twelfth ribs; and the psoas, quadratus lumborum, and transversus abdominis muscles. The subcostal (T12), iliohypogastric, and ilioinguinal nerves (L1) run downward and laterally (Fig. 21-5).

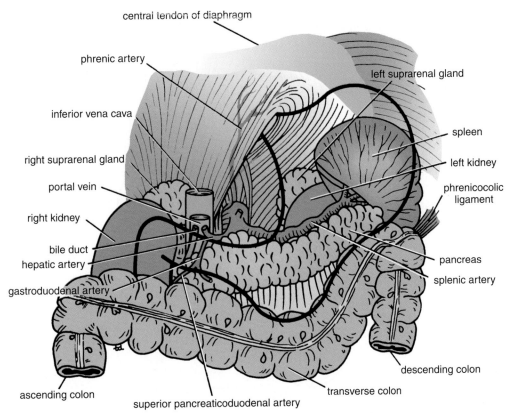

**Figure 21-3**    Structures situated on the posterior abdominal wall behind the stomach. Note the relationships of the kidneys.

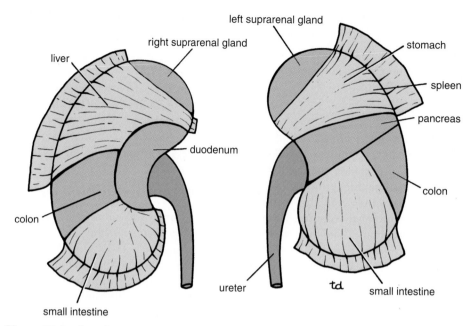

**Figure 21-4**  Anterior relations of both kidneys. Visceral peritoneum covering the kidneys has been left in position. *Shaded areas* indicate where the kidney is in direct contact with the adjacent viscera.

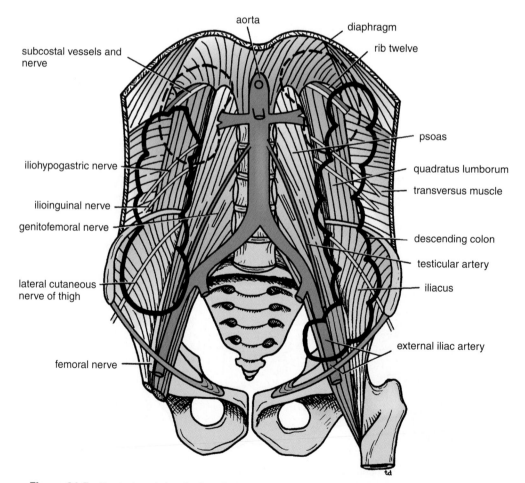

**Figure 21-5**  Posterior abdominal wall showing posterior relations of the kidneys and the colon.

Note that many of the structures are directly in contact with the kidneys, whereas others are separated by visceral layers of peritoneum. For details, see Figure 21-4.

## Blood Supply

### Arteries

The renal artery arises from the aorta at the level of the second lumbar vertebra. Each renal artery usually divides into five **segmental arteries** that enter the hilum of the kidney, four in front and one behind the renal pelvis. They are distributed to different segments or areas of the kidney. **Lobar arteries** arise from each segmental artery, one for each renal pyramid. Before entering the renal substance, each lobar artery gives off two or three **interlobar arteries** (Fig. 21-2). The interlobar arteries run toward the cortex on each side of the renal pyramid. At the junction of the cortex and the medulla, the interlobar arteries give off the **arcuate arteries,** which arch over the bases of the pyramids (Fig. 21-2). The arcuate arteries give off several **interlobular arteries** that ascend in the cortex. The **afferent glomerular arterioles** arise as branches of the interlobular arteries.

### Veins

The renal vein emerges from the hilum in front of the renal artery and drains into the inferior vena cava.

## Lymph Drainage

Lateral aortic lymph nodes around the origin of the renal artery.

## Nerve Supply

Renal sympathetic plexus. The afferent fibers that travel through the renal plexus enter the spinal cord in the tenth, eleventh, and twelfth thoracic nerves.

# RADIOGRAPHIC APPEARANCES OF THE KIDNEY

The radiographic appearances of the kidney are shown in Figures 21-6, 21-7, and 21-8. A CT scan of the kidneys is shown in Figure 21-9.

# Ureter

## Location and Description

The two ureters are muscular tubes that extend from the kidneys to the posterior surface of the urinary bladder (Fig. 21-1). Each ureter measures about 10 in. (25 cm) long and less than ½ in. (1.25 cm) in diameter. It has three constrictions along its course:

- where the renal pelvis joins the ureter
- where it is kinked as it crosses the pelvic brim
- where it pierces the bladder wall (Fig. 21-1)

At its upper end, the ureter is expanded to form a funnel called the **renal pelvis.** It lies within the hilum of the kidney and receives the major calyces (Fig. 21-2). The ureter emerges from the hilum of the kidney and runs vertically downward behind the parietal peritoneum (adherent to it) on the psoas muscle, which separates it from the tips of the transverse processes of the lumbar vertebrae. It enters the pelvis by crossing the bifurcation of the common iliac artery in front of the sacroiliac joint (Fig. 21-1). The ureter then runs down the lateral wall of the pelvis to the region of the ischial spine and turns forward to enter the lateral angle of the bladder.

**PHYSIOLOGIC NOTE**

**Propulsion of Urine**

The urine is propelled along the ureter by peristaltic contractions of the muscle coat, assisted by the filtration pressure of the glomeruli.

## Relations of the Ureters in the Abdomen

### Right Ureter

- **Anteriorly:** The duodenum, the terminal part of the ileum, the right colic and ileocolic vessels, the right testicular or ovarian vessels, and the root of the mesentery of the small intestine (Fig. 21-10).
- **Posteriorly:** The right psoas muscle, which separates it from the lumbar transverse processes, and the bifurcation of the right common iliac artery (Fig. 21-1).

### Left Ureter

- **Anteriorly:** The sigmoid colon and sigmoid mesocolon, the left colic vessels, and the left testicular or ovarian vessels (Fig. 21-10).
- **Posteriorly:** The left psoas muscle, which separates it from the lumbar transverse processes, and the bifurcation of the left common iliac artery (Fig. 21-1).

The inferior mesenteric vein lies along the medial side of the left ureter (Fig. 21-10).

## Relations of the Ureters in the Pelvis

Each ureter runs down the lateral wall of the pelvis to the region of the ischial spine and turns forward to enter the lateral angle of the bladder.

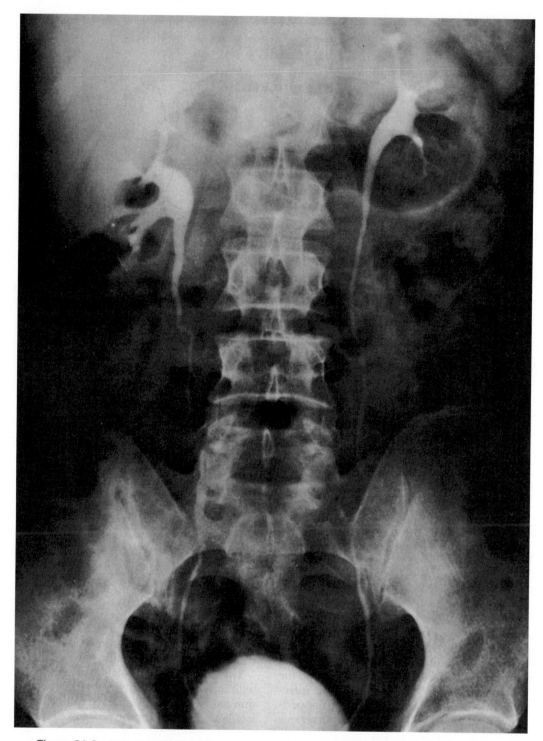

**Figure 21-6**   Anteroposterior radiograph of the ureter and renal pelvis after intravenous injection of an iodine-containing compound, which is excreted by kidney. Major and minor calyces are also shown.

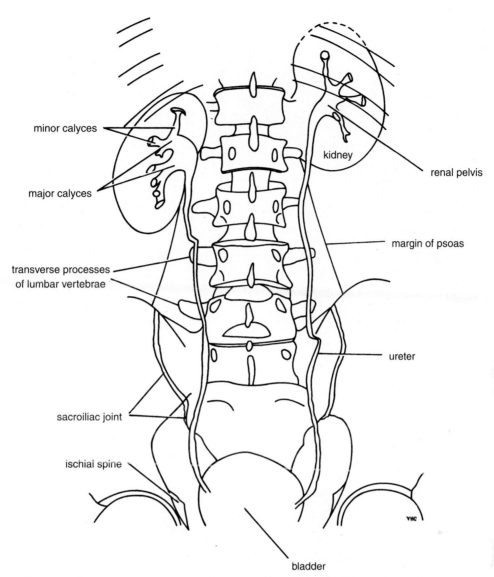

**Figure 21-7**   Representation of the main features seen in the radiograph in Figure 21-6.

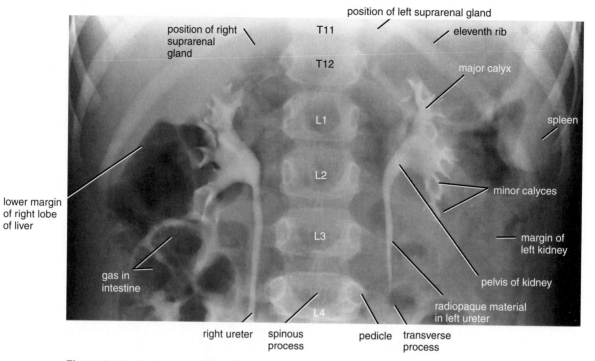

position of left suprarenal gland

position of right suprarenal gland

T11

eleventh rib

T12

major calyx

L1

spleen

L2

minor calyces

lower margin of right lobe of liver

L3

margin of left kidney

gas in intestine

pelvis of kidney

L4

radiopaque material in left ureter

right ureter    spinous process

pedicle    transverse process

**Figure 21-8**   Anteroposterior radiograph of both kidneys 15 min after intravenous injection of an iodine-containing compound. The calyces, the renal pelvis, and the upper parts of the ureters are clearly seen (5-year-old girl).

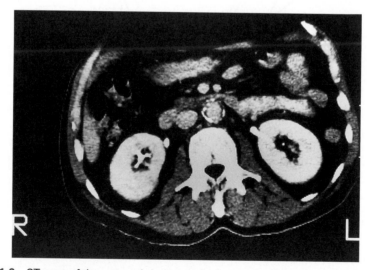

R    L

**Figure 21-9**   CT scan of the upper abdomen at the level of the second lumbar vertebra following an intravenous pyelogram. Note the position of the kidneys and the ureters and the presence of calcification in the aortic wall.

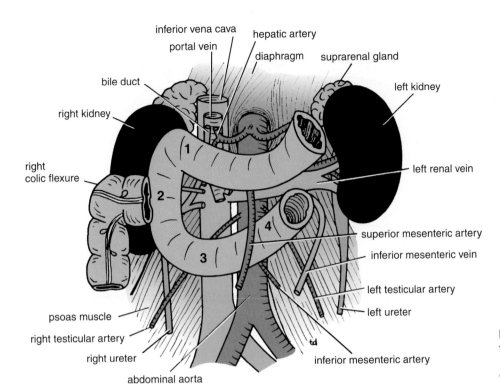

inferior vena cava
hepatic artery
portal vein
diaphragm
suprarenal gland
bile duct
left kidney
right kidney
left renal vein
right
colic flexure
superior mesenteric artery
inferior mesenteric vein
left testicular artery
left ureter
psoas muscle
right testicular artery
right ureter
inferior mesenteric artery
abdominal aorta

**Figure 21-10** Shows the relationships of the kidneys. The *numbers* represent the four parts of the duodenum.

## PHYSIOLOGIC NOTES

### Valvelike Action of the Ureter as It Pierces the Bladder Wall

The ureters pierce the bladder wall obliquely, and this provides a valvelike action that prevents a reverse flow of urine toward the kidneys as the bladder fills.

In the male, the ureter is crossed near its termination by the vas deferens (Fig. 21-11). In the female, the ureter leaves the region of the ischial spine by turning forward and medially beneath the base of the broad ligament (Fig. 21-12); here it is crossed by the uterine artery.

## Blood Supply

### Arteries

The arterial supply to the ureter is as follows:

- Upper end: the renal artery.
- Middle portion: the testicular or ovarian artery.
- The lower end: the superior vesical artery.

### Veins

Venous blood drains into veins that correspond to the arteries.

## Lymph Drainage

The lymph drains to the lateral aortic nodes and the iliac nodes.

## Nerve Supply

Renal, testicular (or ovarian), and hypogastric plexuses (in the pelvis). Afferent fibers travel with the sympathetic nerves and enter the spinal cord in the first and second lumbar segments.

## EMBRYOLOGIC NOTE

### Development of the Kidneys and Ureters

Three sets of structures in the urinary system appear, called the **pronephros, mesonephros,** and **metanephros.** In the human, the metanephros is responsible for the permanent kidney. The metanephros develops from two sources: the ureteric bud from the mesonephric duct and the metanephrogenic cap from the intermediate cell mass of mesenchyme of the lower lumbar and sacral regions.

#### Ureteric Bud

The ureteric bud arises as an outgrowth of the mesonephric duct (Figs. 21-13 and 21-14). It forms the ureter, which dilates at its upper end to form the pelvis of the ureter. The pelvis later gives off branches that form the major calyces, and these, in turn, divide and branch to form the minor calyces and the collecting tubules.

#### Metanephrogenic Cap

The metanephrogenic cap condenses around the ureteric bud (Fig. 21-14) and forms the glomerular cap-

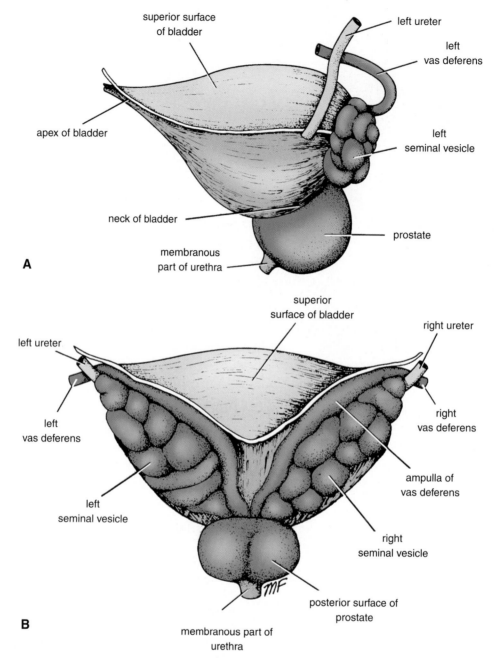

**Figure 21-11   A.** Lateral view of the bladder, prostate, and left seminal vesicle. **B.** Posterior view of the bladder, prostate, vasa deferentia, and seminal vesicles.

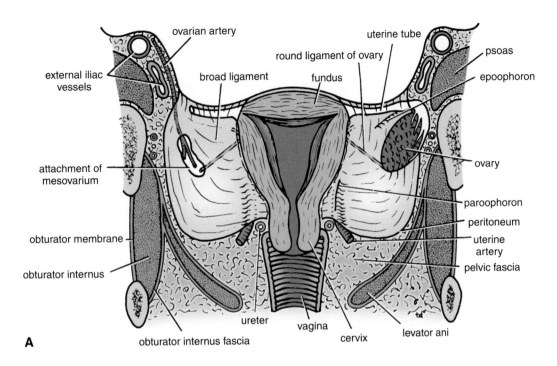

**A**

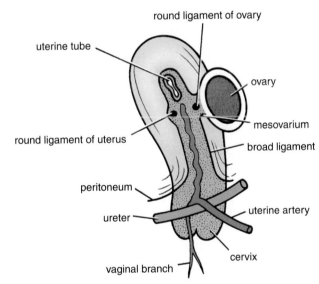

**B**

**Figure 21-12**　**A.** Coronal section of the pelvis showing the uterus, broad ligaments, and right ovary on posterior view. The ureter and uterine artery are seen beneath the base of the broad ligament on each side. **B.** Uterus on lateral view. Note the ureter is crossed by the uterine artery at the base of the broad ligament.

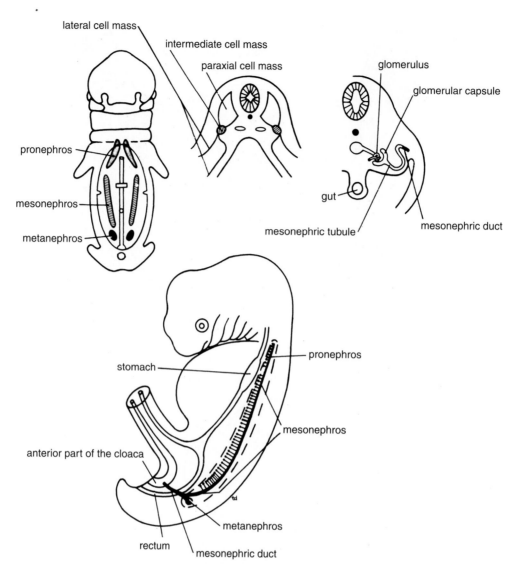

**Figure 21-13** The origins and positions of the pronephros, mesonephros, and metanephros.

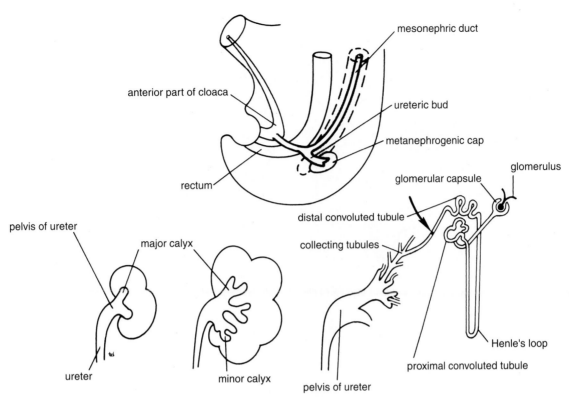

**Figure 21-14** The origin of the ureteric bud from the mesonephric duct and the formation of the major and minor calyces and the collecting tubules. *Arrow* indicates the point of union between the collecting tubules and the convoluted tubules.

sules, the proximal and distal convoluted tubules, and the loops of Henle. The glomerular capsule becomes invaginated by a cluster of capillaries that form the glomerulus. Each distal convoluted tubule formed from the metanephrogenic cap tissue becomes joined to a collecting tubule derived from the ureteric bud. The surface of the kidney is lobulated at first; but after birth, this lobulation usually soon disappears.

The developing kidney is initially a pelvic organ and receives its blood supply from the pelvic continuation of the aorta, the middle sacral artery. Later, the kidneys "ascend" up the posterior abdominal wall. This so-called ascent is caused mainly by the growth of the body in the lumbar and sacral regions and by the straightening of its curvature. The ureter elongates as the ascent continues.

The kidney is vascularized at successively higher levels by successively higher lateral splanchnic arteries, branches of the aorta. The kidneys reach their final position opposite the second lumbar vertebra. Because of the large size of the right lobe of the liver, the right kidney lies at a slightly lower level than the left kidney.

# RADIOGRAPHIC APPEARANCES OF THE URETER

The radiographic appearances of the ureter are shown in Figures 21-6, 21-7, and 21-8.

# Urinary Bladder

## Location and Description

The urinary bladder is situated immediately behind the pubic bones (Figs. 21-15 and 21-16) within the pelvis. In the adult, the bladder has a maximum capacity of about 500 mL. The bladder has a strong muscular wall. Its shape and relations vary according to the amount of urine that it contains. The empty bladder in the adult lies entirely within the pelvis; as the bladder fills, its superior wall rises up into the hypogastric region

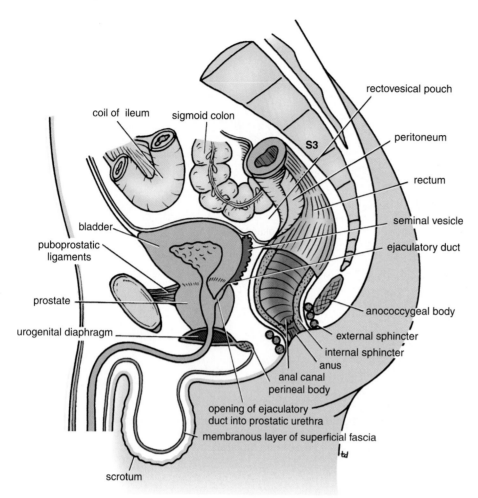

**Figure 21-15** Sagittal section of the male pelvis.

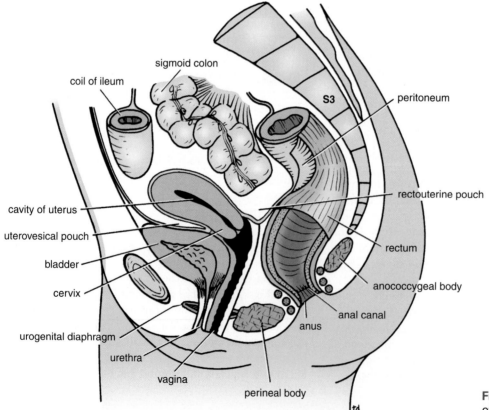

**Figure 21-16** Sagittal section of the female pelvis.

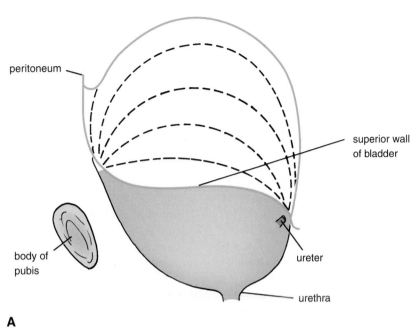

**A**

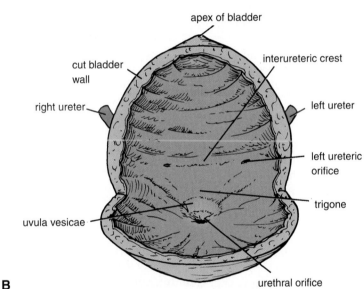

**B**

**Figure 21-17** **A.** Lateral view of the bladder. Note that the superior wall rises as the viscus fills with urine. Note also that the peritoneum covering the superior surface of the bladder is peeled off from the anterior abdominal wall as the bladder fills. **B.** Interior of the bladder in the male as seen from in front.

(Fig. 21-17). In the young child, the empty bladder projects above the pelvic inlet; later, when the pelvic cavity enlarges, the bladder sinks into the pelvis to take up the adult position.

## Shape and Surfaces of the Bladder

The empty bladder is pyramidal (Fig. 21-11), having an apex, a base, and a superior and two inferolateral surfaces; it also has a neck.

The **apex** of the bladder points anteriorly and lies behind the upper margin of the symphysis pubis (Figs. 21-15 and 21-17). It is connected to the umbilicus by the **median umbilical ligament** (remains of urachus).

The **base, or posterior surface,** of the bladder faces posteriorly and is triangular. The superolateral angles are joined by the ureters, and the inferior angle gives rise to the urethra (Fig. 21-11). In the male, the two vasa deferentia lie side by side on the posterior surface of the bladder and separate the seminal vesicles from each other (Fig. 21-11). The upper part of the posterior surface of the bladder is covered by peritoneum, which forms the anterior wall of the rectovesical pouch. The lower part of the posterior surface is separated from the rectum by the vasa deferentia, the seminal vesicles, and the rectovesical fascia (Fig. 21-15). In the female, the uterus and vagina lie against the posterior surface (Fig. 21-16).

The **superior surface** of the bladder is covered with peritoneum and is related to coils of ileum or sigmoid colon (Figs. 21-15 and 21-16). Along the lateral margins of this surface, the peritoneum is reflected onto the lateral pelvic walls.

As the bladder fills, it becomes ovoid, and the superior surface bulges upward into the abdominal cavity (Fig. 21-17). The peritoneal covering is peeled off the lower part of the anterior abdominal wall so that the bladder comes into direct contact with the anterior abdominal wall.

The **inferolateral surfaces** are related in front to the **retropubic pad of fat** and the pubic bones. More posteriorly, they lie in contact with the obturator internus muscle above and the levator ani muscle below.

The **neck** of the bladder lies inferiorly and, in the male, rests on the upper surface of the prostate (Fig. 21-15). Here, the smooth muscle fibers of the bladder wall are continuous with those of the prostate. The neck of the bladder is held in position by the **puboprostatic ligaments** in the male and the **pubovesical ligaments** in the female. These ligaments are thickenings of the pelvic fascia. In the female (Fig. 21-16), because of the absence of the prostate, the bladder neck rests directly on the upper surface of the urogenital diaphragm.

When the bladder fills, the posterior surface and neck remain more or less unchanged in position, but the superior surface rises into the abdomen, as described in the previous paragraphs.

## Interior of the Bladder

The **mucous membrane** of the greater part of the empty bladder is thrown into folds that disappear when the bladder is full. The area of mucous membrane covering the internal surface of the base of the bladder is referred to as the **trigone**. Here, the mucous membrane is always smooth, even when the viscus is empty (Fig. 21-17), because the mucous membrane over the trigone is firmly adherent to the underlying muscular coat.

The superior angles of the trigone correspond to the openings of the ureters, and the inferior angle corresponds to the internal urethral orifice (Fig. 21-17). The ureters pierce the bladder wall obliquely, and this provides a valve-like action, which prevents a reverse flow of urine toward the kidneys as the bladder fills.

The trigone is limited above by a muscular ridge, which runs from the opening of one ureter to that of the other and is known as the **interureteric ridge**. The **uvula vesicae** is a small elevation situated immediately behind the urethral orifice, which is produced by the underlying median lobe of the prostate.

## Muscle Coat of the Bladder

The muscular coat of the bladder is composed of smooth muscle and is arranged as three layers of interlacing bundles known as the **detrusor muscle**. At the neck of the

bladder, the circular component of the muscle coat is thickened to form the **sphincter vesicae.**

## Ligaments of the Bladder

The neck of the bladder is held in position by the **puboprostatic ligaments** in males and by **pubovesical ligaments** in females. These ligaments are formed from pelvic fascia.

## Relations of the Bladder

### In Males (Fig. 21-15)

- **Anteriorly:** Symphysis pubis, retropubic pad of fat, and anterior abdominal wall.
- **Posteriorly:** Rectovesical pouch of peritoneum, vasa deferentia, the seminal vesicles, rectovesical fascia, and the rectum.
- **Laterally:** Obturator internus muscle above and the levator ani muscle below.
- **Superiorly:** Peritoneal cavity, coils of ileum, and sigmoid colon.
- **Inferiorly:** Prostate.

### In Females (Fig. 21-16)

Because of the absence of the prostate, the bladder lies at a lower level in the female pelvis than in the male pelvis, and the neck rests directly on the urogenital diaphragm. The close relationship of the bladder to the uterus and the vagina is of **considerable clinical importance** (Fig. 21-16).

- **Anteriorly:** Symphysis pubis, retropubic pad of fat, and anterior abdominal wall.
- **Posteriorly:** Separated from the rectum by the vagina.
- **Laterally:** Obturator internus muscle above and the levator ani muscle below.
- **Superiorly:** Uterovesical pouch of peritoneum and the body of the uterus.
- **Inferiorly:** Urogenital diaphragm.

## Blood Supply

### Arteries

The superior and inferior vesical arteries, which are branches of the internal iliac arteries, supply the bladder.

### Veins

The veins form the **vesical venous plexus,** which communicates below with the prostatic plexus; it is drained into the internal iliac vein.

## Lymph Drainage

The lymph vessels drain into the internal and external iliac nodes.

## Nerve Supply

The nerve supply to the bladder is from the inferior hypogastric plexuses. The sympathetic postganglionic fibers originate in the first and second lumbar ganglia and descend to the bladder via the hypogastric plexuses. The parasympathetic preganglionic fibers arise as the pelvic splanchnic nerves from the second, third, and fourth sacral nerves; they pass through the inferior hypogastric plexuses to reach the bladder wall, where they synapse with postganglionic neurons. Most afferent sensory fibers arising in the bladder reach the central nervous system via the pelvic splanchnic nerves. Some afferent fibers travel with the sympathetic nerves via the hypogastric plexuses and enter the first and second lumbar segments of the spinal cord.

The sympathetic nerves[*] inhibit contraction of the detrusor muscle of the bladder wall and stimulate closure of the sphincter vesicae. The parasympathetic nerves stimulate contraction of the detrusor muscle of the bladder wall and inhibit the action of the sphincter vesicae.

## PHYSIOLOGIC NOTE

### Micturition

The maximum capacity of the adult bladder is about 500 mL. Micturition is a reflex action that, in the toilet-trained individual, is controlled by higher centers in the brain. The **micturition reflex** is initiated when the volume of urine reaches about 300 mL; stretch receptors in the bladder wall are stimulated and transmit impulses to the central nervous system, and the individual has a conscious desire to micturate. Most afferent impulses pass up the pelvic splanchnic nerves and enter the second, third, and fourth sacral segments of the spinal cord (Fig. 21-18). Some afferent impulses travel with the sympathetic nerves via the hypogastric plexuses and enter the first and second lumbar segments of the spinal cord.

Efferent parasympathetic impulses leave the cord from the second, third, and fourth sacral segments and pass via the parasympathetic preganglionic nerve fibers through the pelvic splanchnic nerves and the inferior hypogastric plexuses to the bladder wall, where they synapse with postganglionic neurons. By means of this

nervous pathway, the smooth muscle of the bladder wall (the detrusor muscle) is made to contract, and the sphincter vesicae is made to relax. Efferent impulses also pass to the urethral sphincter via the pudendal nerve (S2, 3, and 4), and this undergoes relaxation. Once urine enters the urethra, additional afferent impulses pass to the spinal cord from the urethra and reinforce the reflex action. Micturition can be assisted by contraction of the abdominal muscles to raise the intra-abdominal and pelvic pressures and exert external pressure on the bladder.

In young children, micturition is a simple reflex act and takes place whenever the bladder becomes distended. In the adult, this simple stretch reflex is inhibited by the activity of the cerebral cortex until the time and place for micturition are favorable. The inhibitory fibers pass downward with the corticospinal tracts to the second, third, and fourth sacral segments of the cord. Voluntary control of micturition is accomplished by contracting the sphincter urethrae, which closes the urethra; this is assisted by the sphincter vesicae, which compresses the bladder neck.

Voluntary control of micturition is normally developed during the 2nd or 3rd year of life.

# RADIOGRAPHIC APPEARANCES OF THE BLADDER

The radiographic appearances of the bladder are shown in Figures 21-6 and 21-7. A cystourethrogram of the male urethra is shown in Figures 21-19 and 21-20.

## Urethra

The urethra is a small tube leading from the neck of the bladder to the exterior. The opening of the urethra on the surface is called the **urinary meatus.**

### Male Urethra

The male urethra is about 8 in. (20 cm) long and extends from the neck of the bladder to the external meatus on the glans penis (Fig. 21-15). It is divided into three parts: prostatic, membranous, and penile.

The **prostatic urethra** is about 1.25 in. (3 cm) long and passes through the prostate from the base to the apex (Fig. 21-21). It is the widest and most dilatable portion of the urethra. On the posterior wall is a longitudinal ridge called the **urethral crest** (see Fig. 22-9) On each side of this ridge is a groove called the **prostatic sinus** into which

---

[*] The sympathetic nerves to the detrusor muscle are now thought to have little or no action on the smooth muscle of the bladder wall and are distributed mainly to the blood vessels. The sympathetic nerves to the sphincter vesicae are thought to play only a minor role in causing contraction of the sphincter in maintaining urinary continence. However, in males, the sympathetic innervation of the sphincter causes active contraction of the bladder neck during ejaculation (brought about by sympathetic action), thus preventing seminal fluid from entering the bladder.

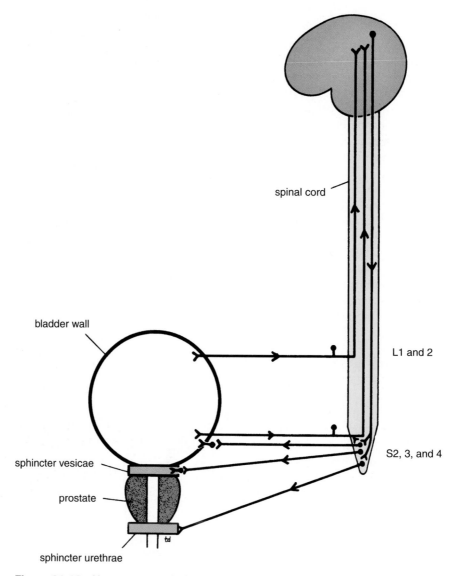

spinal cord

bladder wall

L1 and 2

sphincter vesicae

prostate

S2, 3, and 4

sphincter urethrae

**Figure 21-18** Nervous control of the bladder. Sympathetic fibers have been omitted for simplification.

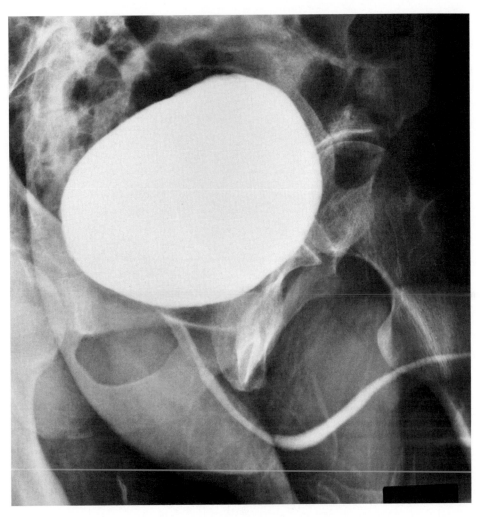

**Figure 21-19**   Cystourethrogram after intravenous injection of contrast medium (28-year-old man).

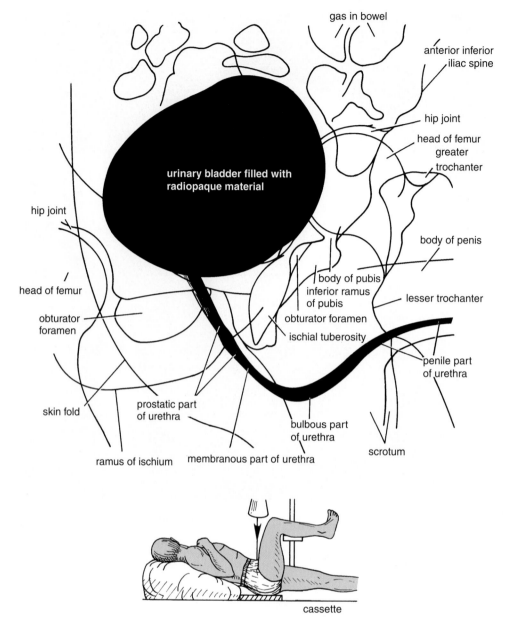

**Figure 21-20**   The main features seen in the cystourethrogram shown in Figure 21-19.

opens the prostatic glands. On the summit of the urethral crest is a depression, the **prostatic utricle,** on the edges of which open the two **ejaculatory ducts.**

The **membranous urethra** is about 0.5 in. (1.25 cm) long and lies within the urogenital diaphragm, surrounded by the sphincter urethrae muscle. It is the shortest and least dilatable part of the urethra (Fig. 21-21).

The **penile urethra** is about 6 in. (15.75 cm) long and is surrounded by the erectile tissue of the bulb and the corpus spongiosum of the penis (Figs. 21-15, 21-21, and 21-22). The external meatus is the **narrowest part of the entire urethra.** The part of the urethra that lies within the glans penis is dilated to form the **fossa terminalis** (navicular fossa). The bulbourethral glands open into the penile urethra below the urogenital diaphragm.

## Female Urethra

The female urethra is about 1.5 in. (3.8 cm) long. It extends from the neck of the bladder to the **external meatus,** where it opens into the vestibule about 1 in. (2.5 cm) below the clitoris (Figs. 21-16 and 21-23). It traverses the sphincter urethrae and lies immediately in front of the vagina. At the sides of the external urethral meatus are the small openings of the ducts of the paraurethral glands. The urethra can be dilated relatively easily.

## Sphincter Urethrae Muscle

The sphincter urethrae muscle surrounds the urethra in the deep perineal pouch. It arises from the pubic arch on the two sides and passes medially to encircle the urethra (Fig. 21-21).

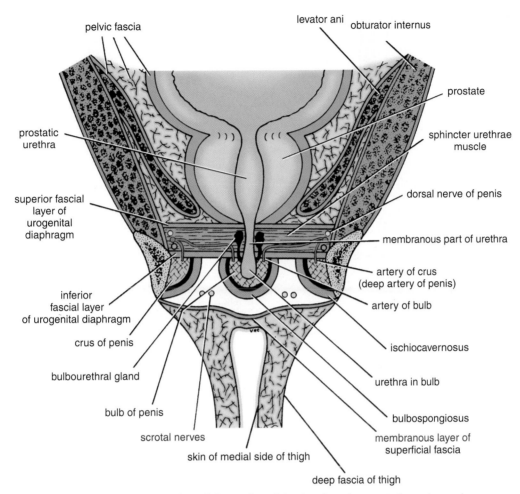

**Figure 21-21** Coronal section of the male pelvis showing the prostatic and membranous parts of the urethra, the urogenital diaphragm, and the contents of the superficial perineal pouch.

<br>

### EMBRYOLOGIC NOTE

## Development of the Bladder, the Fate of the Mesonephric Duct, and the Development of the Urethra in Both Sexes

### Development of the Bladder

The division of the cloaca into anterior and posterior parts by the development of the **urorectal septum** is described on page 759. The posterior portion forms the **anorectal canal** (Fig. 21-24). The entrance of the distal ends of the mesonephric ducts into the anterior part of the cloaca on each side permits one, for purposes of description, to divide the anterior part of the cloaca into an area above the duct entrances called the **primitive bladder** and another area below called the **urogenital sinus**.

The caudal ends of the mesonephric ducts now become absorbed into the lower part of the bladder, so that the ureters and ducts have individual openings in the dorsal wall (Fig. 21-24). With differential growth of the dorsal bladder wall, the ureters come to open through the lateral angles of the bladder, and the mesonephric

ducts open close together in what will be the urethra. That part of the dorsal bladder wall marked off by the openings of these four ducts forms the **trigone** of the bladder (Fig. 21-25). Thus, it is seen that, in the earliest stages, the lining of the bladder over the trigone is mesodermal in origin; later, this mesodermal tissue is thought to be replaced by epithelium of entodermal origin. The smooth muscle of the bladder wall is derived from the splanchnopleuric mesoderm.

The primitive bladder may now be divided into an upper dilated portion, the **bladder**, and a lower narrow portion, the **urethra** (Fig. 21-24). The apex of the bladder is continuous with the **allantois**, which now becomes obliterated and forms a fibrous core, the **urachus**. The urachus persists throughout life as a ligament that runs from the apex of the bladder to the umbilicus and is called the **median umbilical ligament**.

### Fate of the Mesonephric Duct

In both sexes, the mesonephric (or Wolffian) duct gives origin on each side to the **ureteric bud**, which forms the **ureter**, the **pelvis of the ureter**, the **major** and **minor**

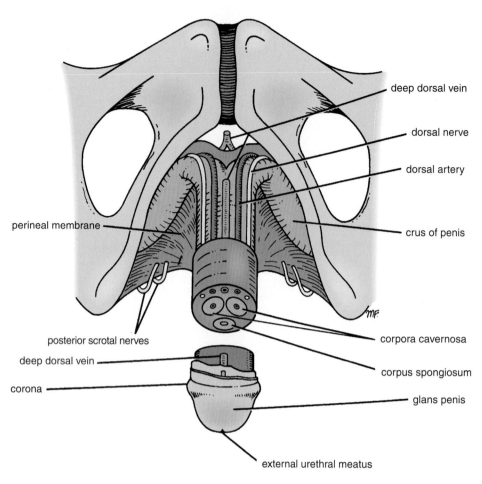

deep dorsal vein

dorsal nerve

dorsal artery

crus of penis

perineal membrane

posterior scrotal nerves

deep dorsal vein

corona

corpora cavernosa

corpus spongiosum

glans penis

external urethral meatus

**Figure 21-22** Root and body of the penis.

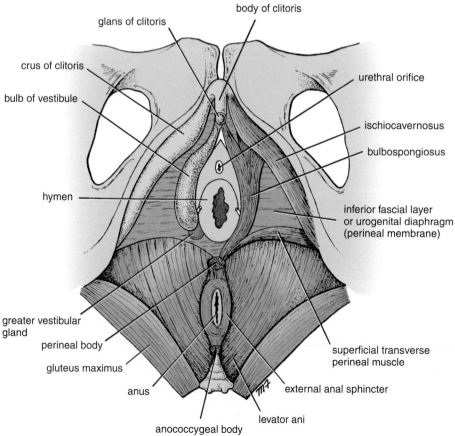

glans of clitoris

body of clitoris

crus of clitoris

bulb of vestibule

urethral orifice

ischiocavernosus

bulbospongiosus

hymen

inferior fascial layer or urogenital diaphragm (perineal membrane)

greater vestibular gland

perineal body

gluteus maximus

anus

anococcygeal body

levator ani

superficial transverse perineal muscle

external anal sphincter

**Figure 21-23** Root and body of the clitoris and the perineal muscles. Shows the urethral orifice.

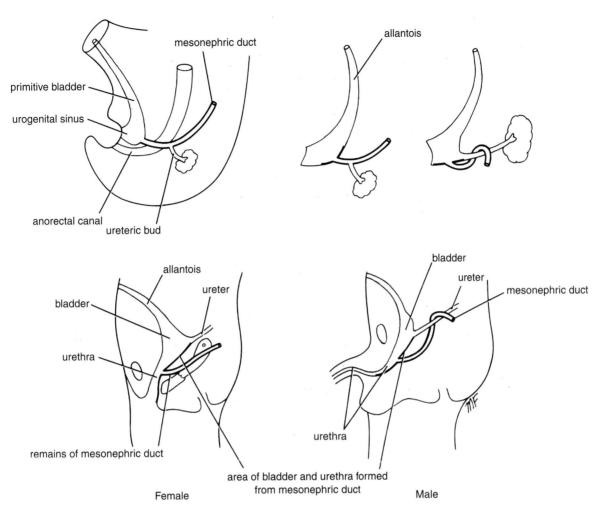

**Figure 21-24** Formation of the urinary bladder from the anterior part of the cloaca and the terminal parts of the mesonephric ducts in both sexes. The mesonephric ducts and the ureteric buds are drawn into the developing bladder.

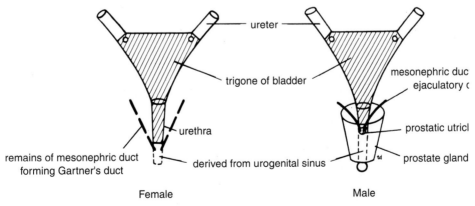

**Figure 21-25** Parts of the bladder and urethra derived from the mesonephric ducts in both sexes (*hatch marks*). The lower end of the urethra in the female and the lower part of the prostatic urethra in the male are formed from the urogenital sinus.

calyces, and the **collecting tubules of the kidney**. Its inferior end is absorbed into the developing bladder and forms the trigone and part of the urethra.

In the male, the upper or cranial end of the mesonephric duct is joined to the developing testis by the efferent ductules of the testis, and so it becomes the **duct of the epididymis**, the **vas deferens**, and the **ejaculatory duct**. From the latter, a small diverticulum arises that forms the **seminal vesicle** (see Fig. 22-5).

In the female, the mesonephric duct largely disappears. Only small remnants persist—as the **duct of the epoophoron** and the **duct of the paroophoron**. The caudal end may persist and extend from the epoophoron to the hymen as **Gartner's duct**.

### Development of the Urethra

In the male, the **prostatic urethra** is formed from two sources. The proximal part, as far as the openings of the ejaculatory ducts, is derived from the mesonephric ducts. The distal part of the prostatic urethra is formed from the urogenital sinus (Fig. 21-25). The **membranous urethra** and the greater part of the **penile urethra** also are formed from the urogenital sinus. The distal end of the penile urethra is derived from an ingrowth of ectodermal cells on the glans penis.

In the female, the upper two thirds of the urethra are derived from the mesonephric ducts. The lower end of the urethra is formed from the urogenital sinus (Fig. 21-25).

# SURFACE ANATOMY OF THE URINARY SYSTEM

## Kidneys

The right kidney lies at a slightly lower level than the left kidney (because of the bulk of the right lobe of the liver), and the lower pole can be palpated in the right lumbar region at the end of deep inspiration in a person with poorly developed abdominal muscles. Each kidney moves about 1 in. (2.5 cm) in a vertical direction during full respiratory movement of the diaphragm. The normal left kidney, which is higher than the right kidney, is not palpable.

On the anterior abdominal wall, the hilum of each kidney lies on the transpyloric plane, about three finger-breadths from the midline (Fig. 21-26). On the back, the kidneys extend from the twelfth thoracic spine to the third lumbar spine, and the hili are opposite the first lumbar vertebra (Fig. 21-26).

## Ureters

On the anterior abdominal wall, the ureter may be indicated by a line drawn downwards from the transpyloric plane at a distance of about 2.5 in. (5 cm) from the midline (Fig. 21-26). At the level of the anterior superior iliac spine, the pelvic portion of the ureter may be indicated by curving the line downward and medially to the pubic tubercle.

On the posterior abdominal wall, the abdominal portion of the ureter may be indicated by a line drawn downward from the level of the first lumbar spine to the posterior inferior iliac spine at a distance of 2 in (5 cm) from the midline.

## Urinary Bladder

In adults, the empty bladder is a pelvic organ and lies posterior to the pubic symphysis. As the bladder fills, it rises up out of the pelvis and comes to lie in the abdomen. The peritoneum covering the distended bladder becomes peeled off from the anterior abdominal wall so that the front of the bladder comes to lie in direct contact with the abdominal wall, as previously discussed.

In children until the age of 6 years, the bladder is an abdominal organ even when empty, since the capacity of the pelvic cavity is not great enough to contain it. The neck of the bladder lies just below the level of the upper border of the symphysis pubis. In infants, suprapubic aspiration is a common procedure to obtain a urine sample in situations of a febrile infant who needs a urine analysis as part of a septic workup.

## Urethra

The **male urethra** is about 8 in (20 cm) long and extends from the neck of the bladder to the external meatus on the glans penis. The prostatic and membranous parts are deeply placed and cannot be palpated directly. The penile part lies within the bulb and shaft of the corpus spongiosum and can be felt throughout its course. The external urethral meatus is the narrowest part of the entire urethra.

The **female urethra** is about 1 ½ in. (3.8 cm) long. It extends from the neck of the bladder to the vestibule of the vulva, where it opens about 1 in. (2.5 cm) below the clitoris.

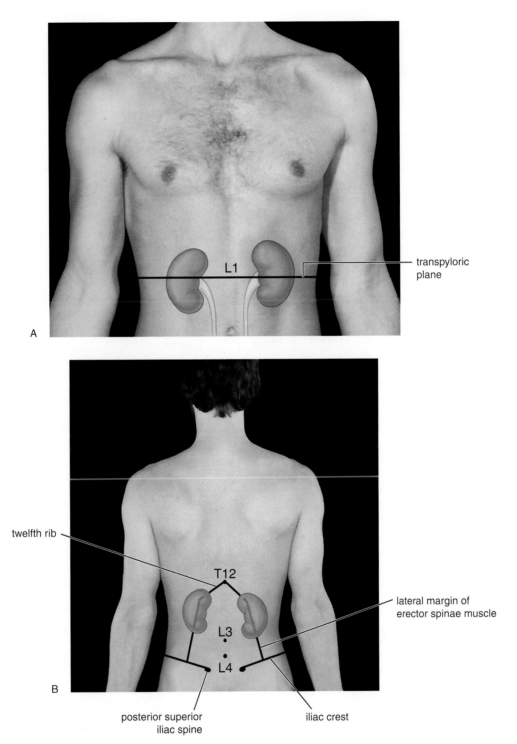

**Figure 21-26   A.** Surface anatomy of the kidneys and ureters on the anterior abdominal wall. Note the relationship of the hilum of each kidney to the transpyloric plane. **B.** Surface anatomy of the kidneys on the posterior abdominal wall.

# Review Questions

## Multiple Choice Questions

### Select the BEST answer for each question

1. The hilum of the right kidney contains the following important structures **except** which?
   A. The renal pelvis
   B. Tributaries of the renal vein
   C. Sympathetic nerve fibers
   D. Part of the right suprarenal gland
   E. Branches of the renal artery

2. The following statements concerning the left kidney are correct **except** which?
   A. The renal papillae open directly into the major calyces.
   B. The left kidney is slightly higher than the right kidney.
   C. The left kidney is related anteriorly to the stomach.
   D. The medulla is composed of approximately 12 renal pyramids.
   E. At the junction of the renal pelvis with the ureter, the lumen of the ureter is narrowed.

3. The right kidney has the following important relationships **except** which?
   A. It is related to the neck of the pancreas.
   B. It is anterior to the right costodiaphragmatic recess.
   C. It is related to the second part of the duodenum.
   D. It is related to the right colic flexure.
   E. It is anterior to the right twelfth rib.

4. The following statements concerning the ureters are correct **except** which?
   A. Both ureters have three anatomic sites that are constricted.
   B. Both ureters receive their blood supply from the testicular or ovarian arteries.
   C. Both ureters are separated from the transverse processes of the lumbar vertebrae by the psoas muscles.
   D. Both ureters pass anterior to the testicular or ovarian vessels.
   E. Both ureters lie anterior to the sacroiliac joints.

5. The following statements concerning the pelvic part of the ureter are correct **except** which?
   A. It enters the pelvis in front of the bifurcation of the common iliac artery.
   B. The ureter enters the bladder by passing directly through its wall, there being no valvular mechanism at its entrance.
   C. It has a close relationship to the ischial spine before it turns medially toward the bladder.

   D. The blood supply of the distal part of the ureter is from the superior vesical artery.
   E. It enters the bladder at the upper lateral angle of the trigone.

6. The following statements concerning the nerve supply to the urinary bladder are correct **except** which?
   A. The sympathetic postganglionic fibers originate in the first and second lumbar ganglia.
   B. The parasympathetic postganglionic fibers originate in the inferior hypogastric plexuses.
   C. The afferent sensory fibers arising in the bladder wall reach the spinal cord via the pelvic splanchnic nerves and also travel with the sympathetic nerves.
   D. The parasympathetic preganglionic fibers arise from the second, third, and fourth sacral segments of the spinal cord.
   E. The parasympathetic postganglionic fibers are responsible for closing the vesical sphincter during ejaculation.

## Completion Questions

### Select the phrase that BEST completes each statement.

7. Pain caused by the passage of a stone down the lower end of the left ureter may be referred to the
   A. umbilical region.
   B. right iliac region.
   C. epigastric region.
   D. penis or clitoris.
   E. none of the above.

8. The sphincter urethrae receives its innervation from the
   A. vagus nerve.
   B. obturator nerve.
   C. pudendal nerve.
   D. inferior rectal nerve.
   E. hypogastric plexus.

9. The narrowest part of the male urethra is the
   A. membranous part.
   B. prostatic part.
   C. external meatus on the glans penis.
   D. penile part.
   E. none of the above

10. The female urethra
    A. is approximately 3 in. (7.62 cm) in length.
    B. is difficult to dilate.
    C. is insensitive to stretching.
    D. opens into the vestibule above the clitoris.
    E. is readily accessible to infection.

## Matching Questions

Match the numbered structures shown on the intravenous pyelogram—obtained 20 min after injection of a suitable contrast medium—with the appropriate lettered structure (5-year-old female).

11. Structure 1

12. Structure 2

13. Structure 3

14. Structure 4

15. Structure 5
    A. Rectum
    B. Pelvis of ureter
    C. Sacrum
    D. Ureter
    E. Urinary bladder

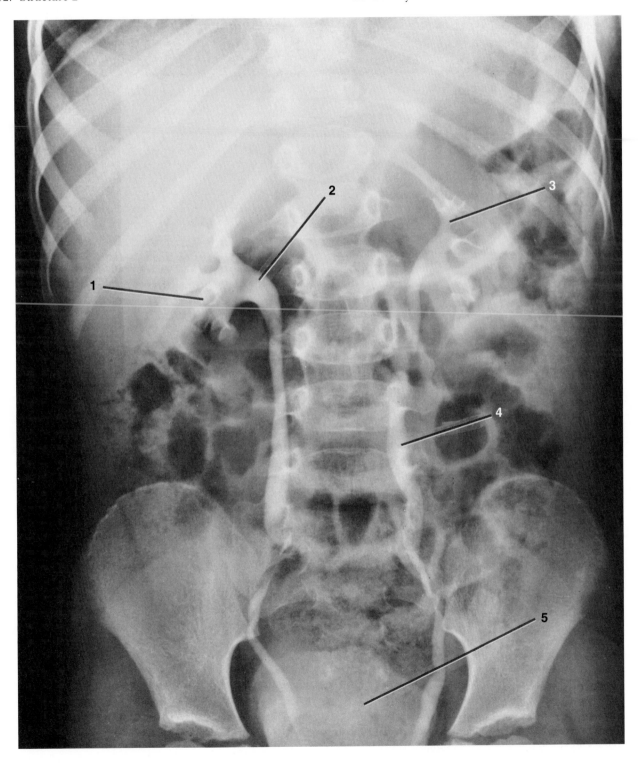

F. Major calyx
G. None of the above

**Match the numbered structures shown on the CT scan of the abdomen at the level of the second lumbar vertebra after intravenous pyelography with the appropriate lettered structures.**

16. Structure 1

17. Structure 2

18. Structure 3

19. Structure 4

20. Structure 5
A. Aorta
B. Vertebral body
C. Gallbladder
D. Pancreas
E. Left ureter
F. Inferior vena cava
G. None of the above

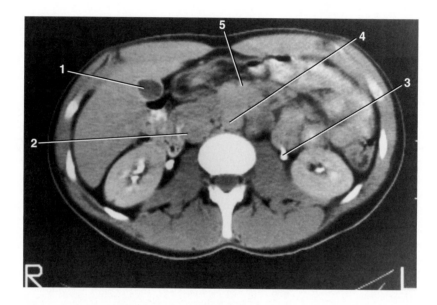

# Answers and Explanations

1. **D** is incorrect. The right suprarenal gland caps the upper pole of the right kidney and does not extend downward to the hilum of the kidney (Fig. 21-1).

2. **A** is incorrect. The renal papillae open directly into the minor calyces of the kidney (Fig. 21-2).

3. **A** is incorrect. The pancreas is not related to the right kidney (Fig. 21-4).

4. **D** is incorrect. The ureters are crossed on their anterior surfaces by the testicular and ovarian vessels (Fig. 21-10).

5. **B** is incorrect. The ureters pierce the bladder wall obliquely, and this provides a valvelike mechanism that prevents urine from reentering the ureter from the bladder cavity.

6. **E** is incorrect. The sympathetic nerves are responsible for the contraction of the sphincter vesicae during ejaculation.

7. **D** is correct. Pain from the upper end of the ureter is referred to the back behind the kidney. Pain from the middle region of the ureter is referred to the inguinal region, and pain from the lower end is referred to the penis or clitoris. This is because the afferent nerves enter the spinal cord at different levels, so the pain is referred along the spinal nerves originating from those spinal cord levels.

8. **C** is correct. The sphincter urethrae muscle is innervated by branches from the pudendal nerve.

9. **C** is correct. The narrowest part of the male urethra is the external meatus on the glans penis.

10. **E** is correct. The female urethra is readily accessible to infection.

11. **G** is correct. Structure 1 is a minor calyx.

12. **B** is correct. Structure 2 is the pelvis of the right ureter.

13. **F** is correct. Structure 3 is a major calyx of the left kidney.

14. **D** is correct. Structure 4 is the left ureter.

15. **E** is correct. Structure 5 is the urinary bladder partially filled with radiopaque material.

16. **C** is correct. Structure 1 is the gallbladder.

17. **F** is correct. Structure 2 is the inferior vena cava.

18. **E** is correct. Structure 3 is the left ureter filled with radiopaque material.

19. **A** is correct. Structure 4 is the aorta.

20. **D** is correct. Structure 5 is the pancreas.

# The Reproductive System

# 22 The Male Genital Organs, the Penis, and the Scrotum

All clinical material relevant to this chapter can be found on the CD-ROM.

# Chapter Outline

The male reproductive system may present the physician with a variety of conditions, from urethral obstruction, to traumatic rupture of the urethra, to infections of the epididymis, testis, or prostate. Benign hypertrophy and carcinoma of the prostate are common clinical conditions. The purpose of this chapter is to review the significant anatomy of the male reproductive system relative to clinical problems.

 # BASIC ANATOMY

The male reproductive system consists of a pair of testes, their excretory ducts and the accessory glands, and the penis (Fig. 22-1). The excretory ducts on each side are the epididymis, the vas deferens, and the ejaculatory duct. The accessory glands are a pair of seminal vesicles, a pair of bulbourethral glands, and the prostate gland.

The external genital organs consist of the penis and the scrotum.

## Scrotum

The scrotum is an outpouching of the lower part of the anterior abdominal wall. It contains the testes, the epididymides, and the lower ends of the spermatic cords (Figs. 22-1, 22-2, and 22-3).

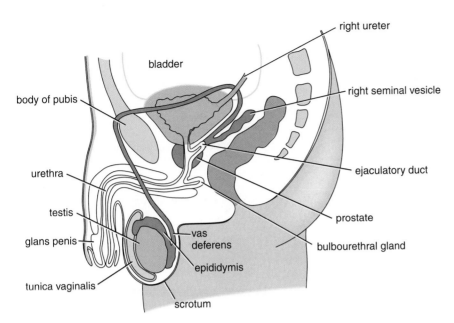

**Figure 22-1** Sagittal section of the male pelvis.

The wall of the scrotum has the following layers:

- **Skin:** The skin of the scrotum is thin, wrinkled, and pigmented and forms a single pouch. A slightly raised ridge in the midline indicates the line of fusion of the two lateral labioscrotal swellings. (In the female, the swellings remain separate and form the labia majora.)

- **Superficial fascia:** This is continuous with the fatty and membranous layers of the anterior abdominal wall; the fat is, however, replaced by smooth muscle called the **dartos muscle** (Fig. 22-2). This is innervated by sympathetic nerve fibers and is responsible for the wrinkling of the overlying skin. The membranous layer of the superficial fascia (Colles' fascia) is continuous in front with the membranous layer of the anterior abdominal wall (Scarpa's fascia), and behind it is attached to the perineal body and the posterior edge of the perineal membrane (see Fig. 22-12). At the sides, it is attached to the ischiopubic rami (see Fig. 22-10). Both layers of superficial fascia contribute to a median partition that crosses the scrotum and separates the testes from each other.

- **Spermatic fasciae:** These three layers lie beneath the superficial fascia and are derived from the three layers of the anterior abdominal wall on each side (Fig. 22-2). The **cremaster muscle** in the cremasteric fascia can be made to contract by stroking the skin on the medial aspect of the thigh. This is called the **cremasteric reflex.** The afferent fibers of this reflex arc travel in the femoral branch of the genitofemoral nerve (L1 and 2), and the efferent motor nerve fibers travel in the genital branch of the genitofemoral nerve.

- **Tunica vaginalis** (Figs. 22-1, 22-2, and 22-3): This lies within the spermatic fasciae and covers the anterior, medial, and lateral surfaces of each testis. It is the lower expanded part of the processus vaginalis; normally, just before birth, it becomes shut off from the upper part of

the processus and the peritoneal cavity. The tunica vaginalis is thus a closed sac, invaginated from behind by the testis.

---

### PHYSIOLOGIC NOTE

#### Spermatogenesis and Temperature

Normal spermatogenesis can only take place at a temperature that is lower than that of the abdominal cavity. When the testes are normally located in the scrotum, they are at a temperature 3°F below that of the abdominal temperature. Should the temperature of the scrotum fall, the dartos muscle in the scrotal wall contracts, causing the surface area of the scrotal skin to be reduced, and, at the same time, the testes move closer to the body for warmth. In addition, the cremaster muscle in the spermatic cord and scrotal wall reflexly contracts, further elevating the testes toward the pelvis. A rise in temperature within the scrotum causes the dartos and cremaster muscles to relax so that the testes move away from the body and cool. It is now recognized that the testicular veins in the spermatic cord that form the pampiniform plexus—together with the branches of the testicular arteries, which lie close to the veins—probably assist in stabilizing the temperature of the testes by a countercurrent heat exchange mechanism. By this means, the hot blood arriving in the artery from the abdomen loses heat to the blood ascending to the abdomen within the veins.

### Lymph Drainage

Lymph from the skin and fascia, including the tunica vaginalis, drains into the superficial inguinal lymph nodes (Fig. 22-4).

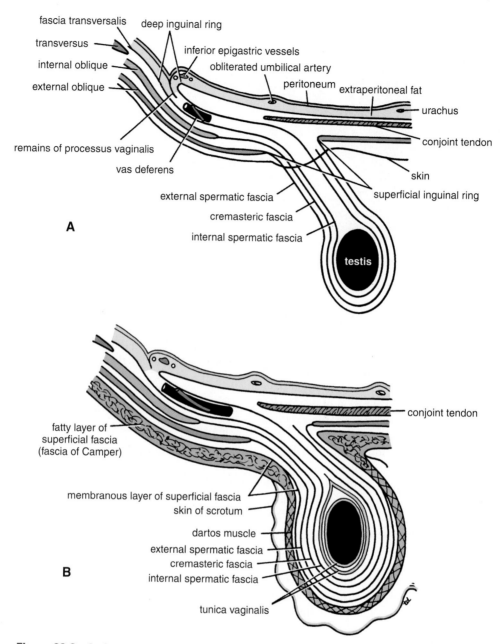

**Figure 22-2** **A.** Continuity of the different layers of the anterior abdominal wall with coverings of the spermatic cord. **B.** The skin and superficial fascia of the abdominal wall and scrotum have been included, and the tunica vaginalis is shown.

# Testis

The testes are paired ovoid organs measuring about 2 in. (5 cm) long and are slightly flattened from side to side (Fig. 22-3). Each testis is a firm, mobile organ lying within the scrotum. The left testis usually lies at a lower level than the right. The upper pole of the gland is tilted slightly forward. Each testis is surrounded by a tough fibrous capsule, the **tunica albuginea.**

Extending from the inner surface of the capsule is a series of fibrous septa that divide the interior of the organ into **lobules.** Lying within each lobule are one to three coiled **seminiferous tubules.** The tubules open into a network of channels called the **rete testis.** Situated within each lobule between the seminiferous tubules are delicate connective tissue and groups of rounded **interstitial cells (Leydig cells)** that produce the male sex hormone **testosterone.** The rete testis is drained by small **efferent**

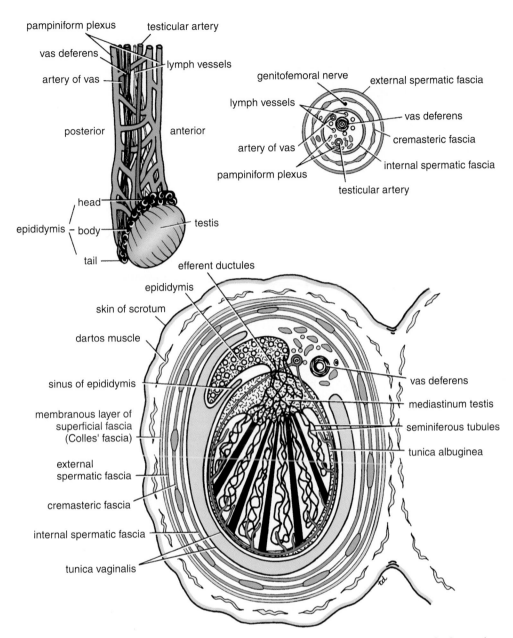

**Figure 22-3**   Testis and epididymis, spermatic cord, and scrotum. Also shown is the testis and epididymis cut across in horizontal section.

**ductules** into the duct at the upper end of the epididymis (Fig. 22-3).

### Function of the Testis

The seminiferous tubules of the testis are responsible for the production of **spermatozoa.** The interstitial cells (Leydig cells) produce the male sex hormone **testosterone.**

# Epididymis

The **epididymis** is a firm structure lying posterior to the testis, with the vas deferens lying on its medial side (Fig. 22-3). It has an expanded upper end, the **head,** a **body,** and a pointed **tail** inferiorly. Laterally, a distinct groove lies between the testis and the epididymis, which is lined with the inner visceral layer of the tunica vaginalis and is called the **sinus of the epididymis** (Fig. 22-3).

The epididymis is a much coiled tube nearly 20 ft (6 m) long, embedded in connective tissue. The tube emerges

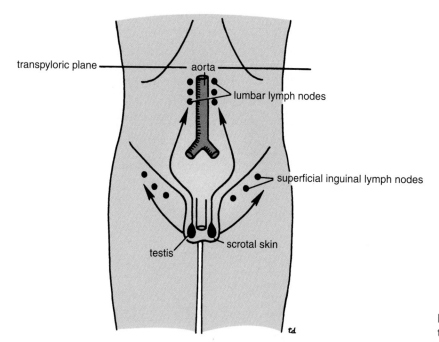

transpyloric plane

aorta

lumbar lymph nodes

superficial inguinal lymph nodes

testis

scrotal skin

**Figure 22-4**  Lymph drainage of the testis and the skin of the scrotum.

from the tail of the epididymis as the **vas deferens,** which enters the spermatic cord.

---

**PHYSIOLOGIC NOTE**

### Function of the Epididymis

The long length of the duct of the epididymis provides storage space for the spermatozoa and allows them to mature. A main function of the epididymis is the absorption of fluid. Another function may be the addition of substances to the seminal fluid to nourish the maturing sperm.

## Blood Supply of the Testis and Epididymis

The testicular artery is a branch of the abdominal aorta. The testicular veins emerge from the testis and the epididymis as a venous network, the **pampiniform plexus** (Fig. 22-3). This becomes reduced to a single vein as it ascends through the inguinal canal. The right testicular vein drains into the inferior vena cava, and the left vein joins the left renal vein.

## Lymph Drainage of the Testis and Epididymis

The lymph vessels (Fig. 22-4) ascend in the spermatic cord and end in the lymph nodes on the side of the aorta (lumbar or para-aortic) nodes at the level of the first lumbar vertebra

(i.e., on the transpyloric plane). This is to be expected because during development the testis has migrated from high up on the posterior abdominal wall, down through the inguinal canal, and into the scrotum, dragging its blood supply and lymph vessels after it.

---

**EMBRYOLOGIC NOTE**

### Development of the Testis

The male sex chromosome causes the genital ridge to secrete testosterone and induces the development of the testis and the other internal and external organs of reproduction.

The **sex cords** of the genital ridge become separated from the coelomic epithelium by the proliferation of the mesenchyme (Fig. 22-5). The outer part of the mesenchyme condenses to form a dense fibrous layer, the **tunica albuginea.** The sex cords become U shaped and form the **seminiferous tubules.** The free ends of the tubules form the **straight tubules,** which join one another in the mediastinum testis to become the **rete testis.** The primordial sex cells in the seminiferous tubules form the **spermatogonia,** and the sex cord cells form the **Sertoli cells.** The mesenchyme in the developing gonad makes up the connective tissue and fibrous septa. The **interstitial cells,** which are already secreting testosterone, are also formed of mesenchyme. The rete testis becomes canalized, and the tubules extend into the mesonephric tissue, where they join the remnants of the mesonephric tubules; the latter tubules become the

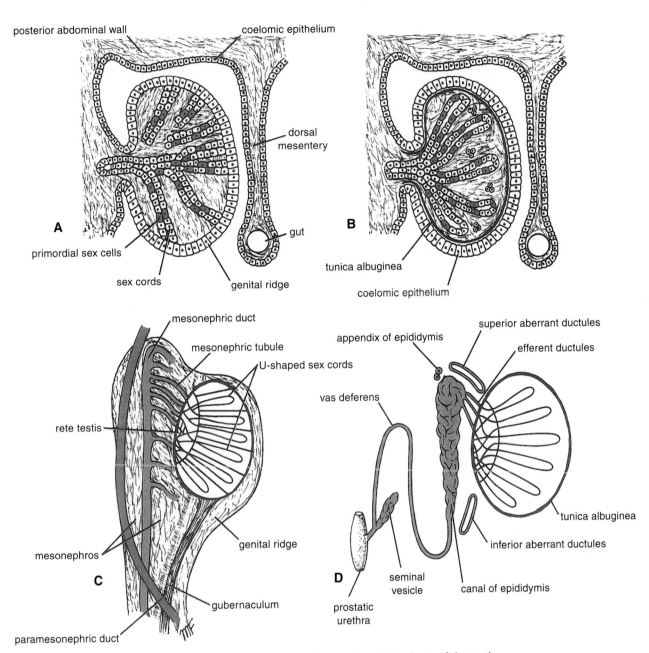

**Figure 22-5** The formation of the testis and the ducts of the testis.

**efferent ductules** of the testis. The **duct of the epididymis,** the **vas deferens,** the **seminal vesicle,** and the **ejaculatory duct** are formed from the mesonephric duct (Fig. 22-5).

### Descent of the Testis

The testis develops high up on the posterior abdominal wall, and in late fetal life, it "descends" behind the peritoneum, dragging its blood supply, nerve supply, and lymphatic drainage after it. The process of the descent of the testis is shown in Figure 22-6.

# Vas Deferens

The vas deferens is a thick-walled tube about 18 in. (45 cm) long that conveys mature sperm from the epididymis to the ejaculatory duct and the urethra. It arises from the lower end or tail of the epididymis and passes through the inguinal canal. It emerges from the deep inguinal ring and passes around the lateral margin of the inferior epigastric artery (Fig. 22-7). It then passes downward and backward on the lateral wall of the pelvis and crosses the ureter in the region of the ischial spine. The vas deferens then runs

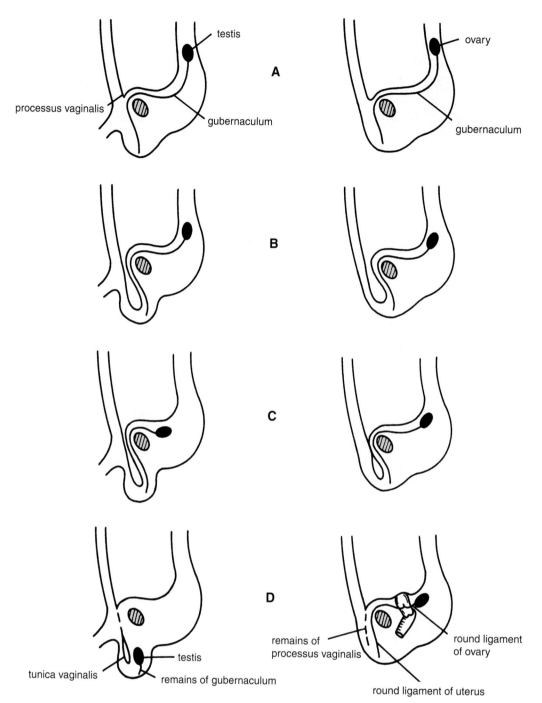

**Figure 22-6**   Origin, development, and fate of the processus vaginalis in the two sexes. Note the descent of the testis into the scrotum and the descent of the ovary into the pelvis.

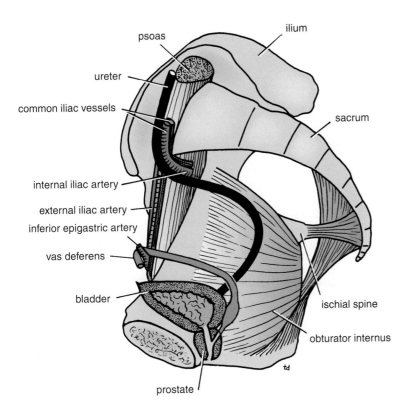

psoas

ureter

common iliac vessels

internal iliac artery

external iliac artery

inferior epigastric artery

vas deferens

bladder

prostate

ilium

sacrum

ischial spine

obturator internus

**Figure 22-7**    Right half of the pelvis showing relations of the ureter and vas deferens.

medially and downward on the posterior surface of the bladder (Fig. 22-7). The terminal part of the vas deferens is dilated to form the **ampulla of the vas deferens** (Fig. 22-8). The inferior end of the ampulla narrows down and joins the duct of the seminal vesicle to form the ejaculatory duct.

# Seminal Vesicles

The seminal vesicles are two lobulated organs about 2 in. (5 cm) long lying on the posterior surface of the bladder (Fig. 22-8). Their upper ends are widely separated, and their lower ends are close together. On the medial side of each vesicle lies the terminal part of the vas deferens. Posteriorly, the seminal vesicles are related to the rectum (see Fig. 22-11). Inferiorly, each seminal vesicle narrows and joins the vas deferens of the same side to form the ejaculatory duct.

Each seminal vesicle consists of a much-coiled tube embedded in connective tissue.

## Blood Supply

### Arteries

Branches of the inferior vesicle and middle rectal arteries supply the seminal vesicles.

### Veins

The veins drain into the internal iliac veins.

## Lymph Drainage

The lymph runs into the internal iliac nodes.

### PHYSIOLOGIC NOTE

#### Function of the Seminal Vesicles

The function of the seminal vesicles is to produce a secretion that is added to the seminal fluid. The secretions contain substances that are essential for the nourishment of the spermatozoa. The walls of the seminal vesicles contract during ejaculation and expel their contents into the ejaculatory ducts, thus washing the spermatozoa out of the urethra.

# Ejaculatory Ducts

The two ejaculatory ducts are each less than 1 in. (2.5 cm) long and are formed by the union of the vas deferens and the duct of the seminal vesicle (Fig. 22-9). The ejaculatory ducts pierce the posterior surface of the prostate and open into the prostatic part of the urethra, close to the margins of the prostatic utricle; their function is to drain the seminal fluid into the prostatic urethra.

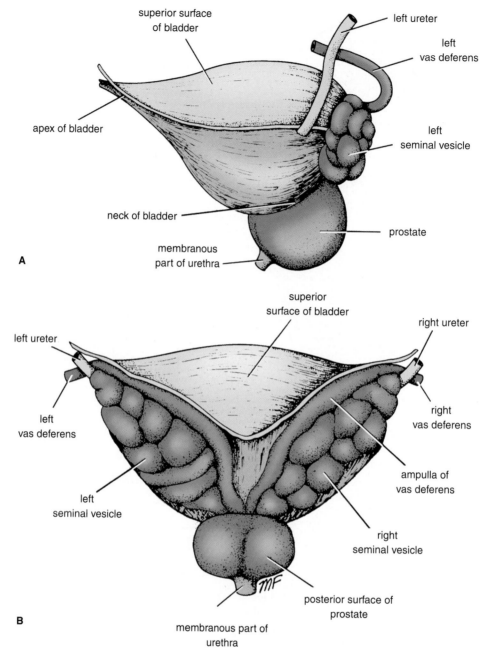

**Figure 22-8  A.** Lateral view of the bladder, prostate, and left seminal vesicle.
**B.** Posterior view of the bladder, prostate, vasa deferentia, and seminal vesicles.

# Prostate

## Location and Description

The prostate is a fibromuscular glandular organ that surrounds the prostatic urethra (Figs. 22-1 and 22-9). It is about 1.25 in. (3 cm) long and lies between the neck of the bladder above and the urogenital diaphragm below (Fig. 22-9).

The prostate is surrounded by a fibrous capsule. Outside the capsule is a fibrous sheath, which is part of the visceral layer of pelvic fascia (Fig. 22-9). The somewhat conical prostate has a base, which superiorly lies against the bladder neck, and an apex, which lies inferiorly against the urogenital diaphragm. The two ejaculatory ducts pierce the upper part of the posterior surface of the prostate to open into the prostatic urethra at the lateral margins of the prostatic utricle (Fig. 22-9).

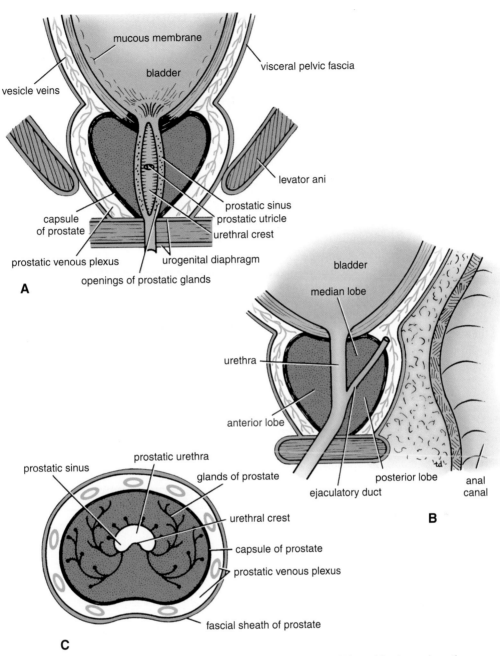

**Figure 22-9**  Prostate in coronal section (**A**), sagittal section (**B**), and horizontal section (**C**). In the coronal section, note the openings of the ejaculatory ducts on the margin of the prostatic utricle.

# Relations

- **Superiorly:** The base of the prostate is continuous with the neck of the bladder, the smooth muscle passing without interruption from one organ to the other. The urethra enters the center of the base of the prostate (Fig. 22-9).
- **Inferiorly:** The apex of the prostate lies on the upper surface of the urogenital diaphragm. The urethra leaves the prostate just above the apex on the anterior surface (Fig. 22-9).

- **Anteriorly:** The anterior surface of the prostate is related to the symphysis pubis, separated from it by the extraperitoneal fat in the retropubic space (cave of Retzius). The fibrous sheath of the prostate is connected to the posterior aspect of the pubic bones by the puboprostatic ligaments. These ligaments lie one on either side of the midline and are condensations of pelvic fascia.
- **Posteriorly:** The posterior surface of the prostate (Figs. 22-1 and 22-9) is closely related to the anterior surface of the rectal ampulla and is separated from it by the

rectovesical septum (fascia of Denonvillier). This septum is formed in fetal life by the fusion of the walls of the lower end of the rectovesical pouch of peritoneum, which originally extended down to the perineal body.

- **Laterally:** The lateral surfaces of the prostate are embraced by the anterior fibers of the levator ani as they run posteriorly from the pubis (Fig. 22-9).

## Structure of the Prostate

The numerous glands of the prostate are embedded in a mixture of smooth muscle and connective tissue, and their ducts open into the prostatic urethra.

The prostate is incompletely divided into five lobes (Fig. 22-9). The anterior lobe lies in front of the urethra and is devoid of glandular tissue. The median, or middle, lobe is the wedge of gland situated between the urethra and the ejaculatory ducts. Its upper surface is related to the trigone of the bladder; it is rich in glands. The posterior lobe is situated behind the urethra and below the ejaculatory ducts and also contains glandular tissue. The right and left lateral lobes lie on either side of the urethra and are separated from one another by a shallow vertical groove on the posterior surface of the prostate. The lateral lobes contain many glands.

### P H Y S I O L O G I C   N O T E

#### Function of the Prostate

The function of the prostate is the production of a thin, milky fluid containing citric acid and acid phosphatase. It is added to the seminal fluid at the time of ejaculation. The smooth muscle in the capsule and stroma contract, and the secretion from the many glands is squeezed into the prostatic urethra. The prostatic secretion is alkaline and helps neutralize the acidity in the vagina.

## Blood Supply

### Arteries

Branches of the inferior vesical and middle rectal arteries supply the prostate.

### Veins

The veins form the prostatic venous plexus, which is between the capsule of the prostate and the fibrous sheath (Fig. 22-9). The prostatic plexus receives the deep dorsal vein of the penis and numerous vesical veins and drains into the internal iliac veins.

### Lymph Drainage

The lymph vessels from the prostate drain into the internal iliac nodes.

## Nerve Supply

The nerve supply to the prostate is from the inferior hypogastric plexuses. The sympathetic nerves stimulate the smooth muscle of the prostate during ejaculation.

## Prostatic Urethra

The prostatic urethra is about 1.25 in (3 cm) long and begins at the neck of the bladder. It passes through the prostate from the base to the apex, where it becomes continuous with the membranous part of the urethra (Fig. 22-10). The prostatic urethra is the widest and most dilatable portion of the entire urethra. On the posterior wall is a longitudinal ridge called the **urethral crest** (Fig. 22-9). On each side of this ridge is a groove called the **prostatic sinus**; the prostatic glands open into these grooves. On the summit of the urethral crest is a depression, the **prostatic utricle**, which is an analog of the uterus and vagina in females. On the edge of the mouth of the utricle are the openings of the two ejaculatory ducts (Fig. 22-9).

## Bulbourethral Glands

The bulbourethral glands are two small glands that lie beneath the sphincter urethrae muscle (Fig. 22-10). Their ducts pierce the perineal membrane (inferior fascial layer of the urogenital diaphragm) and enter the penile portion of the urethra. The secretion is poured into the urethra as a result of erotic stimulation.

### P H Y S I O L O G I C   N O T E

#### Function of the Bulbourethral Glands

The secretion of the glands is added to the seminal fluid. The precise function of the fluid is unknown. The glands are controlled by testosterone, and castration produces atrophy.

## Urogenital Diaphragm

This is a musculofascial diaphragm that fills in the gap of the pubic arch (Fig. 22-10). It is formed by the sphincter urethrae and the deep transverse perineal muscles, which are enclosed between a superior and an inferior layer of fascia of the urogenital diaphragm. The inferior layer of fascia is called the **perineal membrane** (see Fig. 22-14).

Anterior to the urogenital diaphragm is a small gap below the symphysis pubis, through which passes the deep **dorsal vein of the penis** (see Fig. 22-14).

## Deep Perineal Pouch

The closed space that is contained within the urogenital diaphragm, between the superior layer of fascia and the

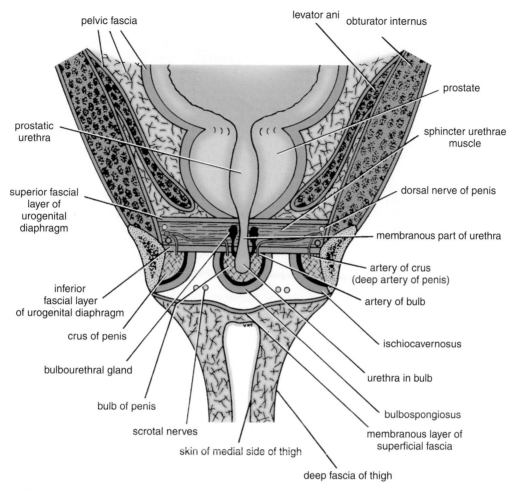

**Figure 22-10** Coronal section of the male pelvis showing the prostate, the urogenital diaphragm, and the contents of the superficial perineal pouch.

perineal membrane, is known as the **deep perineal pouch** (Figs. 22-10 and 22-12).

## Contents of the Deep Perineal Pouch

The deep perineal pouch contains the membranous part of the urethra, the sphincter urethrae, the bulbourethral glands, the deep transverse perineal muscles, the internal pudendal vessels and their branches, and the dorsal nerves of the penis (Fig. 22-10).

# Superficial Perineal Pouch

The superficial perineal pouch is bounded below by the membranous layer of superficial fascia and above by the urogenital diaphragm (Fig. 22-12). It is closed behind by the fusion of its upper and lower walls. Laterally, it is closed by the attachment of the membranous layer of superficial fascia and the urogenital diaphragm to the margins of the pubic arch (Fig. 22-10). Anteriorly, the space communicates freely with the potential space lying between the superficial fascia of the anterior abdominal wall and the anterior abdominal muscles (Fig. 22-10).

## Contents of the Superficial Perineal Pouch

The superficial perineal pouch contains structures forming the **root of the penis**, together with the muscles that cover them—namely, the **bulbospongiosus muscles** and the **ischiocavernosus muscles** (Fig. 22-13). In addition, the **perineal branch of the pudendal nerve** on each side terminates in the pouch by supplying the muscles and the overlying skin.

# Penis

## Location and Description

The penis has a fixed root and a body that hangs free (Figs. 22-14 and 22-15).

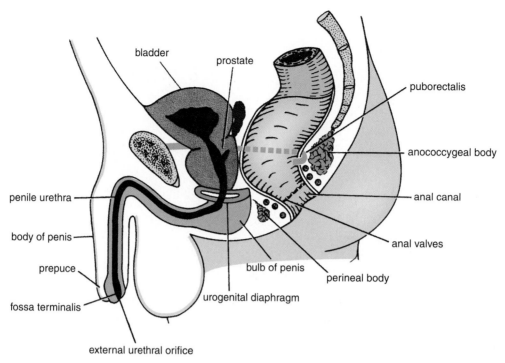

**Figure 22-11**   Sagittal section of the male pelvis.

## Root of the Penis

The root of the penis is made up of three masses of erectile tissue called the **bulb of the penis** and the **right** and **left crura of the penis** (Figs. 22-13, 22-14, and 22-15). The bulb is situated in the midline and is attached to the undersurface of the urogenital diaphragm. It is traversed by the urethra and is covered on its outer surface by the **bulbospongiosus muscles.** Each crus is attached to the side of the pubic arch and is covered on its outer surface by the **ischiocavernosus muscle.** The bulb is continued forward into the body of the penis and forms the **corpus spongiosum** (Fig. 22-13). The two crura converge anteriorly and come to lie side by side in the dorsal part of the body of the penis, forming the **corpora cavernosa** (Figs. 22-14 and 22-15).

## Body of the Penis

The body of the penis is essentially composed of three cylinders of erectile tissue enclosed in a tubular sheath of fascia (**Buck's fascia**). The erectile tissue is made up of two dorsally placed corpora cavernosa (which communicate with each other) and a single corpus spongiosum applied to their ventral surface (Figs. 22-14 and 22-15). At its distal extremity, the corpus spongiosum expands to form the **glans penis,** which covers the distal ends of the corpora cavernosa. On the tip of the

glans penis is the slitlike orifice of the urethra, called the **external urethral meatus.**

The **prepuce** or **foreskin** is a hoodlike fold of skin that covers the glans. It is connected to the glans just below the urethral orifice by a fold called the **frenulum.**

The body of the penis is supported by two condensations of deep fascia that extend downward from the linea alba and symphysis pubis to be attached to the fascia of the penis.

## Muscles of the Penis

### Bulbospongiosus Muscles

The bulbospongiosus muscles, situated one on each side of the midline (Fig. 22-13), cover the bulb of the penis and the posterior portion of the corpus spongiosum. Their function is to compress the penile part of the urethra and empty it of residual urine or semen. The anterior fibers also compress the deep dorsal vein of the penis, thus impeding the venous drainage of the erectile tissue and thereby assisting in the process of erection of the penis.

### Ischiocavernosus Muscles

The ischiocavernosus muscles cover the crus penis on each side (Fig. 22-13). The action of each muscle is to compress

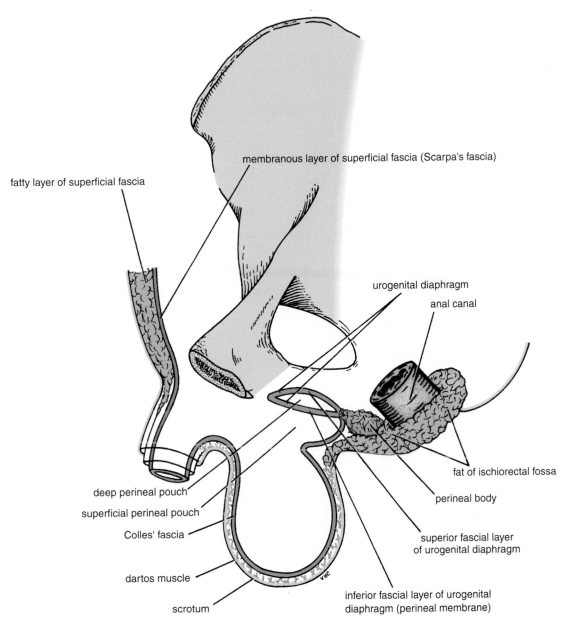

**Figure 22-12**  Arrangement of the superficial fascia in the urogenital triangle. Note the superficial and deep perineal pouches.

the crus penis and assist in the process of erection of the penis.

The muscles of the perineum are summarized in Table 13-8.

## Blood Supply of the Penis

### Arteries

The corpora cavernosa are supplied by the **deep arteries of the penis** (Fig. 22-15); the corpus spongiosum is supplied by the **artery of the bulb.** In addition, there is the **dorsal artery of the penis.** All the above arteries are branches of the internal pudendal artery.

### Veins

The veins drain into the internal pudendal veins.

## Lymph Drainage of the Penis

The skin of the penis is drained into the medial group of superficial inguinal nodes. The deep structures of the penis are drained into the internal iliac nodes.

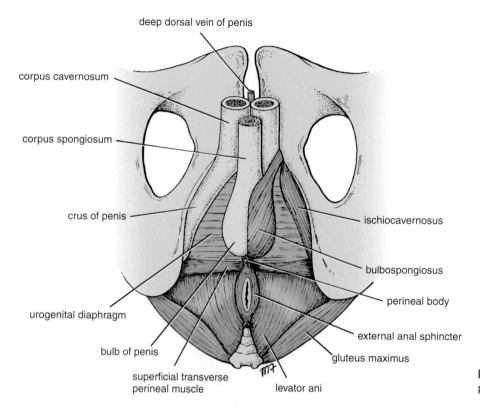

Figure 22-13  Root of penis and perineal muscles.

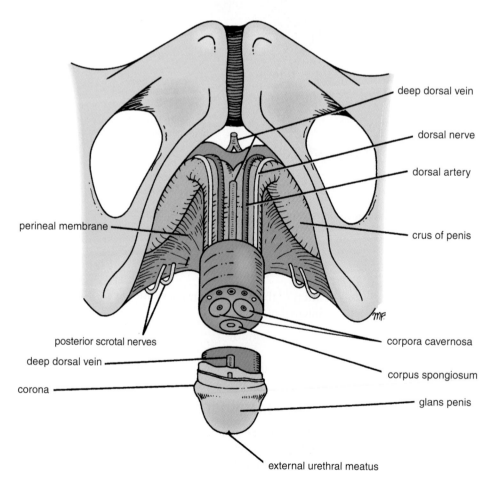

Figure 22-14  Root and body of the penis.

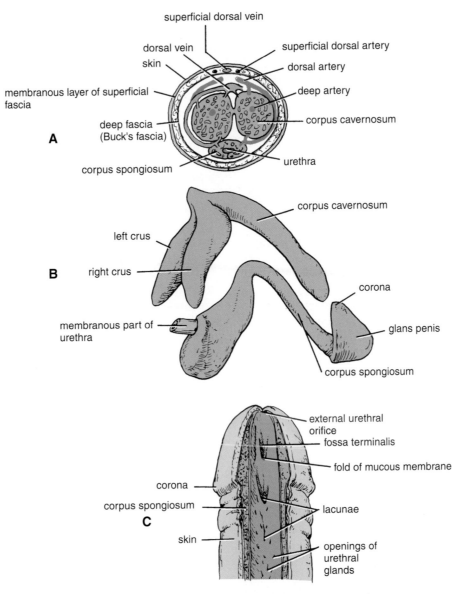

**Figure 22-15**   The penis. **A and B.** The three bodies of erectile tissue, the two corpora cavernosa, and the corpus spongiosum with the glans. **C.** The penile urethra slit open to show the folds of mucous membrane and glandular orifices in the roof of the urethra.

## Nerve Supply of the Penis

The nerve supply is from the pudendal nerve and the pelvic plexuses.

### PHYSIOLOGIC NOTE

#### Erection of the Penis

Erection in the male is gradually built up as a consequence of various sexual stimuli. Pleasurable sight, sound, smell, and other psychic stimuli, fortified later by direct touch sensory stimuli from the general body skin and genital skin, result in a bombardment of the central nervous system by afferent stimuli. Efferent nervous impulses pass down the spinal cord to the parasympathetic outflow in the second, third, and fourth sacral segments. The parasympathetic preganglionic fibers enter the inferior hypogastric plexuses and synapse on the postganglionic neurons. The postganglionic fibers join the internal pudendal arteries and are distributed along their branches, which enter the erectile tissue at the root of the penis. Vasodilatation of the arteries now occurs, producing a great increase in blood flow through the blood spaces of the erectile tissue. The corpora cavernosa and the corpus spongiosum become engorged with blood and expand, compressing their draining veins against the surrounding fascia. By this means, the outflow of blood from the erectile tissue is retarded so that the internal pressure is further accentuated and maintained. The penis thus increases in length and diameter and assumes the erect position.

Once the climax of sexual excitement is reached and ejaculation takes place, or the excitement passes off or is inhibited, the arteries supplying the erectile tissue undergo vasoconstriction. The penis then returns to its flaccid state.

### Ejaculation

During the increasing sexual excitement that occurs during sex play, the external urinary meatus of the glans penis becomes moist as a result of the secretions of the bulbourethral glands.

Friction on the glans penis, reinforced by other afferent nervous impulses, results in a discharge along the sympathetic nerve fibers to the smooth muscle of the duct of the epididymis and the vas deferens on each side, the seminal vesicles, and the prostate. The smooth muscle contracts, and the spermatozoa, together with the secretions of the seminal vesicles and prostate, are discharged into the prostatic urethra. The fluid now joins the secretions of the bulbourethral glands and penile urethral glands and is then ejected from the penile urethra as a result of the rhythmic contractions of the bulbospongiosus muscles, which compress the urethra. Meanwhile, the sphincter of the bladder contracts and prevents a reflux of the spermatozoa into the bladder. The spermatozoa and the secretions of the several accessory glands constitute the **seminal fluid,** or **semen.**

At the climax of male sexual excitement, a mass discharge of nervous impulses takes place in the central nervous system. Impulses pass down the spinal cord to the sympathetic outflow (T1–L2). The nervous impulses that pass to the genital organs are thought to leave the cord at the first and second lumbar segments in the preganglionic sympathetic fibers. Many of these fibers synapse with postganglionic neurons in the first and second lumbar ganglia. Other fibers may synapse in ganglia in the lower lumbar or pelvic parts of the sympathetic trunks. The postganglionic fibers are then distributed to the vas deferens, the seminal vesicles, and the prostate via the inferior hypogastric plexuses.

### Development of the Male External Genitalia

Early in development, the embryonic mesenchyme grows around the cloacal membrane and causes the overlying ectoderm to be raised up to form three swellings. One swelling occurs between the cloacal membrane and the umbilical cord in the midline and is called the **genital tubercle** (Fig. 22-16). On each side of the membrane, another swelling, called the **genital fold,** appears. At the 7th week, the genital tubercle elongates to form the glans. The anterior part of the cloacal membrane, the **urogenital membrane,** now ruptures so that the urogenital sinus opens onto the surface. The entodermal cells of the urogenital sinus proliferate and grow into the root of the phallus, forming a **urethral plate.** Meanwhile, a second pair of lateral swellings, called the **genital swellings,** appears lateral to the genital folds. At this stage of development, the genitalia of the two sexes are identical.

In the male, the phallus now rapidly elongates and pulls the genital folds anteriorly onto its ventral surface so that they form the lateral edges of a groove, the **urethral groove** (Fig. 22-17). The floor of the groove is formed by the entodermal **urethral plate.** The penile urethra develops as the result of the two genital folds fusing together progressively along the shaft of the phallus to the root of the glans penis. During the 4th month, the remainder of the urethra in the glans is developed from a bud of ectodermal cells from the tip of the glans. This cord of cells later becomes canalized so that the penile urethra opens at the tip of the glans.

The **prepuce** or **foreskin** is formed from a fold of skin at the base of the glans (Figs. 22-16 and 22-17). The fold of skin remains tethered to the ventral aspect of the root of the glans to form the **frenulum.** The **erectile tissue**—the corpus spongiosum and the corpora cavernosa—develops within the mesenchymal core of the penis.

# SURFACE ANATOMY OF THE MALE GENITALIA

## Penis

The penis consists of a root, a body, and a glans (Figs. 22-14, 22-15, and 22-18). The **root of the penis** consists of three masses of erectile tissue called the **bulb of the penis** and the **right** and **left crura of the penis.** The bulb can be felt on deep palpation in the midline of the perineum, posterior to the scrotum.

The **body of the penis** is the free portion of the penis, which is suspended from the symphysis pubis. Note that

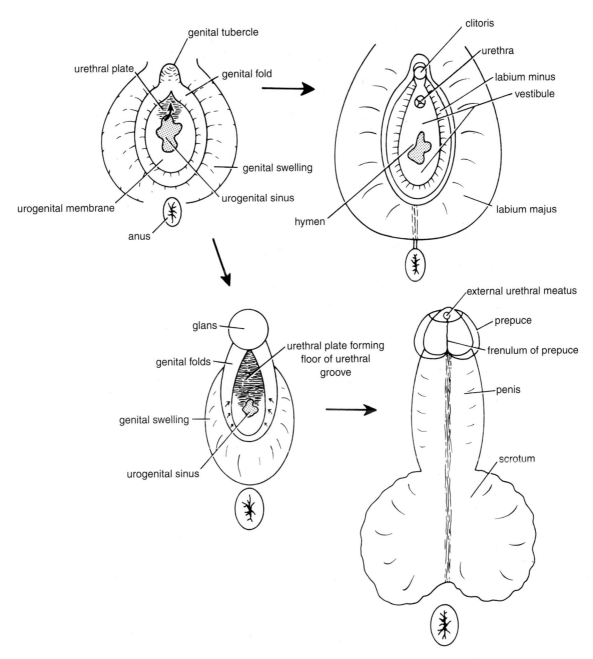

**Figure 22-16**   The development of the external genitalia in the female and male.

the dorsal surface (anterior surface of the flaccid organ) usually possesses a **superficial dorsal vein** in the midline (Fig. 22-15).

The **glans penis** forms the extremity of the body of the penis (Figs. 22-14 and 22-15). At the summit of the glans is the **external urethral meatus**. Extending from the lower margin of the external meatus is a fold connecting the glans to the prepuce called the **frenulum**. The edge of the base of the glans is called the **corona** (Fig. 22-14). The **prepuce** or **foreskin** is formed by a fold of skin attached to

the neck of the penis. The prepuce covers the glans for a variable extent, and it should be possible to retract it over the glans.

# Scrotum

The scrotum is a sac of skin and fascia (Fig. 22-12) containing the testes and the epididymides. The skin of the scrotum is rugose and is covered with sparse hairs. The bilateral origin of the scrotum is indicated by the presence of a dark

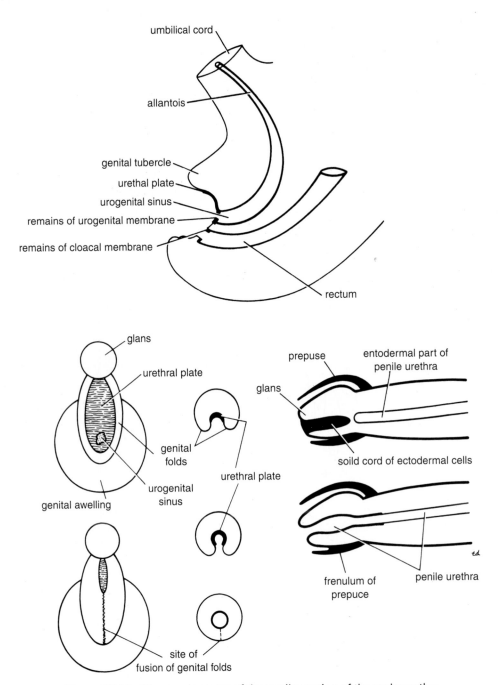

**Figure 22-17** The development of the penile portion of the male urethra.

line in the midline, called the **scrotal raphe**, along the line of fusion.

## Testes

The testes should be palpated. They are oval shaped and have a firm consistency. They lie free within the tunica vaginalis (Fig. 22-3) and are not tethered to the subcutaneous tissue or skin.

## Epididymides

Each epididymis can be palpated on the posterolateral surface of the testis. The epididymis is a long, narrow, firm structure having an expanded upper end or **head**, a **body**, and a pointed **tail** inferiorly (Fig. 22-3). The cordlike **vas deferens** emerges from the tail and ascends medial to the epididymis to enter the spermatic cord at the upper end of the scrotum.

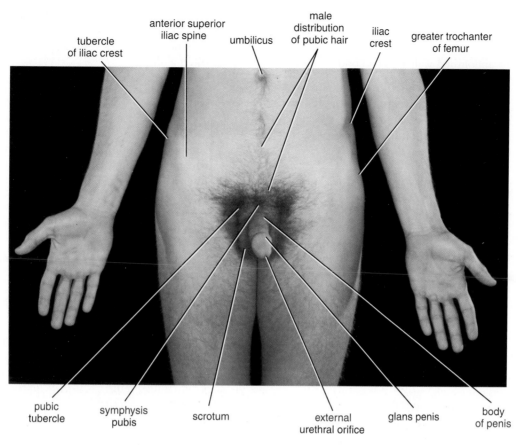

**Figure 22-18**   Anterior view of the pelvis of a 27-year-old man.

# Review Questions

## Multiple Choice Questions

**Select the BEST answer for each question.**

1. The following statements concerning the ductus (vas) deferens are correct **except** which?
   A. It emerges from the deep inguinal ring and passes around the lateral margin of the inferior epigastric artery.
   B. It crosses the ureter in the region of the ischial spine.
   C. The terminal part is dilated to form the ampulla.
   D. It lies on the posterior surface of the prostate but is separated from it by the peritoneum.
   E. It joins the duct of the seminal vesicle to form the ejaculatory duct.

2. The following statements concerning the seminal vesicle are correct **except** which?
   A. The seminal vesicles are related posteriorly to the rectum and can be palpated through the rectal wall.
   B. The seminal vesicles are two lobulated organs that store spermatozoa.
   C. The upper ends of the seminal vesicles are covered by peritoneum.
   D. The function of the seminal vesicles is to produce a secretion that is added to the seminal fluid.
   E. The seminal vesicles are related anteriorly to the bladder, and no peritoneum separates these structures.

3. The following statements concerning the penis are correct **except** which?
   A. Its root is formed in the midline by the bulb of the penis, which continues anteriorly as the corpus spongiosum.
   B. Its roots laterally are formed by the crura, which continue anteriorly as the corpora cavernosa.
   C. The penile urethra lies within the corpus spongiosum.

D. The glans penis is a distal expansion of the fused corpora cavernosa.

E. The penis is suspended from the lower part of the anterior abdominal wall by two condensations of deep fascia.

4. The urogenital diaphragm is formed by the following structures **except** which?
   A. Deep transverse perineal muscle
   B. Perineal membrane
   C. Sphincter urethrae
   D. Colles' fascia (membranous layer of superficial fascia)
   E. Parietal pelvic fascia covering the upper surface of the sphincter urethrae muscle

5. In the male, the following structures can be palpated on rectal examination **except** which?
   A. Bulb of the penis
   B. Prostate
   C. Seminal vesicles
   D. The anterior surface of the sacrum
   E. Ureter

6. The process of ejaculation depends on the following processes **except** which?
   A. The sphincter of the bladder contracts.
   B. The sympathetic preganglionic nerve fibers arising from the first and second lumbar segments of the spinal cord must be intact.
   C. The smooth muscle of the epididymis, ductus (vas) deferens, seminal vesicles, and prostate contracts.
   D. The bulbourethral glands and the urethral glands are active.
   E. The bulbospongiosus muscles relax.

7. A carcinoma of the glans penis is likely to spread eventually into which group of lymph nodes?
   A. External iliac nodes
   B. Internal iliac nodes
   C. Internal and external iliac nodes
   D. Superficial inguinal nodes
   E. Para-aortic nodes at the level of the first lumbar vertebra

8. A carcinoma of the prostate is likely to spread into which group of lymph nodes?
   A. Internal and external iliac nodes
   B. Internal iliac nodes
   C. Para-aortic nodes
   D. Superficial inguinal nodes
   E. Inferior mesenteric nodes

## Completion Questions

**Select the phrase that BEST completes each statement.**

9. The prostatic venous plexus drains into the
   A. internal iliac veins.
   B. inferior vena cava.
   C. external iliac veins.
   D. internal and external iliac veins.
   E. testicular veins.

10. Erection of the penis is a response to the activity of the
    A. sympathetic nerves.
    B. parasympathetic nerves.
    C. sympathetic and parasympathetic nerves.
    D. ilioinguinal nerves.
    E. None of the above

# Answers and Explanations

1. **D** is incorrect. The ductus (vas) deferens lies in direct contact with the posterior surface of the bladder (Fig. 22-1). The inferior end of the ampulla narrows down and joins the duct of the seminal vesicle to form the ejaculatory duct.

2. **B** is incorrect. The seminal vesicles do not store spermatozoa; they produce a secretion that nourishes the spermatozoa.

3. **D** is incorrect. The glans penis is a distal expansion of the corpus spongiosum (Fig. 22-15).

4. **D** is incorrect. Colles' fascia (membranous layer of superficial fascia) takes no part in the formation of the urogenital diaphragm; it is too superficial and lies just beneath the skin.

5. **E** is incorrect. The ureters cannot be felt on rectal examination in both sexes. An abnormal ureter, thickened by disease, can be felt on vaginal examination.

6. **E** is incorrect. During ejaculation, the bulbospongiosus muscles rhythmically contract and compress the urethra, forcing the seminal fluid out of the external meatus.

7. **C** is correct. A carcinoma of the glans penis is likely to spread into the internal and external iliac lymph nodes.

8. **B** is correct. A carcinoma of the prostate is likely to spread into the internal iliac lymph nodes.

9. **A** is correct. The prostatic venous plexus drains into the internal iliac veins.

10. **B** is correct. Erection of the penis is a response to the activity of the parasympathetic nerves.

# 23

# The Perineum, the Female Genital Organs, and Childbirth

All clinical material relevant to this chapter can be found on the CD-ROM.

# Chapter Outline

The female reproductive system may present the medical professional with a variety of conditions, such as infections, unsuspected pregnancy, ectopic pregnancy, spontaneous abortion, abnormal pelvic bleeding, prolapses, or acute pelvic inflammatory disease. The objective of this chapter is to provide an overview of the anatomy relevant to common clinical conditions.

 BASIC ANATOMY

Before studying this chapter, the reader is urged to review the pelvic bones in Chapter 11.

## The Perineum

### Definition of the Perineum

The cavity of the pelvis is divided by the pelvic diaphragm into the main pelvic cavity above and the perineum below

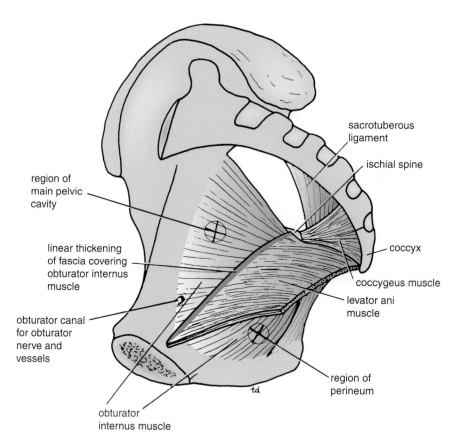

Figure 23-1 Right half of the pelvis showing the muscles forming the pelvic floor. Note that the levator ani and coccygeus muscles and their covering fascia form the pelvic diaphragm. Note also that the region of the main pelvic cavity lies above the pelvic diaphragm, and the region of the perineum lies below the diaphragm.

(Fig. 23-1). When seen from below with the thighs abducted, the perineum is diamond shaped and is bounded anteriorly by the symphysis pubis, posteriorly by the tip of the coccyx, and laterally by the ischial tuberosities (Fig. 23-2).

Anatomically, it is customary to divide the perineum into the anal triangle and the urogenital triangle. The **anal triangle** is bounded behind by the tip of the coccyx and on each side by the ischial tuberosity and the sacrotuberous ligament. The **anus** or lower opening of the anal canal, lies in the midline, and on each side is the **ischiorectal fossa.**

The **urogenital triangle** is bounded in front by the pubic arch and laterally by the ischial tuberosities (Fig. 23-2). In the female, the triangle contains the external genitalia and the orifices of the urethra and the vagina.

## Pelvic Diaphragm

The pelvic diaphragm is formed by the important levatores ani muscles and the small coccygeus muscles and their covering fasciae (Fig. 23-1). It is incomplete anteriorly to allow passage of the urethra and the vagina.

## Levator Ani Muscle

The levator ani muscle is a wide thin sheet that has a linear origin from the back of the body of the pubis, a tendinous arch formed by a thickening of the pelvic fascia covering the obturator internus, and the spine of the ischium (Fig. 23-3).

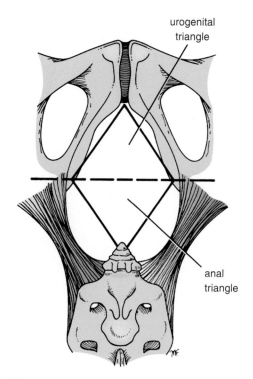

Figure 23-2 Diamond-shaped perineum divided by a *broken line* into the urogenital triangle and the anal triangle.

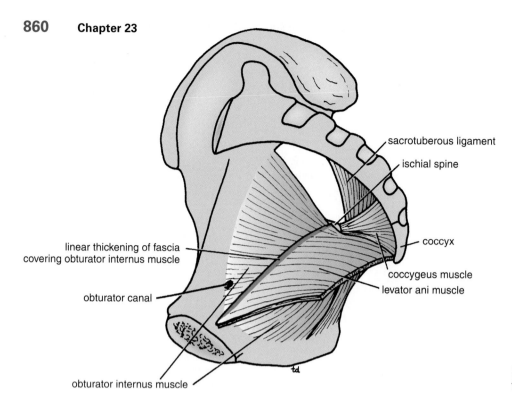

sacrotuberous ligament

ischial spine

linear thickening of fascia
covering obturator internus muscle

coccyx

obturator canal

coccygeus muscle

levator ani muscle

obturator internus muscle

**Figure 23-3**   Inferior wall or
floor of the pelvis.

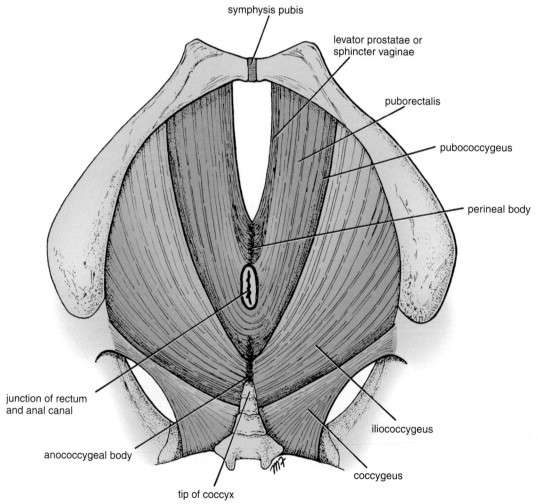

symphysis pubis

levator prostatae or
sphincter vaginae

puborectalis

pubococcygeus

perineal body

junction of rectum
and anal canal

iliococcygeus

anococcygeal body

coccygeus

tip of coccyx

**Figure 23-4**   Levator ani muscle and coccygeus muscle seen on their inferior aspects.
Note that the levator ani is made up of several different muscle groups. The levator ani
and coccygeus muscles with their fascial coverings form a continuous muscular floor to
the pelvis, known as the pelvic diaphragm.

From this extensive origin, groups of fibers sweep downward and medially to their insertion (Fig. 23-4), as follows:

- **Anterior fibers:** The **sphincter vaginae (levator prostatae in the male)** form a sling around the vagina and are inserted into a mass of fibrous tissue, called the **perineal body**, in front of the anal canal. The sphincter vaginae constrict the vagina and stabilize the perineal body.
- **Intermediate fibers:** The **puborectalis** forms a sling around the junction of the rectum and anal canal. The **pubococcygeus** passes posteriorly to be inserted into a small fibrous mass, called the **anococcygeal body**, between the tip of the coccyx and the anal canal.
- **Posterior fibers:** The **iliococcygeus** is inserted into the anococcygeal body and the coccyx.

**Action:** The levatores ani muscles of the two sides form an efficient muscular sling that supports and maintains the pelvic viscera in position. They resist the rise in intrapelvic pressure during the straining and expulsive efforts of the abdominal muscles (as occurs in coughing). They also have an important sphincter action on the anorectal junction, and they serve also as a sphincter of the vagina.

**Nerve supply:** This is from the perineal branch of the fourth sacral nerve and from the perineal branch of the pudendal nerve.

## Coccygeus Muscle

This small triangular muscle arises from the spine of the ischium and is inserted into the lower end of the sacrum and into the coccyx (Figs. 23-3 and 23-4).

**Action:** The two muscles assist the levatores ani in supporting the pelvic viscera.

**Nerve supply:** This is from a branch of the fourth and fifth sacral nerves.

A summary of the attachments of the muscles of the pelvic walls and floor, their nerve supply, and their action is given in Table 13-7.

## Urogenital Diaphragm

The urogenital diaphragm is a triangular musculofascial diaphragm situated in the anterior part of the perineum below the level of the pelvic diaphragm (Fig. 23-5). The urogenital diaphragm fills in the gap of the pubic arch. It is formed by the sphincter urethrae and the deep transverse

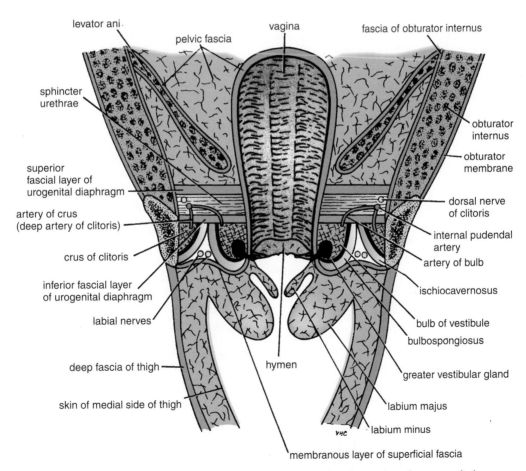

**Figure 23-5**  Coronal section of the female pelvis showing the vagina, the urogenital diaphragm, and the contents of the superficial perineal pouch.

perineal muscles, which are enclosed between a superior and an inferior layer of fascia of the urogenital diaphragm. The inferior layer of fascia is often referred to as the **perineal membrane.**

Anteriorly, the two layers of fascia fuse, leaving a small gap beneath the symphysis pubis. Posteriorly, the two layers of fascia fuse with each other and with the membranous layer of the superficial fascia and the perineal body. Laterally, the layers of fascia are attached to the pubic arch. The closed space that is contained between the superficial and deep layers of fascia is known as the **deep perineal pouch** (Fig. 23-5).

## Contents of the Deep Perineal Pouch

The deep perineal pouch (Fig. 23-5) contains part of the urethra; part of the vagina; the sphincter urethrae, which is pierced by the urethra and the vagina; the deep transverse perineal muscles; the internal pudendal vessels and their branches; and the dorsal nerves of the clitoris.

The urethra and the vagina are described on pages 863 and 876, respectively.

The **sphincter urethrae** is described on page 822. The **internal pudendal vessels** and the **dorsal nerves of the clitoris** have an arrangement similar to the corresponding structures found in the male.

A summary of the muscles of the perineum, their nerve supply, and their action is given in Table 13-8.

# Pelvic Fascia

The pelvic fascia is formed of connective tissue and is continuous above with the fascia lining the abdominal walls. Below, the fascia is continuous with the fascia of the perineum. The pelvic fascia can be divided into parietal and visceral layers.

## Parietal Layer of Pelvic Fascia

The parietal pelvic fascia lines the walls of the pelvis and is named according to the muscle it overlies. For example, over the obturator internus muscle, it is dense and strong and is known as the obturator internus fascia (Fig. 23-5). Over the levator ani and coccygeus muscles, it forms the levator ani and coccygeus fascia or, to describe it more concisely, the superior fascial layer of the pelvic diaphragm. Where the pelvic diaphragm is deficient anteriorly, the parietal pelvic fascia becomes continuous through the opening with the fascia covering the inferior surface of the pelvic diaphragm, in the perineum. In many locations where the parietal fascia comes into contact with bone, it fuses with the periosteum.

Below in the perineum, where the parietal pelvic fascia covers the sphincter urethrae muscle and the perineal membrane, it is known as the perineal layer of the parietal pelvic fascia; that is, it forms the superior fascial layer of the urogenital diaphragm.

## Visceral Layer of Pelvic Fascia

The visceral layer of pelvic fascia is a layer of loose connective tissue that covers and supports all the pelvic viscera. Where a particular viscus comes into contact with the pelvic wall, the visceral layer fuses with the parietal layer. In certain locations, the fascia thickens to form fascial ligaments, which commonly extend from the pelvic walls to a viscus and provide it with additional support. These ligaments are usually named according to their attachments, for example, the pubovesical and the sacrocervical ligaments.

# Pelvic Peritoneum

The parietal peritoneum lines the pelvic walls and is reflected onto the pelvic viscera, where it becomes continuous with the visceral peritoneum (Fig. 23-6). For further details, see page 879.

# The Female Genital Organs

The female reproductive system consists of the external genital organs, a pair of ovaries, a pair of uterine tubes, a uterus, and a vagina.

## External Genital Organs (Vulva)

The external genital organs include the **mons pubis** (hair-bearing skin in front of the pubis) (Fig. 23-7), the labia majora, the labia minora, the clitoris, and the greater vestibular glands (Bartholin's glands).

### Labia Majora

The labia majora are prominent folds of skin extending from the mons pubis to unite in the midline posteriorly (Fig. 23-7). They contain fat and are covered with hair on their outer surfaces. (They are equivalent to the scrotum in the male.)

### Labia Minora

The labia minora are two smaller folds of skin devoid of hair that lie between the labia majora (Fig. 23-7). Their posterior ends are united to form a sharp fold, the fourchette.

Anteriorly, they split to enclose the clitoris, forming an anterior prepuce and a posterior frenulum (Fig. 23-7).

#### Vestibule of the Vagina
The vestibule of the vagina is the space between the labia minora. It has the clitoris at its apex and has the openings of the urethra, the vagina, and ducts of the greater vestibular glands in its floor (Fig. 23-7).

### Clitoris

#### Location and Description
The clitoris, which corresponds to the penis in the male, is situated at the apex of the vestibule anteriorly (Fig. 23-7). It

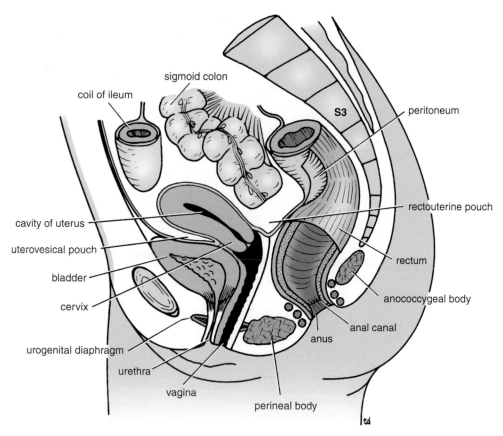

**Figure 23-6**   Sagittal section of the female pelvis.

has a structure similar to the penis. The glans of the clitoris is partly hidden by the prepuce.

### Root of the Clitoris

The root of the clitoris is made up of three masses of erectile tissue called the bulb of the vestibule and the right and left crura of the clitoris (Figs. 23-5 and 23-8).

The **bulb of the vestibule** corresponds to the bulb of the penis, but because of the presence of the vagina, it is divided into two halves (Fig. 23-8). It is attached to the undersurface of the urogenital diaphragm and is covered by the **bulbospongiosus muscles.**

The **crura of the clitoris** correspond to the crura of the penis and become the corpora cavernosa anteriorly. Each remains separate and is covered by an **ischiocavernosus muscle** (Fig. 23-8).

### Body of the Clitoris

The body of the clitoris consists of the two **corpora cavernosa** covered by their **ischiocavernosus muscles.** The corpus spongiosum of the male is represented by a small amount of erectile tissue leading from the vestibular bulbs to the glans.

### Glans of the Clitoris

The glans of the clitoris is a small mass of erectile tissue that caps the body of the clitoris. It is provided with numerous sensory endings. The glans is partly hidden by the **prepuce.**

### Blood Supply, Lymph Drainage, and Nerve Supply

The blood supply, lymph drainage, and nerve supply are similar to those of the penis (see Chapter 22).

## Greater Vestibular Glands (Bartholin's Glands)

The greater vestibular glands are a pair of small mucus-secreting glands that lie under cover of the posterior parts of the bulb of the vestibule and the labia majora (Figs. 23-5 and 23-8). Each drains its secretion into the vestibule by a small duct, which opens into the groove between the hymen and the posterior part of the labium minus (Fig. 23-7). These glands secrete a lubricating mucus during sexual intercourse.

## Urethra

The female urethra is about 1.5 in. (3.8 cm) long. It extends from the neck of the bladder to the **external meatus,** where it opens into the vestibule about 1 in. (2.5 cm) below the clitoris (Figs. 23-6 and 23-8). It traverses the sphincter urethrae and lies immediately in front of the vagina. At the sides of the external urethral meatus are the small openings of the ducts of the paraurethral glands. The urethra can be dilated relatively easily.

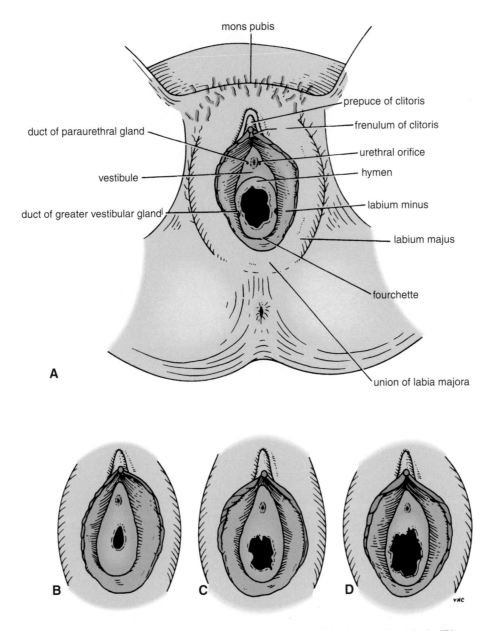

**Figure 23-7    A.** Vulva. Note the different appearances of the hymen in a virgin (**B**), a woman who has had sexual intercourse (**C**), and a multiparous woman (**D**).

## Paraurethral Glands

The paraurethral glands, which correspond to the prostate in the male, open into the vestibule by small ducts on either side of the urethral orifice (Fig. 23-7).

## Vaginal Orifice

The vaginal orifice possesses a thin mucosal fold called the **hymen**, which is perforated at its center. The blood supply to the hymen enters its circumference at the 4 and 8 o'clock positions (this information may be helpful when patients present with vaginal bleeding following tears during sexual intercourse).

While the lower half of the vagina lies within the perineum, the upper half lies above the pelvic floor, and the posterior fornix is related to the rectouterine pouch (pouch of Douglas) in the peritoneal cavity (Fig. 23-6).

## Lymphatic Drainage of the Vulva

The lymphatic drainage of the vulva, including the lower third of the vagina, is into the medial group of superficial inguinal nodes.

The relationship of the various structures of the vulva to one another and to the urogenital diaphragm is shown in Figure 23-8.

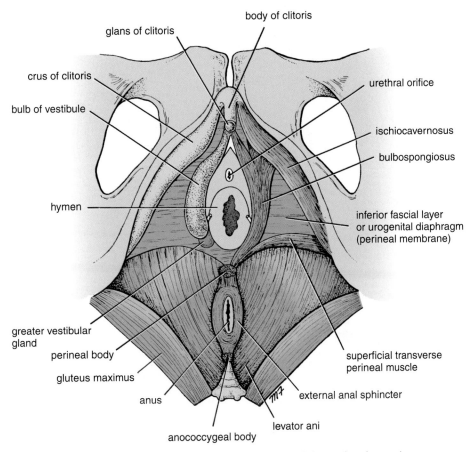

**Figure 23-8**   Root and body of the clitoris and the perineal muscles.

### Erection of the Clitoris

Sexual excitement produces engorgement of the erectile tissue within the clitoris in exactly the same manner as in the male.

#### Orgasm in the Female

As in the male, vision, hearing, smell, touch, and other psychic stimuli gradually build up the intensity of sexual excitement. During this process, the vaginal walls become moist because of transudation of fluid through the congested mucous membrane. In addition, the greater vestibular glands at the vaginal orifice secrete a lubricating mucus.

The upper part of the vagina, which resides in the pelvic cavity, is supplied by the hypogastric plexuses and is sensitive only to stretch. The region of the vaginal orifice, the labia minora, and the clitoris are extremely sensitive to touch and are supplied by the ilioinguinal nerves and the dorsal nerves of the clitoris.

Appropriate sexual stimulation of these sensitive areas, reinforced by afferent nervous impulses from the breasts and other regions, results in a climax of pleasurable sensory impulses reaching the central nervous system. Impulses then pass down the spinal cord to the sympathetic outflow (T1–L2).

The nervous impulses that pass to the genital organs are thought to leave the cord at the first and second lumbar segments in preganglionic sympathetic fibers. Many of these fibers synapse with postganglionic neurons in the first and second lumbar ganglia; other fibers may synapse in ganglia in the lower lumbar or pelvic parts of the sympathetic trunks. The postganglionic fibers are then distributed to the smooth muscle of the vaginal wall, which rhythmically contracts. In addition, nervous impulses travel in the pudendal nerve (S2, 3, and 4) to reach the bulbospongiosus and ischiocavernosus muscles, which also undergo rhythmic contraction. In many women, a single orgasm brings about sexual contentment, but other women require a series of orgasms to feel replete.

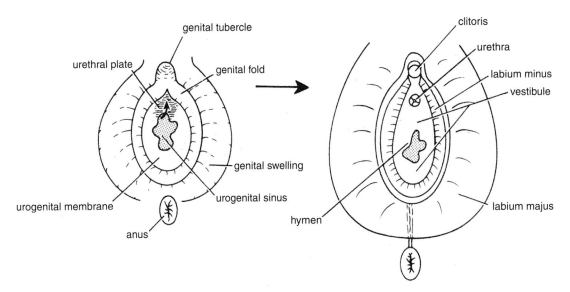

**Figure 23-9**   The development of the external genitalia in the female.

### Development of the Female External Genitalia

Early in development, the embryonic mesenchyme grows around the cloacal membrane and causes the overlying ectoderm to be raised up to form three swellings. One swelling occurs between the cloacal membrane and the umbilical cord in the midline and is called the genital tubercle (Fig. 23-9). On each side of the membrane, another swelling, called the genital fold, appears. At the 7th week, the genital tubercle elongates to form the glans. The anterior part of the cloacal membrane, the urogenital membrane, now ruptures so that the urogenital sinus opens onto the surface. The entodermal cells of the urogenital sinus proliferate and grow into the root of the phallus, forming a urethral plate. Meanwhile, a second pair of lateral swellings, called the genital swellings, appears lateral to the genital folds. At this stage of development, the genitalia of the two sexes are identical.

The changes in the female are less extensive than those in the male. The phallus becomes bent and forms the clitoris (Fig. 23-9). The genital folds do not fuse to form the urethra, as in the male, but develop into the labia minora. The labia majora are formed by the enlargement of the genital swellings.

## Internal Genital Organs

## Ovaries

### Location and Description
Each ovary is oval shaped, measuring 1.5 by 0.75 in. (4 by 2 cm), and is attached to the back of the broad ligament by the **mesovarium** (Fig. 23-10).

That part of the broad ligament extending between the attachment of the mesovarium and the lateral wall of the pelvis is called the **suspensory ligament of the ovary** (Fig. 23-10).

The **round ligament of the ovary**, which represents the remains of the upper part of the gubernaculum, connects the lateral margin of the uterus to the ovary (Figs. 23-10 and 23-11).

The ovary usually lies against the lateral wall of the pelvis in a depression called the **ovarian fossa**, bounded by the external iliac vessels above and by the internal iliac vessels behind (Fig. 23-11). The position of the ovary is, however, extremely variable, and it is often found hanging down in the rectouterine pouch (pouch of Douglas). During pregnancy, the enlarging uterus pulls the ovary up into the abdominal cavity. After childbirth, when the broad ligament is lax, the ovary takes up a variable position in the pelvis.

The ovaries are surrounded by a thin fibrous capsule, the **tunica albuginea**. This capsule is covered externally by a modified area of peritoneum called the **germinal epithelium**. The term germinal epithelium is a misnomer because the layer does not give rise to ova. Oogonia develop before birth from primordial germ cells.

Before puberty, the ovary is smooth; but after puberty, the ovary becomes progressively scarred as successive corpora lutea degenerate. After menopause, the ovary becomes shrunken, and its surface is pitted with scars.

### Function of the Ovaries

The ovaries are the organs responsible for the production of the female germ cells, the ova, and the female sex hormones, estrogen and progesterone, in the sexually mature female.

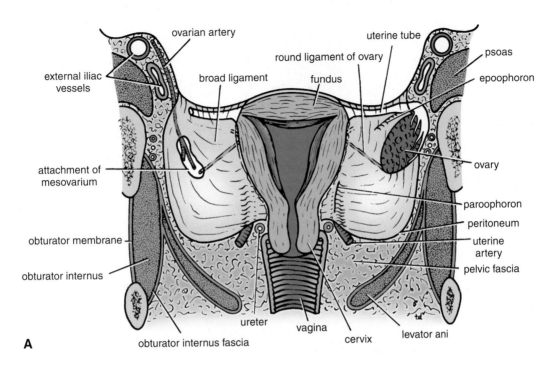

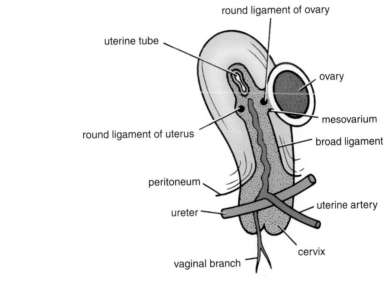

**Figure 23-10**   **A.** Coronal section of the pelvis showing the uterus, broad ligaments, and right ovary on posterior view. The left ovary and part of the left uterine tube have been removed for clarity. **B.** Uterus on lateral view. Note the structures that lie within the broad ligament. Note that the uterus has been retroverted into the plane of the vaginal lumen in both diagrams.

## Blood Supply

### Arteries

The ovarian artery arises from the abdominal aorta at the level of the first lumbar vertebra.

### Veins

The ovarian vein drains into the inferior vena cava on the right side and into the left renal vein on the left side.

## Lymph Drainage

The lymph vessels of the ovary follow the ovarian artery and drain into the para-aortic nodes at the level of the first lumbar vertebra.

## Nerve Supply

The nerve supply to the ovary is derived from the aortic plexus and accompanies the ovarian artery.

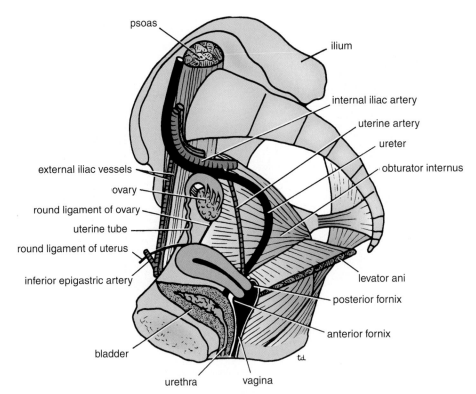

**Figure 23-11** Right half of the pelvis showing the ovary, the uterine tube, and the vagina.

The blood supply, lymph drainage, and nerve supply of the ovary pass over the pelvic inlet and cross the external iliac vessels (Fig. 23-10). They reach the ovary by passing through the lateral end of the broad ligament, the part known as the **suspensory ligament of the ovary**. The vessels and nerves finally enter the hilum of the ovary via the mesovarium. (Compare the blood supply and the lymph drainage of the ovary with those of the testis.)

**EMBRYOLOGIC NOTE**

### Development of the Ovary

The female sex chromosome causes the **genital ridge** on the posterior abdominal wall to secrete estrogens. The presence of estrogen and the absence of testosterone induce the development of the ovary and the other female genital organs.

The **sex cords** contained within the genital ridges contain groups of primordial germ cells. These become broken up into irregular cell clusters by the proliferating mesenchyme (Fig. 23-12). The germ cells differentiate into **oogonia,** and by the 3rd month, they start to undergo a number of mitotic divisions within the cortex of the ovary to form **primary oocytes**. These primary oocytes become surrounded by a single layer of cells derived from the sex cords, called the **granulosa cells**.

Thus, **primordial follicles** have been formed, but later, many degenerate. The mesenchyme that surrounds the follicles provides the ovarian stroma. The relationship of the ovary to the developing uterine tube is shown in Figure 23-13.

## Uterine Tubes

### Location and Description

The two uterine tubes are each about 4 in. (10 cm) long and lie in the upper border of the broad ligament (Figs. 23-10 and 23-11). Each connects the peritoneal cavity in the region of the ovary with the cavity of the uterus. The uterine tube is divided into four parts:

1. The **infundibulum** is the funnel-shaped lateral end that projects beyond the broad ligament and overlies the ovary. The free edge of the funnel has several fingerlike processes, known as **fimbriae,** which are draped over the ovary (Figs. 23-10 and 23-14).

2. The **ampulla** is the widest part of the tube (Fig. 23-14).

3. The **isthmus** is the narrowest part of the tube and lies just lateral to the uterus (Fig. 23-14).

4. The **intramural part** is the segment that pierces the uterine wall (Fig. 23-14).

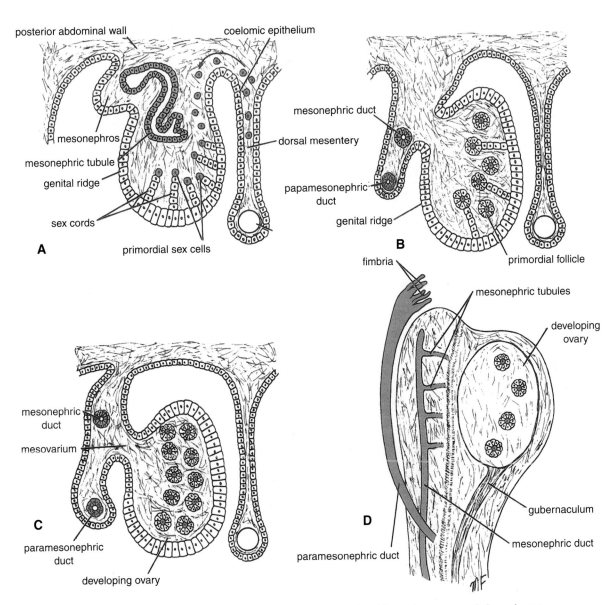

**Figure 23-12** Formation of the ovary and its relationship to the mesonephric and paramesonephric ducts.

---

**P H Y S I O L O G I C   N O T E**

## Function of the Uterine Tube

The uterine tube receives the ovum from the ovary and provides a site where fertilization of the ovum can take place (usually in the ampulla). It provides nourishment for the fertilized ovum and transports it to the cavity of the uterus. The tube serves as a conduit along which the spermatozoa travel to reach the ovum.

### Blood Supply

#### Arteries

The uterine artery from the internal iliac artery and the ovarian artery from the abdominal aorta (Fig. 23-14) supply the uterine tube.

#### Veins

The veins correspond to the arteries.

### Lymph Drainage

The lymph vessels from the uterine tube follow the corresponding arteries and drain into the internal iliac and para-aortic nodes.

### Nerve Supply

Sympathetic and parasympathetic nerves from the inferior hypogastric plexuses supply the uterine tube.

---

**E M B R Y O L O G I C   N O T E**

## Development of the Uterine Tube

Early on in development, the paramesonephric ducts appear on the posterior abdominal wall on the lateral side

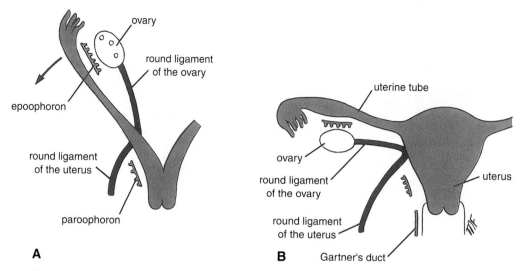

**Figure 23-13** The descent of the ovary and its relationship to the developing uterine tube and uterus.

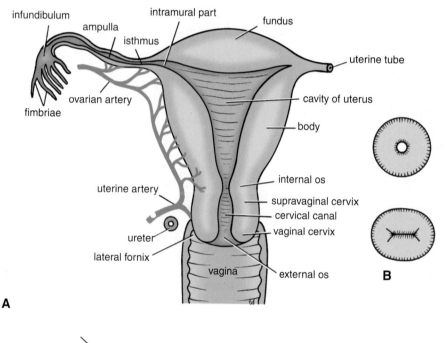

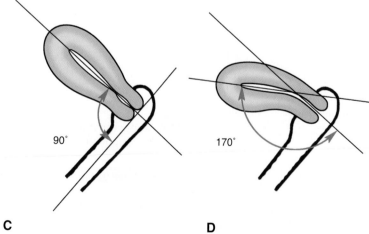

**Figure 23-14** **A.** Different parts of the uterine tube and the uterus. **B.** External os of the cervix: (*above*) nulliparous; (*below*) parous. **C.** Anteverted position of the uterus. **D.** Anteverted and anteflexed position of the uterus.

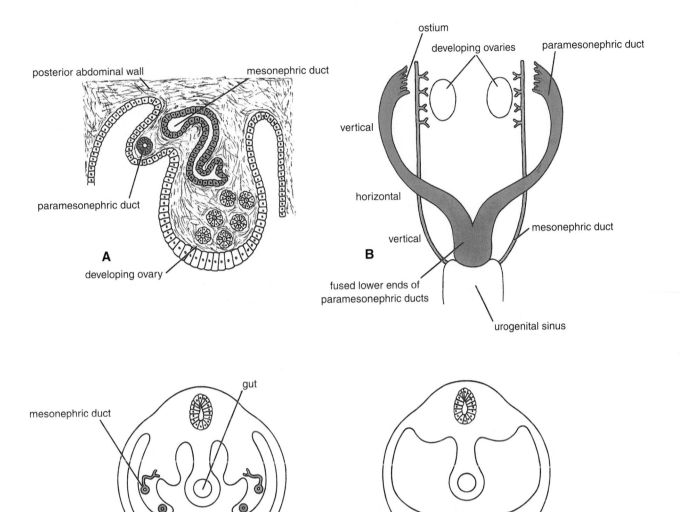

**Figure 23-15**    The relationship of the mesonephric and paramesonephric ducts to the developing ovary. **A.** Cross section of a developing ovary. **B.** Anterior view of ovaries and ducts. **C and D.** Mesonephric and paramesonephric ducts in a cross section of the pelvis. Note the developing broad ligament.

of the mesonephros. The uterine tube on each side is formed from the cranial vertical and middle horizontal parts of the paramesonephric duct (see Fig. 23-15). The tube elongates and becomes coiled; differentiation of the muscle and mucous membrane takes place; the **fimbriae** develop; and the **infundibulum, ampulla,** and **isthmus** are identifiable.

## Uterus

### Location and Description

The uterus is a hollow, pear-shaped organ with thick muscular walls. In the young nulliparous adult, it measures 3 in. (8 cm) long, 2 in. (5 cm) wide, and 1 in. (2.5 cm) thick. It is divided into the fundus, body, and cervix (Fig. 23-14).

The **fundus** is the part of the uterus that lies above the entrance of the uterine tubes.

The **body** is the part of the uterus that lies below the entrance of the uterine tubes. It narrows below, where it becomes continuous with the **cervix.** The cervix pierces the anterior wall of the vagina and is divided into the **supravaginal** and **vaginal parts of the cervix.**

The **cavity** of the uterine body is triangular in coronal section, but it is merely a cleft in the sagittal plane (Fig. 23-14). The cavity of the cervix, the **cervical canal,** communicates with the cavity of the body through the **internal os** and with that of the vagina through the **external os.** Before the birth of the first child, the external os is circular. In a parous woman, the vaginal part of the cervix is larger, and the external os becomes a transverse slit so that it possesses an anterior lip and a posterior lip (Fig. 23-14).

## Relations

- ■ **Anteriorly:** The body of the uterus is related anteriorly to the uterovesical pouch and the superior surface of the bladder (Fig. 23-6). The supravaginal cervix is related to the superior surface of the bladder. The vaginal cervix is related to the anterior fornix of the vagina.
- ■ **Posteriorly:** The body of the uterus is related posteriorly to the rectouterine pouch (pouch of Douglas) with coils of ileum or sigmoid colon within it (Fig. 23-6).
- ■ **Laterally:** The body of the uterus is related laterally to the broad ligament and the uterine artery and vein (Fig. 23-10). The supravaginal cervix is related to the ureter as it passes forward to enter the bladder. The vaginal cervix is related to the lateral fornix of the vagina. The uterine tubes enter the superolateral angles of the uterus, and the round ligaments of the ovary and of the uterus are attached to the uterine wall just below this level.

### PHYSIOLOGIC NOTE

### Function of the Uterus

The uterus serves as a site for the reception, retention, and nutrition of the fertilized ovum.

### Positions of the Uterus

In most women, the long axis of the uterus is bent forward on the long axis of the vagina. This position is referred to as **anteversion of the uterus** (Fig. 23-14). Furthermore, the long axis of the body of the uterus is bent forward at the level of the internal os with the long axis of the cervix. This position is termed **anteflexion of the uterus** (Fig. 23-14). Thus, in the erect position and with the bladder empty, the uterus lies in an almost horizontal plane.

In some women, the fundus and body of the uterus are bent backward on the vagina so that they lie in the rectouterine pouch (pouch of Douglas). In this situation, the uterus is said to be **retroverted.** If the body of the uterus is, in addition, bent backward on the cervix, it is said to be retroflexed.

### Structure of the Uterus

The uterus is covered with peritoneum except anteriorly, below the level of the internal os, where the peritoneum passes forward onto the bladder. Laterally, there is also a space between the attachment of the layers of the broad ligament.

The **muscular wall,** or **myometrium,** is thick and made up of smooth muscle supported by connective tissue.

The **mucous membrane** lining the body of the uterus is known as the **endometrium.** It is continuous above with the mucous membrane lining the uterine tubes and below with the mucous membrane lining the cervix. The endometrium is applied directly to the muscle, there being **no**

**submucosa.** From puberty to menopause, the endometrium undergoes extensive changes during the menstrual cycle in response to the ovarian hormones.

The supravaginal part of the cervix is surrounded by visceral pelvic fascia, which in this region, is often referred to as the **parametrium.** It is in this fascia that the uterine artery crosses the ureter on each side of the cervix.

### Blood Supply

#### Arteries

The arterial supply to the uterus is mainly from the uterine artery, a branch of the internal iliac artery. It reaches the uterus by running medially in the base of the broad ligament (Fig. 23-10). It crosses above the ureter at right angles and reaches the cervix at the level of the internal os (Fig. 23-14). The artery then ascends along the lateral margin of the uterus within the broad ligament and ends by anastomosing with the ovarian artery, which also assists in supplying the uterus. The uterine artery gives off a small descending branch that supplies the cervix and the vagina.

#### Veins

The uterine vein follows the artery and drains into the internal iliac vein.

### Lymph Drainage

The lymph vessels from the fundus of the uterus accompany the ovarian artery and drain into the para-aortic nodes at the level of the first lumbar vertebra. The vessels from the body and cervix drain into the internal and external iliac lymph nodes. A few lymph vessels follow the round ligament of the uterus through the inguinal canal and drain into the superficial inguinal lymph nodes.

### Nerve Supply

Sympathetic and parasympathetic nerves from branches of the inferior hypogastric plexuses.

### Supports of the Uterus

The uterus is supported mainly by the tone of the levatores ani muscles and the condensations of pelvic fascia, which form three important ligaments.

#### The Levatores Ani Muscles and the Perineal Body

The origin and the insertion of the levatores ani muscles are described on page 859. They form a broad muscular sheet stretching across the pelvic cavity, and, together with the pelvic fascia on their upper surface, they effectively support the pelvic viscera and resist the intra-abdominal pressure transmitted downward through the pelvis. The medial edges of the anterior parts of the levatores ani muscles are attached to the cervix of the uterus by the pelvic fascia (Fig. 23-16).

Some of the fibers of levator ani are inserted into a fibromuscular structure called the **perineal body** (Fig. 23-6). This structure is important in maintaining the integrity of the pelvic floor; if the perineal body is damaged during childbirth, prolapse of the pelvic viscera may occur. The perineal

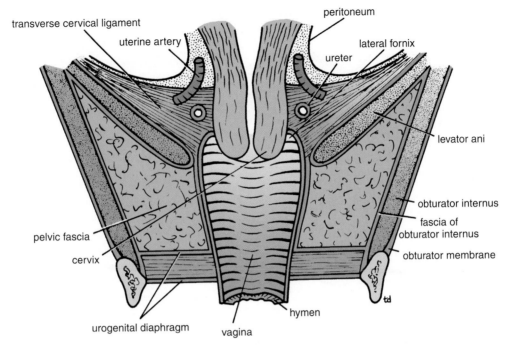

**Figure 23-16**  Coronal section of the pelvis showing relation of the levatores ani muscles and transverse cervical ligaments to the uterus and vagina. Note that the transverse cervical ligaments are formed from a condensation of visceral pelvic fascia.

body lies in the perineum between the vagina and the anal canal. It is slung up to the pelvic walls by the levatores ani and thus supports the vagina and, indirectly, the uterus.

### The Transverse Cervical, Pubocervical, and Sacrocervical Ligaments

These three ligaments are subperitoneal condensations of pelvic fascia on the upper surface of the levatores ani muscles. They are attached to the cervix and the vault of the vagina and play an important part in supporting the uterus and keeping the cervix in its correct position (Figs. 23-16 and 23-17).

■ **Transverse Cervical (Cardinal) Ligaments:** Transverse cervical ligaments are fibromuscular condensations of pelvic fascia that pass to the cervix and the upper end of the vagina from the lateral walls of the pelvis.

■ **Pubocervical Ligaments:** The pubocervical ligaments consist of two firm bands of connective tissue that pass to the cervix from the posterior surface of the pubis. They are positioned on either side of the neck of the bladder, to which they give some support (**pubovesical ligaments**).

■ **Sacrocervical Ligaments:** The sacrocervical ligaments consist of two firm fibromuscular bands of pelvic fascia that pass to the cervix and the upper end of the vagina from the lower end of the sacrum. They form two ridges, one on either side of the rectouterine pouch (pouch of Douglas).

### The Broad Ligaments

The broad ligaments and the round ligaments of the uterus are lax structures, and the uterus can be pulled up or pushed down for a considerable distance before they become taut. Clinically, they are considered to play a minor role in supporting the uterus.

### The Round Ligaments

The round ligament of the uterus, which represents the remains of the lower half of the gubernaculum, extends between the superolateral angle of the uterus, through the deep inguinal ring and inguinal canal, to the subcutaneous tissue of the labium majus (Fig. 23-11). It helps keep the uterus anteverted (tilted forward) and anteflexed (bent forward) but is considerably stretched during pregnancy.

### Uterus in the Child

The fundus and body of the uterus remain small until puberty, when they enlarge greatly in response to the estrogens secreted by the ovaries.

### Uterus after Menopause

After menopause, the uterus atrophies and becomes smaller and less vascular. These changes occur because the ovaries no longer produce estrogens and progesterone.

### EMBRYOLOGIC NOTE

### Development of the Uterus

The uterus is derived from the fused caudal vertical parts of the paramesonephric ducts (Fig. 23-19), and the site of their angular junction becomes a convex dome and forms the **fundus** of the uterus. The fusion between the

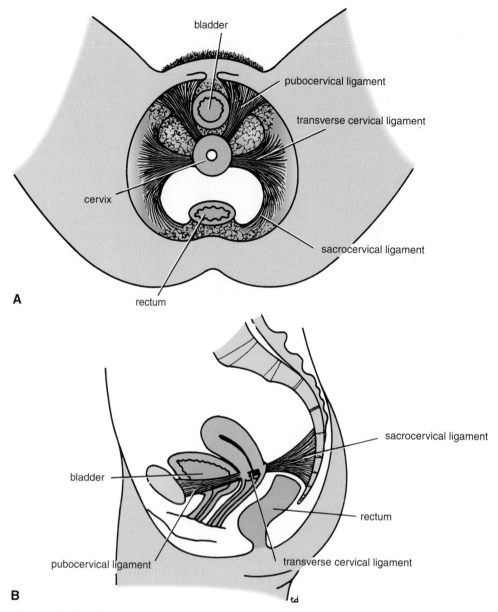

**Figure 23-17**   Ligamentous supports of uterus. **A.** As seen from below. **B.** Lateral view. These ligaments are formed from visceral pelvic fascia.

ducts is incomplete at first, a septum persisting between the lumina. Later, the septum disappears so that a single cavity remains. The upper part of the cavity forms the lumen of the **body** and **cervix** of the uterus. The myometrium is formed from the surrounding mesenchyme.

E M B R Y O L O G I C   N O T E

### The Implantation of the Fertilized Ovum in the Uterus

The blastocyst enters the uterine cavity between the 4th and 9th days after ovulation. Normal implantation takes place in the endometrium of the body of the uterus, most frequently on the upper part of the posterior wall near the midline (Fig. 23-20). As a result of the enzymatic digestion of the uterine epithelium by the trophoblast of the embryo, the blastocyst sinks beneath the surface epithelium and becomes embedded in the stroma by the 11th or 12th day.

E M B R Y O L O G I C   N O T E

### The Formation of the Placenta

The placenta is the organ that carries out respiration, excretion, and nutrition for the embryo, and it is fully

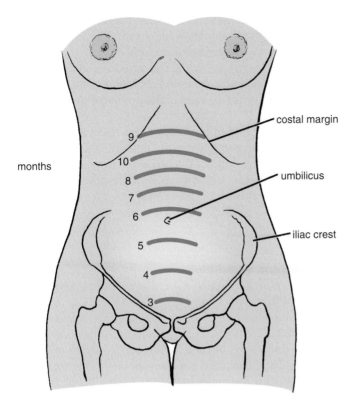

**Figure 23-18** Height of the fundus of the uterus at various months of pregnancy.

formed during the 4th month. The formation of the placenta is complicated and is essentially the development of an organ by mother and child in symbiosis and consists of fetal and maternal parts.

The fetal part develops as follows. The trophoblast becomes a highly developed structure, with villi that continue to erode and penetrate deeper into the endometrium. Large irregular spaces known as **lacunae** appear, which become filled with maternal blood. At the center of each villus is connective tissue containing fetal blood vessels that will eventually anastomose with one another and converge to form the umbilical cord (Fig. 23-21).

The maternal part develops as follows. Under the influence of progesterone, secreted first by the corpus luteum and later by the placenta itself, the endometrium becomes greatly thickened and is known as the **decidua.** Large areas of the decidua become excavated by the invading trophoblastic villi to form the **intervillous spaces.** The maternal blood vessels open into the spaces so that the outer surfaces of the villi of the fetal part of the placenta become bathed in oxygenated blood (Fig. 23-21).

By the 4th month of pregnancy, the placenta is a well-developed organ. As the pregnancy continues, the placenta increases in area and thickness. The placental attachment occupies one third of the internal surface of the uterus.

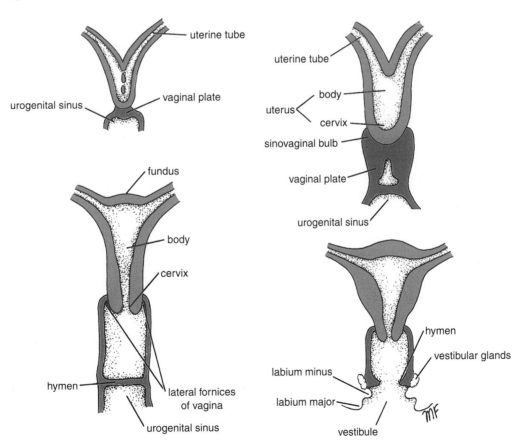

**Figure 23-19** Formation of the uterine tubes, the uterus, and the vagina.

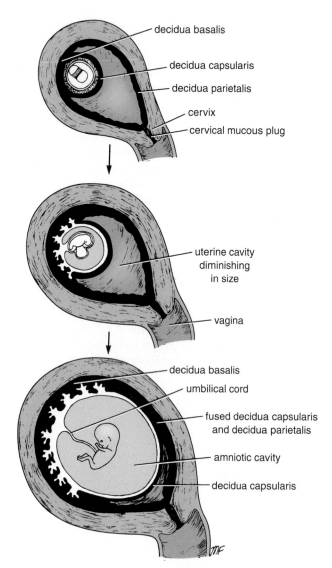

decidua basalis

decidua capsularis

decidua parietalis

cervix

cervical mucous plug

uterine cavity
diminishing
in size

vagina

decidua basalis

umbilical cord

fused decidua capsularis
and decidua parietalis

amniotic cavity

decidua capsularis

**Figure 23-20**   Sagittal section of the uterus showing the developing conceptus expanding into the uterine cavity. The three different regions of the decidua can be recognized. By the 16th week, the uterine cavity is obliterated by the fusion of the decidua capsularis with the decidua parietalis.

At birth, a few minutes after the delivery of the child, the placenta separates from the uterine wall and is expelled from the uterine cavity as the result of the contractions of the uterine musculature. The line of separation occurs through the spongy layer of the decidua (Fig. 23-21).

### Gross Appearance of the Placenta at Birth

At full term, the placenta has a sponge-like consistency. It is flattened and circular, with a diameter of about 8 in. (20 cm) and a thickness of about 1 in. (2.5 cm), and weighs about 1 lb (500 g). It thins out at the edges, where it is continuous with the fetal membranes (Fig. 23-22).

The outer, or maternal, surface of a freshly shed placenta is rough on palpation, dark red, and oozes blood from the torn maternal blood vessels.

The inner, or fetal, surface is smooth and shiny and is raised in ridges by the umbilical blood vessels, which radiate from the attachment of the umbilical cord near its center.

The fetal membranes (see Figs. 19-19 and 19-20), which surround and enclose the amniotic fluid, are continuous with the edge of the placenta. They are the **amnion**, the **chorion**, and a small amount of the adherent **maternal decidua**.

## Vagina

### Location and Description

The vagina is a muscular tube that extends upward and backward from the vulva to the uterus (Fig. 23-6). It measures about 3 in. (8 cm) long and has anterior and posterior walls, which are normally in apposition. At its upper end, the anterior wall is pierced by the cervix, which projects downward and backward into the vagina. It is important to remember that the upper half of the vagina lies above the pelvic floor and the lower half lies within the perineum (Figs. 23-6 and 23-16). The area of the vaginal lumen, which surrounds the cervix, is divided into four regions, or **fornices**: anterior, posterior, right lateral, and left lateral. The vaginal orifice in a virgin possesses a thin mucosal fold called the **hymen,** which is perforated at its center. After childbirth, the hymen usually consists only of tags.

### Relations

- **Anteriorly:** The vagina is closely related to the bladder above and to the urethra below (Fig. 23-6).
- **Posteriorly:** The upper third of the vagina is related to the rectouterine pouch (pouch of Douglas), and its middle third is related to the ampulla of the rectum. The lower third is related to the perineal body, which separates it from the anal canal (Fig. 23-6).
- **Laterally:** In its upper part, the vagina is related to the ureter; its middle part is related to the anterior fibers of the levator ani, as they run backward to reach the perineal body and hook around the anorectal junction (Figs. 23-10 and 23-16). Contraction of the fibers of levator ani compresses the walls of the vagina together. In its lower part, the vagina is related to the urogenital diaphragm and the bulb of the vestibule.

### PHYSIOLOGIC NOTE

### Functions of the Vagina

The vagina not only is the female genital canal but also serves as the excretory duct for the menstrual flow and forms part of the birth canal.

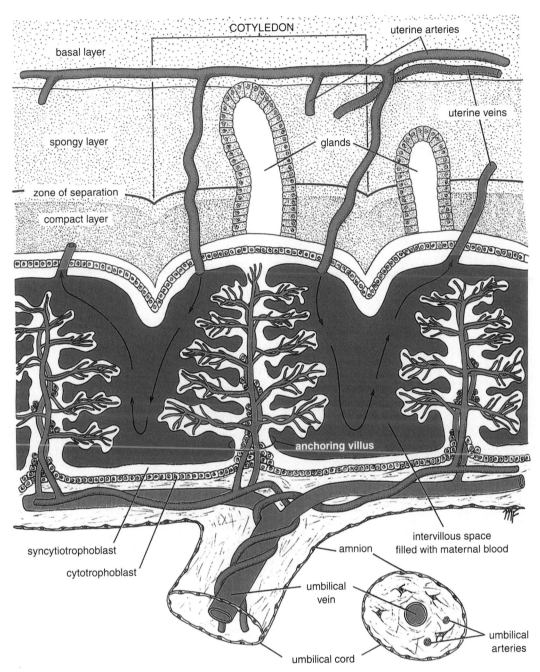

**Figure 23-21** A section through the placenta showing the maternal (*top*) and fetal (*bottom*) parts. Note the maternal part is divided into the basal layer, the spongy layer, and the compact layer. The *heavy solid line* in the spongy layer indicates where separation occurs between the maternal and fetal parts of the placenta during the third stage of labor.

## Blood Supply
### Arteries
The vaginal artery, a branch of the internal iliac artery, and the vaginal branch of the uterine artery supply the vagina.

### Veins
The vaginal veins form a plexus around the vagina that drains into the internal iliac vein.

## Lymph Drainage
The lymph vessels from the upper third of the vagina drain to the external and internal iliac nodes, from the middle third to the internal iliac nodes, and from the lower third to the superficial inguinal nodes.

## Nerve Supply
The nerve supply to the vagina is from the inferior hypogastric plexuses.

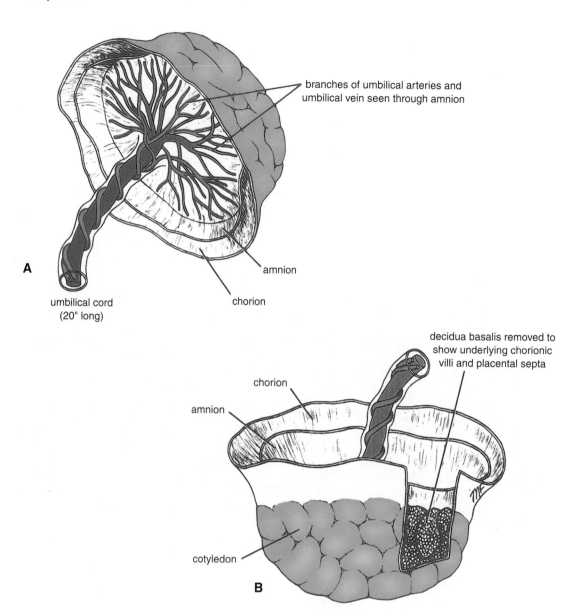

branches of umbilical arteries and
umbilical vein seen through amnion

amnion

**A**

umbilical cord
(20" long)

chorion

decidua basalis removed to
show underlying chorionic
villi and placental septa

chorion

amnion

cotyledon

**B**

**Figure 23-22**   The mature placenta as seen from the fetal surface (**A**) and from the maternal surface (**B**).

## Supports of the Vagina

**The upper part of the vagina** is supported by the levatores ani muscles and the transverse cervical, pubocervical, and sacrocervical ligaments. These structures are attached to the vaginal wall by pelvic fascia (Figs. 23-16 and 23-17).

**The middle part of the vagina** is supported by the urogenital diaphragm (Fig. 23-16).

**The lower part of the vagina,** especially the posterior wall, is supported by the perineal body (Fig. 23-6).

## E M B R Y O L O G I C   N O T E

### Development of the Vagina

The vagina is developed from the wall of the **urogenital sinus** (Fig. 23-19). The fused lower ends of the paramesonephric ducts form the body and cervix of the uterus, and once the solid end of the fused ducts reaches the posterior wall of the urogenital sinus, two outgrowths occur from the sinus, called the **sinovaginal bulbs.** The cells of the sinovaginal bulbs proliferate rapidly and form the **vaginal plate.** The vaginal plate thickens and elongates and extends around the solid end of the fused paramesonephric ducts. Later, the plate is completely canalized and the vaginal fornices are formed.

## Visceral Pelvic Fascia

The visceral pelvic fascia is described on page 862. It covers and supports the pelvic viscera and is condensed to form the pubocervical, transverse cervical, and sacrocervical ligaments of the uterus (Fig. 23-17). The visceral fascia is continuous below with the fascia covering the upper surface of the levatores ani and coccygeus muscles and on the walls of the pelvis with the parietal pelvic fascia.

## Peritoneum

The peritoneum is best understood by tracing it around the pelvis in a sagittal plane (Fig. 23-6).

The peritoneum passes down from the anterior abdominal wall onto the upper surface of the urinary bladder. It then runs directly onto the anterior surface of the uterus, at the level of the internal os. The peritoneum now passes upward over the anterior surface of the body and fundus of the uterus and then downward over the posterior surface. It continues downward and covers the upper part of the posterior surface of the vagina, where it forms the anterior wall of the rectouterine pouch (pouch of Douglas). The peritoneum then passes onto the front of the rectum, as in the male.

In the female, the lowest part of the abdominopelvic peritoneal cavity in the erect position is the rectouterine pouch.

## Broad Ligaments

The broad ligaments are two-layered folds of peritoneum that extend across the pelvic cavity from the lateral margins of the uterus to the lateral pelvic walls (Fig. 23-10). Superiorly, the two layers are continuous and form the upper free edge. Inferiorly, at the base of the ligament, the layers separate to cover the pelvic floor. The ovary is attached to the posterior layer by the **mesovarium.** The part of the broad ligament that lies lateral to the attachment of the mesovarium forms the **suspensory ligament of the ovary.** The part of the broad ligament between the uterine tube and the mesovarium is called the **mesosalpinx.**

At the base of the broad ligament, the uterine artery crosses the ureter (Figs. 23-10 and 23-16).

Each broad ligament contains the following:

▪ The uterine tube in its upper free border.
▪ The round ligament of the ovary and the round ligament of the uterus. They represent the remains of the gubernaculum.
▪ The uterine and ovarian blood vessels, lymph vessels, and nerves.
▪ The epoophoron, a vestigial structure that lies in the broad ligament above the attachment of the mesovar-

ium. It represents the remains of the mesonephros (Fig. 23-10).
▪ The paroophoron is also a vestigial structure that lies in the broad ligament just lateral to the uterus. It is a mesonephric remnant (Fig. 23-10).

# CROSS-SECTIONAL ANATOMY OF THE PELVIS

To assist in the interpretation of CT scans of the pelvis, students should study the labeled cross sections of the pelvis shown in Figures 23-23 and 23-24.

# RADIOGRAPHIC ANATOMY OF THE FEMALE PELVIS

The radiographic appearances of the bony female pelvis are shown in Figures 23-25 and 23-26. The instillation of viscous iodine preparations through the external os of the uterus allows the lumen of the cervical canal, the uterine cavity, and the different parts of the uterine tubes to be visualized (Fig. 23-27). This procedure is known as **hysterosalpingography.** The patency of these structures is demonstrated by the entrance into the peritoneal cavity of some of the opaque medium.

**A sonogram** of the female pelvis shows the uterus and the vagina (Figs. 23-28, 23-29, and 23-30).

# SURFACE ANATOMY

## Bony Landmarks

The perineum, when seen from below with the thighs abducted (Fig. 23-2), is diamond shaped and is bounded anteriorly by the symphysis pubis, posteriorly by the tip of the coccyx, and laterally by the ischial tuberosities.

### Symphysis Pubis

This is the cartilaginous joint that lies in the midline between the bodies of the pubic bones (Fig. 23-31). It is felt as a solid structure beneath the skin in the midline at the lower extremity of the anterior abdominal wall.

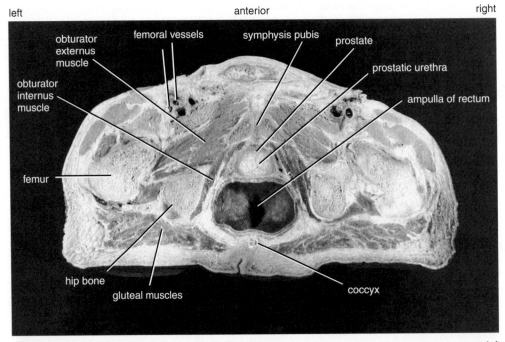

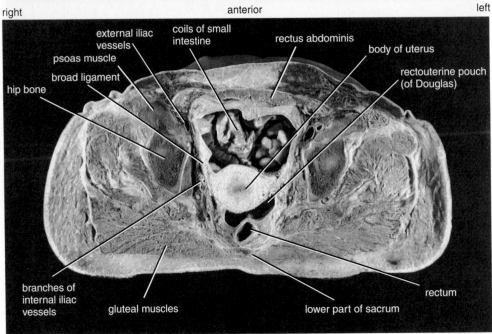

**Figure 23-23   A.** Cross section of the male pelvis as seen from above. **B.** Cross section of the female pelvis as seen from below.

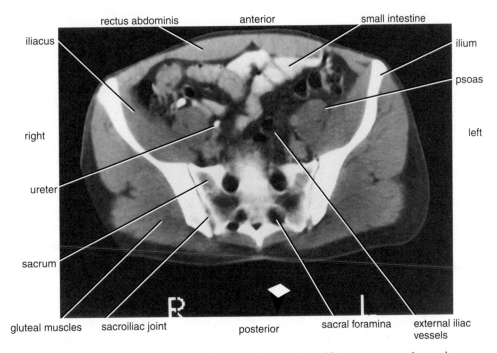

rectus abdominis     anterior     small intestine

iliacus                                                    ilium

                                                           psoas

right                                                      left

ureter

sacrum

gluteal muscles     sacroiliac joint     posterior     sacral foramina     external iliac vessels

**Figure 23-24**   CT scan of the pelvis after a barium meal and intravenous pyelography. Note the presence of the radiopaque material in the small intestine and the right ureter. The section is viewed from below.

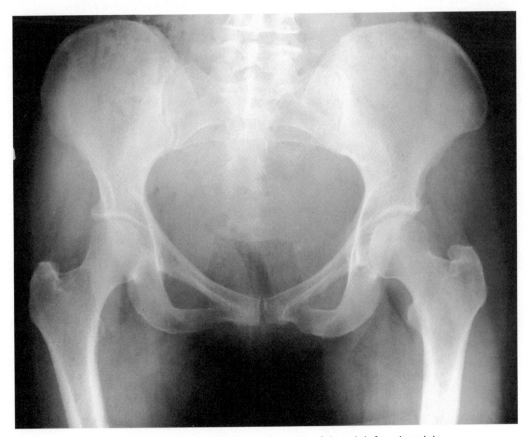

**Figure 23-25**   Anteroposterior radiograph of the adult female pelvis.

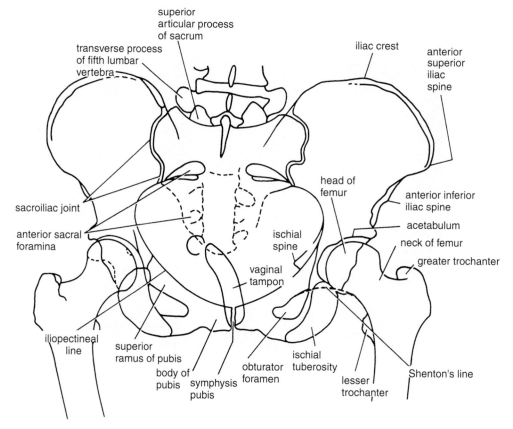

**Figure 23-26** Representation of the radiograph of the pelvis seen in Figure 23-25.

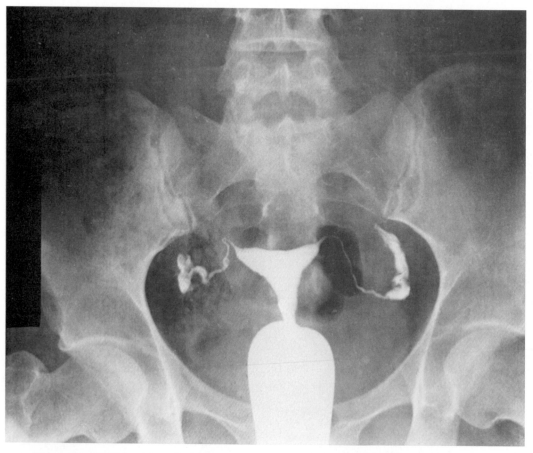

**Figure 23-27** Anteroposterior radiograph of the female pelvis after injection of radiopaque compound into the uterine cavity (hysterosalpingogram).

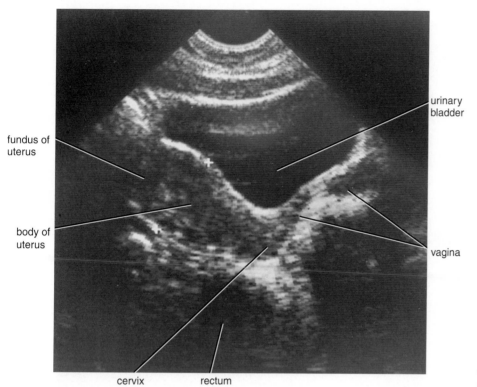

anterior

fundus of
uterus

body of
uterus

urinary
bladder

vagina

cervix          rectum

posterior

**Figure 23-28** Longitudinal
sonogram of the female pelvis
showing the uterus, the vagina,
and the bladder. (Courtesy of
M.C. Hill.)

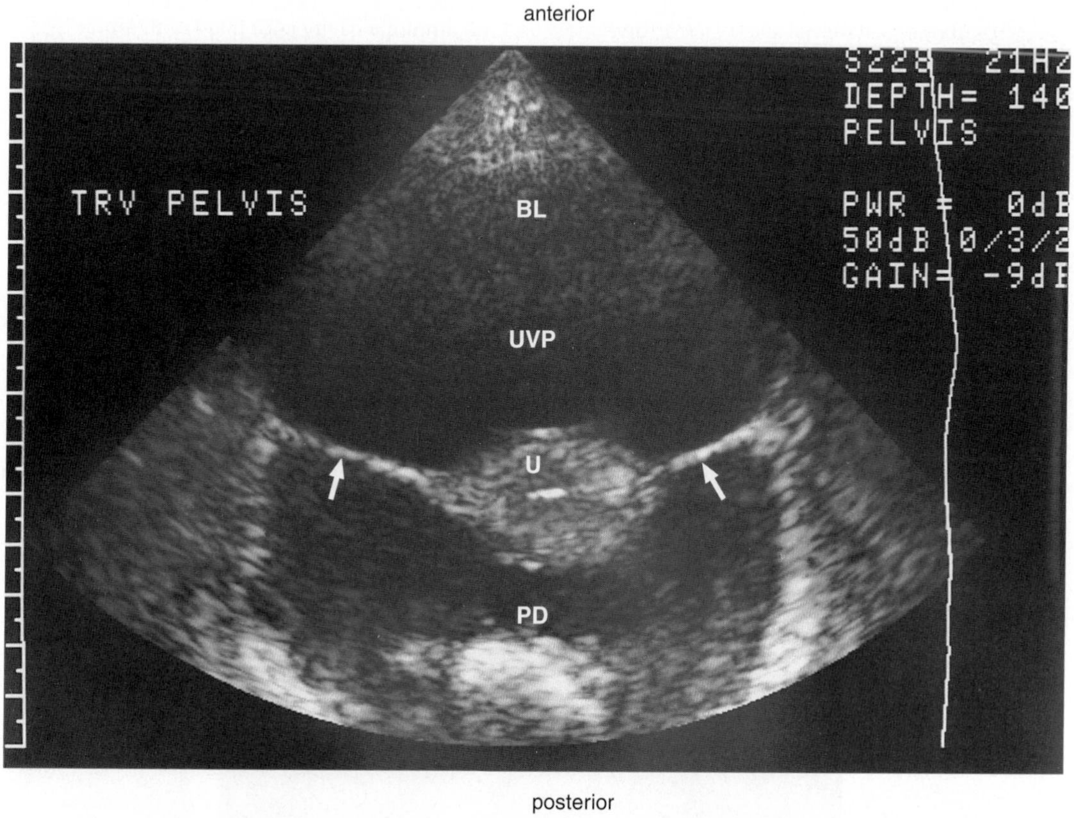

anterior

TRV PELVIS

S228  21Hz
DEPTH= 140
PELVIS

PWR =  0dB
50dB 0/3/2
GAIN= -9dB

BL

UVP

U

PD

posterior

**Figure 23-29** Transverse sonogram of the pelvis in a woman after an automobile accident, in which the liver was lacerated and blood escaped into the peritoneal cavity. The
bladder (*BL*), the body of the uterus (*U*), and the broad ligaments (*white arrows*) are identified. Note the presence of blood (*dark areas*) in the uterovesical pouch (*UVP*) and the
pouch of Douglas (*PD*). (Courtesy of L Scoutt.)

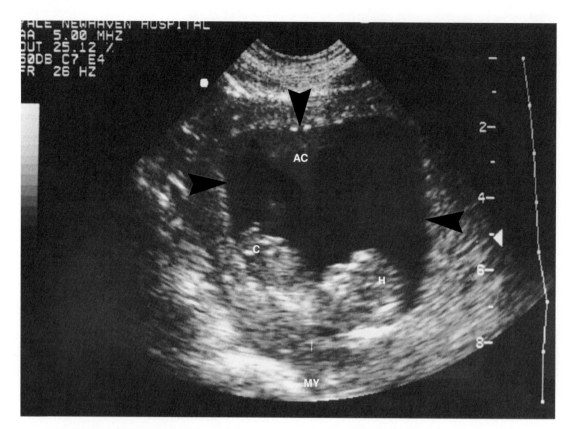

**Figure 23-30** Longitudinal sonogram of a pregnant uterus at 11 weeks showing the intrauterine gestational sac (*black arrowheads*) and the amniotic cavity (*AC*) filled with amniotic fluid; the fetus is seen in longitudinal section with the head (*H*) and coccyx (*C*) well displayed. The myometrium (*MY*) of the uterus can be identified. (Courtesy of L. Scoutt.)

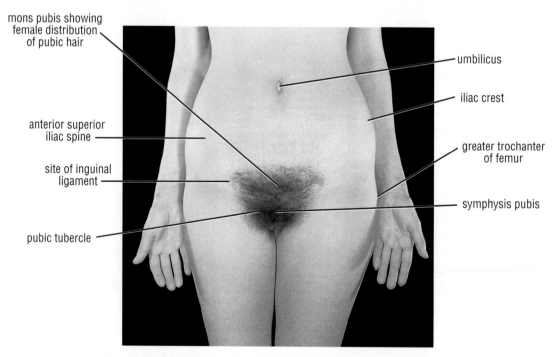

**Figure 23-31** Anterior view of the pelvis of a 29-year-old woman.

## Coccyx

The inferior surface and tip of the coccyx can be palpated in the cleft between the buttocks about 1 in. (2.5 cm) behind the anus (Fig. 23-2).

## Ischial Tuberosity

This can be palpated in the lower part of the buttock (Fig. 23-32). In the standing position, the tuberosity is covered by the gluteus maximus muscle. In the sitting position, the ischial tuberosity emerges from beneath the lower border of the gluteus maximus and supports the weight of the body.

# Vulva

## Mons Pubis

This the rounded, hair-bearing elevation of skin found anterior to the pubis (Fig. 23-32). The pubic hair in the female has an abrupt horizontal superior margin, whereas in the male, it extends upward to the umbilicus.

## Labia Majora

These are prominent, hair-bearing folds of soft skin extending posteriorly from the mons pubis to unite posteriorly in the midline (Fig. 23-32).

## Labia Minora

These two smaller, hairless folds of soft skin lie between the labia majora (Fig. 23-32). Their posterior ends are united to form a sharp fold, the **fourchette**. Anteriorly, they split to enclose the clitoris, forming the anterior **prepuce** and a posterior **frenulum** (Fig. 23-32).

## Vestibule

This is a smooth triangular area bounded laterally by the labia minora, with the clitoris at its apex and the fourchette at its base (Fig. 23-7).

## Vaginal Orifice

This is protected in virgins by a thin mucosal fold called the **hymen**, which is perforated at its center. At the first coitus, the

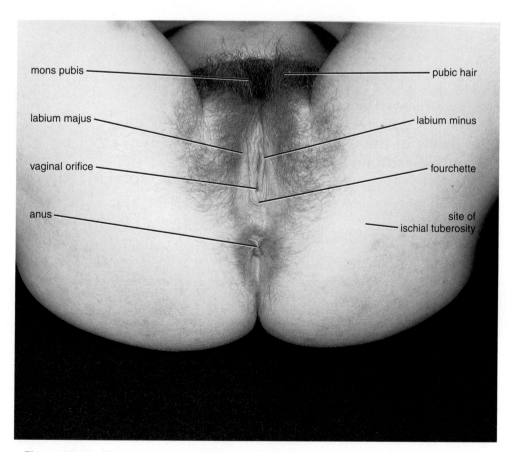

**Figure 23-32** The perineum in a 25-year-old female, inferior view. **A.** With labia together.

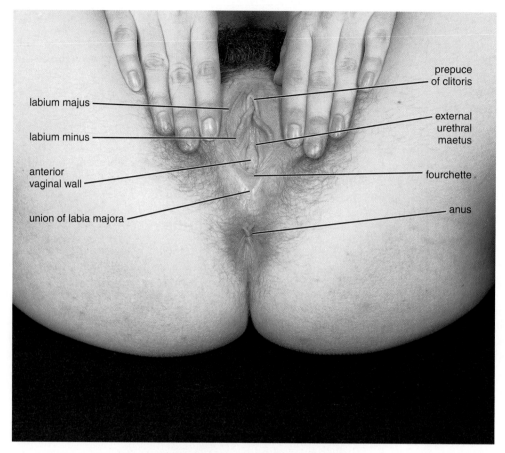

labium majus

labium minus

anterior
vaginal wall

union of labia majora

prepuce
of clitoris

external
urethral
maetus

fourchette

anus

**Figure 23-32** *(Continued)* **B.** With labia separated.

hymen tears, usually posteriorly or posterolaterally, and after childbirth, only a few tags of the hymen remain (Fig. 23-7).

## Orifices of the Ducts of the Greater Vestibular Glands (Bartholin's Glands)

These are small orifices, one on each side, in the groove between the hymen and the posterior part of the labium minus (Fig. 23-7).

## Clitoris

The clitoris is situated at the apex of the vestibule anteriorly (Figs. 23-7 and 23-32). The glans of the clitoris is partly hidden by the **prepuce**.

## External Urethral Meatus

The urethra opens into the vestibule at the **external meatus.** This lies about 1 in. (2.5 cm) below the clitoris (Fig. 23-32).

# Review Questions

## Multiple Choice Questions

### Select the BEST answer for each question.

1. The following statements concerning the uterus are correct **except** which?
   A. The fundus is part of the uterus above the openings of the uterine tubes.
   B. The long axis of the uterus is usually bent anteriorly on the long axis of the vagina (anteversion).
   C. The nerve supply of the uterus is from the inferior hypogastric plexuses.
   D. The anterior surface of the cervix is completely covered with peritoneum.
   E. The uterine veins drain into the internal iliac veins.

2. The following statements concerning the ovary are correct **except** which?
   A. The lymph drainage is into the para-aortic (lumbar) lymph nodes at the level of the first lumbar vertebra.
   B. The round ligament of the ovary extends from the ovary to the upper end of the lateral wall of the body of the uterus.
   C. The ovarian fossa is bounded above by the external iliac vessels and behind by the internal iliac vessels.
   D. The left ovarian artery is a branch of the left internal iliac artery.
   E. The obturator nerve lies lateral to the ovary.

3. The following statements concerning the vagina are correct **except** which?
   A. The area of the vaginal lumen around the cervix is divided into four fornices.
   B. The upper part of the vagina is not covered with peritoneum.
   C. The perineal body lies posterior to and supports the lower part of the vagina.
   D. The upper part of the vagina is supported by the levator ani muscles and the transverse cervical ligaments.
   E. The vaginal wall receives a branch of the uterine artery.

4. The following statements concerning the visceral layer of pelvic fascia in the female are correct **except** which?
   A. In the region of the cervix of the uterus, it is called the parametrium.
   B. It is considered to form the pubocervical, transverse cervical, and sacrocervical ligaments of the uterus.
   C. It covers the obturator internus muscle.
   D. It does not become continuous above with the fascia transversalis.
   E. On the lateral wall of the pelvis, it fuses with the parietal layer of pelvic fascia.

5. The following statements concerning the pelvis are correct **except** which?
   A. The ilium, ischium, and pubis are three separate bones that fuse together to form the hip bone in the 25th year of life.
   B. The platypelloid type of pelvis occurs in about 2% of women.
   C. External pelvic measurements have little practical importance in determining whether a disproportion between the size of the fetal head and the size of the pelvic inlet is likely.
   D. The pelvic outlet is formed by the symphysis pubis anteriorly, the ischial tuberosities laterally, the sacrotuberous ligaments laterally, and the coccyx posteriorly.
   E. The sacrum is shorter, wider, and flatter in the female than in the male.

6. The following statements concerning the muscles and fascia in the pelvis are correct **except** which?
   A. The levator ani muscle is innervated by the perineal branch of the fourth sacral nerve and from the perineal branch of the pudendal nerve.
   B. In the pelvis, the fascia is divided into parietal and visceral layers.
   C. The iliococcygeus muscle arises from a thickening of the obturator internus fascia.
   D. The pelvic diaphragm is strong and has no openings.
   E. The visceral layer of pelvic fascia forms important ligaments that help support the uterus.

7. The following statements concerning the female urethra are correct **except** which?
   A. It lies immediately anterior to the vagina.
   B. Its external orifice lies about 2 in. (5 cm) from the clitoris.
   C. It is about 1.5 in. (3.75 cm) long.
   D. It pierces the urogenital diaphragm.
   E. It is straight, and only minor resistance is felt as a catheter is passed through the urethral sphincter.

8. The following structures can be palpated by a vaginal examination **except** which?
   A. Sigmoid colon
   B. Ureters
   C. Perineal body
   D. Ischial spines
   E. Iliopectineal line

9. The urogenital diaphragm is formed by the following structures **except** which?
   A. Deep transverse perineal muscle
   B. Perineal membrane
   C. Colles' fascia (membranous layer of superficial fascia)
   D. Sphincter urethrae
   E. Parietal pelvic fascia covering the upper surface of the sphincter urethrae muscle

## Completion Questions

**Select the phrase that BEST completes each statement.**

10. A carcinoma of the cervix of the uterus is likely to spread via the lymphatic vessels into the
    A. external iliac nodes.
    B. internal iliac nodes.
    C. superficial inguinal nodes.
    D. internal and external iliac nodes.
    E. presacral lymph nodes.

11. An acute infection of the vaginal orifice is likely to spread via the lymphatic vessels into the
    A. medial group of horizontal superficial inguinal nodes.
    B. internal iliac nodes.
    C. internal and external iliac nodes.
    D. vertical group of superficial inguinal nodes.
    E. None of the above

12. The left ovarian artery originates from the
    A. external iliac artery.
    B. internal iliac artery.
    C. left renal artery.
    D. inferior mesenteric artery.
    E. abdominal part of the aorta.

13. The right ovarian vein drains into the
    A. right internal iliac vein.
    B. inferior vena cava.
    C. inferior mesenteric vein.
    D. right external iliac vein.
    E. right renal vein.

14. In most women, the anatomic position of the uterus when the bladder is empty is
    A. retroverted.
    B. anteverted.
    C. anteflexed.
    D. anteverted and anteflexed.
    E. retroflexed.

15. The uterus receives its blood supply from the
    A. superior vesical artery.
    B. middle rectal artery.
    C. ovarian artery.
    D. uterine artery.
    E. uterine and ovarian arteries.

16. In women with ovarian cancer, it is judicious to examine the
    A. peritoneal cavity for evidence fluid (ascites).
    B. superficial inguinal nodes.
    C. para-aortic nodes at the level of the first lumbar vertebra.
    D. external iliac lymph nodes.
    E. para-aortic lymph nodes and evidence of excessive peritoneal fluid (ascites).

17. The rectouterine pouch (pouch of Douglas) can be most efficiently entered by a surgical incision through the
    A. posterior fornix of the vagina.
    B. anterior fornix of the vagina.
    C. anterior rectal wall.
    D. lateral fornix of the vagina.
    E. posterior wall of the cavity of the uterine body.

18. The pelvic outlet is bounded posteriorly by the coccyx, laterally by the sacrotuberous ligaments and the _____, and anteriorly by the pubic arch.
    A. ischial spines
    B. piriformis muscle
    C. ischial tuberosities
    D. perineal membrane
    E. obturator foramen

19. The pelvic diaphragm is formed by the _____ and coccygeus muscles and their covering fasciae.
    A. piriformis
    B. levator ani

    C. deep transverse perineal muscles
    D. perineal membrane
    E. sphincter urethrae

20. During the second stage of labor, the gutter shape of the pelvic floor tends to
    A. become flat.
    B. cause the baby's head to rotate so that its fronto-occipital diameter assumes the transverse position.
    C. cause the baby's head to rotate so that its fronto-occipital diameter assumes the anteroposterior position with the occipital bone lying posterior.
    D. cause the baby's head to rotate so that its fronto-occipital diameter assumes the anteroposterior position with the frontal bone lying posterior.
    E. interfere with the normal process of labor.

21. The process of orgasm in the female depends in part on the
    A. smooth muscle in the vaginal wall contracting in response to the activity of the parasympathetic innervation.
    B. bulbospongiosus muscles contracting in response to the sympathetic nerve fibers.
    C. ischiocavernosus muscles contracting in response to the activity of the pudendal nerve.
    D. stimulation of the clitoris, which is innervated by the obturator nerve.
    E. stimulation of the labia minora, which are innervated by the obturator nerve.

22. The rectouterine pouch (pouch of Douglas)
    A. is formed by parietal pelvic fascia.
    B. commonly contains coils of jejunum.
    C. lies anterior to the vagina.
    D. lies behind the posterior fornix of the vagina and the body of the uterus.
    E. is not the most dependent part of the female peritoneal cavity when the woman is in the standing position.

## Multiple Choice Questions

**Select the BEST answer for each question.**

23. Support for the uterus, either directly or indirectly, is provided by the following structures **except** which?
    A. The perineal body
    B. The mesosalpinx
    C. The transverse cervical (cardinal) ligaments
    D. The levator ani muscles
    E. The pubocervical ligaments

24. The broad ligaments contain all of the following **except** which?
    A. The round ligament of the ovary
    B. The uterine artery
    C. The round ligament of the uterus
    D. The uterine tube
    E. The ureters

# Answers and Explanations

1. **D** is incorrect. The anterior surface of the cervix lies in direct contact with the posterior surface of the urinary bladder, and there is no peritoneum separating the two structures (Fig. 23-6).

2. **D** is incorrect. The right and the left ovarian arteries are branches of the abdominal aorta at the level of the first lumbar vertebra.

3. **B** is incorrect. The upper third of the posterior wall of the vagina is covered with peritoneum (Fig. 23-6) and is related to the rectouterine pouch (pouch of Douglas).

4. **C** is incorrect. The obturator internus muscle is covered with the parietal layer of pelvic fascia and is called the obturator internus fascia (Fig. 23-5).

5. **A** is incorrect. At puberty, the three separate bones, the ilium, ischium, and pubis, fuse together to form one large irregular bone, the hip bone.

6. **D** is incorrect. The pelvic diaphragm is a gutter-shaped sheet of muscle formed by the levatores ani and coccygeus muscles and their covering fasciae. Its function is to support the pelvic viscera. The pelvic diaphragm is incomplete anteriorly, forming an opening to allow passage of the urethra in males and the urethra and the vagina in females (Fig. 23-4).

7. **B** is incorrect. The female urethra opens into the vestibule at the external meatus about 1 in. (2.5 cm) below the clitoris (Fig. 23-32).

8. **E** is incorrect. The iliopectineal line lies at the brim of the bony pelvis and is far beyond the reach of a vaginal examination.

9. **C** is incorrect. Colles' fascia (membranous layer of superficial fascia) takes no part in the formation of the urogenital diaphragm; it is too superficial and lies just beneath the skin (Fig. 23-5).

10. **D** is correct. A carcinoma of the cervix of the uterus is likely to spread via the lymphatic vessels into the internal and external iliac nodes.

11. **A** is correct. An acute infection of the vaginal orifice is likely to spread via the lymphatic vessels into the medial group of superficial inguinal nodes.

12. **E** is correct. The left and right ovarian arteries originate from the abdominal part of the aorta.

13. **B** is correct. The right ovarian vein drains into the inferior vena cava (the left ovarian vein drains into the left renal vein).

14. **D** is correct. In most women, the anatomic position of the uterus when the bladder is empty is anteverted and anteflexed (Fig. 23-14).

15. **E** is correct. The uterus receives its blood supply from the uterine and ovarian arteries (Fig. 23-14).

16. **E** is correct. In women with ovarian cancer, it is judicious to examine the para-aortic lymph nodes and look for evidence of excessive peritoneal fluid (ascites).

17. **A** is correct. The rectouterine pouch (pouch of Douglas) can be most efficiently entered by a surgical incision through the posterior fornix of the vagina (Fig. 23-6).

18. **C** is correct. The pelvic outlet is bounded posteriorly by the coccyx, laterally by the sacrotuberous ligaments and the ischial tuberosities, and anteriorly by the pubic arch (Fig. 23-2).

19. **B** is correct. The pelvic diaphragm is formed by the levator ani and coccygeus muscles and their covering fascia (Fig. 23-4).

20. **D** is correct. During the second stage of labor, the gutter shape of the pelvic floor tends to cause the baby's head to rotate so that its fronto-occipital diameter assumes the anteroposterior position with the frontal bone lying posterior.

21. **C** is correct. The process of orgasm in the female depends in part on the ischiocavernosus muscles contracting in response to the activity of the pudendal nerve.

22. **D** is correct. The rectouterine pouch (pouch of Douglas) lies behind the posterior fornix of the vagina and the body of the uterus (Fig. 23-6).

23. **B** is incorrect. The mesosalpinx is an area of the broad ligament between the uterine tube and the attachment of the mesovarium. It provides no support of the uterus.

24. **E** is incorrect. The ureters pass forward inferior to the broad ligament (Fig. 23-10).

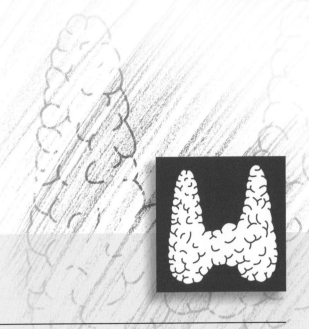

# The Endocrine System

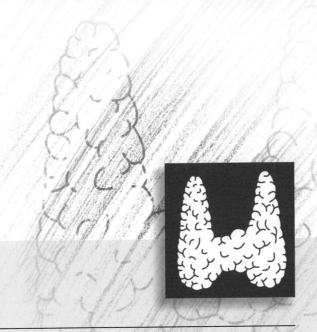

# 24 The Endocrine Glands

All clinical material relevant to this chapter can be found on the CD-ROM.

# Chapter Outline

T he endocrine system is made up of several glands (Fig. 24-1): the pituitary (hypophysis), pineal, thyroid, thymus, parathyroids, suprarenals, islets of Langerhans of the pancreas, testes, ovaries, and, when present, the placenta. In addition, there are groups of cells that form a minor part of the system and are not considered in this chapter: the gastroenteroen-docrine cells, kidney cells, and certain cells of the lung that store and secrete amines.

The endocrine glands have no ducts and consist of masses of cells richly supplied by blood vessels, which pour their secretions (hormones) directly into the blood stream.

The medical professional needs a sound grounding in the structure and function of the endocrine system to be able to apply physiology and hormone therapy in daily clinical practice. Moreover, it must be remembered that disease may affect more than one endocrine gland at the same time in an individual patient, a condition known as multiple endocrine neoplasia. It should also be remembered that patients with advanced malignant disease sometimes produce hormones that are not indigenous to the tissue from which the tumor arose (paraneoplastic syndromes).

This chapter briefly reviews the anatomy of the endocrine system and gives some insight into the different types of hormones produced by the glands.

## PHYSIOLOGIC NOTE

### The Maintenance of Homeostasis

The autonomic nervous system and the endocrine system work closely together to regulate the metabolic activities of the different organs and tissues of the body so as to maintain homeostasis. The autonomic nervous system uses nervous impulses and releases neurotransmitter substances at nerve endings to obtain a rapid and localized response. The endocrine system exerts a slower and more diffuse response by synthesizing and releasing into the bloodstream chemical substances called **hormones**. The specific structure acted on by a hormone is referred to as its **target organ.** The activities of the autonomic nervous system and the endocrine system are integrated and coordinated by the hypothalamus.

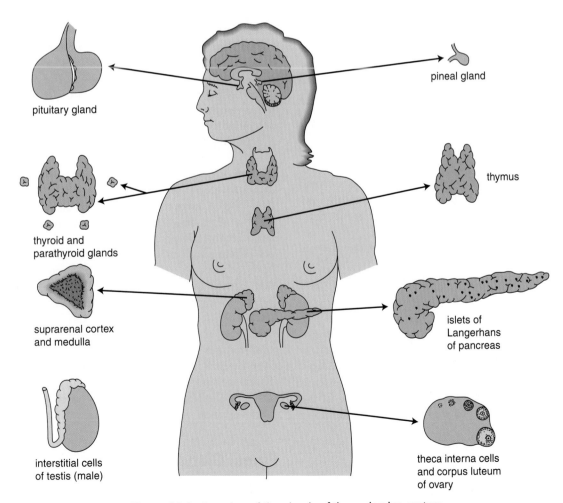

Figure 24-1   Location of the glands of the endocrine system.

BASIC ANATOMY

# Endocrine Glands in the Head and Neck

## Pituitary Gland (Hypophysis Cerebri)

### Location and Description

The pituitary gland is a small, oval structure attached to the undersurface of the brain by the **infundibulum** (Figs. 24-2 and 24-3). During pregnancy, it doubles in size. The gland is well protected by virtue of its location in the sella turcica of the sphenoid bone. Because the hormones produced by the gland influence the activities of many other endocrine glands, the hypophysis cerebri is often referred to as the master endocrine gland. For this reason, it is vital to life.

The pituitary gland is divided into an **anterior lobe,** or **adenohypophysis,** and a **posterior lobe,** or **neurohypophysis.** The anterior lobe is subdivided into the **pars anterior** (sometimes called the pars distalis) and the **pars intermedia,** which may be separated by a cleft that is a remnant of an embryonic pouch. A projection from the pars anterior, the **pars tuberalis,** extends up along the anterior and lateral surfaces of the pituitary stalk (Fig. 24-4).

### Relations

- **Anteriorly:** The sphenoid sinus.
- **Posteriorly:** The dorsum sellae, the basilar artery, and the pons.
- **Superiorly:** The diaphragma sellae, which has a central aperture that allows the passage of the infundibulum. The diaphragma sellae separates the anterior lobe from the optic chiasma.
- **Inferiorly:** The body of the sphenoid, with its sphenoid air sinuses.
- **Laterally:** The cavernous sinus and its contents.

### Blood Supply

The arteries are derived from the **superior** and **inferior hypophyseal arteries,** branches of the internal carotid artery (Fig. 24-4). The veins drain into the intercavernous sinuses. Note the important **hypophyseal portal system** that extends from the median eminence to the anterior lobe of the pituitary and carries releasing and release-inhibiting hormones.

### Hypothalamohypophyseal Tract

This extends from the supraoptic and paraventricular nuclei of the hypothalamus into the posterior lobe of the pituitary (Fig. 24-4). The hormones vasopressin and oxytocin are released at the axon terminals in the posterior lobe of the pituitary.

**PHYSIOLOGIC NOTE**

### Functions of the Pituitary Gland

The pituitary gland influences the activities of many other endocrine glands. The controlling influence is summrized in Figure 24-5. The pituitary gland is, itself, controlled by the hypothalamus, and the activities of the hypothalamus are modified by information received along numerous nervous afferent pathways from different parts of the central nervous system and by the plasma levels of the circulating electrolytes and hormones (Fig. 24-6). The hormonal activities of the pars intermedia of the anterior lobe and the pars posterior (posterior lobe) of the pituitary are summarized in Figure 24-7).

**EMBRYOLOGIC NOTE**

### Development of the Pituitary Gland

The pituitary gland develops from two sources: a small ectodermal diverticulum (**Rathke's pouch**), which grows superiorly from the roof of the stomodeum immediately anterior to the buccopharyngeal membrane, and a small ectodermal diverticulum (the **infundibulum**), which grows inferiorly from the floor of the diencephalon of the brain.

During the 2nd month of development, Rathke's pouch comes into contact with the anterior surface of the infundibulum, and its connection with the oral epithelium elongates, narrows, and finally disappears (Fig. 24-8). Rathke's pouch now is a vesicle that flattens itself around the anterior and lateral surfaces of the infundibulum. The cells of the anterior wall of the vesicle proliferate and form the **pars anterior** of the pituitary; from the vesicle's upper part, there is a cellular extension that grows superiorly and around the stalk of the infundibulum, forming the **pars tuberalis.** The cells of the posterior wall of the vesicle never develop extensively; they form the **pars intermedia.** Some of the cells later migrate anteriorly into the pars anterior. The cavity of the vesicle is reduced to a narrow cleft, which may disappear completely. Meanwhile, the infundibulum has differentiated into the **stalk** and **pars nervosa** of the pituitary gland (Fig. 24-8).

## Pineal Gland

The pineal gland is a small cone-shaped body that projects posteriorly from the posterior end of the roof of the third

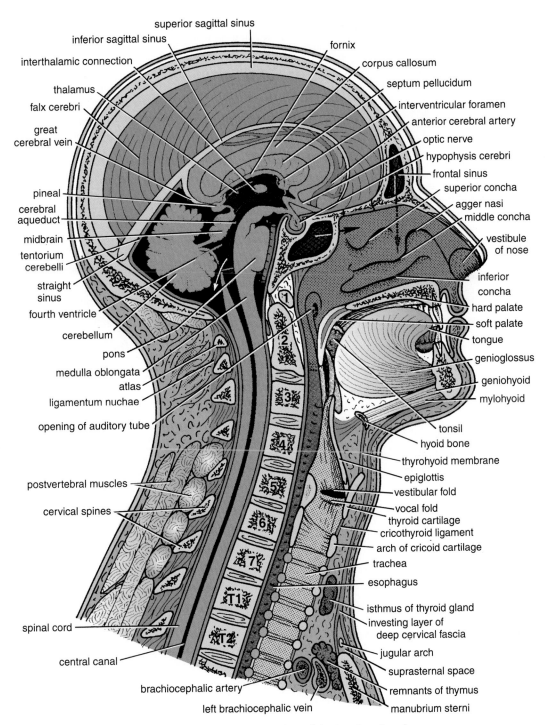

**Figure 24-2**  Sagittal section of the head and neck.

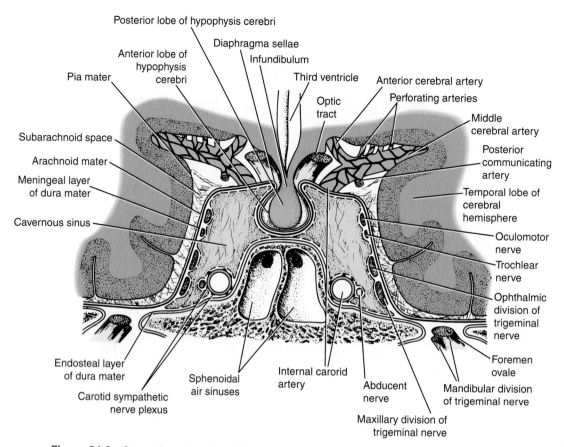

**Figure 24-3**  Coronal section through the body of the sphenoid bone, showing the pituitary gland and cavernous sinuses. Note the position of the internal carotid artery and the cranial nerves.

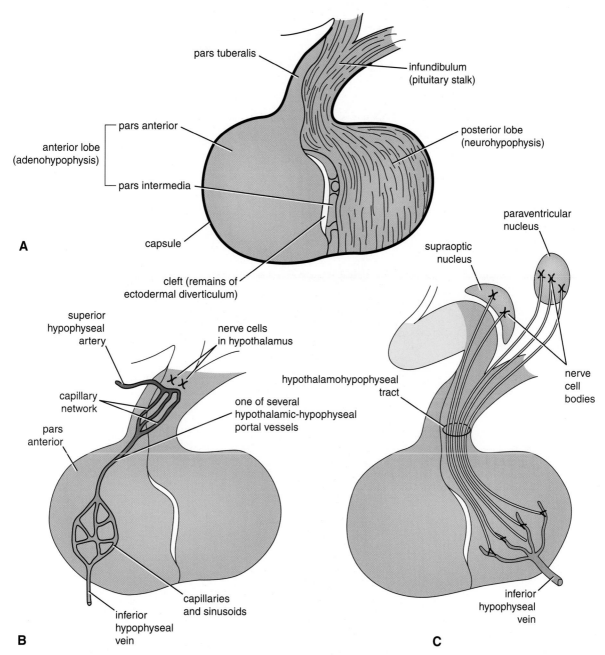

**Figure 24-4   A.** Division of the pituitary gland into the pars anterior, pars intermedia (anterior lobe), and pars nervosa (posterior lobe). **B.** Hypothalamohypophyseal portal vessels. **C.** Hypothalamohypophyseal tract.

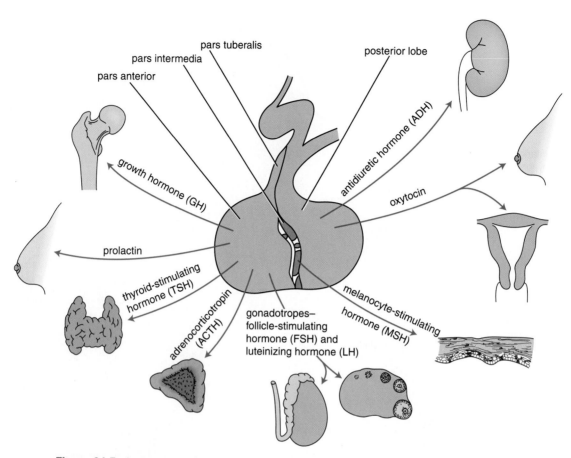

**Figure 24-5** Importance of the pituitary gland in controlling other endocrine glands and tissues in the body. The hormones secreted by the different parts of the pituitary gland in exerting this control are indicated.

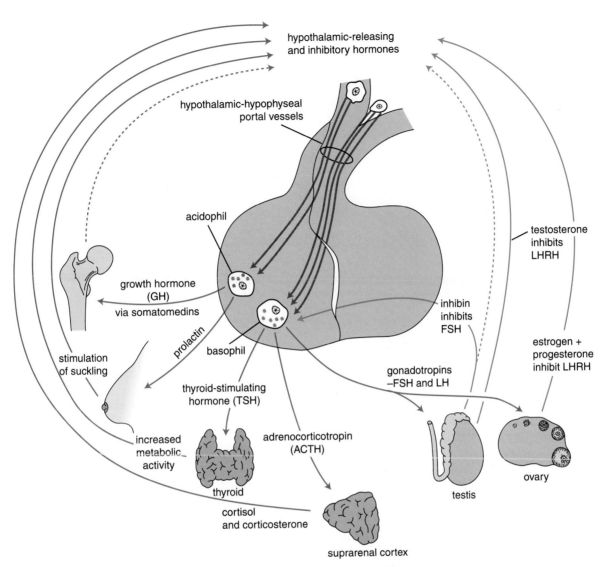

**Figure 24-6**   Control of the secretions of the pars anterior of the pituitary gland by the hypothalamus. Note that the activities of the hypothalamus are modified by information received from other parts of the nervous system and the plasma levels of the circulating hormones.

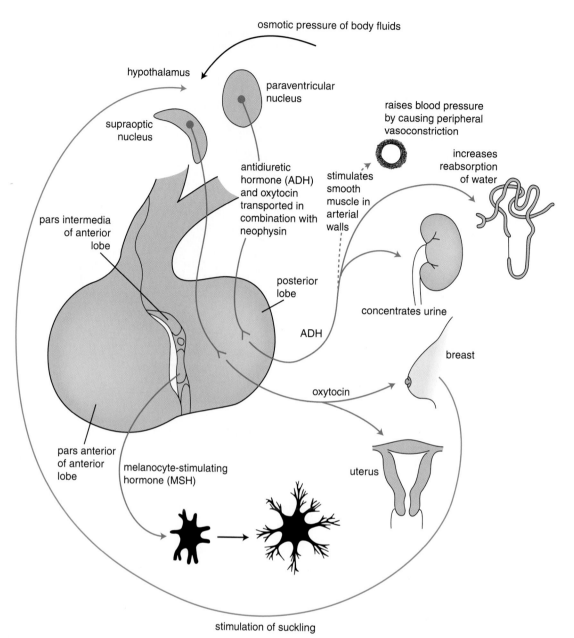

**Figure 24-7** Hormonal activities of the pars intermedia of the anterior lobe and the pars posterior (posterior lobe) of the pituitary gland. Note that the nerve cells of the supraoptic and paraventricular nuclei of the hypothalamus synthesize antidiuretic hormone and oxytocin, which are then transported to the posterior lobe in the hypothalamohypophyseal tract.

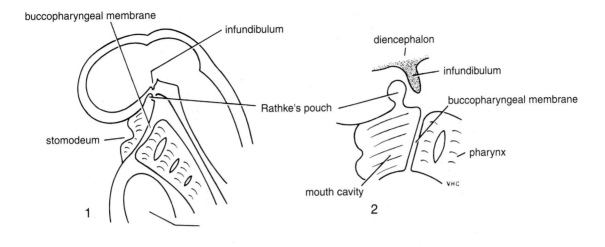

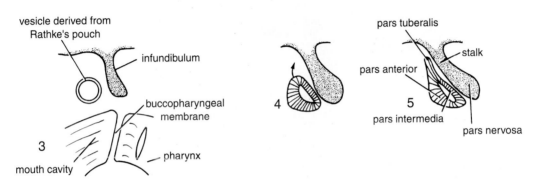

**Figure 24-8** The different stages in the development of the pituitary gland shown in sagittal sections.

ventricle of the brain (Fig. 24-2). The pineal consists essentially of groups of cells, the **pinealocytes,** supported by glial cells. The gland has a rich blood supply and is innervated by postganglionic sympathetic nerve fibers.

### PHYSIOLOGIC NOTE

#### Functions of the Pineal Gland

The pineal gland can influence the activities of the pituitary gland, the islets of Langerhans of the pancreas, the parathyroids, the adrenals, and the gonads. The pineal secretions, produced by the pinealocytes, reach their target organs via the bloodstream or through the cerebrospinal fluid. Their actions are mainly inhibitory and either directly inhibit the production of hormones or - indirectly inhibit the secretion of releasing factors by the hypothalamus. It is interesting to note that the pineal gland does not possess a blood-brain barrier.

**Melatonin** is present in high concentration within the pineal gland. It passes to the anterior lobe of the pituitary and inhibits the release of the gonadotrophic hormone.

The plasma level of melatonin rises in darkness and falls during the day. The pineal gland appears to play an important role in the regulation of the reproductive function.

### EMBRYOLOGIC NOTE

#### Development of the Pineal Gland

The pineal gland develops as a small ectodermal diverticulum in the posterior part of the roof of the diencephalon during the seventh week of development.

## Thyroid Gland

The thyroid gland consists of right and left lobes connected by a narrow isthmus (Fig. 24-9). It is a vascular organ surrounded by a sheath derived from the pretracheal layer of deep fascia. The sheath attaches the gland to the larynx and the trachea.

Each lobe is pear shaped, with its apex being directed upward as far as the oblique line on the lamina of the thy-

**anterior view**

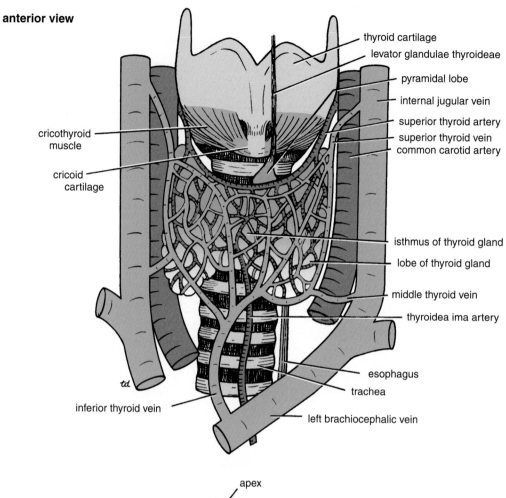

cricothyroid muscle

cricoid cartilage

thyroid cartilage

levator glandulae thyroideae

pyramidal lobe

internal jugular vein

superior thyroid artery

superior thyroid vein

common carotid artery

isthmus of thyroid gland

lobe of thyroid gland

middle thyroid vein

thyroidea ima artery

esophagus

trachea

inferior thyroid vein

left brachiocephalic vein

**lateral view of right lobe**

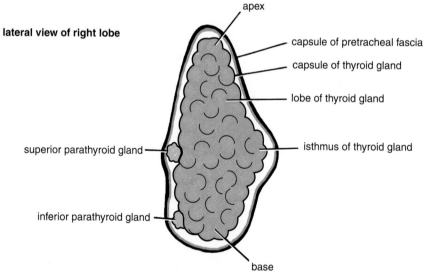

apex

capsule of pretracheal fascia

capsule of thyroid gland

lobe of thyroid gland

isthmus of thyroid gland

superior parathyroid gland

inferior parathyroid gland

base

**Figure 24-9**   The thyroid gland showing its relation to the larynx and trachea and the large blood vessels of the neck. The blood supply and venous drainage of the thyroid gland are also shown. The lower diagram shows the parathyroid glands posterior to the thyroid gland.

roid cartilage; its base lies below at the level of the fourth or fifth tracheal ring.

The **isthmus** extends across the midline in front of the second, third, and fourth tracheal rings (Fig. 24-9). A **pyramidal lobe** is often present, and it projects upward from the isthmus, usually to the left of the midline. A fibrous or muscular band frequently connects the pyramidal lobe to the hyoid bone; if it is muscular, it is referred to as the **levator glandulae thyroideae** (Fig. 24-9).

## Relations of the Lobes

- **Anterolaterally:** The sternothyroid, the superior belly of the omohyoid, the sternohyoid, and the anterior border of the sternocleidomastoid (Fig. 24-10).
- **Posterolaterally:** The carotid sheath with the common carotid artery, the internal jugular vein, and the vagus nerve (Fig. 24-10).
- **Medially:** The larynx, the trachea, the pharynx, and the esophagus. Associated with these structures are the cricothyroid muscle and its nerve supply, the external

laryngeal nerve. In the groove between the esophagus and the trachea is the recurrent laryngeal nerve (Fig. 24-10).

The rounded posterior border of each lobe is related posteriorly to the superior and inferior parathyroid glands (Fig. 24-9) and the anastomosis between the superior and inferior thyroid arteries.

## Relations of the Isthmus

- **Anteriorly:** The sternothyroids, sternohyoids, anterior jugular veins, fascia, and skin.
- **Posteriorly:** The second, third, and fourth rings of the trachea.

The terminal branches of the superior thyroid arteries anastomose along its upper border.

### Blood Supply

The **arteries** to the thyroid gland are the superior thyroid artery, the inferior thyroid artery, and sometimes the thy-

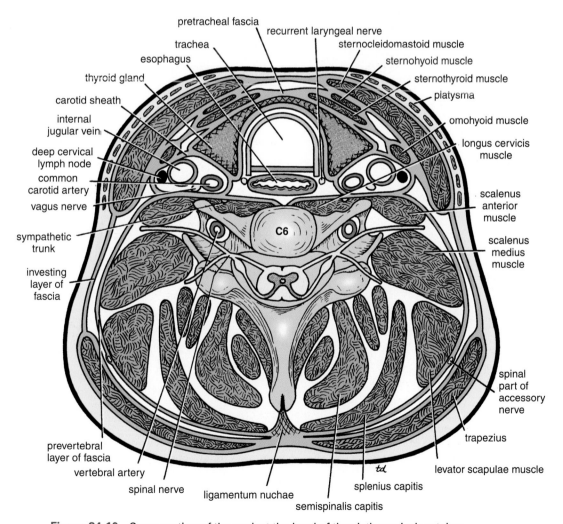

**Figure 24-10**  Cross section of the neck at the level of the sixth cervical vertebra.

roidea ima. The arteries anastomose profusely with one another over the surface of the gland.

The **superior thyroid artery,** a branch of the external carotid artery, descends to the upper pole of each lobe, accompanied by the **external laryngeal nerve** (Fig. 24-9).

The **inferior thyroid artery,** a branch of the thyrocervical trunk, ascends behind the gland to the level of the cricoid cartilage. It then turns medially and downward to reach the posterior border of the gland. The **recurrent laryngeal nerve** crosses either in front of or behind the artery, or it may pass between its branches.

The **thyroidea ima,** if present, may arise from the brachiocephalic artery or the arch of the aorta. It ascends in front of the trachea to the isthmus (Fig. 24-9).

The **veins** from the thyroid gland are the superior thyroid, which drains into the internal jugular vein; the middle thyroid, which drains into the internal jugular vein; and the inferior thyroid (Fig. 24-9). The inferior thyroid veins of the two sides anastomose with one another as they descend **in front of the trachea.** They drain into the left brachiocephalic vein in the thorax.

### Lymph Drainage

The lymph from the thyroid gland drains mainly laterally into the deep cervical lymph nodes. A few lymph vessels descend to the paratracheal nodes.

### Nerve Supply

Superior, middle, and inferior cervical sympathetic ganglia.

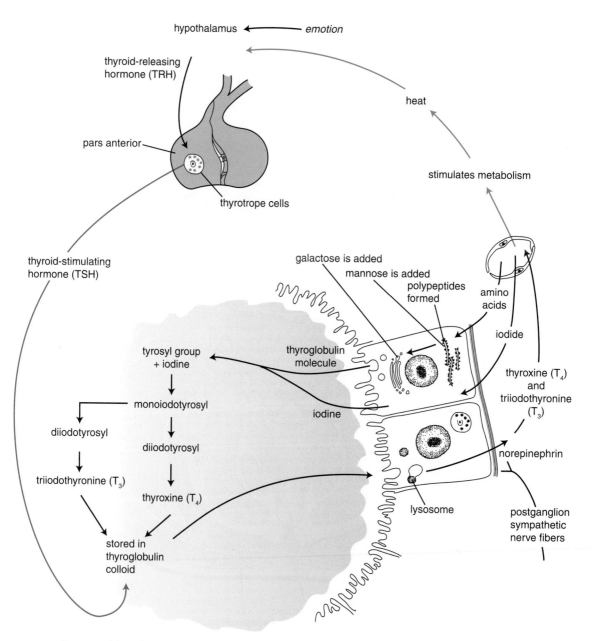

**Figure 24-11**   Formation and control of follicular cell secretion in the thyroid gland. Note the feedback mechanism to the hypothalamus.

## Functions of the Thyroid Gland

In response to the thyroid-stimulating hormone produced by the pars anterior of the pituitary, the hormones **thyroxine** and **triiodothyronine** are liberated from the follicular colloid and enter the blood stream (Fig. 24-11). The thyroid hormones increase the metabolic activity of most cells in the body, increasing oxygen consumption and heat production. The parafollicular cells produce the hormone **thyrocalcitonin,** which lowers the level of blood calcium. The parafollicular cells are stimulated by hypercalcemia and suppressed by hypocalcemia; they are not controlled by the pituitary gland.

## Development of the Thyroid Gland

The thyroid gland begins to develop during the 3rd week as an entodermal thickening in the midline of the floor of the pharynx between the **tuberculum impar** and the **copula** (Fig. 24-12). Later, this thickening becomes a diverticulum that grows inferiorly into the underlying mesenchyme and is called the **thyroglossal duct.** As development continues, the duct elongates, and its distal end

becomes bilobed. Soon, the duct becomes a solid cord of cells, and as a result of epithelial proliferation, the bilobed terminal swellings expand to form the thyroid gland.

The thyroid gland now migrates inferiorly in the neck and passes either anterior to, posterior to, or through the developing body of the hyoid bone. By the 7th week, it reaches its final position in relation to the larynx and trachea. Meanwhile, the solid cord connecting the thyroid gland to the tongue fragments and disappears. The site of origin of the thyroglossal duct on the tongue remains as a pit called the **foramen cecum.** The thyroid gland may now be divided into a small median **isthmus** and two large **lateral lobes** (Fig. 24-12).

In the earliest stages, the thyroid gland consists of a solid mass of cells. Later, as a result of invasion by surrounding vascular mesenchymal tissue, the mass becomes broken up into plates and cords and finally into small clusters of cells. By the 3rd month, colloid starts to accumulate in the center of each cluster so that **follicles** are formed. The fibrous capsule and connective tissue develop from the surrounding mesenchyme.

The **ultimobranchial bodies** (from the fifth pharyngeal pouch) and neural crest cells are believed to be incorporated into the thyroid gland, where they form the **parafollicular cells,** which produce **thyrocalcitonin.** The function of calcitonin is to reduce the plasma concentration of calcium.

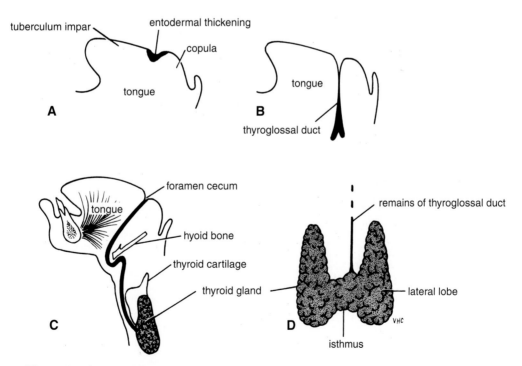

**Figure 24-12**    The different stages in the development of the thyroid gland. **A.** Sagittal section of the tongue showing an entodermal thickening between the tuberculum impar and the copula. **B.** Sagittal section of the tongue showing the development of the thyroglossal duct. **C.** Sagittal section of the tongue and neck showing the path taken by the thyroid gland as it migrates inferiorly. **D.** The fully developed thyroid gland as seen from in front. Note the remains of the thyroglossal duct above the isthmus.

## Parathyroid Glands

The parathyroid glands are ovoid bodies measuring about 6 mm long in their greatest diameter. They are four in number and are closely related to the posterior border of the thyroid gland, lying within its fascial capsule (Fig. 24-9).

The two **superior parathyroid glands** are the more constant in position and lie at the level of the middle of the posterior border of the thyroid gland.

The two **inferior parathyroid glands** usually lie close to the inferior poles of the thyroid gland. They may lie within the fascial sheath, embedded in the thyroid substance, or outside the fascial sheath. Sometimes they are found some distance caudal to the thyroid gland, in association with the inferior thyroid veins; or they may even reside in the superior mediastinum in the thorax.

### Blood Supply

The arterial supply to the parathyroid glands is from the superior and inferior thyroid arteries. The venous drainage is into the superior, middle, and inferior thyroid veins.

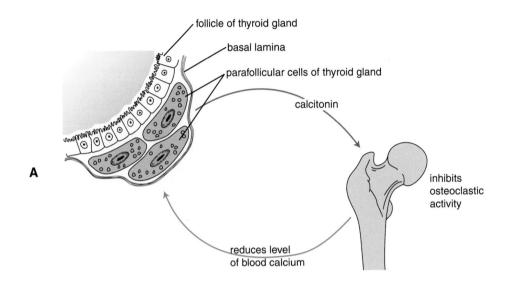

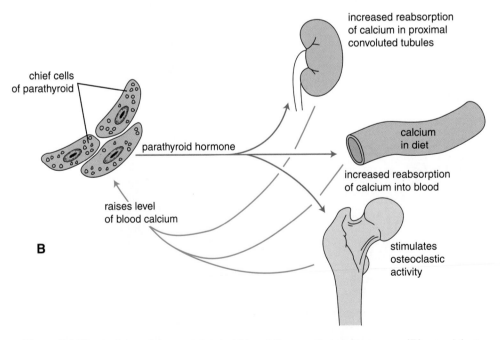

**Figure 24-13** Actions of thyrocalcitonin (**A**) and the parathyroid hormone (**B**) on calcium metabolism.

## Lymph Drainage

Deep cervical and paratracheal lymph nodes.

## Nerve Supply

Superior or middle cervical sympathetic ganglia.

---

### PHYSIOLOGIC NOTE

#### Functions of the Parathyroid Glands

The chief cells produce the parathyroid hormone, which stimulates osteoclastic activity in bones, thus mobilizing the bone calcium and increasing the calcium levels in the blood (Fig. 24-13). The parathyroid hormone also stimulates the absorption of dietary calcium from the small intestine and the reabsorption of calcium in the proximal convoluted tubules of the kidney. It also strongly diminishes the reabsorption of phosphate in the proximal convoluted tubules of the kidney. The secretion of the parathyroid hormone is controlled by the calcium levels in the blood.

---

### EMBRYOLOGIC NOTE

#### Development of the Parathyroid Glands

The pair of **inferior parathyroid glands,** known as **parathyroid III,** develop as the result of proliferation of entodermal cells in the third pharyngeal pouch on each side. As the thymic diverticulum on each side grows in-feriorly in the neck, it pulls the inferior parathyroid with it, so that it finally comes to rest on the posterior surface of the lateral lobe of the thyroid gland near its lower pole and becomes completely separate from the thymus (Fig. 24-14).

The pair of **superior parathyroid glands, parathyroid IV,** develop as a proliferation of entodermal cells in the fourth pharyngeal pouch on each side. These loosen their connection with the pharyngeal wall and take up their final position on the posterior aspect of the lateral lobe of the thyroid gland on each side, at about the level of the isthmus (Fig. 24-14).

In the earlier stages, each gland consists of a solid mass of clear cells, the **chief cells.** In late childhood, acidophilic cells, the **oxyphil cells,** appear. The connective tissue and vascular supply are derived from the surrounding mesenchyme. It is believed that the parathyroid hormone is secreted early in fetal life by the chief cells to regulate calcium metabolism. The oxyphil cells are thought to be nonfunctioning chief cells.

# Endocrine Gland in the Thorax

## Thymus

The thymus is a flattened, bilobed structure (Figs. 24-1 and 24-2) lying between the sternum and the pericardium in the anterior mediastinum. In the newborn infant, the thymus reaches its largest size relative to the size of the body, at which time it may extend up through the superior mediastinum in front of the great vessels into the root of the

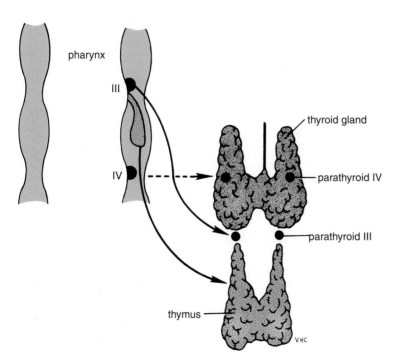

**Figure 24-14**  Parathyroid glands taking up their final positions in the neck.

neck. The thymus continues to grow until puberty but thereafter undergoes involution. It has a pink, lobulated appearance and is the site for the development of thymus processed lymphocytes, T (thymic) lymphocytes, which are distributed to the whole body.

## Blood Supply

The blood supply of the thymus is from the inferior thyroid and internal thoracic arteries.

---

### PHYSIOLOGIC NOTE

### Functions of the Thymus

It is now recognized that the thymus produces a large number of hormones, including the **thymosins,** that influence the maturation and function of lymphocytes within the thymus and elsewhere in the body. It has also been shown that the thymus releases hormones that influence other endocrine glands. Receptors have been identified on thymic cells that indicate that the activity of the thymus can be modified by the hormones of other endocrine glands. The details of these activities are beyond the scope of this book.

---

### EMBRYOLOGIC NOTE

### Development of the Thymus

The thymus arises as an entodermal diverticulum from the third pharyngeal pouch on each side. Each diverticulum grows inferiorly in the neck to reach the anterior aspect of the aorta. The diverticula become solid bars as the result of cellular proliferation. The bars now fuse in the superior mediastinum and lose their connection with the pharyngeal pouches. The entodermal cells form the **corpuscles of Hassall,** and the surrounding mesenchyme forms the connective tissue framework and capsule. The organ becomes invaded by increasing numbers of lymphocytes.

# Endocrine Glands Associated with the Abdomen and Pelvis

## Suprarenal Glands

### Location and Description

The two suprarenal glands are yellowish retroperitoneal organs that lie on the upper poles of the kidneys. They are sur-

rounded by renal fascia (but are separated from the kidneys by the perirenal fat). Each gland has a yellow **cortex** and a dark brown **medulla.**

The **right suprarenal gland** is pyramid shaped and caps the upper pole of the right kidney (Fig. 24-15). It lies behind the right lobe of the liver and extends medially behind the inferior vena cava. It rests posteriorly on the diaphragm.

The **left suprarenal gland** is crescentic in shape and extends along the medial border of the left kidney from the upper pole to the hilus (Fig. 24-15). It lies behind the pancreas, the lesser sac, and the stomach and rests posteriorly on the diaphragm.

---

### PHYSIOLOGIC NOTE

### Functions of the Suprarenal Glands

The cortex of the suprarenal glands secretes hormones that include **mineral corticoids,** which are concerned with the control of fluid and electrolyte balance; **glucocorticoids,** which are concerned with the control of the metabolism of carbohydrates, fats, and proteins; and small amounts of **sex hormones,** which probably play a role in the prepubertal development of the sex organs. The medulla of the suprarenal glands secretes the catecholamines **epinephrine** and **norepinephrine.** The actions of the hormones secreted by the suprarenal cortex are summarized in Figure 24-16. The sympathetic control of the release of catecholamines from the epithelial cells of the suprarenal medulla and the response of the suprarenal cortex and medulla to stress are shown in Figure 24-17.

---

## Blood Supply

### Arteries

The arteries supplying each gland are three in number: inferior phrenic artery, aorta, and renal artery.

### Veins

A single vein emerges from the hilum of each gland and drains into the inferior vena cava on the right and into the renal vein on the left.

## Lymph Drainage

The lymph drains into the lateral aortic nodes.

## Nerve Supply

Preganglionic sympathetic fibers derived from the splanchnic nerves supply the glands. Most of the nerves end in the medulla of the gland.

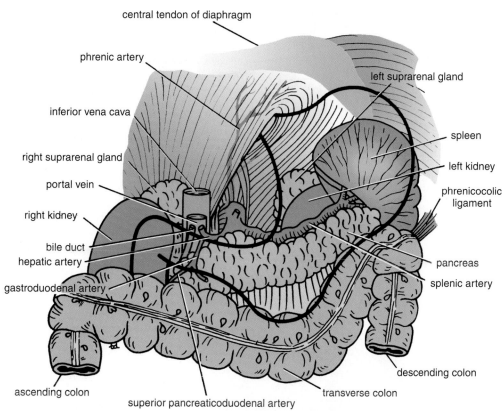

**Figure 24-15** Structures situated on the posterior abdominal wall behind the stomach. Note the position of the suprarenal glands.

### Development of the Suprarenal Glands

The cortex develops from the coelomic mesothelium covering the posterior abdominal wall. At first, a **fetal cortex** is formed; later, it becomes covered by a second **final cortex**. After birth, the fetal cortex retrogresses, and its involution is largely completed in the first few weeks of life.

The **medulla** is formed from the sympathochromaffin cells of the neural crest. These invade the cortex on its medial side. By this means, the medulla comes to occupy a central position and is arranged in cords and clusters. Preganglionic sympathetic nerve fibers grow into the medulla and influence the activity of the medullary cells.

## Islets of Langerhans of the Pancreas

The pancreas is a soft, lobulated organ that lies on the posterior abdominal wall behind the peritoneum. The head lies within the concavity of the duodenum, and the neck, body, and tail extend to the left; the tail lies in contact with the hilus of the spleen (Fig. 24-15). The greater part of the gland produces the exocrine secretion that passes into the duode-num. The endocrine part of the gland is formed of clusters of cells called the **islets of Langerhans**, which are scattered among the exocrine acini. The islets are more numerous in the tail of the pancreas than in its body, neck, or head.

### Functions of the Islets of Langerhans

With special staining techniques, it is possible to identify four cell types in the pancreatic islets. These are **beta, alpha**, and **delta** cells and the **PP cell** (pancreatic polypeptide cell). Specific antibody techniques have enabled researchers to localize specific hormones to individual cells. The beta cells form the majority of the islet cells and secrete **insulin.** The alpha cells secrete glucagon, and the delta cells secrete **somatostatin.** The function of the PP cells in the human is not known. The main action of insulin is to bring about the rapid absorption, storage, and use of glucose in the body (Fig. 24-18). It also has important effects on fat and protein metabolism. Glucagon increases the blood glucose concentration, and therefore, its effect is opposite to that of insulin (Fig. 24-18). The actions of

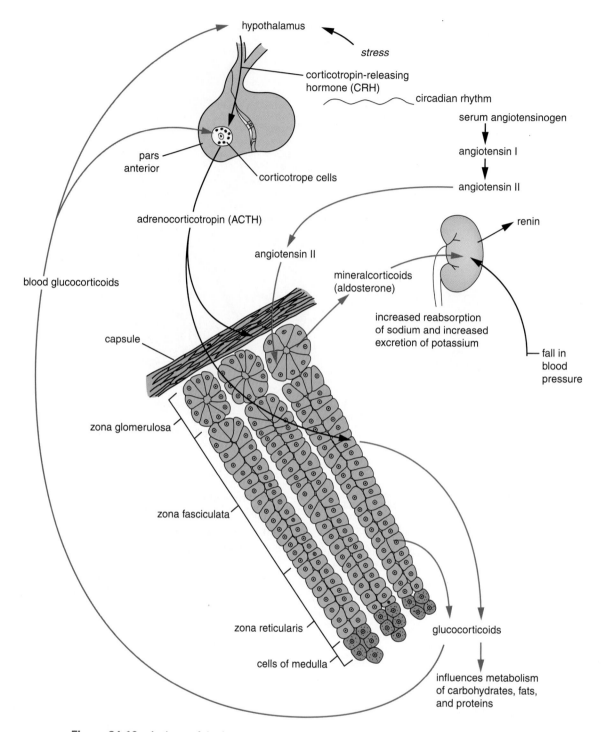

**Figure 24-16**  Actions of the hormones secreted by the suprarenal cortex. The control of the suprarenal cortex by the pituitary gland and the hypothalamus is also shown, as is the renin-angiotensin control of aldosterone secretion.

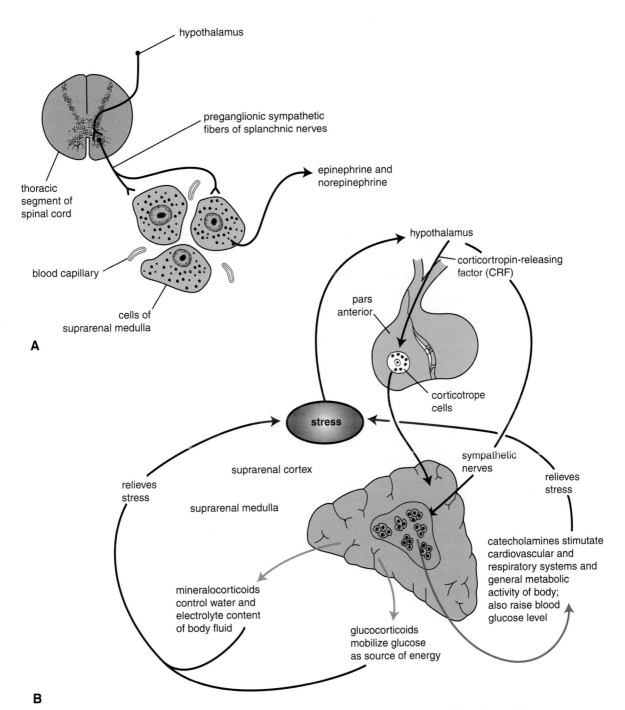

**Figure 24-17   A.** Sympathetic control of the release of catecholamines from the epithelial cells of the suprarenal medulla. **B.** Response of the suprarenal cortex and medulla to stress.

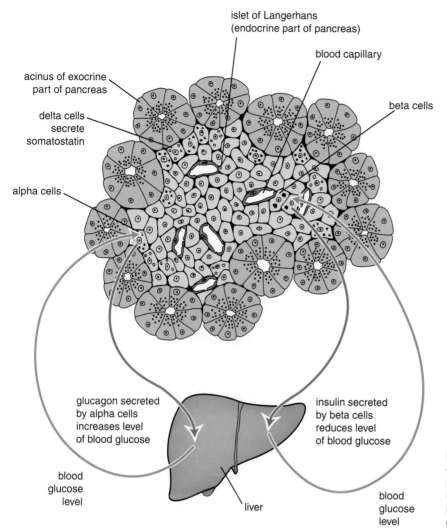

islet of Langerhans
(endocrine part of pancreas)

blood capillary

acinus of exocrine
part of pancreas

beta cells

delta cells
secrete
somatostatin

alpha cells

glucagon secreted
by alpha cells
increases level
of blood glucose

insulin secreted
by beta cells
reduces level
of blood glucose

blood
glucose
level

liver

blood
glucose
level

**Figure 24-18** Effects of insulin and glucagon, secreted by the beta and alpha cells, respectively, of the islets of Langerhans, on the blood glucose level. Note the feedback action of blood glucose on the activities of these cells.

somatostatin are not understood, although the hormone is known to inhibit the secretion of insulin and glucagon.

### Development of the Islets of Langerhans

The development of the Islets of Langerhans is described with the pancreas in Chapter 20 (p. 793).

## Interstitial Cells of the Testis

The testes are paired ovoid organs situated within the scrotum. Each testis is both an exocrine and an endocrine gland. The greater part of each gland is made up of seminiferous tubules whose function is to produce spermatozoa. The spermatozoa constitute the exocrine secretion that passes via ducts into the urethra.

The endocrine part of each testis consists of groups of rounded interstitial cells (Leydig cells) embedded in loose connective tissue between the seminiferous tubules.

### Functions of the Interstitial Cells of the Testis

The function of the interstitial cells is to secrete androgens, which include **testosterone, dihydrotestosterone**, and **androstenedione;** the most important and abundant of these is testosterone. The interstitial cells are present in large numbers in the fetus but atrophy to some extent after birth. They reappear in large numbers at puberty. In the fetus, the chorionic gonadotropin from the placenta stimulates the interstitial cells of the testis to produce testosterone, which is important in the development of the male genitalia and suppressing the formation of the female genitalia;

testosterone also brings about the descent of the testes. The control of the activities of the testis by the hypothalamus and the pars anterior of the pituitary gland are shown in Figure 24-19.

The interstitial cells are less sensitive to heat than are the germinal epithelial cells of the seminiferous tubules. For this reason, testosterone is produced continuously even in individuals with undescended testes.

---

### E M B R Y O L O G I C   N O T E

## Development of the Interstitial Cells of the Testis

The development of the interstitial cells of the testis is described with the development of the testis in Chapter 22 (p. 840).

## Ovaries

Each ovary has an outer cortex and an inner medulla, but the division between the two is ill defined. Embedded in the connective tissue of the cortex are the **ovarian follicles** in different stages of development.

---

### P H Y S I O L O G I C   N O T E

## Functions of the Ovarian Hormones

The ovarian hormones are the steroids estrogen and progesterone.

The estrogens are mainly **B-estradiol** and **estrone.** They are produced by the theca interna cells, which are found in the stroma of the ovary immediately outside the graafian follicle and by the cells of the corpus luteum. At

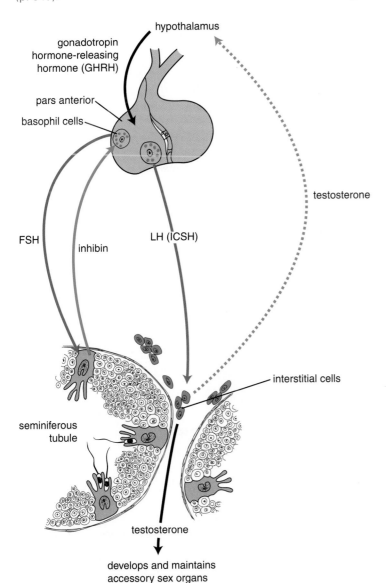

**Figure 24-19**   Control of the activities of the testis by the hypothalamus and the pars anterior of the pituitary gland. Note the possible feedback mechanisms provided by inhibin from the Sertoli cells and testosterone from the interstitial cells (Leydig cells).

puberty, the increased production of estrogen is responsible for the development of the uterus and vagina, the external genitalia, the pelvis, the breasts, and pubic and axillary hair. The estrogens cause the repair and proliferative changes in the endometrium following the menstrual period.

**Progesterone** is produced by the corpus luteum of the ovary. Along with estrogen, it is responsible for the secretory phase in the development of the endometrium during the second half of the menstrual cycle and, in pregnancy, for the development and maintenance of the decidua. Progesterone also causes epithelial enlargement in the duct system of the breasts during the second half of the menstrual cycle, and during pregnancy, progesterone causes extensive alveolar development. The control of the

activities of the ovary by the hypothalamus and the pars anterior of the pituitary gland are shown in Figure 24-20.

# E M B R Y O L O G I C   N O T E

## Development of the Ovary

The development of the ovary is described in Chapter 23, page 868.

## Placenta

The formation and function of the placenta during pregnancy have been described in Chapter 23. One of its main

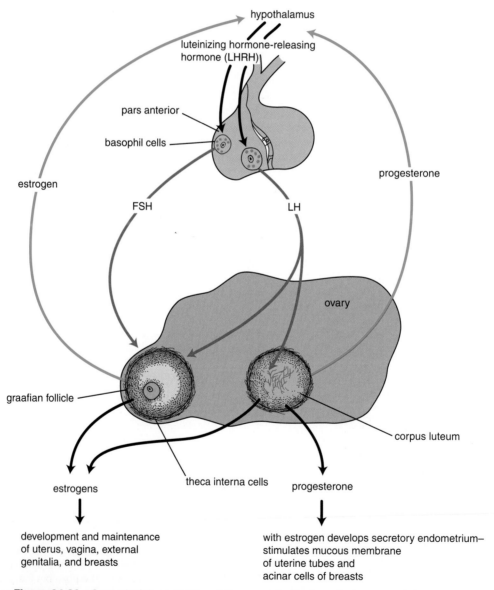

**Figure 24-20**   Control of the activities of the ovary by the hypothalamus and the pars anterior of the pituitary gland. Note the feedback mechanisms provided by estrogen and progesterone.

functions is to serve as an endocrine gland, and the hormones are produced in the fetal part of the placenta in the syncytiotrophoblast.

### Functions of the Placental Hormones

The placenta secretes **estrogens, progesterone, chorionic gonadotropin,** and **chorionic somatomammotropin** (Fig. 24-21).

The placental estrogens are responsible for the continuing enlargement of the uterus, the growth of the glandular ducts in the breast, and the increase in size of the maternal external genitalia. Toward the end of pregnancy, the estrogens bring about relaxation of the sacroiliac joints and the symphysis pubis. The estrogens also increase the contractility of the smooth muscle of the uterine wall and thus promote uterine contractions when labor commences.

Placental progesterone takes over the production of progesterone from the corpus luteum at about the fourth month of pregnancy. It is essential for the maintenance of the decidua throughout pregnancy and is responsible for the further development of the glandular alveoli of the breasts. Toward the end of pregnancy, the amount of

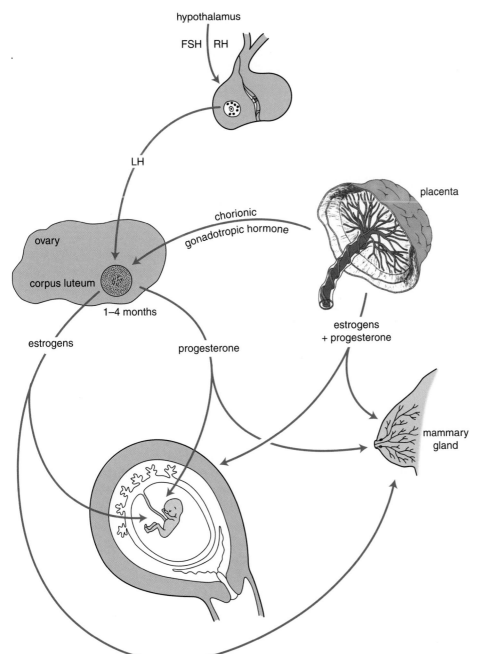

**Figure 24-21** Activities of the placental hormones. The hormones are produced in the syncytiotrophoblast in the fetal part of the placenta.

estrogen produced is increased, while the progesterone level is not increased. The uterine muscle thus becomes more irritable and susceptible to the action of oxytocin from the posterior lobe of the pituitary.

Chorionic gonadotropin appears in the blood stream as the fertilized ovum is being implanted in the endometrium. It is responsible for the continued growth and maintenance of the corpus luteum. The hormone also enters the fetal circulation and, in the male fetus, stimulates the interstitial cells of the testes to produce small amounts of testosterone, thus causing the fetus to develop male genital organs. Chorionic gonadotropin has no effect on the development of the female genital organs.

The presence of chorionic gonadotropin in the urine of a woman 12 days after ovulation is a definite indication of pregnancy.

Chorionic somatomammotropin may influence the carbohydrate and fat metabolism of the mother and make more glucose and fatty acids available for the fetus.

---

**EMBRYOLOGIC NOTE**

### Development of the Placenta

The development of the placenta is described in Chapter 23, page 874.

---

# Review Questions

## Multiple Choice Questions

### Select the BEST answer for each question.

1. The following statements concerning the pituitary gland (hypophysis cerebri) are correct **except** which?
   A. It is separated from the optic chiasma by the diaphragma sellae.
   B. The sphenoid sinus is inferior to it.
   C. It receives its arterial supply from the internal carotid artery.
   D. It is suspended from the floor of the third ventricle by the pars anterior.
   E. It is deeply placed within the sella turcica of the skull.

2. The following statements concerning the hypothalamohypophyseal tract are correct **except** which?
   A. Oxytocin inhibits the contraction of the smooth muscle of the uterus.
   B. The nerve cells of the supraoptic and paraventricular nuclei produce the hormones vasopressin and oxytocin.
   C. The hormones travel in the axons of the tract.
   D. Vasopressin stimulates the distal convoluted tubules and collecting tubules of the kidney, causing increased absorption of water from the urine.
   E. The hormones leave the axons and are absorbed into the bloodstream in the capillaries of the posterior lobe of the pituitary gland.

3. The following statements concerning the hypophyseal portal system are correct **except** which?
   A. The portal system carries releasing hormones and release-inhibiting hormones to the secretory cells of the anterior lobe of the pituitary gland.

B. The production of the releasing hormones and the release-inhibiting hormones can be influenced by the level of the hormone produced by the target organ controlled by the pituitary gland.
   C. The blood vessels commence superiorly in the median eminence and end inferiorly in the vascular sinusoids of the anterior lobe of the pituitary gland.
   D. Afferent nerve fibers entering the hypothalamus influence the production of the releasing hormones by the nerve cells.
   E. The neuroglial cells of the hypothalamus are responsible for the production of the release-inhibiting hormones.

4. The following statements concerning the pineal gland are correct **except** which?
   A. It is small and cone shaped.
   B. It projects from the posterior end of the pons of the brain.
   C. Its activities can influence the functions of the suprarenals and the gonads.
   D. It does not possess a blood-brain barrier.
   E. It produces melatonin, which inhibits the release of the gonadotropic hormone.

5. Which of the following nerves might be injured when tying the inferior thyroid artery during operations on the thyroid gland?
   A. The sympathetic trunk
   B. The internal laryngeal nerve
   C. The descendens cervicalis
   D. The recurrent laryngeal nerve
   E. The superior laryngeal nerve

6. The following statements concerning the parathyroid glands are correct **except** which?
   A. They are four in number.
   B. They are closely related to the posterior border of the thyroid gland within its fascial capsule.
   C. The arterial supply is from the superior and inferior thyroid arteries.
   D. The activity of the cells of the parathyroid glands reduces the calcium levels of the blood.
   E. The inferior parathyroid glands may reside in the superior mediastinum in the thorax.

7. The following statements concerning the thymus are correct **except** which?
   A. In the newborn, it reaches its largest size relative to the size of the body.
   B. In the infant, it may extend up from the anterior mediastinum into the superior mediastinum and even into the root of the neck.
   C. The thymus continues to grow into adult life and, in middle age, starts to involute.
   D. It is the site for the development of the T lymphocytes.
   E. It produces the hormone thymosin, which influences the maturation and function of the T lymphocytes

8. The following statements concerning the left suprarenal gland are correct **except** which?
   A. The gland extends along the medial border of the left kidney from the upper pole to the hilus.
   B. The gland's vein drains into the left renal vein.
   C. The gland is separated from the left kidney by perirenal fat.
   D. The gland lies behind the lesser sac of peritoneum.
   E. The medulla is innervated by postganglionic sympathetic nerve fibers.

9. The suprarenal gland receives its arterial supply from which of the following?
   A. Aorta, inferior phrenic, and renal arteries
   B. Lumbar arteries
   C. Superior phrenic artery
   D. Testicular (ovarian artery)
   E. Subcostal artery

10. The following facts concerning the development of the suprarenal glands are correct **except** which?
    A. The cortex is developed from the coelomic epithelium covering the posterior abdominal wall.
    B. The medulla is developed from the neural crest.
    C. Pre-ganglionic sympathetic nerve fibers grow into the medulla and influence the activity of the medullary cells.
    D. A fetal cortex is first formed that later degenerates and disappears.
    E. The final cortex grows between the fetal cortex and the medulla.

11. The following statements concerning the pancreas are correct **except** which?
    A. The pancreas receives part of the arterial supply from the splenic artery.
    B. The main pancreatic duct opens into the second part of the duodenum.
    C. The islets of Langerhans are more numerous in the head of the pancreas than in its body, neck, or tail.
    D. The bile duct (common bile duct) lies posterior to the head of the pancreas.
    E. The transverse mesocolon is attached to the anterior border of the pancreas.

12. The following statements concerning the interstitial cells of the testis are correct **except** which?
    A. The interstitial cells are embedded in the connective tissue between the seminiferous tubules.
    B. The interstitial cells secrete the hormones dihydrotestosterone, androstenedione, and testosterone.
    C. The interstitial cells are less sensitive to heat than are the germinal epithelial cells of the seminiferous tubules.
    D. The interstitial cells are present in small numbers in the fetus but present in large numbers at puberty.
    E. The interstitial cells receive their blood supply from the testicular artery.

13. The following statements concerning the ovary are correct **except** which?
    A. The lymph drainage is into the para-aortic (lumbar) lymph nodes at the level of the first lumbar vertebra.
    B. The round ligament of the ovary extends from the ovary to the upper end of the lateral wall of the body of the uterus.
    C. The ovarian fossa is bounded above by the external iliac vessels and behind by the internal iliac vessels.
    D. The left ovarian artery is a branch of the left internal iliac artery.
    E. The obturator nerve lies lateral to the ovary.

14. The following statements concerning the ovaries are correct **except** which?
    A. Embedded in the connective tissue of the cortex are the ovarian follicles.
    B. The corpora lutea are restricted to the medulla.
    C. The theca interna cells produce the estrogenic hormones.
    D. The corpus luteum produces the hormone progesterone.
    E. The corpus luteum is controlled by the LH of the pars anterior of the pituitary and, in pregnancy, by the chorionic gonadotropic hormone of the placenta.

# Answers and Explanations

1. **D** is incorrect. The pituitary gland is suspended from the floor of the third ventricle by the infundibulum (Fig. 24-3).

2. **A** is incorrect. Oxytocin stimulates the contraction of the smooth muscle fibers of the uterus (Fig. 24-7).

3. **E** is incorrect. The neurosecretory cells of the hypothalamus are responsible for the production of the releasing hormones and release-inhibitory hormones (Fig. 24-6).

4. **B** is incorrect. The pineal gland projects from the posterior end of the roof of the third ventricle (Fig. 24-2).

5. **D** is correct. The recurrent laryngeal nerve is closely related to the inferior thyroid artery and might be damaged when tying the artery during operations on the thyroid gland.

6. **D** is incorrect. The parathyroid hormone produced by the chief cells of the parathyroid glands raises the calcium levels of the blood.

7. **C** is incorrect. The thymus gland continues to grow until puberty but thereafter undergoes involution.

8. **E** is incorrect. The medulla of the suprarenal gland is innervated by preganglionic sympathetic nerve fibers (Fig. 24-17).

9. **A** is correct. The suprarenal gland receives its arterial supply from the aorta and inferior phrenic and renal arteries.

10. **E** is incorrect. The final cortex of the developing suprarenal gland covers the fetal cortex; the fetal cortex degenerates and disappears during the first few weeks of life.

11. **C** is incorrect. The islets of Langerhans are more numerous in the tail of the pancreas than in its body, neck, or head.

12. **D** is incorrect. The interstitial cells of the testis are present in large numbers in the fetus but atrophy to some extent after birth. They reappear in large numbers at puberty.

13. **D** is incorrect. The right and left ovarian arteries are branches of the abdominal aorta at the level of the first lumbar vertebra.

14. **B** is incorrect. The corpora lutea are derived from the graafian follicles once they have ruptured and shed their ovum. The graafian follicles are found scattered throughout the ovarian cortex, and that is where the corpora lutea are found.

# Appendix

## Notes on Selected Areas of Regional Anatomy of Clinical Importance

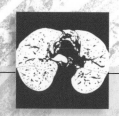

# Appendix Outline

# Face

The face is the window into the individual's personality. Inadequate or untimely delay in the treatment of facial injuries may later lead to irreversible damage and disfigurement. Disfiguring scars may have a devastating effect on the psychosocial relationship of the person with his or her family or peers. This is especially important in the child, since an unsightly injury may result in maladjustment in adulthood.

## Skin of the Face

The skin of the face shows considerable variation in thickness, being thinnest over the eyelids. Wrinkle lines of the face result from the repeated folding of the skin perpendicular to the long axis of the underlying contracting muscles, coupled with the loss of youthful skin elasticity. Thus, we find horizontal wrinkles on the brow, crow's feet wrinkles at the lateral angles of the eyes, and vertical wrinkles above and below both lips.

## Clinical Notes

Surgical incisions of the face heal with less scarring if they are made along wrinkle lines. Sutured lacerations that lie parallel to the tension lines spread less than lacerations that run at right angles to the tension lines.

## Sensory Nerve Supply of the Face

The sensory nerve supply of the skin of the face is supplied by branches of the three divisions of the trigeminal nerve, except for the small area over the angle of the mandible and parotid gland, which is supplied by the great auricular nerve (C2 and C3).

## Arterial Supply of the Face

The face receives a rich arterial supply, with the vessels situated in the superficial fascia. The two main arteries, the facial and the superficial temporal arteries, both branches of the external carotid artery, are supplemented by a number of small arteries that accompany the sensory nerves.

## Clinical Notes

The profuse arterial supply may cause severe bleeding from comparatively small injuries. The great vascularity often permits large flaps of skin resulting from severe injury to be sutured back in position without necrosis occurring.

The arterial pulse can be taken by palpating the superficial temporal artery as it crosses the zygomatic arch in front of the ear. It may also be taken by palpating the facial artery as it winds around the lower margin of the mandible.

## Venous Drainage of the Face

The facial vein runs from the medial angle of the eye, behind the lateral margin of the mouth, and crosses the mandible to drain into the internal jugular vein.

## Clinical Notes

The valveless connection between the facial vein and the cavernous sinus provides a pathway for the spread of infection from the face to the sinus. The central triangular area of skin bounded by the nose, eye, and upper lip is a potentially dangerous area to have a skin infection.

## Lymph Drainage of the Face

The forehead and the anterior part of the face drain into the submandibular lymph nodes. The lateral part of the face, including the lateral parts of the eyelids, drains into the parotid lymph nodes. The central part of the lower lip and the skin of the chin is drained into the submental lymph nodes.

## Development of the Face

Early in development, the face of the embryo is represented by an area bounded cranially by the neural plate, caudally by the pericardium, and laterally by the mandibular process of the first pharyngeal arch on each side (Fig. 1). In the center of this area is a depression in the ectoderm known as the **stomodeum**. In the floor of the depression is the **buccopharyngeal membrane**. By the 4th week, the buccopharyngeal membrane breaks down so that the stomodeum communicates with the foregut.

The further development of the face depends on the coming together and fusion of several important processes, namely, the **frontonasal process**, the **maxillary processes**, and the **mandibular processes** (Fig. 1). The frontonasal process begins as proliferation of mesenchyme on the ventral surface of the developing brain, and this grows toward the stomodeum. Meanwhile, the maxillary process

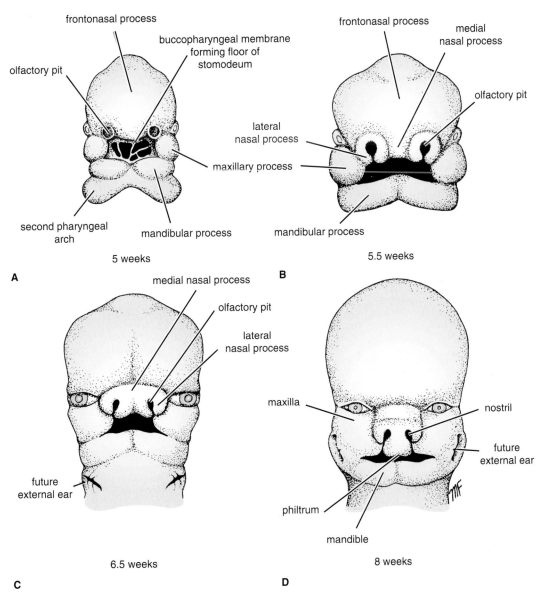

**Figure 1**   Different stages in the development of the face.

grows out from the upper end of each first arch and passes medially, forming the lower border of the developing orbit. The mandibular processes of the first arches now approach one another in the midline below the stomodeum and fuse to form the lower jaw and lower lip (Fig. 1).

The **olfactory pits** appear as depressions in the lower edge of the advancing frontonasal process, dividing it into a **medial nasal process** and two **lateral nasal processes.** With further development, the maxillary processes grow medially and fuse with the lateral nasal processes and with the medial nasal process (Fig. 1). The medial nasal process forms the **philtrum** of the upper lip and the **premaxilla.** The maxillary processes extend medially, forming the upper jaw and the cheek, and finally bury the premaxilla and fuse in the midline. The various processes that ultimately form the face unite during the 2nd month.

The **upper lip** is formed by the growth medially of the maxillary processes of the first pharyngeal arch on each side. Ultimately, the maxillary processes meet in the midline and fuse with each other and with the medial nasal process (Fig. 1). Thus, the lateral parts of the upper lip are formed from the maxillary processes, and the medial part, or philtrum, is formed from the medial nasal process, with contributions from the maxillary processes.

The **lower lip** is formed from the mandibular process of the first pharyngeal arch on each side (Fig. 1). These processes grow medially below the stomodeum and fuse in the midline to form the entire lower lip.

Each lip separates from its respective gum as the result of the appearance of a linear thickening of ectoderm, the **labiogingival lamina,** which grows down into the underlying mesenchyme and later degenerates. A deep groove thus forms between the lips and the gums. In the midline, a short area of the labiogingival lamina remains and tethers each lip to the gum, thus forming the **frenulum.**

At first, the **mouth** has a broad opening, but later, this diminishes in extent because of fusion of the lips at the lateral angles.

## Sensory Nerve Supply to the Skin of the Developing Face

The area of skin overlying the frontonasal process and its derivatives receives its sensory nerve supply from the

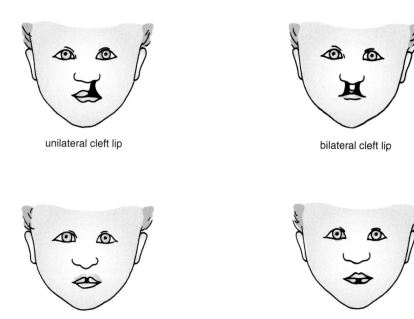

unilateral cleft lip

bilateral cleft lip

median cleft upper lip

median cleft lower lip

oblique facial cleft

**Figure 2**	Various forms of cleft lip.

ophthalmic division of the trigeminal nerve, whereas the maxillary division of the trigeminal nerve supplies the area of skin overlying the maxillary process. The area of skin overlying the mandibular process is supplied by the mandibular division of the trigeminal nerve.

## Muscles of the Developing Face (Muscles of Facial Expression)

The muscles of the face are derived from the mesenchyme of the second pharyngeal arch. The nerve supply of these muscles is the nerve of the second pharyngeal arch—namely, the seventh cranial nerve.

## Congenital Anomalies

### Cleft Upper Lip

Cleft upper lip may be confined to the lip or may be associated with a cleft palate. The anomaly is usually **unilateral cleft lip** and is caused by a failure of the maxillary process to fuse with the medial nasal process (Figs. 2 and 3). **Bilateral cleft lip** is caused by a failure of both maxillary processes to fuse with the medial nasal process, which then remains as a central flap of tissue (Fig. 4). **Median cleft upper lip** is very rare and is caused by the failure of the rounded swellings of the medial nasal process to fuse in the midline.

### Oblique Facial Cleft

Oblique facial cleft is a rare condition in which the cleft lip on one side extends to the medial margin of the orbit (Figs. 2 and 5). This is caused by the failure of the maxillary process to fuse with the lateral and medial nasal processes.

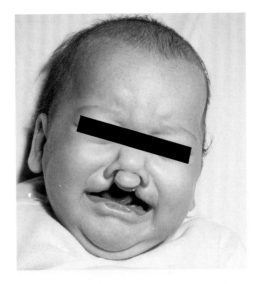

**Figure 4**   Bilateral cleft upper lip and palate. (Courtesy of R. Chase.)

### Cleft Lower Lip

Cleft lower lip is a rare condition. The cleft is exactly central and is caused by incomplete fusion of the mandibular processes (Fig. 2).

### Treatment of Isolated Cleft Lip

The condition of isolated cleft lip usually is treated by plastic surgery no later than 2 months after birth, provided the baby's condition permits. The surgeon strives to approximate the vermilion border and to form a normal-looking lip.

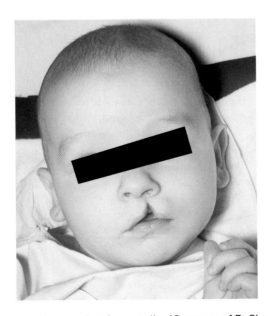

**Figure 3**   Unilateral cleft upper lip. (Courtesy of R. Chase.)

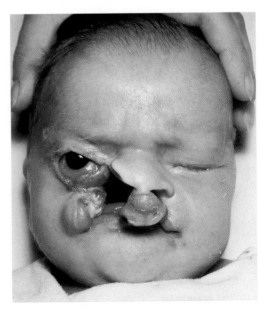

**Figure 5**   Right-sided oblique facial cleft and left-sided cleft upper lip. There also is total bilateral cleft palate. (Courtesy of R. Chase.)

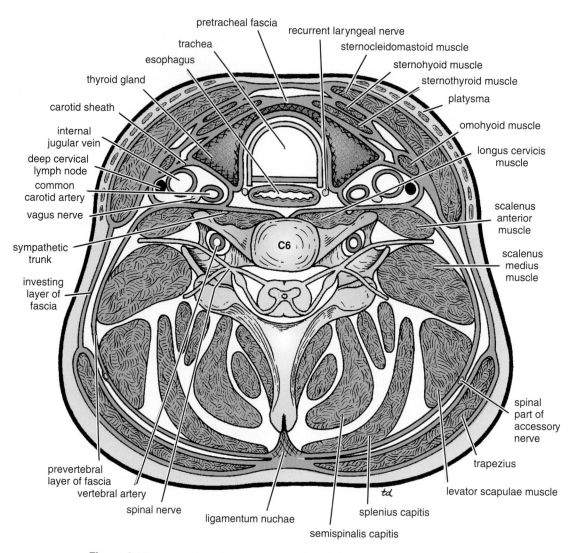

pretracheal fascia
trachea
esophagus
thyroid gland
carotid sheath
internal jugular vein
deep cervical lymph node
common carotid artery
vagus nerve
sympathetic trunk
investing layer of fascia

recurrent laryngeal nerve
sternocleidomastoid muscle
sternohyoid muscle
sternothyroid muscle
platysma
omohyoid muscle
longus cervicis muscle
scalenus anterior muscle
scalenus medius muscle
spinal part of accessory nerve
trapezius
levator scapulae muscle

C6

prevertebral layer of fascia
vertebral artery
spinal nerve
ligamentum nuchae
semispinalis capitis
splenius capitis

**Figure 6** Cross section of the neck at the level of the sixth cervical vertebra.

## Macrostomia and Microstomia

The normal size of the mouth shows considerable individual variation. Rarely, there is incomplete fusion of the maxillary with the mandibular processes, producing an excessively large mouth or macrostomia. Very rarely, there is excessive fusion of these processes, producing a small mouth or microstoma. These conditions can easily be corrected surgically.

# Neck

A large number of vital structures are present in the neck, and blunt and penetrating wounds to the neck are life threatening.

It is suggested that the regional anatomy of the neck as seen on cross section at the level of the sixth cervical vertebra be committed to memory (Fig. 6).

# Wrist and Hand

Wrist and hand injuries occur frequently. The goal of treatment is the preservation of as much function as possible. Particular attention should be paid to the thumb and the pincer action of the thumb and index finger.

It is suggested that the regional anatomy of the wrist, as seen on cross section, be committed to memory and the arrangement of the nerves, blood vessels, and tendons in the palm and fingers be learnt in detail (Figs. 7 and 8).

# Ankle

Ankle injuries are very common. The relevant anatomy of the bones, tendons, nerves, and blood vessels may be reviewed by studying Figure 9.

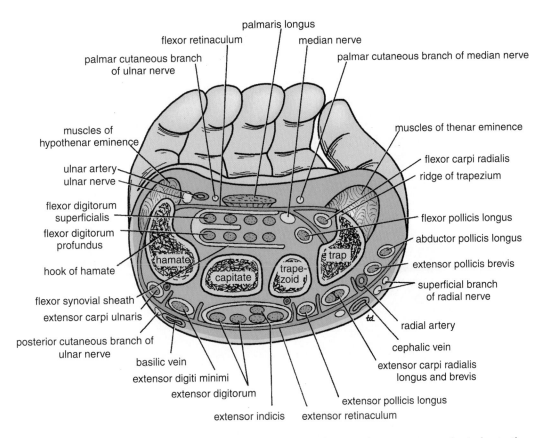

**Figure 7**   Cross section of the hand showing the relation of the tendons, nerves, and arteries to the flexor and extensor retinacula.

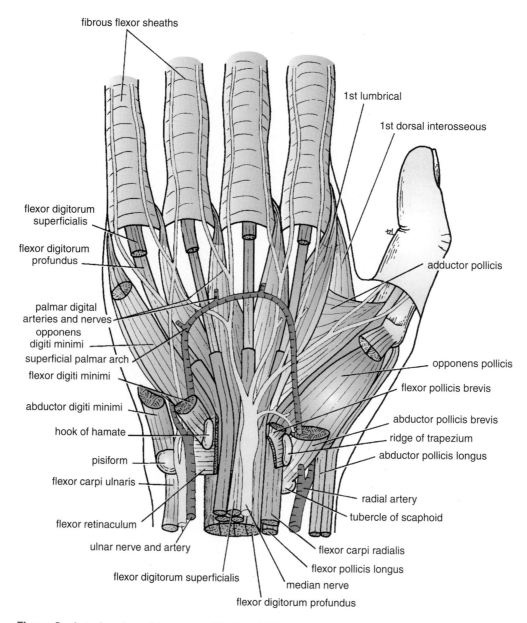

fibrous flexor sheaths

1st lumbrical

1st dorsal interosseous

flexor digitorum superficialis

flexor digitorum profundus

palmar digital arteries and nerves

opponens digiti minimi

superficial palmar arch

flexor digiti minimi

abductor digiti minimi

hook of hamate

pisiform

flexor carpi ulnaris

flexor retinaculum

ulnar nerve and artery

flexor digitorum superficialis

adductor pollicis

opponens pollicis

flexor pollicis brevis

abductor pollicis brevis

ridge of trapezium

abductor pollicis longus

radial artery

tubercle of scaphoid

flexor carpi radialis

flexor pollicis longus

median nerve

flexor digitorum profundus

**Figure 8** Anterior view of the palm of the hand. The palmar aponeurosis and the greater part of the flexor retinaculum have been removed to display the superficial palmar arch, the median nerve, and the long flexor tendons. Segments of the tendons of the flexor digitorum superficialis have been removed to show the underlying tendons of the flexor digitorum profundus.

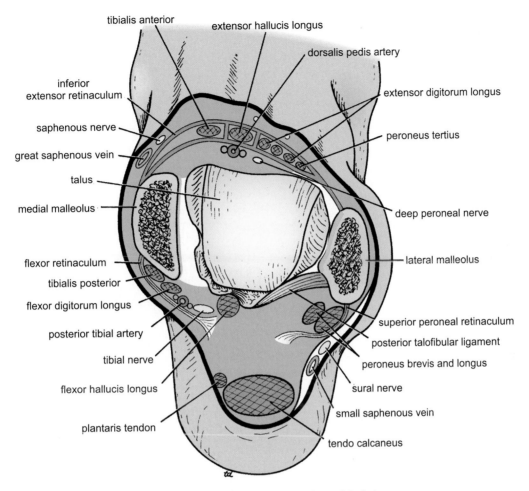

**Figure 9   Relations of the right ankle joint.**

# Index